Exploring the Quantum

Exploring the Quantum

Atoms, Cavities and Photons

Serge Haroche
Professeur, Collège de France

Jean-Michel Raimond
Professeur, Université P. et M. Curie et Institut Universitaire de France

OXFORD
UNIVERSITY PRESS

Great Clarendon Street, Oxford OX2 6DP

Oxford University Press is a department of the University of Oxford.
It furthers the University's objective of excellence in research, scholarship,
and education by publishing worldwide in

Oxford New York

Auckland Cape Town Dar es Salaam Hong Kong Karachi
Kuala Lumpur Madrid Melbourne Mexico City Nairobi
New Delhi Shanghai Taipei Toronto

With offices in

Argentina Austria Brazil Chile Czech Republic France Greece
Guatemala Hungary Italy Japan South Korea Poland Portugal
Singapore Switzerland Thailand Turkey Ukraine Vietnam

Oxford is a registered trade mark of Oxford University Press
in the UK and in certain other countries

Published in the United States
by Oxford University Press Inc., New York

British Library Cataloguing in Publication Data
Data available

Library of Congress Cataloging in Publication Data
Data available

Printed in Great Britain
on acid-free paper by
Biddles Ltd., King's Lynn, Norfolk

ISBN 0–19–850914–6   978–0–19–850914–1 (Hbk)

3 5 7 9 10 8 6 4 2

To

Claudine, Julien and Judith

Fabienne, Yves and Marie

Foreword

The counter-intuitive aspects of quantum physics were illustrated in the early days of the theory by famous thought experiments, from the Einstein and Bohr photon box to Schrödinger's cat. Modern versions of these experiments, involving single particles – electrons, atoms or photons – have now been actually realized in many laboratories around the world. By manipulating these simple systems in a controlled environment, physicists directly unveil the strange features of the quantum. State superpositions, entanglement and complementarity define a novel quantum logic which can be harnessed for information processing, raising great hopes for applications.

This book describes a class of such thought experiments which have come of age. Selecting among a vast and fast expanding domain of research, we have chosen to analyse in detail experiments performed with atoms and photons in high-Q cavities as well as related ones, dealing with ions in traps or cold atoms in optical lattices. In these apparently disparate domains, the same underlying physics is at work: two-level spin-like systems are interacting with quantum harmonic oscillators.

We believe that a description of these real 'spin-spring' experiments provides a more concrete illustration of quantum concepts than that given by abstract discussions about idealized experiments. Although the latter are simpler to analyse and are certainly here to stay in introductory courses of quantum mechanics, real 'thought experiments' should in our view become central in the teaching of modern quantum physics at an intermediate or advanced level. The efforts to carry out these experiments in laboratories have been largely triggered by the hopes placed in the development of quantum information for practical applications in communication and computing. Conversely, this fast expanding field of research is bound to have an increasing influence on the teaching and learning of quantum concepts.

Dealing with real systems necessarily involves a description of the interaction of these systems with their unavoidable environment, in other words a discussion of relaxation and decoherence. These phenomena are described by a formalism (density operator or stochastic Monte Carlo approach) which replaces the simple state description of elementary quantum physics. Mastering this approach and understanding decoherence provide a deep understanding of one important aspect of the quantum, its relation to classical physics. Thought experiments have been invented at the dawn of the quantum age to illustrate the puzzling features of the quantum–classical boundary. It is thus no surprise that an understanding of the modern version of these experiments must also address this important issue.

Starting from the simple goal to describe experiments illustrative of basic quantum laws, we have thus been led to widen our perspective, ending up with a book which presents a comprehensive discussion of many important issues in modern physics. It combines a fundamental approach, based on an analysis of quantum concepts and of useful theoretical tools, with a detailed analysis of experiments, including a brief overview of the various technological developments which have made these experiments possible. In balancing these theoretical and applied points of view, we have tried to convey at the same time the strange beauties of the quantum and the difficulties which had to be overcome to unveil them, and possibly to harness them for achieving useful tasks.

This book is intended for students at the undergraduate or graduate level, with an elementary knowledge of quantum mechanics. We have not assumed that they had been previously exposed to a detailed discussion of concepts such as entanglement, non-locality, decoherence or measurement theory, which we have chosen to expose from scratch. At the same time, we hope that our work will also be useful to teachers in the field of quantum optics and quantum information science. We have attempted to present many examples of physical situations which are computed in detail, and which could be easily turned into instructive problem sets for students.

The many connexions and comparisons we make between atom–cavity, ion trap and cold atom experiments might also be useful to scientists working in these various fields of quantum optics. Reading this book might suggest to them new perspectives for their work, as writing it has helped us to sharpen our understanding and to design new experiments. Finally, theorists in quantum information science might learn here about some of the challenges that experimenters have to address in order to put their bright ideas into practice.

The material of this book is based on the lectures on quantum information we have respectively given at Collège de France from 2001 to 2006, and at Ecole Normale Supérieure (ENS) from 2003 to 2006. The atom–cavity experiments which are at the heart of the book discussions have been carried out by our research team at ENS. We are indebted to all the colleagues, students, postdocs and visitors who have worked with us over the years. We should mention in particular Michel Brune, who has played an essential role in all these experiments and who has provided precious advice to improve the manuscript. Our ENS colleagues Jean Dalibard and Yvan Castin have been consulted on some aspects of cold atom physics and we are glad to acknowledge their illuminating input. Special thanks go also to Luiz Davidovich and Nicim Zagury, from the Federal University of Rio de Janeiro, whose theoretical insights have been precious to design new experiments.

We have devoted the last chapters of the book to the description of experiments performed in other laboratories, mainly in Boulder, Innsbruck, Munich and Mainz. We thank D. Wineland, D. Leibfried, R. Blatt, C. Roos and I. Bloch for helpful discussions and for critical reading of parts of the manuscript. We are of course responsible for any approximation or error remaining in the description of their work. Finally and foremost, we thank Claudine and Fabienne for their constant support and encouragement.

Serge Haroche and Jean-Michel Raimond
Paris, May 2006

Contents

1
Unveiling the quantum

We never experiment with just one electron or atom or (small) molecule. In thought-experiments we sometimes assume that we do; this invariably entails ridiculous consequences...

E. Schrödinger,
British Journal of the Philosophy of Sciences, **3**, 1952.

Thought experiments are at the heart of quantum physics. The fathers of the theory have time and again imagined simple machines manipulating isolated atoms or photons and discussed what quantum rules had to say about the outcome of such idealized experiments. These conceptual constructions have helped them to elaborate a coherent theory. At the same time, they were convinced that such experiments would remain virtual. As late as the middle of the last century, at a time when the quantum ideas were firmly established, Erwin Schrödinger argued that manipulating an isolated atom in the real world would remain forever impossible. And yet, by the end of the twentieth century, those experiments came of age. Physicists can now trap single atoms or photons in a confined region of space, prepare these particles in well-defined states and follow in real time their evolution. These experiments have been made possible by powerful technologies born from quantum concepts, such as modern computers and lasers. Physics has thus come full circle. Thought experiments, considered as pure dreams, have helped to develop the tools which have made these fantasies real.

This history is intellectually satisfying since it displays the deep consistency of our understanding of Nature. By actually performing in the laboratory thought experiments, we are able to observe directly 'in action' the quantum laws that the founding fathers had deduced from remarkable efforts of abstraction. Apart from this intellectual satisfaction, what is the rationale for performing such experiments which remain a technological challenge? Four incentives at least might be invoked.

First, many aspects of the quantum theory are so counter-intuitive that some physicists – including Einstein – have never really accepted them. Entanglement is one of the most intriguing features of the quantum. After interacting, two microscopic quantum systems generally end up in a non-separable state. The properties of each system cannot be described independently of those of the other. This occurs whatever the distance between the components of the entangled state. Entanglement thus naturally leads to non-locality. This notion tells us that physics at one place cannot be described independently of what goes on in another disconnected part of the Universe. This is certainly the quantum feature most difficult to admit by a classical mind. It is only by investigating non-locality through real experiments that one can put Nature to test and find out whether quantum laws give the last word on this intriguing issue.

Our second motivation has to do with the exploration of the connexion between quantum and classical physics. One reason why non-locality is counter-intuitive is that it is only observed in systems made of a few photons, electrons or atoms. Macroscopic systems, directly accessible to our senses, never display non-locality. Nor do they exhibit other strange aspects of the quantum such as state superposition and quantum interference. There seems to be a boundary between the microscopic world, directly displaying quantum laws, and the macro-world where these laws are veiled. By performing thought experiments on systems of increasing size, one might hope to explore this frontier. The phenomenon which tends to blur quantum effects in the macroscopic world, to destroy quantum interference and quantum coherence, is called 'decoherence'. Performing 'thought experiments' is a direct test of our ability to minimize decoherence in real systems and provides a simple and direct way to study this phenomenon.

Our third incentive is more practical. The counter-intuitive features of quantum laws can, in principle, be exploited to process information according to a non-classical logic. Communicating or computing is a quite different state of affairs if one codes information into ensembles of individually addressable atoms or photons. Two quantum states of each of these particles, defined as a quantum bit, or simply a qubit, can be associated to the 0 and 1 values of the usual binary code. These qubits not only exist as 0's and 1's, but can also evolve into entangled superpositions of these states. Secure cryptographic schemes can be elaborated with qubits, enabling two partners to exchange information while being certain that no spy could eavesdrop. Quantum computers solving some problems such as the factoring of large numbers in a time much shorter than classical machines can be envisioned. These hopes have led to the fast development of a new field of research, at the frontier between physics and information science, the 'physics of quantum information'. In this context, simple thought experiments realized with a few atoms or photons are important to demonstrate the feasibility of quantum information and to study its fundamental and technical limitations as well as the possible ways to overcome them.

The last – but not least – incentive for realizing thought experiments is pedagogical. In most textbooks, quantum physics is introduced by discussing early experiments, performed in the first part of the last century, which displayed only indirectly and imperfectly the quantum features. Now that experiments in which quantum laws are clearly apparent have become feasible, it is important to perform them as illustrations of fundamental concepts. Even if quantum computers never come of age, students and researchers studying these experiments will gain a deeper familiarity with quantum laws, which will help them find novel ways to explore Nature and to develop powerful technological tools.

Most of the thought experiments realized recently are made of simple basic elements. Inside a confined region of space, a 'box' as we will call it for short, a few atoms or photons evolve, largely impervious to what happens in the outside world. On this simple stage, the laws of the game are the quantum postulates. State superpositions, quantum interference and entanglement are directly displayed, illustrating as clearly as can be the quantum concepts. In some realizations, the box is a simple configuration of electromagnetic fields trapping atomic particles, without material separation between the inside and the outside. In others, the box has real walls, for example mirrors confining photons.

Simple tools are used to manipulate and detect the particles in the box. If these particles are atoms or ions, laser beams propagating across the box prepare them in well-defined states and manipulate them. Photons scattered off these beams are used to detect the atoms, providing direct sight inside the box. If photons are trapped, atomic beams are used instead to manipulate and probe them. In all cases, light–matter interaction processes are essential to build the box, to manipulate the particles and to detect them.

This book is devoted to these atom–photon-box 'thought experiments' which have become real. Our goal is to describe them and to present the theoretical framework required to understand their physics. Along the way, we will keep in mind our quadruple motivation: the interest of these experiments as tests of the counter-intuitive aspects of the theory, their importance in the exploration of the quantum–classical boundary, their role as elementary steps towards harnessing the quantum for information processing and their utility as a pedagogical approach to quantum phenomena.

We have alluded several times to the strangeness of the quantum world. Saying that the theory is strange, we are in good company. Just recall the famous statement of Richard Feynmann that '*nobody really understands quantum physics*'. This aphorism by one of the fathers of quantum field theory must be taken with a grain of salt. There is a pitfall to avoid when invoking the wonders of the quantum, their manifestations in thought experiments and possible uses in quantum information. The danger is to give the feeling that quantum theory is mysterious or controversial, with promises looming only in the future. In fact, the quantum has already delivered a lot. Most of the fundamental and technological advances of the last century, which make our life so different from our great-grandparents', are due to the deep understanding of Nature brought about by the quantum revolution. If the theory appears strange, it is mainly because we try to describe it with words of our everyday life, which are adapted to the properties of macroscopic objects. Even if quantum concepts are necessary to understand in depth the electric conductivity of metals, the superfluidity of liquid helium or the colour of the sky, these macroscopic phenomena are not 'strange' because they can be described with usual words, which is not the case for an ion in a trap or a photon in a cavity.

A distinction must thus be made between the 'microscopic' quantum strangeness directly displayed in thought experiments and the apparent plainness of the macroscopic physics, which do not violate our common sense in spite of its quantum substrate. The physics of quantum information has the ambition to use the quantum to compute better, but we must not forget that the quantum is already present, albeit in a veiled form, in the physical processes making classical computers so successful. There is thus a 'veiled' and a 'naked' quantum which should be distinguished, even if the opposition between them is sometimes blurred. This book is about the naked quantum, its manifestations in realized thought experiments and its possible applications.

In this introductory chapter, we start (Section 1.1) by recalling some of the great

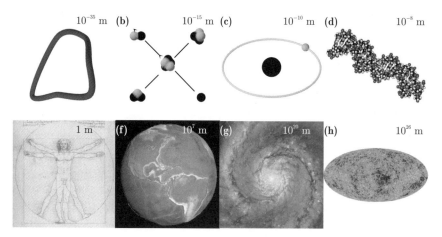

Fig. 1.1 Quantum physics describes Nature at all scales. Pictorial representation of (a) a superstring, (b) a nuclear reaction (fusion of deuterium and tritium), (c) a hydrogen atom, (d) a biological molecule, (e) a human being, (f) a planet (image credit: NASA), (g) the 'whirlpool' galaxy (image credit: NASA/JPL), (h) the microwave blackbody radiation map (image credit: NASA/WMAP Science Team).

advances of quantum theory in the twentieth century. These advances result from the successful work of generations of scientists who have made use of quantum physics, spending only a small fraction of their time puzzling at its wonders and directing their efforts to apply it at explaining Nature. Any meaningful discussion about where the quantum can bring us in the future must be carried on with this background of successes in mind.

We then come back (1.2) to the elaboration of the theory, the long process during which the strange and counter-intuitive manifestations of the quantum have emerged. It is during this process – mainly in the 1920s – that thought experiments were conceived. We recall (1.3) their main features and describe, on a few examples, how they are now realized in the laboratory by exploiting modern technology. We stress that these experiments can be quite generally viewed as manipulations of quantum oscillators and two-level spin-like particles in mutual interaction. We conclude the chapter by presenting the aims and outline of the book (1.4).

1.1   One century of quantum physics

From the very small to the very large, covering about sixty orders of magnitude in dimensions, quantum theory is invoked to describe all phenomena, from the still mysterious vibrations of the microscopic strings which might be the ultimate constituents of Nature, to the fluctuations of the microwave radiation received from the outposts of the known Universe. Between these extremes, we find all the objects of the world around us (Fig. 1.1). Add twenty zeros to the dimensions of the strings and you have the size of the atomic nucleus, the centre of radioactivity and nuclear energy. Five more zeros and here is the atom, a nucleus bound to its swarm of electrons by electromagnetic forces, and simple molecules, small ensembles of atoms which combine according to the laws of chemistry. Two or three orders of magnitude more and we reach bio-molecules, where life appears at the most elementary level. Count six to nine further orders of magnitude and we are in the range of the centimetres to tens of metres, the human scale with its variety of objects, solids, liquids and gases made of huge numbers of atoms. From our scale to the astronomical one, the size of planets and stars where gravitation is dominant, you must still add seven to nine zeros. And you finally need eighteen more orders of magnitude to complete, through the exploration of galaxies, our voyage towards the horizon of our Universe.

1.1.1   Quantum theory as a unifying description of Nature

Along this vertiginous path, physics must account for a boundless variety of phenomena. Some have been empirically known for a long time, others discovered during the last century thanks to the development of powerful methods of investigation. At the very small and very large ends of the dimension scale, others are still raising many questions. Most of these phenomena require us, at least for part of their understanding, to invoke quantum theory whose success has been overwhelming. Theorists will insist on the extreme precision of the description of the interactions between electrons and photons by quantum electrodynamics. They will note the remarkable agreement between theory and experiment in this field: the magnetic moment of the electron, measured in a particle trap experiment, has been found to be, in units of Bohr's magneton $g = 2 \times 1.00159652188$, a value which agrees with theory within a few parts per billion!

Particle physicists will also mention, as a proof of the success of quantum theory, the unification of three out of the four fundamental interactions – the electromagnetic, strong and weak forces – in the Standard Model, which reveals the deep symmetries of Nature. They will also note the promising attempts to include gravitation in a unified theory of strings. And all physicists will certainly emphasize the universality of physics, a remarkable consequence of quantum laws. It is the quantum theory which explains the radiation spectrum of hydrogen in our laboratory lamps, but also in intergalactic space. Quantum chemistry applies to the reactions in laboratory test tubes, but also to the processes in the interstellar dust where molecules are formed and destroyed, producing radiation detected by our telescopes after travelling billions of years through space. Let us finally evoke cosmology and the remarkable link between the infinitely small and large scales, underlined by the quantum theory. It emphasizes the similarity between the phenomena which occurred at the origin of the Universe, in a medium of inconceivably large temperature and density, and those which happen in the violent collisions between particles studied by large accelerators on Earth.

1.1.2   The transistor and the computer revolution

Quantum physics not only describes the structure of matter and the interactions between particles. Through the deep knowledge it gives us about microscopic phenomena, it has provided us with tools to act, communicate, diagnose and compute with a power and a precision which could not be imagined before its advent. Since some of these tools are necessary for realizing the experiments described here, it is appropriate to reflect a little on these achievements of the quantum in our everyday life.

Let us start with the computer. The small desk machines which have become a ubiquitous fixture of offices, homes and laboratories are based on a small processor,

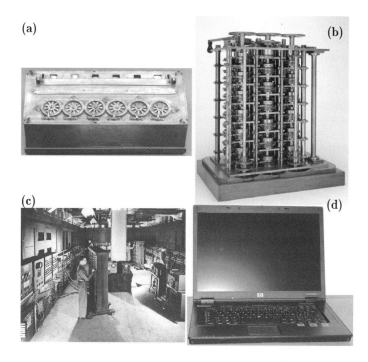

Fig. 1.2 The history of the computer. (a) Pascal machine © Musée des Arts et Métiers, reprinted with permission; (b) Babbage machine © Science and Society Picture Library, reprinted with permission; (c) the ENIAC machine (US Army photo); (d) a modern laptop (photograph by the authors).

made of a silicium chip the size of a stamp, on which a labyrinth of tiny electrical wires is printed. These wires link small layers of semiconductor materials making up microscopic transistors which behave as logical gates. The properties of these materials are ruled by quantum laws (tunnel effect, exclusion principle,...). It is by exploiting these laws that one can achieve extremely complex calculations. The principles of the computer have been known for a long time. The capacity to store information and to compute dates back at least to Pascal's machine in the seventeenth century (Fig. 1.2a). The idea of computer programming can be attributed to Babbage who, in the nineteenth century, had invented a mechanical device able to perform simple computation tasks (Fig. 1.2b).

The development of the electrical industry in the first part of the last century replaced the mechanical parts of Pascal's and Babbage's machines by vacuum tubes, leading to the first modern computers. The machines, which were built during the Second World War and immediately after, under the impulsion of physicists and mathematicians such as Turing, von Neumann and Brillouin, were huge assemblies of cupboard-size boxes, stuffed with tubes and linked by kilometres of wires, requiring the assistance of an army of engineers (Fig. 1.2c). And their poorly reliable performances were by far short of those of the modern laptop computers, a thousand times smaller in weight and volume (Fig. 1.2d)! It is the advent of the transistor replacing the bulky tubes, and the integration of a huge number of them in a chip, which has made all the applications of the modern computer possible, including the delicate manipulation and control of individual atoms and photons described in this book. The logical principles of the computer could have been understood by a nineteenth century scientist, but its practical realization relies on a technology totally inconceivable to a pre-quantum mind.

1.1.3 The laser and the optical revolution

The laser is another example of an invention based on a quantum idea, now ubiquitous in our everyday life. Laser beam shows are today commonplace but, until the 1960s, nobody had seen anything like it. Light had been forever made of random waves, difficult to direct, to focus or to force to oscillate at a well-defined frequency. The laser has changed this state of affairs and has allowed us to tame radiation by exploiting the properties of atomic stimulated emission, discovered by Einstein at the dawn of the quantum era. Lasers now have a huge variety of uses, from the very mundane to the most sophisticated. Laser light travelling through fibres can transport huge amounts of information over very long distances. Laser beams are used to print and read out information on compact disks, with applications for the reproduction of sounds, pictures and movies.

In scientific research, the flexible properties of laser beams have countless applications, some of which are essential to realize the manipulations of single particles which will be our main topic here. Experiments exploit their high intensity to study non-linear optical processes. The extreme monochromaticity of lasers and their time coherence is used for high-resolution spectroscopy of atoms and molecules. Combining monochromaticity and high intensity has proved essential to trap and cool atoms to extremely low temperatures. Laser pulses of femtosecond duration probe very fast processes in molecules and solids and study chemical reactions in real time.

1.1.4 The atomic clock and the measurement of time

The precise measurement of time is another illustration of the importance of quantum processes in science and in society. Atoms emit and absorb radiation at well-defined, never changing frequencies, characteristic of each element. This fundamental quantum property has led physicists to base the measurement of time no longer on the perturbation-prone oscillations of mechanical or electrical pendulums, neither on the fluctuating motions of planets, but on the immutable oscillations of electrons in atoms.

This has led to atomic clocks built by locking a radiosource to the hyperfine radio-frequency transition of caesium atoms propagating in an atomic beam. The atomic resonance is observed by using an interferometric method which we will describe in detail later on. In short, the atoms interact successively with two microwave pulses as they travel along the beam. By detecting downstream one of the hyperfine states and sweeping the microwave frequency, one obtains a modulated signal, called a 'Ramsey fringe' pattern. The centre of this pattern is determined with a precision depending on the fringe spacing, which is inversely proportional to the time interval between the microwave pulses. Recently, laser-cooled atoms have replaced the thermal atoms of former clocks. The interval between the pulses has been considerably increased and

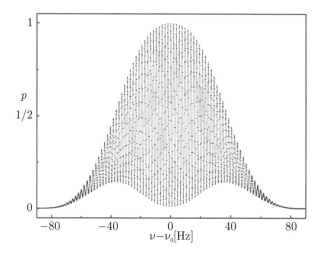

Fig. 1.3 Atomic clock. Ramsey fringe pattern obtained on the hyperfine transition of laser-cooled caesium atoms. From Lemonde *et al.* (2000), reprinted with kind permission of Springer Science and Business Media.

the clock's precision improved by two orders of magnitude.

Figure 1.3 shows the Ramsey fringe pattern of a cold-atom clock, with a fringe spacing of less than 1 Hz, while the microwave frequency is 9.2 GHz. The large signal-to-noise ratio of the fringes makes it possible to determine the atomic frequency within one part in 10^3 of the fringe interval. This clock has accuracy better than a second over a geological time of three million years! The techniques developed for these clocks have much in common with the methods of atomic manipulation described later on.

Time measurements based on atomic clocks and on a related device, the hydrogen maser, have been used in general and special relativity tests. The utility of these clocks is not restricted to fundamental science. Their extraordinary precision is exploited in the Global Positioning System (GPS), which consists of a large set of clocks embarked on satellites. They send signals to small and cheap receivers, not much bigger than a wristwatch. Measuring the arrival times of these signals allows the receiver to determine its location with a precision in the metre range. These devices, combined with navigation computers, have become standard fixtures of planes, boats and cars and we are no longer surprised to be routinely guided from the sky with such an extraordinary precision.

1.1.5 Nuclear magnetic resonance and medical diagnosis

Nuclear Magnetic Resonance (NMR) is another technology based on quantum science which nowadays plays a major role in scientific research and in medical diagnosis. The nuclei of a variety of atoms carry magnetic moments attached to their intrinsic angular momentum or spin. The origin of this magnetism is fundamentally quantum. The quantization of the spin orientation in space, the fact that its projection on any direction can take only discrete values, has been one of the smoking guns of the quantum theory, forcing physicists to renounce their classical views. The evolution of

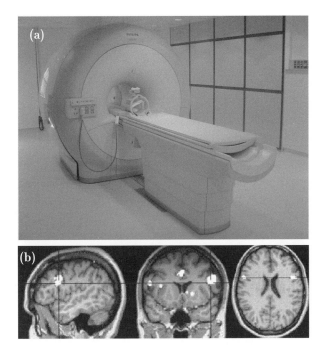

Fig. 1.4 Magnetic resonance imaging. (a) Picture of an MRI imager. Photograph by D. Glutron CIERM–U2R2M. (b) MRI pictures of the brain exhibiting regions concerned by a specific activity (light spots). Reprinted from Rangaswamy *et al.* (2004), with permission from Elsevier.

spins in magnetic fields, both static and time-modulated, requires a quantum approach to be understood in depth. In a solid or liquid sample, the spins are affected by their magnetic environment and their evolution bears witness of their surroundings.

In a typical NMR experiment, one immerses the sample in a large magnetic field which superimposes its effects on the local microscopic field. One furthermore applies to the sample sequences of tailored radio-frequency pulses. These pulses set the nuclei in gyration around the magnetic field and the dance of the spins is detected through the magnetic flux they induce in pick-up coils surrounding the sample. A huge amount of information is gained on the medium, on the local density of spins and on their environment.

This information is widely exploited in chemistry and biology. It has also invaded our lives through the Magnetic Resonance Imaging (MRI), which has become a standard technique of medical diagnosis. The MRI machines used in hospitals (Fig. 1.4a) are the result of three converging quantum technologies carried to a remarkable level of sophistication. The large magnetic field in which the patient is placed is produced by coils cooled to liquid helium temperature and exhibiting superconductivity, a fundamentally quantum effect. Only such coils can produce without heating the large fields required for MRI imaging. The choreography of the spins observed by the NMR method is also ruled by quantum laws. Finally the signals picked up in the detection coils are transformed into images by powerful computers which also exploit quantum effects in their semiconductor circuits. Figure 1.4(b) shows images of the brain made by MRI, exhibiting the regions which are concerned by a specific activity. The change in the magnetic environment produced by the blood flux increase in these regions modifies the NMR signals, and the brain activity – thought or motion command – can thus be directly witnessed.

Information gained in this way is coded into the spins of the protons which make up the hydrogen atoms of our bodies. These spins can be considered, in the language of information, as binary bits which take one value – say 0 – if the spin points along the direction of the field and the other value – say 1 – if it points in the opposite direction. When the spins are rotated by the NMR pulses along a direction normal to the field, quantum laws tell us that they are put in a superposition of the 0 and 1 states. In a way, we can thus say that an NMR machine exploits quantum logic, dealing with qubits in superposition of their ordinary classical binary values.

This simple idea has been carried further in quantum information physics. By performing complex NMR experiments on liquid samples made of organic molecules, simple quantum computing operations have been achieved. The spins of these molecules are manipulated by complex pulse sequences, according to techniques originally developed by chemists, biologists and medical doctors in NMR and MRI. In these experiments, a huge number of molecules contribute to the signal. The situation is thus very different from the manipulation of the simple quantum systems we will be considering. Some of these methods do however apply. We will see that the atoms and ions which are individually manipulated in the modern realizations of thought experiments are also two-level spin-like systems and that one applies to them sequences of pulses similar to those invented for NMR physics.

1.2 Emergence of the microscopic world

When reflecting on the wonderful tools that quantum science has offered us, it is important to remind ourselves that all these inventions have required a deep knowledge of the microscopic structure of matter. The end product is generally a macroscopic contraption whose intimate workings are concealed from the layman enjoying them. It is indeed striking that we are using these devices without wondering about the way they work, most of the time oblivious or plainly unaware of the microscopic effects which are essential to their operation. To a nineteenth century physicist projected in today's world by a time machine, this modern technology would superficially appear familiar, a combination of boxes and keyboards. Its working would, however, appear as pure magic. Up to the beginning of the last century indeed, the microscopic world was a *terra incognita*, whose mere existence was ignored by many – among whom not the least famous scientists of the time.

Even those who – like Lord Kelvin – firmly believed in atoms did not realize the importance of the microscopic world for a deep understanding of Nature. His famous statement that everything was understood about luminous and thermal effects, with the possible exception of two small clouds still obscuring the otherwise clear sky of physics, is typical of the general belief of the end of nineteenth century scientists. Influenced by philosophers such as Ernst Mach, most of them thought that only the directly observable macroscopic quantities – electrical currents, temperature, pressure – really mattered. Once direct relations between these quantities had been found

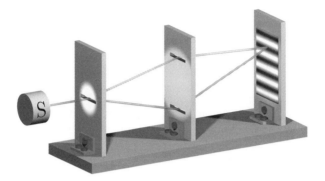

Fig. 1.5 The Young double-slit experiment. A source S of particles collimated by a slit in the left screen illuminates the screen at right through two narrow slits in the central screen. The particle impacts on the detection screen form bright and dark fringes which reveal a quantum interference process.

(Newton equations for mechanics, Maxwell for electrodynamics, Gibbs for thermodynamics), everything was supposed to be understood. The existence of an atomic reality beneath the surface of objects was strongly advocated by some, Boltzmann notably. But many scientists believed that the atomic hypothesis was at best a convenient way to describe things, which should not be given too much objective reality.

Remarkably though, Lord Kelvin's allusion to the two small clouds hinted at some possible trouble in this simple picture. One of these clouds referred to the ether, a hypothetical medium supposed to permeate everything, which Michelson's experiment had just shown to have inconsistent properties. The other cloud pointed to anomalies in the distribution of energy in heated bodies and in the light they radiate (the so-called blackbody problem). In spite of his apparent triumphalism, Lord Kelvin had clearly seen from where the changes were coming. The first cloud announced Einstein's relativity and the second soon led Planck to the idea that the exchanges of energy between matter and radiation occurred in discrete lumps, opening the quantum era with all the consequences which were recalled above.

As we have noted, the main difficulty that the founders of the theory encountered was that the language of everyday life, built on the experience of the macroscopic world, is inadequate to describe microscopic reality. A large object has to be 'here' or 'there'. An atom or a photon, on the contrary, can be 'suspended' between different positions. If it is sent across a double slit in a Young interferometer, as depicted in Fig. 1.5, a drawing inspired by Bohr, it can cross the apparatus in a superposition state, passing in a way *at the same time* through the two openings.

Each state of the superposition is associated to a complex number, its probability amplitude. These amplitudes interfere constructively or destructively depending on their relative phases and the impacts of the particles on a detection screen behind the two slits form a pattern of 'bright' and 'dark' fringes. The latter correspond to points where the particles are less likely to end up when the two slits are open than when only one is. This strange result, which will be discussed in much more detail later on, imparts to the particle, atom or photon, a wavy character. This is strange to a classical mind because, in mundane language, wave and particle aspects are mutually exclusive. Quantum laws however make them in a way 'coexist'. This is what Bohr has called 'complementarity'. This concept is difficult to comprehend and explaining it to the layman does not go without danger. To combine waves and particles in an apparently hybrid entity seems to evoke the monsters, half-man half-beast, which appear in bas-reliefs of Romanesque churches. It may lead to the wrong idea that quantum physics is spooky and ill-defined. This idea may also be reinforced by a misunderstood interpretation of expressions such as 'quantum uncertainties' or 'quantum indeterminacy'. Nothing is more remote from quantum physics which, far from being fuzzy, gives an extraordinarily accurate description of Nature, imparting to the natural entities – nuclei, photons, atoms or molecules – a structural stability and universality which cannot be explained by classical physics.

In order to access this description, one must abandon inadequate images and immerse oneself in the mathematical framework, which is elegant and simple. The notion of state superposition, so fuzzy in classical terms, is merely associated to the linearity which ascribes to each state of a quantum system a vector in an abstract space. These vectors add up and combine according to simple rules of linear algebra. Once these rules have been defined, the theory describes without any ambiguity microscopic phenomena. The concepts of superposition, interference, complementarity and entanglement, which remain vague to the layman, are mere consequences of the mathematical formalism. A new logic, at odds with the classical one, emerges in a perfectly coherent manner. To understand it, one needs to make an effort of abstraction. Moreover, the relationship between theory and observation is less direct than in classical physics. The quantum concepts are generally veiled by decoherence phenomena. The combination of abstraction and apparent remoteness between direct observation and theory makes quantum physics difficult to teach at an elementary level. It explains also the difficulties encountered by the theory when it emerged and the psychological resistance it still has to overcome.

The quantum laws have been revealed through a difficult and long path followed through a lot of turns and pitfalls. To understand the microscopic world, physicists have carried out a complex investigation job, led by a few hints obtained by picking up pieces of a complex puzzle. The spectra of simple atoms, the spatial quantization of the atomic spin, were among the evidence of a veiled quantum world, impossible to interpret as long as physicists stuck to classical concepts. The truth has come progressively, through some brilliant intuitions. Among them are the revelation of matter waves by de Broglie in 1923 and the guess of the exclusion principle by Pauli in 1925. The mathematical formalism evoked above was then elaborated, in 1925–26, by Heisenberg, Schrödinger and Dirac.

But the formalism was not enough. To understand its implications was also essential. To do so, physicists have repeatedly used thought experiments. They imagined simple physical situations in which one important parameter or set of parameters takes an unusual value. It is chosen to make a specific effect, usually negligible because it is masked by the complication of the real world, play a dominant role. Besides this exaggeration, the 'experiment' must be, in all other respects, realistic and obey the rules of physics. In this way, the 'experimenters' try to find out whether the laws of Nature, supposed to be universal, do apply to situations in which the relevant parameters take these unusual values. By analysing the expected behaviour, they hunt for

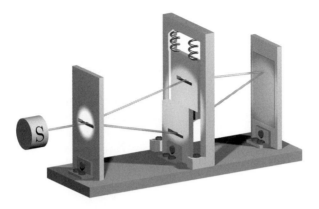

Fig. 1.6 Scheme of Bohr's thought experiment with a moving slit.

possible inconsistencies and contradictions.

Before quantum physics, the thought experiment test had already been used successfully by Einstein in the elaboration of relativity. His 'experiments' with trains going at velocities close to the speed of light, or with elevators falling in a gravitational field, belong to this category. In the fast trains he imagined, all physics was realistic. The yardsticks and the clocks were those of everyday life. Only the velocity of the carriages was unusual. Einstein analysed all the consequences of his theory and concluded that if trains were able to go that fast, strange things would happen. These strange things were, however, perfectly logical and, hence, Einstein concluded that Nature must behave in this odd way.

The quantum thought experiments conceived by Bohr and Einstein during their discussions at the Solvay meetings of 1927 and 1930 were of a different kind. They involved isolated particles, electrons, atoms or photons. The unusual parameter was the sensitivity of the experimental equipment used to manipulate and detect these particles. The sketch in Fig. 1.6, showing an electron crossing a Young double-slit apparatus, is again a drawing inspired by Bohr. The upper slit is here supposed to be light enough to recoil appreciably under the kick of the electron when it is scattered by its edges. And the stiffness of the spring to which the upper slit frame is attached is supposed to be so small that it starts oscillating with observable amplitude when the electron follows the upper path.

This apparatus was conceived for the discussion of complementarity and we will come back to it later. At this stage, let us only note that the issue was to understand the relationship between the fringe visibility and the amount of information acquired about the path of the electron by observing the recoil of the slit. At the time of the Solvay meeting, the experiment was clearly impossible in any imaginable form. However, Bohr was careful to describe the situation in otherwise realistic terms, as shown by the numerous bolts and dials he liked to represent half jokingly in his drawings.

The original blueprint of another Einstein–Bohr experiment is exposed in a window case at the Niels Bohr Institute in Copenhagen, together with a 'semi-realistic' rendition of the apparatus which Gamow, then a postdoctoral visitor in Denmark,

Fig. 1.7 The photon-box thought experiment. Left: sketch inspired by Bohr (1949). Right: model realized by Gamow. The box carries a blackboard with the formulas used by Bohr to interpret the experiment. Note, on top, the spring by which the box is suspended and the horizontal pointer on the left side which, by moving in front of a graduated ruler, yields the box weight. With permission from the Niels Bohr Archive.

made in wood and metal and offered to Bohr as a token of his admiration for his two mentors (Fig. 1.7). Here, the point was to refute a suggestion by Einstein that one could determine with arbitrary precision the emission time of a photon and its energy, in violation of one of the Heisenberg uncertainty relations. Einstein and Bohr envisioned a machine intended to measure the time of escape of a photon from a box and to determine simultaneously the photon energy, through the equivalence principle. This photon box is of course unrealistic because here again, the spring to which it is attached is, in the real world, insensitive to the gravitational mass of a single photon. Making the assumption that such a spring, or the equivalent of it, could exist, Bohr tried to analyse all the consequences in order to show the consistency of quantum theory. We will not analyse here this thought experiment, which is not devoid of difficulties. The fact that Bohr had to invoke general relativity to show the coherence of quantum physics is odd if we remark that gravitation is not yet explained by quantum theory. We mention here this problem only as an example of the semi-realistic considerations leading to the elaboration of thought experiments.

1.3 Thought experiments coming of age

At the end of the twentieth century, the status of thought experiments has changed. Whereas they mainly served as a conceptual aid to theoretical thinking, they have, by becoming real, turned into fundamental tests of the theory. Einstein and Bohr used thought experiments to show that the laws of Nature were logically coherent. Modern experimenters perform realistic versions of these experiments, which up to now have always confirmed that Nature does indeed obey these laws. These experiments have become feasible thanks to the development of modern technologies unimaginable at the time when the quantum concepts were elaborated. This technology is based on

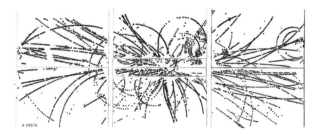

Fig. 1.8 Tracks produced by particles after an energetic collision. This picture is one of the first pieces of evidence of the W boson. © CERN. See also Arnison *et al.* (1983).

our knowledge of relativity and of the quantum, so in a deep way, our understanding of Nature is fully consistent.

Relativity thought experiments were the first to come of age. Of course trains, even the French TGV, still do not rush by platforms at about the speed of light c. Since the 1930s though, accelerators routinely prepare trains of particles travelling close to that speed in high-vacuum tunnels. The natural lifetime of some of these particles provides an accurate clock. That a particle in the accelerator lives longer than at rest has been checked with precision, yielding confirmation of Einstein's idea about the relativity of time. Similar tests have been carried out in low-energy physics too, with particles having velocities a thousand to a million times smaller than c. One takes advantage then of the extreme precision of time measurement in modern atomic physics. The time dilation effect is measured by embarking atomic clocks in planes and satellites. And such experiments could be performed in fast trains too, since the square of the ratio of their velocity to c is of the order of 10^{-12}, well within the sensitivity of atomic clocks. Relativity thought experiments have even become part of routine technology. The correction for time dilation and gravitational effects are essential to the operation of the Global Positioning System (GPS). Satellites also provide modern versions of free-falling elevators of the general relativity thought experiments. Here again relativity tests have been performed or are being planned to check various aspects of Einstein's theory.

Quantum thought experiments have also become real. The first requisite to perform them is to achieve single-particle detection sensitivity. In accelerator experiments the particles are detected by their tracks or by the energy they deposit in bolometers (Fig. 1.8 shows particle tracks following a collision event). A low-energy photon counter also discriminates the arrival of single light quanta, yielding discrete 'clicks'. During these detection events, though, the particles are destroyed as they are being observed. As remarked by Schrödinger, the particles detected in this way are fossil signatures of the past and the physicist works as a palaeontologist reconstructing past history from experimental records.

Genuine atomic or photonic thought experiments have more to them than very sensitive detection. They must be able not only to detect single particles, but to do it without destroying them. The particles should be manipulated with an exquisite sensitivity in such a way that their repeated observation should be possible, as opposed to the static observations mentioned above. Repeating the measurements gives access to correlations of various kinds. Quantum physics, in essence, describes and explains

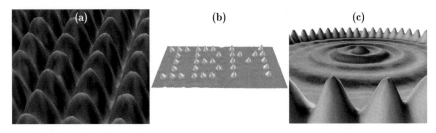

Fig. 1.9 Scanning tunnelling microscopy. (a) a scanning tunnelling microscope image of a nickel surface. Each peak corresponds to a single nickel atom. Reprinted with permission from IBM's Almaden visualization laboratory Web site www.almadem.ibm.com. (b) a famous company's logo written with xenon atoms deposited on a nickel surface. Reprinted by permission from Macmillan Publishers Ltd: Nature, Eigler and Schweizer (1990). (c) a 'quantum corral'. A ring of deposited iron atoms confine the electrons on a copper surface. The electronic stationary wave is clearly visible inside the corral. Reprinted with permission from Crommie *et al.* (1993), © AAAS.

correlations between events. It is thus essential that the thought experiments testing it have the ability to perform repeated measurements on single particles. The physicist should then be compared not to the palaeontologist but rather to the biologist operating *in vivo*.

A first example of single-atom observation and manipulation is provided by Scanning Tunnelling Microscopy (STM). It consists in translating above the surface of a solid a tiny stylus, whose tip is made of a few atoms only. A voltage difference is applied across the small gap between this tip and the surface. When the stylus is close enough, an electron current tunnels through this gap. The current intensity decreases exponentially with the tip-to-surface distance which is controlled by piezoelectrical devices. Moving the stylus across the surface while locking the tunnel current to a fixed value forces it to follow the surface at a fixed distance, visualizing each atom as a bump above the average plane. One thus reconstructs a three-dimensional atomic map resolving single atoms. The tip feels individual atoms in a process similar to Braille reading by the blind, the tiny stylus replacing the finger tip. An example of an atomic map is shown in Fig. 1.9(a).

Large electric fields applied to the tip are used, not only to read out, but to pick up atoms one by one and redeposit them in chosen locations. One can thus write artificial atomic patterns. The three letters of Fig. 1.9(b) have been drawn in this way. Written at the same scale, all the books of the British Library fit on the surface of a stamp. One can also build small enclosures made of a single-atom line, so-called 'quantum corrals'. The electronic density inside the corral area can be recorded by STM methods (Fig. 1.9c). The pattern of electronic waves constrained by the corral boundary is strikingly reminiscent of the water waves at the surface of a pond. The corral is a few tens of angströms across. This is a typical example of the astonishing achievements of nanotechnology. In these experiments, atoms can also be continuously monitored. The motion of atomic lines on surfaces under various physical or chemical processes can be observed in real time. The STM images and the ones based on a related technology, the atomic force microscope, are typical examples of experiments

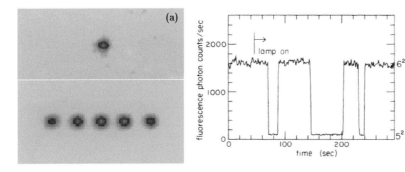

Fig. 1.10 (a) Images of calcium ions in a Paul trap. Top: a single ion. Botton: string of five ions. Reprinted from Nägerl *et al.* (1998a), with kind permission of Springer Science and Business Media and courtesy of R. Blatt. (b) Quantum jumps of a single ion in a trap. Laser-induced fluorescence of a barium ion on the $|6P\rangle \rightarrow |6S\rangle$ transition versus time. The fluorescence is suddenly interrupted by transitions from $|6S\rangle$ to the metastable $|5D\rangle$ level induced by additional lamp irradiation. Reprinted with permission from Nagourney *et al.* (1986). © American Physical Society.

which, until the 1980s, were deemed impossible by most physicists. They rely on extraordinary improvements in tip micro-fabrication and computer control as well as on a deep knowledge of quantum tunnelling properties.

Even though STM manipulations can be performed on single atoms, they are too crude to permit quantum correlation experiments. The atoms on a solid surface are strongly interacting with each other and perturbed by thermal vibrations (phonons) and other collective effects. Under these conditions, decoherence is an effective process and classical concepts can be used to understand most phenomena, even if they occur at the atomic scale. STM lacks one characteristic feature of a quantum thought experiment, an extreme conceptual simplicity. A thought experiment, to be illustrative and demonstrative, should be described by a simple model, emphasizing the physical concepts without the complications coming from the environment, which tend to veil the quantum.

This condition is realized in the quantum optics experiments which emerged at the time of the birth of STM, in the late 1970s. In one class of experiments, single atoms are observed *in vivo* and can be continuously manipulated. Instead of being closely packed and strongly interacting with each other, they are isolated, living so to speak alone in vacuum, only weakly perturbed by neighbouring particles. In another kind of quantum optics experiments, photons instead of atoms are continuously observed and manipulated, again under conditions in which the perturbation from the surrounding is kept minimal.

Single-atom manipulation experiments were first performed in ion traps. These experiments were a direct generalization to charged atoms of the pioneering studies performed by Dehmelt's group in Seattle on a single electron, which had culminated in the extremely precise measurement of the electron magnetic moment recalled above. A configuration of static and oscillating electric fields (Paul trap), or of static electric and magnetic fields (Penning trap) keeps an ion in a confined region of space, away from

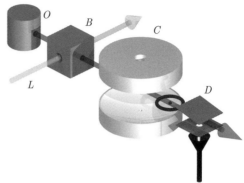

Fig. 1.11 A cavity-QED experiment in the microwave domain. Atoms effusing from the oven O are prepared in a very excited Rydberg state in box B, using laser L. They interact with the photons in the superconducting cavity C before being detected by the state-selective detector D.

all material boundaries. The restoring force acting on the charge of the ion induces its oscillations around equilibrium, with an amplitude which can be reduced down to zero-point fluctuations. Laser beams are used to manipulate the ion, to cool down its motion in the trap, and to detect it.

Observing the ion is a process quite similar to ordinary seeing. The light scattered by the particle is focused by a lens into a photon detector or simply into the eye. It appears as a small diffraction-limited spot, about a micron in diameter, of course much larger than the actual ionic size. The ion has a huge scattering cross-section for the monochromatic laser light, roughly equal to the square of the radiation wavelength. It scatters millions of photons per second from the laser beam, thousands of which are collected by the lens or the naked eye. The first direct 'seeings' of single ions were made in 1980. Figure 1.10(a) shows more recent images of an isolated Ca^+ ion (top) and of a string of five ions (bottom) confined in a trap. When many ions are present the trap confining potential competes with the Coulomb repulsion between the charged particles. If the temperature of the ions is low enough, a quasi-crystalline order is observed.

Not only can the position of the particles be detected in real time with micrometre resolution, but their internal evolution can also be monitored. The ion has electronic levels and the quantum jumps between them can be observed using laser light again, as shown in Fig. 1.10(b). The ion scatters light between two levels separated by a transition resonant with the laser. If a quantum jump induced by a lamp brings the ion from one of these two levels into a third one, the resonant scattering is suddenly interrupted and the ion becomes invisible. The scattered light comes back as suddenly as it vanishes when another spontaneous quantum jump brings the ion back into one of the two levels pertaining to the laser-resonant transition. These jumps appear as a random blinking analogous to the twinkling of a faint star. The telegraphic signals of single ions in traps have become routine in atomic clocks and other high-resolution spectroscopy experiments. These sudden jumps are a direct quantum manifestation, which Schrödinger believed could not be observed. They are a typical example of a thought experiment becoming real.

In another kind of experiment, dual in some way of ion traps, the roles of matter and radiation are exchanged. It is the field instead of the atom which is confined. Photons are stored in a cavity with highly reflecting walls and atoms, sent one by one

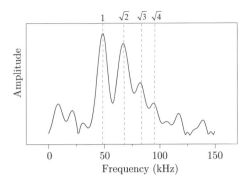

Fig. 1.12 Fourier transform of the time-dependent atomic signal detected after the resonant interaction of a single Rydberg atom with a small field stored in a superconducting cavity. The discrete peaks at frequencies proportional to the square root of successive integers reveal directly the graininess of the photon field. See Chapter 5 for more explanations. Reprinted with permission from Brune *et al.* (1996b). © American Physical Society.

across the cavity, interact with them. These experiments belong to Cavity Quantum Electrodynamics, 'CQED' for short, a subfield of quantum optics which emerged in the 1970s. Its primary goal was to study how the radiative properties of atoms are modified when they radiate close to boundaries. CQED experiments are performed with optical or microwave fields. In one microwave version (Fig. 1.11), photons with wavelength in the millimetre range are bouncing between two mirrors made of highly reflecting superconducting material. A large matter–field coupling is obtained with atoms prepared in very excited Rydberg states, which have a huge electric dipole and interact strongly with microwaves. The Rydberg atoms leaving the cavity are detected by a state-selective field-ionization detector and the atomic signal, a function of the atom–photon interaction time, is used to get information about the cavity field evolution.

Figure 1.12 shows the Fourier transform of such a signal obtained in an experiment described and analysed in a forthcoming chapter. Suffice it to say at this stage that the discrete frequency peaks directly reveal the graininess of the photon field. Each Fourier component corresponds to a photon number, 0, 1, 2 or 3 in this case. The experiment is clear evidence of field quantization in a box, the first quantum effect guessed by Planck, which turned out to be the starting point of the quantum revolution.

These CQED experiments are modern renditions of the Einstein–Bohr photon box. The photons are not directly weighed by a pointer attached to a scale, as was assumed by Bohr, but rather by the atoms which interact with them and carry away, imprinted on their own quantum state, information about the field. In this way, photons can be counted without being destroyed, as one could count marbles in a box. Or the cavity field can be prepared into a strange superposition state, displaying at the same time two different phases or amplitudes. Such a situation reminds us of the famous cat that Schrödinger assumed, in another celebrated thought experiment, to be coherently suspended between life and death.

A striking feature of the ion and photon trap experiments is the simplicity of the model accounting for them. Basically, they can be understood by describing the field and the atoms as interacting quantum oscillators and spins, the most basic quantum systems. This is obvious in the case of CQED. Only the two atomic levels pertaining to the transition resonant with the cavity are relevant and the atom then behaves to a good approximation as a two-level spin-$1/2$-like system. The field in the cavity mode is a quantum oscillator whose elementary excitations are the photons. The coupling between the spin and the oscillator is described by a very simple Hamiltonian, introduced by Jaynes and Cummings in the early days of quantum optics.

The spin and quantum oscillator ingredients are also present in the ion traps. In order to understand the scattering of laser light by an ion, it is convenient to describe it as a two-level system interacting with a classical field. When the trapped ion is moving, its coupling to the laser light is modified by the Doppler shift, modulated by the ion's oscillation. The laser light then couples the electronic degree of freedom of the ion, seen as a two-level spin-like system, to its centre-of-mass motion, viewed as a mechanical quantum oscillator. This coupling is again, under convenient approximations, described by a Jaynes–Cummings Hamiltonian. In both situations, the evolution of spins and oscillators coupled together is fully analytical. The system's evolution is simple enough so that the quantum laws manifest themselves without inessential complications.

We have neglected so far the perturbing effect of the environment, responsible for decoherence. The thought experiments with ions and photons require keeping these perturbations to a minimum. This is achieved by insulating the trapped particles from other atoms or photons with a combination of techniques: high vacuum to reduce the rate of collisions with background molecules, low temperature to minimize thermal photons, highly reflecting cavity mirrors to keep photon loss as small as possible. For a realistic description of the experiment, the residual effect of the environment must be accounted for. Here again, the spin–oscillator model is helpful. The ion spontaneous emission is due to its coupling to a continuum of field modes described as quantum oscillators. Similarly, photons in a cavity can be modelled as escaping in a continuum of outside modes, each of which is also a quantum oscillator.

1.4   Aims and outline of this book

Experiments which display quantum features and can be modelled by spins and oscillators are more than textbook illustration of quantum theory. They are also an ideal testing ground for quantum information science. A two-level atom acts as a qubit, and the same is true of a quantum oscillator if its evolution is restricted to the ground and first excited states (no more than one photon or phonon). Thought experiments with atoms and photons in boxes or with ions in traps have thus naturally evolved into demonstrations of quantum information procedures. Engineering atom–atom or atom–photon entanglement in a deterministic way, realizing quantum logic operations of various kinds and performing simple quantum algorithms has become a flourishing field of research. One of our goals is to describe some of these experiments, with a balanced theoretical and experimental approach.

As a general rule, the simpler the theoretical model the experiment can be reduced to, the more complex the experimental procedure. A kind of 'complementarity' rule can be formulated. The product of the theoretical and practical complexity of thought experiments becoming real has to be greater and, unfortunately, often much greater than some fixed value. If one tries to force the atoms or ions to behave as exact two-level objects and the photons or phonons to be perfect excitations of ideal oscillators, one has to struggle hard with experimental difficulties which, in real life, conspire to blur these simple pictures. If one accepts dealing with complications coming from the environment, the experiment becomes easier, but the theory has to include subtle refinements, and becomes more intricate and less directly related to fundamental concepts.

In practice, a compromise has to be reached and the physicist must deal with some level of complexity, both in theory and in experiment. To do justice to this field of research, it is necessary to detail somewhat these two complementary aspects. This is a challenging task, which must avoid the pitfalls of being either too sketchy on experimental details, giving the impression that one is again merely imagining a virtual experiment, or being too technical, thus burying the physics under engineering precision. We have chosen here the middle ground between these two dangers. In this way, we hope to convey the beauty and the amazement of the physics and to give, at the same time, an understanding of the difficulties which must be overcome to unveil the quantum, and hopefully to harness it for achieving information science goals.

Another challenge is to choose between the many experiments, whose number kept rapidly increasing as the writing of this book progressed. Here again, a decision had to be made between detailing some selected experiments or else making an exhaustive catalogue, at the price of being sketchy on details. In order to be consistent with our middle-ground approach, which requires a precise discussion of both experiment and theory, we have adopted the selective approach.

We have chosen to focus primarily on microwave CQED photon-box experiments. One reason for this choice is of course that these are the ones we know best, since they belong to our own field of research. A more objective argument is that the Jaynes–Cummings model directly corresponds to the microwave CQED situation. At the time it was introduced, it was a gross simplification of practical situations in laser physics. In CQED, it is however the quasi-exact description of the atom–field coupling. The evolution of the atom–field system in a high-Q cavity can thus be described from first principles and with precision. The unavoidable decoherence is also well-understood. It is due to the damping of the field in the cavity and to the spontaneous emission of the atoms. Those are well-known relaxation processes, which can be described accurately and quantitatively.

The understanding of microwave CQED experiments can also be very useful to analyse experiments belonging to other fields of quantum optics. The similarities of the Hamiltonians provide, for instance, a natural link between CQED and ion trap experiments. We use this link to describe some of these experiments too, even though it will be done in a less detailed way. An incursion will finally be made into a novel and very active field, the physics of Bose–Einstein condensates, to which some of the concepts of CQED can also be applied.

This book is also intended to give the reader an opportunity to learn and reflect about the quantum. We will assume that (s)he is familiar with the basics of this physics, but might not have had the opportunity to think too much about all its conceptual aspects. The strange features of the quantum are often presented in negative terms (uncertainty relations, impossibility theorems such as the 'no-cloning' one, non-separability, non-locality and so on). Quantum information is trying to turn these negative points into pluses, showing that, at least in principle, they should allow one to do more than what is possible with classical physics. Learning about thought experiments coming of age should be a good opportunity to think about these somewhat paradoxical issues and to learn about the nascent quantum information science. At a more basic level, the experiments we describe here provide interesting problems for intermediate and advanced quantum physics courses. Tools such as the density operator and its master equation, the Monte Carlo approach to relaxation processes and the Wigner function representation will be described in detail. We hope that this volume will thus be useful to teaching in advanced quantum mechanics.

We start in Chapter 2 with a discussion of the quantum ideas incompatible with a classical description of Nature. This exploration of the strange microscopic world emphasizes the quantum concepts which are illustrated by the experiments described in the following chapters. The principle of state superposition and its consequences are analysed. The ubiquitous phenomenon of quantum interference is described in the light of the complementarity principle. We discuss the symmetrization postulate for the description of identical particles. We then analyse entanglement and non-locality, as it is revealed by Bell's inequality experiments. We next show how decoherence contributes to explaining the quantum–classical boundary and to shed light on the important issue of measurement in quantum physics. The chapter ends by a review of quantum information, which introduces the principles of quantum cryptography, teleportation and quantum computing.

Chapter 3 is a short self-consistent introduction to quantum optics, viewed as a story about field oscillators (or 'springs') coherently coupled with two-level atoms ('spins'). We recall the quantum description of a harmonic oscillator and introduce photon creation and annihilation operators and field quadratures. The description of a coherent state as the closest approximation to a classical field is recalled. We then analyse the coupling of two oscillators, which provides a simple model for the beam-splitter, a basic tool in quantum optics which we will encounter time and again in this book. We next recall the properties of two-level systems seen as pseudo-spins, simple models for atoms interacting with a field mode. We describe how these spins are manipulated and prepared in arbitrary quantum states by classical radiation pulses, according to methods borrowed from NMR. We show how a sequence of pulses realizes a Ramsey interferometer, an essential tool in many of the experiments described later on. After introducing the Jaynes–Cummings Hamiltonian describing the coherent interaction of an atom with a field mode, we analyse the resonant exchange of energy between the spin and the spring described by this Hamiltonian, as well as their dispersive interaction when they are off-resonant. These resonant and dispersive couplings play essential roles in the experiments described in subsequent chapters.

We will have been dealing so far with ideal spins and springs, evolving in splendid isolation. The atoms and fields of real experiments, though, are 'open quantum systems', unavoidably interacting with their environment. This leads to irreversible dissipative processes, which we analyse in Chapter 4. An open system is described by its density operator, a concept which replaces the wave function of elementary quantum physics. Under reasonable assumptions, this operator obeys a 'master equation' which describes the system's damping under the perturbing effect of the environment.

Using a method inspired from quantum information, we derive the master equation by analysing how the open system loses information through its entanglement with its surroundings. This equation can be cast in a form making explicit the elementary 'quantum jumps' experienced by the system as it keeps losing information. We solve the master equation by the powerful Monte Carlo method, imagining thought experiments in which the environment is continuously monitored, thus detecting the quantum jumps undergone by the system. We illustrate these general ideas by describing the spontaneous emission of a two-level atom and the relaxation of a field stored in a cavity. With this master equation formalism at hand, we revisit the spin–spring system introduced in Chapter 3 and analyse the effect of cavity damping on the coherent atom–field evolution. We describe a damped cavity field interacting with a stream of atoms crossing one by one the cavity, a device known as a 'micromaser'. Finally, we consider the collective coupling of many atoms to a damped cavity mode, which leads us to say a few words about the superradiance phenomenon.

Chapter 5 then presents the essentials of cavity QED. After a brief history of the subject, we present the main ingredients of a microwave cavity QED experiment, the Rydberg atoms and the superconducting cavity. We then describe the basic methods used in these experiments. Ramsey interferometry performed by subjecting the atoms to classical pulses before and after they cross the cavity is shown to yield precise information about the atom–cavity coupling. An experiment in which an atom resonantly exchanges energy with the field, undergoing a 'Rabi oscillation', is described. This experiment yields in particular the Fourier transform signals mentioned above (Fig. 1.12), which clearly demonstrate the photon 'graininess' of the field in the cavity. The use of the Rabi oscillation as a tool to entangle atoms to photons and atoms to atoms is described in detail, as well as coherent cavity-assisted collisions in which two atoms are entangled via their common interaction with the cavity field.

Chapter 6 is devoted to experiments performed with a few atoms and photons in a cavity. We show how the photon-box experiment can be turned into a direct test of complementarity, very close in principle to the original Bohr thought experiment with a moving slit (Fig. 1.6). We also show how the cavity photons can be measured and continuously monitored without being destroyed, a striking illustration at the single-particle level of what is called a 'quantum non-demolition' measuring process. We present a simple version of this procedure applied to a field containing at most one photon. The measurement then reduces to a determination of the photon number parity and can be viewed as the operation of a 'quantum gate' coupling conditionally a photonic and an atomic qubit. We show that, in general, this parity measurement can be turned into a direct experimental determination of the field Wigner distribution, yielding a complete description of the non-classical field features.

Chapter 7 is about field state superpositions involving many photons in a cavity. After a brief reminder about the Schrödinger cat problem in quantum optics, we describe how a single atom interacting with a field made of many photons can leave its quantum imprint on this field, bringing it into a superposition of states with distinct classical attributes. We describe the preparation and detection of these 'Schrödinger cat' states. We present simple decoherence models, which account for their extreme fragility. We describe experimental studies of decoherence, which constitute direct explorations of the quantum–classical boundary. We discuss limitations to the size of these cat states and ways to protect them efficiently against decoherence. Finally, we describe proposals to generate and study non-local cats, superpositions of field states delocalised in two cavities.

In the last two chapters, we leave the photon box and analyse other systems bearing strong similarities with it. Chapter 8 is devoted to ion trap physics. We show how ions can be individually confined, manipulated and detected. We then present the Hamiltonian describing the interaction of a trapped ion with laser beams resonant or nearly-resonant with an electronic transition of the ion. We compare it with the Jaynes–Cummings Hamiltonian of CQED. We describe experiments which are the ion-trap version of the Rabi oscillations observed in a photon box. We show how laser beams can be tailored to create artificial environments for the ion vibration, making it possible to generate and study various states of motion. We finally describe quantum logic operations in which ions act as qubits and we show how these operations can be combined to perform simple quantum algorithms and to build entangled superpositions of several ions. We conclude this chapter by discussing proposals to implement logic at a larger scale with trapped ions.

The experiments described so far will have dealt with systems in which the complexity is built 'from the bottom', starting by the control of individual particles and learning how to manipulate more and more of them in a coherent way. One can start instead from a large system and try, by cooling it down, to reduce or suppress decoherence, making the quantum behaviour appear 'from the top'. This is the aim of mesoscopic physics, a very active domain of condensed matter physics. We cannot do justice to this wide field, but Chapter 9, dealing with Bose–Einstein condensates, makes a brief incursion into it. Bose–Einstein condensation is indeed a domain of research in which atomic and condensed matter physics meet. Ultra-cold atomic samples confined in a magnetic trap behave as giant matter waves bearing strong similarities with laser or microwave light fields. The collective behaviour of atoms in these matter waves is also reminiscent of the physics of superfluidity or superconductivity. Many analogies relate the atoms in a Bose–Einstein condensate either to photons in a cavity, or to particles in a liquid or a solid. Some of the ideas of CQED can be generalized and adapted. Quantum logic operations can be performed with Bose–Einstein condensates trapped in optical periodic potentials. Condensates involving two or more different matter wave modes can be combined by using tools reminiscent of the beam-splitter of quantum optics or the Josephson junction of solid state physics. Mesoscopic superpositions of matter waves containing large numbers of atoms could be prepared and studied. These matter Schrödinger cats bear strong similarities with their photonic cousins of CQED.

The book ends with a short conclusion and a technical appendix about the description of quantum states in phase space. In the conclusion, we evoke future prospects and mentioning the challenges ahead. How far will the industry of thought experiments be carried and the quantum classical boundary pushed back? What are the odds to beat decoherence and make quantum information practical? What are the best systems to achieve these goals? Should they be reached from bottom-up, as in atomic physics, or from top-down, as in condensed matter physics? These are some of the open issues we briefly touch on in these final remarks.

2
Strangeness and power of the quantum

No one really understands quantum physics

R. Feynman

The epigraph to this chapter underlines the counter-intuitive nature of the laws which rule the world at the microscopic scale. Paradoxically, these laws are expressed by a mathematical formalism that undergraduate students can easily comprehend. Physicists have been using this formalism successfully without generally attempting to 'understand it' in the way that Feynman had in mind. For a long time it was not well-considered to ponder about the weirdness of the quantum. Those who ventured into interpretations of the quantum postulates were often deemed to be lost for science. This perception has changed with the development of quantum information and the advent of 'thought experiments' manipulating single particles in the laboratory. The strangeness of physics at the microscopic scale plays a central role in this new domain, since its main goal is to take advantage of this strangeness for novel tasks in communication and computing. Rather than trying to 'understand' the quantum weirdness, physicists now work at defining and quantifying it. They study, theoretically and experimentally, entanglement, non-locality and decoherence in systems of increasing complexity. Meanwhile, they acquire a kind of operational familiarity with these concepts. They develop a new kind of intuition, which allows them to guess the result of an experiment before performing it or even simulating it by a calculation. Whether such quantum intuition is different from 'understanding' is a question which we leave to philosophers.

The aim of this chapter is to describe the non-classical aspects of the quantum theory, which are essential in the experiments described later on. It is destined primarily for readers familiar with the quantum formalism, but who might have learned it in textbooks which mainly apply this formalism to solve practical problems, without dwelling too much on conceptual discussions. In order to avoid the pitfalls of everyday language, we will base from the outset our discussion on the supposedly known framework of the theory. Only after the mathematics has been made explicit, will we discuss the concepts in qualitative terms.

The strangeness of the quantum can be traced back to the superposition principle rooted in the linearity of quantum theory. We recall in the first section (2.1) how this principle conditions the properties of the wave function of a quantum particle. We then discuss the ubiquitous quantum interference phenomenon and its relations with complementarity (Section 2.2). The next section (2.3) is about the subtleties of the superposition principle when applied to a system of identical particles. We show that the very notion of identity leads to deep consequences in the quantum realm and we give a brief description of fermionic and bosonic systems. The last three sections relate the superposition principle to entanglement, non-locality and quantum information, our central topics. We show in Section 2.4 that entanglement is a direct consequence of the superposition principle applied to composite systems. We analyse the non-local nature of quantum correlations in an entangled bipartite system and relate entanglement to complementarity. We then discuss the quantum–classical boundary (Section 2.5) and analyse the decoherence phenomenon. This leads us to describe a quantum measurement as an experiment coupling a microscopic system to a macroscopic meter, itself coupled to a large environment. We analyse how decoherence enforces the 'classicality' of the meter by privileging special 'pointer' states impervious to entanglement with the environment. Finally, in Section 2.6, we show how simple quantum systems can be used as elementary qubits for information processing. As we will see, superpositions and entanglement in multi-qubit systems could be exploited to develop novel information processing operations such as quantum key distribution in cryptography, teleportation of the wave function of a particle and quantum computing.

2.1 The superposition principle and the wave function

Let us start by recalling briefly the general framework of quantum theory. Each state of a microscopic system A is represented by a vector in an abstract Hilbert space $\mathcal{H}_A$ and the physical observables of this system are associated to hermitian (self-adjoint) operators in $\mathcal{H}_A$.[1] The linear combination and scalar product of state vectors as well as the operator algebra in $\mathcal{H}_A$ are defined in all quantum mechanics textbooks (Dirac 1958; Messiah 1961; Cohen-Tannoudji et al. 1977; Sakurai 1994). Adopting the compact and elegant Dirac formalism, we represent the state vectors by the standard bra–ket notation, inserting inside the bra $\langle \, |$ or ket $| \, \rangle$ a symbol defining the state.

The description of the most general state $|\psi\rangle$ of A requires the definition of a reference basis $\{|i\rangle\}$ in $\mathcal{H}_A$, obeying the orthogonality and closure relationships:

$$\langle i \, | j \rangle = \delta_{ij} \; ; \qquad \sum_i |i\rangle \langle i| = \mathbb{1} \, , \tag{2.1}$$

where δ_{ij} is the usual Kroneker symbol and $\mathbb{1}$ the unity operator in $\mathcal{H}_A$. The basis states are the eigenstates of a complete ensemble of commuting observables $O_1, O_2, \ldots,$ O_k which define a 'representation' in $\mathcal{H}_A$. Once the representation basis is known, any state $|\psi\rangle$ of A is developed as:[2]

---

[1]We reserve in this section capital letters to describe observables (such as $\mathbf{R}$, $\mathbf{P}$, $\mathbf{S}$, H for position, momentum, spin and energy operators) and lower case letters to describe eigenvalues of observables and classical quantities. Bold symbols represent three-dimensional vectors.

[2]We start for simplicity by considering cases in which the complete set of observables has a discrete spectrum. Extension of the formalism to continuous spectra is recalled below.

$$|\psi\rangle = \mathbb{1} \, |\psi\rangle = \sum_i |i\rangle \, \langle i \, |\psi\rangle \ , \tag{2.2}$$

a linear combination of basis states, entirely defined by the list of C-number coefficients $c_i = \langle i \, |\psi\rangle$.

A measurement of the complete ensemble $\{O_k\}$ randomly projects $|\psi\rangle$ into one of the $|i\rangle$ states with the probability $p_i = |c_i|^2$. The $\langle i \, |\psi\rangle$ scalar product coefficients are thus called 'probability amplitudes'. The normalization of the state ($\langle \psi \, |\psi\rangle = 1$) ensures that the total probability of all measurement outcomes is equal to 1. Immediately after the measurement, the system's state is irreversibly changed, 'jumping' from $|\psi\rangle$ into one of the $|i\rangle$'s. Repeating the measurement immediately afterwards (i.e. before the system has had time to evolve) leaves A with unit probability in the same $|i\rangle$ state. At this stage, we just enunciate the postulates of the quantum theory of measurement. We will come back at many places to the description of measurement procedures, which includes a definition of a measuring apparatus and of its coupling with A. We will then try to sharpen our understanding of the irreversible evolution of a quantum system upon measurement, certainly the most difficult aspect of quantum theory.

In everyday language, eqn. (2.2) can be loosely expressed by saying that *if a system can exist in different configurations (corresponding for example to different classical descriptions), it can also exist in a superposition of these configurations, so to speak 'suspended' between them.* This layman's language is imprecise though, while the mathematical formula (2.2) is unambiguous.

2.1.1 The spin model for a two-level system

As the simplest example of quantum object, let us start by considering a two-state system, which we will describe, without loss of generality, as a 'pseudo-spin' **S**. This analogy will lead us to a geometrical representation of the system, widely used in NMR physics, which will prove very useful. The component of this spin along an arbitrary direction in three-dimensional space can take only one of the two values $\pm\hbar/2$ ($\hbar$ is Planck's constant divided by 2π). The most general observable of this system can be expressed as a linear combination with real coefficients of the 2×2 unity operator $\mathbb{1}$ and of the three Pauli operators $\sigma_i = 2S_i/\hbar$ ($i = X, Y, Z$), which in the basis of the σ_Z eigenstates, are:

$$\sigma_X = \begin{pmatrix} 0 & 1 \\ 1 & 0 \end{pmatrix}; \quad \sigma_Y = \begin{pmatrix} 0 & -i \\ i & 0 \end{pmatrix}; \quad \sigma_Z = \begin{pmatrix} 1 & 0 \\ 0 & -1 \end{pmatrix}. \tag{2.3}$$

These operators satisfy the commutation rules:

$$[\sigma_i, \sigma_j] = 2i\epsilon_{ijk}\sigma_k \ , \tag{2.4}$$

where ϵ_{ijk} is zero if two of the indices are equal and $+1$ or -1, depending on the parity of the permutation of ijk when they are all different.

The eigenvalues of σ_i are ± 1. We will call $|0\rangle$ and $|1\rangle$ the eigenstates of σ_Z with eigenvalues $+1$ and -1 respectively. In the spin language, they correspond to the 'up' and 'down' states along the Z axis. The $0, 1$ notation adopted here is more in the

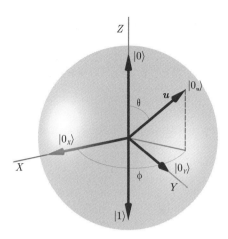

Fig. 2.1 Representation of the Hilbert space of a two-level system on the Bloch sphere. The $|0\rangle$ and $|1\rangle$ states correspond to the 'north' and 'south' poles respectively. The $|0_X\rangle$ and $|0_Y\rangle$ states are mirrored on the 'equator' of the sphere, at its intersection with the X and Y axes. The most general spin state corresponds to a point on the sphere of angular coordinates θ and ϕ.

spirit of quantum information. The eigenstates of σ_X are the symmetric and anti-symmetric linear combinations $|0/1_X\rangle = (|0\rangle \pm |1\rangle)/\sqrt{2}$ whereas the eigenstates of σ_Y are $|0/1_Y\rangle = (|0\rangle \pm i\,|1\rangle)\sqrt{2}$.

The most general traceless observable of the two-level system corresponds to a spin component along the direction defined by the unit vector $\mathbf{u}$ with polar angles θ and ϕ (Fig. 2.1). It is simply expressed in terms of the Pauli matrices as:

$$\sigma_{\mathbf{u}} = \cos\theta\,\sigma_Z + \sin\theta\cos\phi\,\sigma_X + \sin\theta\sin\phi\,\sigma_Y = \begin{pmatrix} \cos\theta & \sin\theta e^{-i\phi} \\ \sin\theta e^{i\phi} & -\cos\theta \end{pmatrix}. \quad (2.5)$$

The observable $\sigma_{\mathbf{u}}$ has also a ± 1 spectrum, with the eigenstates:

$$\begin{aligned} |0_{\mathbf{u}}\rangle &= \cos(\theta/2)\,|0\rangle + \sin(\theta/2)e^{i\phi}\,|1\rangle \;; \\ |1_{\mathbf{u}}\rangle &= \sin(\theta/2)\,|0\rangle - \cos(\theta/2)e^{i\phi}\,|1\rangle \;. \end{aligned} \quad (2.6)$$

When the unit vector $\mathbf{u}$ rotates in space, $|0_{\mathbf{u}}\rangle$ explores the entire Hilbert space of the pseudo-spin. The most general spin state, $c_0\,|0\rangle + c_1\,|1\rangle$, is indeed the eigenstate with eigenvalue $+1$ of the spin component along the direction $\mathbf{u}$ of polar angles θ and ϕ defined by the relation $\tan(\theta/2)e^{i\phi} = c_1/c_0$. The tip of this vector belongs to the sphere of radius unity, called the Bloch sphere in honour of the pioneer of NMR (Fig. 2.1). In other words, the Hilbert space of a two-level system is 'mirrored' onto the Bloch sphere, each of its points representing a possible superposition of $|0\rangle$ and $|1\rangle$. These basis states are respectively imaged on the 'north' and 'south' poles. The $|0_X\rangle$ and $|0_Y\rangle$ states are associated to points on the X and Y axes, on the Bloch sphere's equator. The state $|1_{\mathbf{u}}\rangle$, orthogonal to $|0_{\mathbf{u}}\rangle$ in Hilbert space, corresponds to the point along the direction $-\mathbf{u}$. The identity $|1_{\mathbf{u}}\rangle = |0_{-\mathbf{u}}\rangle$ can easily be checked by inspection of eqn. (2.6). Two orthogonal states in Hilbert space are thus associated to antipodes on the Bloch sphere.

The probability of finding $+1$ when measuring σ_Z on $|0_\mathbf{u}\rangle$ is $\cos^2(\theta/2)$. The projection postulate implies that $\cos^2(\theta/2)$ is also the conditional probability of finding the same result $(+1,+1)$ or $(-1,-1)$ when measuring successively two spin components along directions making the angle θ.

2.1.2 Wave functions in position and momentum spaces

We have so far considered representations associated to observables with a discrete spectrum. The case of continuous representations must be given special attention, because it leads to the particle's wave function, an essential concept in non-relativistic quantum mechanics. The state $|\psi\rangle$ of a point-like particle such as an electron (we disregard here its spin) is naturally expanded onto the basis made of the $|\mathbf{r}\rangle$'s, eigenstates of the particle's position vector operator $\mathbf{R}$ (whose components are the three commuting coordinate operators X, Y and Z). These continuous basis 'states'[3] satisfy relations generalizing (2.1):

$$\langle \mathbf{r}\,|\mathbf{r}'\rangle = \delta(\mathbf{r} - \mathbf{r}') \; ; \qquad \int d^3\,\mathbf{r}\,|\mathbf{r}\rangle\,\langle \mathbf{r}| = \mathbb{1} \; , \qquad (2.7)$$

where $\delta(\mathbf{r} - \mathbf{r}')$ is the three-dimensional Dirac function. Developing $|\psi\rangle$ along the $|\mathbf{r}\rangle$ basis yields:

$$|\psi\rangle = \int d^3\,\mathbf{r}\,|\mathbf{r}\rangle\,\langle \mathbf{r}|\psi\rangle \; , \qquad (2.8)$$

where:

$$\psi(\mathbf{r}) = \langle \mathbf{r}|\psi\rangle \; , \qquad (2.9)$$

is, by definition, the wave function of the particle. The shape of the wave function depends upon the physical situation under consideration. For an electron in an atom, the wave function is localized in a small volume whose dimension is typically of the order of the Bohr radius $a_0 = 0.53$ angströms. For an electron in the periodic potential of a perfect crystal, the wave function extends over the whole macroscopic dimension of the lattice.

In layman's language, we may say that the wave function describes the state of the particle suspended, before measurement, in a continuous superposition of an infinite number of possible positions. Upon measurement of $\mathbf{R}$ performed with a linear precision δr, this superposition 'collapses' into a small wave packet of volume $(\delta r)^3$ around a random position $\mathbf{r}$, with the probability $|\langle \mathbf{r}\,|\psi\rangle|^2(\delta r)^3$. This collapse is quasi-instantaneous, in fact as fast as the physical process measuring the particle position can be. It suddenly transforms a wave function which might be arbitrarily extended

[3]We must use the word 'state' here with caution. The eigenstates of the continuous spectrum of an observable are rigorously defined only as elements of the dual of $\mathcal{H}_A$, i.e. as bras. As in most textbooks, we nevertheless use the ket notation $|\mathbf{r}\rangle$ to represent non-normalized position 'states'. They can be considered as the limit of small wave packets whose amplitude is non-zero in a volume tending towards 0. This mathematical subtlety does not affect the physics discussed here. *In fine*, the particle position appears in physical formulas as an argument of its wave function, in expressions such as $\langle \mathbf{r}\,|\psi\rangle$ where only the unambiguously defined bra $\langle \mathbf{r}|$ is relevant.

into a point-like wave packet. This process already gives us a hint about the non-local character of quantum laws.

The **R** representation is suited for measurements of the particle's position. If one is interested in the particle's momentum **P** or velocity $\mathbf{V} = \mathbf{P}/m$ (m is the particle's mass) it is appropriate to choose the **P** representation and to expand $|\psi\rangle$ over the continuous basis of the momentum eigenstates[4] $|\mathbf{p}\rangle$. Each of these states is itself a superposition of position states:

$$|\mathbf{p}\rangle = \int d^3 \mathbf{r}\, |\mathbf{r}\rangle\, \langle \mathbf{r}|\mathbf{p}\rangle = \frac{1}{(2\pi\hbar)^{3/2}} \int d^3\mathbf{r}\, \exp(i\mathbf{p}\cdot\mathbf{r}/\hbar)\, |\mathbf{r}\rangle \ . \tag{2.10}$$

The momentum eigenstates are plane waves in the **R** representation. This can be considered as a quantum postulate, a direct consequence of de Broglie's hypothesis which associates to a particle of momentum p a plane wave of wavelength $\lambda = h/p$. Alternatively, eqn. (2.10) can be derived from the commutation relations between position and momentum which we will recall shortly. Once the expansion of the $|\mathbf{p}\rangle$'s over the $|\mathbf{r}\rangle$'s is known, going from the **R** to the **P** representation amounts to a basis change. We obtain after a simple exercise on Dirac's notation:

$$|\psi\rangle = \int d^3\mathbf{p}\, |\mathbf{p}\rangle\, \langle \mathbf{p}|\psi\rangle \ , \tag{2.11}$$

with:

$$\widetilde{\psi}(\mathbf{p}) = \langle \mathbf{p}|\psi\rangle = \int d^3\mathbf{r}\langle \mathbf{p}|\mathbf{r}\rangle\langle \mathbf{r}|\psi\rangle = \frac{1}{(2\pi\hbar)^{3/2}} \int d^3\mathbf{r}\, e^{-i\mathbf{p}\cdot\mathbf{r}/\hbar}\psi(\mathbf{r}) \ . \tag{2.12}$$

The probability amplitudes $\psi(\mathbf{r})$ and $\widetilde{\psi}(\mathbf{p})$ are Fourier transforms of each other. The **P**-representation directly yields the density probability $|\widetilde{\psi}(\mathbf{p})|^2$ for a momentum measurement. Immediately after such a measurement, the particle collapses into a momentum eigenstate whose wave function in the **R** representation is the non-normalized plane wave $e^{i\mathbf{p}\mathbf{r}/\hbar}$.

Before A is coupled to any measuring device, eqns. (2.8) and (2.11) are equivalent expressions of $|\psi\rangle$. Using again the language of everyday life, we may say that the particle can be equivalently considered as 'suspended' either among a large number of positions (eqn. 2.8) or among many momentum plane waves (eqn. 2.11). These superpositions thus describe different 'potentialities' which depend upon the kind of measurement we will decide to perform (position or momentum). It is important to understand that these potential outcomes of **R** or **P** measurements have no physical reality before the coupling of A with the corresponding measuring apparatus. We come back to this point later when we have sharpened our definitions of reality and locality in physics.

The probability amplitudes $\psi(\mathbf{r})$ or $\widetilde{\psi}(\mathbf{p})$ define two equivalent tables of C-numbers, each of which containing all information needed to compute the statistics of the outcomes of a specific measurement on $|\psi\rangle$. The wave function is thus a kind of 'quantum

---

[4]The $|\mathbf{p}\rangle$'s, as the $|\mathbf{r}\rangle$'s, are not *stricto sensu* states of $\mathcal{H}_A$, since plane waves are obviously non-normalizable. Only the bras $\langle\mathbf{p}|$ belonging to the dual of $\mathcal{H}_A$ are mathematically defined. It is however possible to consider the $|\mathbf{p}\rangle$ states as the limits of wave packets whose spatial extension goes to infinity. See the related note on page 29 concerning the $|\mathbf{r}\rangle$'s.

identity card' for the system, which can be written in as many different – fully equivalent – languages as there are different representations in $\mathcal{H}_A$.

2.1.3 Heisenberg uncertainty relations

Different components ($i \neq j$) of a particle's position and momentum R_i and P_j commute, while the components of these operators along the same spatial direction are 'conjugate observables' whose commutator is equal to $i\hbar$ times unity:

$$[R_i, P_j] = i\hbar \delta_{ij} \mathbb{1} . \tag{2.13}$$

This special conjugation relation can be considered either as a consequence of the de Broglie matter wave hypothesis or accepted as one of the quantization postulates. In the latter case, it can be used to demonstrate that an eigenstate of $\mathbf{R}$ expands over an infinite number of eigenstates of $\mathbf{P}$, and vice versa (eqn. 2.10). As a direct consequence, a precise measurement of one of the $\mathbf{R}$ coordinates, say X, transforms $|\Psi\rangle$ into a wide superposition of eigenstates of P_x, resulting in large fluctuations in the outcome of a subsequent measurement of the momentum. The more precise the measurement of X, the wider the distribution of the resulting $|\mathbf{p}\rangle$ states, according to the Fourier transformation features. This property can be summarized by the famous uncertainty relation, directly derived from eqn. (2.13):

$$\Delta X \, \Delta P_x \geq \hbar/2 . \tag{2.14}$$

It relates the precisions ΔX and ΔP_x with which conjugate position and momentum components can be determined in a quantum system. The unavoidable perturbations produced by the measurements 'conspire', so to speak, to make it impossible to achieve a product of precisions on the two conjugate observables smaller than $\hbar/2$.

Similar considerations apply to observables with a discrete spectrum, such as the spin components of a particle. The $|0_\mathbf{u}\rangle$ and $|1_\mathbf{u}\rangle$ eigenstates of the $\mathbf{u}$ component of a spin are linear superpositions of eigenstates of spin components in a direction making an angle with $\mathbf{u}$ [such as the eigenstates of σ_Z, see eqn. (2.6)]. A given spin state can be expressed in infinitely many ways depending on the chosen quantization axis. Each is adapted to the prediction of the measurements for a specific orientation of the apparatus measuring the spin (a Stern–Gerlach magnet for instance). As the X and P_x observables, the spin components along two orthogonal directions are non-commuting operators[5] (eqn. 2.4) which cannot be simultaneously measured with precision. If one is perfectly known (with the value $+\hbar/2$ or $-\hbar/2$), the other is fully undetermined (50% probability of finding $+\hbar/2$ and $-\hbar/2$).

2.1.4 Translations of $|\psi\rangle$ in time, space and momentum

Time translation: the Schrödinger equation

We have so far considered a static situation, describing a state $|\psi\rangle$ at a given time. Unless it is an eigenstate of the Hamiltonian H representing the system's energy, $|\psi\rangle$ evolves in time, entailing a continuous change of the probability distribution of the

---

[5]Note, however, that the commutator algebra is different. The commutator of $\mathbf{R}$ and $\mathbf{P}$ is proportional to $\mathbb{1}$, while the commutator of two Pauli operators is proportional to the third one.

outcomes of potential measurements. The state's evolution is ruled by the Schrödinger equation:

$$ i\hbar \frac{d\,|\psi(t)\rangle}{dt} = H\,|\psi(t)\rangle \;. \tag{2.15}$$

The formal solution of this equation is:

$$ |\psi(t)\rangle = U(t,0)\,|\psi(0)\rangle \;, \tag{2.16}$$

where $U(t,0)$ is the system's unitary time-evolution operator, which is simply, when H is time-independent:

$$ U(t,0) = e^{-iHt/\hbar} \;. \tag{2.17}$$

Note that this evolution is always reversible. The inverse unitary operator applied to the system can bring it back to its initial state. This is not the case of the sudden jump upon measurement, in which all information about the prior values of the probability amplitudes is irremediably lost.

For a particle in a potential $V(\mathbf{r})$, eqn. (2.15) translated in the $\mathbf{R}$ representation takes the standard form of a propagation equation for the system's wave function $\psi(\mathbf{r}, t)$:

$$ i\hbar \frac{\partial \psi(\mathbf{r}, t)}{\partial t} = -\frac{\hbar^2}{2m} \Delta \psi(\mathbf{r}, t) + V(\mathbf{r}, t)\psi(\mathbf{r}, t) \;, \tag{2.18}$$

where Δ is the Laplacian operator. The formal solution corresponding to (2.16) becomes:

$$ \psi(\mathbf{r}, t) = \int d^3\mathbf{r}'\,\langle \mathbf{r}|\,U(t,0)\,|\mathbf{r}'\rangle\,\psi(\mathbf{r}', 0) \;, \tag{2.19}$$

where the matrix element $\langle \mathbf{r}|\,U(t,0)\,|\mathbf{r}'\rangle$ is the Green function describing how a localized wave packet centred at $\mathbf{r}'$ at time $t = 0$ propagates at a later time in the potential $V(\mathbf{r})$. We give below an interpretation of eqn. (2.19) in terms of Feynman's space–time diagrams.

The partial differential equation (2.18) is reminiscent of the propagation equation for an acoustic or electromagnetic wave. Schrödinger sought to give a similar interpretation of his equation, viewing it as describing the dynamics of a charge density wave in atoms. This approach turns out to be inconsistent though and we now understand, following the interpretation of Max Born and the Copenhagen school, that the Schrödinger equation does not correspond to the propagation of a real vibration in a material medium or even in a field, but describes the evolution of an abstract amplitude probability wave.

Translations in space and momentum

The Hamiltonian of a system is the generator of its translations in time, meaning that H, acting on $|\psi\rangle$, yields the infinitesimal temporal variation of $|\psi\rangle$. Over a finite time interval, the translation is described by the unitary operator U, an exponential function of the generator H. Similarly, $\mathbf{P}$ is the generator of the system's translation in space. The action of $\mathbf{P}$ on a state yields its infinitesimal change under spatial displacements. By integration, we find that a finite translation by $\Delta\mathbf{r}$ is generated by the exponential unitary operator $\mathcal{T}_R(\Delta\mathbf{r}) = e^{-i\mathbf{P}\cdot\Delta\mathbf{r}/\hbar}$. The action of $\mathcal{T}_R(\Delta\mathbf{r})$ on the basis state $|\mathbf{r}\rangle$ is:

$$\mathcal{T}_R(\Delta\mathbf{r})\,|\mathbf{r}\rangle = e^{-i\mathbf{P}\cdot\Delta\mathbf{r}/\hbar}\,|\mathbf{r}\rangle = |\mathbf{r} + \Delta\mathbf{r}\rangle \;, \tag{2.20}$$

and the effect of the spatial translation on an observable $F(\mathbf{R})$, an arbitrary function of $\mathbf{R}$, is:

$$\mathcal{T}_R^\dagger(\Delta\mathbf{r})F(\mathbf{R})\mathcal{T}_R(\Delta\mathbf{r}) = e^{i\mathbf{P}\cdot\Delta\mathbf{r}/\hbar}F(\mathbf{R})e^{-i\mathbf{P}\cdot\Delta\mathbf{r}/\hbar} = F(\mathbf{R} + \mathbb{1}\Delta\mathbf{r}) \;. \tag{2.21}$$

By exchanging the conjugate operators $\mathbf{R}$ and $\mathbf{P}$, we describe in the same way momentum translations, $\mathbf{R}$ now playing the role of generator. A finite momentum kick $\Delta\mathbf{p}$ is described by the action of $\mathcal{T}_P(\Delta\mathbf{p}) = e^{i\mathbf{R}\cdot\Delta\mathbf{p}/\hbar}$:

$$\mathcal{T}_P(\Delta\mathbf{p})\,|\mathbf{p}\rangle = e^{i\mathbf{R}\cdot\Delta\mathbf{p}/\hbar}\,|\mathbf{p}\rangle = |\mathbf{p} + \Delta\mathbf{p}\rangle \;, \tag{2.22}$$

and:

$$\mathcal{T}_P^\dagger(\Delta\mathbf{p})G(\mathbf{P})\mathcal{T}_P(\Delta\mathbf{p}) = e^{-i\mathbf{R}\cdot\Delta\mathbf{p}/\hbar}G(\mathbf{P})e^{i\mathbf{R}\cdot\Delta\mathbf{p}/\hbar} = G(\mathbf{P} + \mathbb{1}\Delta\mathbf{p}) \;, \tag{2.23}$$

where the observable $G(\mathbf{P})$ is an arbitrary function of $\mathbf{P}$. Space and momentum translations are fundamental symmetry operations playing an essential role in the description of quantum processes. We will encounter them time and again.

2.1.5 Lack of objective reality of $|\psi\rangle$ and no-cloning theorem

The concept of state or wave function is often employed for the description of an ensemble of identical quantum systems, in a statistical context. The expectation value of an observable O measured on a large number of realizations of the same quantum state $|\psi\rangle$ is $\langle O \rangle = \langle\psi|\,O\,|\psi\rangle$. By measurements of non-commuting observables performed on independent realizations of the system, one can acquire complete information on its state. This is the case, for instance, in an NMR experiment in which one studies a sample of many spins, all prepared in the same state $|\psi\rangle = \cos(\theta/2)\,|0\rangle + \sin(\theta/2)e^{i\phi}\,|1\rangle$. Independent measurements of the non-commuting observables σ_i, each on a different subset of this ensemble, can be performed without any back-action of one measurement on another, since they are made on non-interacting subsystems. The quantum averages $\langle\sigma_Z\rangle = \cos\theta$, $\langle\sigma_X\rangle = \sin\theta\cos\phi$ and $\langle\sigma_Y\rangle = \sin\theta\sin\phi$ can be independently determined, yielding the values of θ and ϕ and hence, the complex probability amplitudes which completely define the common state of the spins. As long as experiments are performed on such ensembles, the state or wave function can be considered as an objective concept, since it can be determined by an observer who has no *a priori* knowledge about it.

The situation is very different if – as will be the case in experiments described later on – we deal with an isolated quantum system existing as a *single* copy. The quantum state $|\psi\rangle$ of this isolated system contains all predictive information we can have. This information is encoded into the system by the process which has prepared $|\psi\rangle$. Since a pure state $|\psi\rangle$ can always be considered as an eigenstate of the projector onto $|\psi\rangle$, one can assume that an apparatus has been designed to measure this projector. Using this apparatus on an arbitrary initial state, an experimenter (the 'preparer') measures the value (0 or 1) of the projector and in the case when he finds 1, records that the system is prepared in $|\psi\rangle$. In this way, the preparer knows the system's state and can make all kinds of probabilistic predictions on the outcome of any measurement performed on it.

Can we say, however, that this quantum state, defined on a unique system, has an 'objective reality'? A reasonable criterion of reality is that any other experimenter (a 'measurer', as opposed to the preparer), being given a single copy of this state and not knowing anything about the preparation, should be able to find out what the quantum state is. Clearly, the measurer is, under the conditions we have defined, unable to acquire this information, however clever he is. If he performs a measurement on the system, he obtains partial information, but the state is irreversibly and randomly modified. The probability amplitudes prior to the measurement are irremediably lost and no more information can be acquired about them. Hence, a wave function, or a spin state *existing as a single copy* cannot be determined by someone who has not prepared it (or who has not communicated with the preparer). This 'lack of objective reality' of the wave function of a single quantum system is a fundamental quantum feature (D'Ariano and Yuen 1996) which plays, as we will see, an important role in many quantum information procedures. It has an important corollary, the impossibility of copying exactly an unknown quantum state, a property known as the no-cloning theorem (Wootters and Zurek 1982). If such a copy were possible, quantum mechanics would be inconsistent. By making a large number of identical copies of a given quantum state, one could perform statistical measurements on the copies and deduce from them the wave function of the initial state, in contradiction with the 'lack of objective reality' that we have just defined.

We have restricted our analysis so far to 'pure' quantum states about which the preparer has maximum information. Usually, the situation is less ideal and even the preparer has an imperfect knowledge of the state and of its subsequent evolution, due to random perturbations or to the unavoidable coupling of the system to its environment. In such realistic situations, the system's description in terms of quantum states or wave functions must, as we will see below, be replaced by the density operator formalism.

2.2 Quantum interference and complementarity

Quantum probability amplitudes are C-numbers whose modulus and phase are essential. An amplitude associated to a given measurement outcome often appears as a sum of partial amplitudes corresponding to different system histories. These amplitudes add to each other, or cancel from each other, depending on their relative phases. This leads to interference effects, which are ubiquitous in microscopic physics and illustrate in a dramatic way the wave nature of material particles.

2.2.1 Through which slit does the particle go?

The simplest kind of interference phenomenon is displayed in the famous double-slit experiment, first performed by Young in 1802 with light, then duplicated during the last century with all sorts of material particles. Its principle is recalled in Fig. 1.5 on page 11. The set-up consists of a source S, emitting mono-energetic particles (which can be photons, electrons, atoms, molecules,...) and of three screens interposed on their path. The particles have a momentum p and a de Broglie wavelength $\lambda = h/p$. The

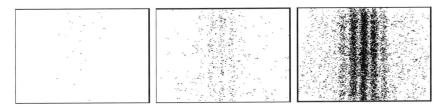

Fig. 2.2 Young experiment with atoms crossing the interferometer one at a time: movie frames showing from left to right the progressive build up of the fringe pattern (note that 'bright fringes' appear here in black and 'dark fringes' in white). Courtesy of F. Shimizu. See also Shimizu *et al.* (1992).

first screen carries a collimating slit restricting the aperture of the incoming particle beam. The second one is pierced by two narrow slits separated by a distance a and placed within the aperture of the primary beam. These slits define two distinct paths for the particles through the apparatus. The third screen is paved by detectors finally recording the impacts of the particles.

After a large number of particles has been detected, one observes on the detection screen a pattern of equidistant lines with a high density of impacts ('bright fringes') separated by stripes where no particles impinge ('dark fringes'). The angular separation of two adjacent bright fringes seen from the second screen is $\theta_f = \lambda/a$. This is easy to explain if one assimilates the beam of particles to a monochromatic wave. The two slits appear as secondary sources radiating forward partial waves whose phases add up at bright fringes and subtract at dark fringes. This explanation has been given, since the nineteenth century, for light interference whose existence was considered as compelling evidence of the wave nature of radiation.

But is it so simple? Refining the experiments, opticians have been able to reproduce it with light so dim that the energy flux corresponded to less than one photon per light transit time across the apparatus. The pattern on the detection screen then builds up in a 'pointillist' way by accumulation of discrete spots, each one registering the arrival of a single photon. Light behaves then as a stream of particles on the detection screen, and yet, the ensemble of detection spots reconstructs the fringes explained by the wave interpretation.

Even more puzzling, at least for a classical mind, is the existence of fringes when similar experiments are performed with electrons (Tonomura *et al.* 1989) or atoms (Carnal and Mlynek 1991; Shimizu *et al.* 1992), whose particle character seems obvious. Figure 2.2 shows the progressive build up of the fringes observed in an experiment performed with laser-cooled neon atoms. The successive frames, taken from a movie recorded during the experiment, show the progressive build up of the fringes when the atomic flux is so low than there is never more than a single atom at a time in the apparatus. The discreteness of the detection events exhibits the corpuscular nature of the atoms. And yet, the final fringe pattern has an unmistakable wavy character. The spooky aspect of this experiment is that each atom crossing the second screen seems to 'know' whether the two slits are simultaneously open or not. If one of the slits is plugged, the pattern of detection points does not show any interference.

The difficulty arises of course when we ask the question: through which slit did a

Fig. 2.3 Principle of the Heisenberg microscope which-path experiment. A light beam (propagating vertically) illuminates the particles behind the slits. The light scattered off the particle allows the observer to pinpoint its position.

given atom cross the second screen? This is a natural question for a classical mind but a meaningless one in the quantum world. If no experiment is performed to measure the position of the atom when it crosses the second screen, this position has no physical reality. Using again inadequate expressions, we may say that the particle crosses the second screen through the two holes at once, *suspended between two classical realities*. The quantum physicist does not resort to such a mysterious and ambiguous statement. He just notices that each particle crosses the apparatus in a linear superposition of two states, corresponding each to a wave packet going through one of the slits. The probability density for recording the particle at a given point is the square of the sum of the two corresponding amplitudes. In short, it is not the particle but its probability amplitude which interferes.

2.2.2 Which-path experiments

What happens if we attempt to acquire information about the particle path? This is the question which puzzled Einstein at the Solvay Conference of 1927 and led Bohr to the notion of complementarity. The discussion of this issue provides an insightful analysis of a position measurement, a process whose effects we have described in our general introduction of the wave function without explaining so far how it can be carried out in practice.

In order to 'see' through which slit the particle goes, let us use a Heisenberg microscope, a device represented on Fig. 2.3. We irradiate a zone behind the central screen of the Young interferometer with a light beam whose photons are scattered off the particle and focused through an appropriate microscope lens system to produce an image of the scattering spot. The optical diffraction laws tell us that the size of the image is of the order of the wavelength of the probing light. In order to pinpoint the slit close to which the scattering event occurred, we must use a radiation whose wavelength is smaller than a. Each probing photon hence carries a momentum larger than h/a. Its scattering, occurring in a random direction, imparts to the particle a recoil whose component normal to the particle beam direction has a fluctuating magnitude $\delta p > h/a$. This random recoil produces an angular fluctuation $\Delta\theta \approx \Delta p/p \geq \hbar/ap = \lambda/a$

of the momentum of the particle. The magnitude of this fluctuation is larger than the angular separation θ_f of the fringes seen from the slits, resulting in a blurring of the interference pattern.

In other words, lifting the quantum ambiguity about the particle's path inside the interferometer perturbs it by at least the amount required to destroy the interference. This qualitative discussion also yields an interpretation of the Heisenberg uncertainty relations by showing on a specific example how acquiring precise information about a particle position suppresses information about its momentum.

This qualitative argument can be made more precise by a few lines of calculation which constitute a simple exercise on the manipulation of the time, position and momentum translation operators of a free particle, respectively $e^{-i\mathbf{P}^2 t/2m\hbar}$, $e^{-i\Delta\mathbf{r}\cdot\mathbf{P}/\hbar}$ and $e^{i\Delta\mathbf{p}\cdot\mathbf{R}/\hbar}$. Let us call $|\psi(0)\rangle$ the wave packet of the particle at the time $t = 0$ when it crosses the double slit. Its evolution up to the time $t = mD/p$ when it reaches the detection plane at a distance D from the slits transforms it, according to Schrödinger's equation, into:

$$|\psi(t)\rangle = e^{-i\mathbf{P}^2 t/2m\hbar} |\psi(0)\rangle \ , \tag{2.24}$$

and the interference pattern is described by the probability density:

$$\Pi(\mathbf{r}) = |\langle \mathbf{r} |\psi(t)\rangle|^2 = |\langle \mathbf{r}| e^{-i\mathbf{P}^2 t/2m\hbar} |\psi(0)\rangle|^2 \ . \tag{2.25}$$

When the particle scatters a photon from the probing beam and undergoes a recoil $\Delta\mathbf{p} = \hbar\Delta\mathbf{k}$, its wave packet at time $t = 0$ is merely translated in momentum, becoming:

$$|\psi_{\mathbf{k}}(0)\rangle = e^{i\Delta\mathbf{k}\cdot\mathbf{R}} |\psi(0)\rangle \ . \tag{2.26}$$

This state, propagating from the slit to the detection screen, is transformed into:

$$|\psi_{\mathbf{k}}(t)\rangle = e^{-i\mathbf{P}^2 t/2m\hbar} e^{i\Delta\mathbf{k}\cdot\mathbf{R}} |\psi(0)\rangle \ , \tag{2.27}$$

and the probability density $\Pi_{\mathbf{k}}(\mathbf{r})$ associated to the perturbed particle is:

$$\Pi_{\mathbf{k}}(\mathbf{r}) = |\langle \mathbf{r} |\psi_{\mathbf{k}}(t)\rangle|^2 = |\langle \mathbf{r}| e^{-i\mathbf{P}^2 t/2m\hbar} e^{i\Delta\mathbf{k}\cdot\mathbf{R}} |\psi(0)\rangle|^2 \ . \tag{2.28}$$

We note the identity, easily deduced from eqn. (2.23):

$$
\begin{aligned}
e^{-i\mathbf{P}^2 t/2m\hbar} e^{i\Delta\mathbf{k}\cdot\mathbf{R}} &= e^{i\Delta\mathbf{k}\cdot\mathbf{R}} [e^{-i\Delta\mathbf{k}\cdot\mathbf{R}} e^{-i\mathbf{P}^2 t/2m\hbar} e^{i\Delta\mathbf{k}\cdot\mathbf{R}}] \\
&= e^{i\Delta\mathbf{k}\cdot\mathbf{R}} e^{-i(\mathbf{P}+\hbar\Delta\mathbf{k})^2 t/2m\hbar} \ ,
\end{aligned}
\tag{2.29}
$$

which, carried in eqn. (2.28) and taking into account eqn. (2.20), leads to:

$$
\begin{aligned}
\Pi_{\mathbf{k}}(\mathbf{r}) &= |\langle \mathbf{r}| e^{-i(\mathbf{P}+\hbar\Delta\mathbf{k})^2 t/2m\hbar} |\psi(0)\rangle|^2 = |\langle \mathbf{r}| e^{-i\Delta\mathbf{k}\cdot\mathbf{P}t/m} e^{-i\mathbf{P}^2 t/2m\hbar} |\psi(0)\rangle|^2 \\
&= |\langle \mathbf{r} - \hbar\Delta\mathbf{k}t/m| e^{-i\mathbf{P}^2 t/2m\hbar} |\psi(0)\rangle|^2 = \Pi(\mathbf{r} - \hbar\Delta\mathbf{k}t/m) \ .
\end{aligned}
\tag{2.30}
$$

Particles which have undergone a momentum kick $\hbar\Delta\mathbf{k}$ thus contribute to an interference pattern $\Pi_{\mathbf{k}}(\mathbf{r})$ deduced from $\Pi(\mathbf{r})$ by a translation $\Delta\mathbf{r} = \hbar\Delta\mathbf{k}t/m = \hbar\Delta kD/p$. Since $\Delta\mathbf{k}$ changes randomly from particle to particle, with a fluctuation larger than $2\pi/a$, the global pattern produced by the ensemble of particles perturbed by the probing light is the superposition of fringes whose angular direction seen from the slits fluctuates by more than $2\pi\hbar/ap = \lambda/a = \theta_f$. This fluctuation, larger than the fringe spacing, washes out the interference pattern, as already deduced from a qualitative argument.

The perturbing effect of the probing beam could be reduced by progressively tuning down its frequency, thus increasing its wavelength above a. We then scatter softer photons off the particles and reduce accordingly the fluctuation of the interference pattern position, which becomes smaller than the fringe separation. The contrast of the pattern then increases. The diffraction spots which could be observed in the microscope are, however, becoming larger than a, so that we progressively lose the ability to distinguish the two interfering paths. An explicit calculation shows that the sum of two dimensionless parameters describing respectively the fringe visibility and the amount of which-path information is bounded by an upper limit. In short, the more visible the interference pattern is, the less distinguishable are the trajectories. More on this problem can be found in Jaeger *et al.* (1993) and Englert (1996).

The quantitative calculation of the fringe translation induced by a momentum kick emphasizes another important aspect of this which-path experiment. By sorting out the particles according to the recoil they have undergone, we get subsets of impacts on the detection screen which exhibit fringes with a full contrast (eqn. 2.30). Finding out the recoil of a given particle can, in principle, be achieved by measuring the momentum of the photon it has scattered. Sorting out the particle recoils thus amounts to performing a coincidence experiment in which we correlate the positions of the particles' impacts on the detection screen with the momenta of the photons they have scattered. These correlations reveal an entanglement between the interfering particles and the probing photons. There is a deep connexion between which-path and entanglement experiments. We will analyse this connexion in more detail later in this chapter, once we have discussed more precisely the concept of entanglement.

The Heisenberg microscope, a virtual device in the early Bohr–Einstein debates, has now been realized in the laboratory with photons, electrons, atoms and molecules in experiments which we will review in Chapter 6. Other situations in which information about the particles' path in an interferometer is acquired via various processes have also been imagined in the early days of the quantum and realized since. An example of such an experiment, also imagined by Bohr, is shown in Fig. 1.6, on page 13. Here, which-path information is acquired through the recoil of the slit itself, supposed to be light enough to keep a record of the particle's crossing. We will analyse in detail this situation in Chapter 6 and describe a modern version of this thought experiment.

To conclude these considerations on which-path experiments, let us remark that information about the trajectory could also be carried by the particle itself, in an internal degree of freedom (spin or polarization). We can think, for example, of a photon interferometer in which the two slits are followed by two polarizers selecting orthogonal polarizations. An analysis of the particle's polarization before detection could tell us through which slit it has travelled. A more mundane explanation for the disappearance of the fringes notes that the electric fields associated to orthogonal polarizations cannot cancel each other. This result – which can be considered as classical when it concerns optical interference – is retrieved in matter wave interference performed with spin-carrying particles. If the spin can be used as a 'which-path' information source, the interference fringes always disappear. This follows from the basic quantum law that the probability amplitudes associated to orthogonal final states do not interfere.

2.2.3 Complementarity according to Bohr

Reflecting on these which-path experiments led Bohr to express his famous 'complementarity principle'. It can be stated in various equivalent ways. We can say that the wave and corpuscular aspects of material or light particles are complementary. The wave nature of particles manifests itself if the experimental set-up does not permit us to measure their positions. On the contrary, their corpuscular nature is revealed if a position measurement is performed or information on the particle path acquired. It is not even necessary to record this information. The mere fact that an interferometer set-up has been modified in order to make this information available (by irradiating in the Young's experiment the second screen with short-wavelength radiation or by coupling the particle to any position-measuring device) is enough to kill the interference and to reveal the corpuscular aspect of the particles.

Complementarity is deeply related to Heisenberg uncertainties.[6] In the Young experiment, the wave-like interference is observed if the particles have a well-defined momentum, incompatible with the determination of their position when they cross the slits. This is often expressed by saying that position and momentum are 'complementary' observables.

Bohr's complementarity is not an independent principle of quantum physics. It is a qualitative statement, or rather an ensemble of equivalent statements, which interpret the content of the mathematical formalism. It emphasizes that a physical quantity has no reality unless it has been effectively measured, or at least unless it could be measured within the context of the experiment (we will come back to the definition of reality when we discuss non-locality). Complementarity insists also on the relativity of classical notions, such as the wave or corpuscular nature of a system. These notions acquire a meaning only in relation to a specific apparatus. It is when the set-up has been precisely defined that the particle in the interferometer 'decides' to behave as an interfering wave or as a particle with a well-defined trajectory. The qualitative notion of complementarity will take a more precise form when we link it to the concept of entanglement, which Bohr did not consider explicitly in his early interpretation of the quantum formalism.

At the transition between the nineteenth and twentieth centuries, the physicist and philosopher Ernst Mach emphasized that science must only describe directly measurable phenomena and avoid unnecessary theoretical considerations. This philosophy has had a mixed influence on modern physics. On the one hand, Mach's disbelief of atoms, on the grounds that they were not directly observed, has had, as we have seen, a negative influence on many scientists. On the other hand, the requirement that all phenomena must be precisely described with reference to a specific experimental context has been a fruitful approach to both relativity and quantum physics. Einstein used a Machian approach for his definition of time and length, always imagining precise measurement devices, sticks and clocks which play a central role in relativity. Heisenberg and Bohr were also, to some extent, driven by Mach's ideas when they analysed

---

[6]For a discussion of subtle aspects of this connexion, see Englert *et al.* (1995) and Storey *et al.* (1995).

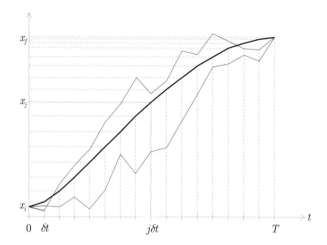

Fig. 2.4 Space–time diagram of the propagation of a particle between two events. The particle goes from x_i at the initial time $t = 0$ to x_f at $t = T$. The thick line is the minimum action trajectory. The thin lines represent two other possible trajectories.

quantum concepts in relation to a specific experimental context. In this respect, the uncertainty relations and the complementarity principle are Machian concepts.

2.2.4   Multiple paths and the recovery of classical dynamics

The phenomenon of interference is very general and plays a fundamental role in all quantum physics. For instance, eqn. (2.19) represents the wave function of a single particle $\psi(\mathbf{r}, t)$ as a sum of partial waves propagating from $\mathbf{r}'$ at time 0 to $\mathbf{r}$ at time t. The probability of finding the particle at $\mathbf{r}$ results from a multiple-path interference process which we analyse now in more detail. This will lead us to a qualitative introduction of Feynman's path-integral approach to quantum physics and will shed interesting light on the quantum–classical boundary.

Let us consider the simple case of a point-like particle of mass m (position and momentum observables X and P) evolving in a potential V in a one-dimensional space. Its Hamiltonian is $H = P^2/2m + V(X)$. The probability amplitude that the particle, initially prepared at point x_i, will pass a time T later at point x_f is the sum of an infinite number of contributions corresponding to all possible trajectories $x(t)$ satisfying the boundary conditions $x(0) = x_i$, $x(T) = x_f$. To show this, we start by expressing this probability amplitude as the matrix element between the initial and the final state of the system's evolution operator:

$$\langle x_f|U(T,0)|x_i\rangle = \langle x_f|\exp\{-i[P^2/2m + V(X)]T/\hbar\}|x_i\rangle \ . \tag{2.31}$$

We expand this expression by slicing the time interval T into infinitesimal intervals δt and by introducing at each of these times a closure relationship on the position eigenstates:

$$\langle x_f|U(T,0)|x_i\rangle \ = \ \int dx_1 dx_2 \ldots dx_n \langle x_f|\exp[-i(P^2/2m + V)\delta t/\hbar]|x_n\rangle \times \ldots$$

$$\times \langle x_{j+1}| \exp[-i(P^2/2m + V)\delta t/\hbar]|x_j\rangle \times \dots$$
$$\times \langle x_1| \exp[-i(P^2/2m + V)\delta t/\hbar]|x_i\rangle \ . \tag{2.32}$$

Each sequence of positions $x_i, x_1, \dots, x_{n-1}, x_n, x_f$ defines a possible trajectory of the particle, associated to a well-defined probability amplitude given by a product of matrix elements of U. Typical trajectories are sketched in Fig. 2.4. This space–time diagram shows that the propagation results from interference between all possible trajectories linking the initial to the final point.

In order to evaluate the amplitude $\langle x_{j+1}| \exp[-i(P^2/2m+V)\delta t/\hbar]|x_j\rangle$ corresponding to an infinitesimal time step δt, we replace in it the exponential operator by the product $\exp[-i(P^2/2m)\delta t/\hbar] \cdot \exp[-iV\delta t/\hbar]$. This replacement is not exact, since P^2 and V do not commute, but the approximation is valid up to negligible terms in δt^2. We then introduce between the two operators the closure expansion on momentum states ($\int dp\,|p\rangle\,\langle p| = \mathbb{1}$) and we express each elementary amplitude as:

$$\langle x_{j+1}| \exp[-i(P^2/2m + V)\delta t/\hbar]|x_j\rangle$$
$$\approx \frac{1}{2\pi\hbar} \int dp \exp[ip(x_{j+1} - x_j)/\hbar - i(p^2/2m)\delta t/\hbar - iV(x_j)\delta t/\hbar] \ . \tag{2.33}$$

The integral over p in this expression is the Fourier transform of a Gaussian of p. It is a Gaussian function of $x_{j+1} - x_j$. The probability amplitude can thus be expressed, within a multiplicative constant, as:

$$\langle x_f|U(T,0)|x_i\rangle \propto$$
$$\int dx_1 dx_2 \dots dx_n \exp\{i[m(x_f - x_n)^2/2\delta t^2 - V(x_n)]\delta t/\hbar\} \dots$$
$$\exp\{i[m(x_{j+1} - x_j)^2/2\delta t^2 - V(x_j)]\delta t/\hbar\} \dots$$
$$\exp\{i[m(x_1 - x_i)^2/2\delta t^2 - V(x_i)]\delta t/\hbar\} \ . \tag{2.34}$$

In the limit $\delta t \to 0$, the ratio $(x_{j+1} - x_j)/\delta t$ identifies with the velocity v_j along a given trajectory and we recognize in each of the exponents in (2.34) the Lagrangian of the system $\mathcal{L} = mv_j^2 - H$. The probability amplitude for the system to go from x_i to x_f in time T is thus a sum of amplitudes – one for each possible classical path – whose phase is the system's action $S = \int \mathcal{L}\delta t$ along the trajectory. This phase is expressed in units of $\hbar$. We have derived this important result by admitting the Schrödinger equation formalism of quantum mechanics. Feynman proceeded the other way around, postulating that a quantum system follows all the classical trajectories with amplitudes having a phase given by the classical action and has derived from there Schrödinger's equation.

At the classical limit $S/\hbar \gg 1$, the phase along a trajectory evolves very fast when the path is slightly modified, by changing for instance one of the x_j. The amplitudes of various neighbouring paths thus interfere destructively, leaving only the contributions of the trajectories for which the phase, hence the action, is stationary (thick line in Fig. 2.4). We retrieve here the classical principle of Maupertuis least action as a limiting case of quantum laws for 'large' systems. The condition $S/\hbar \gg 1$ is easily shown to be equivalent to $L/\lambda \gg 1$, where L is the length of the trajectory and λ the de Broglie wavelength of the quantum particle. The action along a path of length L travelled over a time T is indeed of the order of $mv^2\,T = mvL = 2\pi\hbar L/\lambda$.

The evolution of the particle's wave function is a scattering problem in space–time. If the particle's action in units of $\hbar$ is much larger than 1, the particle follows a classical 'ray'. Suppressing the contributions to $\langle x_f|U(T,0)|x_i\rangle$ coming from trajectories 'far' from the classical one does not appreciably affect this amplitude. In the same way, a short-wavelength optical signal propagating between two points is unaffected by diaphragms blocking regions of space far from the rays of geometrical optics. Quantum mechanics thus plays *vis à vis* classical mechanics the same role as wave optics with respect to ray optics. We have presented here Feynman's path-integral approach in a sketchy way without justifying the subtle mathematical details. This approach sheds light on the issue of the quantum–classical boundary, by explaining how, in a way, classical mechanics is recovered from quantum mechanics when $\hbar$ goes to 0. It does not answer though all the questions raised by this issue. We come back to the quantum–classical boundary problem in Section 2.5.

2.3 Identical particles

2.3.1 Identity in classical and quantum physics

The strangeness of the quantum, stemming from the superposition principle, reveals itself in a striking way in the description of ensembles of electrons, photons or other kinds of identical particles. At first sight, the notion of identity seems familiar to a classical mind. We often say that true twins are identical, or (more recently) that a clone is identical to its 'mother'. The genetic programs which have made them are the same, they look a lot alike, but they are certainly distinguishable by some details. We also say that two objects built according to the same blueprint (two cars of the same model) are identical, but this too is an approximation and we can always, in the end, distinguish between them by some scratches or slight differences. In classical problems involving billiard balls, we often consider the collisions of two identical balls (same mass and volume for example) and deduce general properties from this identity (for example, a collision between two such balls always ends up by two balls rolling away with opposite velocities in their rest frame). But here again, the balls are not identical in every respect. First they can be of different colours, which make their identification easy. And even if their physical differences are unnoticeable to the naked eye, they can always be distinguished by their past histories. Nothing prevents a curious experimentalist from filming the balls as they collide on the billiard table and to distinguish them by their trajectories, labelling them by numbers. In a similar way, classical statistical physicists have always labelled the identical particles of gas they study. Even if following the trajectories of moles of particles is not practical, nothing forbids it being done in principle. We can thus claim that in classical physics identical particles can always be distinguished.

Nothing of this kind is possible in the quantum world where identical particles are fundamentally indistinguishable. This has deep consequences, completely unforeseen in pre-quantum physics. Even if, at some times, the wave packets of different particles are spatially separated and could be identified by continuously following their evolution, this becomes impossible as soon as they interact at close range. The quantum

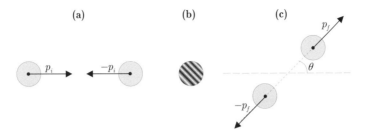

Fig. 2.5 Sketch of a collision of two identical electrons or α particles. The wave packets of the particles are sketched as grey circles.

fuzziness of overlapping wave packets makes it impossible to know which is which after a collision. We could attempt to observe the particles by shining short-wavelength light on them, but we would then disturb their state in an uncontrollable way. The physics of the collisions would be irremediably changed. A quantum physicist does not have in general the opportunity to 'paint' his particles with different colours, or even to shoot a movie to record their trajectories.

Figure 2.5 illustrates as a simple example the collision of two electrons with the same spin orientation or of two spinless α particles (^{4}He nuclei) undergoing a head-on collision. The system evolution is described in its rest frame. We assume that the interaction between the particles depends only on their relative position and is, in the case of electrons, spin-independent. The internal spin variable remains a passive spectator and we can describe the electrons, as the α particles, by their external degrees of freedom only. Initially, the two particles have opposite momenta $\mathbf{p}_i$ and $-\mathbf{p}_i$ (Fig. 2.5a) and their wave packets are separated. When they collide, their wave packets overlap (Fig. 2.5b). They then separate again. We want to describe a scattering event in a direction making the angle θ with the initial one, corresponding to particles having opposite final momenta $\mathbf{p}_f$ and $-\mathbf{p}_f$ ($|\mathbf{p}_f| = |\mathbf{p}_i|$) (Fig. 2.5c).

In order to analyse the process, we need to represent the physical state of the system by a mathematical ket evolving in time. It is natural to write the initial state as $|\mathbf{p}_i, -\mathbf{p}_i\rangle$, assigning the momentum $\mathbf{p}_i$ to the particle which appears first in the ket and the momentum $-\mathbf{p}_i$ to the other. But it is equally possible to express this configuration by the ket $|-\mathbf{p}_i, \mathbf{p}_i\rangle$ obtained by exchanging the states of the 'first' and 'second' particles. Classical physicists would shift from one expression to the other without harm, but in quantum physics, the arbitrariness in the labelling of the particles is not innocuous. If these two mathematical states were legitimate representations of the same physical state, then any superposition $\alpha|\mathbf{p}_i, -\mathbf{p}_i\rangle + \beta|-\mathbf{p}_i, \mathbf{p}_i\rangle$ would also be a possible representation, leading to what is known as 'exchange degeneracy'. The lack of uniqueness in the definition of the system's state cannot be accepted, since it would lead to ambiguous physical predictions, depending upon the choice made about the α and β amplitudes.

2.3.2 The symmetrization postulate

The exchange degeneracy is suppressed by the symmetrization postulate, which selects as mathematical kets the superpositions which are either symmetrical (bosons) or anti-symmetrical (fermions) under exchange of the two particles. More generally, this

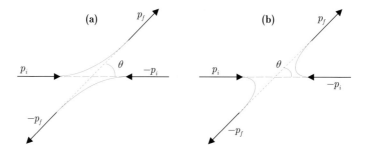

Fig. 2.6 The two interfering amplitudes in a collision between identical particles.

postulate states that an ensemble of $\mathcal{N}$ identical bosons must have a wave function symmetrical under any exchange of two particles, whereas an ensemble of $\mathcal{N}$ fermions must be represented by a wave function anti-symmetrical under the same exchange. The exact formulation of the symmetrization postulate and its application to situations with an arbitrary number of particles is described in quantum physics textbooks. We suppose here that the reader is familiar with this postulate and only discuss some of its deepest consequences.

The electron and all the other spin-$1/2$ fundamental particles making up ordinary and exotic matter (quarks, muons and neutrinos) are fermions. Particles with integer spin which mediate the interactions (photons, W and Z particles,...) are bosons. Composite systems such as nucleons (made of three quarks) or atoms (made of combinations of quarks and electrons) behave in an experiment as bosons or fermions as long as their binding energy is large enough so that they cannot be split into their components. Composite systems with an odd number of quarks plus electrons are fermions; those with an even number are bosons. The proton and the neutron, made of three quarks each, are fermions. The hydrogen atom, a proton bound to an electron, is a boson. So is the α particle, made of twelve quarks, and the ^{4}He atom, made of an α particle and two electrons. The ^{3}He isotope (two protons and one neutron in the nucleus) is a fermion. Cooper pairs of electrons, which interact together via phonon exchange in superconducting metals at very low temperature, behave as bosons.

2.3.3   Interference effects in collision of identical particles

We come back to the particle scattering experiment (Fig. 2.5). Taking symmetrization into account, the initial and final states of the system are:

$$|\psi_i\rangle = (|\mathbf{p}_i, -\mathbf{p}_i\rangle + \epsilon\,|-\mathbf{p}_i, \mathbf{p}_i\rangle)/\sqrt{2}\;; \qquad (2.35)$$

$$|\psi_f\rangle = (|\mathbf{p}_f, -\mathbf{p}_f\rangle + \epsilon\,|-\mathbf{p}_f, \mathbf{p}_f\rangle)/\sqrt{2}\,, \qquad (2.36)$$

with $\epsilon = -1$ for a fermionic collision and $\epsilon = +1$ for bosons. To obtain the probability amplitude of the scattering event transforming $|\psi_i\rangle$ into $|\psi_f\rangle$, we compute the matrix element between these states of the evolution operator U. This matrix element has four terms coming from the expansions (2.35) and (2.36). Since U is symmetrical under particle exchange, these four terms are equal two by two, so that:

$$\langle\psi_f|U|\psi_i\rangle = \langle\mathbf{p}_f, -\mathbf{p}_f|U|\mathbf{p}_i, -\mathbf{p}_i\rangle + \epsilon\langle-\mathbf{p}_f, \mathbf{p}_f|U|\mathbf{p}_i, -\mathbf{p}_i\rangle\,. \qquad (2.37)$$

   The scattering amplitude is the sum of two terms associated to the classical trajectories represented in Fig. 2.6. In the first (Fig. 2.6a), the particle coming from the left with momentum $\mathbf{p}_i$ is deflected by the angle θ and ends up with momentum $\mathbf{p}_f$. The amplitude for this process depends on the interaction potential. Due to the cylindrical symmetry, it is a function of θ only and we write it as $\langle \mathbf{p}_f, -\mathbf{p}_f | U | \mathbf{p}_i, -\mathbf{p}_i \rangle = f_s(\theta)$. In the second trajectory (Fig. 2.6b), the particle coming from the left is deflected by the angle $\pi - \theta$ and its final momentum is $-\mathbf{p}_f$. The corresponding amplitude is $\langle -\mathbf{p}_f, \mathbf{p}_f | U | \mathbf{p}_i, -\mathbf{p}_i \rangle = f_s(\pi - \theta)$. The total scattering amplitude is:

$$\langle \psi_f | U | \psi_i \rangle = f_s(\theta) + \epsilon f_s(\pi - \theta) . \tag{2.38}$$

   The scattering of two identical particles thus exhibits interference between two classical trajectories, a situation reminiscent of Young's experiment. In both cases, the system's evolution is a superposition of two indistinguishable classical histories. The amplitudes associated to these classically distinct but quantum mechanically indistinguishable paths interfere, with a sign depending upon the nature of the particles. For scattering at right angle ($\theta = \pi/2$), we have a fully destructive interference for fermions and constructive interference for bosons. The probabilities of detecting such a scattering event are respectively 0 and $4|f_s(\pi/2)|^2$.

   Special mention should be made of collisions between extremely cold identical particles, whose de Broglie wavelength is much larger than the range of the collision potential. Quantum collision theory shows then that the scattering is isotropic, the outgoing particles flying away in a spherical wave (so called S-wave). The scattering function f_s is θ-independent. There is thus no scattering at all for fermions, which, at very low temperature, fly across each other without interacting. Cold bosons instead are scattered in S-waves more efficiently than distinguishable particles. The isotropic scattering amplitude is twice as large as for distinguishable particles. The total collisional cross-section is obtained by integrating the modulus squared of this scattering amplitude over the $[0, \pi/2]$ interval for θ. For distinguishable particles, this integration would run over the full $[0, \pi]$ interval, since the two processes illustrated in Fig. 2.6 would then be different. The total collisional cross-section of bosons in S-waves is thus twice that of distinguishable particles. The strikingly different properties of cold fermions and bosons collisions have many physical consequences, resulting for example in very different thermal conductivities for bosonic and spin-oriented fermionic isotopes of helium (Lhuillier and Laloe 1982).

   The analogy between identical particle collisions and interferometry can be pushed further and provides us with another example of complementarity. As in the Young double-slit experiment, any modification of the experiment to attempt to pin down which path the system follows during the collision results in a suppression of the interference and restores the classical scattering probability. In the case of electron collisions it is essential that the two particles are initially spin-oriented in the same direction. The spin cannot then be used to distinguish between the paths. The situation is very different if the spins are initially prepared with opposite orientations. The spin then plays the role of a distinctive 'colour' for the two otherwise identical particles. The two paths of Fig. 2.6(a) and (b) become distinguishable, in the same way as the two paths of a Young interferometer can be distinguished by adding spin polarizers behind the two slits. The amplitudes do not interfere and the scattering probability is the sum of the probabilities associated to the two paths.

Elastic two-body collisions between identical particles provided us with a first example of the dramatic differences between fermions and bosons, resulting from the different properties of their quantum states under particle exchange. This also illustrates the importance of removing the exchange degeneracy when describing a physical system. If we had chosen as mathematical kets superpositions such as $(1/\sqrt{2})\,|\mathbf{p}_i, -\mathbf{p}_i\rangle + (e^{i\phi}/\sqrt{2})\,|-\mathbf{p}_i, \mathbf{p}_i\rangle$ with $\phi \neq 0$ or π, we would have found a scattering amplitude different from the fermion and boson values. Only the amplitudes given by eqn. (2.38) with $\epsilon = \pm 1$ have been so far encountered in Nature, but this does not prevent physicists from looking for other kinds of hypothetical particles, called anyons (Frolich and Marchetti 1989), which would obey statistics corresponding to ϕ different from 0 and π.

2.3.4 Fermions and the exclusion principle

The elastic scattering of two cold identical particles has given us a glimpse of an essential concept of quantum physics, the exclusion principle of fermions, first established by Pauli in 1925, which states that two fermions can never be found in the same state. The deep reason why S-wave scattering is impossible for spin-oriented fermions is that the two fermions would end up in the same S-wave. This state cannot be anti-symmetrical under particle exchange, because the two components of the superpositions cancel each other.

This fundamental property of electrons plays an essential role to account for the periodic table of the elements, the structure of atoms and molecules, the laws of chemistry and hence, the principles of biochemistry on which life is based. All the electrons of a system must be distributed among as many different quantum states and this gives to Nature a wealth of structures and chemical combinations. All the properties of solid state systems (metals, semiconductors and insulators) are determined in a fundamental way by the exclusion principle.

For example, the electrons in a metal must fill up energy levels, one particle per energy step, and the electrons at the top of this energy ladder (the so-called Fermi level) are much more energetic that they would be if they were classical particles able to pack in the lowest available energy states. The electric and thermal conductivity of metals are determined by this exclusion property. The mechanical properties of solids are also dependent on it. If two solids repel each other when one tries to push them towards one another, it is because the electrons of their atoms do not want to sit on top of one another. The distinguishable character of solid bodies, the fact that they have clear-cut boundaries, is thus, in a paradoxical way, a consequence of the fundamental identity of the fermions they are made of. The workings of the transistor, based on the properties of semiconducting materials, and all the applications its development have made possible, are also, more or less directly, consequences of the exclusion principle applied to electrons.

We have so far discussed the electron behaviour in atoms or solids without considering their spins. At first sight, this seems a fair approximation, since the main force at play in these systems is the Coulomb interaction, resulting in electronic energies in the range of electron-volts. The magnetic forces coupling together the electron

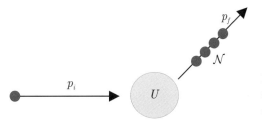

Fig. 2.7 Scattering of a boson in an already occupied state: bosonic amplification of the collision amplitude.

spins are much smaller. They account for small relativistic effects, involving energies in the millielectron-volt range, that one would be inclined to neglect in a qualitative picture. This approach is deceptive. The fermionic wave function anti-symmetrization is responsible for spin correlations involving energies in the electron-volt range. These effects are not due to magnetic forces, but to the action of electric forces in systems whose quantum states must satisfy the exclusion principle.

A simple example is given by the electrons of a helium atom in a configuration where they occupy two different orbital wave functions. Their quantum state is a tensor product of a symmetrical orbital part with an anti-symmetrical spin part, or vice versa. Anti-symmetrical spin states correspond to a total angular momentum $S = 0$, an 'anti-parallel' spin configuration, while a symmetric spin state is a $S = 1$ triplet with parallel spins. In the former situation, the two electrons, in their symmetrical orbital part, have a non-vanishing probability of being at the same point in space, thus repelling each other strongly. In the latter, they have an anti-symmetric wave function and hence cannot be found on top of each other. They thus experience a much weaker Coulomb repulsion. The first state, which corresponds to anti-parallel spins, thus has a larger electrostatic energy than the second, whose spins are parallel.

This looks like a magnetic spin–spin effect imparting different interaction energies to magnetic moments depending upon their relative orientation. But the energies involved are, here, much larger, since the energy difference between the two configurations is of electric, not magnetic origin. In helium and in other atoms, the larger binding of parallel spins than anti-parallel ones corresponds to Hund's rule. A similar effect occurs in ferromagnetic materials and is at the origin of their large magnetization. The strong effective coupling between the spins is reflected in the high phase-transition temperature of these magnetic materials (several hundreds of kelvin), which could not be explained by purely magnetic interactions.

The Pauli principle also accounts for the properties of other fermionic systems such as neutron stars. Basic properties of nuclei, such as their stability against β decay, is also explained by the exclusion principle. Whereas a free neutron decays into a proton, an electron and an anti-neutrino with a lifetime of about 15 minutes, neutrons embedded in nuclei remain stable because the final proton would have to get into an already occupied state. This consequence of the exclusion principle is quite important since it explains the stability of all the ordinary matter around us.

2.3.5 Bosons and bosonic amplification of collision amplitude

While fermions like to be on their own, the symmetry of their wave functions imparts to bosons a gregarious tendency to accumulate in the same quantum state. Once $\mathcal{N}$ bosons are already present in a state, the probability that an additional one will join them is $\mathcal{N}$ times larger than the probability of getting the first boson in this state. To understand this 'stimulated emission' property, let us consider the scattering by an external potential, illustrated in Fig. 2.7. We assume that a particle impinges on the potential in a given state of momentum $\mathbf{p}_i$ in the presence of $\mathcal{N}$ identical bosons in a final state of momentum $\mathbf{p}_f$. We call U the evolution operator of the $\mathcal{N}+1$-particle system in the external potential, which is supposed to act independently on the particles. The probability amplitude that a lonely particle is scattered from $\mathbf{p}_i$ to $\mathbf{p}_f$ is $\langle \mathbf{p}_f|U|\mathbf{p}_i\rangle = U_{fi}$. The potential is weak enough for the perturbative (Born) approximation to be valid, so that $|U_{fi}| \ll 1$ when $\mathbf{p}_i \neq \mathbf{p}_f$ and $\langle \mathbf{p}_i|U|\mathbf{p}_i\rangle = U_{ii} \approx 1$. The mathematical ket representing one particle with momentum $\mathbf{p}_i$ impinging on the potential with $\mathcal{N}$ particles with impulsion $\mathbf{p}_f$ is a fully symmetrized superposition of $\mathcal{N}+1$ orthogonal states in which the label of the particle in the i state is shifted from 1 to $\mathcal{N}+1$:

$$|\psi_i\rangle = \frac{1}{\sqrt{\mathcal{N}+1}}(|\mathbf{p}_i, \mathbf{p}_f, \mathbf{p}_f, \ldots, \mathbf{p}_f\rangle + |\mathbf{p}_f, \mathbf{p}_i, \mathbf{p}_f, \ldots, \mathbf{p}_f\rangle + \cdots$$
$$+ |\mathbf{p}_f, \mathbf{p}_f, \mathbf{p}_f, \ldots, \mathbf{p}_f, \mathbf{p}_i\rangle) , \tag{2.39}$$

whereas the final symmetrical state is:

$$|\psi_f\rangle = |\mathbf{p}_f, \mathbf{p}_f, \mathbf{p}_f, \ldots, \mathbf{p}_f\rangle . \tag{2.40}$$

The scattering amplitude for a boson to join in the final state the $\mathcal{N}$ already present ones is thus a sum of $\mathcal{N}+1$ identical terms, each normalized by $1/\sqrt{\mathcal{N}+1}$, so that:

$$\langle \psi_i|U|\psi_f\rangle = \sqrt{\mathcal{N}+1}U_{fi} . \tag{2.41}$$

The probability for this accumulation to occur is $(\mathcal{N}+1)|U_{fi}|^2$. The 'spontaneous' scattering probability (corresponding to the 1 term in this expression) adds to an $\mathcal{N}$ times larger 'stimulated' probability. The bigger $\mathcal{N}$, the more efficient is the process which tends to accumulate bosons in the same state. Light would not be what it is if photons were not gregarious particles rushing to the same quantum state. When an atom is brought into an excited electronic state, it usually falls back to a less energetic one by spontaneously emitting a photon. If $\mathcal{N}$ photons of the right frequency are already present in a field mode, the stimulated emission process in this mode becomes $\mathcal{N}$ times more efficient. The stimulated emission of light, first understood by Einstein in 1917, gives rise to the laser effect, whose importance in our modern life has already been emphasized.

2.3.6 Bose–Einstein condensation

The quantum concept of particle identity manifests itself spectacularly in the statistical properties of fermions and bosons gases at very low temperatures. We have already briefly discussed electron gases in a metal. The fermionic exclusion principle imparts to these systems an energy much larger than if these particles were distinguishable. For an ensemble of $\mathcal{N}$ weakly interacting bosons of mass m confined in a box of volume V, the effect of statistics is in some respect opposite, and no less spectacular. It leads, at a very low temperature T, to Bose–Einstein condensation (Einstein 1924), a large fraction of the particles accumulating in the single-particle lowest energy state.

In the limit $T \rightarrow 0$, this effect is not surprising. All systems must, according to Boltzmann's law, end up at $T = 0$ K in their ground state, corresponding to absolute order and zero entropy (Nernst law). This situation is an unattainable limit in a classical perfect gas. At the smallest departure from $T = 0$ K, the fraction of particles in the ground state becomes very small, of the order of $1/\mathcal{N}$. Statistical theory shows that the situation is quite different for a gas of bosons. As the temperature goes down, the particle thermal de Broglie wavelength $\lambda_{\text{th}} = \hbar\sqrt{2\pi/mk_bT}$ increases (k_b is Boltzmann's constant). When T reaches a critical value T_{BE} such that λ_{th} becomes – within a numerical factor of the order unity – equal to the average interparticle distance $(V/\mathcal{N})^{1/3}$, a phase transition occurs and a fraction of the particles starts suddenly to accumulate in the ground state. This fraction increases as T is lowered below T_{BE} and becomes very close to 1 at very low, but finite temperature. This situation is in sharp contrast with the classical result.

The explanation of Bose–Einstein condensation is found in statistical mechanics textbooks. Without doing any quantitative calculation here, we can qualitatively understand this phenomenon as a direct consequence of bosonic wave function symmetrization. At thermal equilibrium, according to the general postulate of statistical mechanics, a gas of particles goes randomly through all its available states, exploring them in turn as the particles undergo elastic collisions which change their individual momentum while conserving the system's energy. In this random walk, all the microscopic states corresponding to a given energy are equiprobable. In a classical gas, as soon as you put a bit of energy in the system, the number of available microscopic states becomes huge since all the particle permutations which leave the total energy unchanged are acceptable. The disorder of the classical particles thus increases very fast as T departs from 0 K, greatly favouring the microscopic configurations in which the particles are distributed among different single-particle states. This explains that, in a classical gas at finite temperature, the fraction of particles in the ground state remains exceedingly small.

The situation is quite different for bosons. By admitting only symmetrical wave functions, Nature restricts tremendously the number of available states. The number of microscopic configurations corresponding to a distribution among different single-particle states is strongly decreased, since all permutations between different single-particle states are included in only one allowed symmetrized multi-particle state. This gives a much larger statistical weight to the configurations in which many atoms are in the same state. In the random walk through all available states, it is thus much more probable that many particles will find themselves in the ground single-particle state.

The phenomenon of bosonic stimulated emission discussed above plays an essential role here. Once a sizable number of particles are accumulated in the ground state, the probability that others will fall in it under the effect of collisions increases by a factor proportional to the number of already present particles. An avalanche amplification occurs at the critical temperature, resulting in the Bose condensation. Note that this phenomenon, which requires thermalization induced by elastic collision, does not occur with photons which do not interact with each other.

Bose–Einstein condensation, predicted in the early days of quantum physics, has long remained a theoretical oddity. The discovery in 1936 of superfluidity in liquid

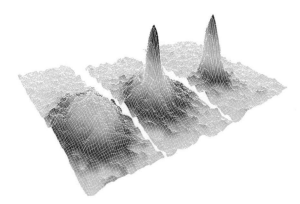

Fig. 2.8 Bose–Einstein condensation of rubidium (Anderson *et al.* 1995). Three snapshots of the velocity distribution of an atomic cloud during evaporative cooling (from left to right: $T > T_{BE}$, $T = T_{BE}$ and $T < T_{BE}$). The sudden appearance of a sharp feature near zero velocity for a low enough temperature is an indication of Bose–Einstein condensation. Courtesy of E. Cornell.

^{4}He has provided a first glimpse at a quantum behaviour attributable to it, but the strong interactions between the atoms in the liquid make it difficult to relate directly the superfluid transition to the simple theory of a perfect gas of bosons. The remarkable fact that superconducting metals carry currents without any resistance at low temperature is related to the bosonic behaviour of the Cooper pairs of electrons, but is only remotely related to Bose–Einstein condensation.

Physicists had to wait until 1995 to observe genuine Bose–Einstein condensation in experiments involving gas samples of weakly-interacting ultra-cold atoms trapped in a magnetic field (Anderson *et al.* 1995; Davis *et al.* 1995; Bradley *et al.* 1995; Leggett 2001). The atoms were brought to a fraction of a microkelvin by a combination of laser and evaporative cooling, using techniques which are briefly described in Chapter 9. Up to a few million atoms can now be condensed in a single wave function, a few micrometres in size, making up a giant matter wave with fascinating properties.

Figure 2.8 shows, as a density plot, the absorption of a probe laser beam by the atomic cloud, after a free-expansion time interval following the suppression of the trap potential. This signal has been recorded in the first condensation experiment (Anderson *et al.* 1995). This density maps – with some distortions (due to atomic interactions) – the atomic velocity distribution as it was at the time the trap was switched off. From left to right in Fig. 2.8, the distributions correspond to three decreasing temperatures: above T_{BE}, at T_{BE} and below T_{BE} respectively. The onset of condensation – with the sharp increase of the number of particles around zero momentum and the disappearance of the broad background of thermal atoms – is spectacular.

Bose–Einstein condensates are now realized with many bosonic isotopes, including hydrogen. Beams of condensed atoms having coherence properties very similar to that of laser beams of photons are obtained by releasing condensed bosons from their trap. Potential applications are many, making the study of Bose–Einstein condensation one of the most promising fields of today's physics. The last chapter is devoted to the description of Bose–Einstein condensate experiments related to quantum information physics.

2.3.7 Identical particles, field theories and second quantization

Exploring the concept of identity in the microscopic world led us to new kinds of strange effects. Particles seem to influence each other merely because they are forced to evolve in a quantum space restricted to symmetric or anti-symmetric wave functions. Spins are aligned in the same direction by this strange influence much more efficiently than by a magnetic force; neutrons are forbidden to decay by the presence of neighbouring protons. Photons or massive bosons are accumulating, as if they were feeling some kind of attraction, into states already occupied by identical partners and, yet, there is no real force pulling them towards each other. These processes, which would be mysterious to a pre-quantum physicist, are deeply rooted in the superposition principle. To understand them, it is essential to recognize that the global wave functions of an ensemble of bosons or fermions are well-defined linear combinations of single-particle product states. These superpositions are involved in quantum interference effects affecting bosons and fermions in different ways.

We have chosen here a pedagogical approach, starting from a single-particle description, and building up a multi-particle wave function satisfying symmetry rules which suppress any exchange degeneracy ambiguity. This approach is rather cumbersome, if conceptually simple. It amounts to introducing first a particle 'labelling', and to erase it in the end by choosing state combinations in which it is irrelevant. Moreover, such a point of view is adequate only for the description of massive particles, for which single-particle wave functions exist at the non-relativistic limit. It is then possible to build by products of such wave functions a multi-particle Hilbert space which is finally symmetrized to account for the properties of bosons and fermions.

This approach is inadequate for photons which are inherently relativistic and for which no complete single-particle wave function exists (even if it is possible to represent by a quantum state the photon polarization). For this reason, we have discussed photons as bosons only in a qualitative way. A more quantitative description will have to await the next chapter in which we present a simple quantum formalism to describe the electromagnetic field. Photons appear as quantum excitations of the field, produced by the action of creation operators on a special state, the field vacuum. This approach recognizes from the outset the deep indistinguishability of photons, without having to label them by non-physical numbers.

The quantum field approach can be generalized to massive bosons and fermions. In fact, at a fundamental level, all the particles in Nature are excitations of specific photon, electron, quark fields etc. (Itzykson and Zuber 1980). The fields are the primary objects of the theory and the indistinguishability of the particles is given from the premises. In this fundamental point of view, the disparate statistical behaviours of fermions and bosons come from the different commutation or anti-commutation rules of their creation and annihilation operators. The quantum field approach also has the merit of allowing for changes in the number of particles, processes which become physically important at high energies and are the stuff studied in particle accelerators. In low-energy physics, only the massless photons can be annihilated or created and the quantum field approach for the massive particles is useful, but not necessary. The alternative simple approach given here is a good approximation of the physics. We present a second quantization description of Bose–Einstein condensates in Chapter 9.

2.4 Entanglement and non-locality

We now analyse the consequences of the superposition principle applied to composite systems. This will lead us to introduce the essential concept of entanglement, which Schrödinger (1935) viewed as being the *'essence of quantum physics'*. Whether a system is composite or not might be at first sight considered as a matter of formal definition. At the ultimate reductionist level, only fundamental particles are elementary and most systems under study are composite. At the other end of the complexity spectrum, one could consider that the whole Universe is a single quantum system described by a wave function in a huge Hilbert space. In this extreme point of view, there is no composite system at all! For most physical problems, these two approaches are too radical. It is convenient to split the complexity at some intermediate level, which usually is neither the level of elementary particles nor that of the Universe as a whole.

We have already seen examples of such pragmatic separations. When we considered an atom crossing an interferometer (Section 2.2), we implicitly assumed that it was a non-composite system in the situation we were describing, even if we know that it is in fact an assembly of quarks and electrons. This non-composite approximation is valid because the binding between the subatomic particles is much stronger than the coupling of the atom to the interferometer. The internal structure of the atom is left unchanged as it travels through the apparatus. Only the global observables such as the position or momentum of the atom evolve and it was thus legitimate to describe each atom as a particle.

But even a simple experiment such as the Young double-slit one requires, when trying to analyse its subtle complementarity aspects, the introduction of composite systems. In fact, as soon as we 'do' something on a system, we have to let it interact with a 'probe' and the global entity we have to describe becomes composite. This is the case when we try to locate an atom inside the Young apparatus by shining photons on it (Fig. 2.3, on page 36). We have then to consider the composite atom–photon system, a main actor in this book. It is also the case when describing an atom carrying a spin through the interferometer. The system is then composite in a slightly different way. We have to consider the interaction of external ($\mathbf{R}$ and $\mathbf{P}$) and internal ($\mathbf{S}$) degrees of freedom of the same particle. And, in the case in which the interferometer itself becomes small enough to be described quantum mechanically, as in Bohr's version of the Young double-slit experiment (Fig. 1.6, on page 13), we have to consider the interaction of the particle with the slit, again a composite system. Finally, in most situations, the systems are unavoidably coupled to their environment – thermal bath of molecules or photons – which, ultimately, must also be dealt with by quantum mechanics. In this way, all systems become composite when we try to describe the effects of the environment, leading to relaxation and decoherence.

At a formal level, a composite system's state is represented in a global Hilbert space, a tensor product of at least two independent smaller spaces. In each of them, distinct observables should be defined and independently measurable. This is the case in all the examples outlined above, in which the component systems have an unambiguous definition. Special mention should be made of systems of identical particles. The Hilbert space is then a tensor product of the individual particle's ones, in which the symmetrization postulate chooses the symmetric or anti-symmetric subspaces. In order to define such a system as composite, one must also identify independent observables belonging to its subparts. For instance, $\mathcal{N}$ photons in a single-mode laser beam, or $\mathcal{N}$ bosons in a single Bose–Einstein condensate state are not composite because all possible observables act globally on the complete system, without distinguishing its parts. To consider such a system as composite, one must design an apparatus able to couple the particles' field into at least two different 'output modes' whose distinct Hilbert spaces are clearly defined in the second quantization frame. Separate observations can then be performed on these modes. This is, for instance, the situation of a laser beam coupled into two output modes by a beam-splitter (Chapter 3) or of a Bose–Einstein condensate split into two different parts by a coherent microwave pulse changing the internal state of the atoms (Chapter 9). In these conditions only, the system of identical particles can be seen as composite.

2.4.1 Bipartite entanglement

As the simplest possible composite object, let us consider a bipartite quantum system S made of two parts A and B, whose states belong to two separate Hilbert spaces $\mathcal{H}_A$ and $\mathcal{H}_B$, in which we define respectively the bases $\{|i_A\rangle\}$ and $\{|\mu_B\rangle\}$[7] (these bases can be either discrete or continuous). If we prepare A and B independently of each other (by performing for instance measurements of observables acting separately in $\mathcal{H}_A$ or $\mathcal{H}_B$) and if we do not couple them together afterwards, S is described by a mere tensor product state of the form $|\psi_S\rangle = |\psi_A\rangle \otimes |\psi_B\rangle$. Each subsystem is described by a well-defined wave function. Any manipulation performed on one part leaves unchanged the measurement results prediction for the other. For all purposes, we can forget B when studying A and vice versa.

The system S can also be prepared by measuring joint observables, acting simultaneously on A and B. Or, even if S has been prepared by measuring separate observables, A and B may be subsequently coupled together by an interaction Hamiltonian. It is then in general impossible to write the global state $|\psi_S\rangle$ as a product of partial states associated to each component of S. Such a situation is referred to as 'quantum entanglement'. The S state, $|\psi_S\rangle$, even if not factorizable, can however, according to the superposition principle, be expressed as a sum of product states $|i_A\rangle \otimes |\mu_B\rangle$. These products make up a basis in the global Hilbert space, $\mathcal{H}_S$, so that an entangled state can be expressed as:

$$|\psi_S\rangle = \sum_{i,\mu} \alpha_{i\mu} |i_A\rangle \otimes |\mu_B\rangle \neq |\psi_A\rangle \otimes |\psi_B\rangle \ , \tag{2.42}$$

where the $\alpha_{i\mu}$ are complex amplitudes. The state $|\psi_S\rangle$ contains information not only about the results of measurements on A and B separately, but also on correlations between these measurements.

---

[7]Since we manipulate subsystems, each associated to its own Hilbert space, we label the kets and the bras with subscripts or superscripts, when necessary to avoid confusion.

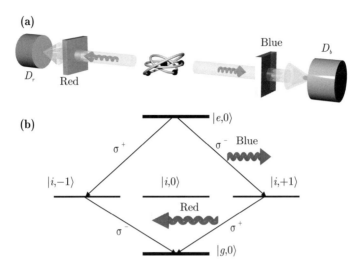

Fig. 2.9 Emission of two polarization-entangled photons in an atomic fluorescence cascade. (a) Scheme of a correlation experiment. An atom emits a blue and a red photon in a cascade process. The red photon is detected by observer D_r, the blue one by D_b. (b) Level scheme depicting the polarization entanglement of the two emitted photons.

Two spin-$1/2$ particles constitute the simplest system in which entanglement occurs. As an example, let us consider the singlet spin state:

$$\left|\psi_{\text{singlet}}\right\rangle = (|0,1\rangle - |1,0\rangle)/\sqrt{2} , \tag{2.43}$$

where the symbols in each ket refer to the spin of particles A and B respectively (0 and 1 being the notation already introduced for the 'up' and 'down' eigenstates of σ_Z). This is the eigenstate with zero total angular momentum, and it is manifestly entangled. It plays an important role in many atomic physics problems. For instance, in the ground level of hydrogen, the electron and nuclear spins are in this state, which has been known to atomic physicists since the early days of quantum physics (Bethe and Salpeter 1977). The entanglement is produced by the hyperfine coupling between the electron and the nuclear spin. The entangled system remains localized in a small region of space, of the order of one Bohr radius. The strong correlation between the two subsystems is a fact which never raised much surprise because the system is microscopic in size.

    In other cases, entanglement survives to the subsystems separation and leads to quantum correlations which have intrigued physicists since the beginning of quantum physics. Figure 2.9(a) describes such a situation: an atom emits two photons propagating in opposite directions by undergoing two cascading transitions from an excited state $|e,0\rangle$ to a ground level $|g,0\rangle$, both with angular momentum 0. The intermediate level $|i\rangle$ of the cascade has an angular momentum 1, corresponding to three Zeeman substates $|i,m\rangle$ ($m = 0, \pm1$). The $|e,0\rangle \rightarrow |i,m\rangle$ transition corresponds to the emission of a blue photon detected by the right detector D_b, while the $|i,m\rangle \rightarrow |g,0\rangle$ transition results in the emission of a red photon into the left detector D_r (the photons of different colours are discriminated by appropriate filters).

The conservation of angular momentum during the cascade entails that the blue and red photons have opposite circular polarizations. This selection rule can be achieved in two different paths, one in which the blue photon has a right circular polarization σ^+ and the red photon a left one σ^-, the other in which these polarizations are reversed. In the first case, the atom cascades through $|i, m = -1\rangle$, in the second case through $|i, m = +1\rangle$. The polarization of the pairs of photons is thus described by an entangled state:

$$|\psi_{\text{photons}}\rangle = \left(|\sigma_b^+\rangle \, |\sigma_r^-\rangle + |\sigma_b^-\rangle \, |\sigma_r^+\rangle\right)/\sqrt{2} \,, \tag{2.44}$$

where the indices indicate the photon colour. The photons propagate over large distances and remain polarization-entangled in the way described by eqn. (2.44). As discussed below, the resulting photon counting correlations have strange properties for a classical mind.

2.4.2   General properties of bipartite pure entangled states

We start by recalling some general properties of bipartite entangled states. We must first emphasize a basic property which looks like a tautology and justifies the name of entanglement: in an entangled state, each part loses its quantum identity. Information about the measurement outcomes is stored in a global wave function (we consider here only the case of pure states for the global system) which cannot be 'torn' apart in two separate partial states. The quantum content of the global state is intricately interwoven between the parts. In many situations, we are, however, interested in performing manipulations or measurements on one part without looking at the other. For example, we might wish to know the probability of finding a given result when measuring an observable O_A attached to subsystem A, without worrying about what happens to B. Predictions about the outcome of such an experiment can be made by introducing the unity operator $\mathbb{1}_B$ in $\mathcal{H}_B$ and extending O_A in $\mathcal{H}_S$, replacing it by $O_A \otimes \mathbb{1}_B$. We can then use the complete wave function $|\psi_S\rangle$ to predict the experimental outcomes of the measurement of $O_A \otimes \mathbb{1}_B$. This approach is cumbersome and can be shortcut by introducing the density operator formalism. The density operator ρ_S of a system described by the quantum state $|\psi_S\rangle$ is the projector onto this state:

$$\rho_S = |\psi_S\rangle \, \langle \psi_S| \,. \tag{2.45}$$

It has the same information content as $|\psi_S\rangle$ itself and, for predictions on the global system S, all quantum rules can be expressed in the density operator formalism. The expectation value of an observable O_S of S is $\text{Tr}(O_S \rho_S) = \langle \psi_S| O_S |\psi_S\rangle$ (Tr is the trace operation). The probability of finding after a measurement the system in a state $|i\rangle$ (corresponding to the projector $P_i = |i\rangle \langle i|$) is $|\langle i|\psi_S\rangle|^2$ in the quantum state description and $\text{Tr}(P_i \rho_S)$ in the density matrix one, two obviously identical expressions.

The density operator approach becomes advantageous for describing a subsystem (say A) without looking at B. Whereas there is no partial quantum state, we can build a partial density operator ρ_A containing all predictive information about A alone, by tracing ρ_S over the subspace of B:

$$\rho_A = \text{Tr}_B(\rho_S) = \sum_{i,i';\mu} \alpha_{i\mu} \alpha_{i'\mu}^* |i_A\rangle \, \langle i_A'| \,. \tag{2.46}$$

From the knowledge of ρ_A, it is straightforward to obtain the probability π_j of finding A in a quantum state $|j_A\rangle$ by computing the expectation value of the corresponding projector $P_j^A = |j_A\rangle\langle j_A|$:

$$\pi_j = \text{Tr}(\rho_A P_j^A) \, . \tag{2.47}$$

We can thus make predictions on A without having to consider B. It is important to recognize however that the information content in ρ_A [and in the equivalent partial density operator $\rho_B = \text{Tr}_A(\rho_S)$] is smaller than in ρ_S since we renounce, by considering A alone, accounting for the correlations between A and B.

In fact, stating that A and B are entangled is equivalent to saying that the partial density operators ρ_A and ρ_B are not projectors on a quantum state. They are, however, hermitian operators which can be expanded as a sum of projectors. There is at least a basis in $\mathcal{H}_A$ in which ρ_A is diagonal (an infinity of bases if one eigenvalue is degenerate). In one of these bases, $\{|j_A\rangle\}$, ρ_A is:

$$\rho_A = \sum_j \lambda_j \, |j_A\rangle\langle j_A| \, , \tag{2.48}$$

where the λ_j are positive or zero eigenvalues whose sum over j is equal to 1. Equation (2.48) appears as a probabilistic expansion. By overlooking what happens to B, we have only a statistical knowledge of the state of A, with a probability λ_j of finding it in $|j_A\rangle$. The expansion (2.48) is quite different from a quantum mechanical superposition of states; here the coefficients are positive probabilities and not complex amplitudes. If A and B are not entangled, ρ_A (and ρ_B) correspond to pure states. One λ_j is equal to 1 and all the others are zero. As soon as at least two λ_j's are non-zero, A and B are entangled.

We have recalled here the essential features of the density operator formalism required to analyse in an elementary way the concepts of entanglement, non-locality and decoherence. Other interesting properties of the density operator are discussed in Chapter 4.

2.4.3 Schmidt expansion and entropy of entanglement

It is instructive to express the state of S in a representation which displays explicitly the entanglement. In general, the superposition in eqn. (2.42) does not tell us at first sight whether the state can be factored. To exhibit clearly this property, it is convenient to choose in $\mathcal{H}_A$ a basis $|j_A\rangle$ in which ρ_A is diagonal. Equation (2.42) then can be written:

$$|\psi_S\rangle = \sum_j |j_A\rangle \, |\tilde{j}_B\rangle \, , \tag{2.49}$$

with:

$$|\tilde{j}_B\rangle = \sum_\mu \alpha_{j\mu} \, |\mu_B\rangle \, . \tag{2.50}$$

The $|\tilde{j}_B\rangle$ are mirroring in $\mathcal{H}_B$ the basis of orthonormal states in $\mathcal{H}_A$ in which ρ_A is diagonal. These mirror states are also orthogonal to each other, as is easily shown by expressing that ρ_A is diagonal:

$$\langle j_A | \, \rho_A \, | j'_A \rangle = \lambda_j \delta_{jj'} = \langle \tilde{j}_B | \tilde{j}'_B \rangle \, . \tag{2.51}$$

Finally, normalizing the mirror state by the transformation $|\hat{j}_B\rangle = |\tilde{j}_B\rangle / \sqrt{\lambda_j}$, we obtain the Schmidt expansion:

$$|\psi_S\rangle = \sum_j \sqrt{\lambda_j} \, |j_A\rangle \, |\hat{j}_B\rangle \; , \tag{2.52}$$

a sum over a basis of product mirror states exhibiting in a transparent way the entanglement between A and B.

The symmetry of this expression shows that ρ_A and ρ_B have the same eigenvalues. Any pure entangled state of a bipartite system can be expressed in this simple form. It applies to systems having discrete as well as continuous bases. For the sake of simplicity, we restrict from now on our analysis to finite-dimensional problems and call n_A and n_B the dimensions of $\mathcal{H}_A$ and $\mathcal{H}_B$, assuming without loss of generality $n_A \leq n_B$. The entanglement between A and B selects in $\mathcal{H}_B$ a 'mirror' subspace spanned by a basis with precisely n_A dimensions. The partial density operator ρ_B must thus, if $n_A \neq n_B$, be completed by zeros. The Schmidt form corresponds to a great simplification over the most general expansion of an entangled state. Whereas the superposition (2.42) may have $n_A n_B$ terms, the Schmidt form has at most n_A terms.

When the eigenvalues of ρ_A and ρ_B are non-degenerate, the $|j_A\rangle$ and $|\hat{j}_B\rangle$ states are unambiguously determined (up to a phase) and the Schmidt expansion is unique. If a non-zero eigenvalue is degenerate, the corresponding states in the expansion are defined up to a unitary transformation within the degeneracy subspace. When changing the basis states, correlated transformations must be performed in $\mathcal{H}_A$ and $\mathcal{H}_B$. The Schmidt expansion is then non-unique, an entangled state $|\psi_S\rangle$ being represented by different equivalent expressions. This ambiguity in the state description has deep physical consequences discussed later.

We now define a measure of the degree of entanglement. It is intuitive that the more the λ_j probabilities are spread out over many non-zero values, more information we lose by focusing separately on one system, disregarding the correlations between A and B. It is reasonable to link this loss of mutual information to the degree of entanglement. A natural way to measure this loss of information is to compute the von Neumann entropy of A or B, defined as:

$$\mathcal{S}_A = \mathcal{S}_B = -\sum_j \lambda_j \log_2 \lambda_j = -\mathrm{Tr}(\rho_A \log_2 \rho_A) = -\mathrm{Tr}(\rho_B \log_2 \rho_B) \; . \tag{2.53}$$

The logarithm function is here in basis 2, well-adapted to quantum information discussions. The entropy of entanglement $\mathcal{S}_e = \mathcal{S}_A = \mathcal{S}_B$, whose form is reminiscent of the Boltzmann statistical concept of entropy, expresses quantitatively the 'degree of disorder' in our knowledge of the partial density matrices of the two parts of the entangled system S.

If the system is separable, one λ_j only is non-zero and $\mathcal{S}_e = 0$. We then have maximum information on the states of both parts. As soon as at least two λ_j are non-zero, $\mathcal{S}_e$ becomes strictly positive and A and B are entangled. The maximum entropy, and, by this definition, maximum entanglement is reached when the λ_j's are equally distributed among the available dimensions of the A or B subspaces, i.e. when

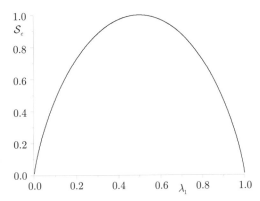

Fig. 2.10 Entropy of entanglement for a bipartite system as a function of the weights of the Schmidt expansion.

ρ_A and ρ_B are proportional to the unity operator in n_A dimensions. Such a unitary operator is invariant in any basis change within the mirror subspaces, meaning that the measurement of any observable of A or B alone restricted to these mirror subspaces yields equiprobable results for all its eigenvalues. In a maximally entangled state, local measurements performed on one part of the system are not predictable *at all*. What can be predicted are the correlations between the measurements performed on both parts. Figure 2.10 shows, as a simple example, the variation of S_e as a function of λ_1 for a pure-state bipartite system in which one part has a dimension 2. There are only two λ values in the Schmidt expansion, related by the normalization $\lambda_1 + \lambda_2 = 1$. The degree of entanglement, equal to zero for $\lambda_1 = 0$ and 1, goes through a maximum equal to 1 for $\lambda_1 = 1/2$.

The degree of entanglement measured by the von Neumann entropy is invariant under local unitary transformations acting on A or B separately. This is a direct consequence of the invariance of the spectrum of the partial density operators ρ_A and ρ_B under local transformations. It is only by acting jointly on A and B or by making non-unitary measurements that one can change the degree of entanglement. This basic property will be illustrated in several situations.

2.4.4 Bell states

Consider now the simple case where both A and B are two-level systems ($n_A = n_B = 2$). We call $\left|0_{A/B}\right\rangle$ and $\left|1_{A/B}\right\rangle$ two orthonormal basis vectors in $\mathcal{H}_A$ and $\mathcal{H}_B$. This labelling applies to any kind of two-level system (spin, photon polarization states or ion hyperfine substates). The global Hilbert space $\mathcal{H}_S$ has four dimensions with an obvious basis made of the maximally entangled 'Bell states' (Braunstein *et al.* 1992):

$$\left|\phi^{\pm}\right\rangle = (\left|0_A 0_B\right\rangle \pm \left|1_A 1_B\right\rangle)/\sqrt{2} ; \qquad \left|\psi^{\pm}\right\rangle = (\left|0_A 1_B\right\rangle \pm \left|1_A 0_B\right\rangle)/\sqrt{2} . \qquad (2.54)$$

They are expressed in the Schmidt form (eqn. 2.52) with $\lambda_1 = \lambda_2 = 1/2$. We should not be concerned by the fact that the Schmidt expansion has positive coefficients whereas $\left|\phi^-\right\rangle$ and $\left|\psi^-\right\rangle$ have negative amplitudes. This can be taken care of by a convenient definition of the Schmidt basis of mirror states whose phase can be freely chosen. From eqn. (2.53) we find that the entropy of entanglement in the four Bell states is maximum, equal to 1. The Bell state $\left|\psi^-\right\rangle$ deserves special mention. It corresponds to the singlet state of a two-spin system and is invariant under a global rotation.

Let us now analyse the strange features of these states. To personalize somewhat the situation, and to follow the playful notation introduced by information scientists, we assume that a system prepared in one of the Bell states is shared by a couple of experimenters called Alice and Bob. They are very far apart and can communicate by classical means (telephone or radio) to compare the results of their measurements. Alice is given subsystem A and Bob B. Each can perform on his (her) part measurements of any spin component $\sigma_{\mathbf{u}}$. He (she) will then find the results 1 and -1 with equal probabilities. This is a direct consequence of the maximum entanglement of Bell states. The partial density operators of A and B are proportional to $\mathbb{1}$. All quantum information is contained in the correlations between their measurements. These correlations are plainly visible if both decide to measure σ_Z. Equation (2.54) shows that they always find the same result if the system is in one of the $|\phi_{\pm}\rangle$ Bell states, and opposite results for $|\psi_{\pm}\rangle$.

Such perfect correlations are not *a priori* surprising. They exist also in classical physics. Think of the following experiment. An ensemble of blue and red otherwise identical balls is sorted in pairs of same colour by a 'preparer' (let us call him Charles). Tossing a dice, he chooses a pair of a given colour (for instance red for 1, 2 or 3 and blue for 4, 5 or 6). He then puts each ball of the pair in two opaque boxes and sends one to Alice and the other to Bob. If the dice tossing has been hidden from them, they do not know which colour they will find until they open their boxes, but they are sure to find the same colour.

2.4.5 The EPR paradox, local realism and hidden variables

The situation is however subtler in the quantum situation, because we can also consider 'superpositions of colours'. For instance, the Bell state $|\phi^+\rangle$ can also be written as:

$$|\phi^+\rangle = \tfrac{1}{2}\left[(|0_A\rangle + |1_A\rangle)(|0_B\rangle + |1_B\rangle) + (|0_A\rangle - |1_A\rangle)(|0_B\rangle - |1_B\rangle)\right] , \qquad (2.55)$$

which makes it explicit that the results of Alice's and Bob's measurements are also always the same if they decide to measure σ_X instead of σ_Z. That the same quantum state can be expressed in different Schmidt forms (compare eqns. 2.54 and 2.55) is – as already noticed above – a consequence of the degeneracy of the partial density operators ρ_A and ρ_B in this case of maximum entanglement.

More generally, the product $\sigma_{\mathbf{u}}^A \sigma_{\mathbf{u}}^B$ of the same observables measured by Alice and Bob satisfy the following relations:

$$
\begin{aligned}
\sigma_{\mathbf{u}}^A \sigma_{\mathbf{u}}^B |\phi^+\rangle &= +|\phi^+\rangle & (\mathbf{u} \text{ in } XOZ) ; \\
\sigma_{\mathbf{u}}^A \sigma_{\mathbf{u}}^B |\phi^-\rangle &= +|\phi^-\rangle & (\mathbf{u} \text{ in } YOZ) ; \\
\sigma_{\mathbf{u}}^A \sigma_{\mathbf{u}}^B |\psi^+\rangle &= +|\psi^+\rangle & (\mathbf{u} \text{ in } XOY) ; \\
\sigma_{\mathbf{u}}^A \sigma_{\mathbf{u}}^B |\psi^-\rangle &= -|\psi^-\rangle & (\text{any } \mathbf{u}) ,
\end{aligned}
\qquad (2.56)
$$

which show that local measurements performed by Alice and Bob on the same observable yield the same result, provided it corresponds to a spin-like component lying in a well-defined plane if the system is prepared initially in $|\phi^+\rangle$, $|\phi^-\rangle$ or $|\psi^+\rangle$. As for the Bell state $|\psi^-\rangle$, it yields perfectly anti-correlated results for any direction $\mathbf{u}$ (following from the rotation-invariance of the singlet state).

Is it possible to understand these perfect correlations (or anti-correlations) by a classical argument of the kind discussed above, involving the tossing of a dice whose result is hidden from Alice and Bob, at the time the system is prepared? Can the reasoning about red and blue balls be generalized in a world where superpositions of colours are possible? These correlations are strange because the results obtained by Alice and Bob are completely random, and yet fully correlated, whatever common measurement choice they decide to make on A and B, even long after they have been spatially separated. This situation was first analysed in a famous paper by Einstein, Podolsky and Rosen (1935) and the experiments in which such correlations are studied are said to be of the 'EPR type'. Although EPR discussed a somewhat different situation (two particles with entangled position and momenta) the physics is basically the same as for a discrete 2×2 system, as has been subsequently shown by Bohm (1951).

The EPR argument goes as follows. Let us assume for the sake of definiteness that Alice and Bob share $|\phi_+\rangle$. Alice, considering her subsystem A and not touching it, can know with certainty whether she will find $+1$ or -1 when measuring an observable $\sigma_{\mathbf{u}}$ (with $\mathbf{u}$ in the XOZ plane). She simply asks Bob to perform the measurement on B and to phone her the result. She knows she would find the same, had she actually performed the experiment. It is then reasonable to assume that the value of $\sigma_{\mathbf{u}}$ in Alice's subsystem does have a well-defined (albeit random) value *before* her measurement, since she has not done anything to her particle and yet has a way to know for sure what value this observable takes. The situation seems similar to the classical experiment with the red and blue balls, in which Alice could also, by asking Bob, know the colour of her ball without opening her box. The value of Alice's spin along any direction of the XOZ plane is what EPR called an 'element of reality' (and so is of course the value of Bob's spin).

The point of view which attributes reality to an unmeasured quantity if its value can be known with certainty without perturbing *locally* the system is known as 'local realism'. The same argument can be made about a continuous infinity of spin-like components, so that Alice and Bob must have, according to local realism, predetermined values of all these quantities 'encoded' in their system. That these values are unknown until one of them has performed a measurement is, according to this point of view, a matter of statistical uncertainty. All happens as if the spin values were determined by some 'hidden variables' (Bell 1966) whose choice has been made, unknown to Alice and Bob, by a virtual Charles at the moment when the Bell state was prepared by a joint operation on A and B, before the system's separation (the equivalent of the classical tossing of a dice). EPR did not refer explicitly to hidden variables but their discussion implicitly contained them. For this reason, local realism is often referred to as 'local hidden variable theory'.

Of course, EPR knew very well that Alice and Bob could not find out all the predetermined values of their observables. Each of them can measure only one spin component, since the measurement perturbs the system's wave function, changing in an unpredictable way the value he (or she) would subsequently find for another component. This fundamental Heisenberg uncertainty is not in contradiction with hidden variable theories, since one could imagine that an actual measurement, which involves the interaction of the system with an external apparatus, changes in an uncontrollable

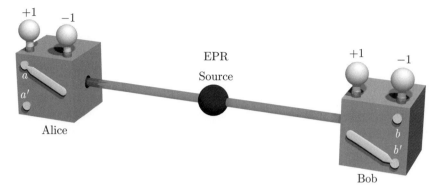

Fig. 2.11 Principle of a CHSH experiment testing Bell's inequalities.

way the value of the hidden variables and affects the results of subsequent measurements. Einstein found this situation quite uncomfortable though. He thought that a satisfactory theory should be able to speak explicitly about all 'real' quantities such as the values of the spin in all directions and he believed that, if quantum mechanics could not do it, it must be incomplete.

Bohr (1935) did not like Einstein's argument based on the explicit consideration of unmeasured quantities. Such a point of view is qualified as 'counterfactual' since it admits the existence of observable values in the absence on any factual measurement of these quantities. This contradicts a basic statement of standard quantum physics which adamantly refuses to consider the value of an observable until it has been measured *'an experiment not done has no result'* as Asher Peres used to say.[8] In the absence of any measuring device, the entangled system is suspended in a superposition of many potential realities. Only an actual interaction with a measuring apparatus can lift the quantum ambiguity and actualize one of the possible results. This point is central to all discussions about complementarity. In short, according to the orthodox quantum theory, as strange as it may seem, Alice's measurements perturb the *whole* entangled system and not only the part she has in her hands and the same is true for Bob. Speaking about local realism and invoking counterfactual quantities is thus, in this point of view, totally heretic.

2.4.6 Bell's inequalities

At this stage of the confrontation between Einstein and Bohr, the debate was rather philosophical. It was not even clear that there were a contradiction between the two approaches, which could, after all, be considered as two equivalent ways of interpreting the laws of Nature. Thirty years after the Bohr–Einstein debate, John Bell's merit has been to show that the hidden variable and orthodox quantum mechanics points of views were irreconcilable and to propose experiments to decide between them. He has shown that Alice and Bob can use the correlations between their measurements to construct statistical quantities, which should satisfy algebraic inequalities if local hidden variable theories are right, whereas these inequalities are violated in some

---

[8]For an illuminating discussion about counterfactual reasoning and Bell inequalities, see Peres (1995).

instances by quantum theory (Bell 1964; 1966; 1987). These so-called Bell inequalities are based on simple logical arguments. They take different forms depending upon the kinds of correlations which are tested on the system.

One of these forms, due to Clauser, Horne, Shimony and Holt (CHSH) (1969), is simple to derive with a classical analogy. To test the CHSH inequality, Alice and Bob proceed in the following way. Sharing a large number of systems, all prepared in the same entangled state, they decide together to measure on their subsystems four spin components corresponding to directions $\mathbf{a}$, $\mathbf{a}'$, $\mathbf{b}$ and $\mathbf{b}'$. Alice measures $\sigma_{\mathbf{a}}^A$ and $\sigma_{\mathbf{a}'}^A$ while Bob detects $\sigma_{\mathbf{b}}^B$ and $\sigma_{\mathbf{b}'}^B$ (Fig. 2.11). Of course, for each realization of the experiment, Alice can only measure one component among the $\mathbf{a}$, $\mathbf{a}'$ polarizations and Bob one among the $\mathbf{b}$, $\mathbf{b}'$. From one realization to the next, they can change the polarizations and thus perform a large number of coincidence measurements on the four couples of polarizations $(\mathbf{a}, \mathbf{b})$, $(\mathbf{a}, \mathbf{b}')$, $(\mathbf{a}', \mathbf{b})$ and $(\mathbf{a}', \mathbf{b}')$. Each of their measurements yields a random value ϵ_a, $\epsilon_{a'}$, ϵ_b or $\epsilon_{b'}$, equal to ± 1. One can think of these values as resulting from a four-coin tossing, with 'heads' for $+1$ and 'tails' for -1. Two of these coins are from one kind (let us say two dimes representing A) whereas the two other are from another denomination (two nickels for B). When trying to find out on which side their coins fall, Alice is able to see only one of the two dimes and Bob one of the two nickels. As soon as they have observed one of their coins, the other is blurred, due to the Heisenberg uncertainties. Nevertheless, by tossing the coins again and again, they are able to build statistical averages over the four-coins configurations.

According to the local realism argument, the four quantities are all defined for each realization of the experiment, even if only two of them can actually be measured. Each of these random sets of values satisfies, in all cases, a simple relationship:

$$(\epsilon_a - \epsilon_{a'})\epsilon_b + (\epsilon_a + \epsilon_{a'})\epsilon_{b'} = \pm 2 . \tag{2.57}$$

An experimentalist can verify this result by himself, tossing again and again two nickels and two dimes and forming the corresponding combination with his $+1$ (head) and -1 (tails) readings. Those more mathematically inclined will simply notice that one of the two $(\epsilon_a + \epsilon_{a'})$ and $(\epsilon_a - \epsilon_{a'})$ expressions is zero whereas the other is equal to ± 2, which demonstrates eqn. (2.57). By accumulating data, Alice and Bob average the four terms in (2.57) and construct the algebraic sum Σ_b of the corresponding quantum mechanical expectation values:

$$\Sigma_b = \langle \sigma_a^A \sigma_b^B \rangle - \langle \sigma_{a'}^A \sigma_b^B \rangle + \langle \sigma_a^A \sigma_{b'}^B \rangle + \langle \sigma_{a'}^A \sigma_{b'}^B \rangle , \tag{2.58}$$

which, being the average of a quantity which can take only the values ± 2, must be bounded by these limits. Hence, according to local realism, we have:

$$-2 \leq \Sigma_b \leq +2 . \tag{2.59}$$

This is the CHSH Bell inequality, established under the very general argument that the spin-like observables of A and B have a contrafactual reality. We have made no assumption about the inner workings of the hidden variables, which are supposed to determine the values of these quantities. We have also made no use of the specific correlation properties of the two spins. This means that these inequalities hold for any system, whether it is in one of the maximally entangled Bell states, in a less entangled

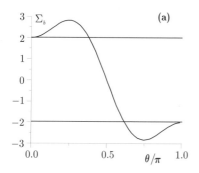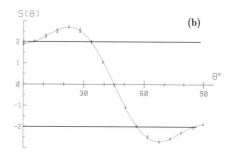

Fig. 2.12 Variations of the Bell sum Σ_b as a function of the angle θ. (a) Theoretical predictions. (b) Experimental data. The points are experimental and the line corresponds to theory, including the measured experimental imperfections. Courtesy of A. Aspect. See also Aspect (2004).

state (including statistical mixtures) or in a separable A–B state. Neither have we made any hypothesis about the physical subsystems A and B, which need not be spins. We only require that we can measure on each of them two bivalued observables which we can always redefine, without loss of generality, as being equal to ± 1 (we will see later other examples of Bell's inequalities applied to correlations of bivalued observables which are not spins).

Let us come back to the spins and describe the same correlation experiment by quantum theory. If we know the initial state (supposed, for the sake of definiteness, to be $|\phi^+\rangle$) and the four polarization directions (in the XOZ plane), the averages $\langle \sigma_a^A \sigma_b^B \rangle$, $\langle \sigma_{a'}^A \sigma_b^B \rangle$, $\langle \sigma_a^A \sigma_{b'}^B \rangle$ and $\langle \sigma_{a'}^A \sigma_{b'}^B \rangle$ are determined. For instance, $\langle \sigma_a^A \sigma_b^B \rangle$ measures the correlation between spin-like components along two directions **a** for A and **b** for B, making an angle θ_{ab}. If Alice measures her spin first and finds it pointing along **a**, Bob's part is immediately projected into the same spin eigenstate (eqn. 2.56). A subsequent measurement by Bob of the spin along **b** yields with a probability $\cos^2(\theta_{ab}/2)$ the same result as Alice's measurement and with probability $\sin^2(\theta_{ab}/2)$ the opposite result (eqn. 2.6). The spin correlation is thus $\langle \sigma_a^A \sigma_b^B \rangle = \cos^2(\theta_{ab}/2) - \sin^2(\theta_{ab}/2) = \cos(\theta_{ab})$. The result depends upon θ_{ab} and is independent of the time-order of the measurements performed by Alice and Bob.

Computing in the same way the other correlations in the Bell sum, we find:

$$\Sigma_b = \cos\theta_{ab} - \cos\theta_{a'b} + \cos\theta_{ab'} + \cos\theta_{a'b'} \ , \tag{2.60}$$

with obvious definitions for the angles $\theta_{a'b}$, $\theta_{ab'}$ and $\theta_{a'b'}$. Let us choose:

$$\theta_{ab} = \theta_{b'a} = \theta_{a'b'} = \theta \ ; \qquad \theta_{a'b} = \theta_{a'b'} + \theta_{b'a} + \theta_{ab} = 3\theta \ . \tag{2.61}$$

The Bell sum now is:

$$\Sigma_b(\theta) = 3\cos\theta - \cos 3\theta \ . \tag{2.62}$$

This function of θ is plotted in Fig. 2.12(a). For a range of θ values, $\Sigma_b(\theta)$ is larger than $+2$, or smaller than -2, clearly violating Bell's inequalities. For $\theta = \pi/4$ the sum takes its maximum value $+2\sqrt{2}$ and for $\theta = 3\pi/4$ its minimum value $-2\sqrt{2}$. The same conclusions hold if Alice and Bob are testing $|\phi^-\rangle$ or $|\psi^+\rangle$ (provided they change accordingly the plane in which the detection polarizations are defined). If they test the rotation-invariant $|\psi^-\rangle$ state, they can choose any plane for the polarizations and they find a Bell sum with an opposite sign with the same $\pm 2\sqrt{2}$ limits, also violating Bell's inequalities.

The CHSH inequalities, or simple variants of them, have been experimentally tested by detecting coincidences on pairs of maximally entangled photons produced, first, by atomic fluorescence (Aspect *et al.* 1981; 1982a;b) and, more recently, by frequency down-conversion processes in non-linear crystals (Ou and Mandel 1988; Rarity and Tapster 1990; Kwiat *et al.* 1995; Zeilinger 1999; White *et al.* 1999).

Figure 2.12 shows the variations of the Bell signal observed in the pioneering experiment of Aspect *et al.* (1982b; 2004). The experimental data is in excellent agreement with the quantum predictions and exhibit a clear violation of Bell's inequalities. This early experiment was performed with photon pairs emitted in a cascade process of the kind depicted in Fig. 2.9, with a photon-polarization Bell state given by eqn. (2.44). It was a technological *tour de force* requiring many hours of photon counting. Since then, the photon sources have greatly improved and similar data can be taken in minutes instead of hours (Kwiat *et al.* 1995).

The distance between Bob and Alice, a few metres in the first experiments, has been increased to kilometres (Tittel *et al.* 1998; Weihs *et al.* 1998) by propagating the entangled photons in optical fibres. In these experiments as in Aspect's one, care has been taken to change the detector's polarizations fast enough to be sure that no signal at the speed of light could propagate between Bob and Alice during each coincidence measurement (Weihs *et al.* 1998). This precludes any explanation of the photon coincidences by a causal disturbance produced by one measurement on the other, which could be explained by a local hidden variable model.[9]

Of course, even the recent photon experiments still have detection noise, and more importantly finite photon detection efficiency. Some have speculated that the imperfect detection could explain the violation of Bell's inequality by some kind of hidden variable 'conspiracy' in which the quantum result could be mimicked (Gisin and Gisin 1999). These speculations do not seem very credible, but they still make it worth improving the detectors so as to close this 'detection loophole' in the chase for elusive hidden variables. Two recent experiments (Rowe *et al.* 2001; Roos *et al.* 2004a), performed with two entangled ions in a trap, are heading in this direction (see Chapter 8).

Bell's inequalities are violated not only by maximally entangled states, but more generally by any bipartite pure state of two spin-like systems exhibiting any degree of entanglement. To prove it, we consider the most general bipartite pure state of two spins which can be written, after a proper definition of the states in the A and B subspaces, as the Schmidt expansion:

$$|\phi\rangle = \sqrt{\lambda_1}\,|0,1\rangle + \sqrt{\lambda_2}\,|1,0\rangle \ , \quad \text{with } \lambda_1 + \lambda_2 = 1 \ . \tag{2.63}$$

---

[9]Earlier experiments, performed in the 1970s, had already shown violations of Bell's inequalities with fixed polarizers in front of Alice's and Bob's detectors (Clauser and Shimony 1978).

If θ_a and θ_b are the angles with OZ of two arbitrary directions **a** and **b** in the XOZ plane, we have (see eqn. 2.6):

$$\langle \sigma_{\mathbf{a}}^A \cdot \sigma_{\mathbf{b}}^B \rangle = -\cos\theta_a \cos\theta_b + 2\sqrt{\lambda_1 \lambda_2} \sin\theta_a \sin\theta_b . \qquad (2.64)$$

We now choose four directions **a**, **a**$'$, **b** and **b**$'$ in the XOZ plane such that their angles with OZ satisfy the relations $\theta_a = \pi - \theta_{a'}$, $\theta_b = 0$ and $\theta_{b'} = -\pi/2$. The Bell sum (2.58) then is:

$$\Sigma_b = -2\left(\cos\theta_a + 2\sqrt{\lambda_1 \lambda_2} \sin\theta_a\right) . \qquad (2.65)$$

It violates the CHSH inequality as soon as $\lambda_1 \lambda_2 \neq 0$ for θ_a arbitrarily close to 0 and π. The CHSH relation (2.59) is satisfied only if $\lambda_1 \lambda_2 = 0$, i.e. if the system is non-entangled. A hidden variable model applied to independent two-level systems is then able to reproduce all the results of quantum physics. In summary, a bipartite spin system in a pure state manifests non-locality if and only if it is entangled.

The detection loophole notwithstanding, there is now very good evidence that the local realism approach of EPR and Bohm, in spite of its reasonable assumptions and tantalizing likeliness, is not the way Nature behaves. Bell's argument provides us with a very strong test enabling us to exclude, on very general grounds, any contrafactual explanation of Nature at the microscopic scale. It seems increasingly more likely, with the development of recent experiments, that there are no hidden variables. Even if this may displease those who do not want to renounce locality, we must admit that this is good news for the consistency and beauty of quantum theory. If Bell's inequalities had been satisfied, it would have meant that the beautiful quantum construction, enabling us to make simple and precise predictions about the probabilities of so many microscopic events, was in some deep way wrong. We would then have had to try to understand how such a flawed theory could have been so successful at explaining and predicting time and again the microscopic world. And we would still have to find out what these mysterious hidden variables were and how they behaved.

Nature is deeply non-local, as Bohr thought it was. What happens in one part of the Universe is strongly correlated to what happens in another part. In an entangled system, the quantum ambiguity inscribed in the superposition principle remains until lifted somewhere by a measurement, which instantaneously actualizes the values of physical quantities that were until then completely undefined everywhere. This is strange indeed, and even appears at first sight to be at odds with causality. We will see however that the relativity principle is not violated by the non-locality of quantum physics, the kind of correlations measured in EPR experiments being unable to carry information faster than light. It should be noted that Einstein, who had understood this point, did not use relativity arguments to fight the quantum concepts he disliked so much. The consistency between the EPR correlations and relativity is by itself also strange, in some way. Quantum laws do ultimately satisfy relativity in their field theory version, but the measurement description we have implicitly invoked to compute the EPR correlations is non-relativistic. If it had violated causality, we could have invoked the necessity to use a relativistic version of measurement theory, dealing with the proper time of detection events. We do not have to do this. Non-relativistic quantum physics is non-local in a way subtle enough not to contradict the inherently relativistic causality principle.

2.4.7   Entanglement, complementarity and the quantum eraser

Entanglement is also deeply related to complementarity. As mentioned in Section 2.2, any which-path experiment finding out through which channel a particle travelled in an interferometer involves the coupling of the particle's translational degrees of freedom to a probe system carrying away information on the path. The particle and the probe evolve into an entangled state which is in the simplest case:

$$|\psi\rangle = (|\psi_a(1)\rangle\,|\psi_p(1)\rangle + |\psi_a(2)\rangle\,|\psi_p(2)\rangle)\,/\sqrt{2}\ , \tag{2.66}$$

where $|\psi_a(1)\rangle$ and $|\psi_a(2)\rangle$ are the orthogonal 'partial' particle states whose amplitudes interfere and $|\psi_p(1)\rangle$ and $|\psi_p(2)\rangle$ are the final states of the probe 'giving away' the path of the particle. The particle–probe ensemble is an EPR bipartite system in which the particle loses its quantum identity. If maximum entanglement is achieved between the particle and the probe, as is the case in eqn. (2.66) when $|\psi_p(1)\rangle$ and $|\psi_p(2)\rangle$ are orthogonal, the partial density operator of the particle becomes an incoherent sum of projectors on the two partial waves. It is then no surprise that the quantum interference vanishes, since there is no longer a quantum state or a coherent wave function for the particle.

Alice cannot observe quantum interference if her particle can be watched by a 'spy' (Bob) having the opportunity to observe a probe entangled to her system. As we will see in Chapter 6, the visibility of Alice's particle interference is directly related to the degree of entanglement between her particle and Bob's probe (Englert 1996). It is not even necessary that Bob actually performs his measurement. The mere fact that his probe is entangled to Alice's particle is enough to kill the interference. The partial density operator of Alice's particle is obtained by tracing over the unread particle states and the coherence of the superposition is lost in this operation.

The discussion of the EPR situation leads us to give another twist to the which-path experiment. Alice, who tries to observe the interference of her particle, and Bob, who observes the which-path probe, can decide to modify the rules of the game. Instead of detecting an observable sensitive to the particle path, Bob can choose to detect a non-commuting observable admitting as eigenstates the superpositions states $(|\psi_p(1)\rangle \pm |\psi_p(2)\rangle)/\sqrt{2}$. To analyse such an experiment, it is convenient to switch bases and to write the particle–probe entangled state in the following form, obviously equivalent to eqn. (2.66):

$$|\psi\rangle = \tfrac{1}{2}[(|\psi_a(1)\rangle + |\psi_a(2)\rangle)(|\psi_p(1)\rangle + |\psi_p(2)\rangle) + (|\psi_a(1)\rangle - |\psi_a(2)\rangle)(|\psi_p(1)\rangle - |\psi_p(2)\rangle)]\ . \tag{2.67}$$

If Bob finds the probe in the state $(|\psi_p(1)\rangle \pm |\psi_p(2)\rangle)/\sqrt{2}$, Alice's particle is found in the correlated superposition state $(|\psi_a(1)\rangle \pm |\psi_a(2)\rangle)/\sqrt{2}$, with the same $\pm$ sign. The coherence of Alice's interference pattern is preserved, provided she detects her particle in coincidence with the outcome of Bob's measurement. By correlating her measurement to Bob's, she finds two interference patterns with opposite phases (corresponding to the $+$ and $-$ signs of the superposition). If she detects her particle irrespective of Bob's finding, or if Bob does not measure anything, she obtains the sum of these two interference patterns which cancel out. We retrieve the washing out produced by tracing over an unread probe.

This restoration of an interference pattern is called a 'quantum eraser' effect (Schwinger *et al.* 1988; Scully and Walther 1989; Scully *et al.* 1989; Kwiat *et al.* 1992; Kim *et al.* 2000). By mixing the which-path states in his measurement, Bob is erasing information about Alice's particle path in the interferometer and this erasure re-establishes the fringes. The interference washing out effect is, in a way, a reversible effect since Bob can decide to 'erase' his path information after Alice's particle has gone through the interferometer, as done in a delayed-choice EPR experiment (Scully and Drühl 1982). We must notice however that quantum erasure requires an active action on Bob's and Alice's parts, whereas the which-path washing out effect can be passive. Bob must definitely detect a superposition of which-path states and Alice must correlate her detections to Bob's results. If Alice and Bob cannot communicate, the entanglement between Alice's particle and Bob's probe always results in a suppression of the interference.

We have described so far a simple situation in which the which-path apparatus is a two-state system. The quantum eraser procedure can be applied to probes which store path information in a larger Hilbert space. The Heisenberg microscope, discussed in Section 2.2, provides an example of such a multistate quantum eraser system. When a particle A crossing the interferometer scatters a photon of momentum $\hbar \mathbf{k}_0$ of the impinging beam, its state undergoes a momentum translation defined by the operator $e^{i\Delta \mathbf{k} \cdot \mathbf{R}}$ (eqn. 2.26). The particle state after this momentum kick, $\left| \psi_{\mathbf{k}}^{(A)}(0) \right\rangle = e^{i\Delta \mathbf{k} \cdot \mathbf{R}} \left| \psi^{(A)}(0) \right\rangle$, is correlated to the photon momentum $\mathbf{k} = \mathbf{k}_0 - \Delta \mathbf{k}$ of the scattered photon. Calling $|1_F : \mathbf{k}\rangle$ the state of the field with one photon of momentum $\mathbf{k}$, we can thus express the state $\left| \Psi^{(A+F)}(t) \right\rangle$ of the 'particle + probe' system at detection time t as the entangled superposition:

$$\left| \Psi^{(A+F)}(t) \right\rangle = \int d^3 k \, c_{\mathbf{k}} \left| \psi_{\mathbf{k}}^{(A)}(t) \right\rangle |1_F : \mathbf{k}\rangle \ , \qquad (2.68)$$

where the $c_{\mathbf{k}}$'s are probability amplitudes describing the angular distribution of the scattered field. This expression generalizes the simple formula (2.66) to the case of a which-path probe belonging to a large Hilbert space spanned by a continuous basis. Tracing over the photon field and using eqn. (2.30), we find that the particle distribution $\Pi_{\mathrm{wp}}(\mathbf{r})$ modified by the which-path detector is:

$$\Pi_{\mathrm{wp}}(\mathbf{r}) = \int d^3 k \, |c_{\mathbf{k}}|^2 \left| \left\langle \mathbf{r} \middle| \psi_{\mathbf{k}}^{(A)}(t) \right\rangle \right|^2 = \int d^3 k \, |c_{\mathbf{k}}|^2 \Pi(\mathbf{r} - \hbar \Delta \mathbf{k} \, t/m) \ , \qquad (2.69)$$

where $\Pi(\mathbf{r})$ is the interference pattern obtained with unperturbed particles. If the photon momentum is large enough to permit a resolution of the particle position better than a, the sum of translated fringe patterns given by eqn. (2.69) cancels out and the quantum interference is washed out, as discussed in Section 2.2. The quantum eraser procedure consists in this case in detecting the momentum of the scattered photon and correlating the observation of the particle distribution to this momentum. In practice, one could collect the scattered photons with a lens located far from the slits. The field is then focused in the focal plane of this lens, each point of this plane corresponding to a different direction of the scattered photon. By sorting out the impacts on the detection screen correlated to a given photon detection point, we obtain a high-contrast interference pattern. The phase of the fringes depends on the chosen photon momentum. We can say that by looking at the far-field of the scattering process, we have blurred the image of the scattering point which would have yielded information incompatible with the observation of the fringes. In this way, the procedure appears as an 'erasure of information'. More to the point, it is, as its simple two-state version, a bipartite correlation experiment of the EPR type.

2.5 The quantum–classical boundary

Quantum physics is essential for the description of phenomena at the atomic or sub-atomic scale. It is also, as seen in Section 2.3, required to understand in-depth collective effects involving very large numbers of particles. But, at the macroscopic scale, the quantum manifestations are usually veiled. The fact that a ball does not go through a screen may well, at the fundamental level, be due to the exclusion principle applied to the electrons of these massive objects, but the net result, the impermeability of solid objects to each other, is not what we would qualify as a weird fact. At the macro-scopic scale all happens as if, by a kind of conspiracy of Nature, the naked quantum was hidden, leaving us with an apparent world described consistently in a classical language. For instance, no one has ever seen a billiard ball going through two holes at once. One is even at pains trying to figure out what such a statement could mean.

One might even think that it is precisely because our instinct has been formed by the observation of macroscopic phenomena in which quantum superpositions are hidden, that such a notion is so hard to comprehend. An example of this attitude is given at the highest conceptual level by Einstein who adamantly refused to see quan-tum physics as the ultimate explanation of Nature, stating that even if this theory is '*logically acceptable without contradiction, it is so contrary to my scientific instincts that I cannot give up the search for a more complete system of concepts*' (Einstein 1936). Our brains, programmed by our personal experience and evolved through count-less generations, can 'understand' the collisions of billiard balls, not the scattering of electrons or atoms.

Stating this leaves unanswered the fundamental question at the heart of the inter-pretation of quantum physics: why do macroscopic objects made of particles which, individually, obey quantum rules behave classically? What is it that makes disappear the overt quantum weirdness at the macroscopic scale? How does the appearance of our classical world emerge from the basic quantum substrate? It is at this fulcrum, at the fuzzy boundary between the quantum and the classical, that the problems of quantum theory interpretation are put forward in the most dramatic way.

We have already given a partial answer to these questions. In the simple case of a single-particle motion, a well-defined trajectory results from quantum interference cancelling all the quantum paths for which the action is not extremum (Section 2.2.4), when the characteristic action of the motion is much greater than $\hbar$. This does not address, however, the case of composite systems. We have considered so far that the system's parts were microscopic. What happens if one of the two partners is a large object? How does the entanglement-free classical situation emerge out of the deeply

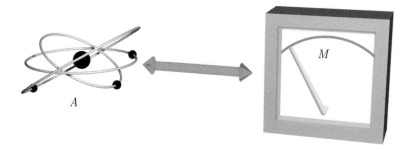

Fig. 2.13 Sketch of an ideal measurement coupling a microscopic object to a meter.

entangled quantum one?[10]

2.5.1 The measurement problem: superpositions in the macro world?

Description of an ideal measurement

The entanglement between microscopic and macroscopic systems is generally encountered when discussing a quantum measurement process. The measured system might be a single particle, an atom A for instance. It is coupled, through some kind of amplification scheme, to a macroscopic object recording information by the position of a 'meter' M on a dial (Fig. 2.13). For the sake of definiteness, we call M the tip of the needle, a point-like massive object evolving along an arc of circle, which we assimilate to a straight Ox axis. Its positions correspond to states denoted $|\xi\rangle$, which we take as Gaussian wave packets of width ΔX. Note that these wave packets are not strictly speaking position eigenstates. Before the measurement, the meter is in the neutral state $|\xi_0\rangle$. The atom is initially in $|\psi\rangle = \sum_i \alpha_i |a_i\rangle$ where the $|a_i\rangle$ are the eigenstates with eigenvalues ϵ_i (assumed to be non-degenerate) of an atomic observable O_A and the α_i are probability amplitudes. A good meter for O_A couples to the atom so that microscopic states $|a_i\rangle$ and $|a_j\rangle$ correlate to macroscopic states $|\xi_i\rangle$ and $|\xi_j\rangle$ verifying the non-overlapping condition $\langle\xi_i|\xi_j\rangle = \delta_{ij}$. The $A+M$ system thus evolves according to:

$$|\psi\rangle\,|\xi_0\rangle \rightarrow \sum_i \alpha_i\,|a_i\rangle\,|\xi_i\rangle \ . \tag{2.70}$$

Such a transformation, in which the evolution of the meter is conditioned to the state of the atom, is easy to implement by Hamiltonian dynamics. One can choose the atom–meter Hamiltonian as:

$$H_{\text{int}} = g\,O_A\,P \ , \tag{2.71}$$

where P is the meter's momentum and g a coupling constant (if O_A is dimensionless, g is a velocity). With this Hamiltonian, the evolution operator $U(t) = \exp(-iH_{\text{int}}t/\hbar)$ is, for each state $|a_i\rangle$, a translation of M along Ox by a distance proportional to the corresponding eigenvalue:

$$U(t)\,|\psi\rangle\,|\xi_0\rangle = \sum_i \alpha_i\,|a_i\rangle\,|\xi_0 + gt\epsilon_i\rangle \ . \tag{2.72}$$

---

[10]For an in-depth discussion of these issues, see for instance Omnès (1994); Giulini *et al.* (1996); Griffiths (2001); and Zurek (2003).

We assume implicitly that, during the measurement, the system's evolution is governed solely by the $A - M$ coupling, with negligible evolution due to the free atom and meter Hamiltonians. This is the case if g is large enough. A good meter should resolve without ambiguity the different atomic eigenvalues. Calling $\delta\epsilon$ the minimum eigenvalue separation, we must have:

$$gt\delta\epsilon \gg \Delta X \,, \tag{2.73}$$

which is again a condition on the strength of the $A - M$ coupling. Equation (2.73) also shows that, for a given apparatus, the resolution is proportional to the duration of the measurement. We have here a special example of an uncertainty relation, relating the precision of a measurement (proportional to the inverse of the minimal detectable difference $\delta\epsilon$) to the duration of the measuring process.

Equation (2.72) shows that information about the atom has been, via the atom–meter coupling, encoded into the macroscopic meter. When looking at it, an observer will find, according to the quantum postulates, one of the $\xi_i = \xi_0 + gt\epsilon_i$ positions (with probability $|\alpha_i|^2$). He infers that the atom is, with the same probability, in the $|a_i\rangle$ state corresponding to the ϵ_i eigenvalue. We do not question the way this macroscopic observation is performed. We just assume that the meter is large enough so that 'looking at it' is a classical process allowing us to determine without ambiguity (and without disturbing it) its position.

In practice, the link between the atomic micro-system and the meter goes through a chain of amplifying stages with intermediate mesoscopic systems getting correlated to each other in cascade. The $|a_i\rangle\,|\xi_i\rangle$ states of the above discussion must then, in all rigour, be replaced by tensor products of the form $|a_i\rangle\,|b_i\rangle\,|c_i\rangle\ldots|\xi_i\rangle$ linking the microscopic atomic states $|a_i\rangle$ to the macroscopic meter states $|\xi_i\rangle$ via a succession of states of the intermediate systems. For example, if one tries to detect an atom in a given energy eigenstate, the first step might be the state-selective ionization of the atom in a large electric field, releasing an electron in state $|b_i\rangle$. The electron is then sped up with an accelerating field, collides with a metallic plate releasing in turn a bunch of secondary electrons in state $|c_i\rangle$... Successive stages of similar amplification follow, leading to larger and larger electron pulses until the resulting macroscopic current, fed into a wire, is able to drive by its magnetic force a meter on a dial, bringing it in the final state $|\xi_i\rangle$. We will skip the chain of intermediate states, which does not fundamentally change the nature of the measurement problem, and consider that the amplification process takes place in a single step, directly linking the microscopic system to a classical meter whose observation is perturbation-free.

The Schrödinger cat paradox

This modelization of an ideal measurement raises however a difficult question linked to the EPR paradox. If the atom is initially in a state superposition, eqn. (2.72) shows that the $A + M$ system evolves into an entangled state. The meter is then involved in a state superposition whose components correspond to macroscopically distinct positions. Before any actual observation of the meter is performed, the weirdness of the quantum thus seems to have invaded the macroscopic world, via the unitary atom–meter evolution. Such a strange entangled state involving a macroscopic system can be related to the fate of the poor cat analysed by Schrödinger (1935) in a famous paper (Fig. 2.14).

CAN THE CAT BE ALIVE AND
DEAD AT THE SAME TIME?

Fig. 2.14 A cartoonist's view of Schrödinger's cat. Reprinted with permission from McEvoy and Zarate (1999). © Icon Books, Web site www.iconbooks.co.uk

To make the situation more dramatic, Schrödinger replaced the inanimate meter by a living animal and assumed that the atom–cat coupling was due to a radiative decay correlating two states of the atom to a 'living' and a 'dead' state of the feline respectively. The details of this (half) deadly contraption are not important. One could imagine for instance that, by decaying from an upper to a lower state, the atom emits a photon whose detection releases a hammer breaking an ampoule of prussic acid. The animal is a measuring device for the atom, in the same way as canary birds were being used, in the not too distant past, to measure the level of carbon monoxide in mine galleries. But, while the density of deadly CO is a classical variable which must assume a well-defined value obeying exclusive logic, a single atom can be suspended between two states according to quantum logic and thus, it seems that Schrödinger's cat can be at the same time *dead and alive*.

In order to discuss this paradox in the simplest possible way, we restrict our analysis to the case of a bivalued spin-like observable O_A with eigenvalues 0 and 1 and eigenstates $|0_A\rangle$ and $|1_A\rangle$, the corresponding states of the meter being $|0_M\rangle$ and $|1_M\rangle$.[11] The translation of the meter is realized by an $A - M$ coupling leaving M unchanged if A is in $|0_A\rangle$ and switching its state if the atom is in $|1_A\rangle$. This operation amounts to an addition modulo 2 (symbolized by the $\oplus$ sign) of the atom and meter eigenvalues ($a = 0, 1$ and $m = 0, 1$):

$$|a_A\rangle \, |m_M\rangle \longrightarrow |a_A\rangle \, |(m \oplus a)_M\rangle \ . \tag{2.74}$$

The conditional dynamics expressed by eqn. (2.74) defines in information theory the operation of a Control-Not (CNOT) gate, the atom and the meter being respectively the control and target 'bits'. The A bit remains unchanged while M undergoes a conditional translation (modulo 2) which amounts to copying information from A into M. As seen in the next section, such gate operations play an essential role in quantum information processing.

Performing a measurement on an atom prepared in the symmetric superposition of the $|0_A\rangle$ and $|1_A\rangle$ states with a meter initially in the $|0_M\rangle$ state brings $A + M$ into the maximally entangled superposition:

---

[11]The observable O_A is $(\mathbb{1} - \sigma_Z)/2$, where σ_Z is the Pauli operator for system A.

$$|\psi_{A+M}\rangle = (|0_A\rangle\,|0_M\rangle + |1_A\rangle\,|1_M\rangle)\,/\sqrt{2}\,, \tag{2.75}$$

which exhibits a cat-like feature since M is macroscopic. According to the EPR argument, such a state can as well be expressed in a different basis exhibiting correlations between the states $|0_A\rangle \pm |1_A\rangle$ and $|0_M\rangle \pm |1_M\rangle$:

$$|\psi_{A+M}\rangle = \tfrac{1}{2}\left[(|0_A\rangle + |1_A\rangle)(|0_M\rangle + |1_M\rangle) + (|0_A\rangle - |1_A\rangle)(|0_M\rangle - |1_M\rangle)\right]\,. \tag{2.76}$$

When looking at the two equivalent expressions (2.75) and (2.76), we recognize a fundamental ambiguity. Is our meter measuring the observable O_A, as eqn. (2.75) implies, or a non-commuting observable with eigenstates $(|0_A\rangle \pm |1_A\rangle)/\sqrt{2}$, as eqn. (2.76) seems to indicate? Are we free to use the same apparatus and to decide, after the atom–meter coupling, which observable we want to measure as in a typical Bell test experiment? This would be strange since the same apparatus could then measure non-commuting observables, in blatant contradiction with complementarity. It would also be strange to 'observe' the two states $(|0_M\rangle \pm |1_M\rangle)/\sqrt{2}$ which represent superposition of meter states at macroscopically distinct positions in space, an odd configuration indeed. In other words, we have the feeling that the expression (2.75) is the natural one, which conveys the essence of the measurement process since it involves the final observation of the position of a macroscopic object (a classical observable), while the expansion in eqn. (2.76) is 'unnatural' since it describes the correlation between atomic states and weird Schrödinger-cat-like meter states that no one has ever observed.

2.5.2 Decoherence: the environment as a measuring apparatus

The EPR paradox of the measurement process is solved by realizing that the meter is coupled to a large environment E. The simplified situation considered so far assumed that the meter is a quantum object in a well-defined state, interacting solely with the measured atom. In a real meter, however, a macroscopic observable such as the position of a pointer on a dial is coupled to a huge number of microscopic variables (positions and momenta of the atoms inside the meter and of the surrounding air molecules, thermal and visible photons, etc). The bipartite situation depicted by Fig. 2.13 is then replaced by the three-party one shown in Fig. 2.15. The meter M is coupled on the one hand to A, and on the other hand to an unavoidable environment E of microscopic variables, which are not detected.

The states of this environment get rapidly entangled with the meter position states. The description of $A + M$ by a wave function is then inadequate. It has to be replaced by a density matrix obtained by tracing the $A + M + E$ global density operator over E. In this density matrix, the coherences between meter states associated to different positions vanish, hence the name of decoherence given to the effect of the environment on the system (Joos and Zeh 1985; Zurek 1991; Giulini *et al.* 1996; Zurek 2003). This decoherence phenomenon suppresses the basis ambiguity and 'forces' the meter to measure the observable O_A, and not any other observable non-commuting with it. We will discuss this point later in more detail, on a specific example. Let us give here only a qualitative account of decoherence and explain how it sheds light on the measurement problem.

The coupling to the environment corresponds to a leakage of information about the position of M. According to complementarity, it destroys quantum coherences

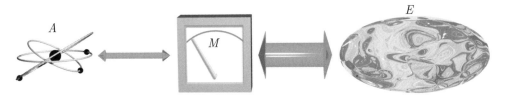

Fig. 2.15 The three partners in a measurement process: the system, the meter and the environment.

between states corresponding to different meter localizations. It is intuitively clear that the presence of a large object at different positions in space perturbs in different ways its microscopic surroundings, by scattering air molecules and thermal photons in disparate final states. In order to simplify the discussion, we start by assuming that the environment is initially in a pure state, of course an idealization of a real situation. Interaction with the meter in states $|0_M\rangle$ or $|1_M\rangle$ very quickly brings E into two final states, $|0_E\rangle$ and $|1_E\rangle$ respectively, with $\langle 0_E|1_E\rangle = 0$. After a very short time, E stores unambiguous information about the position of M. In the case of an atom initially in the symmetric superposition of $|0_A\rangle$ and $|1_A\rangle$, the final $A + M + E$ state is:

$$|\psi_{AME}\rangle = [|0_A\rangle |0_M\rangle |0_E\rangle + |1_A\rangle |1_M\rangle |1_E\rangle]/\sqrt{2} . \tag{2.77}$$

Considering now $A+M+E$ as a bipartite system with $A+M$ and E as components, we see that $|\psi_{AME}\rangle$ is a maximally entangled Bell state. Any measurement on $A+M$ disregards E and is a local operation on $A+M$. All the predictions about its outcomes are contained in the $A+M$ density operator ρ_{A+M}:

$$\rho_{A+M} = \frac{1}{2}\left(|0_A, 0_M\rangle\langle 0_A, 0_M| + |1_A, 1_M\rangle\langle 1_A, 1_M|\right) . \tag{2.78}$$

The coherence between the $|0_A, 0_M\rangle$ and $|1_A, 1_M\rangle$ states has vanished as well as their entanglement, due to the leakage of information about the meter in the environment.

In other words, the environment plays the role of a measuring device which 'observes' M and, according to complementarity, prevents interference effects from building up between the different meter states. By killing the weird coherences between these states, decoherence solves the Schrödinger cat paradox: very quickly the cat is *dead or alive* (a classical alternative) and not in a superposition of *dead and alive* states. The decoherence process also has the essential effect of suppressing the ambiguity of the $A + M$ system description mentioned above. The density matrix of eqn. (2.78) is indeed *not* identical to the sum of the projectors on the states $(|0_A\rangle + |1_A\rangle)(|0_M\rangle + |1_M\rangle)$ and $(|0_A\rangle - |1_A\rangle)(|0_M\rangle - |1_M\rangle)$. This is easily checked by expanding the two density matrices on the basis of the four $|0_A\rangle |0_M\rangle$, $|1_A\rangle |1_M\rangle$, $|0_A\rangle |1_M\rangle$ and $|1_A\rangle |0_M\rangle$ states.

We can extend this discussion to observables with more than two eigenvalues and to arbitrary initial atomic states. The $A + M$ wave function of eqn. (2.72) is then replaced by the density matrix:

$$\rho_{A+M} = \sum_i |\alpha_i|^2 |a_i, \xi_i\rangle\langle a_i, \xi_i| , \tag{2.79}$$

which expresses the measurement postulate of quantum mechanics: if A is initially in a state superposition with amplitudes α_i, a measurement performed with M projects it in one of the states of the superposition with the probability $|\alpha_i|^2$.

We have so far analysed decoherence by assuming that the environment is initially in a pure state. This condition is generally not satisfied and the description of E should involve, from the start, the density matrix formalism. One can however always consider the initial density matrix of E as a sum of projectors on pure states and apply to each of them the above analysis. The $E - M$ coupling quickly brings each of these states into orthogonal states carrying unambiguous information about M. The final result is the $A + M$ density matrix given by eqn. (2.79).

Our analysis has implicitly assumed that the $A - M$ coupling occurs faster than the $M - E$ one. This simplifies the discussion of decoherence, allowing us to define pure meter states before decoherence sets in. Such a situation can indeed be realized in some textbook experiments where great care is taken to protect the meter from the environment (Chapter 7). In most realistic situations, though, the $M - E$ entanglement occurs much faster than the $A - M$ one. One must then describe the $M + E$ system by a density operator even before the meter starts to interact with the atom. In other words, the cat is already in an incoherent mixture of states before we attempt to kill it with our deadly contraption. This does not change the final result, however, and the density matrix of the $A + M$ system is always given, in the end, by eqn. (2.79).

We leave for a subsequent chapter the quantitative estimate of the decoherence time, which is the delay required for the meter states $|0_M\rangle$ and $|1_M\rangle$ to develop orthogonal 'mirror states' in E. In most practical situations, this time is exceedingly short. For example, in a simple model, M is a particle of large mass $\mathcal{M}$ undergoing Brownian motion induced by the kicks of environment molecules at temperature T. The decoherence time is equal to the classical damping time of the mass divided by the dimensionless factor $|(\xi_0 - \xi_1)/\lambda_{\text{th}}|^2$, where $\xi_0 - \xi_1$ is the distance between the two meter positions and $\lambda_{\text{th}} = \hbar\sqrt{2\pi/\mathcal{M}k_bT}$ is its thermal de Broglie wavelength (Unruh and Zurek 1989). We encounter again, in a different context, the classical limit condition expressing that the quantum features of a particle are washed out when the size of the system becomes large compared to its de Broglie wavelength. When the meter positions ξ_0 and ξ_1 are separated by a distance a million times larger than λ_{th}, decoherence is practically instantaneous at the experimental time scale. One can only observe its end product, the classical world with meters pointing in well-defined directions, i.e. cats dead or alive, never in a superposition of the two.

2.5.3   Pointer states

If a set of quantum states can be prepared by a given apparatus, all superpositions of these states are a priori also legitimate, representing possible configurations of the system. We can say that the Hilbert space is a 'democratic' world, with equal opportunity to all states. This is certainly true at the microscopic level. When considering, on the other hand, macroscopic objects such as measuring devices, we have been driven to the conclusion that some states, corresponding to the positions of the meter, are more legitimate than their Schrödinger-cat-like superpositions. More precisely, we have found out that the latter, while they can in principle be prepared via the atom–meter coupling, are much more fragile than the former, their coherence vanishing in a very short time. The selection of the persistent 'position states' among the much more numerous and fragile non-classical Schrödinger-cat-like superpositions is induced by the coupling to the environment. The states which survive environment selection are called 'pointer' states (Zurek 1981; 2003).

The selection of classical pointer states

Let us analyse in more detail the process by which the selection of these states occurs. To keep the discussion simple, we come back to the situation in which the meter is a two-state system. Disregarding now A, we focus on M and E. The main difference between the pointer states $|0_M\rangle$ and $|1_M\rangle$ on the one hand and their superpositions $|0_M\rangle \pm |1_M\rangle$ on the other is that the former couple to E essentially without entanglement, while the latter undergo maximum entanglement. After interaction with the bath of surrounding molecules and photons, the $M + E$ system is in a product state $|0_M\rangle |0_E\rangle$ or $|1_M\rangle |1_E\rangle$ if M is initially in a pointer 'position' state, whereas it is is found in the entangled Bell states $|0_M\rangle |0_E\rangle \pm |1_M\rangle |1_E\rangle$ if M is initially in a Schrödinger-cat superposition. This has, as seen above, deep consequences for measurement. Keeping M in a pointer state or observing its position are classical operations, while conserving it in a Schrödinger-cat superposition, or measuring the phase of this superposition, i.e. distinguishing between the $|0_M\rangle + |1_M\rangle$ and $|0_M\rangle - |1_M\rangle$ states, involve non-classical manipulations which become impossible for large meters. The meter is a good measuring device for O_A whose eigenstates correlate to the pointer states, but is practically unable to measure observables non-commuting with O_A, whose eigenstates are correlated to the fragile superpositions of pointer states.

To deepen our analysis, let us make a slightly more realistic description of E. Since it is made of a huge number of microscopic particles, we model it now as a large ensemble of $\mathcal{N}$ two-state atoms, scattered in state $|0_j\rangle$ and $|1_j\rangle$ respectively by the two pointer states ($j = 1, \ldots, \mathcal{N}$). The $|0_E\rangle$ and $|1_E\rangle$ states are now the tensor products of a large number of two-level atomic states, all in $|0\rangle$ or $|1\rangle$. Coupling a pointer state of the meter to this environment leads to one of the two states of $M + E$:

$$|0_M\rangle |0_E\rangle = |0_M\rangle |0_1, 0_2, \ldots, 0_j, \ldots\rangle \; ; \qquad |1_M\rangle |1_E\rangle = |1_M\rangle |1_1, 1_2, \ldots, 1_j, \ldots\rangle \; . \tag{2.80}$$

This equation describes perfect correlation, without entanglement, between M and environment atoms. Such correlation pleased Einstein because it has a classical interpretation. In both cases, M and E share without conflict information about the meter's position. The meter has classically disseminated information about its position among a large ensemble of independent, non-entangled particles in such a way that many observers can, if they wish, acquire this information without perturbing each other's measurement. It is enough that they measure one of the atoms in E to know what is the position of M. This position is an *objective* property of the meter, since all observers will find the same result by exploiting bits of information scattered in E. The model is of course very simplified, but we recognize the elements of a classical situation: a large object scatters a lot of microscopic particles around itself, and independent observers, each by detecting a subset of these particles, are able to agree about the position of the object.

The situation is quite different if M is initially in the 'cat-like' superposition $|0_M\rangle + |1_M\rangle$ of the two pointer states. According to the linearity of quantum physics, the coupling to the surrounding particles then brings $M+E$ very quickly into the entangled superposition:[12]

$$|\psi_{M+E}\rangle = |0_M\rangle |0_1, 0_2, \ldots, 0_j, \ldots\rangle + |1_M\rangle |1_1, 1_2, \ldots, 1_j, \ldots\rangle . \qquad (2.81)$$

This superposition is very fragile. If we lose track of *any one* of the atoms in E, the quantum coherence vanishes. By tracing over the lost particle, one is left with a density matrix for the rest of the system representing a mere statistical uncertainty between the two terms of the alternative. Moreover, all the observers catching independently particles in E will agree about whether the meter is in the 0 or 1 state. In this way, the position of M after decoherence has occurred can be considered as a classical objective quantity, even if it is *a priori* a completely random one (equal probabilities of finding 0 and 1).

The decoherence approach gives a natural definition of classical states as the ones which, in Hilbert space, are the most immune to entanglement with the environment (Zurek 2003). This definition depends on the properties of E and of its coupling to M. In the case of a mechanical meter, the position states appear as natural pointer states, essentially because the environment particles couple independently to it via interaction potentials depending solely on position. Position eigenstates of M thus remain unentangled in the coupling process. We will see later other examples of pointer states, the so-called coherent states of the field in a cavity, which are also defined by the property of staying immune to entanglement with other environment modes outside the cavity.

Trying to catch Schrödinger's cat by watching the environment

The loss of the Schrödinger cat coherence is due to our inability to track the environment without loss of information. Let us dream, for a moment, that we are almighty and that we can control and observe, with arbitrary precision, *all* the particles of the environment. In this case, no coherence is lost and observing a Schrödinger cat state of the meter becomes possible. We will see, however, that it involves very complex operations. This will give us a visceral feeling of the fragility of mesoscopic coherences.

Suppose that the meter has been prepared in the state $|0_M\rangle + |1_M\rangle$ and coupled to the $\mathcal{N}$-particle environment. Our task is now to measure the meter state after this coupling has occurred. To understand how it can be done, it is convenient to expand the $M + E$ wave function described by eqn. (2.81) in the basis of the Schrödinger cat states $|0_M\rangle \pm |1_M\rangle$:

$$\begin{aligned} |\psi_{M+E}\rangle &= (|0_M\rangle + |1_M\rangle)(|0_1, 0_2, \ldots, 0_j, \ldots\rangle + |1_1, 1_2, \ldots, 1_j, \ldots\rangle) \\ &+ (|0_M\rangle - |1_M\rangle)(|0_1, 0_2, \ldots, 0_j, \ldots\rangle - |1_1, 1_2, \ldots, 1_j, \ldots\rangle) . \end{aligned} \qquad (2.82)$$

This equation shows that the coupling to E has made equiprobable the two phases of the cat, even if it has been initially prepared with the $+$ sign. Tracing over the environment turns the meter state into an incoherent superposition of the two pointer states, which can equivalently be described as an incoherent superposition of the two Schrödinger cat states with the $+$ and $-$ signs. It is thus obvious that a measurement of the cat phase should find the $+$ and $-$ signs with equal probabilities.

---

[12]From now on, we drop in this discussion simple $1/\sqrt{2}$ normalization factors.

Nevertheless quantum laws tell us that, if we detect the observable admitting the cat states as eigenstates, the meter must be found, in each realization of the measurement, in a well-defined cat state with either the + or the − sign (this result being random from one realization to the next of the experiment). Let us analyse how such a measurement could be made by detecting the particles in E. Deciding about the phase of the M cat state requires us to be able to distinguish between the two states $|0_1, 0_2, \ldots, 0_j, \ldots\rangle \pm |1_1, 1_2, \ldots, 1_j, \ldots\rangle$, and this is a tricky operation. One might, at first sight, naively think that it is enough to measure one environment particle in state $|0_j\rangle \pm |1_j\rangle$ to decide in which of the two $|0_M\rangle \pm |1_M\rangle$ states the meter is. This is obviously wrong since the collective state $|0_1, 0_2, \ldots, 0_j, \ldots\rangle \pm |1_1, 1_2, \ldots, 1_j, \ldots\rangle$ is an entangled state, different from the tensor product of the individual $|0_j\rangle \pm |1_j\rangle$ states. Performing such an independent measurement on any particle in E would yield randomly the result + or −, totally uncorrelated to the sign of the entangled superposition. Since any particle in E is maximally entangled to the $\mathcal{N} - 1$ others, we recall from the discussion in Section 2.4 that its partial density matrix is proportional to unity and does not carry information on the global environment's state.

Only a non-local measurement of the environment could enable us to find out in which Schrödinger cat state the meter is in a given realization of the experiment. One way to perform (ideally) this measurement is to pick one of the environment atoms (let say the first one) as a control bit and to couple it through CNOT gate operations to all the others playing the role of target bits. This transforms the entangled environment states according to:

$$|0_1, 0_2, \ldots, 0_j, \ldots\rangle \pm |1_1, 1_2, \ldots, 1_j, \ldots\rangle \longrightarrow (|0_1\rangle \pm |1_1\rangle) |0_2, \ldots, 0_j, \ldots\rangle \ . \qquad (2.83)$$

Through this highly complex operation, all the environment particles are disentangled and the first one carries alone information (± sign) about the phase of the superposition. By finally measuring the control bit, one can find out the sign of the environment superposition and hence, according to eqn. (2.82), the phase of the meter Schrödinger cat. Once this measurement has been made, there is no information left in E, since all the other target particles are in $|0\rangle$. Obviously this scheme is totally impractical since it requires a huge number of CNOT operations and assumes that not a single particle in the environment is lost.

The operation we have outlined in order to force M into one of the two Schrödinger cat superpositions $|0_M\rangle \pm |1_M\rangle$ is a quantum eraser operation of the kind discussed in Section 2.4. The environment is a very efficient which-path detector spying on M. By keeping track of all the environment particles and performing the multiple non-local CNOT operations, we erase information about the meter position and 'revive' one of the two cat-like superpositions. Hence, Schrödinger cat interference can in theory be retrieved in a reversible way by performing correlations between the environment and the meter detections. For all practical purposes though, this kind of operation is impossible when the meter is really macroscopic. The coherence is, in the real world, very quickly lost.

The selection of pointer states: a 'Darwinian' process?

There is thus a great difference between pointer states and their non-classical superpositions. The former disseminate information about themselves in a huge number of independent and redundant copies in E, without getting entangled with it. This makes it easy for many independent observers to agree objectively about the description of the state. This objective agreement is a distinctive mark of classical behaviour. The superpositions of pointer states, on the other hand, copy information in E into a single huge entangled copy, which could be read out only through a complex non-local operation, impossible to achieve in all realistic situations.

W. Zurek has compared the process of decoherence by which the environment selects the robust pointer states and eliminates their fragile superpositions to a Darwinian process ensuring the 'survival of the fittest' (Zurek 2003). In the same way as biological Darwinism selects, out of all the possible living creatures, only a subset of species and eliminates all the 'freaks', the quantum Darwinism chooses among all the possible states of Hilbert space the robust pointer states which behave classically and very efficiently eliminates all the Schrödinger cat 'monsters'. Some ingredients of evolution theory can indeed be recognized here such as the notion of reproduction or the selection rule forced on the system by the environment. The analogy should not be carried out too far though. It is not the quantum state which reproduces itself here, but only information contained in it (more about this later). Moreover the competition for survival in a limited environment, essential in biological Darwinism, is lacking in its decoherence version.

The decoherence picture leads us to give a simple definition of ideal measurement and measuring devices in connexion with the fundamental concept of entanglement. To each physical observable O_A of a quantum system A, there must correspond a macroscopic meter M. This meter is defined by its coupling to A and to a large environment E. Among all the states of M, there is a subset of robust pointer states which couple to E without entanglement and which are thus classical, meaning that independent observers can easily reach a consensus about them. The eigenstates of O_A are the states of A which couple to the M pointer states without entanglement. They are thus stable with respect to their coupling with M, which means that repeated measurements of the same observable leads to identical results, a basic feature of standard quantum measurement theory.

Quantum decoherence versus noise-induced decoherence

The decoherence process described here is a quantum effect, related to entanglement with the environment and to complementarity. The $A + M$ system loses its quantum coherence and has to be described by a density matrix (and no longer by a wave function), because we have to renounce keeping track of the quantum correlations of the system with E. This is not the only way, though, in which quantum coherence might be lost in a quantum system. It can also happen because of statistical uncertainties, of a purely classical nature, about the evolution of the system which results in a randomization of the relative phases of its probability amplitudes. Most generally, the $A + M$ system described above, can be in the state:

$$|\psi_{A+M}\rangle = \sum_i \alpha_i \exp\left[i\phi_i(t)\right]|a_i, \xi_i\rangle , \qquad (2.84)$$

where the $\phi_i(t)$ are classical random functions of t. This *phase noise* could be due to the existence of random classical fields acting on A or M and shifting their energy levels. It could also be due to classical vibrations of the apparatus modifying the length of the arms of an interferometer. These fields or mechanical vibrations constitute a 'classical environmental noise' which affects the system and its evolution, without entangling it to another system. The net effect of this classical noise is the destruction of all interference between the various probability amplitudes. More precisely, in each realization of an experiment in which the wave function (2.84) is prepared, the interference term between two states $|a_i\rangle$ and $|a_j\rangle$ is proportional to $\exp i[\phi_i(t) - \phi_j(t)]$. Since this phase randomly changes from one realization of the experiment to the next, any recording of an interference pattern results in a washing out of the global fringe signal. For all purposes, the system can again be described by the density matrix of eqn. (2.79).

Does this mean however that a quantum environment and classical noise induce decoherence effects of the same nature? There is a great difference between the two – at least at the conceptual level. The classical loss of coherence could, in principle, be corrected by an operation much simpler than the quantum erasure described above. Noise is merely due to an imperfect knowledge of a complex process. It can be measured independently of the interfering system, i.e. without affecting its state in any way. This measurement could for instance be performed by inserting in the apparatus an independent probe recording the stray fields or the mechanical vibrations responsible for the phase diffusion. This could be done in each realization of the experiment and the obtained data could be used to correct the phase of the contribution of this realization to the interference signal. One could also think of correcting actively the stray fields or the vibrations by a feedback action, thus restoring interference directly.

Such correction schemes involve only classical local correlations, since classical noise is inherently reversible. Restoring the coherence is much more complicated in the case of a quantum environment because, contrary to a classical noise source, it is entangled to the system and its different parts are strongly entangled with each other. Information about the loss of coherence is stored in all the quantum correlations in such a way that the loss of information is lethal to the correction process. In this way, the process is inherently irreversible. And even if any loss of information could be avoided, which becomes utterly impractical when the system is truly macroscopic, the quantum noise correction is fundamentally different from the classical one. In the classical noise case, a simple correction allows us to combine all the results of individual realizations of the experiment into a unique interference signal. In the quantum case, we must perform a book-keeping procedure and sort out the data according to the results of the environment measurement. When performing repeatedly the Schrödinger cat measurement experiment discussed above, we must for instance build up two sets of data, one for the + and the other for the − phase of the cat, which should be analysed separately. The sum of these two data sets corresponds to the incoherent situation where nothing is done in the environment. To avoid any ambiguity, we will keep the word 'decoherence' for the fundamental quantum process involving entanglement with the environment and call 'noise induced dephasing' the more mundane classical processes of quantum phase diffusion.

2.5.4 The quantum–classical boundary

Decoherence models versus Copenhagen interpretation

Most quantum mechanics textbooks, following the Copenhagen interpretation, postulate that the evolution of quantum states occurs via two different processes. In the absence of any measurement, the wave function of an isolated system develops in a reversible and unitary way, described by Schrödinger's equation. When one performs at some time an ideal measurement, a very different sort of evolution takes place, resulting in the system collapsing into one of the eigenstates of the measured observable. This latter process is irreversible and essentially random, only its probability being calculable. This irreversible projection is usually not justified by an analysis of the interaction between the system and the apparatus. It is merely postulated as a kind of 'black box' property of the measuring process, an attribute of the classical character of the meter which prevents us from describing it as a quantum entity.

This approach, although sufficient to make all useful predictions about the outcome of measurements, is somewhat unsatisfactory because it splits artificially the description of quantum dynamics into two disparate processes. It gives to the meter a special statute exonerating it from having to obey the 'strange' quantum rules. This is hard to justify since the meter is made of the same kind of particles (quarks and electrons) as the microscopic system it is coupled to. The decoherence model developed in the preceding sections is an alternative approach to measurement theory, which avoids some of the above criticisms. It deals with the micro-system and the meter on an equal footing, both being described by the same quantum laws. The irreversibility of the process is not postulated from the start, but inferred from the unavoidable coupling of the meter to a large environment. The transformation of the entangled atom–meter state into an incoherent density matrix [the transition from eqn. (2.72) to (2.79)] is a general result of quantum relaxation theory. We have so far only presented a qualitative picture of this irreversible behaviour. We will see later (Chapter 4) more precisely that irreversibility and non-unitary behaviour emerge naturally from the usual Schrödinger equation, when looking at the evolution of a small subsystem entangled to a large reservoir. This emerging irreversibility is a quite general feature of statistical quantum mechanics, of the same nature as the irreversibility of classical thermodynamical processes, which results from the cumulative effect of perfectly reversible microscopic events.

The final stage of the decoherence process, described by the density matrix ρ_{A+M}, is strictly equivalent to the predictions of the usual Copenhagen approach, while any mention of a collapse seems to have been avoided. It would be wrong to conclude hastily though that the decoherence point of view allows us to get rid of the projection postulate. Even if in this approach no explicit reference is made to a 'collapse' of the A or M systems, a projection of the wave function does occur in E. By tracing over the environment variables, we consider indeed that unread measurements have taken place in it and ρ_{A+M} is a weighted average of the projectors on the correlated 'collapsed states' of the $A + M$ system.

The notion of collapse associated to measurement is thus deeply rooted in the quantum formalism, whether it is explicitly introduced as a postulate, or implicitly assumed in the rule defining the basic properties of partial density matrices. Should we feel uneasy about this fundamental combination of reversibility and irreversibility in the quantum world? It seems that the uneasiness often comes from a misconception about the nature of the wave function. If we consider it as being the physical system under study, then it is certainly uncomfortable to accept that a measurement appears as an event of a nature so different from any other one, introducing irreversibility in an otherwise reversible world. But the wave function is *not* the system, it is a mathematical representation of the knowledge we have about it at a given time, conditioned to what we know about its initial preparation. This knowledge allows us to make probabilistic predictions about the values of observables which have no reality until they have been determined by a process which excludes the simultaneous knowledge of non-commuting quantities. Once this 'knowledge acquisition' has been performed by a measurement, information about future knowledge must have been modified. The sudden and irreversible change of the wave function appears thus as natural, a mere consequence of the definition of the wave function as representing an amount of information which is changed by getting this information out of the system.

In fact, quantum physics does not deal with events, but with correlations between events. The probability of finding a given result when measuring a system described by a wave function is, in effect, the conditional probability of finding this result, provided that a first measurement, which has prepared the wave function at an earlier time, has yielded a previous result (the eigenvalues corresponding to that wave function, considered as an eigenstate of a set of commuting observables). In other words, quantum mechanical probabilities are *de facto* conditional probabilities and the wave function is only the mathematical representation of these conditional probabilities. In this point of view, no collapse occurs as a physical process.

In the decoherence model, the irreversibility of the measurement and the classical character of the meter are forced upon us by the environment, whose presence cannot be avoided. One can argue that the existence of a complex environment is an essential feature of any imaginable process by which a macroscopic living entity (a human observer) tries to obtain information about a microscopic system. By definition, this observer has, at some level of the regressive chain of amplifier stages discussed above, to be coupled to a complex environment (if nothing else to the thermal bath of molecules and photons essential for his survival). In order to be intelligible and exploitable, the information he seeks to acquire must be separated from this environment. The prescription to perform this separation is to trace over the environment variables. One is then left with the density matrix description of the atom–meter system, which leads to the same predictions as the orthodox collapse approach.

Irreversibility occurs because we, as observers, decide to forget about the environment and renounce keeping track of its correlations with the meter. One can of course, as a pure intellectual game, imagine that one could keep track of all the particles in the environment. By doing so, one could indeed retrieve in principle the quantum coherence of any non-classical superposition. One could even erase through complex operations all which-path information and observe, by performing subtle correlations experiments in the environment, interference effects between live and dead parts of the Schrödinger cat wave function. These thought experiments are utterly impractical though, even more impractical than the one which would consist in waiting long enough to see a glass of water freezing at room temperature.

Is the measurement problem fully solved? Decoherence theory has explained why Schrödinger cats cannot remain suspended coherently between life and death. Exceedingly quickly, they are left in a mundane statistical mixture of the two possibilities, a classical alternative of the hidden variable type. God is still playing dice, but it is not a malicious dice game where the different outcomes can interfere. It has become, owing to decoherence, a perfectly classical game. But it remains irreducibly non-deterministic. Nothing in the theory enables us to predict whether the cat will be found dead or alive, even if we know that one of these possibilities definitely occurs. We can say, according to Einstein terminology, that the death or life of the cat has, even before being recorded by a human mind, become an element of reality (since all entanglement has been destroyed by decoherence), but this element of reality cannot be predicted, only its probability can be estimated. Some physicists find this state of affairs uncomfortable. Others are ready to accept this inherently statistical feature of quantum theory.

Where is the quantum–classical boundary?

Electrons, atoms or photons can be in state superpositions, cats and meters cannot. Decoherence explains why it is so, but we still have an interesting question to answer. Where is the limit between atoms and cats; at which scale do we really encounter the quantum–classical boundary where superpositions of states become impossible? Part of the answer is purely practical. The bigger the object, the more it is coupled to its environment and the faster it loses quantum coherence. Keeping a large object quantum thus seems to be a matter of technology, just protecting it against the environment as much as possible. We will see later examples of this strategy. For instance, by storing a field made of many photons in a cavity with nearly perfectly reflecting walls, one is able to keep a Schrödinger-cat-like state made of several tens of photons alive long enough to be able to observe it. At some point, though, technology reaches fundamental limits. The quality factor of cavities is finite and so, the number of photons you could pack into an observable Schrödinger cat made of photons is also finite, a few tens to a few hundred. Can we do much better with material particles? Can we think of a mesoscopic mass made of billions of atoms, in a superposition of states with different positions? The system would have to be protected against interactions with all kinds of particles which would immediately localize it. It should thus be placed in an exceedingly good vacuum to avoid collisions with background gas. It should also be maintained at a very low temperature in order to suppress the emission of thermal photons. And of course it should be kept in absolute darkness with all microwave, infrared, visible, X and γ ray photons shielded out, as well as cosmic rays of all kinds.

When all this protection is achieved, one kind of environment remains, the gravitational waves acting on everything in the Universe, which cannot, even in principle, be shielded out. These waves, which roam around in all space, have so far never been directly detected because they couple very weakly to matter. They are now actively being hunted by gravitational antennae on Earth and they will soon be investigated by detectors embarked in satellites. The coupling of mesoscopic masses to these fluctuating waves would constitute the ultimate cause of decoherence (Reynaud *et al.* 2002). We should also note that gravitational waves act as a classical perturbation on the masses and the decoherence they induce is thus of the dephasing type, a process in which entanglement plays no role. We will not discuss gravitational decoherence here. In all the situations we will describe, other causes of decoherence, of electromagnetic origin, will be by far dominant.

2.6 Taming the quantum to process information

2.6.1 From classical bits to qubits

Quantum information science was born in the 1980s, out of theoretical considerations about possible ways to harness the strangeness of the quantum world in order to perform useful tasks. The dream of the pioneers of information science, largely inspired by R. Feynman (1985), was to use the weird quantum laws to perform operations impossible with machines obeying classical logic. Their theoretical studies have developed a powerful language to analyse entanglement and non-locality by borrowing useful concepts from information science. Some of these concepts, such as the entropy of information, were imported from physics into computer and communication science at an earlier stage. The two fields have had a long history of collaboration and cross-fertilization. The merit of information scientists has been, in the last years, to induce physicists to think again about the strange aspects of the quantum. The recent coming of age of thought experiments really performed in the laboratory with single atoms, molecules or photons, has given hope that some of the dreams of the information scientists could be realized. The combination of theoretical advances in information science and experimental development in single-particle manipulations has thus given rise to a new domain of research often called the 'physics of quantum information' (Bouwmeester *et al.* 2000; Nielsen and Chuang 2000). This field is thriving now and we briefly review it in this section.

All the signals exchanged by the information machines of our everyday world are coded in the form of discrete entities, called bits, which are to information what atoms are to matter. Each bit can take the values 0 or 1. A bit sequence represents a letter of the alphabet, a piece of text, a piece of music or a picture. Bits can be manipulated, added or multiplied (modulo 2) and the most complex computations can be realized by combining such operations on long bit sequences. In the computers or in the optical fibres of our telecommunication devices, bits are carried by electrical currents or by light beams corresponding to macroscopic fluxes of electrons or photons. Transistor circuits are used to combine bits and manipulate them together according to various conditional dynamics schemes. Even if the physical processes which have made the fast and efficient manipulations of these bits possible are quantum in nature (the transistor and the laser are, as recalled above, born from the knowledge of the quantum laws), the logic to which these bits obey is classical. They are made indeed of large numbers of particles coupled to their environment in an efficient way which destroys all the weird entanglement of the microscopic world.

In a quantum information machine, the bits would be coded in individual quantum systems (atom, electron, photon or possibly a quantum mesoscopic circuit), each of which having two internal states, $|0\rangle$ and $|1\rangle$. This opens the way, in principle, to an extremely dense storage of information. But there is more to it than a simple miniaturization of our computer memories or processors. Quantum bits, or 'qubits' as they are now called, can exist in superpositions of $|0\rangle$ and $|1\rangle$, which is of course forbidden by the classical logic of ordinary bits. The controlled manipulation of atoms and photons opens a new dimension to information science by projecting it into a world where gates can be at the same time open and closed, cats dead and alive, bits assume the value 0 and 1.

2.6.2 Quantum cryptography

Quantum logic promises to be very useful in cryptography, the art of secret communication, as ancient as the history of war and diplomacy (Bennett *et al.* 1992). If Alice and Bob, the two partners already introduced in our discussion about entanglement, want to exchange a message secretly, they can do it, according to the theory of information, by sharing a random key of 0's and 1's, as long as the message itself. Alice, the sender, adds the bits of the message to the successive bits of the key, thus producing a coded sequence that she sends to Bob by a public communication channel (radio-broadcast for instance), without worrying about its possible interception by an eavesdropper (named Eve by theorists in information science). Bob then subtracts bit-wise the key from the scrambled message and reconstructs the original text. If the key is totally random, it is impossible for Eve to decipher the broadcast. The method is secure as long as the key is discarded after each exchange (re-using the same key would jeopardize secrecy because Eve could decipher using correlations between the scrambled messages). The security of the procedure is based on the secret sharing of the key, an operation which must be performed before the exchange of useful information takes place. It is essential, at this crucial stage, to ensure that Eve could not intercept the key.

It is in the process of key sharing that quantum physics can be of a great help. The 'no-go' laws of the quantum world, such as the no-cloning theorem (Wootters and Zurek 1982) or the no-knowledge of a single-particle wave function (D'Ariano and Yuen 1996), can be advantageously exploited by Alice and Bob to make it an impossible task for Eve to steal the key. What seems to be a negative property of the quantum can be turned into a positive feature. Several variants of such procedures have been proposed and realized experimentally [for a review, see Gisin *et al.* (2002)].

Key distribution by exchanging particles in non-orthogonal states

In the scheme invented by Bennett and Brassard in 1984, called BB84 for short and schematized in Fig. 2.16, the key is carried by two-level spin-like particles sent from Alice to Bob. They first decide publicly to prepare and analyse the particles in two bases of eigenstates corresponding to non-commuting observables (which we call, without loss of generality, σ_Z and σ_X). Alice prepares randomly a set of particles in either one of the four eigenstates $|0_Z\rangle, |1_Z\rangle, |0_X\rangle = (|0_Z\rangle+|1_Z\rangle)/\sqrt{2}$ and $|1_X\rangle = (|0_Z\rangle-|1_Z\rangle)/\sqrt{2}$. She keeps a record of her random sequence of 0 and 1's, correlated to the basis (Z or X) to which they correspond and she sends the particles to Bob without information about their states. Bob performs a measurement of each particle. Not knowing in which basis Alice has prepared them, he decides randomly to measure σ_Z or σ_X. Of course, he fails on the average half of the time to use the same basis as Alice. His results will then be totally uncorrelated to those of Alice. For the other half of the particles, he finds the same result as Alice. In a third stage of the procedure, Alice and Bob announce publicly the basis in which they have prepared and detected their particles. They thus learn which are the particles for which their observables coincide

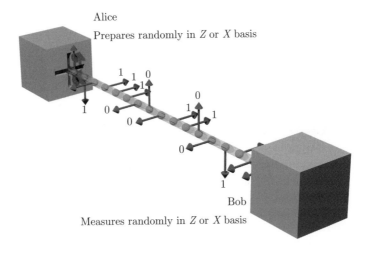

Alice
Prepares randomly in Z or X basis

Bob
Measures randomly in Z or X basis

Fig. 2.16 The Bennett–Brassard key distribution scheme with spin-$1/2$ particles.

and decide to keep the corresponding set of (random) 0 and 1 data as their key.

Assuming that Eve has access to the particle communication channel, what could she do to intercept the key without Alice and Bob noticing? She must try to acquire information on each particle without stopping it. Bob must receive the particles sent by Alice and they should not be aware of any tampering which would make them renounce communicating, which is not the goal pursued by Eve. She can decide to measure σ_Z or σ_X before reinjecting the particle in the Alice-to-Bob channel. She would then be, with respect to Alice, in the same situation as Bob, unaware of the basis in which each particle has been prepared. Half of the time, she will use the wrong basis and her measurement will randomize the eigenvalue in the original basis. This procedure affects the correlations between Bob and Alice in a way they can easily detect. They add to the key preparation a last verification step. They announce publicly the value of the bits for a subset of the key. If Eve had performed measurements, this test will exhibit a large rate of anti-correlated values and Alice and Bob will be able to rule out, with a probability as close to 1 as they wish, any tampering of this kind. If the test is satisfactory, they will discard the publicly announced bits and keep the remainder of the correlated data as their key.

Eve might look for trickier ways to steal the key. If she knows quantum physics, she will however convince herself that it is an impossible task. She might, for example, realize that it is unclever not to use information that Alice and Bob are volunteering by publicly announcing their choice of basis, after they have performed their measurements. For that, Eve might find it better to wait until then before trying to extract information. The possible way to achieve this would be to clone by a unitary reversible operation information contained in each particle on an auxiliary bit (called an 'ancilla' bit in quantum information language) that she would keep aside. But, unfortunately for Eve, quantum cloning of a totally unknown state is forbidden. We have already alluded to this no-go theorem, which we will prove now.

Let us assume *ab absurdo* that Eve had at hand an ideal cloning machine. It would copy an arbitrary input qubit state $|\phi\rangle$ onto an ancilla qubit. The ancilla is

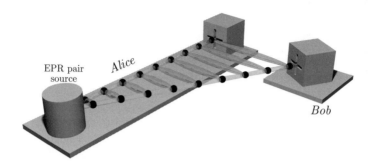

Fig. 2.17 Key distribution by sharing EPR pairs.

initially in the reference state $|0\rangle$ and the machine would perform the transformation $|\phi\rangle\,|0\rangle \to |\phi\rangle\,|\phi\rangle$. In the same way, it would duplicate another input state $|\psi\rangle$ according to $|\psi\rangle\,|0\rangle \to |\psi\rangle\,|\psi\rangle$. Since the transformation is unitary, it preserves the scalar product and hence:

$$\langle\phi|\psi\rangle\langle 0|0\rangle = \langle\phi|\psi\rangle^2 , \qquad (2.85)$$

which shows that the procedure is possible if only $\langle\phi|\psi\rangle = 0$ or 1. A cloning machine is thus limited to the replication of identical or orthogonal states. Since BB84 makes use of non-orthogonal states such as $|0_Z\rangle$ and $|0_X\rangle$, Eve is unable to duplicate perfectly the qubit sent by Alice.

In an effort to overcome this problem, Eve could build an optimal cloning machine (Bruss *et al.* 1998; Buzek and Hillery 1998; Gisin and Massar 1997) which, while imperfect, produces ancilla bits minimizing the anti-coincidence rates in the correlation experiments performed by Alice and Bob. This rate will remain finite however, with a non-zero lower bound, and the verifying procedure outlined above will always allow Alice and Bob to find out whether they are being spied upon or not.

Key distribution by sharing pairs of entangled particles

Quantum physics seems thus to have brought to an end the age-old competition between code makers and code breakers. However smart the code breakers could be, Nature definitely shifts the scale towards the code maker side. In fact, quantum weirdness can be used to realize cryptography schemes different from BB84. One of these schemes, imagined by A. Ekert and schematized in Fig. 2.17, is remarkable because it makes explicit use of entanglement and presents some conceptual advantages (Ekert 1991). Instead of preparing single particles, Alice generates pairs of particles in a maximally entangled Bell state ($|\phi^+\rangle$ for instance, but any of the four Bell states would do). She keeps one particle of each pair for herself and sends the other to Bob. They then decide randomly to measure on their particles either σ_Z or σ_X, performing the measurement of the same observable on half of the pairs on average. As a fundamental feature of entanglement, they must find the same result. The rest of the procedure is the same as in the BB84 scheme. They announce, after the measurement, their choice of basis and keep the results corresponding to the same choice.

This scheme presents many analogies with BB84. As seen above, the time ordering in a Bell correlation experiment is irrelevant. To analyse the correlations, we might as well consider that Alice performs her measurement first on one particle of each pair, before sending the other to Bob. If she chooses randomly to detect either σ_Z or σ_X, she generates with equal probabilities, as in the BB84 scheme, the four states $|0_Z\rangle$, $|1_Z\rangle$, $|0_X\rangle$ and $|1_X\rangle$. She also knows the state of each particle received by Bob. Seen by Alice, the whole procedure is then identical to BB84. Generating an entangled pair and measuring one of its particles is a possible way, if not the simplest, to prepare randomly the other particle in a state (known to her) before sending it to Bob.

The similarity between the Ekert and BB84 schemes entails that the situation is as desperate as it was before for Eve. Whatever she does on the particles travelling from Alice to Bob, she is bound to decrease the amount of correlations initially stored in the Bell states. Alice and Bob will always be able to detect the tampering by performing correlation tests on a subset of the key. The Ekert version of key distribution is somewhat more delicate to implement than the simple BB84 because it is more difficult to prepare entangled states than single-particle states.

The extra effort made to implement the Ekert scheme has a reward though because it makes it possible, at least in principle, to increase the security of the key sharing between Alice and Bob. In the BB84 scheme, since Alice must have performed her state preparation from the start, important information about the key must be kept by her as a record, with all the inherent risks of classical spying by Eve. Once the sequence of the 0's and 1's has become an element of reality by being recorded in a book, it can be copied without perturbation. Alice and Bob cannot have an absolute certainty based on the laws of Nature that Eve has not been looking into their books. This mundane act of spying, disregarded in the above analysis, can be fully avoided in the Ekert scheme. If Alice and Bob store their entangled particles in their safes and decide to measure them just before they need to encode and decode the message, the key is generated only when required, which decreases greatly the risks of mundane spying. The 0 and 1 values do not exist before that time and since there is no 'hidden variables' to peep upon, Eve is at a loss even to perform a classical spying act. This advantage of the Ekert scheme is very hypothetical because it is extremely hard to keep for a long time entangled particles immune from decoherence. We come back to this 'technical' problem later.

Various implementations of the BB84 (Franson and Ilves 1994; Rarity *et al.* 1994; Buttler *et al.* 1998; 2000; Kurtsiefer *et al.* 2002) and Ekert (Ekert *et al.* 1992; Tittel *et al.* 2000) schemes have been realized. The key particles are photons, most often carried in optical fibres. The bits are coded either in the photon polarizations or in their time of arrival. In both cases, it is easy to define observables mathematically equivalent to the Pauli operators of our model. The Ekert-type cryptography experiments are very similar to Bell inequality tests, the difference being a matter of presentation of their results. An experiment showing a violation of Bell's inequalities is *ipso facto* demonstrating entanglement of particles, with an inherent impossibility of tampering secretly with quantum information transmitted between the two partners.

2.6.3   Quantum teleportation

The analysis of quantum cryptography provides a clear illustration of the weirdness of a quantum bit, an object much subtler than its classical cousin. Oddly, its state is in general not known, even fundamentally unknowable. Of course, if it is in one of two possible orthogonal states $|0\rangle$ or $|1\rangle$, it is no more mysterious than a classical bit.

It is easy then to construct an observable admitting these states as eigenstates and a measurement of this observable will yield, without perturbing it, the value of the bit. If, however, the bit exists in a single copy and is suspended coherently between the 0 and 1 values, i.e. if it is in the state $\alpha\,|0\rangle + \beta\,|1\rangle$, there is no way of determining α and β. In spite of this fundamental 'indeterminacy' or rather owing to it, qubits are useful for cryptography, and we will see below that they could also be useful for quantum computing. This is because qubits can be entangled with each other and give rise to interference effects leading to measurements producing useful and exploitable information.

For many potential applications in quantum communication or computing, it might be useful to transmit a qubit at a distance, even if neither the sender nor the receiver knows its state. The simplest way is of course to send the physical bit itself, as is done in a Bell's inequality test or in cryptography experiments. It might be convenient, in other cases, not to send the particle itself, but only quantum information it carries. The classical equivalent of this operation is mundane. It is realized routinely by a fax machine, made of an emitting and a receiving part connected by a phone line. A piece of text written on a sheet of paper is read out and turned into a sequence of bits at the emitting end. The sequence of bits is then sent through the phone line to the receiving end which decodes it and prints it out on another sheet of paper. Incidentally, this classical information has been cloned, since the emitter and the receiver – our friends Alice and Bob – each have in the end a copy of the initial text. Can Alice and Bob achieve the quantum version of this operation; can they transmit to one another all information contained in a qubit?

They will have to proceed in a different way, since there is no emitting machine able to read out the unknown bit in the first place and since it is forbidden in the end that this bit should be cloned. It is possible, though, to copy all information of an initial qubit given to Alice (and unknown to her) into a qubit possessed by Bob. The transmission of this information can, as in the classical fax, be performed at the speed of light. Contrary to its classical counterpart, though, it always erases the initial qubit at Alice's end of the machine. The procedure making this feat possible has been invented by Bennett *et al.* (1993) who have called it 'quantum teleportation'. This name, chosen with obvious reference to science fiction, has certainly played a great role for spreading quantum information among the public, with headlines in newspapers and television programmes about quantum teleportation throughout the world.

Teleportation procedure

Quantum teleportation exploits entanglement and non-locality to circumvent the impossibility of reading out classically the state of a qubit. The principle of the method is illustrated in Fig. 2.18. Before the transmission starts, Alice and Bob share a set of particles labelled $a - b$, entangled in a Bell state (one pair for every qubit they want to transmit). We assume here that this state is $|\phi^+\rangle = (|0_a 0_b\rangle + |1_a 1_b\rangle)/\sqrt{2}$, but any of the other three Bell states would also do. When Alice is given a qubit u in an unknown state $|\psi_u\rangle = \alpha\,|0_u\rangle + \beta\,|1_u\rangle$, she couples it to one of her a particles and performs a joint measurement on the $a - u$ system. Due to the $a - b$ entanglement, this measurement has an immediate effect on the particle b of the same pair belonging to Bob, which is at once projected in a state depending on the initial state of u and on

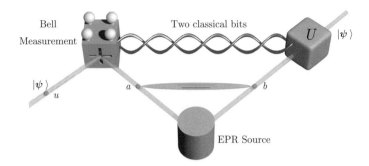

Fig. 2.18 Scheme of a teleportation experiment.

the random result of the measurement performed by Alice. She then sends the result of her measurement (as a set of two classical bits) to Bob, who, using this information, is able to change by a unitary operation the state of his b particle into the state in which u first was.

The mathematics of this procedure is easy. The initial state of the three-particle $a-b-u$ system, a tensor product of the u state with an $a-b$ Bell state, can be equivalently expressed, after a reordering of terms, as an expansion over tensor products of $u-a$ Bell states and b single-particle states:

$$
\begin{aligned}
|\psi_u\rangle \, |\phi_{ab}^+\rangle \;=\;& (\alpha \, |0_u\rangle + \beta \, |1_u\rangle)(|0_a, 0_b\rangle + |1_a, 1_b\rangle)/\sqrt{2} \\
=\;& |\phi_{ua}^+\rangle \, (\alpha \, |0_b\rangle + \beta \, |1_b\rangle)/2 + |\phi_{ua}^-\rangle \, (\alpha \, |0_b\rangle - \beta \, |1_b\rangle)/2 \\
& + |\psi_{ua}^+\rangle \, (\alpha \, |1_b\rangle + \beta \, |0_b\rangle)/2 + |\psi_{ua}^-\rangle \, (\alpha \, |1_b\rangle - \beta \, |0_b\rangle)/2 \; . \quad (2.86)
\end{aligned}
$$

By performing a Bell state measurement on the $u-a$ system, Alice projects with equal probabilities the b particle in one out of four possible superpositions of $|0_b\rangle$ and $|1_b\rangle$. If she finds the $u-a$ pair in $|\phi_{ua}^+\rangle$, b collapses in the state in which u was prior to Alice's measurement and the teleportation is achieved without Bob having to do anything on b. If Alice finds $|\phi_{ua}^-\rangle$, b is projected onto $\alpha \, |0_b\rangle - \beta \, |1_b\rangle$ and Bob must apply the unitary operation σ_Z in order to turn it into the desired state $\alpha \, |0_b\rangle + \beta \, |1_b\rangle$. Likewise, if Alice finds $|\psi_{ua}^+\rangle$ or $|\psi_{ua}^-\rangle$, Bob has to apply the unitary transformations σ_X or $\sigma_Z \sigma_X = i\sigma_Y$ respectively, in order to recover the initial $|\psi_u\rangle$ state.

Hence, to each possible result of Alice's measurement, there corresponds a well-defined unitary operation allowing Bob to put b in the desired quantum state. It is obvious that Bob cannot decide what to do on b before he gets information from Alice about the outcome of her measurement. Teleportation involves a combination of two communication channels: a quantum link provided by the non-local entanglement between a and b, and a classical channel through which Alice sends to Bob two bits of information about the result of her four-valued measurement. Teleportation does not violate causality. Even if the EPR correlations are 'instantaneous' (in fact independent of the time ordering of Alice and Bob measurements), the transmission of information allowing quantum state reconstruction at a distance requires a classical channel which cannot be operated faster than light.

Bell's state measurement

We still have to describe how Alice can perform her Bell state measurement on the $u - a$ pair. We have shown in Section 2.5 that a CNOT gate operating on two qubits amounts to a conditional shift operation (modulo 2) of one of the qubits, determined by the state of the other. We will see later how such a CNOT operation can be implemented in practice with photonic or atomic qubits. A CNOT gate in which u is the control and a the target turns the four Bell states into unentangled product states:

$$
\begin{aligned}
\left|\phi_{ua}^{+}\right\rangle &= (|0_u, 0_a\rangle + |1_u, 1_a\rangle)/\sqrt{2} \to (|0_u\rangle + |1_u\rangle)\,|0_a\rangle\,/\sqrt{2}\;; \\
\left|\phi_{ua}^{-}\right\rangle &= (|0_u, 0_a\rangle - |1_u, 1_a\rangle)/\sqrt{2} \to (|0_u\rangle - |1_u\rangle)\,|0_a\rangle\,/\sqrt{2}\;; \\
\left|\psi_{ua}^{+}\right\rangle &= (|0_u, 1_a\rangle + |1_u, 0_a\rangle)/\sqrt{2} \to (|0_u\rangle + |1_u\rangle)\,|1_a\rangle\,/\sqrt{2}\;; \\
\left|\psi_{ua}^{-}\right\rangle &= (|0_u, 1_a\rangle - |1_u, 0_a\rangle)/\sqrt{2} \to (|0_u\rangle - |1_u\rangle)\,|1_a\rangle\,/\sqrt{2}\;.
\end{aligned}
\tag{2.87}
$$

Applying finally the unitary single-particle transformation:

$$
U_H = (\sigma_X + \sigma_Z)/\sqrt{2}\;,
\tag{2.88}
$$

to the u qubit (Hadamard transform), we complete the transformation of the Bell states into one of four tensor product states, according to the correspondence:

$$
\left|\phi_{ua}^{+}\right\rangle \to |0_u\rangle\,|0_a\rangle\;; \quad \left|\phi_{ua}^{-}\right\rangle \to |1_u\rangle\,|0_a\rangle\;; \quad \left|\psi_{ua}^{+}\right\rangle \to |0_u\rangle\,|1_a\rangle\;; \quad \left|\psi_{ua}^{-}\right\rangle \to |1_u\rangle\,|1_a\rangle\;.
\tag{2.89}
$$

After performing the CNOT gate and the Hadamard one-qubit operation on her $a - u$ system, Alice has merely to read out the two qubits in the OZ basis to get two classical bits yielding complete information about the initial $a - u$ Bell state.

Let us emphasize once more some of the strange aspects of teleportation and the difference from a classical fax operation. Alice and Bob exchange a state that neither of them can determine. After single-particle teleportation, they have no way of checking whether the operation has been successful. They can test the fidelity of the qubit transmission only through statistical measurements involving many u particles prepared in the same quantum state. Non-commuting observables on a subset of particles would be directly measured by Alice, allowing her to determine their state $|\psi_u\rangle$, or rather an approximation of it by a density matrix ρ_u which incorporates the errors she may have made in her measurements. Another subset of particles is teleported to Bob and measured in the same way. If teleportation is imperfect, Bob's measurements yield a density matrix ρ_b slightly different from ρ_u. Comparing their final data, Alice and Bob can estimate the fidelity F of teleportation, generally defined as the expectation value $\mathrm{Tr}(\rho_u \rho_b)$, the limit $F = 1$ corresponding to perfect operation.

Note also that the amplitudes α and β are continuous variables whose precise definition requires an infinite number of binary bits and yet, the transmission of a state superposition with these amplitudes requires only the exchange of two bits of classical information. The combined effects of the quantum and classical channels allow this economy in the required number of classical bits. We should also note that it is not the material particle which has been carried from Alice to Bob, but only quantum information it contains. In this respect, teleportation does appear similar to classical fax operation in which the material sheet of paper is not transmitted, but only information written on it. Bob's particle plays the role of the blank sheet of paper fed into the fax receiver. This blank sheet is quantum, though, since it is entangled to a similar object at Alice's end of the machine. Like the blank sheet of an ordinary fax, it does not contain exploitable information before the operation is completed. Before Alice's manipulations, a measurement by Bob would yield random results. Let us finally recall that the operation has destroyed the original. Contrary to a classical fax transmission which can be repeated if it has failed in a first attempt, one must thus be very confident about the fidelity of quantum teleportation before using it for the transmission of precious information available only in a single copy.

Various teleportation schemes of photon states belonging to a two-dimensional Hilbert space have been demonstrated experimentally using as resource pairs of entangled photons (Bouwmeester *et al.* 1997; Boschi *et al.* 1998; Kim *et al.* 2001; Lombardi *et al.* 2002). In most of these experiments, only part of the teleportation protocol has been implemented. Due to experimental limitations, some of the Bell states could not be distinguished by Alice's measurement, making the procedure fail in a fraction of the attempts. The complete teleportation procedure has also been performed with massive particle qubits in the context of NMR (Nielsen *et al.* 1998) or ion traps (Riebe *et al.* 2004; Barrett *et al.* 2004) over small distances, a few angströms or micrometres. The ion trap experiments are discussed in Chapter 8. Finally, variants of teleportation in which information is encoded in a continuous variable (field amplitude) instead of a discrete one have been theoretically studied (Braunstein and Kimble 1998) and experimentally demonstrated (Furusawa *et al.* 1998).

2.6.4 Quantum computing

By manipulating qubits, it is in principle possible to perform some computations more efficiently (i.e. with fewer elementary steps) than with classical computers operating on ordinary logic. The most famous example is the factorization of large integers. Classically, it requires a number of steps increasing exponentially with the number of bits representing the integer, which makes the operation impractical for numbers having more than about a hundred digits (Lenstra and Lenstra 1993). This difficulty explains why prime factors of huge numbers are used as elements of coding keys in classical cryptography (coding of PINs in credit cards for instance). By contrast, a quantum computer manipulating qubits evolving in state superpositions could factorize in a number of steps increasing only polynomially with the number of digits. A quantum algorithm realizing this feat has been discovered by Shor (1994), triggering great interest in quantum computing and opening the way to many theoretical and experimental studies in quantum information.

In order to build a practical quantum machine, a number of difficult problems have to be addressed. An essential issue is decoherence. How can we manipulate large numbers of qubits and control the coupling to the environment which tends to transform quickly state superpositions into mundane statistical mixtures? Great theoretical progress has been made in this direction by the development of quantum error corrections procedures (Shor 1995; Steane 1996b;a; Ekert and Macchiavello 1996; Laflamme *et al.* 1996) largely based on extrapolation of similar operations in classical computers. These codes, if they could be practically implemented, would allow the operation of quantum logic on a large scale, provided each elementary step could be performed with

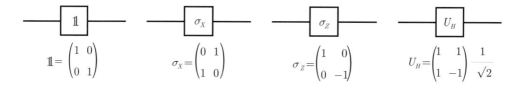

Fig. 2.19 Diagrammatic representation of single-qubit gates, with the corresponding evolution operator: $\mathbb{1}$ for the identity gate, σ_X for the 'not' gate, σ_Z for the 'phase' gate and U_H for the 'Hadamard' gate.

a fidelity exceeding a threshold, very close to 1 (Nielsen and Chuang 2000). Whether practical systems realizing this threshold condition will be discovered remains an open question. It is fair to say that practical quantum computing remains extremely challenging. Many candidates for qubits are being investigated, including photons and atoms which will be our main actors in this book.

Our motivation is here to introduce the general concepts of quantum computation, as an illustration of the ways in which the strange aspects of the quantum could be harnessed. We consider ideal situations, without decoherence, which illustrate most vividly how entanglement and complementarity could be put to practical use. Understanding how a quantum computer works in principle will hopefully motivate the reader to become familiar with the manipulations of atoms and photons described in the following chapters. These manipulations are simple demonstrations of elementary steps which have to be combined to build a complex quantum computing machine.

In a quantum logic process, information is coded in two-level qubits. As in the classical case, the elementary steps of quantum computation can be reduced to the manipulation of individual bits and joint operations on pairs of bits. These manipulations are implemented by unitary operators, to ensure their reversibility. These operators act in the two-dimensional Hilbert space of a single qubit or in the four-dimensional space of a two-qubit system. The devices performing these operations are called quantum gates (single-qubit and two-qubit gates respectively). We have already seen above various examples of such gates.

Single-qubit gates

Single-particle gates implement rotations in the spin-like Hilbert space of the qubit. We represent symbolically these operations by a graph in which the qubit state is associated to a horizontal line and the gate to a square with the symbol of the unitary operation U inside. Figure 2.19 shows various one-qubit gates: the identity (simple transmission of the qubit without change, corresponding to the unity operator, $\mathbb{1}$), the not gate (or 'switch' operation) which exchanges the 0 and 1 bit values, implemented by σ_X, the π-phase gate which leaves the $|0\rangle$ state unaltered and changes the sign of the amplitude associated to $|1\rangle$ (σ_Z) and the already defined Hadamard gate which transforms the $|0\rangle$ and $|1\rangle$ into the symmetric and anti-symmetric state superpositions $(|0\rangle+|1\rangle)/\sqrt{2}$ and $(|0\rangle-|1\rangle))/\sqrt{2}$ respectively $[U_H = (\sigma_X+\sigma_Z)/\sqrt{2}]$. These operations are unitary and hermitian, and are thus their own inverse:

$$\mathbb{1}^2 = \sigma_X^2 = \sigma_Z^2 = U_H^2 = \mathbb{1} \; . \tag{2.90}$$

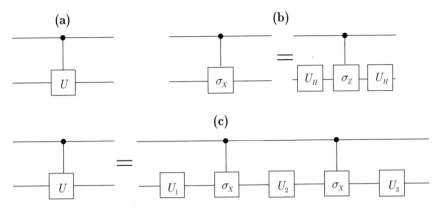

Fig. 2.20 Diagrammatic representation of two-qubit gates. (a) Control-U gate. (b) Link between the control-σ_X (CNOT) and control-σ_Z (controlled phase gate). (c) Control-U as a combination of CNOT and single-qubit gates (see text for definition of U_1, U_2 and U_3).

Other transformations of the qubit state can be realized by single-qubit gates. A rotation by the angle α around the OZ axis implements:

$$U_{\alpha,Z} = \exp(-i\alpha\sigma_Z/2) \,, \qquad (2.91)$$

and its inverse is $U_{-\alpha,Z}$. The most general qubit rotation is defined by its Euler angles α, β and γ and analysed as a (non-commutative) product of rotations around OZ and OY:

$$U_{\alpha,\beta,\gamma} = U_{\alpha,Z} U_{\beta,Y} U_{\gamma,Z} = \exp(-i\alpha\sigma_Z/2) \exp(-i\beta\sigma_Y/2) \exp(-i\gamma\sigma_Z/2) \,. \qquad (2.92)$$

Two-qubit gates

Simple two-qubit gates couple control and target bits according to a conditional dynamics, implementing a logical proposition of the kind *if A is true, then do B*. The B action, performed on the target, can be any unitary operation U. The truth of the A statement corresponds generally to the 1 value of the control, its 'negation' to the 0 value. If the control is initially in the $|1\rangle$ state (A is true) or in the $|0\rangle$ state (A is not true), it remains unaltered in the gate operation and unentangled with the target. We call 'control-U' the gate implementing the 'truth table':

$$|0\rangle \, |\psi\rangle \rightarrow |0\rangle \, |\psi\rangle \;; \qquad |1\rangle \, |\psi\rangle \rightarrow |1\rangle \, U \, |\psi\rangle \,, \qquad (2.93)$$

where the first and the second ket in each expression represents the control and the target bits. Clearly this operation, which transforms the orthogonal basis of two-qubit states $|a, b\rangle$ ($a, b = 0, 1$) into another orthogonal basis is unitary. It can be implemented by a Hamiltonian coupling the two bits reversibly. We give in the following chapters many examples of realizations of such gates.

It is convenient to schematize two-qubit gates by a graph with two horizontal lines representing the control bit (upper line) and the target (lower line). The conditional control is symbolized by a vertical line joining the two qubit lines, with a square vertex

Fig. 2.21 Diagrammatic representation of the quantum computation of a function. The multi-qubit A and B quantum registers are denoted by a slashed horizontal line.

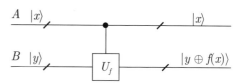

over the target line containing the symbol of the conditional operation (Fig. 2.20a). Special examples of control-U gates are the control-not, already defined, $(U = \sigma_X)$ and the control-phase $(U = \sigma_Z)$. As discussed in Section 2.5, the control-not implements the measurement of the control bit by the target, seen as a meter. When the control is in a state superposition, the output of the gate is entangled. The control-not, and more generally a control-U gate, implement non-local operations on the two qubits, which can change their degree of entanglement.

The control-phase and control-not gates are simply related, one reducing to the other when sandwiched between two Hadamard operations on the target bit. The connexions between the two kinds of gates results from:

$$U_H \sigma_Z U_H = \sigma_X ; \quad U_H \sigma_X U_H = \sigma_Z , \tag{2.94}$$

which are deduced from the definition of the Hadamard transform (eqn. 2.88) and the properties of the Pauli matrices (eqn. 2.3). Figure 2.20(b) illustrates the connexion between the control-not and the control-phase gates. This simple relationship is exploited in many experiments described later on.

The most general control-U gate can be realized by combining the control-not (or the control-phase) gate with single-qubit operations on the target. If the U operation is defined by the Euler angles (α, β, γ), three single-qubit rotations U_1, U_2 and U_3 with Euler angles $[(\gamma - \alpha)/2, 0, 0]$, $[0, -\beta/2, -(\gamma + \alpha)/2]$ and $(\alpha, \beta/2, 0)$ can be combined (Nielsen and Chuang 2000) with two control-not's to realize the control-U operation (Fig. 2.20c).

One-qubit gates and two-qubit control-not are the elementary bricks of quantum computation. They can be used to realize any two-qubit gate and combined to build the most general unitary operation U in the $2^{\mathcal{N}}$ dimensional Hilbert space $\mathcal{H}(\mathcal{N})$ of an $\mathcal{N}$-qubit system. The proof of this general mathematical property is found in Nielsen and Chuang (2000). This theorem is very important, since it entails that one-qubit and two-qubit control-not gates are all that is required to compute any real function $y = f(x)$ with a quantum computer. Let us show indeed that such a computation can always be realized by a unitary operation.

Computing a function $f(x)$ with a quantum machine

Any function $f(x)$ can be approximated by an application of a set of integers $\{x\}$ into another set $\{y\}$. The x variable varies in the $[0, 2^n - 1]$ interval and y in the $[0, 2^p - 1]$ one. By choosing n and p large enough, f can be approximated with arbitrary precision. In order to compute $f(x)$, we need $\mathcal{N} = n + p$ qubits. The x variable is coded into an input 'register' A of n qubits, each integer x being represented by the tensor product $|x_1, x_2, \ldots, x_i, \ldots, x_n\rangle$ with $x_i = 0$ or 1. The x_i's represent the binary notation of the integer x. Likewise, the value y of f is coded in an output register B of p qubits. When there is no ambiguity, we abbreviate the basis states in the A and B Hilbert spaces as

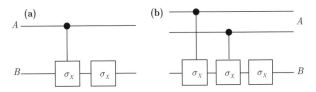

Fig. 2.22 Computation of some simple functions. (a) Single-bit function $f(0) = 1$ and $f(1) = 0$. (b) Two-bit function defined by $f(0) = f(3) = 1$, $f(1) = f(2) = 0$.

$|x_A\rangle$ and $|y_B\rangle$ and we call $|0\rangle$ the state of both registers in which all the bits are 0's. The computation of $f(x)$ is implemented by the transformation $\mathbf{U}_f$ in $\mathcal{H}(\mathcal{N})$, defined in the $|x_A\rangle\,|y_B\rangle$ basis as:

$$\mathbf{U}_f\,|x_A\rangle\,|y_B\rangle = |x_A\rangle\,|[y \oplus_p f(x)]_B\rangle\ , \tag{2.95}$$

where $\oplus_p$ represents addition modulo 2^p. This operation is obviously unitary since it transforms any couple of orthogonal basis states into a pair of orthogonal states (either the x's are different, or the y's). The procedure is symbolized in Fig. 2.21 where the A and B registers are sketched as slashed horizontal lines and the 'hardware' of the computation by a rectangular box. The value of the function for any input variable x is merely obtained by applying $\mathbf{U}_f$ on $|x_A\rangle\,|0_B\rangle$ according to:

$$\mathbf{U}_f\,|x_A\rangle\,|0_B\rangle = |x_A\rangle\,|f(x)_B\rangle\ . \tag{2.96}$$

Measuring the final state of B directly yields the binary bit sequence corresponding to $f(x)$.

According to the universality theorem, the computation can always be realized by a combination of one- and two-qubit gates. As an example, let us consider the simplest possible case where $n = p = 1$, i.e. an application of $\{0,1\}$ into $\{0,1\}$. Figure 2.22(a) shows the gate architecture implementing the function $f(0) = 1$; $f(1) = 0$ with one single-qubit σ_X and one two-qubit control-not gate. A little bit less trivial is the case $n = 2$, $p = 1$ corresponding to functions imaging $\{0,1,2,3\}$ into $\{0,1\}$. The four states of the A register are $|0,0\rangle = |0_A\rangle$, $|0,1\rangle = |1_A\rangle$, $|1,0\rangle = |2_A\rangle$ and $|1,1\rangle = |3_A\rangle$. They are coded into two bits represented as the horizontal upper lines in Fig. 2.22(b). The single-bit of the B register corresponds to the horizontal line at the bottom. The combination of single-qubit not-gates and two-qubit control-not gates shown in the figure implements $f(0) = 1, f(1) = 0, f(2) = 0, f(3) = 1$, as a mere inspection will easily convince the reader.

Quite generally, the quantum computation of $f(x)$ amounts to the measurement by B acting as a meter of the observable:

$$O_f = \sum_x |x_A\rangle\,f(x)\,\langle x_A|\ , \tag{2.97}$$

defined in the Hilbert space of the A register. The unitary transformation $\mathbf{U}_f$ implements the translation by the amount $f(x)$ of the meter pointer initially in the neutral state $|0_B\rangle$. As a general feature of an ideal measurement, the eigenstates $|x_A\rangle$ of O_f are left unaltered by the measurement and unentangled with the pointer states. In other words, the quantum computation conserves the value of the input. This is required to ensure the unitarity, and hence the reversibility of the process.

Quantum parallelism

The principle of quantum computation is, up to a point, similar to that of classical machines, which also operate with gates acting on single bits or pairs of bits (even if classical machines are usually irreversible since they erase input information as the computation proceeds). All similarity ends when we consider the possibility of preparing the A register in a superposition of different $|x_A\rangle$ states. As is also the case for a quantum measurement, entanglement is then produced between A and B. Consider for instance the case in which A is prepared in the fully symmetric superposition of all possible inputs, reached by applying to $|0_A\rangle$ the product $U_{\mathbf{H}} = U_{H_1} U_{H_2} \ldots U_{H_i} \ldots U_{H_n}$ of Hadamard transforms acting independently on the n qubits:

$$
\begin{aligned}
U_{\mathbf{H}} |0_A\rangle &= U_{H_1} U_{H_2} \ldots U_{H_i} \ldots U_{H_n} |0_1, 0_2, \ldots 0_i \ldots 0_n\rangle \\
&= \frac{1}{2^{n/2}} (|0_1\rangle + |1_1\rangle)(|0_2\rangle + |1_2\rangle) \ldots (|0_i\rangle + |1_i\rangle) \ldots (|0_n\rangle + |1_n\rangle) \\
&= \frac{1}{2^{n/2}} \sum_{x=0}^{2^n-1} |x_A\rangle \ .
\end{aligned}
\tag{2.98}
$$

If the B register is initially in $|0_B\rangle$, the $n+p$ qubits evolve under the action of $\mathbf{U_f}$ according to:

$$
\frac{1}{2^{n/2}} \sum_x |x_A\rangle |0_B\rangle \rightarrow \frac{1}{2^{n/2}} \sum_x |x_A\rangle |f(x)_B\rangle \ ,
\tag{2.99}
$$

and the final state is entangled (unless f is a trivial constant function).

Equation (2.99) illustrates a striking feature of quantum computing, distinguishing it fundamentally from its classical counterpart. By superposing the inputs of a computation, one operates the machine 'in parallel', making it compute simultaneously all the values of a function and keeping its state, before any final bit detection is performed, suspended in a coherent superposition of all the possible outcomes.

It is, however, one thing to compute potentially at once all the values of f and quite another matter to be able to exploit this quantum parallelism and extract from it more information than from a mundane classical computation. The final stage of information acquisition must always be a measurement. If one decides to measure the final state of A, one will find randomly any of the x values and B will be projected into the correlated state $|f(x)_B\rangle$ whose measurement yields the value of f corresponding to the (randomly determined) x input. The procedure then amounts to a random computation of the function values, not an obvious improvement over the classical method in which one would systematically compute $f(x)$.

Coding the period of a function into a state superposition

An alternative measurement procedure seems more promising to take advantage of quantum parallelism. It operates when f takes the same value for more than one x and if one seeks to get information about the pattern of x values corresponding to the same y. If one measures first B, a random value y of f is obtained and A is projected into a symmetric superposition of all the states $|x_A\rangle$ such that $f(x) = y$. In the case when f is periodic, this is a superposition of states corresponding to x values separated by a fixed interval, the period of f. If we wish to determine this a *priori* unknown period, it is tantalizing to notice that this information is so to speak 'inscribed' in a superposition which the quantum computing procedure has allowed us to generate in a single computing step. Conversely, the only way to find out classically the period of an unknown function is to compute it for all the values of x successively, a process which involves obviously a much larger number of elementary operations.

We have a first glimpse here at the potential advantages of quantum over classical computing. In fact, Shor's algorithm of factorization (Shor 1994) is based on a theorem stating that the prime factors of a large number can be obtained from the determination of the period of a function depending on this number. We thus see that quantum parallelism may help us in solving such a period problem more efficiently than by a classical algorithm. We must, however, realize that quantum information about the period of the function is still hidden in the superposition and cannot be retrieved by a trivial measurement. Detecting directly the A bits would collapse randomly the superposition into one of the $|x_A\rangle$ states and irreversibly destroy information about the period contained in the superposition created by the measurement of B.

In order to extract this information, we need to be cleverer. Before performing the last measurement on A, we mix again the states of the register and produce in this way an interference effect between the various computing paths. It is from this final interference that we are able to get information about the pattern of x's yielding the same y value. We will not go on discussing Shor's algorithm, which involves technical mathematical points whose analysis would complicate too much the present discussion. A clear description of this algorithm can be found in Ekert and Joasza (1996). To remain at a pedagogical level, we rather analyse a related algorithm, invented by Simon (1994), which also amounts essentially to searching the pattern of inputs corresponding to the same output value of an unknown function.

Simon's algorithm

Simon's algorithm contains basically the same physics as Shor's, in a much simpler context. Its goal is to find, with the smallest possible number of elementary operations, a simple property of an otherwise unknown function $f(x)$ imaging the $[0, 2^n-1]$ interval into $[0, 2^p - 1]$ with $p = n - 1$. We are given a quantum computer 'black box' whose hardware is hidden from us, which computes $f(x)$. We know only the following features of $f(x)$: (i) the variable x is partitioned in pairs (x, x'), for which the function takes the same value $[f(x) = f(x')]$, with no two pairs corresponding to the same output; (ii) the pairing (x, x') is defined by a bit-to-bit translation of the binary sequence describing x:

$$x'_i = x_i \oplus s_i ; \qquad x_i = x'_i \oplus s_i , \qquad (2.100)$$

where $s_i = 0, 1$ is a fixed string of n zeros and ones and $\oplus$ is addition modulo 2. The commutativity of the pairing is a consequence of the relation $s_i \oplus s_i = 0$. We have to find the sequence s_i defining the pairing. The problem looks like the search of an unknown period, but we must notice that the bit to bit addition ($\oplus$ symbol) is different from the sum modulo 2^n ($\oplus_n$ symbol), which corresponds to the definition of an usual period.

Classically, the Simon problem is difficult. It requires a number of operations increasing exponentially with n. If $f(x)$ is computed by a classical machine, the only way to find out s is to calculate randomly $f(x)$ values until, by chance, two outputs

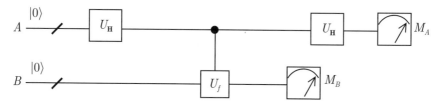

Fig. 2.23 Diagrammatic representation of Simon's algorithm implementation.

are found equal, thus revealing a (x, x') pair. The s sequence is then obtained by $s_i = x_i + x'_i$. The average number of random searches is 2^{n-1}, an exponential function of n.

The quantum method, schematized in Fig. 2.23, is much more economical. We start by generating the symmetrical superposition of all the input states and apply the unknown function to all of them in parallel. We then measure the B register, projecting A into the superposition of two states of a given (random) pair:

$$|\psi_A\rangle = \frac{1}{\sqrt{2}}(|x_1, \ldots, x_i, \ldots, x_n\rangle + |x_1 \oplus s_1, \ldots, x_i \oplus s_i, \ldots, x_n \oplus s_n\rangle) . \qquad (2.101)$$

We then apply Hadamard transforms to the n qubits of A, whose effect is to 'split' the two components of the pair into state superpositions. The superposition coming from each term in eqn. (2.101) is a sum of 2^n states corresponding to all possible strings $z_1, \ldots, z_i, \ldots, z_n$ of 0's and 1's. Each state, in each of these two sums, has an amplitude $\pm[1/2^{(n+1)/2}]$ whose sign depends upon the parity of $\sum_i x_i z_i$ or $\sum_i (x_i \oplus s_i) z_i$ respectively. The final result of this state mixing operation is:

$$
\begin{aligned}
U_{\mathbf{H}}|\psi\rangle_A &= \frac{1}{2^{(n+1)/2}} \sum_z \left[(-1)^{\sum_i x_i z_i} + (-1)^{\sum_i (x_i \oplus s_i) z_i} \right] |z_1, \ldots, z_i, \ldots, z_n\rangle \\
&= \frac{1}{2^{(n+1)/2}} \sum_z (-1)^{\sum_i x_i z_i} \left[1 + (-1)^{\sum_i s_i z_i} \right] |z_1, \ldots, z_i, \ldots, z_n\rangle .
\end{aligned}
$$

$$(2.102)$$

Equation (2.102) exhibits an interference effect. The probability amplitude of a given final qubit sequence $z_1, \ldots, z_i, \ldots, z_n$ is the sum of two partial amplitudes coming from the splitting of each of the two states of the pair randomly selected by the B measurement. These partial amplitudes interfere constructively or destructively, depending on the parity of $\sum_i s_i z_i$. The 2^{n-1} z values for which $\sum_i s_i z_i = 0$ (modulo 2) correspond to equiprobable states, while the 2^{n-1} ones corresponding to $\sum_i s_i z_i = 1$ (modulo 2) have a zero probability and are missing ('dark fringes' in the interference vocabulary). This interference effect is revealed by a final measurement of A. Detecting the n qubits yields a sequence of 0 and 1 values $z_1^1, \ldots, z_i^1, \ldots, z_n^1$, which satisfy the constructive interference condition:

$$\sum_i s_i z_i^1 = 0 \quad (\text{modulo } 2) . \qquad (2.103)$$

We thus get a linear relation constraining the unknown s_i's. By repeating n times the sequence of operations, we obtain n independent relations of the same kind: $\sum_i s_i z_i^k =$

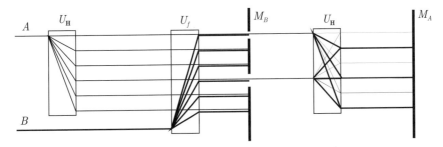

Fig. 2.24 Quantum computation viewed as multi-particle interference for the Simon algorithm. The first Hadamard gates on A split the initial state into a superposition of paths, which get entangled with register B by the computation of the function f. The measurement (M_B) on B acts as a double slit selecting only two paths for A. The final Hadamard gates produce an interference pattern between these paths, from which the result is finally extracted by the measurement of A (M_A). The final states corresponding to constructive interference are shown as thick lines.

0 (modulo 2) with $k = 1, 2, \ldots, n$. This set of n independent linear relations allows us to find out the n unknown parameters and to solve the Simon enigma. If, by a stroke of (un)luck, two measurements yield dependent equations (which is very unlikely for large n), we would merely have to perform the operation once more.

As a final remark, we should add that we have been over-careful in our implementation of this algorithm. Recording the measurement of the B register is superfluous. It is enough that the 'black box' computing the f function has entangled A and B, even if the f value is not read. This entanglement is sufficient to reduce A to a statistical mixture of superposition pairs of the form $|x_A\rangle + |(x \oplus s)_A\rangle$. Their subsequent splitting by the Hadamard transforms yields, whatever x is, to a constructive interference only for the z values which satisfy eqn. (2.103). In any case, the answer to Simon's problem is obtained in about n computation steps, much smaller than the 2^{n-1} steps required in the classical version.

General features of quantum computation: multi-particle interference

Simon's algorithm illustrates on a simple example the main features of a quantum computer, which make it potentially more efficient than a classical machine. A collective quantum state superposition is induced by an application of single-bit Hadamard transforms mixing coherently the $|0\rangle$ and $|1\rangle$ states of each qubit of the input register A. This ensures massive parallelism of the computation with, in general, entanglement of the bits. Useful information is coded into a state superposition in a form which cannot be read out directly, but only after a second mixing of the bits. The successive application of two sequences of Hadamard transforms amounts to the operation of a kind of multi-particle interferometer, splitting the qubit 'trajectories' and recombining them, before a final measurement which exploits a constructive interference effect to get useful information. The operation of this interferometer is schematized in Fig. 2.24.

As a multi-particle interference process, the operation of a quantum computer is very delicate and jeopardized by all kinds of decoherence effects. The machine prepares a large multi-state Schrödinger cat which must remain coherent throughout the computation. We will not dwell here on this important challenge to quantum computation, just keep in mind the fact that decoherence is the big enemy which would have to be efficiently controlled.

Simon's algorithm is one of several examples of games which quantum logic is able to solve advantageously. Others are the Deutsch and Josza (1992) and Grover search algorithms (Grover 1997; 1998) whose principle is described in Nielsen and Chuang (2000). The Deutsch–Josza algorithm (Deutsch and Josza 1992) corresponds also to a situation in which an exponentially difficult problem becomes polynomial. In the Grover algorithm, a classically difficult problem becomes only slightly easier when quantum logic is used (division by 2 of the exponent in the exponential law giving the number of steps).

Simple experimental demonstrations of the Deutsch–Josza (Gulde *et al.* 2003) and Grover (Vandersypen *et al.* 1999) problems have been made on various systems, in quantum optics and Nuclear Magnetic Resonance (NMR) experiments. In the quantum optics case, the qubits are carried by ions in a trap (Gulde *et al.* 2003). In the NMR experiments (Vandersypen *et al.* 1999), the qubits are nuclear spins of organic molecules in a liquid sample. Performing NMR pulse sequences on these spins amounts to operating together a huge number of elementary computers, one for each molecule in the sample. The quantum coherence is realized within each molecule, not between the molecules. An experiment implementing Shor's algorithm to factorize 15 has been realized in NMR (Vandersypen *et al.* 2001).

3

Of spins and springs

The sciences do not try to explain, they hardly even try to interpret, they mainly make models. By a model, is meant a mathematical construct which, with the addition of certain verbal interpretation describes observed phenomena. The justification of such a mathematical construct is solely and precisely that it is expected to work

John von Neumann

Cavity quantum electrodynamics and ion trap experiments provide striking illustrations of quantum concepts. The appeal of their physics comes from the basic simplicity of the systems they manipulate, which can essentially be modelled as two-state particles, so called 'spins', coupled to quantum field oscillators or 'springs'. In fact spins and springs are ubiquitous in physics. Before considering them in the CQED and ion trap contexts, it is instructive to review the roles they play in a wide variety of phenomena.

Two-state systems are simplified representations of real systems whose configuration can be idealized as a dichotomic quantity (a spin pointing up or down, a photon with vertical or horizontal polarization, a particle passing through one slit or the other in a Young interferometer). Once the complications of the real world have been eliminated and the essence of the physics reduced to the two-state configuration, calculations become simple and the quantum concepts are revealed in phenomena which become analytically tractable.

The two-state approach is systematically used in some textbooks such as the *Feynman lectures in physics* (Feynman 1965). The essence of the tunnel effect, for instance, is captured in a simplified model representing a particle trapped in two potential wells separated by a barrier. The particle evolves in a Hilbert space spanned by two states, corresponding to wave packets localized in each of the wells. These states are coupled by an interaction term describing the tunnelling amplitude across the barrier. This coupling produces a splitting of the particle's energy levels into a symmetrical and an anti-symmetrical superposition of the two wave packet configurations. If the particle is initially localized in one well, it subsequently oscillates across the barrier at a rate equal to the transition frequency between the two energy eigenstates. This model accounts for the Josephson effect, the inversion of the ammonia molecule and other tunnelling phenomena we will encounter later on.

In light–matter interaction physics, the representation of atoms as two-level entities is often a very good approximation. When an atom interacts with a field resonant or

(a)   (b)

(c)   (d)

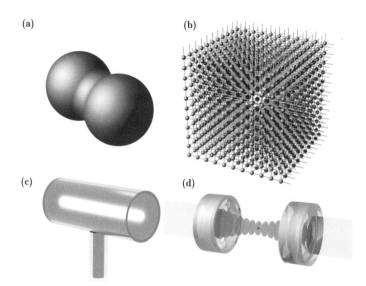

Fig. 3.1 Artist's view of some systems more or less faithfully described by the quantum oscillator model: (a) diatomic molecule, (b) crystal, (c) atomic sample in a glass cell scattering non-resonant light and (d) single mode of an optical cavity.

nearly resonant with an atomic transition between two non-degenerate energy states, it is legitimate to neglect all the other levels. The atom can then be treated as a spin represented by a vector evolving on the Bloch sphere. The dynamics of this vector is ruled by the famous Bloch equations, widely used in the description of magnetic and optical resonance phenomena.

As ubiquitous as they are, spins alone are however unable to capture the essence of many physical processes. Spring models are better suited to represent systems which, contrary to spins, are able to give or receive many quanta of energy. A harmonic oscillator with its Hamiltonian quadratic with respect to the system coordinates describes all kinds of systems close to equilibrium, in a wide range of situations (Fig. 3.1). Atoms bound in a molecule, whose vibration modes are described as an ensemble of oscillators, provide a typical example of such a near-equilibrium situation. The spring model is also well-suited for analysing the motion of atoms or ions in a crystal. The phonons, quantum vibrations of crystalline media, result directly from the quantization of the classical collective oscillation modes. Of course, the spring model for particle vibrations in matter is only an approximation. For molecules, the model breaks down when the vibration amplitude is of the order of the extension of the binding potential well. The potential is then far from being quadratic and is more adequately described by Lennard–Jones functions. For crystals, the phonon model obviously loses its validity when the vibration amplitude is so high that the solid eventually melts!

Atoms interacting with light can also, in some situations, be described by a spring model. If the atom remains with a high probability in its ground state, modelling the valence electron in the atom as a harmonically bound particle excited by the light field provides a remarkably good description of the physics. This is the Thomson–Lorentz model of the atom, which represents an atomic ensemble as a collection of oscillators. The fraction of oscillators with a given frequency defines the 'oscillator strength' of the corresponding transition in the atomic spectrum. This model gives a very precise description of the refractive index of non-resonant atomic media. It also accounts well for the scattering of light by density fluctuations in gases. The blue colour of clear skies results directly from the fourth-power frequency dependence of the scattering cross-section of light by an elastically bound electron.

Here again, the quantum oscillator model is only an approximation to a more complex reality. It breaks down when the atom is driven too strongly by the light field. Non-linear phenomena, which cannot be explained by the Thomson model, then come into play. These non-linear effects become important when the atoms are subjected to very intense non-resonant radiation, or even to a moderate field resonant with an atomic transition. The spin model is then a much better description of the atom than the spring one.

In all the situations discussed so far, the oscillator description is only an approximation. For the description of radiation, however, the oscillator picture is more than a near-equilibrium simplifying assumption. An electromagnetic field can quite generally be described as a collection of independent oscillators, the validity of this model breaking down only for exceedingly large field intensities such that particle-pair creations become significant. These are extreme situations which we will not consider here.

The intertwined history of field and oscillator physics can be traced back to the early days of Maxwell's theory. It was then recognized that the electromagnetic field can be conveniently expanded along an ensemble of independent, orthogonal field modes. A possible mode basis is made of the plane electromagnetic waves defined by their wave vector $\mathbf{k}$, frequency $\omega = c|\mathbf{k}|$, and transverse polarization $\boldsymbol{\epsilon}$ normal to $\mathbf{k}$. The Gaussian modes, used in laser physics, are another possible set of basis modes. For a given mode, the field geometry is completely defined. The only free parameter is the 'normal mode coordinate', a complex number, α, describing the field amplitude. Maxwell's equations, written for the α's, are identical to those describing the dynamics of a harmonic oscillator complex amplitude $\alpha = x + ip$ in phase space (x and p being the oscillator's position and momentum): $d\alpha/dt = -i\omega\alpha$. This analogy results from the similarity of the field and oscillator Hamiltonians. The field energy stored in each mode is the sum of two contributions, quadratic functions of the electric and magnetic field amplitudes, while the energy of a mechanical oscillator is the sum of potential and kinetic energies, quadratic in its position and momentum coordinates.

The field mode expansion is at the heart of the crisis of classical physics at the end of the nineteenth century, which we have mentioned in Chapter 1. Calculating the spectrum of the field radiated by heated bodies amounts to finding the thermal equilibrium of the field oscillators. This calculation leads to a striking absurdity when one applies to these oscillators Boltzmann's rules of classical statistical physics. These rules attribute to each oscillator at temperature T an average thermal energy k_bT, where k_b is Boltzmann's constant. The mode density being proportional to ω^2, the blackbody spectrum diverges at high frequency and the total energy is infinite. This 'ultraviolet' puzzle, not the last one in the history of physics, was solved by Planck (1900). Making the bold assumption that the energy exchanged by field oscillators with heated bodies is quantized in multiples of $E = \hbar\omega$, he showed that this quantization renders the average thermal equilibrium energy per mode much smaller at high frequencies than the classical value $k_b T$. Planck got rid of the ultraviolet divergence and found the experimentally observed blackbody spectrum. Five years later, Einstein postulated that light itself was made of lumps of energy and introduced the concept of photons. Quantum electrodynamics was thus born from what appeared as an *ad hoc* modification of the book-keeping rules of the harmonic oscillator energies. It became gradually, as we have recalled, the most precise theory in physics.

This development took some time, though. The ultraviolet divergence, which had been eliminated in the blackbody spectrum by the quantum rules, came back with a vengeance under a different guise. The origin of this new difficulty was in the 'vacuum quantum fluctuations'. The ground state of a mechanical oscillator has a zero average position and momentum, but presents non-vanishing mean square values, imposed by the Heisenberg uncertainty relations. Similarly, the vacuum state of each field mode has electric and magnetic fields with zero mean values, but finite energy densities. In a simple approach, these vacuum fluctuations, whose sum over modes diverges at high frequencies, have an infinite effect on charged particles.

This second ultraviolet catastrophe was cured in the 1940s by the renormalization theory. In qualitative terms, this theory recognizes that the measurable parameters of all particles – mass, charge, magnetic moment – are always 'dressed' by vacuum fluctuations which cannot be switched off. The observed finite quantities result from a subtraction of infinities, the infinitely large parameters of the unphysical 'naked' particles being balanced by the infinite corrections produced by the vacuum fluctuations. Only the differential effect of these fluctuations, which affect in slightly different ways a free particle and a particle interacting with an external field, can be observed. This differential effect produces small corrections of the energy levels of charged particles in an electric potential. This is the radiative shift of atomic energy states, first observed by Lamb in the hydrogen spectrum. The precession frequency of the spin of a charged particle in an external magnetic field is also slightly affected by these corrections, resulting in a so-called 'anomalous magnetic moment'. This qualitative discussion gives a deceptively naive view of the sophisticated renormalization theory (Itzykson and Zuber 1980). We have already mentioned its astounding precision in the prediction of the magnetic moment of the electron.

Near mirrors imposing their boundary conditions, the mode structure of the electromagnetic field is not invariant under translation. The Lamb shift becomes position-dependent. The spatial derivative of this energy shift represents a force produced by the boundary on the atom. This is the Casimir and Polder (1948) effect, which may also affect macroscopic objects in the vicinity of metallic surfaces. Its most striking manifestation is that two metallic plates attract each other in vacuum (Casimir 1948). This effect results, in a very qualitative picture, from the differential radiation pressure between the unconfined quantum fluctuations outside the plates and the fluctuations of the confined field between them. The precise calculation of these effects is rather cumbersome, since the continuum of field modes has to be taken into account. We will deal below with a much simpler situation, in which a single field mode (the resonant cavity one) plays the dominant role. The perturbing effect of all the other modes is taken into account in the definition of the atomic energies. We will thus be able to describe safely our systems with a simple minded, one-dimensional harmonic oscillator approach.

Springs thus play an essential role in electrodynamics, both classical and quantum. They are also important players in statistical physics, as our brief incursion into the thermal radiation problem has already shown. More generally, oscillators are useful elements to build an explicit model for the environment of a quantum system undergoing a relaxation process. Describing quantum relaxation amounts to coupling a quantum system with a large thermodynamic 'bath' (Louisell 1973; Haake 1973). A convenient bath, amenable to a simple description, is made of an infinite set of quantum oscillators, spanning a wide frequency range, each being weakly coupled with the system under study. In many situations, this model is more than an abstraction. The spontaneous emission of an atom results from its coupling to the environment made of the initially empty modes of the electromagnetic field around it, which can be described as an ensemble of quantum oscillators in their ground states. The damping of a cavity mode is also generally due to its coupling to an environment of oscillators corresponding to the free-space propagating modes outside the cavity. These modes are coupled to the cavity field either by finite transmission across the mirrors or by scattering processes due to imperfections of the mirror surfaces.[1]

This chapter describes the properties of quantum oscillators and spins essential for the analysis of the experiments to come. It can be viewed as a short self-consistent course in quantum optics. We first recall (Section 3.1) the main features of a quantum field mode, using the analogy with the mechanical oscillators we have just recalled. We introduce the concept of photon and describe various field states, among which the coherent states which are the best approximation of a classical field with well-defined amplitude and phase. We also describe in simple terms the coupling of fields with electric charges, which yields simple models to describe classical sources of radiation and photon counters. We next consider the coupling of two modes (Section 3.2) and introduce a simple model of a linear beam-splitter, a very useful quantum optics component. We show how, by combining two beam-splitters, one can separate, then recombine a field mode, realizing a Mach–Zehnder interferometer, a device widely employed in quantum optics. We also describe a homodyne field measurement in which a beam-splitter is used to mix a quantum field with a known reference of the same frequency, the resulting beating yielding information about the field coherence properties. The beam-splitter also provides a simple model for photon losses in a cavity storing a field mode. We briefly analyse such a model, which leads us to an elementary discussion of cavity field damping.

We then turn (in Section 3.3) to the other player in the spin–spring game and describe an atomic system as a spin-$1/2$. We recall how this spin can be manipulated with classical fields, preparing and analysing arbitrary spin state superpositions. We show that a combination of two classical field pulses realizes an atomic Ramsey interferometer, analogous to the optical Mach–Zehnder device.

We finally describe in Section 3.4 the coupling of a spin to a spring oscillator, ruled by a very simple Hamiltonian, first introduced in quantum optics by Jaynes and Cummings (1963). We discuss the eigenstates of the complete atom–field Hamilto-

---

[1]Choosing explicitly a model for the environment is not the only approach to relaxation studies. As we will see in Chapter 4, it is possible to derive from general principles a damping equation for a system, without any detailed description of the environment.

nian, called the 'dressed states', and, as an immediate application, the 'vacuum Rabi oscillation' performed by an atom repeatedly emitting and re-absorbing a photon in a cavity. We extend this study to the quantum Rabi oscillations of an atom in a small coherent field, which exhibit the striking phenomenon of 'collapse and revivals', a direct manifestation of field quantization. We conclude this section by analysing the non-resonant interaction of an atom with a field mode in a cavity. The two interacting systems experience reciprocal frequency shifts. The atomic shift depends upon the number of photons in the field, making the effect useful to measure light intensity at the quantum level. The cavity field shift, on the other hand, can be viewed as a single-atom index effect which depends upon the state of the atom. These dispersive quantum effects are essential ingredients in experiments described later on.

The interested reader can find more information about oscillators, spins and their interaction in quantum optics textbooks.[2]

3.1 The field oscillator

3.1.1 Mechanical and field oscillators in quantum physics

One-dimensional mechanical oscillator

A particle of mass m moving in one dimension in the quadratic potential $V(x_c) = m\omega^2 x_c^2/2$ is the paradigm of a harmonic oscillator (Fig. 3.2). Clasically, its state is defined by its position and momentum coordinates x_c and p_c. The motion of the particle in phase space, identified with the complex plane of the variable $x_c + ip_c$, is a rotation at angular frequency ω along an ellipse centred at the origin.

In quantum physics, the position and momentum become operators, X and P, satisfying the commutation relation $[X, P] = i\hbar$ and the oscillator's Hamiltonian H_x is:

$$H_x = \frac{P^2}{2m} + \frac{m\omega^2 X^2}{2} . \tag{3.1}$$

Introducing natural units for X and P:

$$x_0 = \sqrt{\frac{\hbar}{2m\omega}} \quad \text{and} \quad p_0 = \sqrt{\frac{m\omega\hbar}{2}} , \tag{3.2}$$

and dimensionless position and momentum operators by:

$$X_0 = \frac{X}{2x_0} \quad \text{and} \quad P_0 = \frac{P}{2p_0} , \tag{3.3}$$

the Hamiltonian is:

$$H_x = \hbar\omega \left[P_0^2 + X_0^2 \right] . \tag{3.4}$$

It is convenient to define the non-hermitian operator $a = X_0 + iP_0$ and its conjugate $a^\dagger$. These operators correspond, in quantum physics, to the normalized amplitude

---

[2]Field quantization is covered in detail in Cohen-Tannoudji *et al.* (1992a). A good general introduction to quantum optics and atom–field coupling can be found in a great variety of textbooks such as Glauber (1965); Loudon (1983); Knight and Allen (1983); Meystre and Scully (1983); Perina (1991); Cohen-Tannoudji *et al.* (1992b); Carmichael (1993); Walls and Milburn (1995); Barnett and Radmore (1997); Scully and Zubairy (1997); Leonhardt (1997); Meystre and Sargent (1999); Vogel *et al.* (2001); and Schleich (2001).

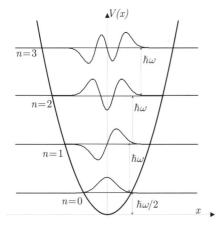

Fig. 3.2 A mechanical quantum oscillator. The parabola represents the confining potential, the horizontal lines the Fock states ladder. We have also indicated the first wave functions. The oscillator spatial coordinate is expressed in reduced units ($x = x_c/2x_0$).

$x_c/2x_0 + ip_c/2p_0$ of the classical oscillator and its complex conjugate. The commutation relation between X and P leads to:

$$[a, a^\dagger] = \mathbb{1} . \tag{3.5}$$

Expressing the position and momentum operators in terms of a and $a^\dagger$:

$$X_0 = \frac{a + a^\dagger}{2} \quad \text{and} \quad P_0 = i\frac{a^\dagger - a}{2} , \tag{3.6}$$

we get a canonical expression for the Hamiltonian:

$$H_x = \hbar\omega \left[a^\dagger a + 1/2\right] . \tag{3.7}$$

The constant term $\hbar\omega/2$ in H_x represents the vacuum fluctuation energy, which can be eliminated by redefining the energy origin. We then get the equivalent expression:

$$H'_x = \hbar\omega a^\dagger a . \tag{3.8}$$

The spectrum and all the properties of the quantum oscillator directly derive from this simple Hamiltonian and from the commutation relations of a and $a^\dagger$. Let us only recall here the main results of this elementary exercise. The Hamiltonian is proportional to the number operator $N = a^\dagger a$ satisfying the following commutation relations:

$$[a, N] = a \quad \text{and} \quad [a^\dagger, N] = -a^\dagger . \tag{3.9}$$

The non-degenerate spectrum of N is made of the non-negative integers. We denote by $|n\rangle$ the associated eigenstates, the 'Fock states', such that $N|n\rangle = n|n\rangle$. They form an orthogonal basis of the Hilbert space: $\langle n|p\rangle = \delta_{n,p}$ and $\sum_n |n\rangle \langle n| = \mathbb{1}$. The energy of $|n\rangle$ (eigenvalue of H'_x) is $E_n = n\hbar\omega$. It is the energy of n quanta $\hbar\omega$. These quanta will be called here 'phonons'.[3]

---

[3] Usually, phonons are defined as quanta of collective vibration modes in solid. We use this term here for a single oscillator.

The operators a and $a^\dagger$ have a simple action on the Fock states:

$$a\,|n\rangle = \sqrt{n}\,|n-1\rangle \quad \text{and} \quad a^\dagger\,|n\rangle = \sqrt{n+1}\,|n+1\rangle \ . \tag{3.10}$$

The operator a, which removes an elementary excitation, is called the phonon 'annihilation operator', whereas $a^\dagger$ is the 'creation' operator. All Fock states can be generated from the ground state $|0\rangle$ by repeated action of the creation operator:

$$|n\rangle = \frac{a^{\dagger n}}{\sqrt{n!}}\,|0\rangle \ . \tag{3.11}$$

It is convenient to express the position and the momentum in dimensionless units, as eigenvalues x and p of the operators X_0 and P_0. This amounts to measuring X in units of $2x_0$ and P in units of $2p_0$. With this notation, the wave function of the oscillator ground state $|0\rangle$, a Gaussian centred at the origin, is:

$$\Psi_0(x) = \langle x\,|0\rangle = (2/\pi)^{1/4}\,e^{-x^2} \ . \tag{3.12}$$

The standard deviations of X_0 and P_0 in this minimum uncertainty state are equal to $1/2$. More generally, the wave functions of the $|n\rangle$ state is:

$$\Psi_n(x) = \langle x\,|n\rangle = (2/\pi)^{1/4}\,\frac{1}{\sqrt{2^n n!}}e^{-x^2}H_n(x\sqrt{2}) \ , \tag{3.13}$$

where:

$$H_n(u) = (-1)^n e^{u^2}\frac{d^n}{du^n}e^{-u^2} \ , \tag{3.14}$$

is the Hermite polynomial of order n. The wave function of the nth excited state is thus a Gaussian of x modulated by a polynomial function presenting n nodes. Its parity with respect to $x \to -x$ reflection is $(-1)^n$, a property which will be useful to recall in Chapters 6 and 7. The wave functions of the $n = 0$ to 3 Fock states are represented in Fig. 3.2.

The Fock states are the stationary states of the oscillator in the Schrödinger point of view. Any state expands on this basis with probability amplitudes presenting time-dependent phases. In the Heisenberg point of view, all oscillator states become stationary and the operators evolve in time. The dynamical equation for $\tilde{a}$ is:[4]

$$\frac{d\tilde{a}}{dt} = \frac{1}{i\hbar}\left[\tilde{a}, \tilde{H}\right] = -i\omega\,[\tilde{a}, N] = -i\omega\tilde{a} \ . \tag{3.15}$$

This is formally equivalent to the evolution equation of the classical oscillator complex amplitude with the trivial solution:

$$\tilde{a}(t) = a(0)e^{-i\omega t} \ . \tag{3.16}$$

---

[4]We use systematically, in the following, a tilde to denote the operators or state vectors in an interaction representation with respect to a part or to the totality of the system's Hamiltonian. This convention applies to the Heisenberg point of view, which corresponds to the interaction representation with respect to the total Hamiltonian.

Field mode oscillator

A field mode in a cavity is a one-dimensional oscillator (Loudon 1983; Cohen-Tannoudji *et al.* 1992a) which can be described by the formalism we have just recalled. The operators a and $a^\dagger$ are now the photon annihilation and creation operators, and N is the photon number operator. The Hamiltonian is $H_c = \hbar\omega_c(N + 1/2)$, ω_c being the cavity mode angular frequency. This Hamiltonian is replaced by $H'_c = \hbar\omega_c N$ when more convenient.

The electric field at point $\mathbf{r}$ in the cavity mode is a linear combination of the creation and annihilation operators. It can be written, in the Schrödinger picture:

$$\mathbf{E}_c = i\mathcal{E}_0 \left[\mathbf{f}(\mathbf{r})a - \mathbf{f}^*(\mathbf{r})a^\dagger \right] \;, \tag{3.17}$$

where $\mathcal{E}_0$ is a normalization factor, with the dimension of an electric field. The dimensionless vector function $\mathbf{f}$ describes the spatial structure of the field mode (relative field amplitude and polarization), as given by the solution of Maxwell equations. It obeys the Helmholtz equation $\Delta\mathbf{f} + (\omega_c^2/c^2)\mathbf{f} = 0$ and satisfies the boundary conditions for the electric field on the cavity mirrors and at infinity. We use a simple normalization, with $|\mathbf{f}| = 1$ at the points where the field mode amplitude is maximum. One of these points is taken as the origin. The function $\mathbf{f}$ can also be written as $\mathbf{f}(\mathbf{r}) = \boldsymbol{\epsilon}_c(\mathbf{r})f(\mathbf{r})$, where $\boldsymbol{\epsilon}_c$ is a unit vector (possibly complex) describing the field polarization at point $\mathbf{r}$, and f a scalar function with maximum value 1.

It is convenient to use also the Heisenberg picture, which provides a more direct link between quantum and classical physics. In this point of view, the electric field operator has a harmonic time dependency:

$$\widetilde{\mathbf{E}}_c(\mathbf{r}, t) = i\mathcal{E}_0 \left[\boldsymbol{\epsilon}_c f(\mathbf{r})ae^{-i\omega_c t} - \boldsymbol{\epsilon}_c^* f^*(\mathbf{r})a^\dagger e^{i\omega_c t} \right] \;. \tag{3.18}$$

The term involving the time-dependent annihilation operator $\tilde{a} = a\exp(-i\omega_c t)$ is called the positive frequency part of the field. Classically, it corresponds to the field amplitude evolving in phase space (also called the Fresnel plane in optics).

The electric field is related to the potential vector $\mathbf{A}(\mathbf{r}, t)$ by $\mathbf{E} = -\partial\mathbf{A}/\partial t$. We thus have, in the Heisenberg picture:

$$\widetilde{\mathbf{A}}(\mathbf{r}, t) = \frac{\mathcal{E}_0}{\omega_c} \left[\boldsymbol{\epsilon}_c f(\mathbf{r})ae^{-i\omega_c t} + \boldsymbol{\epsilon}_c^* f^*(\mathbf{r})a^\dagger e^{i\omega_c t} \right] \;, \tag{3.19}$$

and hence, in the Schrödinger representation:

$$\mathbf{A}(\mathbf{r}) = \frac{\mathcal{E}_0}{\omega_c} \left[\boldsymbol{\epsilon}_c f(\mathbf{r})a + \boldsymbol{\epsilon}_c^* f^*(\mathbf{r})a^\dagger \right] \;. \tag{3.20}$$

Photon number states and field per photon

The Fock states $|n\rangle$ of the field oscillator are now also called 'photon number states'. The ground state $|0\rangle$ is referred to as the 'vacuum'. The field and vector potential being linear functions of a and $a^\dagger$, have zero expectation values in photon number states, including the vacuum. These states present however field fluctuations. The electromagnetic energy density $\varepsilon_0|\mathbf{E}_c|^2$, where ε_0 is the vacuum permittivity, involves terms like a^2 or $a^{\dagger 2}$, which average to zero in a Fock state, but also terms proportional to $a^\dagger a = N$, which do not cancel. As expected, the average energy density is non-vanishing in the Fock states. The special case of the vacuum $|0\rangle$ is particularly striking. It has a zero average electric field but a finite energy density, as had been recalled in the qualitative discussion of 'vacuum fluctuations' in the introduction of this chapter.

A simple energy argument can be used to determine the normalization factor $\mathcal{E}_0$ (Haroche 1992). The electromagnetic energy in the n-photon state $|n\rangle$ is $\hbar\omega_c(n+1/2)$, a result which implies:

$$\left\langle n \right| \int \varepsilon_0 |\mathbf{E}_c|^2 \, d^3\mathbf{r} \left| n \right\rangle = \hbar\omega_c(n + 1/2) \;, \tag{3.21}$$

where the integral is over the whole space. Expanding the electric field modulus and removing the operator products averaging to zero, the normalization condition can be written as:

$$\langle n | \varepsilon_0 \mathcal{E}_0^2 \mathcal{V}(2N + 1) | n \rangle = \hbar\omega_c(n + 1/2) \;, \tag{3.22}$$

where we define the effective mode volume $\mathcal{V}$ by:

$$\mathcal{V} = \int |\mathbf{f}(\mathbf{r})|^2 \, d^3\mathbf{r} \;. \tag{3.23}$$

The field normalization is thus:

$$\mathcal{E}_0 = \sqrt{\frac{\hbar\omega_c}{2\varepsilon_0 \mathcal{V}}} \;. \tag{3.24}$$

It represents the r.m.s. electric field amplitude of the vacuum in the mode and depends only on the frequency and on the cavity geometry. This analysis, valid for a field stored in a finite volume can be straightforwardly extended to a single mode representing a freely propagating field, such as a plane wave or laser beam. In this case, one can imagine a fictitious single-mode cavity ('quantization box'), such as that of a ring laser, whose linear dimensions are much larger than the characteristic size of the experiment under consideration. The vacuum field r.m.s. amplitude is then extremely small. When a single-mode description is inadequate (e.g. for spontaneous emission), the field is expanded over a continuum of modes quantized in a large fictitious box whose volume is taken to infinity at the end of the calculation (Cohen-Tannoudji *et al.* 1992a;b).

Field quadratures

The electric field operator is a function of $\mathbf{r}$, which depends upon the mode geometry. It is often convenient to introduce simpler position-independent field operators. In analogy with the mechanical oscillator case, we define the dimensionless 'field quadratures operators', X_ϕ, by:[5]

---

[5]The normalization used here is arbitrary. Another convention uses the normalization factor $\mathcal{E}_0$ instead of $1/2$. The quadratures then have the dimension of an electric field. We could also define X_ϕ as $(ae^{-i\phi}+a^\dagger e^{i\phi})/\sqrt{2}$, which would multiply by 2 the commutator $[X_\phi, X_{\phi+\pi/2}]$ given by eqn. (3.26). The factors of 2 would then be removed from some of the following formulas, without modifying the physics.

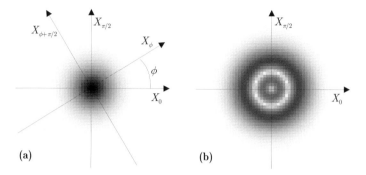

(a)   **(b)**

Fig. 3.3 The phase space for the field in terms of quadratures. (a) Graphical illustration of the X_0, $X_{\pi/2}$, X_ϕ and $X_{\phi+\pi/2}$ quadratures together with a representation of the vacuum state wave function (see text). (b) Graphical representation of a three-photon Fock state in the quadrature phase space.

$$X_\phi = \frac{ae^{-i\phi} + a^\dagger e^{i\phi}}{2} .$$ (3.25)

The operator X_0 corresponds, in the case of a mechanical oscillator, to the dimensionless position operator while $X_{\pi/2}$ is associated to P_0.

The two-dimensional phase space coordinates (Schleich 2001) can thus be taken as $\{X_0, X_{\pi/2}\}$ or, more generally, as $\{X_\phi, X_{\phi+\pi/2}\}$, as shown in Fig. 3.3. Any two orthogonal field quadratures X_ϕ and $X_{\phi+\pi/2}$ satisfy the commutation rule and Heisenberg uncertainty relation:

$$\left[X_\phi, X_{\phi+\pi/2}\right] = i/2 \quad \text{and} \quad \Delta X_\phi \Delta X_{\phi+\pi/2} \geq 1/4 .$$ (3.26)

The eigenstates $|x_\phi\rangle$ of the quadrature X_ϕ are, as the position and momentum eigenstates of the mechanical oscillator, non-normalizable (they have an infinite average energy).[6] They can be considered as the limit when ΔX_ϕ goes to zero of 'quadrature wave packets' which have a well-defined quadrature component, at the expense of a large indeterminacy in the conjugate quantity $X_{\phi+\pi/2}$. Such states are said to be 'squeezed' (Kimble 1992). Eigenstates of the quadrature operator correspond to states with infinite squeezing. The eigenstates of orthogonal quadratures are linked by a Fourier transform relation:[7]

$$\left|x_{\phi+\pi/2}\right\rangle = \frac{1}{\sqrt{\pi}} \int dy\, e^{2ixy} |y_\phi\rangle .$$ (3.27)

Any field state can be expanded on the X_ϕ eigenstates. The expansions of Fock states are independent of ϕ. They are given by the wave functions (3.13), represented on Fig. 3.2. The Fock states carry no phase information and have zero quadrature (and zero electric field) averages. Their wave functions can be used to give a simple

---

[6]See notes on pages 29 and 30 about states with infinite norm.

[7]The factor of 2 in the argument of the exponential in eqn. (3.27) is due to the peculiar normalization of X_ϕ. This factor would be lacking with the definition $X_\phi = (ae^{-i\phi} + a^\dagger e^{i\phi})/\sqrt{2}$ (see note on page 110). It is replaced by $1/\hbar$ for conjugate observable satisfying the canonical relation $[X, P] = i\hbar$.

picture of the Fock states in phase space. Using a polar representation, we plot along each direction ϕ a density of shade proportional to the square modulus $|\langle x_\phi | n\rangle|^2$ of the wave function at distance r from the origin. The vacuum state $|0\rangle$ appears as a circular shape centred at the origin (Fig. 3.3a). Other Fock states appear as sets of rings, as shown in Fig. 3.3(b) for a three-photon state. This representation is only a visual aid. More quantitative pictures are given below and in the appendix.

3.1.2   Coupling field to charges

We have considered so far a field mode evolving freely. Its coupling to matter must also be considered to account for radiation generation and detection. The description of the matter–field interaction is an essential topic of quantum electrodynamics which we touch on only briefly here, restricting ourselves to the simple results which will be used in the following. Let us first consider a free electron of charge q, mass m, whose position and momentum operators are $\mathbf{R}$ and $\mathbf{P}$. It is coupled to a field mode described in the Schrödinger picture by its potential vector $\mathbf{A}(\mathbf{r})$. Neglecting the small magnetic effects linked to the electron spin, the Hamiltonian of this system is in the Coulomb gauge:[8]

$$
\begin{aligned}
H_{qf} &= \hbar\omega_c \left(a^\dagger a + 1/2\right) + \frac{1}{2m}\left[\mathbf{P} - q\mathbf{A}(\mathbf{R})\right]^2 \\
&= \hbar\omega_c \left(a^\dagger a + 1/2\right) + \frac{\mathbf{P}^2}{2m} - \frac{q}{2m}\left[\mathbf{P}\cdot\mathbf{A}(\mathbf{R}) + \mathbf{A}(\mathbf{R})\cdot\mathbf{P}\right] + \frac{q^2}{2m}\mathbf{A}^2(\mathbf{R}) \ .
\end{aligned}
\tag{3.28}
$$

In this full quantum treatment, the electron position is an operator in the expression of the vector potential. If in H_{qf} we replace $\mathbf{P}$ and $\mathbf{R}$ by the momentum and position $\mathbf{p}$ and $\mathbf{r}$ of a classical charge, we obtain by differentiating the Hamilton equations:

$$
\frac{\partial H_{qf}}{\partial p_i} = \frac{dr_i}{dt} \ ,
\tag{3.29}
$$

and:

$$
\frac{\partial H_{qf}}{\partial r_i} = -\frac{dp_i}{dt} \ ,
\tag{3.30}
$$

where r_i and p_i stand for x, y, z and p_x, p_y and p_z. Equation (3.29) shows that the velocity $\mathbf{v} = d\mathbf{r}/dt$ of the classical charge in the presence of radiation is no longer $\mathbf{p}/m$ but becomes $(\mathbf{p} - q\mathbf{A})/m$. The force $\mathbf{f} = m d\mathbf{v}/dt$ exerted by the field on the particle is obtained by combining eqns. (3.29) and (3.30), leading to:

$$
\mathbf{f} = q\mathbf{v} \times (\nabla \times \mathbf{A}) - q\frac{\partial \mathbf{A}}{\partial t} \ .
\tag{3.31}
$$

We recognize here the classical Lorentz force experienced by the charge, with its magnetic $(\nabla \times \mathbf{A})$ and electric $(-\partial\mathbf{A}/\partial t)$ components. That we recover this force can be taken as a justification of the charge–field Hamiltonian (3.28) which describes the

---

[8]In the Coulomb gauge, the field, far from its sources, is described by a divergence-free potential vector $\mathbf{A}(\mathbf{r})$ $(\nabla \cdot \mathbf{A} = 0)$. The scalar potential of this field is negligibly small.

coupled evolution of the charge and the field in the quantum as well as in the classical framework. Examining the developed expression of this Hamiltonian in eqn. (3.28), we distinguish the free field and particle contributions $[\hbar\omega_c(a^\dagger a + 1/2) + \mathbf{P}^2/2m]$ from the q-depending terms which combine observables of both subsystems and describe the coupling between them.

We next consider an electron bound in a hydrogen-like atom, which moves in a small region of space with linear dimensions of the order of the Bohr radius a_0.[9] The electron Hamiltonian then becomes $\mathbf{P}^2/2m + V(\mathbf{R})$ where $V(\mathbf{r})$ is the Coulomb potential produced at point $\mathbf{r}$ by the nucleus sitting at the origin. The coupling of the electron with the field is still described by the q-dependent terms in eqn. (3.28). The momentum of the electron is then of the order of $\hbar/a_0$ and the ratio between the expectation values of the terms proportional to q and q^2 in eqn. (3.28) is of the order of $\xi = \hbar/(qa_0A) = \hbar\omega_c/(qa_0E)$ where A and E are typical values for the vector potential and electric field respectively. In a resonant or nearly resonant photon absorption or emission process, ξ is equal to the ratio between the atomic transition energy $\hbar\omega_c$ and the polarization energy qEa_0 of the atom in the field of the light wave. This ratio is very small, unless the field is so intense that it strongly perturbs the atomic energy levels, a situation we will not consider here. We thus neglect the $\mathbf{A}^2$ term in eqn. (3.28) and write the charge–field interaction Hamiltonian as:

$$H_{qf}^{\text{int}} \approx -\frac{q}{2m}\left[\mathbf{P}\cdot\mathbf{A}(\mathbf{R}) + \mathbf{A}(\mathbf{R})\cdot\mathbf{P}\right] = -\frac{q}{m}\mathbf{P}\cdot\mathbf{A}(\mathbf{R}) , \qquad (3.32)$$

where we have made use of the commutation identity $\mathbf{P}\cdot\mathbf{A} = \mathbf{A}\cdot\mathbf{P}$ resulting from the relation $\nabla\cdot\mathbf{A} = 0$ satisfied by the vector potential of the field (see note on page 112).

The electron–field interaction can also be written as:

$$H_{qf}^{\text{dip}} = -q\mathbf{R}\cdot\mathbf{E}(0) , \qquad (3.33)$$

an expression which involves the scalar product of the electric field at the nucleus with the atomic dipole operator:

$$\mathbf{D} = q\mathbf{R} . \qquad (3.34)$$

Equation (3.33), valid in the limit $a_0 \ll \lambda$ (electric-dipole approximation), is deduced from eqn. (3.28) by a gauge transformation (Cohen-Tannoudji et al. 1992a). Let us first analyse the case of a classical field. In general, a gauge transformation replaces the vector potential $\mathbf{A}$ and the scalar potential $\mathcal{U} = 0$ defining the field in the Coulomb gauge by new potentials $\mathbf{A}' = \mathbf{A} + \nabla\chi(\mathbf{r}, t)$ and $\mathcal{U}' = -\partial\chi(\mathbf{r}, t)/\partial t$ where $\chi(\mathbf{r}, t)$ is an arbitrary scalar function. This transformation leaves the electric and magnetic fields invariant and thus does not change the physics of the charge–field interaction process. Making the specific gauge choice:

$$\chi(\mathbf{r}, t) = -\mathbf{r}\cdot\mathbf{A}(0, t) , \qquad (3.35)$$

has the effect to cancel the new vector potential up to first order in (a_0/λ). The electron kinetic energy, equal to $(\mathbf{P} - q\mathbf{A})^2/2m$ in the old gauge, becomes $\mathbf{P}^2/2m$ in the new one,

---

[9]We assume for simplicity that the positively charged nucleus has an infinite mass, so that its coordinates do not appear in the dynamical equation of the system. A more careful analysis would separate the centre of mass from the relative motion of the atomic charges and introduce a reduced mass for the electron. The important point is that the dynamics of the charges can be described in terms of a single particle which we call 'the electron' in the following.

as for a free particle. The leading contribution to the coupling between the field and the electron is no longer contained in the kinetic term. It becomes an electric potential contribution: $q\mathcal{U}'(\mathbf{R}, t) = -q\partial\chi(\mathbf{R}, t)/\partial t = q\mathbf{R}\cdot\partial\mathbf{A}(0, t)/\partial t = -q\mathbf{R}\cdot\mathbf{E}(0, t)$. Replacing the Coulomb gauge by the new one amounts to applying to the electron wave function and operators the transformation $\exp[iq\chi(\mathbf{R}, t)/\hbar] = \exp[-iq\mathbf{R}\cdot\mathbf{A}(0, t)/\hbar]$, a simple translation in momentum space. This change of representation does not alter the physical predictions. In a classical description of the field, the $-q\mathbf{R}\cdot\mathbf{E}(0, t)$ interaction is time-dependent. When the field is described as a quantum system, the coupling becomes time-independent in the Schrödinger point of view, and we get the result expressed by eqn. (3.33).

The interaction Hamiltonian (3.33) is the first term in the multipole expansion of the charge–field interaction (Jackson 1975; Cohen-Tannoudji *et al.* 1992a). If the atomic dipole operator has a vanishing matrix element on the transition resonant with the field, the expansion must be carried out to higher orders in a_0/λ. The next term is the sum of two contributions. One describes the interaction of the charge quadrupole with the spatial derivatives of the electric field at the origin and the other the coupling of the magnetic dipole of the charge current with the magnetic field at $\mathbf{r} = 0$ (the interaction of the spin with the magnetic field cannot of course be neglected at this order). Selection rules determine whether one of these terms has a non-zero matrix element between the levels of the relevant transition. We will see in Chapter 8 an example of an electric quadrupole transition. If the interaction remains zero at this order, the expansion must be carried out further (the next order involves electric octupole and magnetic quadrupole transitions).

This formalism, which describes in very general terms the evolution of coupled quantized charge and field systems, can be simply adapted to account for the coupling of matter and radiation when one of the two systems is classical. If it is the field, we have just seen that we can consider $\mathbf{A}$ and $\mathbf{E}$ in eqns. (3.32) and (3.33) as classical time-varying quantities. We use this result in Section 3.3. If a classical charge is interacting with a quantum field, we can consider it as a very massive particle whose evolution is not affected by its coupling to radiation. We merely take the limit of eqn. (3.28) when m goes to infinity, while the (now classical) quantity $\mathbf{p}/m$ remains finite and represents the time-dependent velocity $\mathbf{v}$ of the charge. The quadratic term $q^2 A^2/2m$ in eqn. (3.28) vanishes and the Hamiltonian of the field driven by the classical charge is:

$$H_{qf}^{\text{class}} = \hbar\omega_c(a^\dagger a + 1/2) - q\mathbf{v}\cdot\mathbf{A}(\mathbf{r}) , \qquad (3.36)$$

this equation being valid even if the charge moves in an extended region of space (no electric-dipole approximation is made here). For a distribution of classical point charges q_j located at points $\mathbf{r}_j$, we define the current:

$$\mathbf{j}(\mathbf{r}, t) = \sum_j q_j\mathbf{v}_j\delta(\mathbf{r} - \mathbf{r}_j) , \qquad (3.37)$$

which is, through the $\mathbf{v}_j$'s and the $\mathbf{r}_j$'s, a function of time. The field Hamiltonian then becomes:

$$H_{qf}^{\mathbf{j}} = \hbar\omega_c\left(a^\dagger a + 1/2\right) - \int d^3\mathbf{r}\,\mathbf{j}(\mathbf{r}, t)\cdot\mathbf{A}(\mathbf{r}) , \qquad (3.38)$$

an expression which we will use in the next section.

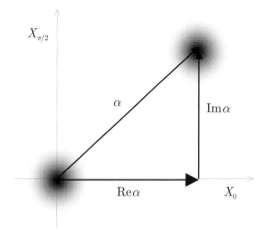

Fig. 3.4 Pictorial representation of the action of the displacement operator on the vacuum state. The displacement by a complex amplitude α amounts to a displacement by $\mathrm{Re}\,\alpha$ along the X_0 quadrature axis, followed by a displacement by $\mathrm{Im}\,\alpha$ along the $X_{\pi/2}$ quadrature.

3.1.3 Coherent states

We will now study a family of oscillator states, the so-called coherent or Glauber states, which play a very important role in quantum optics. The Fock states we have described above, with their zero average electric field and indefinite phase, clearly do not correspond to the classical picture of an excited oscillator. Intuitively, we rather visualize it as having a well-defined amplitude and phase, described by a complex amplitude. In dimensionless reduced units, we write this amplitude $\alpha = \alpha' + i\alpha'' = x + ip$. For a mechanical oscillator, such classical states can be obtained, as shown in Fig. 3.4, by displacing suddenly the oscillator from its rest position, shifting its position by x while imparting to it a momentum kick p. The resulting states, written $|\alpha\rangle$, are the coherent states.

Displacement operator

The displacement turning $|0\rangle$ into $|\alpha\rangle$ is described by the product of the translation operators in phase space $\mathcal{T}_x = e^{-i2x_0 x P/\hbar} = e^{-2i\alpha' X_{\pi/2}}$ and $\mathcal{T}_p = e^{i2p_0 p X/\hbar} = e^{2i\alpha'' X_0}$ [see eqns. (2.20) and (2.22) for a general definition of these operators, as well as (3.2) and (3.3) for the connexion between the eigenvalues of X, P and X_0, $P_0 = X_{\pi/2}$]. Manipulating functions of operators which do not commute requires some care. We can take advantage here of the Glauber identities (Barnett and Radmore 1997):

$$e^A e^B = e^{A+B} e^{[A,B]/2} \quad \text{or} \quad e^{A+B} = e^A e^B e^{-[A,B]/2} \,, \tag{3.39}$$

valid when both A and B commute with $[A,B]$. By applying them to $A = 2i\alpha'' X_0$ and $B = -2i\alpha' X_{\pi/2}$, we transform $\mathcal{T}_p \mathcal{T}_x$ according to:

$$e^{2i\alpha'' X_0} e^{-2i\alpha' X_{\pi/2}} = e^{i\alpha'\alpha''} e^{-2i(\alpha' X_{\pi/2} - \alpha'' X_0)} = e^{i\alpha'\alpha''} e^{(\alpha a^\dagger - \alpha^* a)} \,. \tag{3.40}$$

Up to a phase factor,[10] $\mathcal{T}_p \mathcal{T}_x$ is equal to the unitary displacement operator:

$$D(\alpha) = e^{(\alpha a^\dagger - \alpha^* a)} \,, \tag{3.41}$$

[10]This phase factor depends upon the order of the x and p translations: $\mathcal{T}_p \mathcal{T}_x = e^{2i\alpha'\alpha''} D(\alpha)$, while $\mathcal{T}_x \mathcal{T}_p = e^{-2i\alpha'\alpha''} D(\alpha)$.

which satisfies:
$$D^{-1}(\alpha) = D^{\dagger}(\alpha) = D(-\alpha) ; \qquad D(0) = \mathbb{1} . \tag{3.42}$$

Choosing the simplest phase convention, we define the coherent state $|\alpha\rangle$ as:
$$|\alpha\rangle = D(\alpha) |0\rangle , \tag{3.43}$$

the vacuum $|0\rangle$ being the coherent state corresponding to a zero displacement amplitude.

Coherent states and classical currents

In the case of a mechanical oscillator, the physical processes turning the ground state into coherent states are mere translations in phase space, which can be implemented by manipulating the particle directly (see Chapter 8). There is also a very simple, even if slightly less intuitive way to generate these states for a field. The displacement operator $D(\alpha)$ is then the evolution operator describing the action of a monochromatic classical current coupled to the mode. For the sake of simplicity, we assume that this current is localized at the origin. The current density is $\mathbf{j}(\mathbf{r}, t) = \mathbf{J}(t)\delta(\mathbf{r})$ where $\mathbf{J}(t) = \mathbf{J}_0 \exp(-i\omega_c t) + \mathbf{J}_0^* \exp(i\omega_c t)$ is a real vector oscillating at the cavity frequency ω_c and δ is the Dirac distribution. This current density is a classical parameter whose dynamical evolution is not modified by its coupling to the cavity. It is produced by a large 'source' whose quantum fluctuations are negligible.

The source–cavity coupling is given by eqn. (3.38):
$$V_{sc}^{\mathbf{j}} = -\int d^3\mathbf{r}\, \mathbf{j}(\mathbf{r}, t) \cdot \mathbf{A}(\mathbf{r}) = -\mathbf{J}(t) \cdot \mathbf{A}(0) . \tag{3.44}$$

In the interaction representation with respect to the free cavity Hamiltonian H_c, the complete Hamiltonian reduces to $\widetilde{V}_{sc}^{\mathbf{j}}$ in which the vector potential has to be expressed in the Heisenberg picture (eqn. 3.19). The scalar product $\mathbf{J} \cdot \widetilde{\mathbf{A}}(0, t)$ involves two terms oscillating at $2\omega_c$ and two time-independent terms. The field state evolves on a time scale much longer than ω_c^{-1}. The fast 'counter-rotating' terms at $2\omega_c$ have a negligible effect on the evolution. Dropping them is known as the 'secular approximation',[11] called 'rotating wave approximation' (RWA) in quantum optics (Cohen-Tannoudji *et al.* 1992b). It retains, in the real current $\mathbf{J}$, the terms which rotate in the complex plane in the same direction as the field observables.

Defining $J_1 = \mathbf{J}_0 \cdot \boldsymbol{\epsilon}_c^*$, we get:
$$\widetilde{V}_{sc}^{\text{sec}} = -\frac{\mathcal{E}_0}{\omega_c}(J_1 a^{\dagger} + J_1^* a) . \tag{3.45}$$

The state of the field mode evolves from the vacuum $|0\rangle$ at $t = 0$ into:
$$\left| \widetilde{\Psi}(t) \right\rangle = \exp(-i\widetilde{V}_{sc}^{\text{sec}} t/\hbar) |0\rangle . \tag{3.46}$$

We recognize in the evolution operator the displacement $D(\alpha)$ for the amplitude $\alpha = iJ_1\mathcal{E}_0 t/\hbar\omega_c$, proportional to the source–cavity interaction time t and to the projection,

---

[11]Such approximations were first used in astronomy, when fast processes (like the rotation of the moon) are neglected for the calculation of very long-term evolutions, on the scale of centuries (*secula* in Latin).

J_1, of the current along the field polarization. This α value is proportional to the amplitude of the classical field injected in the cavity during the same time by the same current.

Action of the displacement and annihilation operators on coherent states

The displacement operator has a simple action on a, merely adding to it a multiple of the unity operator. To show this, we let $A = \alpha a^\dagger - \alpha^* a$, and we write the displaced operator $D(\alpha)aD(-\alpha)$ as:

$$e^A a e^{-A} = a + [A, a] + \frac{1}{2!}[A, [A, a]] + \dots \, , \tag{3.47}$$

where we use the Baker–Hausdorf lemma (Barnett and Radmore 1997). Since $[A, a] = -\alpha \mathbb{1}$, the first two terms only are non-vanishing and:

$$D(\alpha)aD(-\alpha) = a - \alpha \mathbb{1} \, . \tag{3.48}$$

Displacements in phase space can be combined. The action of the classical current [displacement $D(\alpha)$] on a coherent state $|\beta\rangle$ produces a coherent state with an amplitude $\alpha + \beta$. More precisely, using the Glauber operators identity (eqn. 3.39), we get:

$$D(\alpha)D(\beta) = D(\alpha + \beta)e^{(\alpha\beta^* - \alpha^*\beta)/2} \, , \tag{3.49}$$

which shows that:

$$D(\alpha)|\beta\rangle = e^{(\alpha\beta^* - \alpha^*\beta)/2}|\alpha + \beta\rangle \, . \tag{3.50}$$

Within a global phase, the coherent state amplitudes are additive, like the amplitudes of classical fields. The addition of coherent fields plays an important role in quantum optics as we will see later when describing homodyne detection techniques.

The additivity rule entails the obvious relation $D(-\alpha)|\alpha\rangle = |0\rangle$. Combined with eqn. (3.48), it leads to the identity:

$$(a - \alpha \mathbb{1})|\alpha\rangle = 0 \, , \tag{3.51}$$

which means that the coherent state $|\alpha\rangle$ is an eigenstate of the annihilation operator a with the eigenvalue α. Similarly, the adjoint equation:

$$\langle\alpha|a^\dagger = \alpha^*\langle\alpha| \, , \tag{3.52}$$

expresses that $\langle\alpha|$ is a left eigenvector of $a^\dagger$ with the eigenvalue α^*. Note that the annihilation and creation operators being non-hermitian, their spectrum is non-real (it is in fact the complete C-numbers set).

From eqns. (3.51) and (3.52), we deduce that the average of a $(a^\dagger)$ in the coherent state $|\alpha\rangle$ is α (α^*). The electric field, a linear combination of a and $a^\dagger$, has a non-zero average in a coherent state. In the Schrödinger picture:

$$\langle\alpha|\mathbf{E}_c|\alpha\rangle = i\mathcal{E}_0\left[\mathbf{f}(\mathbf{r})\alpha - \mathbf{f}^*(\mathbf{r})\alpha^*\right] \, . \tag{3.53}$$

This expectation value coincides with the expression of a classical electric field with amplitude α in units of $\mathcal{E}_0$.

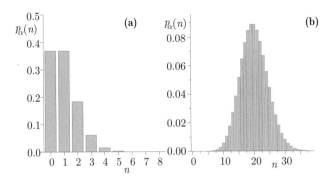

Fig. 3.5 Photon number statistical distributions. (a) Coherent field with $\bar{n} = 1$ photons on the average. (b) Coherent field with $\bar{n} = 20$.

More generally, the expectation in the coherent state $|\alpha\rangle$ of any normally ordered[12] product $a^{\dagger p}a^q$ of a and $a^{\dagger}$ is $\alpha^{*p}\alpha^q$. For instance, the expectation value of the photon number $N = a^{\dagger}a$ in $|\alpha\rangle$ is:

$$\bar{n} = \langle \alpha | a^{\dagger}a | \alpha \rangle = |\alpha|^2 , \tag{3.54}$$

while the average of the squared photon number is easily calculated by expressing $N^2 = a^{\dagger}aa^{\dagger}a$ in normal order as $N^2 = a^{\dagger 2}a^2 + a^{\dagger}a$:

$$\langle \alpha | N^2 | \alpha \rangle = |\alpha|^4 + |\alpha|^2 . \tag{3.55}$$

Fock state representation of coherent states

The expansion of coherent states on the Fock state basis is deduced from a simple expression of $D(\alpha)$, obtained with the help of the Glauber operator identity:

$$D(\alpha) = e^{\alpha a^{\dagger} - \alpha^* a} = e^{-|\alpha|^2/2} e^{\alpha a^{\dagger}} e^{-\alpha^* a} . \tag{3.56}$$

Expanding the exponentials in power series, we see that $\exp(-\alpha^* a)$ leaves $|0\rangle$ invariant. Using the expression of $a^{\dagger n}|0\rangle$, we get:

$$|\alpha\rangle = \sum_n c_n |n\rangle = e^{-|\alpha|^2/2} \sum_n \frac{\alpha^n}{\sqrt{n!}} |n\rangle . \tag{3.57}$$

The photon number probability distribution, $p_{\alpha}(n) = |c_n|^2$, obeys a Poisson law:

$$p_{\alpha}(n) = e^{-|\alpha|^2} \frac{|\alpha|^{2n}}{n!} , \tag{3.58}$$

with an average photon number $\bar{n} = |\alpha|^2$ and a photon number mean square-root deviation:

$$\Delta N = \sqrt{\langle \alpha | N^2 | \alpha \rangle - \langle \alpha | N | \alpha \rangle^2} = |\alpha| = \sqrt{\bar{n}} . \tag{3.59}$$

Figure 3.5(a) and (b) shows $p_{\alpha}(n)$ for 1 and 20 photons on the average. The relative photon number mean square-root deviation, $\Delta N / \bar{n} = 1/\sqrt{\bar{n}} = 1/|\alpha|$, is inversely

---

[12] A normally ordered product of a and $a^{\dagger}$ has all the creation operators to the left of the annihilation operators (see the appendix).

proportional to the field amplitude. The energy of a coherent state is ill-defined when its average photon number is of the order of unity or smaller (Fig. 3.5a). For $\bar{n}$ larger (Fig. 3.5b), $\Delta N/\bar{n}$ gets progressively smaller. This can be viewed as a manifestation of the correspondence principle. As the number of quanta increases, the coherent state becomes more and more classical, with smaller and smaller relative energy fluctuations.

Scalar product of coherent states
The scalar product of two coherent states is:

$$\langle \alpha\,|\beta\rangle = e^{-(|\alpha|^2+|\beta|^2)/2}\sum_{n,p}\frac{\alpha^{*n}}{\sqrt{n!}}\frac{\beta^p}{\sqrt{p!}}\,\langle n\,|p\rangle\;. \tag{3.60}$$

Using the orthogonality of Fock states, the double sum reduces to:

$$\langle \alpha\,|\beta\rangle = e^{-(|\alpha|^2+|\beta|^2)/2}\sum_{n}\frac{(\alpha^*\beta)^n}{n!} = e^{-|\alpha|^2/2}e^{-|\beta|^2/2}e^{\alpha^*\beta}\;. \tag{3.61}$$

The modulus squared of this scalar product:

$$|\langle \alpha\,|\beta\rangle|^2 = e^{-|\alpha-\beta|^2}\;, \tag{3.62}$$

is a Gaussian versus the distance between the two states, with a width 1. Two coherent states are never exactly orthogonal. They become approximately so when the difference between their amplitudes is much larger than one.

The coherent states form a non-orthogonal over-complete basis of the field mode Hilbert space, with the closure relationship:

$$\mathbb{1} = \frac{1}{\pi}\int |\alpha\rangle\langle\alpha|\,d^2\alpha\;, \tag{3.63}$$

where the integral is over the whole complex plane ($d^2\alpha = d\alpha' d\alpha''$). This is demonstrated by expanding the projector $|\alpha\rangle\langle\alpha|$ over the $|n\rangle\langle p|$ dyadic operators and showing that all terms with $n \neq p$ cancel while the $n = p$ terms reduce to the Fock states closure relation. Any state can be written as a superposition of coherent states. The $|n\rangle$ Fock states, for instance, expand as:

$$|n\rangle = \frac{1}{\pi\sqrt{n!}}\int e^{-|\alpha|^2/2}(\alpha^*)^n\,|\alpha\rangle\,d^2\alpha\;. \tag{3.64}$$

Since the coherent state basis is over-complete, the $|\alpha\rangle$ expansion is not unique. The vacuum $|0\rangle$, a particular coherent state, can also be expressed as a Gaussian sum of $|\alpha\rangle$'s with all α values:

$$|0\rangle = \frac{1}{\pi}\int e^{-|\alpha|^2/2}|\alpha\rangle\,d^2\alpha\;. \tag{3.65}$$

Coherent state wave functions
It is instructive to compute the coherent state wave function in the X_0 and $X_{\pi/2}$ representations. The distribution of the X_0 quadrature is, using (3.40):

$$\langle x\,|\alpha\rangle = \langle x|\,D(\alpha)\,|0\rangle = e^{-i\alpha'\alpha''}\langle x|\,e^{2i\alpha''X_0}e^{-2i\alpha'X_{\pi/2}}\,|0\rangle\;, \tag{3.66}$$

where, for the sake of compactness, we have omitted the $_0$ indices on the eigenvectors of X_0. We use $\langle x|\,X_0 = \langle x|\,x$ and inject, between the two exponential operators, the

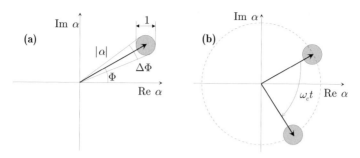

Fig. 3.6 (a) Pictorial representation of a coherent state in the Fresnel plane. (b) Time evolution of a coherent state.

closure relation for the eigenvectors $|y\rangle$ of $X_{\pi/2}$ (we omit here also the $\pi/2$ index since no confusion is possible). We then get:

$$\langle x\,|\alpha\rangle = \frac{1}{\sqrt{\pi}}e^{-i\alpha'\alpha''}e^{2i\alpha''x}\int dy\,\langle x\,|y\rangle\,e^{-2i\alpha'y}\,\langle y\,|0\rangle\ . \tag{3.67}$$

Introducing the scalar products expressions [note that $\langle y\,|0\rangle$ is the wave function of the vacuum in $X_{\pi/2}$ representation, $(2/\pi)^{1/4}\exp(-y^2)$], and performing the Gaussian integration, we get finally:

$$\langle x\,|\alpha\rangle = (2/\pi)^{1/4}\,e^{-i\alpha'\alpha''}e^{2i\alpha''x}e^{-(x-\alpha')^2}\ . \tag{3.68}$$

This scalar product is a Gaussian centred on $x = \alpha'$, modulated by a phase factor $\exp(2i\alpha''x)$, corresponding to translation by α'' along the $X_{\pi/2}$ axis.

The probability distribution for the X_0 quadrature:

$$\Pi_0(x) = (2/\pi)^{1/2}\,e^{-2(x-\alpha')^2}\ , \tag{3.69}$$

is a translated Gaussian with a width $1/\sqrt{2}$. The probability distribution for $X_{\pi/2}$ is a similar Gaussian centred at α''. The coherent state is thus a minimal uncertainty state satisfying $\Delta X_\phi\Delta X_{\phi+\pi/2} = 1/4$. This wave function supports our description of the coherent state as a translated vacuum. Moreover, the real and imaginary parts of α appear as convenient quantities to define coordinates in phase space.

The Q function of a coherent state

The modulus squared of the scalar product of two coherent states is a natural measure of their overlap in phase space. This notion can be generalized to the overlap of a coherent state $|\beta\rangle$ with an arbitrary field state $|\Psi\rangle$. We are thus led to define the Husimi–Q function, $Q^{[|\Psi\rangle\langle\Psi|]}(\beta)$ (Schleich 2001; Vogel *et al.* 2001):

$$Q^{[|\Psi\rangle\langle\Psi|]}(\beta) = \frac{1}{\pi}|\langle\beta\,|\Psi\rangle\,|^2\ , \tag{3.70}$$

computed by expanding $|\Psi\rangle$ over the coherent states basis.

The $Q^{[|\Psi\rangle\langle\Psi|]}$ function is a regular, continuous function of β, everywhere positive. From the over-complete basis relation (eqn. 3.63), we get:

$$\int Q^{[|\Psi\rangle\langle\Psi|]} \, d^2\beta = 1 \ . \tag{3.71}$$

The $Q^{[|\Psi\rangle\langle\Psi|]}$ distribution looks like a probability distribution in classical phase space. It belongs to a family of phase space distributions for quantum oscillators. Another very useful member of this family is the Wigner function W. Both Q and W are described in detail in the appendix, where we show that their exact knowledge fully determines the quantum state $|\Psi\rangle$. Although mathematically equivalent, Q and W display however complementary features of the field state.

The Q function for the coherent state α is:

$$Q^{[|\alpha\rangle\langle\alpha|]}(\beta) = \frac{1}{\pi} e^{-|\alpha-\beta|^2} \ , \tag{3.72}$$

a Gaussian with width 1 centred at the amplitude α. This leads to a simple pictorial representation of coherent states, displayed in Fig. 3.6(a). We plot, in the Fresnel plane, the region where $Q^{[|\alpha\rangle\langle\alpha|]}$ is above $1/\pi e$. This region is a circular disk, with radius unity, centred at α. The finite size of this region reflects the quadratures uncertainty relations. The coherent state can thus be understood qualitatively as a classical field with amplitude fluctuating inside this 'uncertainty disk' due to quantum fluctuations. We use this simple pictorial representation of coherent states thoroughly in the following.

The direction of the vector joining the origin of phase space to a point inside this uncertainty disk exhibits an angular fluctuation $\Delta\Phi \approx 1/\sqrt{n}$ which represents the phase dispersion of the coherent field. As the mean photon number increases, this fluctuation becomes smaller and the phase better defined. This feature is often described as a phase–photon number uncertainty relation.

Time evolution

The time evolution of a coherent state is obtained from its expansion over the Fock stationary states basis. Each Fock state evolves as $|n(t)\rangle = \exp[-i(n+1/2)\omega_c t]\,|n\rangle$. The evolution of $|\alpha\rangle$ is thus given by:

$$e^{-|\alpha|^2/2} \sum_n \frac{\alpha^n}{\sqrt{n!}} e^{-i(n+1/2)\omega_c t} |n\rangle = e^{-i\omega_c t/2} \left|\alpha e^{-i\omega_c t}\right\rangle \ . \tag{3.73}$$

Within an irrelevant global phase factor,[13] the state remains coherent. The complex amplitude rotates at ω_c around the origin. It has the same dynamics as the classical amplitude. The average electric field evolves exactly as the classical field.

In a pictorial representation (Fig. 3.6b), the disk representing the state's uncertainty radius rotates at constant angular velocity ω_c around the origin, in the clockwise direction. The time evolution of the coherent state wave function in the quadrature

---

[13]This extra factor can be removed by redefining the energy origin and using H'_c instead of H_c.

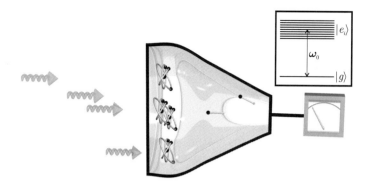

Fig. 3.7 Scheme of a photon counter. The photocathode is paved with atoms converting photons into electrons by the photo-electric effect. The relevant atomic energy level diagram is shown in the inset.

representation provides additional insight into this evolution. The probability distribution Π_0 in the X_0 representation, given at initial time by eqn. (3.69), evolves as:

$$\Pi_0(x,t) = (2/\pi)^{1/2} \, e^{-2(x-\alpha' \cos \omega_c t - \alpha'' \sin \omega_c t)^2} , \qquad (3.74)$$

which describes a translated vacuum state wave function, with a displacement rotating at ω_c around the phase space's origin.

3.1.4   Photon counting

How can we distinguish a coherent state from a Fock state? Or a coherent state from a squeezed state of light? To answer these questions, we need to describe briefly the methods used to detect the field in quantum optics. The prime devices for this task are photon counters. Either inserted directly in the light beam, or used as detectors at the output ports of interferometers, they extract information from the field by registering clicks corresponding to the absorption of photons. The click rate yields the field intensity at the detector's location, and click correlations can be directly related to the spatio-temporal fluctuations of field observables.[14]

Simple photon counting model

Figure 3.7 presents a simple model of photodetector. A photocathode, of area S, is paved with atoms converting impinging photons into electrons by photo-ionization. The elementary photocounter (Glauber 1963) is thus a single atom which we will describe as having a ground state $|g\rangle$ and a continuum of excited states $|e_i\rangle$, corresponding to electronic states above the ionization threshold (inset in Fig. 3.7).[15] The atom, initially in $|g\rangle$, is irradiated by the light field and we collect the electron

---

[14]We must realize that photon counting is far from being an ideal measurement, since it is destructive for field quanta. We will see in Chapter 6 that quantum non-destructive (QND) methods of field measurements are also possible, which satisfy the ideal postulates of a quantum measurement, counting photons without destroying them. These methods can, however, be implemented only in very special situations. Destructive photon counting is by far the general method of field detection.

[15]We assume that this atom has a single electron for simplicity, but this is not essential in this treatment.

produced by photo-ionization. The field with a given polarization ϵ is in an initial state involving a set of modes labelled by the index j. The absorption is governed by the atom–field coupling, deduced from eqn. (3.32) and written, within the secular approximation neglecting fast evolving terms (see note, on page 116) as:

$$H_i^d \approx -\frac{q}{m}\mathbf{p}\cdot\sum_j \mathbf{A}_j(\mathbf{r}) \approx -\frac{q}{m}\sum_i(\mathbf{p}\cdot\epsilon)_{ig}\,|e_i\rangle\,\langle g|\sum_j \frac{\mathcal{E}_j}{\omega_j}a_j e^{i\mathbf{k}_j\cdot\mathbf{r}} + \text{h.c.} , \quad (3.75)$$

where $\mathcal{E}_j$ is the field per photon in mode j (angular frequency ω_j, wave vector $\mathbf{k}_j$) and $\mathbf{r}$ is, in the electric dipole approximation, the atomic position (a classical quantity). We assume that the excited atomic continuum extends over a relatively narrow absorption band whose centre defines an average transition frequency ω_0. The field modes relevant for the absorption are those whose frequencies ω_j are close to ω_0, so that:

$$H_i^d \approx -\sum_i \frac{q}{im\omega_0}(\mathbf{p}\cdot\epsilon)_{ig}\,|e_i\rangle\,\langle g|\sum_j i\mathcal{E}_j a_j e^{i\mathbf{k}_j\cdot\mathbf{r}} + \text{h.c.} \quad (3.76)$$

The sums over j are the 'positive' and 'negative' frequency parts of the electric field operator, which respectively annihilate and create a photon:

$$E^+(\mathbf{r}) = i\sum_j \mathcal{E}_j a_j e^{i\mathbf{k}_j\cdot\mathbf{r}} \; ; \qquad E^-(\mathbf{r}) = -i\sum_j \mathcal{E}_j a_j^\dagger e^{-i\mathbf{k}_j\cdot\mathbf{r}} . \quad (3.77)$$

Grouping atomic parameters in a single coupling constant κ_{ig}, we express the interaction Hamiltonian as:

$$H_i^d = -\sum_i \left[\kappa_{ig}\,|e_i\rangle\,\langle g|\,E^+(\mathbf{r}) - \kappa_{ig}^*\,|g\rangle\,\langle e_i|\,E^-(\mathbf{r})\right] . \quad (3.78)$$

We finally go into the interaction representation with respect to the atom and field Hamiltonians, H_a and H_f:

$$\widetilde{H}_i^d = e^{i(H_a+H_f)t/\hbar}\,H_i^d\,e^{-i(H_a+H_f)t/\hbar} . \quad (3.79)$$

The atom–field coupling then involves the electric field operators in the Heisenberg point of view, as time-dependent operators:

$$\widetilde{H}_i^d = -\sum_i \left[\kappa_{ig}e^{i\omega_{ig}t}\,|e_i\rangle\,\langle g|\,\widetilde{E}^+(\mathbf{r},t) - \kappa_{ig}^* e^{-i\omega_{ig}t}\,|g\rangle\,\langle e_i|\,\widetilde{E}^-(\mathbf{r},t)\right] . \quad (3.80)$$

Assuming that, at $t = 0$, the system is in the state $|\widetilde{\Psi}(0)\rangle = |g\rangle|\widetilde{\Psi}_f\rangle$ we obtain, at first order of perturbation theory, the state at time t:

$$\left|\widetilde{\Psi}(t)\right\rangle \approx |g\rangle\left|\widetilde{\Psi}_f\right\rangle + \frac{1}{i\hbar}\int_0^t dt'\,\widetilde{H}_i^d(t')\,|g\rangle\left|\widetilde{\Psi}_f\right\rangle$$

$$= |g\rangle \left|\widetilde{\Psi}_f\right\rangle - \frac{1}{i\hbar} \sum_i \kappa_{ig} |e_i\rangle \int_0^t dt' \, e^{i\omega_{ig}t'} \widetilde{E}^+(\mathbf{r}, t') \left|\widetilde{\Psi}_f\right\rangle . \quad (3.81)$$

We finally get the total photo-ionization probability, $P_e(t)$, at time t by summing the contributions of all final states:

$$P_e(t) = \sum_i \left\langle \widetilde{\Psi}(t) \,\middle|\, e_i \right\rangle \left\langle e_i \,\middle|\, \widetilde{\Psi}(t) \right\rangle , \quad (3.82)$$

which directly yields the total atomic excitation probability:

$$P_e(t) = \frac{1}{\hbar^2} \sum_i |\kappa_{ig}|^2 \int_0^t \int_0^t dt' dt'' e^{i\omega_{ig}(t'-t'')} \left\langle \widetilde{\Psi}_f \,\middle|\, \widetilde{E}^-(\mathbf{r}, t'') \widetilde{E}^+(\mathbf{r}, t') \,\middle|\, \widetilde{\Psi}_f \right\rangle . \quad (3.83)$$

Summing over the final atomic states $|e_i\rangle$ is equivalent to an integration over a continuum. We assume that the coupling $|\kappa_{ig}|^2$ and the continuum mode density are varying slowly in the relevant frequency range, and we replace them by constants. The exponential integration over ω_{ig} introduces a Dirac distribution of $t' - t''$. Hence:

$$P_e(t) = \xi \int_0^t dt' \left\langle \widetilde{\Psi}_f \,\middle|\, \widetilde{E}^-(\mathbf{r}, t') \widetilde{E}^+(\mathbf{r}, t') \,\middle|\, \widetilde{\Psi}_f \right\rangle , \quad (3.84)$$

where ξ, proportional to the product of the average coupling and mode density, defines the efficiency of the atomic detector. The derivative $dP_e(t)/dt$ represents the probability of detecting a photo-ionization process per unit time at time t, with an atom at $\mathbf{r}$. It is the single-photon counting rate $w_1(\mathbf{r}, t)$:

$$w_1(\mathbf{r}, t) = \frac{dP_e(t)}{dt} = \xi \left\langle \widetilde{\Psi}_f \,\middle|\, \widetilde{E}^-(\mathbf{r}, t) \widetilde{E}^+(\mathbf{r}, t) \,\middle|\, \widetilde{\Psi}_f \right\rangle . \quad (3.85)$$

The average photon counting rate is proportional to the mean value of the hermitian operator $\widetilde{E}^- \widetilde{E}^+$, a normally-ordered product of the negative and positive frequency parts of the electric field (the positive frequency part involving photon annihilation operators is at the right). If the field is defined by its density operator $\tilde{\rho}$ in the interaction representation, the counting rate is:

$$w_1(\mathbf{r}, t) = \xi \text{Tr}\left[\tilde{\rho} \widetilde{E}^-(\mathbf{r}, t) \widetilde{E}^+(\mathbf{r}, t)\right] . \quad (3.86)$$

We similarly define the double counting rate describing the probability of detecting a photo-ionization event at $\mathbf{r}_1$, in a unit time interval around t_1 and a photo-ionization event at $\mathbf{r}_2$, in a unit time interval around t_2:

$$w_2(\mathbf{r}_1, t_1; \mathbf{r}_2, t_2) = \xi_1 \xi_2 \text{Tr}\left[\tilde{\rho} \widetilde{E}^-(\mathbf{r}_1, t_1) \widetilde{E}^-(\mathbf{r}_2, t_2) \widetilde{E}^+(\mathbf{r}_2, t_2) \widetilde{E}^+(\mathbf{r}_1, t_1)\right] , \quad (3.87)$$

where ξ_1 and ξ_2 are constants given by expressions analogous to ξ characterizing the sensitivity to the field of the two detector atoms at $\mathbf{r}_1$ and $\mathbf{r}_2$. Formulas (3.86) and (3.87) are essential to describe field measurements in a wide variety of quantum optics experiments. Higher-order field correlation functions are obtained in a similar way, by defining the p-photon counting rates as the expectation values of properly ordered products of $2p$ field operators taken at different times.

Real detectors have of course, as shown in Fig. 3.7, sensitive cathodes made of more than one atom. Their counting rates are given by expressions (3.86) and (3.87) in which the constants ξ and $\xi_1\xi_2$ are replaced by $\mathcal{N}(\mathbf{r})\xi$ and $\mathcal{N}(\mathbf{r}_1)\mathcal{N}(\mathbf{r}_2)\xi_1\xi_2$, where $\mathcal{N}(\mathbf{r})$ is the number of elementary field absorbers in a detection 'pixel' at point $\mathbf{r}$.

The many-mode expansion (3.77) is required to describe precisely the field's temporal and spatial variations. A real counting process lasting a finite time Δt involves generally a finite-size wave packet which expands over a continuum of modes with different frequencies and spatial distributions. In some cases, a simplified description in terms of a single mode is possible, which gives a more transparent meaning to the photon-counting formulas. Detecting, during a time interval Δt, the field of a running wave light beam of wave vector $\mathbf{k}$ impinging normally on a photocathode of area S is equivalent to counting all the photons of a single-mode field quantized in a fictitious slab of volume $\mathcal{V} = Sc\Delta t$. Calling a the field operator of this 'equivalent mode', the positive and negative frequency parts of the field operators E^+ and E^- are then proportional to $ae^{i\mathbf{kr}}/\sqrt{Sc\Delta t}$ and $a^\dagger e^{-i\mathbf{kr}}/\sqrt{Sc\Delta t}$. If the detector's quantum efficiency is unity, the average number of photons detected during time Δt is then:

$$\langle n_c \rangle = w_1 \Delta t = \langle a^\dagger a \rangle = \sum_n n\tilde{\rho}_{nn} , \qquad (3.88)$$

a quite intuitive result.

The distribution $p_c(n)$ of the photon number count around its average also contains, as we will see in the next section, interesting information about the field. It seems natural to guess that, for a perfect detector with unit quantum efficiency, this distribution reflects that of the photon number in the incident field:

$$p_c(n) = \langle n| \tilde{\rho} |n \rangle = Tr[\tilde{\rho} |n\rangle \langle n|] . \qquad (3.89)$$

This formula is in fact far from being obvious, since two distributions $[p_c(n)$ and $\tilde{\rho}_{nn}]$ which have the same average, are not necessarily identical. It is however justified by an analysis of the statistics of the random number of absorption processes in the $\mathcal{N}$ atom detector irradiated by a single-mode field (Vogel *et al.* 2001).

Imperfect detection

For a detector having a single-photon counting efficiency $\eta < 1$, the photon counting distribution becomes:

$$p_c(n, \eta) = \sum_{m \geq n} \binom{m}{n} \eta^n (1 - \eta)^{(m-n)} \tilde{\rho}_{mm} , \qquad (3.90)$$

in which $\binom{m}{n} = m!/n!(m-n)!$ is the binomial coefficient. The imperfect photon count distribution is, quite intuitively, the convolution of the ideal detection probability (ρ_{mm}) by the probability $\binom{m}{n} \eta^n (1-\eta)^{(m-n)}$ that $m-n$ photons have been missed by the detector. The mean photon count then becomes $\langle n_c(\eta) \rangle = \sum_n n p_c(n, \eta) = \eta \langle n \rangle$, an obvious result. We leave as an exercise to show that the variance of the photo-counting rate $\Delta_c^2(\eta)$ is:

$$\Delta_c^2(\eta) = \eta^2 \Delta N^2 + \eta(1-\eta)\langle n \rangle \ . \tag{3.91}$$

This formula shows that, for an imperfect counting ($\eta < 1$), the variance of the number of clicks depends not only on the variance of the photon number, ΔN^2, but also on its mean $\langle n \rangle$. To the quantum field state uncertainty scaled down by the detection efficiency ($\eta^2 \Delta N^2$), we must add the classical uncertainty [$\eta(1-\eta)\langle n \rangle$] due to the random partition of the photons between those which are detected (with probability η) and those which are missed (with probability $1 - \eta$).

For a coherent state [$\Delta N^2 = \langle n \rangle$ and $\Delta_c^2(\eta) = \eta\langle n \rangle = \langle n_c(\eta) \rangle$], the photon count distribution is Poissonian, as the photon number distribution in the incident field, with an average reduced by the detection efficiency η. For a Fock state field ($\Delta N^2 = 0$), we have $\Delta_c^2(\eta) = \eta(1-\eta)\langle n \rangle < \eta\langle n \rangle$. The photon count fluctuations then entirely come from the classical randomness of the detection events. These fluctuations vanish when the detection is perfect ($\eta = 1$). For imperfect detection, the variance of the number of photon clicks is in general sub-Poissonian, with a maximum for $\eta = 1/2$. It is then half its value in a Poissonian distribution with the same average. The sub-Poissonian character fades away as the detection efficiency tends towards 0, the counting distribution becoming, at this limit, Poissonian. This behaviour illustrates the importance of good detection efficiency for the observation of non-classical field features. The same is true for the detection of squeezed states of light, which also require a high detection efficiency.

The photon counting probabilities (3.89) and (3.90) are defined over a statistical ensemble of counting sequences realized on identical fields. In most cases, however, the same field is probed again and again during successive time intervals Δt, in a single experimental run. It is generally assumed that the statistics of these photo-counts are identical to those obtained on an ensemble of field realizations (see Section 3.2.4). This ergodicity hypothesis is justified if the field is in a steady state, being for instance continuously fed by a source as it gets depleted by the counting process and coupling to the environment. We will not discuss in any detail the validity of this quantum ergodic assumption. It seems quite natural if the sampling occurs over intervals much longer than all the relevant field correlation times. The field in each time sequence has then 'forgotten' its previous state and can be considered as an independent realization in an ensemble of experimental 'runs'.

3.2   Coupled field modes

We now explore the quantum dynamics of coupled oscillators. This will give us useful insights into the physics of beam-splitters and interferometers, which are essential tools in the experiments presented in following chapters. We start by analysing the dynamics of two mechanical oscillators coupled by a harmonic interaction potential. We then present a model of the beam-splitter, a device coupling linearly two modes of the field and we describe the Mach–Zehnder optical interferometer, realized by combining two beam-splitters. Finally, we present field damping in a cavity as resulting from the coupling to an environment of 'bath oscillators' and analyse in these terms the relaxation of a coherent state.

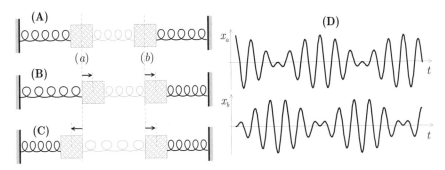

Fig. 3.8 Two coupled mechanical oscillators. (A) Pictorial representation: two identical oscillators coupled by a spring. (B) Symmetrical eigenmode. (C) Anti-symmetrical eigenmode. (D) Classical beat signal for $g/\omega_0 = 1/6$, when the oscillator (a) alone is initially excited. Upper and lower traces: amplitudes for (a) and (b) respectively.

3.2.1 Coupled mechanical oscillators

Let us consider two mechanical oscillators, labelled (a) and (b), with the same mass and spring stiffness and hence a common oscillation frequency ω_0. They are linked together by an additional coupling spring (Fig. 3.8A). The system's Hamiltonian is the sum of the (a) and (b) oscillators energies plus the coupling spring potential energy:

$$V_{ab} = m\omega_0 \frac{g}{2}(X_a - X_b)^2 \ , \tag{3.92}$$

where X_a and X_b ($[X_a, X_b] = 0$) are the two oscillator positions and g is a coupling constant depending upon the stiffness of the coupling spring. For the sake of simplicity, we assume $g \ll \omega_0$ and we express all quantities up to first order in g/ω_0.

Expanding the coupling energy V_{ab}, we include the terms proportional to X_a^2 and X_b^2 in the free-oscillator Hamiltonians, which amounts to redefining the oscillator frequencies. They both become $\omega \approx \omega_0 + g/2$.

We are then left with two oscillators at frequency ω coupled by the Hamiltonian $-m\omega g X_a X_b$ (we can neglect, to first order, the difference between ω and ω_0). Calling a and b the annihilation operators associated to the renormalized oscillators, we can replace $X_a X_b$ by $(\hbar/2m\omega)(a + a^\dagger)(b + b^\dagger)$ in the coupling term, and the total Hamiltonian is:

$$H = H_a + H_b + H_{ab} \ , \tag{3.93}$$

with:

$$H_a = \hbar\omega(a^\dagger a + 1/2) \ ; \quad H_b = \hbar\omega(b^\dagger b + 1/2) \ ; \quad H_{ab} = -\hbar g(a + a^\dagger)(b + b^\dagger)/2 \ . \tag{3.94}$$

The coupling Hamiltonian H_{ab} involves four terms. Two of them, proportional to ab and $a^\dagger b^\dagger$, correspond to transitions in which both oscillators simultaneously lose or gain an excitation quantum. These processes do not conserve the total energy. We can neglect these anti-resonant terms and keep only the two others, again a secular approximation, very similar to the one performed above for the coupling of a classical current to a field mode. The oscillator coupling then becomes:

$$H_{ab} = -\hbar\frac{g}{2}(ab^\dagger + a^\dagger b) \, . \tag{3.95}$$

The two remaining terms, which combine an annihilation and a creation operator, describe a single-phonon exchange between (a) and (b). Any linear coupling between two oscillators, whatever their nature, is described by an interaction term assuming this simple form. Our discussion is thus, as we will see shortly, not limited to the mechanical case.

To describe the dynamics of this system, it is convenient to separate it into independent parts by defining new eigenmodes combining the dynamical variables of (a) and (b). Let us define the symmetrical and anti-symmetrical superpositions of a and b as:

$$a_1 = \frac{a+b}{\sqrt{2}} \; ; \quad a_2 = \frac{a-b}{\sqrt{2}} \, . \tag{3.96}$$

These superpositions are genuine annihilation operators, which obey the commutation relations of independent oscillators. They commute with each other and satisfy $\left[a_i, a_i^\dagger\right] = 1$ for $i = 1, 2$. The total Hamiltonian becomes:

$$H = \hbar(\omega - g/2)\left(a_1^\dagger a_1 + 1/2\right) + \hbar(\omega + g/2)\left(a_2^\dagger a_2 + 1/2\right) \, . \tag{3.97}$$

The coupled oscillators evolve as two independent systems with frequencies $\omega - g/2 = \omega_0$ and $\omega + g/2 = \omega_0 + g$. The symmetrical mode (a_1) corresponds to an oscillation of the 'centre of mass' in which the two masses remain separated by a constant distance (Fig. 3.8B). The coupling spring does not exert any force and the oscillation frequency is thus ω_0. The 'anti-symmetrical' mode (a_2) corresponds to an oscillation of the mass separation, leaving their centre of mass motionless (Fig. 3.8C). The effect of the coupling is then maximum and the frequency higher than ω_0 by an amount g. We will encounter a similar mode analysis when treating the coupling of an atom to a field mode in Section 3.4.1.

The superposition of two eigenmodes with different frequencies gives rise to beats in the evolution of the system. An arbitrary initial condition generally corresponds to a superposition of eigenmodes. The position of each mass versus time is thus a sum of two oscillating functions, at frequencies ω_0 and $\omega_0 + g$. This results in an oscillation at ω_0, with a modulated amplitude. This 'beat signal' is represented in Fig. 3.8(D), in the simple case where (a) is at $t = 0$ out of its equilibrium position and (b) at equilibrium (both being initially motionless). In this beating process, the two oscillators periodically exchange their energy. After half a beat period, π/g, (a) comes to rest, while (b) is taking over all the energy.

3.2.2   Coupled field modes

In order to couple two field oscillators, we need a beam-splitter playing the role of the coupling spring of the mechanical model. Two possible realizations of this device are shown in Fig. 3.9. On the left, two modes (a) and (b) of the radiation field with the same frequency propagate at right angles. At their intersection, there is a partially transmitting, partially reflecting dielectric slab inclined at 45 degrees with respect to the beam directions. We assume that the modes are geometrically matched: they have

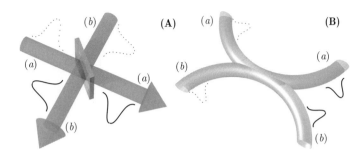

Fig. 3.9 Two beam-splitter models. (A) Two freely propagating modes coupled by a semi-transparent mirror. (B) Two modes guided in single-mode fibres coupled in a fibre coupler. Incoming wave packets (dotted lines) and outgoing ones (solid lines) are represented pictorially in both cases.

the same polarization (orthogonal to the incidence plane) and the spatial structure of mode (b) reflected by the slab exactly matches the transmitted mode (a). Figure 3.9(B) represents an optical fibre coupler realizing the same mode mixing. The upper fibre, (a), and the lower fibre, (b), are placed side by side over some length,[16] so that light is coupled from one mode into the other.

In both cases, the two field-mode amplitudes are classically mixed. In the case of a lossless dielectric slab, for instance, the emerging electric field amplitudes, E'_a and E'_b, are linked to the incoming ones, E_a and E_b, by a unitary and symmetrical[17] matrix relation (Born and Wolf 1980):

$$\begin{pmatrix} E'_a \\ E'_b \end{pmatrix} = U_c \begin{pmatrix} E_a \\ E_b \end{pmatrix} = \begin{pmatrix} t(\omega) & r(\omega) \\ r(\omega) & t(\omega) \end{pmatrix} \begin{pmatrix} E_a \\ E_b \end{pmatrix} , \qquad (3.98)$$

where $t(\omega)$ and $r(\omega)$ are complex transmission and reflection coefficients, functions of the frequency ω of the field modes, which satisfy the unitarity conditions $|t(\omega)|^2 + |r(\omega)|^2 = 1$ and $t(\omega)r^*(\omega) + t^*(\omega)r(\omega) = 0$. The general relation (3.98) holds for any kind of lossless symmetrical semi-reflecting device, whether dielectric or metallic, and for a fibre coupler as well. By choosing properly the phase origin of the transmitted field, one can always assume that $t(\omega)$ is real positive, which entails that $r(\omega)$ is purely imaginary. We then introduce a mixing angle θ characterizing the beam-splitter operation at frequency ω, set $t(\omega) = \cos\theta/2$, $r(\omega) = i\sin\theta/2$, and for a monochromatic field U_c is:

$$U_c(\theta) = \begin{pmatrix} \cos(\theta/2) & i\sin(\theta/2) \\ i\sin(\theta/2) & \cos(\theta/2) \end{pmatrix} . \qquad (3.99)$$

By combining a beam-splitter with phase shifters, we can realize more general mode mixing transformations. A possible set-up is sketched in Fig. 3.10. Two phase-retarding plates are sandwiching the beam-splitter on the path of the (a) beam, inducing on the

---

[16]In practical couplers, the cladding around the fibre core is partially removed. The modes' evanescent waves around both fibre cores are thus superposed and directly coupled.

[17]The unitarity is imposed by energy conservation, while the symmetry of the matrix reflects the invariance of optics laws when propagation directions are reversed.

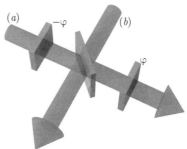

Fig. 3.10 Principle of a generalized beam-splitter with adjustable phase φ. A semi-reflecting plate, inclined at $45°$ with the (a) and (b) beams, is sandwiched between two retarding plates inducing opposite phase shifts on beam (a).

field opposite phase shifts $\pm\varphi$. This combination of optical elements is a generalized beam-splitter whose action is described by the product of matrices:

$$
\begin{aligned}
U_c(\theta,\varphi) &= \begin{pmatrix} e^{i\varphi} & 0 \\ 0 & 1 \end{pmatrix} \begin{pmatrix} \cos(\theta/2) & i\sin(\theta/2) \\ i\sin(\theta/2) & \cos(\theta/2) \end{pmatrix} \begin{pmatrix} e^{-i\varphi} & 0 \\ 0 & 1 \end{pmatrix} \\
&= \begin{pmatrix} \cos(\theta/2) & ie^{i\varphi}\sin(\theta/2) \\ ie^{-i\varphi}\sin(\theta/2) & \cos(\theta/2) \end{pmatrix} .
\end{aligned}
\tag{3.100}
$$

This description of classical mode mixing is valid for monochromatic fields interacting continuously with the beam-splitter. In general, a realistic experiment involves time-limited wave packets propagating in modes (a) and (b) and 'colliding' on the beam-splitter during a time τ, as illustrated on Fig. 3.9. These transient fields are developed over a collection of field modes spanning a small frequency interval $\Delta\omega = 1/\tau$. To avoid complications arising from the spectral analysis of this time-dependent problem, we take the limit in which $\Delta\omega$ is very small and τ very long yet finite, so that we can define a 'before' and an 'after' for the collision. We assume that the time envelopes of the wave packets are identical and thus overlap perfectly on the beam-splitter. We consider also that the r, t and φ parameters do not vary over the narrow spectrum of the wave packets, so that we can define a single matrix (3.99) or (3.100) describing the same mixing process for all the frequency components of the light field.

Operation of a quantum beam-splitter

What is the action of this beam-splitter on the quantum states and observables of the field? We now show that it is described by an interaction Hamiltonian H_{ab}, which, at the classical limit of large numbers of photons, has the same effect on the field operators as the transformation (3.100) on the classical fields. This Hamiltonian acts transiently on the fields, coupling the two quantum oscillators during the finite time τ corresponding to the simultaneous passage of the wave packets on the beam-splitter. Since it describes a linear interaction, H_{ab} is made of two terms associated to single-photon exchange between the two modes:

$$
H_{ab}(t) = -\hbar\frac{g(t)}{2}\left(e^{-i\varphi}ab^{\dagger} + e^{i\varphi}a^{\dagger}b\right) ,
\tag{3.101}
$$

where a and b are the annihilation operators in the modes and $g(t)$ is a slowly varying real function of time, switching adiabatically the coupling on and off for the time interval τ around $t = 0$. This Hamiltonian is an obvious generalization of the linear coupling between mechanical oscillators described above. In the following, we consider the action of H_{ab} alone. This implicitly assumes that we use an interaction representation with respect to the Hamiltonians of the free modes.[18]

The operation of the beam-splitter can be analysed either in the Schrödinger or Heisenberg point of views. In the first case, the quantum states of both modes evolve due to the coupling Hamiltonian H_{ab}. This approach is used in the next section to describe the evolution of some special quantum states.

We focus here on the Heisenberg point of view, which lends itself to a direct comparison between the classical and quantum descriptions. The quantum states are constant and the dynamical variables are the field operators. In the interaction picture with respect to the free mode Hamiltonians, these operators are time-independent before and after the 'collision' with the beam-splitter. We can thus identify them, for $t \ll -\tau$ and $t \gg \tau$, with the a and b operators of the Schrödinger picture. The beam-splitter transforms the input operator a into the output operator a':

$$a' = U^\dagger a U \ , \tag{3.102}$$

where U is the evolution operator:

$$U = e^{-(i/\hbar)\int H_{ab}(t)\,dt} \ , \tag{3.103}$$

the integral being carried over the whole duration τ of the coupling. The evolution operator can be rewritten as:

$$U(\theta, \phi) = e^{-iG(\varphi)\theta/2} \ , \tag{3.104}$$

where:

$$G(\varphi) = -(e^{-i\varphi}ab^\dagger + e^{i\varphi}a^\dagger b) \quad \text{and} \quad \theta = \int g(t)\,dt \ . \tag{3.105}$$

Note that we omit the θ and φ arguments in U when no confusion is possible. We now have:

$$\begin{aligned} a' &= U^\dagger a U = e^{iG\theta/2} a e^{-iG\theta/2} = a + \frac{i\theta}{2}[G,a] \\ &+ \frac{i^2\theta^2}{2!2^2}[G,[G,a]] + \cdots + \frac{i^n\theta^n}{n!2^n}[G,[G,[\cdots,[G,a]]]] + \cdots \ , \end{aligned} \tag{3.106}$$

where we have used again the Baker–Hausdorff lemma (eqn. 3.47) to develop the exponentials. The nested commutators on the right-hand side can be easily computed: $[G,a] = e^{i\varphi}b$ and $[G,[G,a]] = a$. All the odd-order terms in the development are proportional to $e^{i\varphi}b$, while the even-order ones are proportional to a. The coefficients of a and $ie^{i\varphi}b$ thus coincide with the series expansion of $\cos\theta/2$ and $\sin\theta/2$ respectively:

$$a' = U^\dagger a U = \cos(\theta/2)\,a + ie^{i\varphi}\sin(\theta/2)\,b \ , \tag{3.107}$$

and similarly:

$$b' = U^\dagger b U = ie^{-i\varphi}\sin(\theta/2)\,a + \cos(\theta/2)\,b \ . \tag{3.108}$$

The output mode operators are linear combinations of the input ones. These relations are formally identical to the input–output matrix relations (3.100) for the classical

---

[18]To simplify the notation, we will omit the tilde that we should use in an interaction representation.

electric field amplitudes. This identity, which justifies the expression (3.101) of H_{ab}, expresses the correspondence principle. When the fields involve many photons, the annihilation operators behave as the classical field complex amplitudes. The mixing parameter θ is directly related to the transmission and reflection coefficients of the beam-splitter, which can be measured in a classical optics experiment. These results are independent of the precise details of the model.

Taking the hermitian conjugate of the previous relations and noting that $U^\dagger(\theta, \varphi) = U(-\theta, \varphi)$, we find similarly:

$$Ua^\dagger U^\dagger = \cos(\theta/2)\, a^\dagger + ie^{-i\varphi} \sin(\theta/2)\, b^\dagger \; ; \qquad Ub^\dagger U^\dagger = ie^{i\varphi} \sin(\theta/2)\, a^\dagger + \cos(\theta/2)\, b^\dagger \; . \tag{3.109}$$

Action of the beam-splitter on some input states

We now analyse the action of the beam-splitter on selected initial states in the Schrödinger point of view. The final state $|\Psi'\rangle$ is given in terms of the input state $|\Psi\rangle$ by $|\Psi'\rangle = U\,|\Psi\rangle$. The first trivial case corresponds to the vacuum impinging in both modes, $|\Psi\rangle = |0,0\rangle$. This state is invariant under the action of U. A beam-splitter has no effect in the dark.

Let us consider now a single photon impinging in mode (a), the initial state being $|1,0\rangle$ [in the following, the first and second labels in the kets refer to (a) and (b) respectively]. The sequence of equalities:

$$U\,|1,0\rangle = Ua^\dagger\,|0,0\rangle = Ua^\dagger U^\dagger U\,|0,0\rangle = Ua^\dagger U^\dagger\,|0,0\rangle \; , \tag{3.110}$$

leads, using eqn. (3.109), to:

$$U\,|1,0\rangle = \left[\cos(\theta/2)\, a^\dagger + ie^{-i\varphi} \sin(\theta/2)\, b^\dagger\right]|0,0\rangle = \cos(\theta/2)\,|1,0\rangle + ie^{-i\varphi} \sin(\theta/2)\,|0,1\rangle \; . \tag{3.111}$$

For a single photon impinging on the beam-splitter in mode (b), we get:

$$U\,|0,1\rangle = \left[ie^{i\varphi} \sin(\theta/2)\, a^\dagger + \cos(\theta/2)\, b^\dagger\right]|0,0\rangle = ie^{i\varphi} \sin(\theta/2)\,|1,0\rangle + \cos(\theta/2)\,|0,1\rangle \; . \tag{3.112}$$

A single photon is thus in general dispatched in a coherent superposition between the two modes, with a final probability distribution corresponding to the classical light intensity transmission and reflection coefficients $\cos^2(\theta/2)$ and $\sin^2(\theta/2)$. In particular, for $\theta = \pi/2$, we have:

$$U(\pi/2, \varphi)\,|1,0\rangle = (|1,0\rangle + ie^{-i\varphi}\,|0,1\rangle)/\sqrt{2} \; . \tag{3.113}$$

Is this an entangled state? If we define the system as two oscillators (a) and (b) coupled by the beam-splitter, the answer is definitely 'yes'. The two parts of this bipartite entity are entangled by sharing in an ambiguous way a quantum of excitation. Correlations between measurements performed independently on the two modes could reveal this entanglement. If we renounce to perform such independent measurements, though, we can define two new modes $(\pm)$, linear combinations of (a) and (b) corresponding to the photon creation operators $(a^\dagger \pm ie^{-i\varphi}b^\dagger)/\sqrt{2}$. The state (3.113) then represents a single photon in the $(+)$ mode with the vacuum in the $(-)$ mode and, in this point of view, it is not entangled. This example shows that the notion of entanglement depends upon the basis in which we choose to describe the system (this choice being in turn determined by the kind of measurements we intend to perform on it). The unitary transformation changing the tensor products of Fock states in modes (a) and (b) into those of photon number states in modes ($\pm$) is non-local with respect to the (a)-(b) partition since it mixes the two modes. Such a non-local operation can result in different degrees of entanglement in the two representations, turning an entangled state into a separate one.[19]

The single-photon state (3.111) explores, when θ and φ are tuned, the whole two-dimensional Hilbert space spanned by $|1,0\rangle$ and $|0,1\rangle$. We note the formal analogy between a single photon 'suspended' between the two modes and a spin on the Bloch sphere. The states (3.111) and (3.112) are indeed identical to those defined by eqn. (2.6), provided one makes the change $\phi \to \pi/2 - \varphi$ and identifies $|1,0\rangle$ and $|0,1\rangle$ with the spin states $|0\rangle$ and $|1\rangle$ respectively. It has been suggested (Chuang and Yamamoto 1996; Knill *et al.* 2001; Fattal *et al.* 2004) to exploit this analogy and to code information into the path of single-photon bimodal states, $|1,0\rangle$ and $|0,1\rangle$, playing the roles of logical $|0\rangle$ and $|1\rangle$ qubit states. In this context, the beam-splitter which transforms the logical qubit states into their most general superpositions realizes a one-qubit quantum gate (see Section 2.6).

Let us now consider the action of the beam-splitter on fields containing many photons. Assume first that n-photons impinge in (a), the initial state being $|n,0\rangle$. A calculation similar to the single-photon one yields:

$$U\,|n,0\rangle = U\frac{(a^\dagger)^n}{\sqrt{n!}}\,|0,0\rangle = \frac{1}{\sqrt{n!}}U(a^\dagger)^n U^\dagger U\,|0,0\rangle \ . \tag{3.114}$$

Noting that $U(a^\dagger)^n U^\dagger = (Ua^\dagger U^\dagger)^n$, and using again eqn. (3.109):

$$U\,|n,0\rangle = \frac{1}{\sqrt{n!}}\left[\cos\frac{\theta}{2}a^\dagger + ie^{-i\varphi}\sin\frac{\theta}{2}b^\dagger\right]^n |0,0\rangle \ . \tag{3.115}$$

The right-hand side can be expanded according to the binomial law:

$$U\,|n,0\rangle = \sum_{p=0}^{n}\binom{n}{p}^{1/2}[\cos(\theta/2)]^{n-p}\left[ie^{-i\varphi}\sin(\theta/2)\right]^{p}|n-p,p\rangle \ . \tag{3.116}$$

The final quantum state is a coherent superposition of terms corresponding to all possible partitions of the n incoming photons between the two modes. If we describe the system as a bipartite (a)-(b) entity, the final state is thus in general massively entangled. When θ is an even multiple of π, the output state is disentangled, in fact identical to the initial one [all photons in mode (a)]. When θ is an odd multiple of π, all photons are channelled in mode (b) (again a disentangled situation).

To be specific, let us consider in more detail a balanced beam-splitter ($\theta = \pi/2, \varphi = 0$). The output state is:

$$U\,(\pi/2,0)\,|n,0\rangle = \frac{1}{\sqrt{2^n}}\sum_{p=0}^{n}\binom{n}{p}^{1/2}(i)^{p}|n-p,p\rangle \ . \tag{3.117}$$

---

[19]Note that the notion of entanglement is here independent of the number of particles in the system. A single photon can be in an entangled state or not, depending upon the basis choice.

The probability of getting $n-p$ photons in (a) and p in (b) is proportional to $\binom{n}{p}$, the number of partitions of the incoming photons into the two outputs. From the photon counting point of view, the quantum beam-splitter behaves as a random device. Each photon 'freely chooses' to emerge either in (a) or (b), with equal probabilities. The average photon number in each output mode is thus $n/2$. The photon number in one mode obeys, in the limit of large incoming photon numbers $n \gg 1$, Gaussian statistics with a width $\sqrt{n/2}$. The relative fluctuations, proportional to $1/\sqrt{n}$, are negligible for large fields and the two output energies are balanced according to the correspondence principle. This simple picture, treating photons as billiard balls, is valid because we consider only the output energies. It is not sufficient to account for the massive entanglement between the two modes, which would be revealed by more complex photon coincidence experiments.

Suppose now that the impinging field in mode (a) is in a coherent state $|\alpha\rangle$ with $\bar{n} = |\alpha|^2$ photons on the average, while mode (b) is in vacuum. The output state of the beam-splitter is now:

$$U|\alpha, 0\rangle = UD_a(\alpha)U^\dagger |0, 0\rangle \;, \tag{3.118}$$

where $D_a(\alpha)$ is the displacement operator in mode (a). Using the operator identity $Uf(A)U^\dagger = f(UAU^\dagger)$, which holds whenever f can be expanded in power series, we get:

$$UD(\alpha)U^\dagger = e^{\alpha Ua^\dagger U^\dagger - \alpha^* UaU^\dagger} \;. \tag{3.119}$$

Using eqn. (3.109) and its hermitian conjugate, and separating the commuting contributions of a and b in the exponentials, we get:

$$U|\alpha, 0\rangle = D_a\left[\alpha \cos(\theta/2)\right] D_b\left[ie^{-i\varphi}\alpha \sin(\theta/2)\right]|0, 0\rangle \;, \tag{3.120}$$

where D_b is the displacement operator acting on mode b. The output state is thus:

$$U|\alpha, 0\rangle = \left|\alpha \cos(\theta/2), ie^{-i\varphi}\alpha \sin(\theta/2)\right\rangle \;. \tag{3.121}$$

It is an unentangled tensor product of coherent states in the two modes, with amplitudes obeying the same relations as the classical fields (eqn. 3.100). The coherent state dynamics, entirely classical even for small photon numbers, is thus very different from that of Fock states. The non-entanglement of coherent states involved in linear coupling operations is a very basic feature of these states, with far-reaching consequences, as we will see below when discussing relaxation and decoherence.

Let us note also that the two modes exchange periodically their energies as a function of the mixing angle θ or, equivalently, as a function of the coupling time. This periodic energy exchange is the counterpart of the classical beating phenomenon described in the previous section.

A beam-splitter can also be used to mix two excited modes. As a simple example, suppose that (a) and (b) initially contain a single photon. The output state is then:

$$U|1, 1\rangle = Ua^\dagger b^\dagger |0, 0\rangle = Ua^\dagger U^\dagger Ub^\dagger U^\dagger |0, 0\rangle \;. \tag{3.122}$$

Using eqn. (3.109), we get:

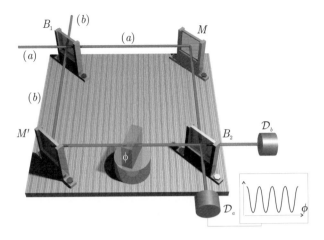

Fig. 3.11 A Mach–Zehnder interferometer.

$$U\,|1,1\rangle = \frac{i\sin\theta}{\sqrt{2}}\left[e^{+i\varphi}\,|2,0\rangle + e^{-i\varphi}\,|0,2\rangle\right] + \cos\theta\,|1,1\rangle \;, \qquad (3.123)$$

which is, in general, an entangled state. In the case of a balanced beam-splitter ($\theta = \pi/2, \varphi = 0$):

$$U(\pi/2, 0)\,|1,1\rangle = (|2,0\rangle + |0,2\rangle)\,/\sqrt{2} \;, \qquad (3.124)$$

which is a superposition of two photons propagating together in mode (a) or (b). The two photons are bunched in the same mode after the beam-splitter (Hong *et al.* 1987). This can be seen as a consequence of the bosonic nature of the photons. It is also a quantum interference effect. The probability amplitude of getting one photon in each mode cancels, because of the exact destructive interference between two indistinguishable quantum paths. In one of the paths, both photons are transmitted, with probability amplitude $t \times t = 1/2$. In the other, they are both reflected, with the quantum amplitude $ir \times ir = -1/2$.

This process and the resulting state are non-classical. A superposition of two photons either in mode (a) or in mode (b) is the simplest example of a multi-particle quantum state superposition. We will show later how non-linear optical elements could be used to generalize this situation to larger photon numbers, preparing a superposition of n photons either in (a) or (b).

This photon 'bunching' effect is expected here because the polarizations of the two photons are the same. Photons with orthogonal polarizations behave differently. If they are initially in an entangled polarization state, anti-symmetric versus photon exchange, they emerge 'anti-bunched' in different spatial modes. This is required to preserve the overall state symmetry versus photon exchange. Photon entanglement and Bell states analysis techniques are based on this feature. Let us add that these photon coincidence experiments are technically difficult, since the wave packets must be well-matched, both in space (mode matching) and in time (coherence length overlap).

3.2.3   The Mach–Zehnder interferometer

The beam-splitter is a key element in most optical interferometers. We choose to describe here the Mach–Zehnder device because of its symmetry. The two interfering modes correspond to well-identified photon trajectories, with a simpler geometry than in the Young design. A Mach–Zehnder interferometer is sketched in Fig. 3.11. It has two input modes, (a) and (b), which are first mixed by the beam-splitter B_1. The field is supposed to impinge in mode (a) while (b) is initially in vacuum. It is dispatched by B_1 into two different paths, folded by the mirrors M and M'. A tuneable dephasing element is inserted in mode (b). The two modes are mixed again by a second beam-splitter B_2 and finally impinge on photon counters $\mathcal{D}_a$ and $\mathcal{D}_b$. We consider here only the case of balanced beam-splitters with $\theta = \pi/2$. As is well-known in classical optics, this situation provides maximum fringe visibility. We also assume for the sake of simplicity (but this is not an essential condition) that B_1 and B_2 are identical, implementing both the same transformation $U(\pi/2, 0)$.

We have shown (eqn. 3.121) that coherent state amplitudes are transformed as classical fields by beam-splitters. We thus consider first the case of a coherent state input $|\alpha\rangle$, containing $\bar{n} = |\alpha|^2$ photons on the average. Let us show that we retrieve then, through a quantum calculation, the well-known classical behaviour of this device. Following the evolution of the field wave packets, the interferometer operation consists in two successive applications of the beam-splitter transformation U, sandwiching the action of the dephasing element onto (b).

The input state, $|\alpha, 0\rangle$, is first transformed by B_1 into $|\alpha/\sqrt{2}, i\alpha/\sqrt{2}\rangle$ (eqn. 3.121 with $\varphi = 0$). The phase-shifter, acting on mode (b), changes the classical phase by an amount ϕ. In the Schrödinger point of view, the corresponding field transformation is $|i\alpha/\sqrt{2}\rangle \rightarrow |ie^{i\phi}\alpha/\sqrt{2}\rangle$. In the same way, any Fock state $|n\rangle$ would be transformed into $e^{in\phi}|n\rangle$ (the vacuum state being unchanged). Finally, B_2 again mixes the modes, resulting in the final state $U|\alpha/\sqrt{2}, ie^{i\phi}\alpha/\sqrt{2}\rangle$ expressed as:

$$e^{[(\alpha/\sqrt{2})Ua^\dagger U^\dagger - (\alpha^*/\sqrt{2})UaU^\dagger + (i\alpha e^{i\phi}/\sqrt{2})Ub^\dagger U^\dagger + (i\alpha^* e^{-i\phi}/\sqrt{2})UbU^\dagger]}|0,0\rangle$$

$$= \left| \frac{\alpha}{2}\left(1 - e^{i\phi}\right), \frac{i\alpha}{2}\left(1 + e^{i\phi}\right) \right\rangle , \tag{3.125}$$

and the mean photon numbers n_a and n_b recorded by $\mathcal{D}_a$ and $\mathcal{D}_b$ are:

$$n_{\substack{a \\ b}} = \bar{n}(1 \mp \cos\phi)/2 . \tag{3.126}$$

When ϕ is tuned, these signals exhibit phase-opposite oscillations with a 100% contrast. All photons end up in $\mathcal{D}_b$ for $\phi = 0$ and in $\mathcal{D}_a$ for $\phi = \pi$. This is obviously the classical result. It is also important to remark that (a) and (b) are always disentangled when the interferometer is fed with coherent states.

Contrary to a widespread misconception, the phase of the input field plays no role in the optical interference signal obtained here. The phase of α does not appear in the photon counting rates (eqn. 3.126) and must not be confused with the adjustable phase difference ϕ between the two interferometer paths. To stress this point, it is instructive to consider an input field with n photons in a Fock state, carrying no phase information at all. We get after a straightforward calculation using eqns. (3.107)–(3.109) the final state as:

Fig. 3.12 Single-photon interference in a Mach–Zehnder interferometer. Number of detected photons in one of the output ports as a function of phase ϕ. The three traces (a) to (c) correspond to increasing data accumulation times and increasing photon numbers in each phase bin. Traces (a): courtesy of P. Grangier; (b) and (c) reprinted with permission from Grangier *et al.* (1986).

$$|\Psi\rangle = \frac{1}{\sqrt{n!}} \left[\frac{1}{2} \left(1 - e^{i\phi}\right) a^\dagger + \frac{i}{2} \left(1 + e^{i\phi}\right) b^\dagger \right]^n |0,0\rangle \ . \tag{3.127}$$

This equation describes a binomial partition of the photons in the two output ports, with the probabilities:

$$p_{\ b}^{\ a} = (1 \mp \cos\phi)/2 \ . \tag{3.128}$$

The average photon counting signal in (a) and (b) is exactly the same as for an initial coherent state with the same mean photon number. Note, however, that the two modes, seen as distinct entities, are in general massively entangled, a feature which is not reflected in the single-photon counting signals but which could be revealed by higher order photon correlation experiments. Before we perform independent measurements on the (a) and (b) modes, we can as well view the state (3.127) as representing n photons all in the same state, linear superposition of the two output modes of the first beam-splitter. In this point of view, there is no entanglement [see discussion following eqn. (3.113), on page 132].

In the situation described so far, many particles cross the interferometer at the same time. This feature is irrelevant to the operation of the device. The photon partition probabilities between modes (a) and (b) is n-independent. Thus, if only one photon at a time impinges on the interferometer, the average signals are the same as for a coherent or n-photons input. Single-photon interference has been observed in a variety of situations. One of the most striking experiments, illustrating directly the present discussion, has been performed by Grangier and co-workers (Grangier *et al.* 1986). They used a Mach–Zehnder interferometer and a single-photon source based on an atomic cascade. The mean time between single-photon emissions by the source was much larger than the photon's transit time through the interferometer. Figure 3.12 presents the detected photon number in one of the output ports.

When the number of detected photons is very small (Fig. 3.12a), the events are randomly distributed over the phase bins. When the number of recorded photons increases (Fig. 3.12b) the sinusoidal fringe signal shows up. The signal-to-noise ratio increases with the number of detected photons (Fig. 3.12c). Even though this interference is a single-particle effect, its observation clearly requires many realizations of the experiment with different phase settings. The final noise reflects the statistical nature of the signal. The signal-to-noise ratio, proportional to the square root of the number of experiments per phase bin, can be increased to arbitrarily high values by lengthening the data acquisition time.

This experiment illustrates essential quantum features. First, the interference is not a collective effect. It involves a quantum interference process, associated to the superposition of two different and indistinguishable quantum paths for a single particle propagating from the source to the detector. In this experiment, the photon 'always interferes with itself', following a famous statement by Dirac.[20] Putting it in a slightly provocative way, we could say that a Mach–Zehnder or a Young double-slit experiment, even performed with large light intensities, is a fundamentally quantum effect, directly illustrating the superposition principle. This experiment is also a beautiful illustration of the wave–particle duality. The photons behave as particles at the detection time, giving well-defined 'clicks' on the detector. However, each of these individual particles has undergone the quantum interference process, going through the interferometer through the two paths 'at the same time', as a wave (we use here a pictorial and largely inadequate language). Of course, this is true as long as the two interfering paths are left indistinguishable, in the absence of decoherence (see Chapter 2).

In most cases, interferometers are not designed to illustrate quantum mechanics, but to measure with precision optical phases. It is thus relevant to discuss the elementary limits set by quantum mechanics to their sensitivity. Since we count light quanta, it seems that the input photon fluctuations should be minimized to get the best sensitivity. An input Fock state with n photons thus appears *a priori* as an ideal choice. The statistics of the detected photons obeys a binomial law with the average photon number n_a in $\mathcal{D}_a$ being np_a ($n_b = np_b$ in $\mathcal{D}_b$) (eqn. 3.128). The mean square-root deviation of the photon number in mode (a), ΔN_a, measuring the statistical partition noise, is:

$$\Delta N_a = \sqrt{np_a p_b} = \sqrt{n}\frac{|\sin\phi|}{2} . \tag{3.129}$$

It cancels for $\phi = 0$ [in which case no photon exits in mode (a)] and $\phi = \pi$, when all photons are detected in $\mathcal{D}_a$. The photon number variance is maximum for $\phi = \pi/2$, when the interferometer as a whole behaves as a balanced beam-splitter. This variance, associated to the partition of the photons, is directly linked to the granular structure of the light field and is thus called 'shot noise'.

The sensitivity of the interferometer, η_a, for a detection of mode (a), can be defined as the inverse of the smallest phase shift, $\delta\phi_m$, which produces a signal change δn_a in $\mathcal{D}_a$ larger than the statistical noise ΔN_a. With $dn_a/d\phi = (n/2)\sin\phi$, we get $\delta\phi_m = 1/\sqrt{n}$ and:

$$\eta_a = \sqrt{n} . \tag{3.130}$$

Note that the same result is obtained for η_b, corresponding to a detection of mode (b). The sensitivity of the interferometer is proportional to the square root of the incoming photon number. This result is the same as for a one-photon device when the experiment is repeated n time per phase bin. What is perhaps more counter-intuitive is that the sensitivity is independent of the phase setting. The partition noise and the slope of the fringes vary both as $\sin\phi$ and their ratio is phase-independent. At $\phi = 0$,

---

[20]This statement is only valid, however, for linear beam-splitting elements and for a single-port photon counting detection. Experiments with non-linear beam-splitters or measuring higher order correlations of the output fields involve interference between paths implying many photons at once (see Chapter 7).

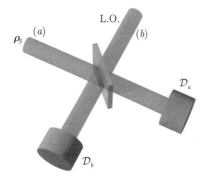

Fig. 3.13 Scheme of the homodyne detection of an unknown field described by the density operator ρ_S.

for instance, there is no noise, but the fringe slope is also zero, the ratio of these two quantities keeping a finite value.

Let us discuss briefly the case of a coherent input, with $\bar{n}$ photons on the average (which is far easier to produce than an n-photon Fock state). The mean photon numbers in each detection channel are given by eqn. (3.126). Since the outputs are two independent coherent fields, the standard deviations of the signals are $\Delta N_a = \sqrt{\bar{n}}|\sin(\phi/2)|$ and $\Delta N_b = \sqrt{\bar{n}}|\cos(\phi/2)|$. For $\phi = \pi/2$, $\Delta N_a = \Delta N_b = \sqrt{\bar{n}/2}$. This fluctuation is $\sqrt{2}$ larger than for a Fock state with the same average photon number (eqn. 3.129). This increased noise results from the r.m.s. addition of the incoming photon number fluctuations of the coherent state with the partition noise.

The sensitivity of the interferometer with a coherent input, obtained by the same argument as above, is now phase- and mode-dependent:

$$\eta_a = \sqrt{\bar{n}}|\cos(\phi/2)| \qquad \text{and} \qquad \eta_b = \sqrt{\bar{n}}|\sin(\phi/2)| . \qquad (3.131)$$

The optimum sensitivity, $\sqrt{\bar{n}}$, which is achieved for $\phi = 0$ in $\mathcal{D}_a$ and $\phi = \pi$ in $\mathcal{D}_b$, is the same as in the Fock state input case. It corresponds now to a dark fringe situation. For $\phi = 0$, for instance, no photon is detected in $\mathcal{D}_a$ and there is thus no noise in this port. The fringe slope is also going to zero, but the ratio of these two quantities remains finite. Note that this dark fringe setting is well-adapted to interferometric measurement with strong light intensities, since it reduces the detector saturation problems. This is the case for interferometric gravitational waves sensors. For other applications, it might be convenient to combine the two output port signals and to compute their difference. The optimal sensitivity, again $\sqrt{\bar{n}}$, is obtained for $\phi = \pi/2$ and corresponds now to the maximum fringe slope.

These results can be substantially modified if mode (b), instead of being empty, contains initially a non-classical squeezed state (Kimble 1992). The sensitivity can then be increased beyond the standard quantum limit set by shot noise as discussed above. Another way to increase the sensitivity beyond the shot noise limit is to use multi-particle interferometers, based on non-linear beam-splitting elements (see Chapter 7).

3.2.4 Homodyne field measurement

The experiments described so far are not sensitive to the phase of the field. The same amount of information is extracted by a Mach–Zehnder interferometer from a coherent and from a Fock state. To get information about the phase, one must resort to another kind of interference method which mixes the field, not with itself, but with a reference beam, called the local oscillator (L.O.). If this reference has the same frequency as the field to be measured, the method is referred to as 'homodyne' detection.

The mixing is performed again with a beam-splitter according to the scheme shown in Fig. 3.13. The ingoing 'signal' field in (a) is described by its density operator ρ_s. The L.O., impinging in (b), is in a coherent state $|\beta\rangle = |\beta_0 e^{i\phi}\rangle$. The beating between the reflected L.O. beam and the transmitted signal is measured by counting the mean number of photons n_a received by the detector $\mathcal{D}_a$ (assumed to be ideal with unit quantum efficiency), placed in the transmitted signal output beam. At the same time, the transmitted L.O. beam beating with the reflected signal is received by another perfect detector $\mathcal{D}_b$, which counts the resulting mean number of photons n_b. We adopt here the Heisenberg point of view: the field operators are transformed by the mode mixing and the field states are invariant. Taking into account eqns. (3.107) and (3.108), and setting $\theta = \pi/2$ (balanced homodyne detection) and $\varphi = 0$ we obtain:

$$n_{\substack{a\\b}} = \frac{1}{2}\mathrm{Tr}\left[\rho_s |\beta\rangle\langle\beta| (a^\dagger \mp ib^\dagger)(a \pm ib)\right]$$

$$= \tfrac{1}{2}[\langle\beta| b^\dagger b |\beta\rangle + \mathrm{Tr}(\rho_s a^\dagger a)] \pm \frac{i}{2}\mathrm{Tr}\left[\rho |\beta\rangle\langle\beta| (a^\dagger b - ab^\dagger)\right] . \qquad (3.132)$$

The contributions in n_a and n_b which are proportional to the L.O. and signal intensities are equal, while the contributions corresponding to the beating between these two fields are opposite. Subtracting the outputs of $\mathcal{D}_a$ and $\mathcal{D}_b$, we finally get the difference counting signal:

$$n_a - n_b = i\beta_0\mathrm{Tr}\rho_s\left(a^\dagger e^{i\phi} - ae^{-i\phi}\right) = 2\beta_0\langle X_{\phi+\pi/2}\rangle_s , \qquad (3.133)$$

which measures the expectation value in the field state described by ρ_s of the signal field quadrature $\pi/2$-out-of-phase with the L.O.[21] By sweeping the phase of the local oscillator, we can measure the expectation value of any field quadrature.

In fact, the method has the potential to extract even more information. The distribution of the difference of photocounts in the two output channels observed on an ensemble of identically prepared fields (or on a single steady-state realization on which the measurement is repeated many times) reconstructs the probability distribution of the field quadrature around its average value. This is shown to be true at the limit where the amplitude of the L.O. oscillator is very large, so that it can be considered as a classical field whose fluctuations are negligible (Smithey *et al.* 1993).

Repeating the measurement for a large set of quadrature phases, we get information which can be exploited to reconstruct the field Wigner function, and hence its density operator. We show in the appendix that the probability density of a given quadrature is the integral of the W function along a line in phase space orthogonal to the direction of this quadrature. Measuring the quadrature fluctuations for all possible phases thus amounts to determining the integrals of W along all possible directions in the phase plane. A procedure known as the Radon transform can then be used to find out W from

---

[21]This phase offset between the L.O. and the measured quadrature is due to the phase choice made for the beam-splitter ($\varphi = 0$). Setting $\varphi = \pi/2$ yields a homodyne detection of the quadrature in phase with the local oscillator.

these integrals (Smithey *et al.* 1993). This transform is employed in a different context in medical tomography. Here, the integrated optical density of an inhomogeneous medium irradiated by X rays is measured along different directions in a plane and the Radon inversion reconstructs the density of the absorbing medium in this plane. In this way, pictures of the inner parts of the body are made. Quantum tomography is analogous to medical tomography. The field to be measured is mixed via a beam-splitter with L.O. references of various phases and the fluctuations of the homodyne beat signal are measured. A Radon inversion procedure reconstructs W and hence the field density operator.

3.2.5 Beam-splitters as couplers to the environment: damping of a coherent state

Beyond its practical importance in interferometers, the linear beam-splitter is a useful model to study various fundamental effects in quantum optics. We have seen in particular that this linear device leaves coherent states immune to entanglement, preserving their classical character. We show here that this immunity to entanglement survives if the coherent state is coupled not to one single mode, but to a continuum of modes initially in vacuum. This situation describes in general terms field damping, the mode continuum being a model for a reservoir. For example, a field stored in a cavity made of mirrors facing each other is coupled to reservoir modes by the scattering of light on mirror imperfections. This scattering process can be modelled as a linear coupling between the cavity and the reservoir modes similar to the coupling performed by an ensemble of beam-splitters. We will understand, with this simple model, that coherent states are the natural 'pointer states' in quantum optics.

We consider a field mode (a) (annihilation operator a, frequency ω) coupled to a large set of other modes at frequencies ω_i described by their annihilation operators b_i. In the interaction picture with respect to the free field Hamiltonian, the coupling of (a) with an environment mode (b_i) is:

$$\widetilde{H}_i = -\hbar \frac{g_i}{2} (a b_i^\dagger e^{-i\delta_i t} + a^\dagger b_i e^{+i\delta_i t}) , \qquad (3.134)$$

where $\delta_i = \omega - \omega_i$ and the g_i are coupling constants depending smoothly on i, which we can take as real by a proper definition of the (b_i) modes. The environment modes span a wide frequency range. The relaxation of (a) is mainly due to those modes with a frequency very close to ω (as required by energy conservation). Over a short interval $\delta\tau$ around a given time t, we can 'freeze' $\widetilde{H}_i$ and write it as the beam-splitter Hamiltonian (3.101), with $\varphi_i = \delta_i t$.

The mode (a) contains initially a coherent state $|\alpha\rangle$. Assuming that the reservoir is at zero temperature, all modes b_i are initially in vacuum. The action of the coupling Hamiltonian $\sum_i \widetilde{H}_i$ during the time interval $\delta\tau$ (much shorter than the characteristic relaxation time, but much longer than the field period) is thus equivalent to the coupling of (a) to the reservoir modes (b_i) by a set of beam-splitters each having an amplitude transmission $\cos(\theta_i/2) \approx 1 - (g_i \delta\tau)^2/8$ very close to 1. Since (a) is not appreciably depleted during time $\delta\tau$, we can sum up independently the actions of these beam-splitters acting 'in parallel'. Coupling to mode (b_i) alone transforms $|\alpha\rangle|0\rangle$ into $|\alpha \cos(\theta_i/2)\rangle|i\alpha \exp(-i\varphi) \sin(\theta_i/2)\rangle$ (eqn. 3.121). Expanding the transmission and reflection coefficients in powers of $g_i \delta\tau \ll 1$ and summing up all relevant modes, we get the global state of the field at time $\delta\tau$:

$$\left| \widetilde{\psi}(\delta\tau) \right\rangle \approx \left| \alpha \left(1 - \sum_i \frac{g_i^2 \delta\tau^2}{8} \right) \right\rangle \prod_i \left| i\alpha \exp(-i\varphi_i) g_i \delta\tau/2 \right\rangle . \qquad (3.135)$$

Mode (a) still contains a coherent state whose amplitude is slightly reduced, but which remains unentangled with the reservoir [the situation is much more complex if (a) is initially in a Fock state]. The amplitude reduction corresponds to an energy transfer into the environment modes. In order to estimate the amplitude loss, we must count the number of environment modes participating in the process. During the short time interval $\delta\tau$, all modes in a frequency interval of the order of $1/\delta\tau$ around ω are appreciably coupled to (a), as can be guessed from a simple time–energy uncertainty relation argument. The sum over i in eqn. (3.135) thus contains a number of terms scaling as $1/\delta\tau$. Each of these terms being proportional to $\delta\tau^2$, the total amplitude decrease is, during this short time interval, quasi-linear in time:

$$\left| \widetilde{\psi}(\delta\tau) \right\rangle \approx \left| \alpha \left(1 - \frac{\kappa}{2} \delta\tau \right) \right\rangle \prod_i \left| i\alpha \exp(-i\varphi_i) g_i \delta\tau/2 \right\rangle , \qquad (3.136)$$

where κ is a constant depending upon the mode density in the environment and the distribution of the coupling constants g_i. Note that the argument developed here is analogous to that used to derive the Fermi Golden Rule in standard perturbation theory.

For describing the system's evolution during the next time interval $\delta\tau$, the initial state of the environment is *a priori* slightly different. Since the leaking amplitude is diluted among a large number of modes, the process in which some amplitude would leak back from the environment into (a) is however very unlikely. It is thus safe, for the computation of the (a) mode evolution, to assume that all modes (b_i) remain practically empty all the time. We can thus consider independently the amplitude reductions resulting from successive time intervals. At an arbitrary time t, the state of (a) is thus still coherent, with the amplitude:

$$\alpha(t) \approx \alpha \left(1 - \frac{\kappa \delta\tau}{2} \right)^{t/\delta\tau} \approx \alpha \, e^{-\kappa t/2} . \qquad (3.137)$$

At any time, (a) still contains a coherent state, unentangled with the mode reservoir. The coherent states, impervious to entanglement when they are coupled to a single beam-splitter, keep this remarkable property when a large set of beam-splitters couples them to a big environment. The field-plus-reservoir system remains factorized throughout its evolution. According to the definition introduced in Chapter 2, the coherent states are thus the 'pointer states' of field decoherence. The amplitude of these states is exponentially damped with the rate $\kappa/2$ and the field energy decays with the time constant $T_c = 1/\kappa$, which can be identified with the experimental cavity damping time. In a pictorial representation, the disk representing the coherent state follows a logarithmic spiral in phase space, its centre reaching the origin after an infinite time (Fig. 3.14).

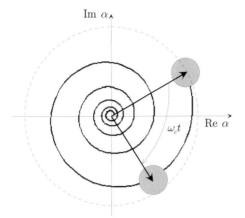

Fig. 3.14 Pictorial representation of the time evolution of a relaxing coherent state.

Note that the environment modes also contain at any time coherent states resulting from the accumulation of tiny coherent amplitudes along the successive time intervals. The global mode–environment state can be written as:

$$|\alpha e^{-\kappa t/2}\rangle \prod_i |\beta_i\rangle \, , \tag{3.138}$$

where the partial amplitudes β_i are such that:

$$\sum_i |\beta_i|^2 = \bar{n}(1 - e^{-t/T_c}) \, , \tag{3.139}$$

a relation resulting simply from the total energy conservation.

The derivation presented here is heuristic and qualitative. Its conclusions are correct though, as a much more rigorous description of coherent state damping will prove in Section 4.4.4. We will come back to the analysis of field relaxation in terms of linear coupling with a large set of oscillators in Chapter 7, when dealing with the decoherence of mesoscopic superpositions of coherent states.[22]

3.3 The spin system

We have, in the previous sections, described springs, either isolated or in mutual interaction. Although this physics is rich, these models are not sufficient to describe the variety of effects encountered in quantum optics. We must introduce another player in the game, the spin system. This section briefly recalls useful notation, only partly introduced in Chapter 2, and describes how classical fields provide tools to manipulate individual spin states using techniques borrowing amply from nuclear magnetic resonance.

3.3.1 A two-level atom

We consider a two-level atom whose upper level $|e\rangle$ is connected to level $|g\rangle$ by an electric dipole transition at angular frequency ω_{eg}. This system is equivalent to a spin-$1/2$ evolving in an abstract space, with a magnetic field oriented along the 'vertical' Z

---

[22]See also the appendix, which describes field damping using the formalism of the field characteristic function.

axis accounting for the energy difference between $|e\rangle$ and $|g\rangle$. These states correspond to the eigenstates of the spin along Z, which we denote, as in Chapter 2, by $|0\rangle$ and $|1\rangle$. The assignment of $|e\rangle$ and $|g\rangle$ with qubit states is of course arbitrary. Here we will make the correspondence $|e\rangle \rightarrow |0\rangle$ and $|g\rangle \rightarrow |1\rangle$, which, with the conventions of quantum information, make $|e\rangle$ and $|g\rangle$ eigenstates of σ_Z with eigenvalue $+1$ and -1 respectively, the atomic Hamiltonian being:

$$H_a = \frac{\hbar\omega_{eg}}{2}\sigma_Z , \tag{3.140}$$

where we have set the zero of energy half-way between the two levels. Note that the chosen qubit assignment means that the state $|1\rangle$ is less excited than the state $|0\rangle$, which can be surprising. In some cases, it might be more convenient to use the opposite qubit choice and we will in the following shift freely from one qubit definition to the other.

Let us introduce also the atomic raising and lowering operators $\sigma_{\pm}$:

$$\sigma_{\pm} = \tfrac{1}{2}(\sigma_X \pm i\sigma_Y) . \tag{3.141}$$

In terms of the spin eigenstates along the Z axis, these operators are:

$$\sigma_+ = |0\rangle\langle 1| ; \qquad \sigma_- = \sigma_+^\dagger = |1\rangle\langle 0| . \tag{3.142}$$

It follows that:

$$\sigma_+|1\rangle = |0\rangle ; \qquad \sigma_-|0\rangle = |1\rangle ; \qquad \sigma_+|0\rangle = 0 ; \qquad \sigma_-|1\rangle = 0 . \tag{3.143}$$

These atomic excitation creation/annihilation operators have a fermionic commutation relation:

$$[\sigma_-, \sigma_+]_+ = 1 , \tag{3.144}$$

where $[,]_+$ denotes an anti-commutator. This relation results from the fact that the atom carries at most one excitation. There is a clear analogy between $\sigma_{\pm}$ and the photon creation and annihilation operators defined in Section 3.1.1. The atomic Hamiltonian is in terms of these operators:

$$H_a = \hbar\omega_{eg}(\sigma_+\sigma_- - \sigma_-\sigma_+)/2 = \hbar\omega_{eg}(\sigma_+\sigma_- - \mathbb{1}/2) . \tag{3.145}$$

3.3.2 Manipulating a spin-$1/2$ with a classical field

Being able to prepare or analyse the spin in a state corresponding to an arbitrary point on the Bloch sphere is essential in the quantum optics experiments described below. This is achieved by realizing rotations on the Bloch sphere induced by classical fields, resonant or quasi-resonant with the atomic transition. According to eqns. (3.33) and (3.34), these unitary operations result from the action of the atom–field Hamiltonian:

$$H = H_a + H_r , \tag{3.146}$$

where:

$$H_r = -\mathbf{D} \cdot \mathbf{E}_r , \tag{3.147}$$

describes the coupling of the atomic dipole operator, $\mathbf{D} = q\mathbf{R}$, with the classical time-dependent electric field:

$$\mathbf{E}_r = i\mathcal{E}_r \left[\boldsymbol{\epsilon}_r e^{-i\omega_r t} e^{-i\varphi_0} e^{-i\varphi} - \boldsymbol{\epsilon}_r^* e^{+i\omega_r t} e^{i\varphi_0} e^{i\varphi} \right] . \qquad (3.148)$$

In this equation, $\mathcal{E}_r$ is the real amplitude of the classical field, ω_r its angular frequency, $\boldsymbol{\epsilon}_r$ the complex unit vector describing its polarization and $\varphi + \varphi_0$ its phase (its splitting in two parts is explained below).

Assuming that $|0\rangle$ and $|1\rangle$ are levels of opposite parities, the odd-parity $q\mathbf{R}$ dipole operator is purely non-diagonal in the Hilbert space spanned by $|0\rangle$ and $|1\rangle$ and develops along the $\sigma_\pm$ matrices according to:

$$\mathbf{D} = d(\boldsymbol{\epsilon}_a \sigma_- + \boldsymbol{\epsilon}_a^* \sigma_+) , \qquad (3.149)$$

where we have introduced the notation:

$$q \langle g | \mathbf{R} | e \rangle = d\boldsymbol{\epsilon}_a , \qquad (3.150)$$

with d being the dipole matrix element of the atomic transition (assumed to be real without loss of generality) and $\boldsymbol{\epsilon}_a$ the unit vector describing the atomic transition polarization. Calling $\mathbf{u}_x$, $\mathbf{u}_y$ and $\mathbf{u}_z$ the unit vectors along the axes Ox, Oy and Oz in real space, we have $\boldsymbol{\epsilon}_a = (\mathbf{u}_x \pm i\mathbf{u}_y)/\sqrt{2}$ for a $\sigma^\pm$-circularly polarized transition. A π-polarized transition corresponds to $\boldsymbol{\epsilon}_a = \mathbf{u}_z$.

The atom–field interaction can be described by a time-independent Hamiltonian, after a simple representation change and a secular approximation. We write $H_a = \hbar\omega_r \sigma_Z/2 + \hbar\Delta_r \sigma_Z/2$, introducing the atom–field detuning:

$$\Delta_r = \omega_{eg} - \omega_r . \qquad (3.151)$$

We then use an interaction representation with respect to the first part of the atomic Hamiltonian, $\hbar\omega_r \sigma_Z/2$. In the atom–field interaction Hamiltonian, $\sigma_\pm$ are accordingly replaced by $\sigma_\pm \exp(\pm i\omega_r t)$. Among the four terms in the expansion of the $\mathbf{D} \cdot \mathbf{E}_r$ scalar product, two are now time-independent. The two others oscillate at angular frequencies $\pm 2\omega_r$. We can perform the same approximation as the one used in Section 3.1.3 for the coupling of a classical current to a quantum cavity field and neglect the fast oscillating terms. The total Hamiltonian is then:

$$\tilde{H} = \frac{\hbar\Delta_r}{2}\sigma_Z - i\hbar\frac{\Omega_r}{2} \left[e^{-i\varphi}\sigma_+ - e^{i\varphi}\sigma_- \right] , \qquad (3.152)$$

where:

$$\Omega_r = \frac{2d}{\hbar}\mathcal{E}_r \boldsymbol{\epsilon}_a^* \cdot \boldsymbol{\epsilon}_r e^{i\varphi_0} , \qquad (3.153)$$

is the classical Rabi frequency.[23] The phase φ_0 is adjusted to make Ω_r real positive. The phase φ depends on the phase offset between the classical field and the atomic transition dipole. It can be tuned by sweeping the phase of the classical field.

---

[23]We have considered here the case of an electric dipole transition. A similar two-level Hamiltonian is obtained for a quadrupole transition between $|e\rangle$ and $|g\rangle$ (see Chapter 8), with a different definition for the Rabi frequency Ω_r.

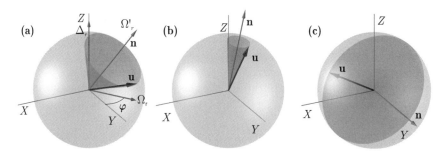

Fig. 3.15 Evolution on the Bloch sphere, viewed in the rotating frame, of the atomic pseudo-spin in an oscillating electric field. The Bloch vector **u** precesses at angular frequency Ω'_r around the effective field along **n**. (a) $\Delta_r \approx \Omega_r$ and arbitray φ. (b) Far off-resonant case. (c) Atom–field resonance ($\Delta_r = 0$) and $\varphi = 0$.

Note that similar equations describe the evolution of a spin subjected to a static magnetic field along OZ and to a time-dependent magnetic field, oscillating along a direction in the equatorial plane of the Bloch sphere. This field can be expanded as the sum of two components rotating in opposite directions in the equatorial plane. The interaction representation used above describes the physics in a frame rotating with the field component quasi-synchronous with the Larmor precession of the spin in the static field. The other component is seen in this frame as a fast rotating field, whose effect can be neglected (hence the name of Rotating Wave Approximation – RWA – given to the secular approximation in this context).

Adopting the nuclear magnetic resonance language, we can write $\widetilde{H}$ in terms of the Pauli matrices:

$$\widetilde{H} = \hbar \frac{\Omega'_r}{2} \boldsymbol{\sigma} \cdot \mathbf{n} \ , \tag{3.154}$$

with $\boldsymbol{\sigma}$ being the formal vector of components σ_X, σ_Y and σ_Z:

$$\Omega'_r = \sqrt{\Delta_r^2 + \Omega_r^2} \ , \tag{3.155}$$

and:

$$\mathbf{n} = \frac{\Delta_r \mathbf{u}_Z - \Omega_r \sin \varphi \, \mathbf{u}_X + \Omega_r \cos \varphi \, \mathbf{u}_Y}{\Omega'_r} \ . \tag{3.156}$$

Here $\mathbf{u}_X$, $\mathbf{u}_Y$ and $\mathbf{u}_Z$ are the unit vectors in the Bloch sphere space, with axes OX, OY and OZ, not to be confused with the real space unit vectors $\mathbf{u}_x$, $\mathbf{u}_y$ and $\mathbf{u}_z$. The Hamiltonian of eqn. (3.154) describes the Larmor precession at the angular frequency Ω'_r of the spin around an effective magnetic field along **n**. This precession, observed in the frame rotating at angular frequency ω_r around OZ is known in atomic and NMR physics as the 'Rabi oscillation'. We have represented it in figure 3.15 in three different cases: $\Delta_r \approx \Omega_r$ (Fig. 3.15a), $\Delta_r \gg \Omega_r$ (Fig. 3.15b) and $\Delta_r = 0$ (Fig. 3.15c). The vertical component of **n** is proportional to Δ_r. Its horizontal component, proportional to the electric field amplitude, makes an angle $\varphi + \pi/2$ with OX.

For a very large atom–field detuning, the effective magnetic field points in a direction close to one of the poles of the Bloch sphere. A spin initially prepared in $|0\rangle$, i.e. at the north pole, undergoes in the frame rotating at ω_r around OZ, a precession of very small amplitude, always remaining close to its initial position (Fig. 3.15b). The spin projection along OZ is nearly constant and there is thus no energy transfer between the field and the atom. A far off-resonant field does not affect appreciably the atomic state. As one tunes the classical field closer to resonance, the direction $\mathbf{n}$ of the effective magnetic field swings away from the polar direction. The circular trajectory of the tip of the Bloch vector widens, bringing it periodically farther and farther from its initial position. For a zero atom–field detuning (resonant case), the Bloch vector rotates at frequency Ω_r around an axis in the equatorial plane (Fig. 3.15c). We choose here for definiteness $\varphi=0$. The vector $\mathbf{n}$ is then the unit vector of the OY axis, $\mathbf{u}_Y$. A Bloch vector initially along OZ subsequently precesses along the Bloch sphere's great circle in the XOZ plane. It goes periodically from $|0\rangle$ to $|1\rangle$ and back.

If the Bloch vector is initially parallel or anti-parallel to $\mathbf{u}_Y$, it does not evolve at all. There are thus, in the frame rotating at ω_r around OZ, two stationary states of the atom field coupling, $|0/1_Y\rangle$, eigenstates of the atom–classical field Hamiltonian (see eqn. 3.154) with eigenenergies $\pm\hbar\Omega_r/2$. An arbitrary atomic state can be expressed as a superposition of the $\{|0/1_Y\rangle\}$ eigenstates, with probability amplitudes evolving at frequencies $\pm\Omega_r/2$. The classical Rabi oscillation at Ω_r can thus be interpreted as a quantum interference between these probability amplitudes.

Being merely a spin rotation, the resonant Rabi oscillation provides a simple way to prepare a state with arbitrary polar angles on the Bloch sphere. The interaction during a time $t = \theta/\Omega_r$ with a resonant field having a phase φ performs the rotation:

$$\exp(-i\theta\boldsymbol{\sigma}.\mathbf{n}/2) = \cos(\theta/2)\mathbb{1} - i\sin(\theta/2)\boldsymbol{\sigma}.\mathbf{n} = \begin{pmatrix} \cos(\theta/2) & -\sin(\theta/2)e^{-i\varphi} \\ \sin(\theta/2)e^{i\varphi} & \cos(\theta/2) \end{pmatrix},$$

(3.157)

the first identity resulting from a power series expansion of the exponential, with $(\boldsymbol{\sigma}.\mathbf{n})^2 = \mathbb{1}$. This rotation transforms $|e\rangle = |0\rangle$ into $\cos(\theta/2)|0\rangle + \sin(\theta/2)e^{i\varphi}|1\rangle$, which is the state whose Bloch vector points in the direction of polar angles θ, φ (see eqn. 2.6, on page 28). In quantum information terms, any single-qubit gate mapping $|0\rangle$ into an arbitrary state can thus be performed by a resonant Rabi oscillation with conveniently chosen θ and φ parameters.

In particular, a $\pi/2$-pulse ($\Omega_r t = \pi/2$) produces the atomic state evolution:

$$|e\rangle \longrightarrow |0_{\pi/2,\varphi}\rangle = \left[|e\rangle + e^{i\varphi}|g\rangle\right]/\sqrt{2} \; ; \qquad |g\rangle \longrightarrow \left[-e^{-i\varphi}|e\rangle + |g\rangle\right]/\sqrt{2} . \quad (3.158)$$

This transformation mixes $|e\rangle$ and $|g\rangle$ with equal weights. It can be used to prepare an atomic state superposition or to analyse it. In the latter case, the Rabi oscillation is used to map the states $|0/1_{\pi/2,\varphi}\rangle$ back onto $|e\rangle$ and $|g\rangle$, which can be directly detected. The $\pi/2$-pulse plays a very important role in the Ramsey atomic interferometer (see next paragraph). It behaves as a balanced beam-splitter for the atomic state. It transforms one input state into a coherent quantum superposition, as the optical beam-splitter in Section 3.2.2 transformed an incoming wave packet into a coherent superposition of two wave packets propagating along different paths.[24]

---

[24]Compare the rotation described by eqn. (3.157) with the unitary beam-splitter operation given by eqn. (3.100) and notice that they are identical provided one makes the notation change $\varphi \to -\varphi+\pi/2$.

A π-pulse ($\Omega_r t = \pi$) exchanges $|g\rangle$ and $|e\rangle$ according to the transformations:

$$|e\rangle \longrightarrow e^{i\varphi} |g\rangle \; ; \qquad |g\rangle \longrightarrow -e^{-i\varphi} |e\rangle \; , \tag{3.159}$$

which also play an important role in quantum information operations. The $|g\rangle \leftrightarrow |e\rangle$ state exchange can also be performed by an 'adiabatic rapid passage sequence', a procedure again borrowed from nuclear magnetic resonance. The atom, initially in $|g\rangle$ or $|e\rangle$, is subjected to an oscillating field with a strong amplitude and a detuning Δ_r initially set to a very large value. The atomic spin precesses around an effective magnetic field oriented at a very small angle from the Z axis. The detuning Δ_r is then slowly varied, going across resonance and finally reaching a very large value again. The effective magnetic field thus undergoes a slow rotation by an angle π around an axis in the horizontal plane. During this evolution, the Bloch vector continuously precesses around this field. Provided the precession frequency is, at any time, much faster than the field rotation, the spin adiabatically follows the magnetic field and remains nearly aligned with it. It thus ends up pointing along Z, in a direction opposite to the initial one. To an excellent approximation, $|e\rangle$ and $|g\rangle$ are exchanged. The whole sequence duration must be short compared to the atomic relaxation time. This explains the apparent oxymoron ('adiabatic rapid') in the name coined for this process. The adiabatic passage is largely insensitive to the precise field frequency and amplitude. It is much more robust to experimental imperfections than a standard Rabi π-pulse.

Let us finally consider the 2π-pulse ($\Omega_r t = 2\pi$), leading to:

$$|e\rangle \longrightarrow - |e\rangle \; ; \qquad |g\rangle \longrightarrow - |g\rangle \; . \tag{3.160}$$

As expected, the atomic energy is not affected by this 2π-rotation around the effective magnetic field. However, the atomic state experiences a global phase shift by π. This phase shift is a well-known property of a 2π-spin rotation. It will prove very useful in the following.

3.3.3   The Ramsey interferometer

The Ramsey (1985) interferometer is realized by combining two $\pi/2$-Rabi pulses and is closely related to the Mach–Zehnder device, discussed in Section 3.2.3. The first pulse R_1 creates a coherent superposition of atomic states. Before the second pulse R_2 probes it, a tuneable phase shift is applied to the atomic coherence. The operation of the device is sketched in Fig. 3.16(a), using a representation mixing time along the horizontal axis with atomic internal state along the vertical axis. A comparison with Fig. 3.16(b) showing a Mach–Zehnder device makes conspicuous the analogy between the two interferometers.

Let us follow the atomic state transformations for an atom initially in $|e\rangle$. The pulse R_1, with phase $\varphi = 0$, produces the quantum superposition $(|e\rangle + |g\rangle)/\sqrt{2}$. The dephasing element changes it into $(|e\rangle + e^{i\phi} |g\rangle)/\sqrt{2}$. This operation can be achieved by transiently modifying the atomic transition frequency, applying a transient electric or magnetic field. Within an irrelevant global phase factor, this operation produces a phase shift of the atomic coherence, which can be tuned by changing the transient field amplitude or duration. The second pulse R_2, again with phase $\varphi = 0$, then produces the state:

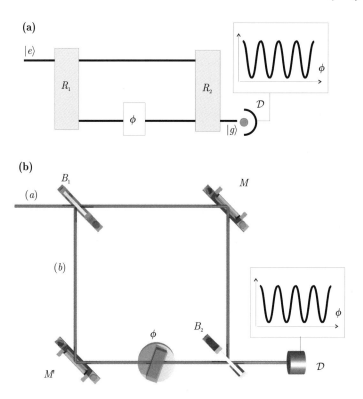

Fig. 3.16 (a) Ramsey set-up as an atomic interferometer. Two indistinguishable paths lead the atom from the initial level $|e\rangle$ to the final level $|g\rangle$. (b) Scheme of a Mach–Zehnder optical interferometer. The topological analogy with the Ramsey interferometer is conspicuous.

$$|\Psi\rangle = \tfrac{1}{2}\left[(1 - e^{i\phi})\,|e\rangle + (1 + e^{i\phi})\,|g\rangle\right] \; . \tag{3.161}$$

The probability of finding finally the atom in $|e\rangle$ or $|g\rangle$ when it is initially prepared in $|e\rangle$ is:

$$P_{e,g|e} = (1 \mp \cos\phi)/2 \; , \tag{3.162}$$

a result formally identical to the Mach–Zehnder photon counting signal. Detection in $|e\rangle$ or $|g\rangle$ is equivalent to the (a) and (b) output port signals in eqn. (3.126). The interpretation of the interference fringes is here quite clear. When $\phi = 0$, the two Ramsey pulses, performing identical transformations without any evolution of the spin in between, merely add their effects. The combination of these two $\pi/2$-Rabi pulses amounts to a π-pulse and the atom finally flips from $|e\rangle$ to $|g\rangle$. When $\phi = \pi$, the action of R_2 on the phase-shifted atomic coherence undoes the effect of R_1 and the atom finally comes back to $|e\rangle$.

These oscillations in the detection probability can also be seen as an atomic interference process. The transition from $|e\rangle$ to $|g\rangle$, for instance, may occur by stimulated emission either during R_1 or during R_2. Two indistinguishable quantum paths, illustrated in Fig. 3.16(a), lead the atom from the initial to the final state. The corresponding amplitudes must be summed to obtain the final transition probability and Ramsey fringes correspond to the interference between these amplitudes,

As in the Mach–Zehnder case, the Ramsey set-up is a very sensitive probe of any phase disturbance acting on the system while it crosses the interferometer. If an external perturbation produces an additional phase shift, Φ, of the atomic coherence between R_1 and R_2, the Ramsey fringes are shifted accordingly. The detection probabilities become $P_{e,g|e} = [1 \mp \cos(\phi + \Phi)]/2$.

Instead of tuning the phase of the atomic coherence between the two Ramsey pulses, it is possible to leave this phase ϕ constant and to scan the fringes by changing the relative phase of the two pulses R_1 and R_2.[25] Assuming that R_1 and R_2 perform the transformations (3.158) with $\varphi = 0$ and $\varphi = \phi$ respectively, we get the same final detection probabilities as above.

In practice, it is not convenient to use two separate field sources for R_1 and R_2, with a tuneable phase difference between them. It is more convenient to use the same source twice. The relative phase ϕ can then be adjusted by changing the source frequency ω_r and thus the atom–field detuning $\Delta_r = \omega_{eg} - \omega_r$. The atom undergoes two interactions with the source, from $t = 0$ to $t = t_r$ and again from $t = T_r$ to $t = T_r + t_r$. We assume that $T_r \gg t_r$, a very good approximation in most practical cases. Since the atom is free during the time interval $T_r - t_r$ between the two pulses, it is convenient to describe the action of the two pulses in an interaction representation with respect to the atomic free evolution. The atom–field Hamiltonian is then:

$$\widetilde{H} = -i\hbar \frac{\Omega_r}{2} g(t) \left[e^{i\Delta_r t}\sigma_+ - e^{-i\Delta_r t}\sigma_- \right] , \qquad (3.163)$$

where $g(t) = 1$ during the pulses and $g(t) = 0$ otherwise.

Consider, first, that Δ_r is much smaller than the spectral width $1/t_r$ of the pulses. We can then neglect the variation of $\Delta_r t$ in eqn. (3.163) during each interaction. This phase can be taken as zero for the first pulse and as $\Delta_r T_r$ for the second. Comparing with the standard Hamiltonian of eqn. (3.152), we see that the second pulse has a phase:

$$\phi_r = -\Delta_r T_r = (\omega_r - \omega_{eg})T_r . \qquad (3.164)$$

We recover the Ramsey fringe signal (eqn. 3.162), with ϕ being replaced by ϕ_r. The spacing of these fringes, when the source frequency ω_r is swept, is $1/T_r$. The spectroscopic resolution of the Ramsey interferometer is thus proportional to the total 'interrogation time', T_r, and not to the short duration, t_r, of each pulse.

These features explain the importance of the Ramsey interferometer for high-resolution spectroscopy experiments. During most of the interrogation time, the atom can be efficiently shielded from the outside world and from the action of the source, thus minimizing light shifts and power broadening. Most modern atomic clocks (Bordé 1999) are based on Ramsey interferometry, as already discussed in Chapter 1. Figure 1.3, on page 8 presents Ramsey fringes observed in a cold-atom clock. The transfer probability between the two ground state hyperfine levels is plotted as a function of the frequency offset Δ_r. The modulation due to the Ramsey atomic interference process

---

[25]Similarly, in a Mach–Zehnder device, it is also possible to scan the fringes by setting the phase of B_1 to $\varphi_1 = 0$ and tuning the phase φ_2 of B_2.

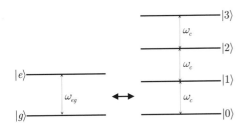

Fig. 3.17 A two-level atom coupled to a single field mode treated as an harmonic oscillator.

has an excellent contrast demonstrating the quality of the atomic coherence preservation during the whole interaction time ($T_r = 520$ ms). For a very large atom–source detuning Δ_r, of the order of $1/t_r$, the atom–field detuning is not negligible compared to the spectral width of the pulses, which no longer realize exact $\pi/2$-pulses. In an optical language, the interferometer operates with unbalanced beam-splitters and the contrast is accordingly reduced. The typical width of the fringes envelope is thus $1/t_r$ and the number of fringes observed with a high contrast is of the order of T_r/t_r.

To summarize, Ramsey fringes can be scanned in three different ways. One can either use two identical pulses with fixed phase and frequency and induce a variable phase accumulation of the atomic coherence between the pulses, or leave the atomic coherence unperturbed and tune either the relative phase or the common frequency of the pulses. We will have, in the following chapters, many opportunities to use Ramsey interferometry, under one form or another, in a context very different from metrology. We will not be interested in the absolute spectral resolution it permits, but rather in the sensitivity of the method to measure small phase shifts on an atomic coherence signal. These phase shifts, induced by quantum fields interacting with the atom, will reveal various subtle aspects of the atom–radiation interaction processes.

3.4 Coupling a spin and a spring: the Jaynes–Cummings model

We now consider the coupling of a two-level system, modelled as a spin-$1/2$, with a quantum harmonic oscillator (Fig. 3.17). This is the typical situation in cavity quantum electrodynamics, when a single atom is coupled to a cavity mode. The same model describes as well a single ion in a trap, when two internal ionic states are coupled by laser beams to the harmonic oscillator motional states in the trap (see Chapter 8). The dynamics of a neutral atom in a laser trap can also be described in this way. We discuss here the CQED case, but our conclusions also apply with minor modifications to many other situations. Historically, this model was introduced by Jaynes and Cummings (1963) as an idealization of the matter–field coupling in free space. The advent of CQED techniques turned it into a very precise description of actual experimental situations.

3.4.1 The spin–oscillator Hamiltonian and the dressed states

The complete Hamiltonian of the atom–cavity system is the quantum version of eqn. (3.146) and can be expressed as:

$$H = H_a + H_c' + H_{ac} , \qquad (3.165)$$

where H_a and $H'_c = \hbar\omega_c N$ are the atom and cavity Hamiltonians.[26] The coupling Hamiltonian, H_{ac}, is $-\mathbf{D} \cdot \mathbf{E}_c$, where $\mathbf{D}$ is the atomic dipole operator introduced above and $\mathbf{E}_c$ the cavity electric field operator at the atomic location. In order to avoid unnecessary complications at this stage, we assume that the atom is sitting at the cavity centre, with $f(\mathbf{r}) = 1$.

The atom–field coupling Hamiltonian is thus:

$$H_{ac} = -d\left[\boldsymbol{\epsilon}_a \sigma_- + \boldsymbol{\epsilon}_a^* \sigma_+\right] \cdot i\mathcal{E}_0 \left[\boldsymbol{\epsilon}_c a - \boldsymbol{\epsilon}_c^* a^\dagger\right] . \tag{3.166}$$

The expansion of the scalar product involves four terms. Two of them are proportional to $\sigma_- a$ and $\sigma_+ a^\dagger$. The first corresponds to a transition from the upper level $|e\rangle$ to the lower level $|g\rangle$, together with the annihilation of a photon. The second describes the reverse process: emission of a photon by an atom performing a transition from $|g\rangle$ to $|e\rangle$. When the cavity mode and atomic transition frequencies, ω_c and ω_{eg}, are close, these terms correspond to highly non-resonant processes. They play a minor role in the evolution, as compared to the two other terms, proportional to $\sigma_+ a$ and $\sigma_- a^\dagger$, corresponding to the usual processes of photon absorption or emission. Neglecting the two anti-resonant terms amounts to performing the rotating wave approximation, already encountered above in a classical context.

With this approximation, the atom–cavity coupling reduces to:

$$H_{ac} = -i\hbar \frac{\Omega_0}{2} \left[a\sigma_+ - a^\dagger \sigma_-\right] , \tag{3.167}$$

where we introduce the 'vacuum Rabi frequency' Ω_0:

$$\Omega_0 = 2 \frac{d\mathcal{E}_0 \boldsymbol{\epsilon}_a^* \cdot \boldsymbol{\epsilon}_c}{\hbar} . \tag{3.168}$$

We assume, for the sake of simplicity, that $\boldsymbol{\epsilon}_a^* \cdot \boldsymbol{\epsilon}_c$ is real and positive, hence Ω_0.[27] The frequency Ω_0 measures the strength of the atom–field coupling. It is proportional to the interaction energy of the atomic dipole with a classical field having the r.m.s. amplitude of the vacuum in the cavity. The analogy between the atom–field and the two-oscillator couplings is obvious (compare with eqn. 3.95). Both interactions involve two terms describing a single quantum exchange between the coupled systems. Equation (3.167) describes a special case of non-linear beam-splitter in which one of the modes (the field) can store an arbitrary number of quanta, while the other (the atom) saturates with a single excitation. This atomic non-linearity gives rise to a wealth of new effects.

Let us examine briefly the 'uncoupled' eigenstates of $H_a + H'_c$. They are the tensor products $|e, n\rangle$ and $|g, n\rangle$ of atomic and cavity energy states. Their energies are $\hbar(\omega_{eg}/2 + n\omega_c)$ and $\hbar(-\omega_{eg}/2 + n\omega_c)$. When the atom–cavity detuning:

$$\Delta_c = \omega_{eg} - \omega_c , \tag{3.169}$$

is equal to 0 or small compared to ω_c, the uncoupled states $|e, n\rangle$ and $|g, n + 1\rangle$ are degenerate or nearly degenerate. The excited states of $H_a + H'_c$ are thus organized as

---

[26]We take the vacuum as the zero energy for the field, getting rid of the $\hbar\omega_c/2$ offset.

[27]If this is not the case, we make Ω_0 real positive by multiplying it by a proper phase term $e^{i\varphi}$. The two terms of H_{ac} are then, as in eqn. (3.152), multiplied by $e^{\pm i\varphi}$.

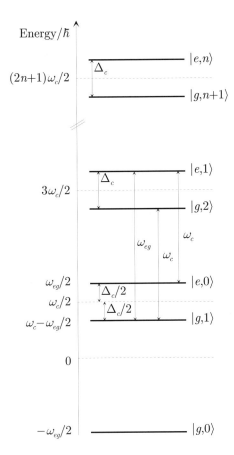

Fig. 3.18 Uncoupled atom–cavity energy states in the case $\Delta_c > 0$. Apart from the ground state $|g, 0\rangle$, they form an infinite ladder of two-level manifolds separated by the cavity frequency ω_c. In each manifold, the states $|e, n\rangle$ and $|g, n + 1\rangle$ are separated by $\hbar \Delta_c$.

a ladder of doublets, separated from each other by the energy $\hbar \omega_c$ (Fig. 3.18). The ground state $|g, 0\rangle$, representing the atom in its lower state in the cavity vacuum, is the only unpaired state at the bottom of this ladder.

The nth excited doublet stores $n + 1$ elementary excitations either as $n + 1$ field quanta ($|g, n + 1\rangle$) or as the sum of n photons added to one atomic excitation ($|e, n\rangle$). The operator $M = a^\dagger a + \sigma_+ \sigma_-$, representing the total number of atomic and field excitations, commutes with H and is a constant of motion. The excitation number being conserved, the atom–field coupling H_{ac} connects only states inside each doublet. The diagonalization of the full Hamiltonian thus amounts to solving separate two-level problems.

Eigenenergies and eigenvectors

Let us call H_n the restriction of the total Jaynes–Cummings Hamiltonian to the nth doublet. Introducing the 'n-photon Rabi frequency':

$$\Omega_n = \Omega_0 \sqrt{n + 1} \,, \tag{3.170}$$

we can write H_n in matrix form as:

$$H_n = \hbar \omega_c \left(n + 1/2 \right) \mathbb{1} + V_n \,, \tag{3.171}$$

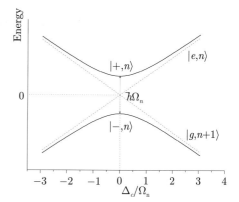

Fig. 3.19 Dressed state energies as a function of the atom–cavity detuning Δ_c. The uncoupled state energies are represented as dotted lines.

with:

$$V_n = \frac{\hbar}{2} \begin{pmatrix} \Delta_c & -i\Omega_n \\ i\Omega_n & -\Delta_c \end{pmatrix} = \frac{\hbar}{2}[\Delta_c \sigma_Z + \Omega_n \sigma_Y] \ . \tag{3.172}$$

The diagonalization of H_n is thus formally identical to the determination of the eigenstates of a spin placed in a magnetic field whose components along OZ and OY are proportional to Δ_c and Ω_n. This fictitious field makes with Z the angle θ_n defined by:

$$\tan \theta_n = \Omega_n / \Delta_c \ . \tag{3.173}$$

This 'mixing angle', varying between 0 and π, is useful to express in a simple form the eigenstates of the atom–field system, which are represented in the $\{|e, n\rangle, |g, n + 1\rangle\}$ basis by the same amplitudes as the $|0, 1\rangle_{\theta_n, \varphi}$ spin states whose Bloch vectors point along the θ_n, $\varphi = \pi/2$ direction on the Bloch sphere (eqn. 2.6). From this simple analogy, we deduce immediately that the eigenvalues of H_n are:

$$E_n^\pm = (n + 1/2) \hbar\omega_c \pm \frac{\hbar}{2}\sqrt{\Delta_c^2 + \Omega_n^2} \ , \tag{3.174}$$

with the corresponding eigenvectors:

$$\begin{aligned} |+, n\rangle &= \cos(\theta_n/2) |e, n\rangle + i \sin(\theta_n/2) |g, n + 1\rangle \ ; \\ |-, n\rangle &= \sin(\theta_n/2) |e, n\rangle - i \cos(\theta_n/2) |g, n + 1\rangle \ . \end{aligned} \tag{3.175}$$

These states, which are generally entangled, are called the 'dressed states' of the atom–field system.

The variation of the dressed energies as a function of Δ_c is represented in Fig. 3.19. For very large detunings, the dressed energies almost coincide with the uncoupled state energies $(n+1/2)\hbar\omega_c \pm \hbar\Delta_c/2$. At zero detuning, the uncoupled energy levels cross. The atom–cavity coupling transforms this crossing into an avoided crossing, the minimum distance between the dressed states being the coupling energy $\hbar\Omega_n$. We examine now in more detail two interesting limiting cases: the resonant case $(\Delta_c = 0)$ and the 'large detuning case' $(\Delta_c \gg \Omega_n)$.

3.4.2 Resonant case: Rabi oscillation induced by n photons

In the resonant case, the mixing angle is $\theta_n = \pi/2$ for all n values. The dressed states, separated by $\hbar\Omega_n$, are:

$$|\pm, n\rangle = [|e, n\rangle \pm i\,|g, n+1\rangle]/\sqrt{2} \,. \tag{3.176}$$

We have here a situation similar to two coupled degenerate oscillators, whose eigenmodes are the symmetric and anti-symmetric superpositions of the independent free modes. The eigenmode frequency difference corresponds then to the beating note in the correlated motion of the two oscillators (Fig. 3.8D). Here, the separation of the dressed states at resonance corresponds to the frequency of the reversible energy exchange between the atom and the field.

To be more specific, let us consider the simple case of an atom initially $(t = 0)$ in state $|e\rangle$, inside a cavity containing n photons. The initial state, $|\Psi_e(0)\rangle = |e, n\rangle$, expanded on the dressed states basis is:

$$|\Psi_e(0)\rangle = [|+, n\rangle + |-, n\rangle]/\sqrt{2} \,. \tag{3.177}$$

We will describe the evolution of this state in the interaction representation with respect to the constant term $\hbar\omega_c(n+1/2)\mathbb{1}$ in eqn. (3.171). At time t, $|\Psi_e(0)\rangle$ becomes:

$$\left|\widetilde{\Psi}_e(t)\right\rangle = \left[|+, n\rangle\, e^{-i\Omega_n t/2} + |-, n\rangle\, e^{i\Omega_n t/2}\right]/\sqrt{2} \,. \tag{3.178}$$

The probabilities of finding the atom in $|e\rangle$ or $|g\rangle$ are obtained by reverting to the uncoupled basis:

$$\left|\widetilde{\Psi}_e(t)\right\rangle = \cos\frac{\Omega_n t}{2}\,|e, n\rangle + \sin\frac{\Omega_n t}{2}\,|g, n+1\rangle \,. \tag{3.179}$$

Similarly, if the atom is initially in state $|g\rangle$ in a cavity containing $n + 1$ photons,[28] the combined atom–field state at time t is:

$$\left|\widetilde{\Psi}_g(t)\right\rangle = -\sin\frac{\Omega_n t}{2}\,|e, n\rangle + \cos\frac{\Omega_n t}{2}\,|g, n+1\rangle \,. \tag{3.180}$$

The atom–field coupling results in a reversible energy exchange between $|e, n\rangle$ and $|g, n+1\rangle$ at frequency Ω_n. This is the quantum version of the Rabi oscillation phenomenon which can be understood as a 'quantum beat' between the two dressed states (see eqn. 3.178).

For the initial state $|e, 0\rangle$ (or $|g, 1\rangle$), this oscillation occurs at the frequency Ω_0, hence the name of vacuum Rabi frequency coined for this parameter. The oscillation between $|e, 0\rangle$ and $|g, 1\rangle$ can also be viewed as an 'oscillatory spontaneous emission'. An atom initially in $|e\rangle$ emits a photon while undergoing a transition to $|g\rangle$. In free space, the photon escapes and the atom remains in state $|g\rangle$. This is the usual irreversible spontaneous emission process. In the cavity, on the contrary, the photon emitted by the atom remains trapped. It can be absorbed, and then emitted again and so on. In

---

[28] Note that an atom in $|g\rangle$ does not evolve in the cavity vacuum since the state $|g, 0\rangle$ is an eigenstate of the full atom–field Hamiltonian.

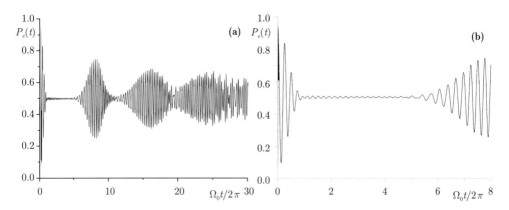

Fig. 3.20 Quantum collapse and revivals: computed probability $P_e(t)$ of finding the atom in state $|e\rangle$ versus the interaction time t in units of $2\pi/\Omega_0$. The cavity contains initially an $\bar{n} = 15$ photon coherent field. (a) Behaviour on a long time scale. (b) Zoom on the collapse and the onset of the first revival.

more mathematical terms, the coupling to a continuum of modes (free-space case) is replaced here by the coupling to a single quantum oscillator. The irreversible damping process is thus replaced by a coherent oscillation. The observation of this coherent atom–field coupling requires atomic and cavity field lifetimes much longer than the Rabi period ('strong coupling regime' of cavity QED).

As the coupling of two field modes by a beam-splitter (Section 3.2.2), the Rabi oscillation produces in general an entanglement (eqns. 3.179 and 3.180). We defer the discussion of this physics to Section 5.4, in which we present experiments demonstrating the generation via Rabi oscillation of atom–cavity and atom–atom entanglement.

3.4.3   Collapse and revival of Rabi oscillations induced by a coherent field

Suppose now that the atom, initially in $|e\rangle$, interacts with a coherent field $|\alpha\rangle = \sum_n c_n |n\rangle$, stored in the cavity. The average photon number is $\bar{n} = |\alpha|^2$. The c_n coefficients and the photon number distribution $p_\alpha(n)$ are given by eqns. (3.57) and (3.58). Using eqn. (3.179), we get the atom–field state at time t:

$$|\Psi_e(t)\rangle = \sum_n c_n \cos\frac{\Omega_0\sqrt{n+1}t}{2} \, |e, n\rangle + c_n \sin\frac{\Omega_0\sqrt{n+1}t}{2} \, |g, n+1\rangle \;, \qquad (3.181)$$

and the probability $P_e(t)$ of finding the atom in $|e\rangle$:

$$P_e(t) = \sum_n p_\alpha(n)\frac{1 + \cos\Omega_0\sqrt{n+1}t}{2} \;. \qquad (3.182)$$

The Rabi oscillation in a coherent field is thus a sum of terms oscillating at the frequencies $\Omega_0\sqrt{n+1}$, weighted by the probability of finding the corresponding photon number n in the initial coherent state.

Figure 3.20 shows $P_e(t)$ for $\bar{n} = 15$. This signal presents remarkable features, which attracted a lot of interest in the early days of quantum optics and cavity QED

(Faist *et al.* 1972; Eberly *et al.* 1980). At short times, the Rabi oscillation occurs at a frequency close to $\Omega_0\sqrt{\bar{n}+1}$, proportional to the classical field amplitude when $\bar{n} \gg 1$. These oscillations are, however, quite rapidly damped away and P_e reaches a stationary value of $1/2$. This 'collapse' of the Rabi oscillation is due to the dispersion of Rabi frequencies reflecting the uncertainty of the photon number in the coherent field. We can qualitatively estimate the collapse time t_c by expressing that the phase variation of the oscillation over the width $\sqrt{\bar{n}}$ of the Poisson law is of the order of 2π. A simple calculation leads to:

$$t_c \approx \pi/\Omega_0 \ . \tag{3.183}$$

The collapse time is thus of the order of the vacuum Rabi period. About $\sqrt{\bar{n}}$ Rabi oscillations are observed before the collapse, as can be checked on Fig. 3.20(b).

A similar collapse of the Rabi oscillation would also occur for an atom interacting with a noisy classical field, presenting a continuous distribution of amplitudes. There is thus a possible classical interpretation for the collapse, but none for the other conspicuous phenomenon exhibited by Fig. 3.20. After an idling period, the Rabi oscillations 'revive'. The contributions of the different photon numbers come back into phase and lead again to large-amplitude oscillations in P_e. Let us give an order of magnitude estimate of the revival time, t_r. Two oscillations corresponding to consecutive photon numbers, n and $n+1$, come back in phase, for $n \approx \bar{n}$, at a time t_r such that $t_r\Omega_0/2\sqrt{\bar{n}} \approx 2\pi$, or:

$$t_r \approx \frac{4\pi}{\Omega_0}\sqrt{\bar{n}} \ . \tag{3.184}$$

This time being independent of n, all the oscillatory components in the Rabi signal then rephase, producing large-amplitude oscillations in $P_e(t)$. The revival time is thus $\sqrt{\bar{n}}$ times greater than the collapse time, $\bar{n}$ times greater than the Rabi period.

After an additional time interval of the order of t_c, the Fock state contributions get out of phase again and the revival ends. A second revival occurs later on, after another delay of the order of t_r. The Rabi oscillation signal thus exhibits a series of revivals, separated by idling periods. The duration of the revivals increases progressively (Fig. 3.20a). They end up overlapping and, at long times, the signal becomes an oscillation with an average frequency $\Omega_0\sqrt{\bar{n}}$ and an apparently chaotic amplitude.[29]

The quantum revival phenomenon is insensitive to the phase relations between the Fock state contributions and hence does not depend upon the field coherence. It does not even require a Poissonian photon number distribution. Similar phenomena are observed for an atom coupled to a thermal field (Rempe *et al.* 1987). To observe them, it is only necessary that a finite number of irrational frequency components contribute to the signal.

The quantum revivals are thus directly linked to the quantization of the spectrum of the Rabi oscillation and, hence, to the field energy quantization itself. This explains the theoretical interest in this phenomenon (Knight and Radmore 1982; Gea-Banacloche 1991; Fleischhauer and Schleich 1993; Buzek and Knight 1995). We describe in Chapter 5 an observation of the quantum collapse and revival effect in cavity QED. Related effects in ion trap physics will be discussed in Chapter 8. In analysing

---

[29]It should be emphasized that the Rabi signal at long times is not chaotic. It is only extremely complex, due to the superposition of very many irrational frequencies.

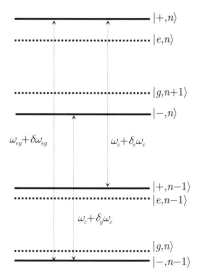

Fig. 3.21 Dressed state energies in the far-detuned case for two adjacent manifolds. The effect of the non-resonant interaction is a slight shift of the atom–cavity energy levels. This results in atomic transition and cavity mode frequency shifts, illustrated by the vertical arrows.

here this phenomenon, we have focused our attention on the atom alone. In fact, the atom and field systems get entangled during the collapse phase, leading to the appearance of mesoscopic field state superpositions. We come back to this remarkable feature in Chapter 7.

For very large $\bar{n}$ values, we must, according to the correspondence principle, recover the evolution of a spin in a classical resonant field (Section 3.3.2). This limit corresponds to $\bar{n}$ going to infinity while $\Omega_0 \sqrt{\bar{n}}$ becomes the classical Rabi frequency Ω_r. The vacuum Rabi frequency, Ω_0, must accordingly go to zero as $1/\sqrt{\bar{n}}$.[30] The collapse time, π/Ω_0, is thus rejected to infinity. The atom evolves between $|e\rangle$ and $|g\rangle$ with a steady oscillation amplitude. This is, as expected, the classical result

3.4.4 Non-resonant spin–spring coupling and single-atom index effects

Remarkable phenomena, which can be interpreted as single-atom refractive index effects, are also expected when the atom and the field are not at resonance. Exact expressions of the atom–field eigenstates and energies are given by eqns. (3.174) and (3.175) for an arbitrary detuning Δ_c. Rather than discussing these general expressions, we focus here on the large detuning case, for which a perturbative treatment leads to simple formulas with a clear physical interpretation. A mere inspection of Fig. 3.19 shows that for $|\Delta_c| \gg \Omega_n$ the dressed levels tend towards the uncoupled ones. When $\Delta_c \to +\infty$, we deduce from eqn. (3.173) that $\theta_n \to 0$, entailing that $|+,n\rangle \to |e,n\rangle$ and $|-,n\rangle \to -i|g,n+1\rangle$ (eqn. 3.175). These conclusions are reversed for large negative detunings: when $\theta_n \to \pi$, $|+,n\rangle \to i|g,n+1\rangle$ and $|-,n\rangle \to |e,n\rangle$.

Since the dressed states almost coincide with the uncoupled ones, the probability of an atomic transition accompanied by photon absorption or emission is negligible, as expected from an energy conservation argument in this off-resonant situation. The energies of the dressed states are however slightly different from the uncoupled ones,

---

[30]This can be achieved, for instance, by increasing the size of the cavity, thus 'diluting' the vacuum field $\mathcal{E}_0$, while feeding more and more quanta in the cavity to keep the Rabi frequency constant.

as seen on Fig. 3.19. These energies can be expressed by expanding eqn. (3.174) to lowest order in powers of the small parameter Ω_n/Δ_c (for the sake of definiteness, we assume from now on $\Delta_c > 0$):

$$E_n^{\pm} = (n + 1/2)\,\hbar\omega_c \pm \hbar \left(\frac{\Delta_c}{2} + \frac{\Omega_n^2}{4\Delta_c}\right) . \tag{3.185}$$

Figure 3.21 shows the effect of this energy shift on two adjacent doublets of the atom–cavity system. In each doublet the level spacing is slightly increased, the effect being larger in the upper one. More quantitatively, the shifts $\Delta_{e,n}$ and $\Delta_{g,n}$ of the states $|e, n\rangle$ and $|g, n\rangle$ are:

$$\Delta_{e,n} = \hbar(n + 1)s_0 \; ; \qquad \Delta_{g,n} = -\hbar n s_0 \; , \tag{3.186}$$

with:

$$s_0 = \frac{\Omega_0^2}{4\Delta_c} . \tag{3.187}$$

These shifts, which affect globally the atom–cavity states, can be given complementary physical interpretations, depending upon which of the two interacting subsystems we are observing. They correspond to reciprocal changes of the atomic transition and cavity mode frequencies. Let us consider these two frequencies in turn.

Quantized light shift and Lamb shift in a cavity

The atomic transition frequency measured in a cavity containing n photons is equal to $1/\hbar$ times the energy splitting between levels $|g, n\rangle$ and $|e, n\rangle$ (Fig. 3.21). It is thus shifted from its free-space value by:

$$\delta\omega_{eg} = (2n + 1)s_0 . \tag{3.188}$$

This shift is the sum of two contributions. The first one, $2ns_0$, is proportional to the photon number n. It corresponds to the light shift experienced by the atom in the detuned quantum cavity field. It is of the same nature as the usual light shifts induced on atoms by macroscopic laser beams. As we will see in Chapter 6, the quantization of this shift is observable at the limit of very small photon numbers, provided Δ_c is not too large. The second contribution in eqn. (3.188), equivalent to the light shift produced by 'half a photon', is observed in the cavity vacuum. It is a Lamb shift effect, due to the coupling of the excited atomic level to the vacuum fluctuations in the cavity mode. The strong atom–cavity coupling makes this effect non-negligible when Δ_c is of the order of a few Ω_0. We describe in Chapter 5 an experiment in which this shift has been observed. Note that level $|g\rangle$ is not affected by this vacuum shift.

Single-atom index shift

The same shift, viewed from the field perspective, corresponds to a change of the cavity mode frequency. When an atom in level $|e\rangle$ interacts non-resonantly with the cavity, the $|e, n\rangle$ levels corresponding to different n values are equidistant, with a spacing $\hbar(\omega_c + s_0)$ (Fig. 3.21). This means that the cavity frequency is offset from ω_c by:

$$\delta_e\omega_c = s_0 . \tag{3.189}$$

For an atom in level $|g\rangle$, the $|g, n\rangle$ to $|g, n+1\rangle$ spacings are $\hbar(\omega_c - s_0)$, independent of n. The mode frequency is thus displaced by an opposite amount:

$$\delta_g \omega_c = -\delta_e \omega_c = -s_0 \ . \tag{3.190}$$

These cavity shifts have a transparent interpretation in terms of refractive index. The atom, which cannot absorb or emit light in the detuned mode, behaves as a piece of transparent dielectric. It plays the same role as the dielectric plungers used to tune cavities in microwave equipment. The atomic index changes the effective cavity length and hence the mode frequency. This effect is dispersive (frequency dependent) and s_0 has the same sign as Δ_c. A single atom can thus control the frequency of a field containing many photons. We will see in Chapter 7 that this leads to spectacular effects when this single-atom index shift becomes larger than the cavity mode spectral width.

Mechanical effects

It is instructive to analyse how energy is conserved in these non-resonant interaction processes. An atom in state $|e\rangle$ entering a cavity tuned below the atomic transition ($\Delta_c > 0$) increases the field frequency. If the cavity contains n photons, the field energy is augmented by $n\hbar s_0$. In addition, the internal atomic energy is increased by $\hbar s_0$, as a result of the cavity-induced Lamb shift. Where does the total energy increase $(n + 1)\hbar s_0$ come from? The atom behaves in this case as a small speck of dielectric material whose index is greater than 1. As a piece of glass in a light beam, it experiences when entering the cavity a repulsive dipole force, proportional to the gradients of the n-photon field and of the vacuum fluctuations in the cavity. This force tends to expel it from the centre of the cavity. To overcome it, the atom loses some of its kinetic energy, precisely balancing the gain in energy of the cavity field and its own gain of 'Lamb shift energy'. The residual force remaining when $n = 0$ is a kind of cavity-induced Casimir effect (Haroche *et al.* 1991; Englert *et al.* 1991).

If the atom is in level $|g\rangle$, there is no vacuum contribution and the sign of the energy exchange with the n-photon field is reversed. The cavity field frequency is shifted down (for $\Delta_c > 0$) and the atom, which is attracted by the light force towards the cavity centre gains some kinetic energy. In this case, the mechanical effect requires at least the presence of one photon in the cavity. For negative detunings, the frequency shifts as well as the forces change sign, but the same energy conservation arguments apply. The manifestations of these atom–cavity forces will be briefly described in Chapter 5.

Measuring one subsystem with the other

The reciprocal shifts experienced by the atom and the field in this dispersive regime can be exploited to perform quantum measurements on one of the two subsystems, using the other as a meter, according to the general procedure outlined in Chapter 2. To understand this point, it is convenient to describe the dispersive atom–cavity shift effects by the action of an effective Hamiltonian of the combined system. The states $|e, n\rangle$ and $|g, n\rangle$ and their shifted energies, $E^s_{e,n} = \hbar\omega_{eg}/2 + n\hbar\omega_c + (n + 1)\hbar s_0$ and $E^s_{g,n} = -\hbar\omega_{eg}/2 + n\hbar\omega_c - n\hbar s_0$, are the eigenstates and eigenvalues of an effective Hamiltonian H_{eff}:

$$H_{\text{eff}} = \hbar\omega_c a^\dagger a + \frac{\hbar}{2}\sigma_Z \omega_{eg} + \hbar s_0 \sigma_Z a^\dagger a + \hbar s_0 \sigma_+ \sigma_- \ . \tag{3.191}$$

In this formula, one can either group the two terms containing σ_Z as $V_1 = (\hbar/2)\sigma_Z(\omega_{eg} + 2s_0 a^\dagger a)$ or the two terms proportional to $a^\dagger a$ as $V_2 = \hbar(\omega_c + s_0 \sigma_Z)a^\dagger a$. The first expression V_1 appears as a pseudo-spin Hamiltonian whose frequency is shifted by an amount $2s_0 a^\dagger a$ proportional to the field photon number operator. In this form, the atomic spin whose precession is affected by the number of photons, appears as a meter for the measurement of $a^\dagger a$. We will describe in detail experiments illustrating this idea in Chapter 6. The second expression V_2 appears as a field Hamiltonian whose frequency is shifted by a quantity $s_0 \sigma_Z$ proportional to the atomic energy operator. In this form, it is the field whose frequency depends upon the orientation of the spin which is a meter fit for the measurement of the atomic energy. We analyse in detail this situation in Chapter 7.

The linear refractive index approach is valid only when the condition $\Delta_c \gg \Omega_n$ is fulfilled. This means that the atomic index is linear as long as the photon number is not too large:

$$n \ll (\Delta_c/\Omega_0)^2 \ . \tag{3.192}$$

When this condition is not fulfilled, the spacing between levels $|+, n\rangle$ and $|+, n+1\rangle$ becomes n-dependent. It is no longer possible to define an intensity-independent cavity mode frequency. Describing the atom in terms of a refractive index requires the introduction of non-linear and imaginary terms, making this picture of little use, since exact expressions are available. We say more in Chapter 7 about the saturation of the single-atom index effect.

The simple spin–oscillator model analysed here can be adapted, as we will see later, to describe related situations in which a field mode is resonant or nearly resonant with an atomic transition between $|e\rangle$ and $|g\rangle$, with a third level $|i\rangle$ separated from $|e\rangle$ and $|g\rangle$ by transitions whose frequencies are widely different from ω_c. This third level then remains unperturbed and can be used as a reference to observe the effects of the atom–field coupling on $|e\rangle$ and $|g\rangle$.

The dance of the spin and the spring obeys the simple Jaynes and Cummings choreography, and yet, in spite of its simplicity, it involves, as we have seen, subtle steps illustrating various facets of the quantum concepts. Observing this dance will occupy us in the following chapters. But before turning to the description of experiments, we must realize that we are not the only spectators. As we look at the atom and the field in the cavity, the outside world, the environment, is also watching them. The presence of this unavoidable spy alters the show in ways that we analyse in the next chapter.

4

The environment is watching

On each landing, opposite the lift-shaft, the poster with the enormous face gazed from the wall. It was one of those pictures which are so contrived that the eyes follow you about when you move. BIG BROTHER IS WATCHING YOU, the caption beneath it ran.

G. Orwell, 1984.

Whatever care is exercised to protect it, a quantum system A interacts, if only slightly, with its environment E. Information about the system's state is always leaking into the correlations building up between A and E. In other words, the environment, as a quantum version of Orwell's Big Brother, is continuously 'watching' the system and collecting information about it. The mere possibility that this information could be retrieved, really or virtually, is enough to destroy the coherences of the system's wave function in the same way as looking at the path of a particle in a Young double-slit set-up washes out the interference fringes (see Section 2.2). The concept of a wave function then becomes inadequate and A must in general be described not by a state vector, but by a density matrix. Only the special 'pointer states' of A – such as coherent states of the field in a cavity – remain unentangled pure states throughout the system's evolution.

We intend here to go deeper into this analysis and present the general relaxation theory which describes, under very general assumptions valid in quantum optics, the dynamical evolution of an open system A coupled to a large environment E.[1] The system A could be, for instance, the field in a cavity QED experiment (Chapters 5, 6 and 7) or an ion in a trap (Chapter 8). In the first case, E is made of the modes of the field to which the cavity is coupled via transmission through mirrors or scattering on cavity walls imperfections. The environment of a trapped ion comprises the modes of the field into which it can spontaneously emit photons. In both cases, E may also include other parts, more difficult to describe in detail (absorbing atoms in the cavity walls in the first case, charges in the electrodes perturbing the motion of the ion in the second).

---

[1]In NMR or atomic physics, the environment is traditionally called a 'reservoir' or 'bath' and the system's evolution is referred to as a 'relaxation' process. In the context of quantum information – especially when dealing with the evolution of a mesoscopic system at the border between the quantum and the classical worlds – the name 'environment' is preferred and relaxation is called 'decoherence'.

The complexity of E makes it impossible to follow the detailed evolution of the $A + E$ system. The same limitation is encountered in classical statistical mechanics, when a system evolves in an irreversible way, coupled to a thermodynamical reservoir. Nevertheless, as we will see, the dynamics of A alone can be described by a differential master equation for its reduced density operator, ρ_A. Analysing the dynamics of an open quantum system ruled by this equation will shed light on the deep links – already evoked in Chapter 2 – between entanglement, complementarity and decoherence.

The first section (4.1) of this chapter is devoted to a review of the main properties of the density matrix. We introduce the useful concept of 'density matrix purification', which provides an insight into the nature of statistical mixtures of states and we give a definition of a generalized measurement procedure, which we relate to the projective measurements of elementary quantum theory. We also apply these general considerations to a two-state system, whose density operator is entirely defined by a three-dimensional real polarization vector. Pure states correspond to polarizations whose tips are on the Bloch sphere introduced in the previous chapter, while statistical mixtures are associated to points inside this sphere.

The usual approach to relaxation, discussed in a variety of textbooks, introduces a simple model for the environment (a bath of harmonic oscillators spanning a wide resonance frequency interval, for instance) and assumes a small set of reasonable approximations to derive the differential master equation for ρ_A. We use here a different approach, more in the spirit of quantum information physics. We first describe, in Section 4.2, a general representation of a 'quantum process', mapping a density matrix onto another one. This quantum map describes quite generally a Hamiltonian evolution, an unread measurement or the effect of a relaxation mechanism. The properties of these quantum maps can be deduced from very simple arguments describing the evolution of A as resulting from its coupling with a virtual 'environment simulator' B, smaller and much simpler to describe than the real environment E. The system's dynamics can then be viewed as a kind of generalized EPR experiment in which correlated measurements are virtually performed on the A and B parts. These results are quite general and do not depend on the precise nature of E.

With a few simple and reasonable hypotheses on E, we then cast (Section 4.3) the master equation under the generic Lindblad form. It is expressed in terms of a small number of operators acting on A alone, which describe the various kinds of 'quantum jumps' experienced by the system. In the case of a cavity field, for instance, the quantum jumps correspond to the loss or gain of a photon exchanged with the environment. Simple considerations about the nature of these jumps make it possible to infer all the terms of the master equation. The only free parameters left are damping rates, which cannot be guessed from this approach, but which turn out to be easily measurable quantities.

We introduce then in Section 4.4 the Monte Carlo description of quantum relaxation. We assume that A is at any time in a pure state. Under the influence of E, this wave function undergoes random jumps, whose statistics can be computed. Between jumps, the wave function is also modified, illustrating that information can be acquired on a system's state even in the case of a negative measurement. We analyse in detail this somewhat counter-intuitive result and relate it to examples of everyday life, in which changes of our knowledge of a classical system do occur under the effect of negative measurements. The density matrix ρ_A is recovered by averaging many quantum trajectories of the wave function, corresponding to different random choices of the quantum jumps. This approach is very efficient for numerical simulations. It also provides an insightful description of quantum jumps observed in real experiments in which single quantum systems are continuously monitored.

As simple examples, we use the Monte Carlo approach to describe the spontaneous emission of a two-level atom in a superposition of its excited and ground state and to cavity field relaxation at zero and finite temperature T. We unveil interesting differences between the relaxation of Fock and coherent states at $T = 0$ K. The non-entanglement of coherent states with their environment as they relax towards vacuum (see Chapter 3) finds a simple explanation in this Monte Carlo description. More generally, we precise the conditions that a Lindblad equation with a single jump operator must fulfil for the eigenstates of this operator to be 'pointer states' impervious to entanglement with the environment.

We then illustrate these general considerations by applying the relaxation formalism to our favourite system, a spring and a spin in interaction. We consider in Section 4.5 a cavity mode resonantly coupled to a single two-level atom and describe how the Rabi oscillation phenomenon is altered by atomic and cavity relaxation. We finally turn to the description of a damped field mode interacting with a collection of negligibly relaxing spins. In Section 4.6, the two-level atoms are coupled sequentially to the spring, providing a simple model of a 'micromaser', a device in which atoms passing one by one through a cavity sustain a steady-state field presenting remarkable non-classical features. In the last section (4.7), the spins are simultaneously coupled to the spring, realizing a simple model for co-operative emission phenomena (collective Rabi oscillations and superradiance). Choosing alternatively the master equation and the Monte Carlo approach, we describe the dynamical features of these collective spin–spring systems as they appear either on a single realization, or on averages performed on many independent 'runs' of the experiment.

4.1 Quantum description of open systems

4.1.1 Main properties of the density operator

We have already introduced in Chapter 2 the density operator as an essential tool for the description of an open system. If A interacts with an environment E, its density operator ρ_A is the partial trace of the global $A + E$ density operator over E. It contains all information needed to predict the statistics of measurements performed on A, without having to describe E explicitly. The probability of finding the eigenvalue ϵ_j of an observable O_A of A is $\pi_j = \text{Tr}(\rho_A P_j^A)$ where P_j^A is the projector on the corresponding eigenspace and the expectation value of O_A is:

$$\langle O_A \rangle = \sum_j \pi_j \epsilon_j = \text{Tr}_A(\rho_A O_A) \, . \tag{4.1}$$

The probabilistic interpretation implies that ρ_A must be a positive hermitian operator of trace one, with non-negative eigenvalues λ_i. The associated eigenvectors, $\left| \lambda_i^{(A)} \right\rangle$, form an orthonormal basis in the Hilbert space $\mathcal{H}_A$ of A (here assumed to have a finite dimension N_A). In this basis, ρ_A is in diagonal form:

$$\rho_A = \sum_i \lambda_i \left| \lambda_i^{(A)} \right\rangle \left\langle \lambda_i^{(A)} \right| , \tag{4.2}$$

with the normalization condition:

$$\sum_i \lambda_i = 1 . \tag{4.3}$$

When only one of the eigenvalues, λ_k, is non-zero, the system is in the pure quantum state $\left| \lambda_k^{(A)} \right\rangle$. When two or more λ_i are non-zero, they can be viewed as probabilities of exclusive events summing up to 1, and ρ_A is interpreted as a statistical mixture of quantum states. Since quantum mechanics only computes probabilities and averages over statistical ensembles, the predictions made with ρ_A are recovered by imagining repeated experiments in which A is prepared in one of the states $\left| \lambda_i^{(A)} \right\rangle$ with probability λ_i, the random choice of the state resulting from a classical dice tossing. Discarding the environment by the partial trace operation thus introduces a classical uncertainty on the quantum state of A. Note that the density matrix approach can also be used to describe imperfect system preparation, leading to a fuzzy knowledge of the system's initial state, even when no environment is involved.

Equation (4.2) shows that ρ_A is generally not a projector, unless all probabilities λ_i vanish but one. Hence, ρ_A^2 is generally different from ρ_A and:

$$\mathrm{Tr}_A(\rho_A^2) = \sum_i \lambda_i^2 \leq 1 , \tag{4.4}$$

the equality being realized only for pure states. The quantity $1 - \mathrm{Tr}_A(\rho_A^2)$, called 'linear entropy', is often used as a definition of the 'distance' between ρ_A and a pure state. The difference between $\mathrm{Tr}\rho_A^2$ and $\mathrm{Tr}\rho_A$ is also a measure of the degree of entanglement between A and E (the greater the entanglement, the more mixed the reduced density matrix).

The amount of 'mixedness' of a state can also be evaluated with the von Neumann entropy, introduced in Chapter 2:

$$\mathcal{S}_e = -\mathrm{Tr}_A(\rho_A \log_2 \rho_A) = -\sum_i \lambda_i \log_2(\lambda_i) , \tag{4.5}$$

where the logarithm is expressed in base 2 in the quantum information context. The entropy is zero for a pure state (a single eigenvalue is one, all others are zero). It is maximal, equal to $\log_2 N_A$, when ρ_A is proportional to the unit matrix: $\rho_A = \mathbb{1}_A/N_A$. This density matrix describes then a statistical mixture of N_A orthogonal states occurring with equal probabilities. This is the situation in which we have minimum information about the system's state. The 'disorder' is maximal.

Let us conclude this brief reminder by recalling that the unitary evolution of ρ_A under the action of the Hamiltonian H_A of A is described by a differential equation directly deduced from the Schrödinger equation for the wave function. The linearity of quantum mechanics indicates that ρ_A, initially a weighted sum of projectors, evolves, under H_A, as the projector on a pure state $|\phi\rangle \langle\phi|$:

$$\frac{d\,|\phi\rangle\,\langle\phi|}{dt} = \frac{d\,|\phi\rangle}{dt}\,\langle\phi| + |\phi\rangle\,\frac{d\,\langle\phi|}{dt} \; , \tag{4.6}$$

with:

$$i\hbar\frac{d\,|\phi\rangle}{dt} = H_A\,|\phi\rangle \; . \tag{4.7}$$

The Schrödinger equation for ρ_A is thus:[2]

$$i\hbar\frac{d\rho_A}{dt} = [H, \rho_A] \; . \tag{4.8}$$

The evolution from the initial time $t = 0$ to the final time t is obtained by introducing the evolution operator $U(t)$:

$$\rho_A(t) = U(t)\rho_A(0)U^\dagger(t) \; , \tag{4.9}$$

this relation resulting from the state evolution: $|\phi(t)\rangle = U(t)\,|\phi(0)\rangle$.

4.1.2 Density matrix of a qubit: Bloch sphere representation

As a simple example of the density matrix representation, let us consider the case of a two-level system, a qubit with states $|0\rangle$ and $|1\rangle$. The density matrix is then a 2×2 hermitian matrix:

$$\rho_A = \begin{pmatrix} \rho_{00} & \rho_{01} \\ \rho_{10} & \rho_{11} \end{pmatrix} \; . \tag{4.10}$$

Its real positive diagonal terms ρ_{00} and ρ_{11}, called the 'populations', are the probabilities of finding the qubit in $|0\rangle$ or $|1\rangle$. They sum up to unity: $\rho_{00} + \rho_{11} = 1$. The non-diagonal terms $\rho_{01} = \rho_{10}^*$, called the 'coherences', are zero for a statistical mixture of $|0\rangle$ and $|1\rangle$. Since ρ_A is a positive operator, they satisfy the inequality:

$$|\rho_{10}| = |\rho_{01}| \leq \sqrt{\rho_{00}\rho_{11}} \; , \tag{4.11}$$

the upper bound being reached for pure states.

The density matrix ρ_A can be expanded with real coefficients onto the operator basis made of the Pauli operators plus $\mathbb{1}$. The coefficient of $\mathbb{1}$ is $^1/_2$, since the Pauli components are traceless. We thus write:

$$\rho_A = \tfrac{1}{2}(\mathbb{1} + \mathbf{P} \cdot \boldsymbol{\sigma}) \; , \tag{4.12}$$

where $\mathbf{P}$ is a three-dimensional vector, $\mathbf{P} = (u, v, w)$, and $\boldsymbol{\sigma}$ is the formal vector made up of σ_X, σ_Y and σ_Z. The components of $\mathbf{P}$ are linked to populations and coherences by:

$$u = \rho_{01} + \rho_{10} \; ; \qquad v = i(\rho_{01} - \rho_{10}) \; ; \qquad w = \rho_{00} - \rho_{11} \; . \tag{4.13}$$

The vector $\mathbf{P}$ is called the 'polarization' of the qubit. Its modulus P is bounded by one: $P \leq 1$, the equality being realized for pure states only. This result follows from $\mathrm{Tr}(\rho_A^2) \leq 1$, after a few lines of operator algebra. The von Neumann entropy of ρ is:

---

[2]This equation, in the Schrödinger point of view, is different from the one describing the motion of observables in the Heisenberg picture (the order of the terms in the commutator is reversed).

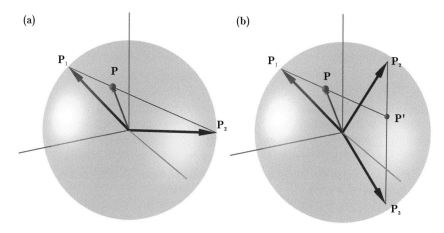

Fig. 4.1 Representation of an arbitrary mixed state of a qubit inside the Bloch sphere as a weighted sum of two (a) and three (b) pure states.

$$S = -\frac{1+P}{2}\log_2\left(\frac{1+P}{2}\right) - \frac{1-P}{2}\log_2\left(\frac{1-P}{2}\right) . \qquad (4.14)$$

To each ρ_A we can associate a point, the tip of the corresponding $\mathbf{P}$ vector, which is located in or on the Bloch sphere introduced in Chapter 2. The surface of the sphere ($P = 1$) is the locus of the pure states (with $S = 0$), whose expressions are given as a function of the polar angles θ and φ of $\mathbf{P}$ by the first line in eqn. (2.6). The statistical mixtures correspond to the inside of the sphere ($P < 1$). The closer the point to the centre, the larger the von Neumann entropy. The centre of the sphere ($\mathbf{P} = 0$) corresponds to the totally unpolarized maximum entropy state whose density operator is $\mathbb{1}/2$.

Any qubit mixed state can be represented in an infinite number of ways as a statistical mixture of two pure states, since a vector $\mathbf{P}$ inside the sphere can be expressed as a linear combination of two vectors, $\mathbf{P}_1$ and $\mathbf{P}_2$, whose tips are at the intersection of the Bloch sphere surface with an arbitrary line passing by the extremity of $\mathbf{P}$: $\mathbf{P} = \lambda\mathbf{P}_1 + (1 - \lambda)\mathbf{P}_2$, λ being a real coefficient between 0 and 1 (Fig. 4.1a). The density matrix, being a linear function of the Bloch vector $\mathbf{P}$, is then a weighted sum of the projectors on the pure states $|u_1\rangle$ and $|u_2\rangle$ corresponding to $\mathbf{P}_1$ and $\mathbf{P}_2$:

$$\rho_A = \frac{1}{2}[\mathbb{1} + \lambda\mathbf{P}_1 \cdot \boldsymbol{\sigma} + (1 - \lambda)\mathbf{P}_2 \cdot \boldsymbol{\sigma}] = \lambda|u_1\rangle\langle u_1| + (1 - \lambda)|u_2\rangle\langle u_2| . \qquad (4.15)$$

There is thus an ambiguity of representation of the density operator of a qubit, a feature that we will analyse below in a more general situation. The ambiguity is removed (provided $P \neq 0$) if we add the condition that $|u_1\rangle$ and $|u_2\rangle$ should be orthogonal, i.e. that $\mathbf{P}_1$ and $\mathbf{P}_2$ should be aligned along a diameter of the sphere. The decomposition corresponds then to the diagonalization of ρ_A in an orthogonal basis.

The density matrix ρ_A can also be cast in infinitely many ways as a mixture of three pure states. We first express it as a mixture of two density matrices, represented by vectors $\mathbf{P}'$ and $\mathbf{P}_1$. The vector $\mathbf{P}_1$ is on the sphere surface, corresponding to a pure state and $\mathbf{P}'$ is inside it, corresponding to a mixed state ($\mathbf{P}$ is on the line joining

$\mathbf{P}_1$ and $\mathbf{P}'$, see Fig. 4.1b). In turn, $\mathbf{P}'$ is expressed as a mixture of two pure states, $\mathbf{P}_2$ and $\mathbf{P}_3$. Cascading the process, ρ_A can be expressed as an arbitrary mixture of non-orthogonal projectors, with an arbitrary number of terms which may exceed by far the dimension of the Hilbert space.

4.1.3   Freedom of preparation and purification of the density operator

We now come back to a system A whose Hilbert space $\mathcal{H}_A$ has a finite dimension N_A. The expression of ρ_A as a sum of N_A *orthogonal* projectors is unique, up to an obvious freedom of bases choice when one of the λ_i is degenerate. There are, however, as for a qubit, *infinitely many ways* to cast a statistical mixture as a sum of an arbitrary number of *non-orthogonal* projectors. Sorting out two of these non-orthogonal representations, we have for instance:

$$\rho_A = \sum_{i=1}^{N_A} \lambda_i \left| \lambda_i^{(A)} \right\rangle \left\langle \lambda_i^{(A)} \right| = \sum_{j=1}^{N_1} p_j \left| \phi_j^{(A)} \right\rangle \left\langle \phi_j^{(A)} \right| = \sum_{k=1}^{N_2} q_k \left| \xi_k^{(A)} \right\rangle \left\langle \xi_k^{(A)} \right| , \qquad (4.16)$$

where the $\left| \phi_j^{(A)} \right\rangle$ and $\left| \xi_k^{(A)} \right\rangle$ are non-orthogonal normalized vectors ensembles, and the p_j and q_k are corresponding sets of positive probabilities summing up to one. The number of terms N_1 and N_2 in these expressions is arbitrary, and can be larger than the dimension N_A of $\mathcal{H}_A$.

These equivalent expressions correspond to different scenarios for the preparation of the *same* statistical mixture ρ_A. One can imagine that, instead of being randomly cast in one of the mutually exclusive states $\left| \lambda_i^{(A)} \right\rangle$, A is prepared repeatedly in one of the $\left| \phi_j^{(A)} \right\rangle$ with probability p_j, or else in one of the $\left| \xi_k^{(A)} \right\rangle$ with probability q_k. Any observer, being given one of these statistical ensembles of realizations and allowed to make arbitrary measurements on them, will be fundamentally unable to tell which preparation procedure has been used, since they all correspond to the same matrix ρ_A. This 'freedom of preparation' gives to an experiment performed on a statistical mixture an infinite number of equivalent interpretations. We now show that these interpretations are related to each other in a simple way.

The density matrix ρ_A generally results from the trace of the formidably complex $A+E$ density operator over the environment's degrees of freedom. It is often fruitful to view it as resulting from the partial trace of a much simpler entangled state involving A and an 'environment simulator' B. Setting the dimension, N_B, of its Hilbert space $\mathcal{H}_B$ to be larger than the number of terms, N_1 or N_2, appearing in eqn. (4.16), we can write:

$$\rho_A = \sum_{j=1}^{N_1} p_j \left| \phi_j^{(A)} \right\rangle \left\langle \phi_j^{(A)} \right| = \text{Tr}_B \left| \Phi^{(AB)} \right\rangle \left\langle \Phi^{(AB)} \right| , \qquad (4.17)$$

if we define the entangled state $\left| \Phi^{(AB)} \right\rangle$ as:

$$\left| \Phi^{(AB)} \right\rangle = \sum_{j=1}^{N_1} \sqrt{p_j} \left| \phi_j^{(A)} \right\rangle \otimes \left| \phi_j^{(B)} \right\rangle , \qquad (4.18)$$

where the ensemble $\left\{\left|\phi_j^{(B)}\right\rangle\right\}$ is an orthonormal set of vectors in $\mathcal{H}_B$. We call $\left|\Phi^{(AB)}\right\rangle$ a 'purification' of ρ_A. In a similar way, we can define N_2 orthogonal vectors $\left\{\left|\xi_k^{(B)}\right\rangle\right\}$ in $\mathcal{H}_B$ so that the entangled state:

$$\left|\Xi^{(AB)}\right\rangle = \sum_{k=1}^{N_2} \sqrt{q_k}\, \left|\xi_k^{(A)}\right\rangle \otimes \left|\xi_k^{(B)}\right\rangle\, , \tag{4.19}$$

when traced over B, leads to:

$$\mathrm{Tr}_B \left|\Xi^{(AB)}\right\rangle\left\langle\Xi^{(AB)}\right| = \sum_{k=1}^{N_2} q_k \left|\xi_k^{(A)}\right\rangle\left\langle\xi_k^{(A)}\right| = \rho_A\, . \tag{4.20}$$

More generally, any expansion of ρ_A over a set of less than N_B projectors can be written as the partial trace over B of a well-chosen purification. This procedure leads to a very compact representation of the density matrix. When the expansion of ρ_A involves only two terms, B can be simply a qubit. As a result, any density matrix for a single qubit can be written as the partial trace of an entangled two-qubit state.

The simplicity of the environment simulator B allows us to describe a physical procedure implementing the partial trace operation. The 'owners' of A and B, Alice and Bob as they are usually called in quantum information, initially share the entangled state $\left|\Phi^{(AB)}\right\rangle$. Bob performs on B the measurement of an observable O_B, having the $\left\{\left|\phi_j^{(B)}\right\rangle\right\}$ as eigenvectors, with the non-degenerate eigenvalues ϵ_j. When Bob obtains ϵ_j (with a probability p_j), the state owned by Alice is projected onto $\left|\phi_j^{(A)}\right\rangle$. We will use later this simple procedure for the conditional preparation of a pure state of A. When Bob does not record his result (an 'unread' measurement), Alice gets instead a statistical mixture of the states $\left|\phi_j^{(A)}\right\rangle$, with probabilities p_j. This is precisely ρ_A. The partial trace operation leading to ρ_A thus corresponds to an unread measurement performed on B. The same considerations apply obviously to any purification of ρ_A, with a proper choice of the measured observable.

Let us show now that *any* representation of ρ_A in terms of non-orthogonal projectors with less than N_B terms can be obtained by an unread measurement of a proper observable, performed on the *same* purification $\left|\Phi^{(AB)}\right\rangle$. This result is called the 'GHJW theorem' [for Gisin (1989), Hughston, Josza and Wooters]. Let us consider the two purifications $\left|\Phi^{(AB)}\right\rangle$ and $\left|\Xi^{(AB)}\right\rangle$ of ρ_A and cast them in the Schmidt form (see eqn. 2.52, on page 57):

$$\left|\Phi^{(AB)}\right\rangle = \sum_i \sqrt{\lambda_i}\left|\lambda_i^{(A)}\right\rangle\otimes\left|i^{(B)}\right\rangle\, ; \qquad \left|\Xi^{(AB)}\right\rangle = \sum_i \sqrt{\lambda_i}\left|\lambda_i^{(A)}\right\rangle\otimes\left|i'^{(B)}\right\rangle\, . \tag{4.21}$$

The kets $\left|\lambda_i^{(A)}\right\rangle$ and the probabilities λ_i are the eigenvectors and eigenvalues of ρ_A. They are thus common to the two Schmidt expansions, which have the same Schmidt number N_s ($N_s \leq N_B$). The sets $\{|i^{(B)}\rangle\}$ and $\{|i'^{(B)}\rangle\}$ are two orthonormal sets of vectors in $\mathcal{H}_B$. There is thus (at least) one unitary transformation, U_B, connecting them: $\left|i^{(B)}\right\rangle = U_B \left|i'^{(B)}\right\rangle$ (note that U_B is defined without ambiguity only when $N_s = N_B$). Hence, the two purifications are linked by the unitary transformation $\mathbb{1}_A \otimes U_B$:

$$\left|\Phi^{(AB)}\right\rangle = (\mathbb{1}_A \otimes U_B) \left|\Xi^{(AB)}\right\rangle . \qquad (4.22)$$

We can thus cast $\left|\Phi^{(AB)}\right\rangle$ in terms of the $\left|\xi_k^{(A)}\right\rangle$:

$$\left|\Phi^{(AB)}\right\rangle = (\mathbb{1}_A \otimes U_B) \sum_{k=1}^{N_2} \sqrt{q_k} \left|\xi_k^{(A)}\right\rangle \otimes \left|\xi_k^{(B)}\right\rangle = \sum_{k=1}^{N_2} \sqrt{q_k} \left|\xi_k^{(A)}\right\rangle \otimes \left|\xi_k'^{(B)}\right\rangle , \qquad (4.23)$$

with $\left|\xi_k'^{(B)}\right\rangle = U_B \left|\xi_k^{(B)}\right\rangle$.

Alice and Bob initially share the purification $\left|\Phi^{(AB)}\right\rangle$. If Bob performs an unread measurement of O_B, he projects A onto $\sum_{j=1}^{N_1} p_j \left|\phi_j^{(A)}\right\rangle \left\langle \phi_j^{(A)}\right| = \rho_A$. If he decides instead to measure an observable O_B' with eigenvectors $\left|\xi_k'^{(B)}\right\rangle$, he projects A onto $\sum_{k=1}^{N_2} q_k \left|\xi_k^{(A)}\right\rangle \left\langle \xi_k^{(A)}\right| = \rho_A$. By obvious generalization, all the different representations of the same density matrix can thus be viewed as resulting from unread measurements of different observables of B, performed on an unique entangled A/B purification. The density matrix purification is a powerful tool to unite apparently unrelated representations.

4.1.4 Generalized and projective measurements

We shall now describe the effect of a measurement on the density operator of a system A. Let us define a *generalized measurement* $\mathcal{M}$ by a set of (non-necessarily hermitian) operators M_i of $\mathcal{H}_A$, in arbitrary number, each of them being associated to a possible measurement result m_i. We will show that it is possible to design a procedure which, applied to A prepared in the mixed state ρ_A, yields the results m_i with the probabilities π_i given by:

$$\pi_i = \text{Tr}(M_i \rho_A M_i^\dagger) , \qquad (4.24)$$

while projecting A conditionally in the mixed state described by the final density operator:

$$\rho_{A|i} = \frac{M_i \rho_A M_i^\dagger}{\pi_i} . \qquad (4.25)$$

Since the probabilities of all the measurement outcomes must sum up to 1 for an arbitrary ρ_A, the operators M_i must fulfil the normalization condition:

$$\sum_i M_i^\dagger M_i = \mathbb{1} . \qquad (4.26)$$

When the M_i's are orthogonal projectors P_i, satisfying the conditions $P_i^\dagger = P_i$, $P_i^\dagger P_i = P_i^2 = P_i$ and $\sum_i P_i = \mathbb{1}$, we recognize in eqns. (4.24)–(4.26) the expression of the projective measurement postulates, as defined in elementary textbooks. What can we say about $\mathcal{M}$ when the M_i are not projectors and how can we relate it to a standard measurement?

The generalized procedure $\mathcal{M}$ performed on A can be viewed as resulting from a projective measurement on an environment simulator B, after it has been entangled to A by a proper unitary operation $U_\mathcal{M}$. To show this, let us assume that the Hilbert space $\mathcal{H}_B$ of B has a dimension at least equal to the number of M_i operators and let us associate to each M_i a state $\left|u_i^{(B)}\right\rangle$ of an orthonormal basis of $\mathcal{H}_B$. The entangling operation $U_\mathcal{M}$ is defined by its action on the tensor product of a generic state $\left|\phi^{(A)}\right\rangle$ of A by a reference state $\left|0^{(B)}\right\rangle$ of B as:[3]

$$U_\mathcal{M}\left|\phi^{(A)}\right\rangle \otimes \left|0^{(B)}\right\rangle = \sum_i M_i \left|\phi^{(A)}\right\rangle \otimes \left|u_i^{(B)}\right\rangle . \tag{4.27}$$

If A is in a statistical mixture $\sum_i p_i \left|\phi_i^{(A)}\right\rangle \left\langle\phi_i^{(A)}\right|$, each $\left|\phi_i^{(A)}\right\rangle$ is linearly transformed according to eqn. (4.27) and the entangling operation $U_\mathcal{M}$ transforms $\rho_A \otimes \left|0^{(B)}\right\rangle\left\langle 0^{(B)}\right|$ according to:

$$U_\mathcal{M}\left(\rho_A \otimes \left|0^{(B)}\right\rangle\left\langle 0^{(B)}\right|\right)U_\mathcal{M}^\dagger = \sum_{i,j} M_i\rho_A M_j^\dagger \otimes \left|u_i^{(B)}\right\rangle\left\langle u_j^{(B)}\right| . \tag{4.28}$$

Suppose that we perform, after this entanglement, a projective measurement of the observable O_B of B admitting the $\left|u_i^{(B)}\right\rangle$ as eigenstates with eigenvalues m_i. It is straightforward to show that obtaining the result m_i projects A, with the probability π_i given by eqn. (4.24), in the statistical mixture $\rho_{A|i}$ defined by eqn. (4.25). We have thus designed a procedure which coincides with the definition of $\mathcal{M}$ given above. The generalized measurement rules (4.24)–(4.26) do not imply, thus, any additional quantum postulate and we might wonder why we need to consider such processes at all.

The reasons are practical and theoretical. We must first recognize that in real laboratory life, most measurements are not projective. We will discuss this point in more detail in Chapter 6. Let us only consider here, as a simple example, a photo-detection performed with a unit efficiency detector (Section 3.1.4). A field in a mixture of the $|1\rangle$ and $|0\rangle$ Fock states: $\rho_A = p_1 |1\rangle\langle 1| + p_0 |0\rangle\langle 0|$ ($p_1 = 1 - p_0$ being the probability of detecting the photon) always ends up, whether a click has been recorded or not, in the vacuum state $|0\rangle$. This reduction is not described by the standard measurement postulate. It corresponds to a generalized measurement, defined by the operators $M_1 = |0\rangle\langle 1|$ (one photon detected) and $M_0 = |0\rangle\langle 0|$ (no photon detected). At a more fundamental level, considering generalized measurements as resulting from projective measurements on an auxiliary system will allow us to give a very simple and insightful description of the irreversible evolution of open systems. Note that the number of operators describing a generalized measurement on a system A evolving in a Hilbert space of N_A dimensions is not bounded, because the size of the auxiliary system B can be arbitrarily large. It is thus possible to describe in these terms complex measurements yielding non-exclusive results.

There is a fundamental difference between the unitary Hamiltonian evolution described by eqn. (4.9) and a generalized (or projective) measurement transforming the

---

[3]It is easy to check that $U_\mathcal{M}$ is unitary, as a consequence of eqn. (4.26).

system's state according to eqn. (4.25). During free evolution, the linearity of the process entails that each pure state projector belonging to the development of the density matrix can be treated separately, the partial contributions to the evolution being finally summed. This is not generally the case for the state change due to a measurement. The normalization term in the denominator of eqn. (4.25) makes the transformation from ρ_A to $\rho_{A|i}$ non-linear. The density matrix must then be considered as a whole and is altered in a global way. By acquiring information on the system through measurement, we change drastically its description and hence its density matrix. As for pure states, there is something really special about quantum measurements performed on statistical mixtures.

What happens, finally, if the generalized measurement is not recorded? It leaves then the density matrix in a mixture $\rho_{A|u}$ of the $\rho_{A|i}$, weighted by the probabilities π_i of the different measurement outcomes:

$$\rho_{A|u} = \sum_i \pi_i \rho_{A|i} = \sum_i M_i \rho_A M_1^\dagger . \qquad (4.29)$$

This transformation is free of normalization and is linear. An unread measurement, which does not provide direct information on the system, modifies ρ_A in a way quite different from a 'read-out' measurement. The transformations of the components of the density operator can be computed separately, as for a Hamiltonian evolution. In the case of a projective procedure, the unread measurement simply suppresses the off-diagonal elements of the density operator between the states associated to different measurement results, while leaving unchanged the probabilities of finding the system in any of these states. Note also the formal analogy between eqns. (4.29) and (4.9) describing the unitary evolution of ρ_A. We show in the next section that *any* linear quantum process which does not imply a read-out measurement can be cast in the general form of eqn. (4.29).

4.2 Quantum maps: the Kraus sum representation

A linear quantum process transforming a density matrix into another can result from a unitary evolution, from an unread measurement, or, in the most general case, from the coupling of the system with the environment. In all cases, the initial density matrix of A, ρ_A, is transformed into a new one, $\mathcal{L}_A(\rho_A)$. The process $\mathcal{L}_A$ is called a 'quantum map'. It is described by a linear operator in the super-Hilbert space of the operators acting in $\mathcal{H}_A$ (a 'super-operator'). We show in this section, inspired by J. Preskill (2005), that quantum maps can always be cast in a simple, finite form, the 'Kraus sum representation' (Kraus 1983).

4.2.1 Complete positivity

What are the minimal conditions that a quantum map $\mathcal{L}_A$ must fulfil? First, it should be a linear operation preserving hermiticity, conserving the trace of ρ_A and its positivity:

$$\mathcal{L}_A(p\rho_A + q\rho'_A) = p\mathcal{L}_A(\rho_A) + q\mathcal{L}_A(\rho'_A) \quad \text{with } p + q = 1 ; \qquad (4.30)$$

$$[\mathcal{L}_A(\rho_A)]^\dagger = \mathcal{L}_A(\rho_A) ; \qquad (4.31)$$

$$\text{Tr}\left[\mathcal{L}_A(\rho_A)\right] = 1 \; ; \tag{4.32}$$

$$\left\langle \phi^{(A)} \middle| \mathcal{L}_A(\rho_A) \middle| \phi^{(A)} \right\rangle \geq 0 \qquad \text{for all } \left| \phi^{(A)} \right\rangle \text{ in } \mathcal{H}_A . \tag{4.33}$$

These conditions are necessary to ensure the probabilistic interpretation of ρ_A. There is, in addition, a more subtle requirement that $\mathcal{L}_A$ must also fulfil. Imagine that A has been entangled in the past, through some unspecified interaction, with an arbitrary system B. The global system is described by the joint density operator ρ_{AB}. The superoperator $\mathcal{L}_A$ affects only A. The system B remains unaffected, since it no longer interacts with A. The superoperator acting on ρ_{AB} is thus $\mathcal{L}_A \otimes \mathbb{1}_B$, where $\mathbb{1}_B$ is the identity superoperator in $\mathcal{H}_B$. The complete quantum map on $\mathcal{H}_A \otimes \mathcal{H}_B$ must preserve the positivity of ρ_{AB}:

$$\left\langle \phi^{(AB)} \middle| \mathcal{L}_A \otimes \mathbb{1}_B(\rho_{AB}) \middle| \phi^{(AB)} \right\rangle \geq 0 \quad \text{for all } \left| \phi^{(AB)} \right\rangle \text{ in } \mathcal{H}_{AB} . \tag{4.34}$$

This condition must apply for any system B and any ρ_{AB}. The super-operator $\mathcal{L}_A$ is then said to be 'completely positive'. In simple terms, we require that $\mathcal{L}_A$ is a physically acceptable process, even when A is entangled with any other part of the universe.

There are processes which are not completely positive, even if they fulfil criteria (4.30)–(4.33). As a simple example, let us consider two qubits A and B, and build the quantum map $\mathcal{L}_T$ performing a transposition of ρ_A alone:

$$\mathcal{L}_T\left[\begin{pmatrix} \rho_{A;00} & \rho_{A;01} \\ \rho_{A;10} & \rho_{A;11} \end{pmatrix}\right] = \begin{pmatrix} \rho_{A;00} & \rho_{A;10} \\ \rho_{A;01} & \rho_{A;11} \end{pmatrix} . \tag{4.35}$$

This map satisfies criteria (4.30)–(4.33). It fails, however, to pass the complete positivity test. Suffice it to consider the pure joint state $\left| \Psi^{(AB)} \right\rangle = (|0,0\rangle + |1,1\rangle)/\sqrt{2}$, one of the maximally entangled Bell states. The corresponding partially transposed density matrix is, in the $\{|0,0\rangle, |0,1\rangle, |1,0\rangle, |1,1\rangle\}$ basis:

$$\mathcal{L}_T \otimes \mathbb{1}_B \left(\left| \Psi^{(AB)} \right\rangle \left\langle \Psi^{(AB)} \right| \right) = \begin{pmatrix} 1 & 0 & 0 & 0 \\ 0 & 0 & 1 & 0 \\ 0 & 1 & 0 & 0 \\ 0 & 0 & 0 & 1 \end{pmatrix} , \tag{4.36}$$

which has a -1 eigenvalue and is therefore not an acceptable density matrix. Let us note that this 'partial transpose' operation can be used as a test of entanglement between two qubits, the final density matrix being positive only when the two systems are separable (Horodecki *et al.* 1996; Peres 1996).

4.2.2   The Kraus sum

In this paragraph, we show that any quantum map can be expressed as a finite sum of operator products, whatever the nature of the physical process producing this map. The map $\mathcal{L}_A$ transforms the initial density matrix ρ_A into $\mathcal{L}_A(\rho_A)$. Since it is a linear operation, we can assume, without loss of generality, that ρ_A is the projector onto a pure state $\left| \phi_1^{(A)} \right\rangle$:

$$\rho_A = \left| \phi_1^{(A)} \right\rangle \left\langle \phi_1^{(A)} \right| . \tag{4.37}$$

When ρ_A is a more general mixture, each term will be treated separately.

We use a trick reminiscent of the density matrix purification. We assume that A, owned by Alice, is initially entangled with an ancillary system B, owned by Bob, in the state:

$$\left|\Psi^{(AB)}\right\rangle = \frac{1}{\sqrt{N_A}} \sum_i \left|i^{(A)}\right\rangle \otimes \left|i^{(B)}\right\rangle , \qquad (4.38)$$

where the set of vectors $\{|i^{(A)}\rangle\}$ is an orthogonal basis of $\mathcal{H}_A$ (with N_A elements) and $\{|i^{(B)}\rangle\}$ a set of N_A orthogonal vectors of $\mathcal{H}_B$ (whose dimension, N_B, is assumed to be larger than N_A). The latter set can be completed by $N_B - N_A$ vectors to provide a basis in $\mathcal{H}_B$. A condition, essential in this demonstration, requires that A is not initially entangled with another system. In particular, when the quantum map is induced by the coupling between A and the environment E, it is essential that A is not initially entangled with E.

The maximally entangled state (4.38) is a generalization of the Bell states studied in Chapter 2 to systems with N_A and $N_B > 2$. This state contains no information on each subsystem (whose partial density matrix is proportional to unity), but predicts strong correlations between the results of measurements performed on A and B. By measuring a proper observable of B, Bob can project his part on any state of $\mathcal{H}_B$ and prepare *ipso facto* for Alice a correlated state of $\mathcal{H}_A$, in particular $\left|\phi_1^{(A)}\right\rangle$ which is correlated to a well-defined state $\left|\phi_1^{(B)}\right\rangle$ of $\mathcal{H}_B$.

In order to prepare the final state $\mathcal{L}_A(\rho_A)$ corresponding to an initial density matrix $\rho_A = \left|\phi_1^{(A)}\right\rangle \left\langle\phi_1^{(A)}\right|$, Alice and Bob can follow two equivalent procedures. In the first, Alice asks Bob to measure a proper observable of B, admitting $\left|\phi_1^{(B)}\right\rangle$ as an eigenstate. When Bob obtains the desired result (and tells it to Alice), she knows that her system is projected onto the density matrix ρ_A. She then lets her system A undergo the quantum process (for instance the coupling with an environment E) and she gets, in the end, the state $\mathcal{L}_A(\rho_A)$.

Alternatively, Alice can first submit A to the quantum process, and then ask Bob to project his subsystem into $\left|\phi_1^{(B)}\right\rangle$. When Bob tells her that he has got the desired state, Alice can convince herself that her system has been prepared in the same quantum state as in the first procedure. Since $\mathcal{L}_A$ does not act in $\mathcal{H}_B$, the system under Bob's control is indeed in no way affected by the quantum process. Hence, the transformations induced by the quantum map acting on A and the measurement performed on B commute. We retrieve the essential property that the quantum EPR correlations are independent of the timing of the independent manipulations performed on the two subsystems.[4] Now that we have analysed the physics of these two equivalent procedures, let us get quickly through the maths.

First route

Let us first focus on the measurement initially performed by Bob. The entangled state $\left|\Psi^{(AB)}\right\rangle$ can be expanded over terms containing the state $\left|\phi_1^{(A)}\right\rangle$. We introduce

---

[4]If the two procedures outlined above yielded different results, Bob separated from Alice by an arbitrarily large distance could, by deciding when he performs his measurement, send to Alice information faster than light!

$N_A - 1$ states $\left|\phi_j^{(A)}\right\rangle$ $(j = 2, \ldots, N_A)$ so that the set $\left\{\left|\phi_j^{(A)}\right\rangle\right\}$ $(j = 1, \ldots, N_A)$ is an orthonormal basis of $\mathcal{H}_A$. The basis change from $\{|i^{(A)}\rangle\}$ to $\left\{\left|\phi_j^{(A)}\right\rangle\right\}$ is defined by the unitary matrix U_A:

$$\left|\phi_j^{(A)}\right\rangle = \sum_i U_{A,ji} \left|i^{(A)}\right\rangle , \qquad (4.39)$$

where $U_{A,ji}$ is a matrix element of U_A. We perform simultaneously in B the transformation from $|i^{(B)}\rangle$ to $\left|\phi_j^{(B)}\right\rangle$ defined by:

$$\left|\phi_j^{(B)}\right\rangle = \sum_i U_{A,ji}^* \left|i^{(B)}\right\rangle , \qquad (4.40)$$

where $U_{A,ji}^*$ is the complex conjugate of $U_{A,ji}$. This second vector transformation is the restriction of a unitary transformation in $\mathcal{H}_B$ to the subspace generated by the N_A vectors $\{|i^{(B)}\rangle\}$. After a bit of algebra and using the unitarity of U_A, we get:

$$\left|\Psi^{(AB)}\right\rangle = \frac{1}{\sqrt{N_A}} \sum_i \left|i^{(A)}\right\rangle \otimes \left|i^{(B)}\right\rangle = \frac{1}{\sqrt{N_A}} \sum_j \left|\phi_j^{(A)}\right\rangle \otimes \left|\phi_j^{(B)}\right\rangle . \qquad (4.41)$$

Any pure state of $\mathcal{H}_A$, $\left|\phi_1^{(A)}\right\rangle$, can thus be viewed as a belonging to a term in the expansion of $\left|\Psi^{(AB)}\right\rangle$ over a proper basis.

In order to prepare this state for Alice, Bob performs a measurement of an observable having the $\left\{\left|\phi_j^{(B)}\right\rangle\right\}$ as non-degenerate eigenvectors. When Bob does not get the result corresponding to the required state, $\left|\phi_1^{(B)}\right\rangle$, he and Alice (who share a classical communication channel) discard the attempt and prepare again the entangled state $\left|\Psi^{(AB)}\right\rangle$ afresh. After about N_A trials, Bob will obtain the result corresponding to $\left|\phi_1^{(B)}\right\rangle$. The state of A, handled by Alice, is accordingly projected onto $\left|\phi_1^{(A)}\right\rangle$. Bob then gets out of stage and Alice lets the quantum map $\mathcal{L}_A$ act over $\mathcal{H}_A$ alone, resulting in the final density matrix, $\mathcal{L}_A\left(\left|\phi_1^{(A)}\right\rangle\left\langle\phi_1^{(A)}\right|\right)$.

Second route

In the alternate route to the final state, Alice and Bob first let the extended map $\mathcal{L}_A \otimes \mathbb{1}_B$ act on the initial entangled state $\left|\Psi^{(AB)}\right\rangle$. Bob then performs the selection measurement.

Before Bob's measurement, the global system is described by:

$$\rho_{AB} = \mathcal{L}_A \otimes \mathbb{1}_B \left(\left|\Psi^{(AB)}\right\rangle\left\langle\Psi^{(AB)}\right|\right) . \qquad (4.42)$$

Since $\mathcal{L}_A \otimes \mathbb{1}_B$ preserves hermiticity, trace and positivity, $\mathcal{L}_A$ being completely positive (we see here the importance of this hypothesis), ρ_{AB} is a *bona fide* density matrix. It can thus be diagonalized and expanded as a weighted sum of orthogonal projectors:

$$\rho_{AB} = \sum_\mu q_\mu \left|\Phi_\mu^{(AB)}\right\rangle\left\langle\Phi_\mu^{(AB)}\right| , \qquad (4.43)$$

where the $\left|\Phi_\mu^{(AB)}\right\rangle$ are orthonormal states of $\mathcal{H}_{AB}$ and the q_μ are positive numbers summing up to 1. All the terms in this expression are independent of the choice of $\left|\phi_1^{(A)}\right\rangle$, since Bob has not yet decided to perform his measurement. The subspace spanned in $\mathcal{H}_B$ by the vectors $\{|i^{(B)}\rangle\}$ has a dimension equal to N_A. The effective dimension explored in $\mathcal{H}_{AB}$ is thus N_A^2 and there are thus at most N_A^2 terms in this expansion, whatever the actual dimension of $\mathcal{H}_B$.

Bob then performs his projective measurement, defined in $\mathcal{H}_A \otimes \mathcal{H}_B$ by the projectors $P_j = \mathbb{1}_A \otimes \left|\phi_j^{(B)}\right\rangle\left\langle\phi_j^{(B)}\right|$. When he does not get the result corresponding to $\left|\phi_1^{(B)}\right\rangle$, he and Alice decide to restart the whole process. When, after N_A attempts on the average, Bob gets the right result, A is projected by P_1 onto the final density matrix $\mathcal{L}_A(\rho_A)$ which can be written:

$$\mathcal{L}_A\left(\left|\phi_1^{(A)}\right\rangle\left\langle\phi_1^{(A)}\right|\right) = N_A \sum_\mu q_\mu \left\langle\phi_1^{(B)}\middle|\Phi_\mu^{(AB)}\right\rangle\left\langle\Phi_\mu^{(AB)}\middle|\phi_1^{(B)}\right\rangle , \qquad (4.44)$$

where the N_A factor comes from the normalization after the measurement, the result corresponding to $\left|\phi_1^{(A)}\right\rangle$ being obtained with a $1/N_A$ probability.

The scalar products in the right-hand side of eqn. (4.44) are kets and bras in $\mathcal{H}_A$. The transformation $\left|\phi_1^{(A)}\right\rangle \rightarrow \sqrt{N_A q_\mu}\left\langle\phi_1^{(B)}\middle|\Phi_\mu^{(AB)}\right\rangle$ is a linear mapping of $\mathcal{H}_A$ onto $\mathcal{H}_A$ itself, since both the transformations $\left|\phi_1^{(A)}\right\rangle \rightarrow \left|\phi_j^{(B)}\right\rangle$ and $\left|\phi_j^{(B)}\right\rangle \rightarrow \left\langle\phi_j^{(B)}\middle|\Phi_\mu^{(AB)}\right\rangle$ are anti-linear[5] [the anti-linearity of the first appears clearly when comparing eqns. (4.39) and (4.40), while the second is anti-linear by definition of the relationship between bras and kets].

There are thus at most N_A^2 linear operators M_μ in $\mathcal{H}_A$ such that:

$$M_\mu\left|\phi_1^{(A)}\right\rangle = \sqrt{N_A q_\mu}\left\langle\phi_1^{(B)}\middle|\Phi_\mu^{(AB)}\right\rangle . \qquad (4.45)$$

We can rewrite in terms of these operators the final expression of the density matrix and we get, by comparing the two routes:

$$\mathcal{L}_A\left(\left|\phi_1^{(A)}\right\rangle\left\langle\phi_1^{(A)}\right|\right) = \sum_\mu M_\mu\left|\phi_1^{(A)}\right\rangle\left\langle\phi_1^{(A)}\right|M_\mu^\dagger . \qquad (4.46)$$

This relation, derived here for an initial pure state of A remains valid for any initial density matrix ρ_A.

We obtain the essential result that *any* physically sound quantum map can be expressed as a sum of $N_K \leq N_A^2$ terms:

$$\mathcal{L}_A(\rho_A) = \sum_\mu M_\mu\rho_A M_\mu^\dagger . \qquad (4.47)$$

---

[5]This means that a phase $\exp(i\varphi)$ in the argument reflects in a phase $\exp(-i\varphi)$ in the result.

We leave it as an exercise to check that the 'Kraus operators' M_μ obey the normalization condition:[6]

$$\sum_\mu M_\mu^\dagger M_\mu = \mathbb{1} \; , \tag{4.48}$$

ensuring that the trace of the density matrix is preserved.

The unitary evolution under the Hamiltonian H_A or the unread generalized measurements (see previous section) were already cast in this form, which appears now as fully general. In fact, eqn. (4.47) shows that *any* linear process affecting an open system A can be viewed as an unread generalized measurement, i.e. as resulting from an entanglement of A with an environment simulator B, followed by an unread projective measurement in B. We will come back below to this very important property of quantum maps.

The essential result of the Kraus sum formulation is that all quantum maps can be expressed in terms of a finite number of operators, not exceeding the square of the dimension of the system under study (four Kraus operators at most for a qubit). This number is independent of the size of the environment, which may be huge and have an infinite number of states and degrees of freedom.

The Kraus sum representation is not uniquely determined. Replacing M_μ by $N_\nu = \sum_\mu V_{\nu\mu} M_\mu$, where the $V_{\nu\mu}$ are the elements of a unitary matrix, we obtain the same quantum map, as can be checked with a few lines of algebra. A quantum map may also be expressed with more than N_A^2 operators. Generalized unread measurements, for instance, may have an unlimited number of possible results and hence of terms in the sum. However, all quantum maps, including these complex measurements, can always be reformulated in terms of N_A^2 operators at most.

Let us stress one important condition for the existence of a quantum map describing the effect of an environment E on a system A. It is required that A and E are not initially entangled. This is necessary to define the generalized EPR state (4.38), which was the starting point of our argument. If A and E have initial quantum correlations, their state is defined by a density operator ρ_{AE} which contains, in the $A - E$ correlations, information often essential to determine the subsequent evolution of A. It is then, in general, impossible to define a linear map deducing the state of A after its interaction with E from the knowledge of ρ_A alone. We will see shortly, however, that when E is a large system with a short memory time, the effect of these correlations is rapidly washed out, and a quantum map description for the evolution of ρ_A can be, after all, retrieved.

4.3 The Lindblad master equation

We now apply the results of the previous section to describe the evolution of a system A coupled to an environment E in thermal equilibrium. The operator sum representation summarizes in a compact way the evolution of ρ_A from the initial time $t = 0$ (when A and E are uncorrelated) to the final time $t = t_f$. The M_μ operators are functions

---

[6]Hint: show that the matrix representing $\sum_\mu M_\mu^\dagger M_\mu$ in the $\left|\phi_j^{(A)}\right\rangle$ basis is identical to N_A times the reduced density matrix representing $\rho_B = \mathrm{Tr}_A \rho_{AB}$ in the $\left|\phi_j^{(B)}\right\rangle$ basis (ρ_{AB} is given by eqn. 4.43). Use then a simple argument to show that $\rho_B = \mathbb{1}/N_A$.

of t_f, so that the Kraus expression is in general not directly helpful. Our goal is to replace this time-integrated expression by a first-order differential master equation with time-independent coefficients for the density matrix, $\rho_A(t)$.

We first discuss, in physical terms, the conditions that the environment E must fulfil for such an equation to exist. Basically, E should be a 'large' system, a 'sink' with many degrees of freedom, whose evolution is not appreciably affected by its interaction with A. It is then possible to make a Markov approximation which amounts to neglecting memory effects in the $A + E$ system evolution. We obtain an equation in which $d\rho_A(t)/dt$ depends only on $\rho_A(t)$, and not on the state at prior times. The conditions for this approximation are satisfied by a wide class of physical environments.

We then write down a generic expression of the master equation known as the Lindblad form. This equation is expressed in terms of a small number of operators acting on A, which are associated to the various kinds of quantum jumps that the system may experience. We interpret the Lindblad equation and show how its simple structure allows us to guess its form for systems of interest. We finally discuss important features of the relaxation of a two-level atom and of a field mode, essential processes in the experiments described in later chapters.

The deductive approach chosen here to derive the master equation is based on simple quantum information principles and does not invoke any specific model for the environment (Preskill 2005). In usual quantum optics books (Cohen-Tannoudji *et al.* 1992b; Carmichael 1993; Barnett and Radmore 1997), the same equation is arrived at through quite a different route. The environment is precisely described (it is often a continuum of quantum oscillators, as discussed in Section 3.2.5) and its coupling with A is explicitly analysed. This approach has the advantage of yielding the exact expression of the master equation coefficients, while the deductive route taken here requires some reasonable physical guesses to get the explicit form of the Lindblad operators.

The quantum information perspective has, on the other hand, the merit of pointing out quite general features of the master equation, which are independent of the nature of the environment, provided it satisfies the general conditions of the Markov approximation. The physical picture which emerges from the Lindblad approach is more transparent and more general than the one suggested by the traditional point of view on relaxation phenomena.

4.3.1 Markov approximation and coarse-grained evolution

Assuming that a first-order differential equation for ρ_A exists, we usually deal with it numerically by dividing the time into small 'slices' of duration τ. During each, the evolution of ρ_A must be incremental, $\rho_A(t + \tau) - \rho_A(t)$ being a first-order quantity in τ, which we identify with $\tau d\rho_A/dt$. We thus require τ to be very short compared to the characteristic time T_r of evolution of ρ_A. Solving the differential equation amounts to computing ρ_A values at successive time steps, thus realizing a stroboscopic analysis of the system's dynamics. Mathematically, the 'exact' evolution is recovered by going to the limit $\tau = 0$.

Physically, though, we will see that the equation describing the evolution of a system coupled to a large environment cannot be taken to this mathematical limit. The time slicing must remain coarse-grained, allowing a quasi-continuous following of the system's observables expectation values, but renouncing a too precise description of its evolution at very short times. The spacing of the successive 'stroboscopic snapshots' will have to be much larger than a minimum step, measuring the memory time of the environment.

To make the connexion with the formalism developed in the previous section, we would like to express the transformation from $\rho_A(t)$ to $\rho_A(t+\tau)$ as an 'infinitesimal' quantum map, $\mathcal{L}_\tau$, developed as a Kraus sum. If at time t the environment is in a steady state described by the density operator $\bar{\rho}_E$ and the $A+E$ system in a tensor product described by the density operator:

$$\rho_{AE}(t) = \rho_A(t) \otimes \bar{\rho}_E , \tag{4.49}$$

the Kraus formalism can be directly applied, leading to the density matrix for A at time $t+\tau$:

$$\rho_A(t+\tau) = \mathcal{L}_\tau[\rho_A(t)] = \sum_{\mu=0}^{N_K-1} M_\mu(\tau)\rho_A(t)M_\mu^\dagger(\tau) , \tag{4.50}$$

where the $M_\mu(\tau)$ are $N_K \leq N_A^2$ Kraus operators depending upon τ, the interaction between A and E and upon $\bar{\rho}_E$. This formula, after simple transformations, will lead us directly to the differential master equation we are seeking. Before that, we need to discuss the physical assumptions made to derive it and the limitations they imply for the stroboscopic step τ.

We have supposed that $A+E$ is, at time t, in a separable state, an important *a priori* condition to get a quantum map according to the discussion of the last section. We must note, however, that we have in eqn. (4.49) overlooked two terms which result from important physical processes. First, the environment density operator presents at time t a fluctuation $\delta\rho_E(t)$ around its steady-state value, due to its prior interaction with A. Second, A and E are generally entangled, again a consequence of their coupling in the past of t. This entanglement adds a contribution $\delta\rho_{AE}(t)$ to the global density operator. Equation (4.49) should thus be replaced by:

$$\rho_{AE}(t) = \rho_A(t) \otimes [\bar{\rho}_E + \delta\rho_E(t)] + \delta\rho_{AE}(t) . \tag{4.51}$$

There are two difficulties implied by this exact expression. First, and foremost, the existence of correlations between A and E at time t makes it impossible to follow rigorously the line of reasoning developed above to get a quantum map linking linearly $\rho_A(t)$ to $\rho_A(t+\tau)$. Second, even if such a map exists, it would *a priori* depend upon the fluctuating initial state of the environment and the M_μ would be functions of t, leading to unwanted time-varying coefficients in the master equation.

It can be shown however that, if E is large enough (in a way we will shortly precise), the evolution over not too small a time interval τ can be computed as if A and E were not entangled at the beginning of this interval, E then being in its steady state. In other words, when the conditions of a large environment are met, all happens for the incremental evolution of ρ_A over a finite time interval τ as if the $A+E$ system was initially described by eqn. (4.49).

This important result is based on the Markov approximation. We will not justify it in detail, which is done in quantum optics textbooks following the traditional approach to relaxation (Cohen-Tannoudji *et al.* 1992b; Carmichael 1993; Barnett and Radmore 1997), but only discuss it in qualitative terms, insisting on its physical interpretation.

Usually, the environment, or 'reservoir' as it is called in statistical physics, is a very large system with levels spanning a wide energy range $\hbar\Delta\omega$. The correlation time of the observables R_E of E, defined loosely as the damping time of $\mathrm{Tr}(\rho_E R_E(t)R_E(t+\tau))$, is very short, of the order of $\tau_c = \hbar/\Delta\omega$. For time intervals smaller than τ_c, the observables of E appearing in the coupling Hamiltonian H_{AE} remain nearly constant, and A and E undergo a coherent evolution. During the next τ_c time interval, the phase relations between A and E are lost and a new coherent evolution starts again. As a result, the environment fluctuation $\delta\rho_E(t)$ and the A/E correlation $\delta\rho_{AE}(t)$ have a very short correlation time, of the order of τ_c, meaning that products of two matrix elements of these quantities taken at times differing by more than τ_c vanish on average.

These arguments provide the order of magnitude of the time scale T_r for the evolution of ρ_A. The system, coupled to the amnesic environment, undergoes a random walk. Each step has a τ_c duration, and corresponds to a typical $V\tau_c/\hbar$ variation for the quantum phases of a generic matrix element of ρ_A. Here, V is the order of magnitude of the H_{AE} matrix elements. The steps corresponding to successive τ_c intervals add quadratically. The quantum phase dispersion accumulated over the total time t, i.e. after t/τ_c steps, $\Delta\Phi(t)$, is thus given by:

$$[\Delta\Phi(t)]^2 = \frac{V^2\tau_c^2}{\hbar^2}\frac{t}{\tau_c} = \frac{t}{T_r} , \tag{4.52}$$

where we define the evolution time scale T_r by:

$$T_r = \frac{\hbar^2}{V^2\tau_c^2}\tau_c . \tag{4.53}$$

The Markov condition of a very short memory time of the environment:

$$\tau_c \ll T_r , \tag{4.54}$$

is thus satisfied when:

$$V\tau_c/\hbar \ll 1 . \tag{4.55}$$

The evolution of the system's quantum phases is then negligible during the memory time of the environment, a joint condition relating the environment's spectrum to the characteristic $A - E$ coupling V. This condition is usually satisfied by the reservoirs encountered in quantum optics (ensemble of field modes in which atoms decay, atoms in cavity mirrors exchanging photons with the field inside, etc.). The width of the reservoir in these cases is of the order of the atom or field frequency, while $V/\hbar$ is typically of the order of a Rabi frequency, many orders of magnitude smaller. When the Markov condition is met, the system's evolution can be analysed with two different time 'units', the very short memory time τ_c setting the time scale of its fluctuations and correlations, and the long time T_r, measuring the dynamical evolution of its observables.

The short memory time implies that the initial conditions defined by $\delta\rho_E(t)$ and $\delta\rho_{AE}(t)$ influence $\rho_A(t')$ up to $t' \approx t+\tau_c$ only. Since $\delta\rho_E(t)$ depends on the state of A

slightly prior to t, it is also clear that, on this short time scale, $\rho_A(t')$ depends not only on $\rho_A(t)$ but also, through $\delta\rho_E(t)$, on $\rho_A(t'')$ at times t'' comprised between $t - \tau_c$ and t. If we choose a time slice τ much larger than τ_c, the transient evolution of ρ_A over the very small fraction τ_c/τ at the beginning of the slice is entirely negligible compared to that which is produced by the fluctuations and correlations appearing in the system from $t' \approx t + \tau_c$ up to $t + \tau$. In other words, neglecting the initial reservoir fluctuations and system–environment correlations at time t does not appreciably change the value of $\rho_A(t+\tau)$. It is thus legitimate, under these Markov conditions, to replace the initial state (4.51) by (4.49). The incremental change of A is then well-approximated by the quantum map transformation (4.50).

We have thus justified the stroboscopic description of the system's evolution outlined at the beginning of this section. Choosing a finite time interval $\tau \gg \tau_c$ and applying repetitively infinitesimal quantum maps to the successive values of ρ_A at multiples of τ, we obtain a 'coarse-grained description' of the system's dynamics. On the one hand, the slicing must be coarse enough to be insensitive to the fast fluctuations affecting the system's evolution at the τ_c time scale. It should, on the other hand, be fine enough to provide a near-continuous description of the evolution of the density matrix over times of the order of T_r. Hence the double inequality:

$$\tau_c \ll \tau \ll T_r \ , \tag{4.56}$$

essential for the validity of this procedure, which leaves ample space to choose a convenient τ value when the Markov condition (4.55) is fulfilled.

4.3.2   Master equation in Lindblad form

Keeping all these conditions in mind, we now define the derivative of ρ_A as:

$$\frac{d\rho_A(t)}{dt} = \frac{\mathcal{L}_\tau[\rho_A(t)] - \rho_A(t)}{\tau} \ , \tag{4.57}$$

with the quantum map $\mathcal{L}_\tau$ defined by (4.50) and τ obeying (4.56). Since $\mathcal{L}_\tau[\rho_A(t)] = \rho_A(t + \tau) = \rho_A(t) + O(\tau)$, where $O(\tau)$ is a first-order contribution in τ, one of the operators M_μ should be of the order of unity. We write it, without loss of generality, as:

$$M_0 = \mathbb{1} - iK\tau + O(\tau^2) \ , \tag{4.58}$$

where K is an operator independent of τ. We isolate hermitian and anti-hermitian parts in K and define two hermitian operators by:

$$H = \hbar\frac{K + K^\dagger}{2} \quad \text{and} \quad J = i\frac{K - K^\dagger}{2} \ , \tag{4.59}$$

so that:

$$K = \frac{H}{\hbar} - iJ \ . \tag{4.60}$$

Up to the first order in τ, we have:

$$M_0(\tau)\rho_A M_0^\dagger(\tau) = \rho_A - \frac{i\tau}{\hbar}[H, \rho_A] - \tau(J\rho_A + \rho_A J) \ . \tag{4.61}$$

We recognize, in the right-hand side, a commutator describing a unitary evolution of ρ_A induced by the Hamiltonian-like term H. This term is the sum of the free

Hamiltonian H_A of A plus relaxation-induced energy shift contributions, due to the coupling of A with E. They describe a *renormalization* of the energy levels of A.[7] The general analysis developed here does not provide an explicit expression for these shifts. We will simply assume in the following that they have been included in the 'naked' energy levels of A and we will replace formally H by H_A in eqn. (4.61).

All other terms, M_μ with $\mu \neq 0$, in the Kraus sum are of order τ. We can thus write:

$$M_\mu(\tau) = \sqrt{\tau} L_\mu ,\tag{4.62}$$

where the $N_K - 1$ operators L_μ are of order unity, independent of τ. The Kraus operator normalization condition is then:

$$\sum_{\mu=0}^{N_K-1} M_\mu^\dagger(\tau) M_\mu(\tau) = \mathbb{1} - 2J\tau + \sum_{\mu\neq0} \tau L_\mu^\dagger L_\mu = \mathbb{1} .\tag{4.63}$$

Note that the Hamiltonian H_A does not appear in this normalization (it induces its own a map which always preserves the trace of the density matrix). We can thus express J in term of the L_μ:

$$J = \frac{1}{2} \sum_{\mu\neq0} L_\mu^\dagger L_\mu .\tag{4.64}$$

Arranging all terms, we finally get the master equation for ρ_A, in a generic form, the Lindblad (1976) expansion:

$$\frac{d\rho_A}{dt} = -\frac{i}{\hbar} [H_A, \rho] + \sum_{\mu\neq0} \left(L_\mu \rho_A L_\mu^\dagger - \frac{1}{2} L_\mu^\dagger L_\mu \rho_A - \frac{1}{2} \rho_A L_\mu^\dagger L_\mu \right) .\tag{4.65}$$

Our derivation of this master equation is based on very general arguments, stemming from the Kraus sum formulation and the Markov approximation. We show now that the physical interpretation of the master equation in the Lindblad form is transparent and makes it possible, in many cases, to guess the L_μ operators by a simple description of the relaxation events which affect A.

4.3.3   Physical interpretation of the master equation

Inspired by the definition of a general measurement (Section 4.1.4), let us imagine a process, much simpler than the A/E coupling, which reproduces the evolution of the density matrix ρ_A from t to $t + \tau$. This process involves the Hamiltonian coupling of A with a simple 'environment simulator' B during the time interval τ, followed by an unread measurement performed on B at time $t + \tau$.

We assume that A is at time t in the pure state $|\phi^{(A)}\rangle$ (this is not a restrictive hypothesis, since all projectors composing an arbitrary ρ_A can be handled separately). We want to recover the final density matrix $\mathcal{L}_\tau \left(|\phi^{(A)}\rangle \langle \phi^{(A)}| \right)$. The system A, owned by Alice, is now decoupled from E, but it interacts with a much smaller system B, owned by Bob. The dimension of $\mathcal{H}_B$, the Hilbert space of B, is equal to the number

---

[7] When A is an atom undergoing spontaneous emission due to its coupling to the vacuum modes of the free radiation field, these additional terms describe the Lamb shifts of the A states.

N_K of Kraus operators (it is generally much smaller than the dimension of E). The system B is initially (at time t) in a reference state $|0^{(B)}\rangle$ and evolves, during the time interval τ, together with A, under a unitary process defined by the evolution operator U_{AB}:

$$U_{AB} |\phi^{(A)}\rangle \otimes |0^{(B)}\rangle = \left(M_0 |\phi^{(A)}\rangle \right) \otimes |0^{(B)}\rangle + \sum_{\mu \neq 0} \left(M_\mu |\phi^{(A)}\rangle \right) \otimes |\mu^{(B)}\rangle$$

$$= \left[1 - \frac{i}{\hbar} H_A \tau - J\tau \right] |\phi^{(A)}\rangle \otimes |0^{(B)}\rangle + \sqrt{\tau} \sum_{\mu \neq 0} \left(L_\mu |\phi^{(A)}\rangle \right) \otimes |\mu^{(B)}\rangle . \quad (4.66)$$

This transformation corresponds to the $U_\mathcal{M}$ mapping defined in eqn. (4.27), the M_μ merely replacing the M_i's. The $\{|\mu^{(B)}\rangle\}$ and $|0^{(B)}\rangle$ form an orthonormal basis of $\mathcal{H}_B$. The final kets in $\mathcal{H}_B$ 'label' the Kraus operators acting on A. It is easy to show that the transformation defined by eqn. (4.66) is unitary.

An unread measurement performed by Bob of an observable O_B, with the $|\mu^{(B)}\rangle$ as non-degenerate eigenvectors, projects A onto $\mathcal{L}_\tau \left(|\phi^{(A)}\rangle \langle \phi^{(A)}| \right)$. There is thus no difference, for Alice, between the quantum map induced by the coupling to the complex environment E and the much simpler unitary process induced by the environment simulator B. This picture provides a very economical description of the relaxation process. Imagine for instance that there are only two Kraus operators M_0 (of order unity) and M_1. The system B is then a simple qubit, initially in state $|0\rangle$, staying in $|0\rangle$ when M_0 acts on A, and performing a transition to $|1\rangle$ when M_1 is involved.

The environment simulator picture also stresses the importance of entanglement in the relaxation process. The state of A and B at the end of the time interval is entangled, with a Schmidt number equal to the number of Kraus operators. Bob thus has access to 'which-path' information about A. The unread measurement performed by Bob erases this entanglement, resulting into an increased mixedness for the state of A. The relaxation is, finally, due to information leaking into the environment.

To be equivalent with the action of E, Bob should perform an unread measurement of O_B. What if he performs a complete measurement of O_B whose result is known to him and Alice? The system A is then projected in a well-defined pure state, depending upon the read-out result. Obviously, the density operator corresponding to the unread measurement case is recovered by summing, with their respective probabilities, the projectors on the pure states associated to all possible measurement outcomes.

Even if this measurement is not actually performed, imagining that it is done will lead us to a very insightful interpretation of the system's evolution. Let us examine separately two cases: either B is found in its initial state $|0^{(B)}\rangle$, or it is found in another one, $|\mu^{(B)}\rangle$. The probability, p_0, of the first outcome is:

$$p_0 = \langle \phi^{(A)} | M_0^\dagger M_0 | \phi^{(A)} \rangle = 1 - \tau \sum_{\mu \neq 0} \langle \phi^{(A)} | L_\mu^\dagger L_\mu | \phi^{(A)} \rangle . \quad (4.67)$$

The state of A is then projected onto:

$$|\phi_0^{(A)}\rangle = \frac{M_0}{\sqrt{p_0}} |\phi^{(A)}\rangle = \frac{1 - i H_A \tau / \hbar - J\tau}{\sqrt{p_0}} |\phi^{(A)}\rangle . \quad (4.68)$$

When this event occurs, with a probability close to unity, A undergoes an infinitesimal change, which can be viewed as an elementary time step in a continuous, non-unitary evolution described by an effective non-hermitian Hamiltonian $H_e = H_A - i\hbar J$.

The environment simulator B may also be found in one of the other states $\left|\mu^{(B)}\right\rangle$. The corresponding probability p_μ is:

$$p_\mu = \left\langle \phi^{(A)}\left| M_\mu^\dagger M_\mu \right| \phi^{(A)}\right\rangle = \tau \left\langle \phi^{(A)}\left| L_\mu^\dagger L_\mu \right| \phi^{(A)}\right\rangle . \tag{4.69}$$

Note that $\sum_\mu p_\mu = 1$ (where the sum extends from $\mu = 0$), as expected. The state of A after this measurement is:

$$\left|\phi_\mu^{(A)}\right\rangle = \frac{L_\mu \left|\phi^{(A)}\right\rangle}{\sqrt{p_\mu/\tau}} . \tag{4.70}$$

When such an event occurs, with a small probability of order τ, the state of A undergoes a drastic change. It 'jumps' from $\left|\phi^{(A)}\right\rangle$ to the completely different state $L_\mu \left|\phi^{(A)}\right\rangle$ (within a normalization). The evolution of A is thus mostly a slow non-unitary evolution, combined with 'quantum jumps', occurring with a small probability, induced by the $N_K - 1$ 'quantum jump operators' L_μ.

Let us conclude this section by an important remark. As with the M_μ's, the L_μ operators which derive from them are not uniquely defined. The same master equation can be expressed in terms of different sets of jump operators. Even the number of these operators can be different in different representations of the same relaxation process. This means that the environment simulators B are not real systems. They are merely defined in *gedanken* experiments which allow us to compute, after proper averaging, the real evolution of A. Once the trace over B has been made, which amounts to imagining that Bob has performed an unread measurement on his simulator, the same evolution must be recovered for ρ_A, independently of the specific choice made for the L_μ's (see Section 7.5 for a description of a field relaxation process by different L_μ representations).

4.3.4 Master equations for a spin or a spring

Spontaneous emission of a two-level atom

If we are able to identify which quantum jumps may affect the state of A, we can infer the jump operators L_μ (within numerical factors) and hence guess the structure of the Lindblad master equation. Let us apply this strategy to the elementary case of a two-level atom emitting a spontaneous photon in free space on the $|e\rangle \rightarrow |g\rangle$ transition.[8] The environment E is made of all modes quasi-resonant with the atomic transition, which we assume here to be initially in vacuum ($T = 0$ K temperature of radiation).

We imagine a *gedanken* experiment in which the environment is monitored. The atom is fully surrounded by unit-efficiency photodetectors, so that the escaping photons cannot fail to be recorded. During the time interval τ, two situations may occur.

---

[8]We assume here that the recoil of the atom when it emits a photon is negligible, so that the emission process is entirely disconnected from the external degrees of freedom of the atom. The effect of photon recoil in spontaneous emission is discussed in Chapter 8.

Either no photon is recorded, or one of the detectors clicks. The environment simulator is then a two-state system, the state $|0^{(B)}\rangle$ corresponding to no detection and $|1^{(B)}\rangle$ to the detection of a photon. From energy conservation, the detection of B in $|1^{(B)}\rangle$ corresponds to a projection of the atom onto its lower level $|g\rangle$. There is thus only one jump operator, proportional to the lowering operator $\sigma_- = |g\rangle\langle e|$: $L_1 = \sqrt{\Gamma}\sigma_-$ and the master equation is:

$$\frac{d\rho_A}{dt} = -i\frac{\omega_{eg}}{2}[\sigma_z, \rho_A] - \frac{\Gamma}{2}(\sigma_+\sigma_-\rho_A + \rho_A\sigma_+\sigma_- - 2\sigma_-\rho_A\sigma_+) \;. \tag{4.71}$$

To make it more expicit, let us express it in terms of the atomic populations (ρ_{Aee}, ρ_{Agg}) and coherences ($\rho_{Aeg} = \rho_{Age}^*$):

$$\frac{d\rho_{Aee}}{dt} = -\Gamma\rho_{Aee} \;; \qquad \frac{d\rho_{Agg}}{dt} = \Gamma\rho_{Aee} \;; \qquad \frac{d\rho_{Aeg}}{dt} = -i\omega_0\rho_{Aeg} - \frac{\Gamma}{2}\rho_{Aeg} \;. \tag{4.72}$$

These equations describe the well-known statistical features of the spontaneous emission: an exponential decay of the excited state probability with the rate Γ and of the atomic coherence with the rate $\Gamma/2$. Note that Γ cannot be obtained through the simple and general argument leading to the Lindblad form of the master equation. It can, however, be determined by a Fermi golden-rule argument describing explicitly the coupling of the atomic transition to the continuum of field modes. We recall that, for a two-level atom with an electric dipole d between $|e\rangle$ and $|g\rangle$, this rate is (Cohen-Tannoudji *et al.* 1992b):

$$\Gamma = \frac{d^2\omega_{eg}^3}{3\pi\varepsilon_0\hbar c^3} \;, \tag{4.73}$$

a formula which will be useful to remember shortly.

Relaxation of a cavity field in a thermal environment

Let us now discuss an important example in the CQED context, the relaxation of a cavity field. The system A is a field mode stored in a cavity, coupled to the environment E (other freely propagating field modes coupled by diffraction on mirror defects, or electrons and Cooper pairs in the mirrors, etc). What are the possible jump operators? Assuming, reasonably, that the coupling with the environment is linear in the cavity field amplitude, there are only two possible jumps. The mode may lose a photon in the environment or gain a photon from it.[9]

The environment simulator B has, accordingly, three different states, the initial one, $|0^{(B)}\rangle$, and $|-^{(B)}\rangle$ and $|+^{(B)}\rangle$, corresponding respectively to the loss or to the gain of a photon by the cavity field. These latter states are associated to the jump operators L_- and L_+. The first, L_-, is proportional to the annihilation operator a: $L_- = \sqrt{\kappa_-}a$. The second, L_+, is proportional to the photon creation operator: $L_+ = \sqrt{\kappa_+}a^\dagger$.

---

[9]Other more exotic kinds of environments, which can absorb or emit several quanta at a time, are considered in Section 8.3. The system is then a mechanical oscillator (ion oscillating in a trap) and the photons of the present analysis are replaced by phonons.

We can relate the rates κ_- and κ_+ by a general thermodynamical argument, independent of the specifics of E. Any process in which A loses (or gains) a quantum must be accompanied by a correlated jump of the environment in the opposite direction, a transition upwards (or downwards) on the energy ladder of E. The probabilities for these transitions to occur are proportional to those for finding in E an initial state linked to a final state having with it an energy difference $\pm\hbar\omega_c$, necessary to conserve the energy of the $A + E$ system. A given transition in the environment will thus contribute to the gain and loss of photons in a ratio equal to the probability of finding E in the upper state of this transition divided by the probability of finding it in its lower state. Assuming that the environment is in thermal equilibrium at temperature T, this ratio is simply equal to $\exp(-\hbar\omega_c/k_bT)$ (Boltzmann law). We thus have:

$$\kappa_+ = \kappa_- e^{-\hbar\omega_c/k_bT} \ . \tag{4.74}$$

The relation between the loss and gain rates can be expressed in a more transparent form, by introducing the average number $n_{\rm th}$ of thermal photons per mode at frequency ω_c, given by Planck's law:

$$n_{\rm th} = \frac{1}{e^{\hbar\omega_c/k_bT} - 1} \ . \tag{4.75}$$

Equation (4.74) can then be rewritten as:

$$\frac{\kappa_-}{\kappa_+} = \frac{1 + n_{\rm th}}{n_{\rm th}} \ , \tag{4.76}$$

which leads us to define κ_- and κ_+ in term of a unique cavity rate κ as:

$$\kappa_- = \kappa(1 + n_{\rm th}) \ ; \qquad \kappa_+ = \kappa n_{\rm th} \ , \tag{4.77}$$

and to write the Lindblad equation for the cavity field as:

$$\begin{aligned}
\frac{d\rho_A}{dt} &= -i\omega_c \left[a^\dagger a, \rho_A \right] - \frac{\kappa(1 + n_{\rm th})}{2} \left(a^\dagger a\rho_A + \rho_A a^\dagger a - 2a\rho_A a^\dagger \right) \\
&\quad - \frac{\kappa n_{\rm th}}{2} \left(aa^\dagger \rho_A + \rho_A aa^\dagger - 2a^\dagger \rho_A a \right) \ .
\end{aligned} \tag{4.78}$$

The commutator term in this equation can be removed by switching to an interaction representation with respect to the field Hamiltonian $\hbar\omega_c a^\dagger a$. The relaxation terms are not modified, since a is changed into $a\exp(-i\omega_c t)$ while $a^\dagger$ becomes $a^\dagger \exp(i\omega_c t)$.

The field master equation turns into a simple rate equation for the photon number distribution $p(n) = \rho_{Ann} = \langle n| \rho_A |n\rangle$. Using eqn. (4.78) and the action of a and $a^\dagger$ on Fock states, we obtain a closed set of equations for $p(n)$:

$$\frac{dp(n)}{dt} = \kappa(1 + n_{\rm th})(n+1)p(n+1) + \kappa n_{\rm th} n p(n-1) - [\kappa(1 + n_{\rm th})n + \kappa n_{\rm th}(n+1)]p(n) \ . \tag{4.79}$$

The first two terms on the right-hand side describe the transitions towards the n-photons state $|n\rangle$, increasing its occupation probability. They originate either from the upper state $|n + 1\rangle$, at a rate $\kappa(1 + n_{\rm th})(n + 1)$ or from the lower level $|n - 1\rangle$, with a rate $\kappa n_{\rm th} n$. The last term describes the transition out of state $|n\rangle$, towards

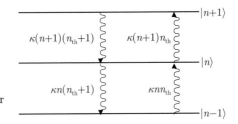

Fig. 4.2 Transition rates in the Fock state ladder under cavity relaxation.

$|n+1\rangle$ or $|n-1\rangle$. The transitions between the Fock states and the corresponding rates are summarized in Fig. 4.2.

These rates can be used to determine the photon statistics in the steady-state solution of the master equation. We invoke a 'detailed balance argument', stating that the transfer rates from $|n\rangle$ to $|n-1\rangle$ and from $|n-1\rangle$ to $|n\rangle$ are equal when the system A ceases to evolve:

$$\kappa(1+n_{\text{th}})np(n) = \kappa n_{\text{th}}np(n-1) , \qquad (4.80)$$

which leads immediately to:

$$\frac{p(n)}{p(n-1)} = \frac{n_{\text{th}}}{1+n_{\text{th}}} = e^{-\hbar\omega_c/k_bT} . \qquad (4.81)$$

We thus find, not unexpectedly, that the mode A ends up in the Boltzman equilibrium distribution at the temperature T. The average steady-state photon number is then given by Planck's law (4.75).

It is also instructive to determine how the average photon number $\langle N \rangle = \sum_n np(n)$ reaches this equilibrium. Calculating the rate $d\langle N \rangle /dt = \sum_n n\, dp(n)/dt$ with the help of eqn. (4.79) leads to:

$$\frac{d\langle N \rangle}{dt} = -\kappa(\langle N \rangle - n_{\text{th}}) . \qquad (4.82)$$

The κ coefficient is the exponential decay rate of the field mean energy towards its thermal equilibrium, a quantity which is classically denoted as $\omega_c/Q = 1/T_c$, where Q is the dimensionless mode quality factor. In spite of the obvious environment's complexity, cavity relaxation depends only upon two simple dimensionless parameters, Q and n_{th}. They both generally depend upon T (when the cavity is superconducting, the mirror losses, and hence Q, vary rapidly with temperature, see Section 5.2.2).

Atom relaxing in a thermal field: spontaneous and stimulated processes

Let us now briefly come back to the 'spin' relaxation. We have considered above its damping due to spontaneous emission at $T = 0$ K. The master equation (4.71) can be simply modified to describe the coupling of the two-level atom with a field in thermal equilibrium at a finite temperature. In fact, the arguments developed above for a cavity field are quite general and can be extended immediately to the atomic relaxation, merely replacing a and $a^\dagger$ by σ_- and σ_+ and κ by Γ. In the presence of a thermal environment, the jump operator $\sqrt{\Gamma}\sigma_-$ thus becomes $\sqrt{\Gamma(1+n_{\text{th}})}\sigma_-$ and a new Lindblad operator, $\sqrt{\Gamma n_{\text{th}}}\sigma_+$ describing the thermal excitation of the atom needs to be added. The two-level atom master equation thus becomes:

$$\frac{d\rho_A}{dt} = -i\frac{\omega_{eg}}{2}[\sigma_z, \rho_A] - \frac{\Gamma(1+n_{\text{th}})}{2}(\sigma_+\sigma_-\rho_A + \rho_A\sigma_+\sigma_- - 2\sigma_-\rho_A\sigma_+)$$

$$-\frac{\Gamma n_{\text{th}}}{2} \left(\sigma_- \sigma_+ \rho_A + \rho_A \sigma_- \sigma_+ - 2\sigma_+ \rho_A \sigma_-\right) . \tag{4.83}$$

The first line in the right-hand side describes the processes in which the atom loses photons into the continuum of field modes. These are described by the same term as in the $T = 0$ K case, with a mere enhancement of the emission rate by the factor $1 + n_{\text{th}}$. The '1' term describes spontaneous emission and the n_{th} one is due to stimulated processes in which the atom emits in modes already containing n_{th} thermal photons on average. The second line describes the symmetrical processes in which the atom absorbs thermal photons from the field. The corresponding rate is proportional to n_{th}, without the 1 term, because there is of course no 'spontaneous absorption' of energy from the vacuum. The atomic relaxation is fully described by two constants, the spontaneous emission rate Γ and the average thermal photon number n_{th}. By developing similar arguments to describe the thermalization of atoms with radiation, Einstein has, in 1917, understood the necessity to consider stimulated emission as an essential process of atom–field energy exchanges.

4.4 Quantum Monte Carlo trajectories

Most often, the Lindblad master equation which describes the relaxation process in a complete way cannot be solved analytically. One has then to resort to numerical solutions, which are difficult to obtain when A is large. The number of elements of the density matrix, ρ_A, is the square of the Hilbert space dimension, N_A^2. The calculations rapidly become too cumbersome to be performed, even with supercomputers. In comparison, it is much simpler to follow the Hamiltonian evolution of a wave function, since it involves only N_A complex variables.

We have seen above that the environment simulator description allows us to understand the system's relaxation from t to $t + \tau$ as an elementary unitary evolution entangling A to an external system B having as many states as the number of Lindblad operators plus one. The infinitesimal evolution of the density matrix is recovered by tracing the quantum state at $t + \tau$ over all results of measurements virtually performed on B. Computing the quantum state resulting from the elementary unitary $A + B$ interaction is a simple as following an ordinary Hamiltonian evolution.

By extending this procedure to the whole time interval, we now introduce the notion of random evolutions of the system's state vectors, 'quantum Monte Carlo trajectories', whose average over many realizations allow us to recover the full density matrix (Dalibard *et al.* 1992; Plenio and Knight 1998). This is a very efficient scheme for numerical simulations. More fundamentally, the quantum Monte Carlo point of view provides an illuminating insight into relaxation mechanisms. We start by presenting the principle of the method. We next analyse the deep difference between single trajectories and averages described by the density matrix. We then use the Monte Carlo approach to describe the quantum trajectories followed by some simple systems undergoing relaxation: a two-level atom in a superposition of its excited and ground state, a Fock state and a coherent state of radiation in a cavity. We comment on the counter-intuitive features of some of these trajectories and reach interesting conclusions concerning the different ways in which a quantum system's wave function is changed via information acquisition. Other examples of Monte Carlo simulations applied to the evolution of Schrödinger cat states of the field are given in Chapter 7.

4.4.1   The Monte Carlo recipe

We assume again, without loss of generality, that A is initially, at time $t = 0$, in a pure state $|\phi^{(A)}(0)\rangle$. The evolution is divided in small time steps, each with a duration τ. During each step, the action of the environment E is mimicked by an environment simulator B. At the end of the first step (at time τ), Bob performs a measurement of the observable O_B indicating which quantum jump occurred, if any. Instead of tracing over all possible measurement results, we assume that Bob keeps a record of the obtained value. This measurement thus projects A onto a new state vector, which is taken as the initial condition for the next step. The environment simulator B is then re-prepared in its neutral state $|0^{(B)}\rangle$. This periodic initialization of B reflects the fact that the short-memory real environment E is not modified during the relaxation process and that its entanglement with A at the beginning of each time interval can be neglected to compute the system's state at the end of this interval. The total evolution is followed in this way, step after step. At any stage, A is in a pure state, which depends upon the previous measurement results stored by Bob. The exact density matrix evolution is then recovered by averaging the projector on the state of A over very many realizations of such evolutions.

In a numerical simulation, the measurements performed by Bob are conditioned to classical dice tossings reproducing the probabilities of outcomes of the various jumps undergone by the system. The recursive algorithm to simulate the evolution of A is as follows:

- From the initial state of A at time t, $|\phi^{(A)}(t)\rangle$, compute the no-jump probability p_0 (eqn. 4.67) and the $N_K - 1$ jump probabilities p_μ associated to the jump operators L_μ (eqn. 4.69).
- Choose whether a jump occurs or not. If it does occur, choose randomly the corresponding jump operator L_ν. Both choices are performed by drawing a single random value r, chosen between 0 and 1. No jump occurs if $r < p_0$. If a jump takes place, its index ν is the smallest integer such that $\sum_{\mu=0}^{\nu} p_\mu > r$.
- In the no-jump case, compute the elementary evolution of $|\phi^{(A)}\rangle$ under the effective non-hermitian Hamiltonian $H_e = H_A - i\hbar J$. Renormalize the final result using the value of p_0 (eqn. 4.68) to get the new state in this case.
- In the event of a jump, compute the new state and renormalize it with p_ν (eqn. 4.70).
- Repeat the sequence of steps with the resulting state $|\phi^A(t + \tau)\rangle$ as a new initial state.

In many cases, it is not necessary to apply the complete algorithm to all time steps. The probability distribution for the time of the next jump can often be computed explicitly. The jump time is chosen randomly according to this distribution. The non-hermitian effective Hamiltonian evolution is directly integrated up to this jump time.

A large number of individual quantum trajectories, $|\phi_j^{(A)}(t)\rangle$, all starting from the same initial state, are constructed in this way. From the definition of the procedure, the time-dependent density matrix $\rho_A(t)$ is obtained by an average of the projectors $\Pi_j = |\phi_j^{(A)}(t)\rangle\langle\phi_j^{(A)}(t)|$. An additional average should be performed on the initial state of A when it corresponds to a statistical mixture.

As a consistency check, it is easy to verify that the average of Π_j over very many trajectories, $\overline{\Pi}$, obeys the same Lindblad equation as the density matrix ρ_A. After a time interval τ, $\overline{\Pi}(t+\tau)$ is a mixture of the no-jump and jump evolutions, with weights p_0 and $1-p_0$. Replacing in this mixture the jump and no-jump wave functions by their explicit expressions (eqns. 4.68 and 4.70), and setting $d\overline{\Pi}/dt = [\overline{\Pi}(t+\tau) - \overline{\Pi}(t)]/\tau$, we get finally:

$$\frac{d\overline{\Pi}}{dt} = -\frac{i}{\hbar}\left[H_A, \overline{\Pi}\right] + \sum_\mu \left(L_\mu \overline{\Pi} L_\mu^\dagger - \tfrac{1}{2}\overline{\Pi} L_\mu^\dagger L_\mu - \tfrac{1}{2} L_\mu^\dagger L_\mu \overline{\Pi}\right) , \qquad (4.84)$$

which is precisely the Lindblad form of the master equation. Since $\overline{\Pi}(0)$ coincides, by construction, with the initial density matrix, $\rho_A(t)$ and $\overline{\Pi}(t)$ coincide at all times. The Lindblad and Monte Carlo approaches are fully equivalent.

The picture which emerges from this Monte Carlo analysis for the evolution of A under Alice control, is that of a stroboscopic measurement of the environment by Bob, performed with a period $\tau \gg \tau_c$. Since the results of these measurements are unknown to Alice, they correspond for her to tracing $A + E$ over E to obtain $\rho_A(t)$. Whether these stroboscopic measurements or trace operations are periodically performed or not, the final density matrix of A is the same. The Markov condition $\tau \gg \tau_c$ means, however, that successive observations of the environment should not be performed within its memory time. This restrictive condition makes sense. If the environment was interrogated at a rate larger than $1/\tau_c$, it would be repeatedly projected on the same eigenstate of the measured observable, resulting in a 'freezing' of the evolution of the $A + E$ system. This inhibition of a system's evolution under the effect of observations repeated fast enough is known as the quantum Zeno effect (Misra and Sudarshan 1977; Spiller 1994). The irreversible Markovian evolution of an open quantum system implies that its large environment is effectively monitored at a rate too slow for the quantum Zeno effect to occur.

The quantum Monte Carlo procedure is very efficient for numerical simulations. As mentioned above, it only requires the computation of a wave vector, with N_A coefficients. Many trajectories have to be computed, in principle. However, a reasonable approximation of ρ_A is rapidly obtained. Moreover, the simulations of different trajectories are independent calculations, well-adapted to parallel or distributed processing. As soon as the Hilbert space dimension is larger than a few tens, the Monte Carlo method is overwhelmingly more efficient than the direct solution of the master equation. This is one of the reasons which explain the success of this approach in quantum optics. The other reason has to do with the recent development of real experiments in which quantum trajectories can actually be observed.

4.4.2 Quantum trajectories for spontaneous emission

Beyond its mere computational convenience, the Monte Carlo method sheds light on the difference between a quantum average and an individual quantum trajectory. The density matrix describes an ensemble average over very many identical realizations of an experiment. The individual trajectories of the Monte Carlo situation correspond to a single realization of one of those experiments, in which the environment is continuously monitored to witness the quantum jumps.

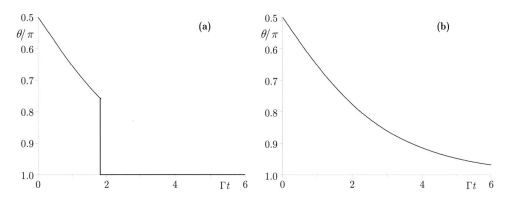

Fig. 4.3 Two typical Monte Carlo trajectories describing the spontaneous emission of a two-level atom initially in an equal probability superposition of its excited and ground states. Polar angle $\theta(t)$ of the atomic Bloch vector, in units of π, as a function of Γt (the θ axis is oriented downwards, to represent an evolution towards the south pole of the Bloch sphere). (a) Discontinuous trajectory corresponding to the emission of a photon. (b) Continuous trajectory corresponding to no emission.

Two-level atom spontaneously emitting a photon

Let us illustrate this fundamental difference by coming back to the simple thought experiment considered above, in which the spontaneous emission of a single atom, initially in its upper state $|e\rangle$, surrounded at $T = 0$ K by photodetectors is monitored. The corresponding master equation (4.71) involves a single jump operator, $L_1 = \sqrt{\Gamma}\sigma_-$. It predicts an exponential decay of the excited state population ρ_{Aee}, with a rate Γ. A single realization of the experiment exhibits a very different evolution. At a random time, one of the detectors clicks, indicating that the atom has experienced a jump. Just before the click (we neglect propagation delays), the atom was in $|e\rangle$. Immediately after, it is in $|g\rangle$. On such a trajectory, the population of $|e\rangle$ evolves as a step function. The time of the emission is, however, random. The average of these step functions over many trajectories with different switching times yields the exponential decay.

In order to understand more deeply the meaning of the non-hermitian evolution in the no-jump case, assume now that the atom has been initially prepared in the superposition $|\Psi_0\rangle = (|e\rangle + |g\rangle)/\sqrt{2}$. It has a $1/2$ probability of emitting at some time a photon and the same probability of never emitting one. Half of the Monte Carlo trajectories will thus exhibit a jump, and the other half not. In all cases, if we wait long enough, the atom will however be found in the ground state $|g\rangle$. How is this possible? In the case that a jump occurred, the explanation is simple. The emission of a photon has suddenly projected the atom in its ground state.

If no jump is detected, the evolution into the final state is less easy to interpret. The atomic state has changed continuously from $(|e\rangle + |g\rangle)/\sqrt{2}$ into $|g\rangle$ under the effect of the non-hermitian term $-i\hbar J = -(i\hbar/2)\Gamma\sigma_+\sigma_-$. After a time t without photon click, the state has become:

$$\left|\Psi^{\text{no click}}(t)\right\rangle = \frac{1}{\sqrt{\mathcal{N}}}e^{-\Gamma\sigma_+\sigma_- t/2}|\Psi_0\rangle = \frac{1}{\sqrt{2\mathcal{N}}}\left[e^{-\Gamma t/2}|e\rangle + |g\rangle\right], \qquad (4.85)$$

where $\mathcal{N} = 1 + e^{-\Gamma t}$. The longer the time without photodetection, the larger is the probability amplitude of finding the atom in $|g\rangle$. The system's wave function then evolves not because a physical emission process occurred, but because the knowledge about the system has changed. If no photon is detected during a time interval, it becomes logically more likely that the atom is in the state which cannot emit light. The non-unitary part of the Monte Carlo wave function evolution thus accounts for an acquisition of information due to a negative measurement ('no click' in the environment). We give in the next section another example of such a continuous evolution of a quantum state induced by the observation of an unchanged environment.

This discussion is illustrated in Fig. 4.3 which represents two typical Monte Carlo trajectories, one corresponding to photon emission (Fig. 4.3a) and the other to the absence of a click (Fig. 4.3b). In both cases, we plot the polar angle $\theta(t)$ defined by:

$$|\Psi(t)\rangle = \cos \frac{\theta(t)}{2} |e\rangle + \sin \frac{\theta(t)}{2} |g\rangle \ , \tag{4.86}$$

as a function of time $[\theta(0) = \pi/2, |g\rangle$ corresponds to $\theta = \pi]$. The Bloch vector representing the atom evolves from the equatorial plane to the south pole of the Bloch sphere. In the trajectory where a photon is recorded, the evolution is a continuous rotation down the south hemisphere, followed by a sudden jump projecting the Bloch vector onto its final position. In the trajectory where no click occurs, the vector rotates continuously until it reaches the south pole.

In experiments involving large atomic ensembles, the observed fluorescence signal is the average over very many individual emissions and is well-described by the density matrix formalism. The quantum trajectory approach is then a useful computational trick, but is not essential to account for the observed signals. It is in recent experiments in which single quantum systems are really monitored (see Chapters 1, 2, 7, and 8), that the Monte Carlo approach shows all its power. More than a convenient procedure, it becomes then an essential theoretical tool to account for the observations. These single systems (atoms or ions in traps, fields in cavities) do exhibit detectable quantum jumps. In some cases (ions in traps) these jumps are visible with the naked eye. In the density matrix approach, this conspicuous jump behaviour is somewhat 'hidden', and the existence of the jumps is recovered only through the calculation of complex correlations functions (Cohen-Tannoudji and Dalibard 1986). The Monte Carlo approach makes them much more accessible. Of course, real environments are more complex than the ideal one considered above, but the Monte Carlo method can easily be modified to account for these complications (by adding for instance an extra dice tossing to account for the finite detection efficiency of photodetectors). Realistic simulations of 'typical' observed quantum trajectories can be obtained in this way.

Spontaneous emission by a driven two-level atom: photon anti-bunching

As an interesting example of such single-atom experiments, let us consider a two-level atom coupled by spontaneous emission to the vacuum modes of the field and driven by a classical field inducing a Rabi oscillation between $|e\rangle$ and $|g\rangle$. In this case, the atom is periodically re-excited and can emit a large number of light quanta. This situation corresponds, for instance, to an atom crossing an intense resonant laser beam and resonantly scattering photons from this beam in all directions.

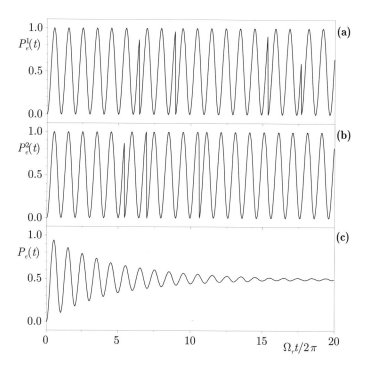

Fig. 4.4 Spontaneous emission by a driven atom. (a) and (b) Probabilities $P_e^j(t)$ of finding at time t in level $|e\rangle$ an atom which has been reset at time $t = 0$ in level $|g\rangle$, corresponding to two typical quantum trajectories ($j = 1, 2$). The classical Rabi frequency is $\Omega_r = 21\Gamma$. (c) Average $P_e(t)$ of 10^4 trajectories. $P_e(t)$ is proportional to the photon correlation function $g^{(2)}(t)$ which presents a zero minimum at $t = 0$ (photon anti-bunching).

The master equation for this system is obtained by a simple modification of eqn. (4.71), adding in the commutator term the time-dependent contribution $H_r = -D\,E_r$, which describes the effect of the classical resonant field E_r coupled to the atomic dipole D (see eqn. 3.147). Solving this master equation by the Monte Carlo method yields quantum trajectories which simulate an experiment in which detectors placed around the atom would detect the sequence of photons randomly scattered. Figure 4.4(a) and (b) shows, on different trajectories labelled by the index j, the probability $P_e^{(j)}(t)$ of finding at time t in state $|e\rangle$ an atom initially prepared in state $|g\rangle$. These simulations correspond to a classical Rabi frequency $\Omega_r = 21\,\Gamma$. On each realization, the atomic probability oscillates for some time, then undergoes a sudden jump which resets the atom in state $|g\rangle$, before resuming its oscillation. Each jump corresponds to the detection of a photon by one of the virtual detectors surrounding the atom. Recording the sequence of these jumps on individual trajectories simulates the photon click sequence as it would be observed in an experiment with perfect detectors.

Summing the contributions of many such trajectories yields the quantum average $P_e(t)$, shown in Fig. 4.4(c). It exhibits an oscillation at frequency $\Omega_r/2\pi$, whose amplitude decays exponentially with the rate $\Gamma/2$. This damped signal, starting from $P_e(0) = 0$ at $t = 0$, can be given a very simple physical interpretation. The quantity

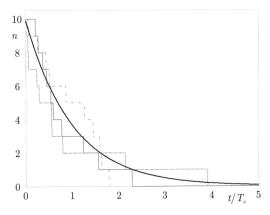

Fig. 4.5 Quantum Monte Carlo simulation of a 10-photon Fock state relaxation in a $T = 0$ K environment. Photon number n as a function of time in units of the cavity damping time T_c. Thin lines (solid, dashed and dotted): three individual staircase trajectories. Thick line: average over 10 000 trajectories corresponding to the usual exponential decay.

$\Gamma P_e(t)\,dt$ is the conditional probability that, in a perfect experiment, a scattered photon will be detected in the time interval $[t, t+dt]$, following detection of a previous photon at time $t = 0$. The counting of a photon by a detector located at an arbitrary position $\mathbf{r}_1$ resets indeed the atom in level $|g\rangle$, so that the quantum trajectories shown in Fig. 4.4(a) and (b) describe the histories of atoms which have yielded a photon click in one of the detectors at $t = 0$, before being subsequently re-excited by the classical field. The conditional probability $\Gamma P_e(t)$ is proportional to $\int d^3\mathbf{r}_1 d^3\mathbf{r}_2\, g^{(2)}(\mathbf{r}_1, 0; \mathbf{r}_2, t)$, where $g^{(2)}(\mathbf{r}_1, 0; \mathbf{r}_2, t) = \langle E^-(\mathbf{r}_1, 0)E^-(\mathbf{r}_2, t)E^+(\mathbf{r}_2, t)E^+(\mathbf{r}_1, 0)\rangle$ is the second-order correlation function of the steady-state field scattered by the atom, proportional to the two-photon counting rate defined by eqn. (3.87), on page 124.

This correlation function is zero at $t = 0$, meaning that two photons are never detected immediately after one another. This photon anti-bunching effect, characteristic of a non-classical field state, has been observed in a pioneering experiment (Kimble *et al.* 1977). It is one of the first examples of single quantum system observation. The physical explanation for this effect is very clear in the Monte Carlo point of view. The atom being projected into its ground state by photon emission needs a time of the order of $1/\Omega_r$ to be re-excited before it has a chance to emit another photon.

4.4.3 Monte Carlo trajectories of a relaxing Fock state

It is very instructive to apply the quantum Monte Carlo approach to cavity mode relaxation. We start by considering the evolution from an initial Fock state $|n\rangle$. Consider first an environment at zero temperature. The single jump operator, $L_1 = \sqrt{\kappa}a$, describes then the loss of a photon by the cavity mode. The non-hermitian part of the Hamiltonian describing the 'no-jump' evolution reduces to $H_e = -i\hbar J = -i\hbar\kappa a^\dagger a/2$. The effective Hamiltonian is proportional to the cavity one $H'_c = \hbar\omega_c a^\dagger a$. The no-jump evolution is thus formally obtained by setting the mode frequency at the complex value $\omega_c - i\kappa/2$. During the first time interval τ, the probability of a jump is $p_1 = \tau\kappa\langle n|a^\dagger a|n\rangle = \kappa\tau n$, an intuitive result, which could have been guessed from eqn. (4.79). Being an eigenvector of H'_c, $|n\rangle$ is also an eigenvector of the non-hermitian 'no-jump' Hamiltonian. Besides an irrelevant phase factor, the state thus does not evolve in the no-jump case. When a jump occurs, $|n\rangle$ is transformed into $|n-1\rangle$. Throughout the evolution, the cavity thus remains in a Fock state, the photon number decreasing by one unit at each jump. The field energy is a staircase function of

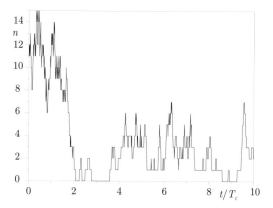

Fig. 4.6 Quantum Monte Carlo simulation of a single trajectory corresponding to a 10-photon Fock state relaxing in an environment at a finite temperature with $n_{\text{th}} = 3$.

time. The jump rate is the largest at the start, when the photon number is maximum. The last photon is lost at a rate κ, which is also the classical energy damping rate.

Figure 4.5 presents simulations of a 10-photon Fock state decay. The staircase curves correspond to random individual quantum trajectories. All eventually end up in the vacuum state $|0\rangle$. The evolution of the standard deviation, ΔN, of the photon number can be guessed from these curves. Zero in the initial and final states, it reaches a maximum during the evolution. The solid line presents the evolution of the average photon number obtained from 10 000 such trajectories. It is in agreement with the exponential decay with a rate κ predicted by the density matrix approach, which cannot be distinguished from the statistical average at this scale.

For a finite environment temperature T, L_1 is multiplied by $\sqrt{1 + n_{\text{th}}}$ and another jump operator $L_2 = \sqrt{n_{\text{th}}\kappa}a^\dagger$ must be considered. The two possible jumps lower or increase the photon number by one. The field is again at any time in a Fock state, but there is no stationary final state for any trajectory. After a transient regime lasting a time $\sim 1/\kappa$ during which the average field energy adjusts to its equilibrium value, the photon number keeps jumping up and down randomly around n_{th} (Fig. 4.6).

4.4.4 Paradoxical quantum trajectory of a damped coherent state

Let us now describe the $T = 0$ K Monte Carlo trajectory starting from an initial coherent state $|\beta\rangle$. The properties of this trajectory, which may seem paradoxical at first sight, will help us explain the important features of coherent state relaxation, which we have already discussed in Chapter 3 (no entanglement with the environment and exponential decay of the field amplitude).

The probability of counting a photon in the first time interval is $p_1 = \kappa\tau\langle\beta|a^\dagger a|\beta\rangle = \kappa\tau|\beta|^2 = \kappa\tau\bar{n}$. If a photon is counted, the state is unchanged since $|\beta\rangle$ is an eigenstate of the jump operator $\sqrt{\kappa}a$. An evolution occurs, on the contrary, if no photon is recorded in this time bin. The imaginary frequency contribution to the effective Hamiltonian produces a decrease of the state complex amplitude, the field evolving, in the interaction picture with respect to its free Hamiltonian, according to:

$$|\beta\rangle \rightarrow \left|\beta e^{-\kappa\tau/2}\right\rangle . \tag{4.87}$$

This situation is quite paradoxical, since the field's state amplitude decreases only if no photon is recorded! How is this possible? First of all, how come that the state does not change when we are sure that one photon has been lost? We have here the combination of two effects, which tend to change the photon number in opposite directions. On the one hand, the loss of a quantum reduces the average energy in the cavity. On the other hand, the registration of a photon click modifies our knowledge about the state in which the field was just before the photon was emitted, corresponding to an increase by precisely one unit of the photon number expectation at this time.

To understand this point, let us ask a question whose answer is a simple exercise in conditional probabilities: knowing the *a priori* Poissonian photon number distribution $p(n) = e^{-\bar{n}} \bar{n}^n / n!$ (centred at $n = \bar{n}$ at time t) *and* that a photon has been lost and detected as a 'click' c in the time interval $(t, t + \delta t)$, what is the *conditional* probability $p(n|c)$ that the field mode contained n photons at time t? If the photon emission probability were n-independent, $p(n|c)$ would merely be equal to $p(n)$. The probability of losing a photon is however proportional to the expectation value of $a^\dagger a$, hence to n. Detecting a photon has thus the effect of shifting the probability distribution towards larger $n's$, since the corresponding states are more likely to lose quanta.

This result can be made quantitative by invoking Bayes law on conditional probabilities. The probability we seek, $p(n|c)$, is related to the reverse conditional probability $p(c|n)$ of detecting a click during the same time interval, *conditioned* to the prior existence of n photons. This probability is obviously $p(c|n) = \kappa n \delta t$. It is slightly different from the *a priori* probability $p_c = \kappa \bar{n} \delta t$ of detecting a click when the field is in the coherent state, a superposition of n states around the mean $\bar{n}$. We can then relate $p(n|c)$ and $p(c|n)$ by expressing in two different ways the *joint probability* $p(n, c)$ that a click is detected during the considered time interval *and* that the field contains n photons:

$$p(n, c) = p(c|n)p(n) = p(n|c)p_c , \qquad (4.88)$$

which yields immediately the Bayes law result:

$$p(n|c) = p(n)\frac{p(c|n)}{p_c} = \frac{n}{\bar{n}}p(n) = e^{-\bar{n}}\frac{\bar{n}^{n-1}}{(n-1)!} = p(n-1) . \qquad (4.89)$$

This formula shows that $p(n|c)$ is precisely equal to the *a priori* probability that the initial field contains $n - 1$ photons. The maximum of $p(n - 1)$, hence of $p(n|c)$, occurs for $n = \bar{n} + 1$. The knowledge that a click has occurred biases the photon number distribution before this click towards larger n values, with a photon number expectation exceeding the *a priori* average value by exactly one unit. The loss of the photon signalled by the click brings this number back down to $\bar{n}$. The two effects exactly cancel each other and the state of the field has not changed!

This is a very peculiar property of coherent states, a consequence of their Poissonian photon number distribution. For non-Poissonian fields with larger fluctuations, this effect could be even more counter-intuitive and result in an overall increase of the photon expectation number, just after the loss of a photon! We leave the reader to construct a photon number distribution leading to such a situation. Note finally that this counter-intuitive increase of the field energy upon loss of photons requires the existence of fluctuations in the initial energy distribution. The effect does not occur for a Fock state which jumps down the energy ladder at each click and loses its energy in a perfectly intuitive way.

The apparent strangeness of the coherent state behaviour as photons get lost is not a quantum effect. It is rather a general property of conditional probabilities, which present analogies with classical situations in everyday life. The distribution of the wealth among the population of a town presents large fluctuations. If you pick up a lady leaving the cashier of an elegant department store, it is likely that – in spite that she has just given out some money – she is still wealthier than the average person in town.

We have now to understand the other side of the coherent state evolution paradox. Why are they decreasing in amplitude precisely during the time intervals when no photon clicks are registered? We have already encountered the same effect (continuous quantum state evolution without jump) when analysing the spontaneous emission of a single atom in a superposition of its emitting and non-emitting states. It is here again a matter of conditional probability. If no photon is detected in a time interval, it is more likely that the photon distribution has less photons that was *a priori* assumed.

To consider an extreme situation, a coherent field has a small probability, equal to $e^{-\overline{n}}$, to contain no photon at all. This is the probability that the detector will never click, however long one waits. The longer the time interval t without click, the more likely it becomes that the field is effectively in vacuum. Its wave function evolves continuously without jump under the effect of the non-hermitian operator $1 - i\kappa a^\dagger a/2$. If t becomes very long, the field ends up effectively in $|0\rangle$. This seems counter-intuitive, but we must not forget that this event has a very small probability of occurring. This example is once again a striking illustration of an important aspect of measurement theory. Even the absence of a click is information whose knowledge affects the wave function of the field.

In Monte Carlo simulations, the randomness is generated only by the jumps. The evolution between the jumps is continuous and deterministic. The insensitivity of a coherent state to jumps makes its Monte Carlo trajectory certain. To determine the state at time t of a field initially in state $|\beta\rangle$, we have to piece together its partial evolutions during the successive intervals between jumps $t_1, t_2, \ldots, t_i, \ldots, t_N$ adding up to t. Whatever the distribution of these intervals, the final state is uniquely defined, in the interaction picture, as $\left|\beta e^{-\kappa(t_1+t_2+\ldots+t_N)/2}\right\rangle = \left|\beta e^{-\kappa t/2}\right\rangle$. The field state remains pure, as already indicated by eqn. (4.87). We retrieve here that coherent states are impervious to entanglement with the environment under the effect of photon loss, a property which we have derived in Chapter 3 from a different argument. This basic feature is a direct consequence of their counter-intuitive evolution during and between photon clicks.

4.4.5   Monte Carlo simulations, jump operators and pointer states

Coherent states of a field A, which do not get entangled with their environment E and stay pure throughout the system's evolution, are 'pointer states' of the damping process, in the way defined in Chapter 2. We can also assume without loss of generality that the environment interacting with this coherent field remains in a time-dependent pure state. If the initial state of E is a mixture, it is always possible to consider an auxiliary environment E_a, uncoupled to A, and to define in the combined $E + E_a$ system, viewed as a new environment E', a purification $\rho_{E'}(0) = \left|\Psi^{(E')}(0)\right\rangle\left\langle\Psi^{(E')}(0)\right|$ whose partial trace over E_a reproduces the initial density operator of E. Obviously, the effect of E' on A is identical to that of the real environment E. The extended environment E' thus starts in a pure state and remains pure throughout the evolution since it remains unentangled with the field. From the initial $A + E'$ tensor product state, the combined system evolves while remaining an unentangled tensor product of the field state $|\alpha e^{-\kappa t/2}\rangle$ and an environment state $|\Psi^{(E')}(t)\rangle$ according to:

$$|\alpha\rangle \left|\Psi^{(E')}(0)\right\rangle \to \left|\alpha e^{-(\kappa/2+i\omega_c)t}\right\rangle \left|\Psi^{(E')}(t)\right\rangle . \tag{4.90}$$

Note that we include here the free field evolution. This very special evolution distinguishes coherent states from all other states of the field which, being superpositions of coherent states, get entangled with the environment and are quickly transformed into statistical mixtures. We devote Chapter 7 to the study of the Schrödinger cat states, superpositions of two coherent states with different phases, and analyse in detail their decoherence properties.

Exact pointer states

The notion of pointer state can be extended to other damping mechanisms corresponding to different Lindblad operators. The special behaviour of coherent states in quantum optics is related to their being eigenstates of the photon annihilation operator a. More generally, there are two necessary conditions that a state $|\psi^{(A)}\rangle$ should satisfy to be a pointer state of a relaxation process defined by a single Lindblad operator L:

- It should be, before the first jump at time $t = 0$, an eigenstate of L. This is required for this jump not to introduce any randomness in the Monte Carlo trajectory of the system:

$$L \left|\psi^{(A)}(0)\right\rangle = \lambda \left|\psi^{(A)}(0)\right\rangle . \tag{4.91}$$

- The non-unitary evolution between jumps should leave the state as an eigenstate of L. If the evolution mixes eigenstates with different eigenvalues, the first jump would be deterministic, but subsequent ones would result in a random projection of the system wave function on different states and the ensemble of quantum trajectories would correspond to a statistical mixture. Condition (4.91) should thus be supplemented by:

$$L \left|\psi^{(A)}(\tau)\right\rangle = \lambda' \left|\psi^{(A)}(\tau)\right\rangle , \tag{4.92}$$

with, up to first-order in τ:

$$\left|\psi^{(A)}(\tau)\right\rangle = \left[(1+\tau|\lambda|^2/2)\mathbb{1} - \frac{i\tau}{\hbar}H_A - \frac{\tau}{2}L^\dagger L\right] \left|\psi^{(A)}(0)\right\rangle , \tag{4.93}$$

where the coefficient in front of the unity operator accounts for the state normalization after the non-unitary evolution. We thus get:

$$\begin{aligned} L \left|\psi^{(A)}(\tau)\right\rangle &= (1+\tau|\lambda|^2/2)L \left|\psi^{(A)}(0)\right\rangle - \frac{i\tau}{\hbar}LH_A \left|\psi^{(A)}(0)\right\rangle - \frac{\tau}{2}LL^\dagger L \left|\psi^{(A)}(0)\right\rangle \\ &= \lambda \left|\psi^{(A)}(\tau)\right\rangle - \frac{i\tau}{\hbar}[L, H_A] \left|\psi^{(A)}(0)\right\rangle - \frac{\tau\lambda}{2}[L, L^\dagger] \left|\psi^{(A)}(0)\right\rangle . \end{aligned} \tag{4.94}$$

The second line is obtained from the first by replacing, in the first term of the right-hand side, $|\psi^{(A)}(0)\rangle$ by its expression in terms of $|\psi^{(A)}(\tau)\rangle$ (consistently to first order in τ). Note that the normalization factor is absorbed in this operation. A mere inspection of eqn. (4.94) shows that it implies eqn. (4.92) if the eigenstates of L are also eigenstates of $[L, H_A]$ and $[L, L^\dagger]$. This is in particular the case if L, $L^\dagger$ and H_A satisfy the following commutation relations:

$$[L, H_A] = f_1(L) ; \quad [L, L^\dagger] = f_2(L) , \tag{4.95}$$

where f_1 and f_2 are arbitray functions. Equation (4.94) then reduces – up to first order in τ – to eqn. (4.92) with:

$$\lambda' = \lambda - i(\tau/\hbar)f_1(\lambda) - (\tau\lambda/2)f_2(\lambda) . \tag{4.96}$$

The system, initially in an arbitrary eigenstate of L, remains in a pure state after the second application of L and, by iteration, after an arbitrary number of jumps. The commutation relations (4.95) are thus sufficient conditions to ensure the existence of pointer states of a decoherence process described by a single jump operator L.

Clearly, these conditions are satisfied for a coherent state $|\alpha\rangle$ undergoing photon jumps in a $T = 0$ K environment.[10] It is an eigenstate of the jump operator $L = a\sqrt{\kappa}$ with the eigenvalue $\alpha\sqrt{\kappa}$. The commutators of $a\sqrt{\kappa}$ with $H = H'_c$ and $a^\dagger\sqrt{\kappa}$ are $\sqrt{\kappa}\left[a, \hbar\omega_c a^\dagger a\right] = \sqrt{\kappa}\hbar\omega_c a$ and $\kappa\left[a, a^\dagger\right] = \kappa\mathbb{1}$, which satisfy eqn. (4.95) with $f_1(\lambda) = \hbar\omega_c\lambda$ (linear function) and $f_2(\lambda) = \kappa$ (constant function). Inserting these functions in eqn. (4.96) and replacing λ and λ' by α and α' yields $\alpha' = \alpha[1 - i\tau(\omega_c - i\kappa/2)]$. Iterating the procedure, we find that the coherent state evolves at time t into $\left|\alpha e^{-i\omega_c t}e^{-\kappa t/2}\right\rangle$, which corresponds to the result derived above (eqn. 4.90).

Decoherence-free subspaces

The eigenstates of L with the eigenvalue $\lambda = 0$ satisfying in addition the condition $LH_A\left|\Psi^{(A)}\right\rangle = 0$ are peculiar pointer states. Not only do they remain unentangled with E, but they also undergo a unitary evolution. They stay forever in this L-eigenspace as it can easily be deduced from eqn. (4.94). Once the system has fallen into this eigenspace, it remains trapped in a pure state evolving under the effect of H_A, without any influence from the environment. Since the probability of quantum jumps has vanished, no information is leaking form A towards E. Such an eigenspace of L is called a *decoherence-free subspace* (Lidar et al. 1998). If this eigenspace is non-degenerate, the corresponding pointer state is in addition time-invariant. For a field relaxing into a zero-temperature environment, the vacuum is a stationary decoherence-free state.

Hermitian jump operator: link with measurement-induced decoherence

The case of an hermitian Lindblad operator L_h commuting with the free Hamiltonian H_A deserves special attention. The commutator $\left[L_h, L_h^\dagger\right]$ then obviously vanishes and a simple inspection of eqn. (4.94) shows that if A is in an eigenstate $\left|\phi_{\lambda,\nu}^{(A)}\right\rangle$ of L_h at time t with eigenvalue λ, it remains an eigenstate with the *same* eigenvalue at time $t + \tau$, and hence, by iteration, at any subsequent time (λ is now a real quantity and

---

[10]In the $T \neq 0$ case, coherent states are no longer exact pointer states. Among all states, they correspond, however, to the minimum growth rate of entanglement with the environment (Zurek et al. 1993).

the ν index in $\left| \phi_{\lambda,\nu}^{(A)} \right\rangle$ allows for possible degeneracy of the L_h eigenspaces). It is easy to show that the master equation (in the interaction picture with respect to H_A), expressed in the $\left| \phi_{\lambda,\nu}^{(A)} \right\rangle$ basis of common eigenstates of H_A and L_h is:

$$\frac{d}{dt} \left\langle \phi_{\lambda,\nu}^{(A)} \middle| \rho_A \middle| \phi_{\lambda',\nu'}^{(A)} \right\rangle = -\frac{(\lambda - \lambda')^2}{2} \left\langle \phi_{\lambda,\nu}^{(A)} \middle| \rho_A \middle| \phi_{\lambda',\nu'}^{(A)} \right\rangle . \tag{4.97}$$

A coherence between two states with λ values differing by an amount $\delta\lambda$ decays at a rate proportional to $\delta\lambda^2$ (note that λ has the dimension of the square root of a frequency). On the other hand, a $\left| \phi_{\lambda,\nu}^{(A)} \right\rangle$ state, or any superposition of such states with the *same* λ, remains stationary (in the interaction picture). An eigenspace of an hermitian jump operator commuting with H_A is thus decoherence-free, even if its eigenvalue is different from zero. Equation (4.97) has an interesting connexion with measurement theory. A hermitian jump operator L_h is a physical observable of A. The irreversible evolution described by eqn. (4.97), which transforms any superposition of eigenstates of L_h corresponding to different eigenvalues into a statistical mixture, has the same final state as an unread ideal measurement of L_h (see Section 2.5.1).

Approximate pointer states

We have considered so far ideal situations where the master equation had perfect pointer or decoherence-free states. What can we say about the eigenstates of L when the additional conditions which we have imposed on the commutators $[L, H_A]$ and $[L, L^\dagger]$ are not fulfilled? An eigenstate of L is then generally contaminated by other eigenstates by the non-unitary evolution between successive jumps and A gets eventually entangled with its environment. If the evolution starts from an eigenstate of L, this entanglement is slow, however, and can be neglected at the beginning of the process, as long as the mixing with other states remains small. This argument can be made quantitative by computing the rate of variation of the trace of the square of the system's density operator:

$$\frac{d}{dt} \mathrm{Tr} \rho_A^2 \bigg|_{t=0} = 2 \mathrm{Tr} \rho_A \frac{d\rho_A}{dt} = 2 \mathrm{Tr}(\rho_A L \rho_A L^\dagger - \rho_A L^\dagger L \rho_A) . \tag{4.98}$$

Note that the Hamiltonian term does not contribute to this rate, due to the invariance of the trace by permutation of operators. Assuming that, at $t = 0$, A is in a pure state $\left| \phi^{(A)} \right\rangle$ not entangled with E, the rate of variation of $\mathrm{Tr} \rho_A^2$ is at this time:

$$\frac{d}{dt} \mathrm{Tr} \rho_A^2 \bigg|_{t=0} = 2 \left(\left\langle \phi^{(A)} \middle| L^\dagger \middle| \phi^{(A)} \right\rangle \left\langle \phi^{(A)} \middle| L \middle| \phi^{(A)} \right\rangle - \left\langle \phi^{(A)} \middle| L^\dagger L \middle| \phi^{(A)} \right\rangle \right) . \tag{4.99}$$

It is equal to zero if $\left| \phi^{(A)} \right\rangle$ is an eigenstate of L. The rate of variation of the linear entropy $1 - \mathrm{Tr}(\rho_A^2)$ is then zero and the system is relatively impervious to entanglement with E. Of course higher-order derivatives of $\mathrm{Tr} \rho_A^2$ are non-zero, leading to a slow entanglement process. States which are initially eigenstates of L and lose this property as time evolves will be called 'approximate pointer states'.

We will illustrate these general considerations in Chapter 8, devoted to ion trap physics. We will see then that it is possible to engineer environments which produce unusual damping mechanisms on an oscillator. For instance, the jump operator of the form a^q may annihilate q phonons at once. Or it may project the oscillator in a coherent superposition of states having lost different numbers of phonons. We analyse in detail in Section 8.3.2 these exotic decoherence processes and describe their approximate and exact pointer states and decoherence-free states.

Pointer states of the Brownian motion

Let us conclude this discussion by considering decoherence in a one-dimensional Brownian motion process. It can be modelled by the evolution in an external potential of a macroscopic particle of mass m, interacting with a bath of oscillators in thermal equilibrium at temperature T (Dekker 1977; Caldeira and Leggett 1983; 1985; Unruh and Zurek 1989; Zurek 1991). The corresponding master equation can be expressed as (Giulini *et al.* 1996, p.71):

$$\frac{d\rho_A}{dt} = \frac{1}{i\hbar}[H,\rho_A] + \frac{\gamma}{2i\hbar}[X,P\rho_A + \rho_A P] - \frac{m\gamma k_b T}{\hbar^2}[X^2\rho_A + \rho_A X^2 - 2X\rho_A X] , \quad (4.100)$$

where X and P are the particle's position and momentum operators and γ a friction rate which describes the damping of the particle velocity. The first term in the right-hand side of eqn. (4.100) accounts for the particle's free motion and the second for the slow friction. The third term, which primarily interests us here, describes a fast diffusion of the particle position and is responsible for its decoherence. Note that this term takes a Lindblad form, with a jump operator proportional to X. At very short times, the diffusion term is dominant. Developing the master equation on the matrix elements $\langle x| \rho_A |x'\rangle = \rho_{Axx'}$, we express this term as:

$$\frac{d\rho_{Axx'}}{dt} = -\frac{m\gamma k_b T}{\hbar^2}(x-x')^2 \rho_{Axx'} . \quad (4.101)$$

This is a special case of the general solution (4.97) applied to a hermitian Lindblad operator proportional to X. The coherence $\rho_{Axx'}$ between two positions of the particle separated by $\delta x = |x - x'|$ thus decays at a rate:

$$\frac{1}{T_D} = \frac{m\gamma k_B T \delta x^2}{\hbar^2} = 2\pi^2\gamma \left(\frac{\delta x}{\lambda_{th}}\right)^2 , \quad (4.102)$$

where $\lambda_{th} = \hbar\sqrt{2\pi/mk_bT}$ is the thermal de Broglie wavelength of the particle at temperature T. This rate is proportional to the square of the distance between the two positions, measured in units of λ_{th}, a huge quantity even for a speck of dust. For a typical Brownian particle of 10 μm^3 with a mass of 10^{-10} g at $T = 300$ K, $\lambda_{th} \sim 10^{-18}$ m. The decoherence rate of $\rho_{Axx'}$ for $|x - x'| = 10$ μm is $1/T_D = 2 \cdot 10^{14}\gamma$. This is a hundred trillion times larger than the rate of the particle's energy damping![11] Any superposition state of the form $|x_0 + \delta x/2\rangle + |x_0 - \delta x/2\rangle$ loses its quantum coherence in a time exceedingly short compared to the characteristic time of

---

[11] For such short decoherence times, the validity of the Markov approximation is questionable. A more rigorous approach (Braun *et al.* 2001) is then required, which leads qualitatively to the same result: in an extremely short time, the coherence between $|x\rangle$ and $|x'\rangle$ vanishes.

the viscous damping. Even such a small object is classical for all practical purposes, since superposition states corresponding to separations larger than its size ($\sim 1\ \mu$m) decohere much too fast to make their practical observation possible.

Among the position state superpositions, those in which δx tends to 0 live the longest and are thus privileged by decoherence. The pointer states of the Brownian diffusion correspond, at the limit $\delta x = 0$, to the eigenstates $|x\rangle$ of the position operator X. Since this operator is (within a numerical factor) the Lindblad operator of the diffusion process, we retrieve on this special example the general result established above. The molecules of the gas constitute the particle's environment which constantly measures its position, forcing it to 'choose' a well-defined x value instead of being coherently suspended between several. The position eigenstates are here approximate pointer states since they do not remain stable under the effect of the other (much slower) processes contributing to the particle's evolution. This approximation is very good, though, because of the huge difference between the time scales of decoherence and velocity (or energy) damping.

Discussing the existence of pointer states remains relatively simple for relaxation processes involving a single Lindblad operator (or commuting hermitian Lindblad operators, sharing a set of common eigenvectors). When several non-commuting operators correspond to different kinds of jumps, the conditions for the existence of pointer states cannot in general be satisfied. The system then evolves into a mixture of states and never comes back to a pure state. This is for example the case for the thermalization of a field in a cavity, which undergoes two kinds of jumps (photon emission and absorption) whose operators, proportional to a and $a^\dagger$, do not admit common eigenstates. The thermal equilibrium is a statistical mixture and each Monte Carlo trajectory corresponds to a random, never ending succession of jumps between Fock states (see Fig.4.6).

4.5 Damped spin–spring system: from Rabi to Purcell

We can now revisit the atom–single-mode system described in Chapter 3 and analyse how it is affected by its coupling to an environment at temperature T. The master equation accounting for the evolution of the spin–spring density operator ρ consists of a commutator term associated to the coherent Jaynes–Cummings evolution and of Lindblad terms describing the atom and field relaxation processes. Piecing together eqns. (3.167), (4.78) and (4.83), we write this equation as:

$$\frac{d\rho}{dt} = \frac{1}{i\hbar}[H_a + H'_c + H_{ac}, \rho] + \sum_i \left[L_i\rho L_i^\dagger - \tfrac{1}{2}(L_i^\dagger L_i\rho + \rho L_i^\dagger L_i) \right], \qquad (4.103)$$

where H_a and H'_c are the atom and cavity Hamiltonians. The coupling Hamiltonian H_{ac} is given by eqn. (3.167) and the L_i jump operators ($i = 1$ to 4) are $\sqrt{\Gamma(1 + n_{\text{th}})}\sigma_-$, $\sqrt{\Gamma n_{\text{th}}}\sigma_+$, $\sqrt{\kappa(1 + n_{\text{th}})}a$ and $\sqrt{\kappa n_{\text{th}}}a^\dagger$. The infinitesimal evolution of this system is reproduced by coupling it in a unitary way to a simple environment simulator restricted to five states (one for the no-jump event and one for each L_i jump). The evolution is expected to be coherent, i.e. dominated by the Hamiltonian contribution, when the vacuum Rabi frequency Ω_0 is much larger than the irreversible decay rates $\Gamma(1 + n_{\text{th}})$ and $\kappa(1 + n_{\text{th}})$. This expresses the condition for the strong coupling regime of cavity QED. We will see in the next chapter how these conditions are realized in practice.

Let us remark that, in a specific experiment, the situation is often simpler than what is described by the general master equation (4.103). In optical experiments with $\hbar\omega_c \ll k_b T$, the average number of thermal photons is negligible and only two jump operators remain, $\sqrt{\Gamma}\sigma_-$ and $\sqrt{\kappa}a$. The system's damping is then described by the atomic spontaneous emission rate and the cavity Q factor. In microwave experiments, on the other hand, the atomic emission rate is usually negligible compared to cavity damping and thermal photons become a relevant factor, unless T is extremely small. There are then again two dominant Lindblad operators, $\sqrt{\kappa(1 + n_{\mathrm{th}})}a$ and $\sqrt{\kappa n_{\mathrm{th}}}a^\dagger$. Finally, for microwave experiments at extremely low temperatures, $n_{\mathrm{th}} \sim 0$ and the master equation reduces to a single jump operator, $\sqrt{\kappa}a$. We restrict our discussion in the remainder of this chapter to microwave experiments, with or without thermal photons.

4.5.1   The damped vacuum Rabi oscillation

The simplest situation to be examined in this context is the vacuum Rabi oscillation. Qualitatively, we expect competition between the reversible spin–spring energy exchange at frequency Ω_0 and the photon loss at a rate $\kappa = 1/T_c = \omega_c/Q$. If $\Omega_0 \gg \kappa$, many Rabi oscillations occur before the photon is eventually lost in the environment. The probability $P_e(t)$ of observing the atom in the upper state $|e\rangle$ exhibits damped oscillations. At very long times, P_e tends towards zero since the atom ends up in level $|g\rangle$ and the cavity in vacuum. If, on the other hand, $\Omega_0 \ll \kappa$, we expect an overdamped evolution for $P_e(t)$, which tends exponentially towards zero. The atom, even if not directly coupled to the environment, is irreversibly damped via its coupling with the strongly relaxing spring. The cavity appears then as an environment for the atom. This regime is reminiscent of spontaneous emission in free space, but occurs, as shown below, at a much faster rate.

The above qualitative discussion can be made quantitative by solving the master equation (4.103) with $\Gamma = n_{\mathrm{th}} = 0$. Since there is at most one excitation in the system, ρ has non-vanishing matrix elements restricted to the three-dimensional Hilbert space spanned by the states $|1\rangle = |e, 0\rangle$, $|2\rangle = |g, 1\rangle$ and $|3\rangle = |g, 0\rangle$. States $|1\rangle$ and $|2\rangle$ are coherently coupled by the Jaynes–Cummings Hamiltonian, while $|2\rangle$ decays towards $|3\rangle$ at the rate κ. Adding these coherent and incoherent evolutions, the master equation reduced to the non-zero matrix elements of ρ is easily obtained:

$$\frac{d\rho_{11}}{dt} = -\frac{\Omega_0}{2}(\rho_{12} + \rho_{21}) \; ; \tag{4.104}$$

$$\frac{d\rho_{22}}{dt} = \frac{\Omega_0}{2}(\rho_{12} + \rho_{21}) - \kappa\rho_{22} \; ; \tag{4.105}$$

$$\frac{d}{dt}(\rho_{12} + \rho_{21}) = \Omega_0(\rho_{11} - \rho_{22}) - \frac{\kappa}{2}(\rho_{12} + \rho_{21}) \; ; \tag{4.106}$$

$$\frac{d\rho_{33}}{dt} = \kappa\rho_{22} \; , \tag{4.107}$$

with an obvious physical interpretation. The coherent terms, proportional to Ω_0, couple the 'populations' ρ_{11} and ρ_{22} to the 'coherences' ρ_{12} and ρ_{21}. The damping term, proportional to κ, does not affect ρ_{11}. It contributes to the decay of ρ_{22} and to the

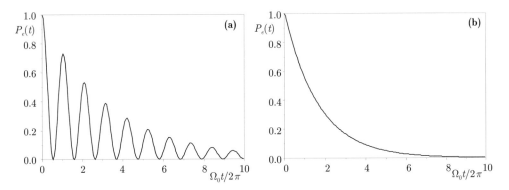

Fig. 4.7 Spontaneous emission in a damped cavity. Probability of finding the atom in the upper level, P_e, as a function of time t, in units of the vacuum Rabi period $2\pi/\Omega_0$. (a) Oscillatory regime for $\kappa = 0.1\,\Omega_0$. (b) Overdamped regime for $\kappa = 10\,\Omega_0$.

corresponding increase of ρ_{33}. Finally, the coherences ρ_{12} and ρ_{21}, which connect two states of different relaxation rates, κ and 0, are damped with the average rate $\kappa/2$.

There are only three independent variables in eqns. (4.104)–(4.107): $x = \rho_{11}$, $y = \rho_{22}$ and $z = \rho_{12} + \rho_{21}$. With this notation, the master equation takes the matrix form:

$$\frac{d}{dt}\begin{pmatrix} x \\ y \\ z \end{pmatrix} = \begin{pmatrix} 0 & 0 & -\Omega_0/2 \\ 0 & -\kappa & \Omega_0/2 \\ \Omega_0 & -\Omega_0 & -\kappa/2 \end{pmatrix}\begin{pmatrix} x \\ y \\ z \end{pmatrix}. \tag{4.108}$$

The eigenvalues λ_0, $\lambda_\pm$ of the 3×3 matrix in eqn. (4.108) are the solutions of:

$$(\kappa/2 + \lambda)\left(\lambda^2 + \kappa\lambda + \Omega_0^2\right) = 0 \; ;$$

$$\lambda_0 = -\frac{\kappa}{2} \; ; \quad \lambda_\pm = -\frac{\kappa}{2} \pm \frac{\kappa}{2}\left(1 - 4\Omega_0^2/\kappa^2\right)^{1/2}. \tag{4.109}$$

The solutions $\lambda_\pm$ can be real or complex, leading to the two different regimes qualitatively analysed above.

If the cavity damping rate is weak:

$$\kappa < 2\Omega_0 , \tag{4.110}$$

the $\lambda_\pm$ eigenvalues are complex and the evolution is described by damped oscillating functions. Figure 4.7(a) shows $P_e(t) = \rho_{11}(t) = x(t)$ for $\kappa = 0.1\,\Omega_0$. The atom undergoes a Rabi oscillation in the field of the single photon that it emits and reabsorbs. The excitation eventually decays away, with half the cavity damping rate. The system excitation is, half of the time, carried by the atom which does not decay in this model.

4.5.2 Overdamped regime: the Purcell effect

When the cavity damping rate is strong and obeys:

$$\kappa > 2\Omega_0 , \tag{4.111}$$

all the eigenvalues in eqn. (4.109) are real and $P_e(t)$ exhibits an irreversible damping (Fig. 4.7b, corresponding to $\kappa = 10\,\Omega_0$). In the limit of large cavity damping, $\kappa \gg$

Ω_0, the eigenvalue with the smallest modulus, λ_+, corresponding to the largest time constant, essentially determines the atomic decay, which is quasi-exponential with a rate:

$$\Gamma_c = -\lambda_+ \approx \Omega_0^2/\kappa . \tag{4.112}$$

This rate can also be retrieved by performing a direct approximation on eqns. (4.104)–(4.107), justified by the $\kappa \gg \Omega_0$ condition. The probability ρ_{22} of finding a photon in the cavity decays then at a rate much larger than the rate at which it is fed by the coherent Rabi oscillation. This probability thus remains very small, as well as the coherences ρ_{12} and ρ_{21}. The cavity is at all times practically in its vacuum state and can be seen as a genuine environment for the atomic system. The atom is irreversibly damped via its indirect coupling to the cavity environment. The very small coherence $\rho_{12}+\rho_{21}$ is estimated by assuming that it follows adiabatically the variations of the excited state population. Cancelling the slow evolution rate of the coherence in eqn. (4.106), we get a simple algebraic equation yielding $\rho_{12} + \rho_{21} \approx (2\Omega_0/\kappa)\rho_{11}$. Injecting this expression in eqn. (4.104), we finally get directly $d\rho_{11}/dt = -(\Omega_0^2/\kappa)\rho_{11}$, in agreement with eqn. (4.112).

It is instructive to express this cavity-assisted damping rate in terms of the basic atomic and cavity parameters. Taking eqns. (3.24) and (3.168) into account, we get:

$$\Gamma_c = \frac{2d^2}{\varepsilon_0 \hbar} \frac{Q}{\mathcal{V}} , \tag{4.113}$$

(we assume here, for the sake of simplicity, that the atomic transition and cavity mode polarizations are matched: $|\epsilon_a^* \cdot \epsilon_c| = 1$). This damping rate can be compared to the spontaneous emission rate of the same atom in free space, given by eqn. (4.73). The ratio η of these two rates, independent of the atomic dipole d, is a function of cavity parameters only. A simple calculation yields:

$$\Gamma_c = \eta\Gamma , \tag{4.114}$$

with:

$$\eta = \frac{3}{4\pi^2} \frac{Q\lambda_c^3}{\mathcal{V}} , \tag{4.115}$$

where $\lambda_c = 2\pi c/\omega_c$ is the wavelength of the cavity field, resonant with the atomic transition. The factor η increases linearly with Q and is inversely proportional to $\mathcal{V}/\lambda_c^3$, the cavity volume measured in units of λ_c^3. Large spontaneous emission enhancement factors require cavities with small losses and small volumes. In the range of Q and $\mathcal{V}$ values satisfying $\eta \gg 1$ and $Q \ll \omega_c/2\Omega_0$, the atom decays exponentially much faster than in free space. When Q approaches the value corresponding to the transition to the strong coupling regime ($Q = \omega_c/2\Omega_0$), eqn. (4.113) is no longer valid and the irreversible overdamped regime transforms continuously into the oscillatory one.

The spontaneous emission rate enhancement in a cavity with a moderate Q has been discussed in the context of radio-frequency magnetic resonance experiments by Purcell (1946). His qualitative reasoning was based on a Fermi Golden Rule argument, describing the coupling of a two-level atom to a continuum of modes. The spontaneous emission rate in free space is proportional to the spectral density of modes per unit

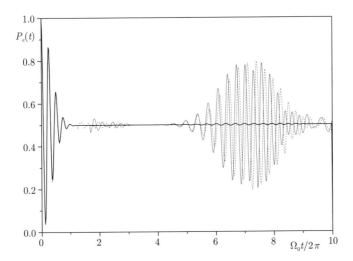

Fig. 4.8 Rabi oscillation collapse and revival in a damped cavity. Probability $P_e(t)$ of finding the atom in $|e\rangle$ at time t (measured in units of $2\pi/\Omega_0$). The initial coherent field has on average $\bar{n} = 15$ photons and $\kappa = 8 \cdot 10^{-3}\Omega_0$. The thin solid and dotted lines correspond to two typical quantum trajectories (six and four jumps respectively) and the thick solid line to the average over 10 000 such histories (six quantum jumps on the average).

volume at the atomic frequency w_{eg}, proportional to w_{eg}^2/c^3. A strongly damped cavity at frequency $w_c = w_{eg}$ is also seen by the atom as a continuum. Its spectral mode density is proportional to $Q/w_c\mathcal{V}$ (one mode over the volume $\mathcal{V}$ in the spectral bandwidth Q/w_c). The ratio of these spectral densities is, when all π factors are taken into account, precisely the Purcell enhancement η. This seminal remark by Purcell, who noted for the first time that the emission rate of an atom may depend on its surrounding, was the starting point of cavity quantum electrodynamics, as we will recall in Section 5.1.

4.5.3 Damping of Rabi oscillation revivals

Let us now consider the effect of cavity damping on the atomic evolution induced by a quantum field containing initially many photons in the cavity. The Rabi oscillation of an atom interacting with a coherent field in state $|\alpha\rangle$ with $|\alpha| \gg 1$ exhibits the collapse and revival phenomenon described in Chapter 3. The Rabi frequencies $\Omega_0\sqrt{n+1}$ corresponding to the different $|n\rangle$ states beat together, leading first to a disappearance of the oscillation after a time of the order of π/Ω_0, then to a re-phasing of the various components after a time $4\pi\sqrt{\bar{n}}/\Omega_0$.

The oscillation revival phenomenon is very sensitive to cavity relaxation, as a numerical simulation based on a Monte Carlo approach demonstrates. Figure 4.8 shows the probability $P_e(t)$ of finding the atom in state $|e\rangle$ at time t, for a coherent field containing initially $\bar{n} = 15$ photons on the average. The atom is supposed to be undamped ($\Gamma = 0$) and the cavity environment is at $T = 0$ K ($n_{\text{th}} = 0$). The cavity damping rate is $\kappa = 8 \cdot 10^{-3}\Omega_0$. It corresponds to the loss of one photon on average every six Rabi periods $2\pi/(\Omega_0\sqrt{\bar{n}+1})$. The probability $P_e(t)$ is obtained from the calculation of individual atom–field quantum trajectories in which photon losses are registered at random times. The thin solid and dotted lines correspond to two such trajectories, in which, respectively, six and four photons have been lost.

The revivals are clearly apparent on each trajectory, but they have different phases, due to the diverse frequency chirps experienced by the Rabi oscillation during the two field histories. Each time a photon is lost, the Rabi frequency, proportional to $\sqrt{n+1}$, makes a jump. This jump affects the phase of the oscillations at later times. Summing over a large number of such trajectories, we get the thick solid line representing the average Rabi signal. Not surprisingly, the sum over oscillations with random phases results in a near-complete suppression of the revival signal. No more than two or three photons should be lost on average on each trajectory to get a good contrast of the average revival. This defines an upper limit for κ ($\sim 0.01\,\Omega_0$ for $\bar{n} = 15$) and, accordingly, a lower limit for the cavity Q factor compatible with the observation of the revival phenomenon. These limits become very severe and difficult to fulfil for larger fields, since the revival time increases linearly with $\sqrt{\bar{n}}$.

When studying the Rabi oscillation, we focus our attention on the atomic evolution, tracing over the field variables. Without discussing the field dynamics here, we can nevertheless conclude from the trajectories shown in Fig. 4.8 that the coherent field must undergo some non-trivial evolution. A naive guess would be that the field stays in a coherent state, with an amplitude decaying exponentially, as is the case when no atom is present. In such a case, however, the field would follow, as described in Section 4.4.4, a single perfectly determined quantum trajectory and the Rabi revival signal would be the same in all individual 'runs'. The fact that the atomic histories do exhibit randomness indicates that the field does not remain in a coherent state.

We will see in Chapter 7 that during the collapse and revival phenomenon the field experiences a very interesting dynamical evolution, getting entangled with the atom and evolving continuously into a superposition of coherent states with different phases. The damping of the atomic Rabi revival signal turns out to be intimately related to the decoherence of these field state superpositions, which is analysed in detail in Section 7.4.

4.6   Kicking a spring with spins: the micromaser

The Jaynes–Cummings model of cavity QED had initially been proposed as an idealization of a laser or maser. This model, dealing with one atom at a time, is a gross approximation of the usual laser device, in which many atoms collectively interact with a single field mode. There is, however, a context in which the Jaynes–Cummings Hamiltonian exactly describes the operation of a laser system. If the atoms are sent one by one across the cavity, the laser/maser field builds up by successive interactions of the Jaynes–Cummings type. This device has effectively been realized in the microwave domain. It is called a 'micromaser' (Gallas *et al.* 1985; Walther 1988; Raithel *et al.* 1994).

Its operation is described by a theory which is an instructive illustration of the master equation formalism applied to a quantum oscillator coupled to two environments exerting competing effects. On the one hand, the field stored in a cavity with a finite Q factor is in contact with a large reservoir which tends to thermalize it at its own temperature. On the other hand, it interacts with a stream of excited atoms which

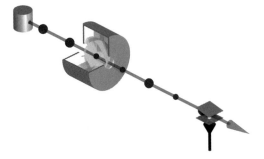

Fig. 4.9 Principle of an ideal micromaser. Atoms, initially in $|e\rangle$, create a large field in the cavity by cumulative emissions. The final state of the atoms is detected downstream.

emit photons in the mode and tend to increase its excitation. Ideally, the coupling with the atoms is coherent and deterministic. In practice, though, unavoidable fluctuations in the atomic flux and atom–cavity interaction parameters make the atomic ensemble act as a 'large' environment for the field, which must be described statistically. The atoms are detected after cavity crossing and the state of the field is inferred from the observation of this atomic environment.

We will start by a 'naive' analysis neglecting relaxation and describing the atom–field system by a simple quantum state. We will next turn to more realistic situations in which field damping, thermal photons and atomic fluctuations play an important role. The master equation and single quantum trajectory descriptions inspired by the Monte Carlo approach will be essential for the understanding of the micromaser dynamics.

4.6.1 Ideal micromaser

Let us first consider a very simple thought experiment depicted in Fig. 4.9. An initially empty lossless cavity is excited by a stream of two-level atoms, prepared in $|e\rangle$. The atoms, separated by regular time intervals t_a, interact resonantly with the field one at a time on the $|e\rangle \rightarrow |g\rangle$ transition. For each atom, the interaction time t_i, much shorter than t_a, is supposed to be adjustable (e.g. by controlling individual atomic velocities). The final atomic state, $|e\rangle$ or $|g\rangle$, is measured downstream.

What is the most efficient way to pump energy into the cavity field? Obviously, optimum efficiency is achieved when each atom releases one photon. The first atom prepares by spontaneous emission a single-photon state. The second atom then, stimulated by the already present photon, emits a second one and so on. This optimal situation is realized (Krause *et al.* 1987; 1989) by adjusting the interaction time of the $(n+1)$-th atom, t_{n+1}, in such a way that it experiences an exact π-quantum Rabi pulse in the already present n photon Fock state:[12]

$$t_{n+1} = \frac{\pi}{\Omega_0 \sqrt{n+1}} \; . \tag{4.116}$$

In this ideal situation, the cavity field jumps up the ladder of Fock states, by one step in each t_a interval. The field 'trajectory' is deterministic. It can be monitored by checking that all the exiting atoms are detected in $|g\rangle$. The detection of each outgoing atom is then the signature of a field quantum jump.

---

[12]We assume that the atom–field coupling is constant during cavity crossing.

This ideal situation requires a precise tuning of the atom–field interaction time, difficult to achieve in practice. In a more realistic experiment, successive atoms cross the cavity with a given velocity and the atom–field interaction times are equal. Fock states can still be generated in this case, provided the atomic detection is ideally efficient. Counting the number n of atoms exiting in $|g\rangle$ directly projects the field state onto the Fock state $|n\rangle$.

The climb of the field up the Fock states ladder is now a random process. At each time interval t_a, the field has a probability of remaining in the same Fock state (corresponding to the atom being detected in $|e\rangle$) and a complementary probability of jumping up one step in the ladder (atomic detection in $|g\rangle$). The timing of the field jumps is then non-deterministic. In each realization of the experiment, the field follows a different trajectory, monitored by detecting the atoms.

Assume now, still as a thought experiment, that the atomic detector is located very far from the cavity. Before the atoms have reached it, the field is entangled with all the atoms which have crossed the cavity. It is then described by a density operator, obtained by tracing out the atomic states. This density matrix results from the average of all possible field trajectories in the Fock states ladder, weighted by their probabilities. These individual trajectories can be simulated by a dice-tossing procedure very similar to the Monte Carlo calculation analysed above. Not detecting the atoms at all obviously produces the same field density matrix. At a given time, the photon probability distribution variance reflects the randomness of the number of jumps having occurred so far.

4.6.2   Trapping states

An interesting situation occurs when the interaction time t_i is tuned so that the probability of an atom to emit a photon is strictly zero when the cavity already contains n_0 photons. This condition can be written:

$$\Omega_0\sqrt{n_0 + 1}t_i = 2p\pi , \tag{4.117}$$

where p is an integer. On each quantum trajectory, the photon number increases until it reaches n_0. All subsequent atoms are then detected in $|e\rangle$ and the cavity field remains 'trapped' in this Fock state (Meystre *et al.* 1988). This deterministic trapping state is reached at a random time, differing from one quantum trajectory to the next.

Figure 4.10(a) present a numerical simulation of such trajectories (thin lines) for a trapping state with $n_0 = 10$ and $p = 1$. All trajectories converge to the 10-photon Fock state. The histograms in Fig. 4.10(b)–(d) and the thick line in Fig. 4.10(a) show the evolution of the photon number distribution $p(n)$ and of its average. Note that a pure trapping state is asymptotically reached even if the atoms are not detected. At intermediate times, the large photon number variance reflects the transient statistical nature of the field.

4.6.3   The real micromaser

The thought experiments described so far assume a lossless cavity and perfect control of the atoms. A more complete description of a realistic micromaser must take into account cavity damping, finite atomic detection efficiency, statistical fluctuations in the atom–cavity interaction time and thermal photons in equilibrium with the cavity

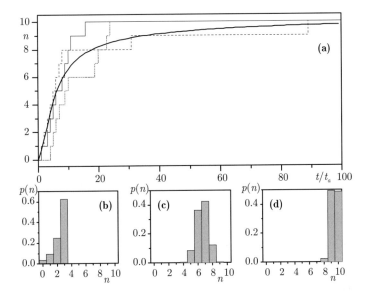

Fig. 4.10 Monte Carlo simulation of a trapping state situation for $n_0 = 10$ and $p = 1$. (a) Thin lines (solid, dashed and dotted): photon number versus time for individual field trajectories. The time unit is t_a. Thick solid line: photon number versus time averaged over 10 000 Monte Carlo trajectories. (b), (c) and (d) Histograms of the photon number distribution $p(n)$ at times $t = 3$, 10 and $60\,t_a$ respectively.

walls at finite temperature. Remarkably, when these complications are taken into account, some of the features of the simple model survive. A real micromaser can produce non-classical light (Rempe *et al.* 1990) and can exhibit properties reminiscent of the trapping states (Weidinger *et al.* 1999).

Time scales

The operation of a realistic micromaser depends upon three time parameters. The first one is the atom–cavity interaction time t_i. The second is the mean time interval $\overline{t_a}$ between atoms, the reciprocal of the maser pumping rate $r_a = 1/\overline{t_a}$. In a realistic situation, the statistics of atomic arrival times is Poissonian. The third parameter is the cavity damping time, $T_c = Q/\omega_c = 1/\kappa$. For a genuine micromaser operation, the following double inequality must be fulfilled:

$$t_i \ll \overline{t_a} \ll T_c \ .$$ (4.118)

The left inequality means that one atom at most is present at any time inside the cavity. The right one means that many atoms have a chance to deposit a photon in the cavity during its damping time, allowing for the build-up of a field containing many photons. Each atom leaving at most one photon, which survives an average time T_c, the total photon number in the field is bounded by:

$$N_e = T_c/\overline{t_a} = r_a Q/\omega_c \ ,$$ (4.119)

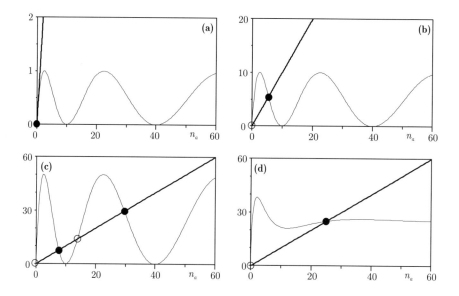

Fig. 4.11 Graphical solution of the semi-classical rate equation for the micromaser. The loss (thick line) and gain (thin line) curves are presented for three pumping rates, increasing from (a) to (c) ($r_a = 1$, 10 and 50 in units of T_c^{-1} respectively). The interaction time t_i is such that $\Omega_0 t_i/2 = 2\pi/\sqrt{40}$. The stable (unstable) solutions are depicted by solid (open) circles. Note that the zero-intensity point is stable below threshold, unstable above. (d) The gain curve is averaged over an atomic Maxwellian velocity distribution for $r_a = 50$ [the average velocity corresponds to the interaction time used for panels (a)–(c)]. There is now only one stable operating point above threshold.

a quantity much larger than one when the inequalities (4.118) are fulfilled. The mean number of active atoms in the cavity, $t_i/\overline{t_a}$, is much smaller than one, meaning that there is no atom at all most of the time. This maser operates in a remarkable regime, with a mesoscopic field sustained by less than one active atom at a time. Its dynamics, in which quantum fluctuations play an essential role, is very different from that of an ordinary laser or maser oscillator (Filipowicz *et al.* 1986; Lugiato *et al.* 1987; Bergou *et al.* 1989a; Rempe *et al.* 1991; Davidovich 1996).

Semi-classical approach

The maser steady-state results from a competition between the gain produced by repeated kicks from the emitting atoms and the photon losses. A simple gain/loss model (Haroche 1992) yields the evolution of the average photon number in the cavity, n_a, provided one can neglect the fluctuations of the photon number around its average. This approximation holds when the cavity contains a large field. The photon number n_a can then be treated as a continuous classical variable, which evolves over time scales much longer than $\overline{t_a}$.

The gain is the number of photons released per unit time by the atoms undergoing stimulated emission in the n_a photon field. It is the product of the atomic pumping rate, r_a, by the probability that each atom emits an extra photon. The gain is an oscillating function of n_a:

$$G(n_a) = r_a \sin^2\left(\frac{\Omega_0}{2}t_i\sqrt{n_a+1}\right) \approx r_a \sin^2\left(\Theta\sqrt{n_a/N_e}\right) , \tag{4.120}$$

where we introduce the dimensionless interaction time Θ defined by:

$$\Theta = \frac{\Omega_0}{2}t_i\sqrt{N_e} . \tag{4.121}$$

The loss rate being proportional to the photon number, the rate equation balancing gain and losses is:

$$\dot{n}_a = -\frac{n_a}{T_c} + r_a \sin^2\left(\Theta\sqrt{n_a/N_e}\right) . \tag{4.122}$$

The steady states ($\dot{n}_a = 0$) are obtained graphically as the intersections of the curves giving the losses and gain versus n_a. The loss curve is a straight line with slope $1/T_c$, running through the origin. The gain is given by an oscillating curve reflecting the atomic Rabi oscillations. The graphs in Fig. 4.11(a)–(c) represent the loss line for a given T_c value along with three gain curves corresponding to increasing pumping rates r_a.

The curve corresponding to the smallest r_a value (Fig. 4.11a) intersects the loss line only at $n_a = 0$. The field cannot build up, corresponding to a maser below threshold. The maser threshold is reached when the gain and loss curves have the same slope at the origin, i.e. when:

$$r_a > \frac{N_e}{T_c\Theta^2} \quad \text{or} \quad r_a > \frac{4}{T_c(\Omega_0 t_i)^2} . \tag{4.123}$$

For the pumping rate corresponding to Fig. 4.11(b), the gain and loss curves intersect at two points. The first, $n_a = 0$ (open circle), corresponds now to an unstable solution. Any small departure from $n_a = 0$ increases the gain, which becomes larger than the losses. The photon number thus increases until it reaches a value corresponding to the second intersection (solid circle). This solution is stable since the slope of the gain curve is, at the corresponding intersection, smaller than $1/T_c$. When the photon number fluctuates above the equilibrium value, the losses overcome the gain, and the photon number is pulled back to its original value. For the higher pumping rate, corresponding to Fig. 4.11(c), two stable solutions coexist (solid circles) with an unstable one between them (open circle). Which stable point is actually reached depends upon the initial condition and the maser behaviour is, in this semi-classical model, hysteretic.

This is radically different from the behaviour of an ordinary maser, which presents only one stable operating point above threshold. The standard situation is recovered when the atoms are not velocity-selected (Fig. 4.11d). The dispersion of the interaction times t_i washes out the Rabi oscillations and the gain G reaches rapidly an asymptotic value $r_a/2$. We recover the ordinary gain saturation of a standard maser. There is, above threshold, a single stable operating point. When the intersection of the loss and gain curves occurs in the saturation region, the steady-state photon number is proportional to r_a.

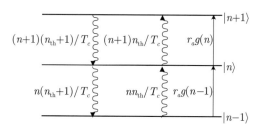

Fig. 4.12 Micromaser jumps in the Fock state ladder. The solid arrows correspond to atomic emission and the wavy ones to relaxation processes (photon losses or thermal photons creation).

Quantum model

The semi-classical model neglects the graininess of the photon number and the quantum fluctuations around its mean value. These features are, as we see now, essential for a full understanding of the micromaser dynamics. In order to take them into account, the field must be described by its density operator ρ which yields predictions for an ensemble of identical micromasers, each of them corresponding to a quantum trajectory. There is no preferred phase in this ensemble, since the cavity is initially empty and since the atoms, injected in $|e\rangle$, carry no phase information. All information about ρ is contained in its diagonal elements in the Fock state basis, $\langle n| \rho |n\rangle = p(n)$, the photon number probability distribution. The off-diagonal elements of ρ remain zero throughout the evolution.

The evolution equation for $p(n)$ is:

$$
\begin{aligned}
\dot{p}(n) = {} & \frac{1 + n_{\text{th}}}{T_c} \left[(n+1)p(n+1) - np(n)\right] + \frac{n_{\text{th}}}{T_c} \left[np(n-1) - (n+1)p(n)\right] \\
& - r_a g(n)p(n) + r_a g(n-1)p(n-1) \, ,
\end{aligned}
\tag{4.124}
$$

where $g(n) = \sin^2(\Omega_0 t_i \sqrt{n+1}/2)$ is the emission probability of an atom interacting with the $|n\rangle$ photon number state and n_{th} is the mean thermal photon number. The first line in the right-hand side of eqn. (4.124) describes cavity damping (see eqn. 4.79) and the last line accounts for the field amplification produced by the atoms. This equation is valid provided that $p(n)$ evolves over a characteristic time scale much longer than $\overline{t_a}$, so that each atom has only an infinitesimal effect on the field (Filipowicz et al. 1986; Haroche 1992; Davidovich 1996). The discrete cumulative emissions can then be described by a continuous differential equation. Note that this 'coarse graining' procedure, similar to the one introduced above in the derivation of the master equation, is essential to describe the effect of the atoms as a relaxation process. Equation (4.124) also assumes that the atomic pumping process obeys Poissonian statistics. Quite different results are obtained when the atoms are injected at regular time intervals (Bergou et al. 1989a).

The master equation (4.124) has a simple intuitive interpretation. It describes a diffusion process in the Fock states ladder (Fig. 4.12). The first line has already been commented on in Section 4.3.4. The first term in its right-hand side describes the field downwards jump from $|n\rangle$ to $|n-1\rangle$, resulting in a decrease of $p(n)$ with a rate $n(n_{\text{th}} + 1)/T_c$, and the downwards jump from $|n+1\rangle$ to $|n\rangle$, leading to an increase of $p(n)$. The second term describes jumps from $|n-1\rangle$ to $|n\rangle$ and $|n\rangle$ to $|n+1\rangle$ corresponding to the creation of thermal photons. Proportional to n_{th}, they vanish at zero temperature.

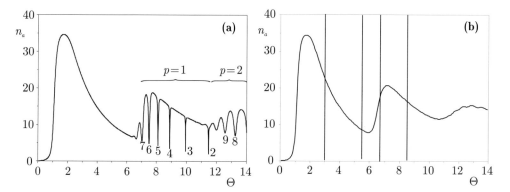

Fig. 4.13 Average photon number n_a in the quantum micromaser model as a function of the dimensionless interaction time Θ for $N_e = 40$. (a) $n_{\text{th}} = 0$ (zero temperature case). The trapping state dips are labelled according to the p and n_0 values. (b) $n_{\text{th}} = 0.1$. The small thermal field blurs the trapping states features. The vertical lines correspond to the four Θ values analysed in Fig. 4.14.

The last line in eqn. (4.124) describes the effect of the atomic gain, which counteracts cavity losses. It has the same structure as the loss terms in the first line. The two parts now correspond to the upward jumps from $|n\rangle$ to $|n+1\rangle$ and $|n-1\rangle$ to $|n\rangle$. They occur with rates $r_a g(n)$ and $r_a g(n-1)$, proportional to the micromaser pumping rate and to the atomic emission probabilities in the initial state of the jump.

The micromaser steady-state corresponds to $\dot{p}(n) = 0$. The photon number distribution is then obtained by balancing the gain and the losses in eqn. (4.124). Equating the up and down jump rates, we find the recurrence relation:

$$\frac{n_{\text{th}} + 1}{T_c} n p(n) = \left[r_a g(n-1) + \frac{n_{\text{th}}}{T_c} n \right] p(n-1) , \qquad (4.125)$$

which is immediately solved, assuming $p(0) \neq 0$, as:

$$p(n) = p(0) \prod_{k=1}^{n} \left[\frac{n_{\text{th}}}{n_{\text{th}} + 1} + \frac{N_e}{n_{\text{th}} + 1} \frac{g(k-1)}{k} \right] , \qquad (4.126)$$

where $p(0)$ is obtained by normalizing $p(n)$.

The average photon number $n_a = \sum_n n p(n)$ is plotted in Fig. 4.13(a) for $N_e = 40$ and $n_{\text{th}} = 0$ as a function of the dimensionless interaction time Θ. This parameter can be simply varied by changing t_i, keeping r_a and, hence, N_e constant.

For very low Θ values, the micromaser is below the semi-classical threshold and the average photon number close to zero. Around the semi-classical threshold, the photon number increases sharply. It reaches a maximum value close to N_e for $\Theta \approx \pi/2$ when each atom has a probability close to one of emitting a photon. Above this value, the atomic gain and hence n_a decrease. A second increase in n_a occurs around $\Theta = 2\pi$. It corresponds to the apparition of a second stable solution in the semi-classical model.

In addition to these slow variations, the evolution of n_a also presents narrow dips for precise values of Θ above 2π (Fig. 4.13a). These dips correspond to the trapping

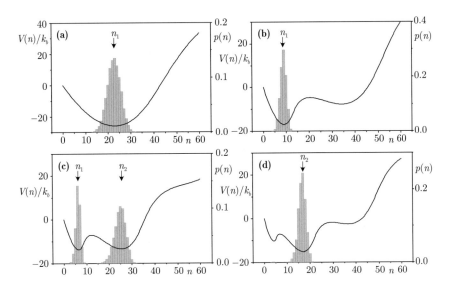

Fig. 4.14 Effective potential $V(n)$ (in units of k_b, thick solid line) and photon number distribution $p(n)$ (histogram) for $N_e = 40$, $n_{\rm th} = 0.1$ and four different Θ values (indicated by vertical lines in Fig. 4.14b): $\Theta = 3$, 5.5, 6.7 and 8.5 for panels (a)–(d) respectively.

states predicted by the ideal lossless model. They occur for $\Theta = p\pi\sqrt{N_e/(n_0+1)}$, where n_0 and p are integers, and correspond to cancellations of the atomic emission probability $g(n_0)$. The dips in Fig. 4.13(a) are labelled according to the n_0 and p values. When the trapping state condition is met, $p(n)$ cancels exactly for $n > n_0$ as clearly expressed by eqn. (4.126), taken for $n_{\rm th} = 0$. The diffusion process up the Fock state ladder experiences an abrupt cut-off so that the steady-state photon number n_a, bounded by n_0, is much smaller than for a slightly different Θ value. At the exact trapping state condition, $p(n)$ is very close to δ_{n,n_0} and the micromaser naturally generates a good approximation of a Fock state, with a very small photon number variance.

These trapping state features are very sensitive to residual thermal fields. Thermal fluctuations tend to drive the micromaser above the '$g = 0$ barrier', and completely wash out the trapping state dips as soon as the thermal photon number $n_{\rm th}$ is larger than 0.1 (Meystre *et al.* 1988; Weidinger *et al.* 1999). This washing out effect is shown in Fig. 4.13(b), which presents the average photon number as a function of Θ, for $N_e = 40$ and $n_{\rm th} = 0.1$.

Effective potential model

It is possible to get a simple intuitive picture of the steady-state micromaser operation when both the average and the variance of the photon number are large (Haroche 1992). The logarithm of $p(n)$, is, from eqn. (4.126), a sum of n terms. Treating n as a continuous variable, this sum can be replaced by an integral:

$$p(n) = p(0)e^{-V(n)/k_b} , \qquad (4.127)$$

where:

$$V(n) = -k_b \int_0^n \log \left[\frac{n_{\text{th}}}{n_{\text{th}} + 1} + \frac{N_e g(x-1)}{(n_{\text{th}} + 1)x} \right] dx \ . \qquad (4.128)$$

The photon number distribution appears as the equilibrium thermal distribution of a fictitious particle moving in a one-dimensional potential $V(n)$ (at a 1 kelvin temperature). The extrema of this potential correspond to $N_e g(n-1)/n = 1$, i.e. to the intersections of the loss and gain curves in the semi-classical model (within the approximation $n_{\text{th}} \ll 1$ and $n - 1 \approx n$). The minima of $V(n)$ correspond to stable semi-classical operating points and the maxima to unstable ones.

Figure 4.14 presents the effective potentials and photon number distributions for $N_e = 40$, $n_{\text{th}} = 0.1$ and four Θ values ($\Theta = 3$, 5.5, 6.7 and 8.5) corresponding to the vertical lines in Fig. 4.13. The small thermal photon number suppresses the trapping state features, not essential for the present discussion. The qualitative variations of the micromaser steady state with Θ can be understood from the change in shape of this potential. For a micromaser above threshold and Θ small (Fig. 4.14a), $V(n)$ has a single minimum at n_1 and $p(n)$ is peaked around n_1, which is a slowly varying function of Θ. For a larger Θ value, the potential develops a second minimum at $n_2 > n_1$, corresponding to another semi-classical stable operating point. As long as $V(n_2) > V(n_1)$, $p(n)$ remains peaked at n_1, with negligible values around n_2 (Fig. 4.14b). For a still larger Θ (Fig. 4.14c), the two minima have an equal depth. The photon number distribution is then doubly peaked around n_1 and n_2. Finally, for the largest Θ value in Fig. 4.14(d), the second minimum becomes the deeper one and $p(n)$ is peaked around n_2. When Θ is further increased, another minimum appears for $n_3 > n_2$ and so on. We understand in this way the main features of the variations of the average photon number versus Θ (Fig. 4.13). In particular, the fast variation of N around $\Theta = 2\pi$ reflects the switching of the maser operating point between n_1 and n_2.

The quantum model of the micromaser shows that the steady-state photon number distribution is, in most cases, peaked around a single n value corresponding to the deepest minimum of the effective potential. The other semi-classical 'stable' operating points are in fact metastable. If the micromaser transiently operates around one of these secondary minima, statistical fluctuations will eventually activate its passage above the potential barriers separating this secondary potential minimum from the deepest one. This process can be described by a dynamical diffusion equation for $p(n)$ obtained from eqn. (4.124) by making a continuous variable approximation for n (Bergou *et al.* 1989b; Benkert *et al.* 1990; Haroche 1992). It is reminiscent of the thermal activation of a chemical reaction whose progression is often described by a potential with minima separated by potential barriers (Kramers process). Note that, in the micromaser case, in the limit $n_{\text{th}} \ll 1$, the fluctuations are not of thermal origin. They originate in quantum field fluctuations and in the Poissonian fluctuations of the atomic pumping.

The existence of a single steady-state solution seems to be at odds with the semi-classical hysteretic model, which predicts in some cases several stable operating points. We must keep in mind that $p(n)$ describes in fact an ensemble of masers. In a single realization, the photon number switches between two different potential minima at a rate which decreases exponentially with the height of the barrier separating them. It spends random time intervals close to each minimum. The global time spent in each

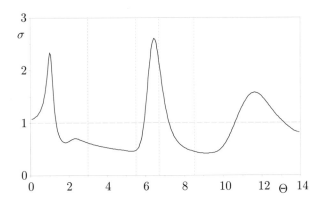

Fig. 4.15 Reduced photon number dispersion σ in the quantum micromaser model as a function of the reduced interaction time Θ for $N_e = 40$ and $n_{\text{th}} = 0.1$. The dashed horizontal line corresponds to the coherent state value, $\sigma = 1$. The vertical dotted lines correspond to the Θ values used for Fig. 4.14.

well averages out to a value proportional to the corresponding $p(n)$ (Carvalho *et al.* 1989; Buchleitner and Mantegna 1998; Benson *et al.* 1994). The single-trajectory description thus recovers the features of the statistical analysis. The connexion with the semi-classical approach is also very clear. When N_e becomes large, the potential barriers become so high that the switching time between the wells diverges. The metastable operating points then become stable for all practical purposes.

Photon number variance and sub-Poissonian state generation

Most standard laser oscillators produce fields with a Poissonian photon number distribution (coherent states) and a unit normalized dispersion: $\sigma = \sqrt{\Delta N^2/n_a} = 1$. The micromaser is an exception to this general rule and can generate fields with a normalized dispersion markedly different from one. We have already shown above that this is the case when a trapping state condition is satisfied in the absence of thermal photons. Strongly sub-Poissonian quasi-Fock states are then generated (Rempe and Walther 1990; Rempe *et al.* 1990; 1991; Davidovich 1996). Even when all trapping state features are washed out by a small thermal field, non-Poissonian statistics can still survive.

Figure 4.15 presents σ versus Θ in trapping-state-free conditions ($N_e = 40$ and $n_{\text{th}} = 0.1$). The $p(n)$ distribution, and hence its normalized dispersion, are inferred from the effective potential (see eqn. 4.127) (Haroche 1992). The normalized dispersion exhibits sharp maxima, larger than one, for Θ around one (threshold) and around the values $(2\pi, 4\pi, ...)$ corresponding to the switching of the maser between stable semi-classical operating points. The field has, in these ranges of Θ values, super-Poissonian statistics. Outside these domains, σ is smaller than one (sub-Poissonian statistics). The super-Poissonian behaviour around $\Theta = 2\pi, 4\pi, ...$ has an obvious interpretation. Two minima of the effective potential then have the same depth and the maser thus 'hesitates' between two different n values, resulting in large fluctuations of the photon number.

We now concentrate on the description of the field fluctuations when $p(n)$ is a single-peak distribution. We develop $V(n)$ near its minimum at $n = n_1$, up to second order, as:

$$V(n) = V(n_1) + \tfrac{1}{2}(n - n_1)^2 V''(n_1) \, . \tag{4.129}$$

Plugging this expression into eqn. (4.127), we obtain a Gaussian approximation for $p(n)$ around n_1:

$$p(n) = p(0)e^{-V(n_1)/k_b} \exp\left[-\frac{1}{2k_b}(n - n_1)^2 V''(n_1)\right] \, , \tag{4.130}$$

with a variance given by:

$$(\Delta N)^2 = k_b/V''(n_1) \, . \tag{4.131}$$

With the help of eqn. (4.128), we then get:

$$\sigma = \sqrt{\Delta N^2/n_1} = \frac{\sqrt{n_{\mathrm{th}} + 1}}{\sqrt{1 - N_e g'(n_1)}} \approx \frac{1}{\sqrt{1 - N_e g'(n_1)}} \, , \tag{4.132}$$

where we assume, in the last expression, $n_{\mathrm{th}} \ll 1$. Above threshold, $1 - N_e g'(n_1)$ is always positive (the threshold condition is $1 - N_e g'(n_1) = 0$). The super- or sub-Poissonian character of the field is directly related to the sign of $g'(n_1)$, i.e. to the sign of the differential gain around the operating point. Near threshold, the differential gain is positive and the field is super-Poissonian ($\sigma > 1$). Well above threshold, the maser operates usually around points where the differential gain is negative (Fig. 4.11). There, the field is sub-Poissonian.

The result expressed by eqn. (4.132) is quite general for a steady-state laser with Poissonian atomic injection. As a particular case, let us consider a maser operating at a point where the differential gain is zero. This is the case of an ordinary laser, well above threshold (strongly saturated gain). We then have $\sigma = 1$, a well-known result for a standard laser. The origin of the Poissonian character of the field is then easy to understand. The photon number fluctuations merely reflect those of the injected atoms. A small change in the atom rate integrated over the cavity damping time results in a proportional change in the photon number.

Assume now that the system operates with a negative differential gain. A positive fluctuation of the atom number is counter-balanced by the decrease of the gain and the Poissonian fluctuations of the atomic beam are squeezed in the photon output. These conclusions are reversed when the differential gain is positive. The Poissonian fluctuations of the pumping process are then enhanced by the gain variations.

There is another possible way of controlling the field photon statistics in this device. Instead of squeezing the atomic Poisson noise by a negative differential gain, it is possible to drive the maser with a sub-Poissonian atomic beam. The photon distribution is then sub-Poissonian, even if the gain is saturated (Bergou *et al.* 1989a; Benkert *et al.* 1990; Bergou *et al.* 1989b).

Due to its conceptual simplicity, the micromaser is often used as a paradigm of a laser or maser device. Theorists have used it as a 'tool' for thought experiments illustrating various aspects of complementarity (Scully *et al.* 1991). The micromaser is also a real device, which has led to many interesting experimental achievements. We discuss briefly in the next chapter its realization with Rydberg atoms and superconducting cavities.

4.7   Collective coupling of $\mathcal{N}$ spins to a spring: superradiance

In the micromaser, the field oscillator is excited by a collection of atoms, coupled to the cavity one at a time. Alternatively, we consider now an ensemble of atoms interacting simultaneously with a cavity field (Haroche 1984). The atoms then emit and absorb photons collectively, resulting in the build-up of strong interatomic correlations. The dynamics of the atom–field evolution is strongly influenced by these correlations, with characteristic time constants depending upon the number $\mathcal{N}$ of atoms in the sample. Here again, the master equation approach is very useful. The field mode is damped due to the finite cavity Q and the atoms are indirectly coupled to the cavity reservoir. Solving the resulting master equation by the Monte Carlo method will shed interesting light on the collective atomic emission processes.

Collective atom–field coupling has been studied extensively since the beginnings of quantum optics. One of the first problems in this context was the collective spontaneous emission of an ensemble of two-level atoms, all initially excited and radiating in the free-space continuum of modes. The sample then radiates much faster than a single atom and produces a short and intense radiation burst, within a time inversely proportional to $\mathcal{N}$. This is the 'superradiance' effect (Dicke 1954). For a review, see Gross and Haroche (1982) and Haroche (1984). In most experiments, superradiance is a complex phenomenon, since it involves all field modes and depends upon the spatial distribution of the emitting atoms (McGillivray and Feld 1976).

The situation we are considering here is much simpler, since only one field mode is involved. Moreover, we assume that all atoms 'see' the same field, being confined in a volume small compared to the field wavelength, or located at points with equal field amplitude in the mode's standing wave pattern. The single mode and the symmetrical coupling condition greatly simplify the theory, while retaining the main features of collective radiative effects. These conditions are realized in realistic cavity QED situations (see the next chapter).

4.7.1   The $\mathcal{N}$-spin system

Assuming that all atoms are motionless and identically coupled to the maximum field amplitude in the cavity, the atom–field Hamiltonian is (Tavis and Cummings 1969):

$$H = \omega_{eg} J_Z + \hbar \omega_c a^\dagger a - i \frac{\hbar \Omega_0}{2} \left(a J_+ - a^\dagger J_- \right) . \tag{4.133}$$

The operators J_Z, $J_\pm$ are defined as the sums of the corresponding individual spins:

$$J_Z = \tfrac{1}{2} \sum_j \sigma_{Z,j} ; \qquad J_\pm = \sum_j \sigma_{\pm,j} , \tag{4.134}$$

where the index j labels the atoms. They satisfy the standard commutation relations of angular momentum $\mathbf{J} = (J_X, J_Y, J_Z)$, with $J_\pm = (J_X \pm i J_Y)$. The Hamiltonian commuting with $\mathbf{J}^2$, the total angular momentum J is a good quantum number.

We consider now that all atoms are initially in the same state, $|e\rangle$ or $|g\rangle$. This initial condition corresponds to a maximum value of the angular momentum: $J = \mathcal{N}/2$. The

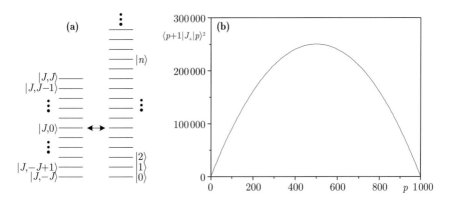

Fig. 4.16 (a) Finite Dicke states ladder and infinite Fock states ladder. (b) Variation of $\langle p+1|\,J_+\,|p\rangle^2$ versus p for $\mathcal{N} = 1\,000$.

atomic system subsequently evolves in the $(\mathcal{N}+1)$-dimensional eigenspace of $\mathbf{J}^2$ with the eigenvalue $J(J+1)$. These states are fully symmetrical by atomic permutation, which results from the symmetry of the coupling and of the initial condition.

The permutation-invariant eigenstates of the collective atomic Hamiltonian are the $\mathcal{N}+1$ symmetric 'Dicke' states $|J,M\rangle$ with $J_Z\,|J,M\rangle = M\,|J,M\rangle$ (Dicke 1954). They form a non-degenerate ladder of equidistant levels. The highest one, $|J,J\rangle$, is $|e,e,\dots,e\rangle$ and the lowest one is $|J,-J\rangle = |g,g,\dots,g\rangle$. All other states are generated by repeated actions of $J_\pm$, according to:

$$J_\pm\,|J,M\rangle = \sqrt{(J\pm M+1)(J\mp M)}\,|J,M\pm 1\rangle \ . \tag{4.135}$$

We retrieve the single-atom case by letting $J = 1/2$ in these equations. The atom–field Hamiltonian H describes the joint evolution of two systems along ladders of equidistant states (Fig. 4.16a). The field ladder is infinite, while the atomic one is bounded.

There is a clear analogy between the field creation and annihilation operators, $a^\dagger$ and a, and the atomic raising and lowering operators, J_+ and J_-. To make it more explicit, we introduce the number of atomic excitations, $p = J + M$, and we make the notation change $|p\rangle = |J, p - J\rangle$. Figure 4.16(b) presents the parabolic variation of $\langle p+1|\,J_+\,|p\rangle^2 = (p+1)(\mathcal{N}-p)$ as a function of p between 0 and $\mathcal{N}$. Around $p = 0$, this quantity is close to $\mathcal{N}(p+1)$, which is reminiscent of the linear variation of $\langle n+1|\,a^\dagger\,|n\rangle^2 = (n+1)$ versus n. Near the bottom of its energy ladder, the collective atomic system behaves as a harmonic oscillator. When the atom–field system is weakly excited, it can be described as two coupled springs, with a coupling frequency $\Omega_0\sqrt{\mathcal{N}}/2$. This approximation breaks down when the number of atomic excitations becomes of the order of $\mathcal{N}$. The square of the matrix element of J_+ then saturates for $p = \mathcal{N}/2$ and decreases again. It cancels for $p = \mathcal{N}$.

A similar oscillator analogy can be made near the top of the ladder. Defining now the number of *lost* atomic excitations $q = J - M$ and letting $|q\rangle = |J, J - q\rangle$, we write:

$$J_-\,|q\rangle = \sqrt{(\mathcal{N}-q)(q+1)}\,|q+1\rangle \ , \tag{4.136}$$

which, near $q = 0$, approximates as $J_- |q\rangle = \sqrt{\mathcal{N}} \sqrt{q+1} |q+1\rangle$. The operator J_- then plays the role of the creation operator for an inverted harmonic oscillator, gaining a negative energy each time a quantum is added. When the atomic system is initially inverted, the evolution starts with a coupled oscillators dynamics, but one of the oscillators is unstable, with a propensity to run away from its vacuum state. This instability is at the origin of the superradiance phenomenon, as discussed in the next section.

The $|J, M\rangle$ Dicke states describe situations in which strong correlations exist between the dipoles of different atoms. The expectation value of any $\sigma_{+,i}\sigma_{-,j}$ product, with $i \neq j$, does not depend upon the couple of atoms considered:

$$\langle J, M| J_+ J_- |J, M\rangle = \langle J, M| \left(\sum_i \sigma_{+,i}\right)\left(\sum_j \sigma_{-,j}\right) |J, M\rangle$$

$$= \mathcal{N}(\mathcal{N} - 1) \langle J, M| \sigma_{+,i}\sigma_{-,j} |J, M\rangle + \langle J, M| \sum_i \sigma_{+,i}\sigma_{-,i} |J, M\rangle \ . \quad (4.137)$$

The last term in this equation represents the number of excited atoms, equal to $J + M$ (note that $\sigma_{+,i}\sigma_{-,i} = |e_i\rangle\langle e_i|$). Taking into account eqn. (4.135), we get:

$$\langle J, M| \sigma_{+,i}\sigma_{-,j} |J, M\rangle = \frac{J^2 - M^2}{\mathcal{N}(\mathcal{N} - 1)} \ . \quad (4.138)$$

This matrix element measures the correlations between the dipoles belonging to two arbitrary atoms in the symmetric state. It has a parabolic variation with M, with a maximum value $\approx 1/4$ for $M = 0$, when $\mathcal{N}/2$ excitations are symmetrically shared among the atoms. The correlations vanish at the two ends of the atomic energy ladder. These correlations reveal that the symmetric atomic states are massively entangled. In the simple case $\mathcal{N} = 2$, the $M = 0$ state is the maximally entangled $(|e, g\rangle + |g, e\rangle)/\sqrt{2}$ Bell state (Chapter 2). The entanglement results here from a fundamental ambiguity: when a photon is emitted or absorbed, there is no way to assign this process to a specific atom.

The build-up of these atomic correlations plays an important role in the system's evolution. Let us couple, at the initial time $t = 0$, $\mathcal{N}$ atoms, all in $|e\rangle$, with a resonant cavity at $T = 0$ K. One expects, as in the single-atom case, that the emission regime depends upon the relative magnitudes of κ and Ω_0. The system is expected to exhibit an oscillatory behaviour when κ is very small and an irreversible, over-damped emission process when κ is large. We consider now briefly these two situations.

4.7.2   Collective emission in a cavity: oscillatory regime

Let us first consider an ideal cavity ($\kappa = 0$). The evolution, ruled by H (eqn. 4.133), is unitary. The resonant coupling conserves the global number of excitations (total number of photons and excited atoms). The evolution thus occurs in a $(\mathcal{N}+1)$-dimensional Hilbert space spanned by the states $|q, q\rangle$ describing q lost atomic excitations along with q photons in the cavity.

The Bohr frequencies describing the system's evolution are given by the eigenvalues of H, a simple tridiagonal matrix in the $|q, q\rangle$ basis. Simple analytical solutions are

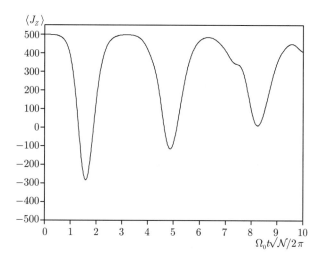

Fig. 4.17 Collective emission in the oscillatory regime. Average atomic energy $\hbar\omega_{eg}\langle J_Z\rangle$ in units of $\hbar\omega_{eg}$ for $\mathcal{N}=1\,000$ atoms emitting in an ideal cavity. The time unit is $2\pi/\Omega_0\sqrt{\mathcal{N}}$.

obtained for small $\mathcal{N}$ values ($\mathcal{N}\leq 4$) (Haroche 1984). For large $\mathcal{N}$ values, there is an asymptotic solution, predicting a quasi-periodic evolution of the average atomic energy, with a frequency $\approx \Omega_0\sqrt{\mathcal{N}}/2(\pi+\log\mathcal{N})$ (Bonifacio and Preparata 1970; Scharf 1970; Haroche 1984). The atomic system and the cavity periodically exchange their energy, but at a faster rate than in the single-atom–cavity case.

Figure 4.17 presents, for $\mathcal{N}=1\,000$, the evolution of $\langle J_Z\rangle$, proportional to the average atomic energy, $\hbar\omega_{eg}\langle J_Z\rangle$, obtained by a numerical integration of the Schrödinger equation. The oscillation is anharmonic, the system spending longer times around the inverted state $q=0$. As shown above, the atomic system is then equivalent to an inverted harmonic oscillator, coupled to the cavity spring with an energy exchange frequency $\Omega_0\sqrt{\mathcal{N}}/2$. This simple picture is certainly not valid when $q\approx\mathcal{N}$, due to the non-linear behaviour of the collective atomic system. The atom–field coupling is then much stronger, the interaction Hamiltonian matrix elements reaching values of the order of $\Omega_0\mathcal{N}\sqrt{\mathcal{N}}/2$ for $q\approx\mathcal{N}/2$. This explains the sharp cusp features near the minima of $\langle J_Z\rangle(t)$. The time spent around these cusps being very small, the characteristic oscillation time is determined by the slow processes occurring near $q=0$, explaining that the period is, up to a logarithmic correction, of the order of $2\pi/\Omega_0\sqrt{\mathcal{N}}$.

The average atomic energy remains, at all times, larger than the minimum value corresponding to all atoms in $|g\rangle$. This is directly related to the existence of large fluctuations in the system. At a given time, a distribution of symmetric atomic states with different M values is populated. The centre of gravity of this distribution yields the average atomic energy. At the minimum, the centre of gravity of the distribution has an energy excess over $-\mathcal{N}\hbar\omega_{eg}/2$ of the order of half the distribution dispersion. On the other hand, after a whole period of the oscillation, the average energy comes back close to $\mathcal{N}\hbar\omega_{eg}/2$. At this time, the fluctuations almost vanish. In other words, the atomic energy fluctuations oscillate at the same frequency as the average energy.

These fluctuations can be understood by assimilating the atomic angular momentum to a classical pendulum, generalizing the Bloch vector model to angular momentums larger than $1/2$. The collective emission process is equivalent to the oscillation of this pendulum in a gravitational field (Bonifacio *et al.* 1971a). The potential and kinetic energies of the pendulum play the role of the atomic and field energies respectively. At $t = 0$, the atomic pendulum is pointing upwards in the fictitious gravitational field.

Classically, this state is metastable. The pendulum possesses, however, quantum fluctuations ($\langle J_x^2 \rangle = \langle J_y^2 \rangle \neq 0$). They trigger the evolution, i.e. the emission process around $t = 0$. They can be mimicked by imparting a slight initial random 'tipping angle' to the pendulum (Gross and Haroche 1982). All quantum mechanical expectation values can be recovered by computing classical trajectories and averaging them over the known statistics of the random initial tipping angle.

The characteristic departure time of a pendulum from unstable equilibrium varies rapidly with the initial tipping angle. It is then clear that the pendulums pass through their minimum potential energy at various times, whose dispersion reflects the initial distribution of tipping angles. This results in large fluctuations of the atomic energy at this time, representing a macroscopic amplification of the initial tipping angle distribution. This model also explains the suppression of atomic fluctuations over a full period. The fastest pendulums (larger tipping angle) stay long enough around the upward position to be caught up by the slowest ones. Note, however, that the dephasing processes between the pendulums gradually blur the oscillations, as seen in Fig. 4.17.

Cavity losses are described by a master equation, which can be numerically solved. In the semi-classical pendulum picture, these losses are represented by a viscous damping of the pendulum's velocity (friction proportional to κ). For $\kappa < \Omega_0 \sqrt{\mathcal{N}}$, damped oscillations are observed, with a final state in which the atomic system and the cavity are both in their ground state (pendulum at rest in its stable equilibrium position). For larger damping, $\kappa > \Omega_0 \sqrt{\mathcal{N}}$, the pendulum motion is overdamped. The atomic energy decays monotonically towards zero. We devote the next paragraph to a more detailed discussion of this situation, reminiscent of the Purcell regime in the spin-$1/2$ case.

4.7.3   The overdamped regime: superradiance in a cavity

In a very strongly damped cavity ($\kappa \gg \Omega_0 \sqrt{\mathcal{N}}$) at $T = 0$ K, the photons emitted by the collective atomic sample are lost so fast that the cavity stays, at all times, practically empty and its entanglement with the atomic system remains very small. It then plays the role of an environment for the atomic system. Even though the atoms are not directly coupled to the outside world (the cavity environment), they 'see' it indirectly, through their coupling to the rapidly relaxing cavity mode. The rigorous description of this situation is given by a master equation for the coupled atom–field system, relaxing into the cavity's environment. The very fast time constant for the cavity field makes it possible to eliminate the field variables from the dynamics. We then obtain a master equation for the atomic density matrix ρ_a alone, which describes an irreversible radiative cascade down the ladder of the symmetric atomic states $|q\rangle = |J, J - q\rangle$ (Bonifacio *et al.* 1971a;b; Agarwal 1974).

If there is no initial coherence between different $|q\rangle$ states, the off-diagonal density

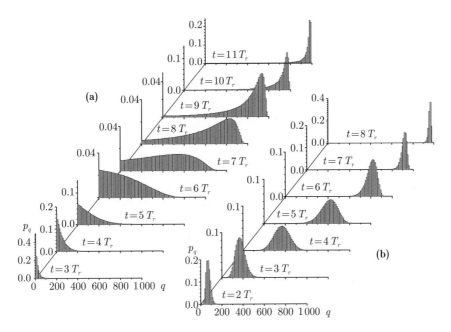

Fig. 4.18 Calculated $p_q(t)$ distributions for $\mathcal{N} = 1\,000$ atoms. (a) Initial state $|q_0 = 0\rangle = |J, J\rangle$. (b) Initial state $|q_0 = 10\rangle = |J, J - 10\rangle$. The times are in units of $T_r = 1/(\mathcal{N}\Gamma_c)$. Note the vertical scale changes.

matrix elements remain zero and the process is entirely described by rate equations involving the states populations $p_q = \rho_{a;q,q}$:

$$\frac{dp_q}{dt} = -\Gamma_{q \to q+1} p_q + \Gamma_{q-1 \to q} p_{q-1} , \tag{4.139}$$

where:

$$\Gamma_{q \to q+1} = \Gamma_c (\mathcal{N} - q)(q + 1) . \tag{4.140}$$

The first term in eqn. (4.139) describes the departure from $|q\rangle$ due to the loss of one photon under the action of J_-, the second the arrival in $|q\rangle$ from $|q - 1\rangle$. The transition rates (eqn. 4.140) can be intuitively understood. For a single atom ($J = 1/2$), they reduce to $\Gamma_{0 \to 1} = \Gamma_c$, the cavity-enhanced spontaneous emission rate of the Purcell regime (Section 4.5). For $\mathcal{N} > 1$, this rate is multiplied by the square of the matrix element of J_- between $|q\rangle$ and $|q + 1\rangle$. Noting that this square can also be written as $\langle q | J_+ J_- | q \rangle$, we see that the rate equation derives from a compact Lindblad expression involving the single jump operator $\sqrt{\Gamma_c} J_-$. In the interaction picture, this Lindblad equation is:

$$\frac{d\rho_a}{dt} = \Gamma_c J_- \rho_a J_+ - \frac{\Gamma_c}{2} \left[J_+ J_- \rho_a + \rho_a J_+ J_- \right] . \tag{4.141}$$

This equation could have been guessed directly from the single-atom spontaneous emission master equation (4.71) (expressed in the interaction picture) by replacing in it Γ by Γ_c and σ_- by J_-.

Figure 4.18(a) shows the $p_q(t)$ distribution, resulting from a numerical solution of eqn. (4.141) with $\mathcal{N} = 1\,000$. It is computed at increasing times, for a system initially in state $|0\rangle = |J, J\rangle$. The time is defined in units of $T_r = 1/(\mathcal{N}\Gamma_c)$. Starting from the $\delta_{q,0}$ distribution, $p_q(t)$ spreads out very quickly over a wide range of q values as the atomic system cascades down. At early times $(t = 3T_r)$, $p_q(t)$ has an exponential shape versus q. At later times $(t \approx 7T_r)$, when the average q value gets close to $\mathcal{N}/2$, $p_q(t)$ is a bell-shaped distribution practically covering all q values (Bonifacio et al. 1971b). At still later times $(t \approx 10T_r)$, towards the end of the cascade, $p_q(t)$ gets narrower and ends up as the $\delta_{q,\mathcal{N}}$ distribution.

At the middle of the emission, the wide distribution of $p_q(t)$ correspond to large fluctuations of the atomic energy. These fluctuations, as in the oscillatory regime, result from an amplification of fluctuations occurring in the initial stage of the process, when the atomic system is close to the top of the states ladder. These fluctuations are greatly suppressed if the emission starts from the state $|q_0\rangle$, with $0 \ll q_0 \ll J$. We have represented in Fig. 4.18(b) $p_q(t)$ for $q_0 = 10$, computed for the same times as in Fig. 4.18(a). The system still cascades down irreversibly, with a similar time constant for the evolution of the average energy, but the fluctuations are now much smaller.

In order to understand the dynamics of these fluctuations, it is fruitful to analyse this problem from a different perspective and to follow the evolution of the system, not on a quantum mechanical average, but along single quantum trajectories (Haake et al. 1979). One can imagine that the photons escaping from the cavity are detected with unit efficiency, recording in this way the successive jumps of the atomic system. These jumps occur at random times and the process can be mimicked by a Monte Carlo simulation. Each trajectory is a random unidirectional walk down the energy states ladder. The p_q distribution is recovered by averaging over a large number of such trajectories.

The decay of the atomic energy versus time on a typical trajectory is shown in Fig. 4.19(a). We choose as a natural time unit $T_r = 1/\mathcal{N}\Gamma_c$. The excitation first remains, for a few T_r, nearly constant. During this time, a few photons are emitted and the interatomic correlations build up. The atomic dipoles, once correlated, behave as an ensemble of phased antennas radiating very fast, much more rapidly than a single one. This is a classical interference effect. The initial phasing of the dipoles, though, is an intrinsically quantum process. This acceleration of the collective emission is a characteristic feature of superradiance (Gross and Haroche 1982). A similar effect occurs in free space. It is, however, much more complicated then, due to the coupling of the atoms with a continuum of field modes. The single-mode situation retains, in a much simpler context, the essence of superradiance.

The whole process can be described by two time constants defined in Fig. 4.19(a): a delay time, T_d, corresponding to the emission of $\mathcal{N}/2$ photons and a collective radiation pulse duration, T_w, during which the bulk of the emission occurs (width at half maximum of the time-derivative of the atomic energy).

An ensemble of such trajectories (thin lines) along with their average (thick line) is shown in Fig. 4.19(b). All trajectories have a similar shape, with the same T_w values, and differ only by the duration of the early phasing stage, resulting in a spread of the delay T_d over a range of the order of T_w. A cut at the average delay time, $\overline{T_d}$, shows that the dispersion of atomic energies is then of the order of $\mathcal{N}\hbar\omega_{eg}$. This explains the

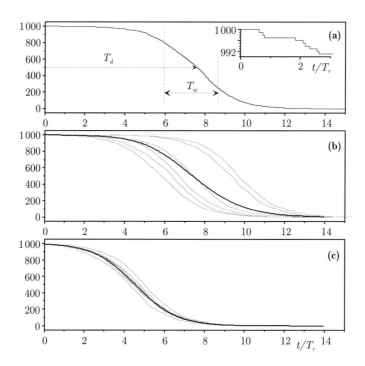

Fig. 4.19 Monte Carlo trajectories for the collective emission of $\mathcal{N} = 1\,000$ atoms. Atomic energy in units of $\hbar\omega_{eg}$ versus time, in units of T_r. (a) A typical trajectory. The initial state is $|q = 0\rangle = |J, J\rangle$. The arrows indicate the delay time T_d and the emission pulse duration T_w. The inset displays the first quantum jumps at the beginning of the evolution. (b) Six trajectories (thin lines) for the same initial condition as in (a) and atomic energy averaged over 10 000 trajectories (thick line) (c) Six trajectories starting from $|q_0 = 10\rangle = |J, J - 10\rangle$ (thin lines) and average over 10 000 such trajectories (thick line).

large width of $p_q(t)$ exhibited in Fig. 4.18(a) around $t = \overline{T_d}$.

The Monte Carlo approach also allows us to estimate T_d, T_w and their fluctuations (Haroche 1984). On a typical trajectory starting from the fully inverted state $|0\rangle = |J, J\rangle$, the successive steps take an average time inversely proportional to the $\Gamma_{q \to q+1}$ rates. The average delay time $\overline{T_d}$ is thus of the order of:

$$
\begin{aligned}
\overline{T_d} &= \frac{1}{\Gamma_{0 \to 1}} + \frac{1}{\Gamma_{1 \to 2}} + \cdots + \frac{1}{\Gamma_{q \to q+1}} + \cdots \\
&= \frac{1}{\Gamma_c} \left[\frac{1}{\mathcal{N}} + \frac{1}{2(\mathcal{N} - 1)} + \frac{1}{3(\mathcal{N} - 2)} + \cdots \right] \\
&\approx T_r \left[1 + \frac{1}{2} + \frac{1}{3} + \cdots \right] \approx T_r \log \mathcal{N} \, .
\end{aligned}
\tag{4.142}
$$

For $\mathcal{N} = 1\,000$ atoms, $\overline{T_d} = 6.9\,T_r$.

The radiation time, T_w, corresponds to the bulk of the emission, when the jump rate is the fastest, each jump taking a time of the order of $4/\mathcal{N}^2\Gamma_c$. The emission of half the photon number $\mathcal{N}/2$ at this rate would take a time $2/\mathcal{N}\Gamma_c = 2T_r$. This is a lower bound for T_w, which is in fact slightly longer, due to the decrease of the jump rate when q departs from $\mathcal{N}/2$. The actual value, as seen in Fig. 4.19(a), is $T_w \approx 2.6\,T_r$.

The Monte Carlo trajectory description can also be used to estimate the fluctuations of the superradiance delay time. Coming back to a single quantum trajectory, we remark that the uncertainties about the intervals between jumps are of the same order as their average. Assuming that the jump events are not correlated, one can quadratically add their contributions and get the standard deviation of T_d:

$$\Delta T_d = T_r \left[1 + \frac{1}{2^2} + \frac{1}{3^2} + \cdots \right]^{1/2} \approx 1.3\,T_r \;. \tag{4.143}$$

It is of the order of T_w, as noticed above (Fig. 4.19b).

Finally, the Monte Carlo trajectories also explain the strong suppression of fluctuations when the atomic system starts from an already correlated state $|q_0\rangle$ (Haake and Glauber 1972). Figure 4.19(c) presents a set of trajectories for $\mathcal{N} = 1\,000$ and $q_0 = 10$, corresponding to the conditions of Fig. 4.18(b). The trajectories have the same general shape as for the $q_0 = 0$ initial condition. The bulk of the emission occurs, however, sooner and the dispersion of the delay times is much smaller. These features explain the reduced fluctuations in the atomic energy reflected by the narrow $p_q(t)$ distribution in Fig. 4.18(b).

The characteristic time $T_d(q_0)$ and its fluctuations are now given by the truncated series:

$$\overline{T_d}(q_0) = T_r \left[\frac{1}{q_0 + 1} + \frac{1}{q_0 + 2} + \cdots \right] \approx T_r \log \frac{\mathcal{N}}{q_0 + 1} \;, \tag{4.144}$$

and:

$$\Delta T_d(q_0) = T_r \left[\frac{1}{(q_0 + 1)^2} + \frac{1}{(q_0 + 2)^2} + \cdots \right]^{1/2} \approx \frac{T_r}{\sqrt{q_0 + 1}} \;. \tag{4.145}$$

For $\mathcal{N} = 1\,000$ and $q_0 = 10$, these parameters become $\overline{T_d} = 4.5\,T_r$ and $\Delta T_d = T_r/3.3$. The relative dispersion of the delay, $\Delta T_d/\overline{T_d}$ is now 0.07, down from 0.2. This reduction of the delay fluctuations comes from the initial phasing of the dipoles.

The Monte Carlo trajectory description of the overdamped superradiance regime is very analogous to the pendulum description of the oscillatory collective emission. Each quantum trajectory in Fig. 4.19 can be viewed as the evolution of the potential energy of a classical over-damped pendulum in a gravitational field. The various trajectories correspond to different initial tipping angles.

We have analysed so far superradiance at $T = 0$ K. A simple modification of the master equation (4.141) can take into account the presence of an initial thermal field and describe the collective emission at a finite temperature T. We consider here only the case of an initially fully inverted system. The thermal effects are particularly simple to account for when the average number of blackbody photons in the mode, n_{th}, is much smaller than the atom number $\mathcal{N}$. The thermal fluctuations merely add up to the vacuum one and have only the effect of speeding up the phasing time of the atomic dipoles. This results in a shortening of the emission delay time T_d from $T_r \log \mathcal{N}$ to $T_r \log \left[\mathcal{N}/(1 + n_{\text{th}}) \right]$ (Gross and Haroche 1982). All the features of the system's evolution remain qualitatively unchanged.

The superradiance phenomenon was the focus of intense theoretical and experimental activity in the early days of quantum optics. The simple model described in this section has been experimentally realized with Rydberg atoms in microwave cavities. We briefly describe these experiments in the next chapter.

We got acquainted in the last two chapters with the theoretical tools needed to analyse the behaviour of open systems in quantum optics. We know how to describe the coherent coupling of quantum springs and spins, and how to account for the irreversible effects related to their interactions with unavoidable environments. We need now to learn how to keep these dissipative effects minimal in an experiment, so that the juggling between the spins and the springs will reveal, without unnecessary complications, all the beauty and the strangeness of the quantum.

5

Photons in a box

Cavity quantum electrodynamics (CQED) can be defined, in a nutshell, as the physics of a spin and an oscillator in interaction. This system has lent itself to the realization of thought experiments illustrating the basic laws of quantum physics. The fundamental ideas behind these experiments are simple, but implementing them in the laboratory has not been easy. All the perturbations which in real life tend to blur the conceptual simplicity of the physics must be eliminated or at least reduced to a minimum. The goal is to isolate an atom in a cavity with highly reflecting walls, the equivalent of a 'box' separating it efficiently from the external world. The field is quantized in this box as a set of harmonic oscillators, one of them being resonant or nearly resonant with a transition between two levels of the atom which thus behaves as a spin. The field oscillator cannot be considered independently of the cavity walls which define the mode structure, so that the atom is directly interacting with a macroscopic object, the box which surrounds it. This basic feature of an ideal thought experiment – the coherent and controlled coupling of a small quantum system with a large classical one – had for a long time been considered as impossible to achieve in practice.

In order to observe many of the effects theoretically described in Chapter 3, the atom–field coupling Ω_0 must be much larger than the atom and cavity relaxation rates T_a^{-1} and T_c^{-1}. Moreover, Ω_0 must be larger than the reciprocal of the atom–cavity interaction time t_i. These conditions, characterizing the 'strong coupling regime' of CQED, are simple to state but difficult to fulfil. Basically, they mean that a single atom in a cavity is enough to sustain a laser oscillation. They also imply that a single photon in the cavity is able to induce strong non-linear effects on the atom. Lasers have started with amplifying media made up of moles of atoms, and usual non-linear optics involve intense fields with billions of photons. These figures underline the huge gap which has been bridged to reach the single-photon–single-atom strong coupling regime in CQED.

The aim of this chapter is to present the experimental methods developed in this field, which will also be useful to describe experiments analysed in the next chapters. We will depart from the simplicity of the quantum concepts to dig into the complexity of an experimenter's real life. We think that it is indeed important to describe the sophistication of the techniques which are required to unveil the direct manifestations of the quantum.

The initial objective of CQED was to analyse the radiative properties of atoms in a vacuum 'modified' by mirrors. In the early days of this field of research, the observed effects resulted in mere alterations of the atomic radiative parameters – spontaneous emission rate enhancement or inhibition and Lamb shift modifications – while the dynamical behaviour of the atom–field system remained qualitatively unchanged. This 'perturbative regime' of CQED dealt with atoms near mirrors or in cavities with moderately high quality factors Q. With the better resonators which were subsequently developed, the coupling of the atom to one mode of the field has become a dominant effect in the system's evolution. The radiative properties in this strong coupling regime radically differ from what is observed on an atom in free space. Spontaneous emission, for instance, is replaced by quantum Rabi oscillation and becomes a reversible process. The qualitative change in the dynamics leads to new quantum effects, making it possible to entangle in a controllable way atoms and photons and opening the way to quantum information applications.

Modern CQED techniques result from fifty years of incremental improvements in cavity design and atom manipulations, which have largely benefited from the technological development of lasers, quantum optics and solid state devices. We review these developments in Section 5.1 in their historical perspective. We start by analysing the perturbative regime of CQED and show how it has gradually evolved into strong coupling studies. We follow in this discussion the parallel developments of microwave and optical experiments and we say a few words about possible applications to optronics and communication technologies.

We have chosen, after this brief historical introduction, to analyse in more detail the experiments performed with superconducting microwave cavities and Rydberg atoms with maximum electronic angular momentum ('circular' states). This system constitutes a quasi-ideal realization of the spin–spring model. In Section 5.2, we describe the circular Rydberg atoms and the superconducting cavities which are the two 'partners' in these experiments and we analyse the coupling between them. As first examples, we present in Section 5.3 two basic experiments unveiling in a dramatic way the quantumness of the atom–field system. We end (Section 5.4) by a description of entanglement procedures in CQED. We show how the resonant Rabi oscillation is used to entangle one atom and the field, two atoms, or two field modes. We conclude by the description of non-resonant atom–cavity experiments which provide an alternative recipe for atom–atom entanglement.

5.1   A short history of cavity QED

The radiative properties of an atom in free space are determined by its coupling to the continuum of radiation modes in vacuum. Irreversible spontaneous emission arises from the coupling of atoms in excited states with resonant field modes. The dispersive coupling to all other modes is the source of the Lamb shift effect, a modification of the atomic energy levels. These fundamental phenomena were considered, in the early days of atomic physics, as unalterable features of the atom–field coupling.

While it is true that the quantum fluctuations of the vacuum in a given mode cannot be changed, it has however gradually been realized that the spatial and frequency distributions of the modes depend upon physical limiting conditions imposed to the field, which can be modified and controlled. Disposing mirrors or conductors around the atom changes the distribution of modes with which it interacts, and hence the density of vacuum fluctuations that it experiences at a given frequency. This density can be increased or reduced with respect to its free-space value, leading to inhibited or enhanced atomic radiative properties.

5.1.1   Early theories

The first discussion of a CQED concept is found in the short note by Purcell (1946) mentioned in Chapter 3, in which he predicted that the spontaneous emission rate of a spin in a magnetic resonance set-up, exceedingly low in free space, should be considerably enhanced in a resonant structure, such as a resonant $R - L - C$ circuit used to filter the free-induction decay NMR signals. To explain this effect, Purcell invoked a simple Fermi Golden Rule argument, recalled in the last chapter. When the quality factor Q of the circuit is not too high, the oscillator's resonance width is much larger than the spontaneous emission rate. The circuit then behaves, from the spin point of view, as a continuum. The mode density is, however, much larger than its value in free space, entailing a multiplication of the spontaneous emission rate by a factor $\eta = (3/4\pi^2)Q\lambda^3/\mathcal{V}$, where $\mathcal{V}$ is the resonator mode's volume and λ the wavelength of the transition (eqn. 4.115).

This argument was applied to a spin in an NMR context, but the same effect is expected for an atom in an optical cavity. Spontaneous emission can thus be enhanced at will by imposing boundary conditions to the field radiated by the atom. The factor η reaches high values in small-volume, high-quality cavities. The Purcell argument applies only as long as the damped cavity mode can be considered as a continuum. For very high quality factors, the mode width is comparable to the spontaneous emission rate and the perturbative CQED approach breaks down as the atom–cavity system enters the strong coupling regime.

Purcell's note was followed by a series of papers dealing with spontaneous emission rates and level shifts near mirrors. An extensive bibliography is given by Hinds (1994). Let us mention here the paper by Casimir and Polder (1948), calculating the force between an atom and a conducting plane. This force is the gradient of the position-dependent level shift experienced by the atom near the plane and relates thus directly to a CQED effect. This work was soon extended to the case of two parallel metallic plates (Casimir 1948), leading to the prediction of the 'Casimir effect', an attraction of the plates in free space due to the 'radiation pressure' of vacuum fluctuations. This 'CQED effect without atoms' has attracted a lot of theoretical interest (Barton 1994). It has also led to detailed analyses of the forces and frictions experienced by mirrors moving in the vacuum (Jaekel and Reynaud 1997; Lambrecht *et al.* 1996).

The remark by Kleppner (1981) that the spontaneous emission can be inhibited as well as enhanced has been another important milestone in CQED theory. The density of modes in which an atom can radiate is suppressed when it is located in a waveguide below cut-off for its transition frequency and its spontaneous emission rate becomes accordingly much smaller than in free space. That an atom can be kept, in principle,

(a)   (b)   (c)

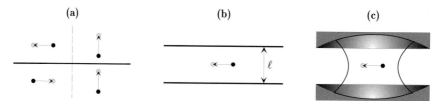

Fig. 5.1 Spontaneous emission with three different boundary conditions. (a) Atom, represented as a radiating dipole, in front of a plane mirror (thick horizontal solid line). The interference with the electric image (below the mirror) is destructive when the atomic dipole is parallel to the mirror (left side) and constructive when it is perpendicular (right side). (b) Spontaneous emission is completely suppressed between two mirrors with $\ell < \lambda/2$ when the atomic dipole is parallel to the mirrors. (c) Spontaneous emission is strongly enhanced in a resonant Fabry–Perot cavity.

for an arbitrarily long time in one of its radiative excited states is a very striking illustration of CQED concepts.

These effects have been discussed in situations corresponding to various boundaries around the atom. The simplest configuration corresponds to an atom located at a small distance d from a plane conductor, with a transition wavelength $\lambda \gg d$ (Fig. 5.1a). The electric field parallel to the mirror nearly vanishes at the atomic location for all modes with wavelength λ. As a result, an atom with a dipole polarized along the surface finds no mode to emit a photon in and its spontaneous emission is inhibited. If the dipole is normal to the surface instead, the atom remains coupled to modes with a non-vanishing electric field normal to the surface. The spontaneous emission rate is then twice that in free space.

These spontaneous rate alterations are easily understood in terms of 'electric images' (Jackson 1975), which account for the field boundary conditions (Haroche 1992). The field emitted at large distance by an atom located in the vicinity of a mirror results from the interference between the direct radiation and the field reflected from the surface. This latter component is the same as if it were radiated by a charge-conjugate image of the dipole in the surface (Fig. 5.1a). For a dipole parallel to the surface, the image dipole is opposite and the interference destructive. For a dipole normal to the surfaces, the image is identical to the dipole. The interference is constructive, explaining the doubled emission rate.[1]

These effects become even more spectacular when the atom is surrounded by conducting walls. Multiple electric images then reinforce the interference effect. In the case of two parallel mirrors (Fig. 5.1b) with a gap ℓ, all field modes with $\lambda > 2\ell$ and an electric field parallel to the plates are excluded. An atom with a dipole aligned along the mirrors is unable to radiate if its wavelength is above the cut-off $\lambda_c = 2\ell$. This leads to a total spontaneous emission inhibition, wherever the atom is inside the gap. There is a completely destructive interference between the directly emitted field and the components radiated by the set of images resulting from multiple reflections.

---

[1]The combined atom + image dipole is twice the 'naked' atomic dipole, leading to a quadrupled field intensity, but this field is radiated only in the half-space above the mirror, hence the factor of 2 of the overall radiation rate.

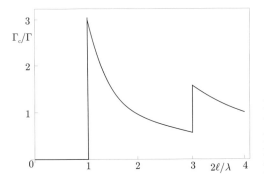

Fig. 5.2 Spontaneous emission rate Γ_c (in units of the free-space rate Γ) in an ideal plane-parallel waveguide as a function of $2\ell/\lambda$, for a dipole parallel to the mirrors and located at mid-gap.

If the mirror separation is increased to $\ell = \lambda/2$, modes with an electric field parallel to the mirrors suddenly appear and the spontaneous emission rate Γ_c jumps from zero to thrice its free-space value. The rate Γ_c is plotted in Fig. 5.2 versus $2\ell/\lambda$ for a dipole located at mid-gap and parallel to the mirrors, in units of the free-space rate Γ. It exhibits sharp rises for $\ell = (2p + 1)\lambda/2$, where p is an integer. This resonant enhancement can be made much larger by a better confinement of the field in a resonant cavity. If plane mirrors are replaced by spherical conducting surfaces, the field modes are focused in a small region of volume $\mathcal{V}$ between the mirrors (Fig. 5.1c). The spontaneous emission rate of a resonant atom located in this region is then considerably enhanced by the Purcell effect (eqn. 4.115).

The electric image picture explains the radiative shifts as well as the spontaneous emission rate modifications. The image field has one quadrature $\pi/2$-out-of-phase with the atomic dipole, which either adds to or subtracts from the self-reaction field, leading to spontaneous emission enhancement or inhibition. It has also a quadrature in-phase with the atomic dipole, which modifies its oscillation frequency and accounts for the cavity shifts.

The perturbative regime of cavity QED explains the radiative properties of a single atom interacting with its images in conducting walls. This situation is comparable to the radiation of an atom interacting with its neighbours in an atomic sample. Single-atom rates are then modified by collective radiation reaction and transition frequencies are shifted by van der Waals interactions between the atoms. These effects have also been extensively studied since the early work of Dicke (1954) on superradiance. In fact, CQED and superradiance theory have developed in parallel, borrowing ideas and concepts from each other. The emission rate enhancement produced by a cavity and by collective emission processes can indeed be combined together by placing $\mathcal{N}$ two-level atoms in a cavity, as was discussed in Section 4.7. In the overdamped regime, the collective emission is then faster than in free space by a combination of two enhancing factors. Each of the $\mathcal{N}$ atoms in the sample radiates along with η images, the overall emission rate being multiplied by $\mathcal{N}\eta$.

5.1.2 Perturbative regime: single-atom experiments

The first single-emitter experiments have been performed in the optical domain, with dye molecules embedded in a Langmuir–Blodgett film structure, deposited on a mirror. The film acted as a spacer between the dye and the mirror. Modifications of the spontaneous emission patterns and limited alterations of the rates have been observed

(Drexhage 1974).

A clear-cut test of Purcell's predictions was made in the 1980s in the microwave domain (Goy *et al.* 1983). This experiment was performed with sodium Rydberg atoms, crossing one at a time a Fabry–Perot microwave cavity. Rydberg atoms are superb tools for matter–field coupling studies, because of their huge electric dipole on millimetre-wave transitions and because they can be efficiently and selectively detected by field-ionization. These unique features will be described in detail in Section 5.2, when we turn to the strong coupling regime experiments which are the prime topic of this book. We present here only a sketchy review of this early Rydberg CQED study. Sodium atoms prepared in the state $|23S\rangle$ (principal quantum number 23 and zero orbital momentum) were injected one by one in the cavity tuned to resonance with the $|23S\rangle \rightarrow |23P\rangle$ transition at 340 GHz. The partial spontaneous emission rate on this transition is 150 s^{-1}. The cavity quality factor was $Q \approx 10^6$ and the mode volume $\mathcal{V} = 70$ mm^3. The Purcell enhancement factor, $\eta = 530$, increased the free-space emission rate to $\Gamma_c = 8\,10^4$ s^{-1}. The Purcell effect was demonstrated by sweeping the distance between the mirrors and observing the sharp increase of the $|23P\rangle$ level population when the cavity was at resonance.

The first evidence of spontaneous emission inhibition was obtained in an electron-trap experiment (Gabrielse and Dehmet 1985). The emitter was a single electron in a Penning trap (Ghosh 1995). The radiative damping of the cyclotron motion at millimetre-wave frequency was inhibited when the microwave cavity around the electron, made up of the trap electrodes, was non-resonant. This experiment has been considerably improved since then (Peil and Gabrielse 1999). Soon after this first observation, spontaneous emission inhibition was demonstrated in an atomic system (Hulet *et al.* 1985). This experiment made again use of Rydberg atoms. They were prepared in a 'circular' state $|22C\rangle$, with principal quantum number $n = 22$ and maximum angular momentum. This state has a very long spontaneous emission lifetime (460 μs), and decays only by a millimetre-wave transition at $\lambda = 0.45$ mm towards the circular state $|21C\rangle$ (we say much more on circular states in Section 5.2.1). The atoms were sent in a waveguide made up of two parallel metallic plates. The $|22C\rangle \rightarrow |21C\rangle$ transition was tuned via the Stark effect by applying a static electric field across the mirrors. When the radiated wavelength was made larger than twice the intermirror gap, the number of atoms surviving the transit in $|22C\rangle$ was noticeably increased, revealing the inhibition of the spontaneous emission process. A closely related effect, the inhibition of blackbody radiation absorption, had been observed previously in a similar experiment (Vaidyanathan *et al.* 1981).

Observations of spontaneous emission modifications have also been reported since then in the near-infrared and optical domains.[2] An experiment (Jhe *et al.* 1987a) has been performed with caesium atoms prepared in the $|5D_{5/2}\rangle$ state, which can decay only to $|6P_{3/2}\rangle$ by a 3.49 μm infrared transition. The atoms crossed, at thermal velocity, a 8 mm-long 1.1 μm-wide gap between two parallel metallic mirrors. A static magnetic field **B** was applied, inducing the precession of the atomic optical dipole in a plane perpendicular to it. When **B** was perpendicular to the mirror plane, the atomic

---

[2]For spontaneous emission enhancement and inhibition, see for instance Heinzen *et al.* (1987); Jhe *et al.* (1987a); De Martini *et al.* (1987); Yablonovich *et al.* (1988); Yokoyama *et al.* (1990); Martorell and Lawandy (1990); Yokoyama *et al.* (1991); Björk *et al.* (1991); and Morin *et al.* (1994).

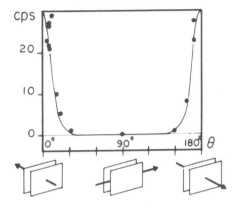

Fig. 5.3 Spontaneous emission inhibition in the infrared. Number of caesium atoms in the $|5D\rangle$ level at the exit of a gap between parallel plane mirrors, as a function of the angle θ between the normal to the mirrors and the applied magnetic field **B**. The arrows in the bottom represent the magnetic field for $\theta = 0$, 90 and 180°. Spontaneous emission inhibition is observed when the atomic dipole precessing around **B** remains parallel to the mirrors ($\theta = 0$ and 180°). Reprinted with permission from Jhe *et al.* (1987a). © American Physical Society.

dipole remained in this plane and the spontaneous emission rate was considerably reduced, down to 4% of its free-space value (Jhe *et al.* 1987b). Almost all atoms survived the 8-mm travel through the gap, whereas the free-space spontaneous emission rate would only correspond to a 0.6 mm propagation. When **B** was parallel to the mirrors, the precessing optical dipole periodically acquired a component normal to the mirrors, which strongly radiated. The overall spontaneous emission rate was then enhanced instead of being inhibited. Figure 5.3 presents the number of atoms detected per second in $|5D_{5/2}\rangle$ at the exit of the gap, for a constant input flux, as a function of the angle of **B** with the direction normal to the mirrors. The spontaneous emission inhibition, revealed by the strong transmission of the atoms in the excited $|5D_{5/2}\rangle$ state, is only observed in narrow domains close to 0 and 180°.

Radiative shifts of atomic levels near a mirror or in cavity-like structures have also been explored experimentally.[3] These shifts depend upon the atomic position. At a distance $d \ll \lambda$ from the cavity wall (λ being the wavelength of relevant atomic transitions), they are of a van der Waals type, resulting from the instantaneous electrostatic interaction of the atom with its closest electric image (Sandoghdar *et al.* 1992). The shift then varies as $1/d^3$. At larger distances, $d \approx \lambda$, retardation effects come into play. The van der Waals interaction evolves into the Casimir–Polder interaction, which decreases as $1/d^4$ (Sukenik *et al.* 1993). In a high-Q cavity, far from its walls, the atoms experience shifts which vary dispersively when a cavity mode is tuned around atomic resonance: the shift cancels and changes sign at resonance (Heinzen and Feld 1987; Brune *et al.* 1994).

Note that all these shifts are position-dependent, meaning that the energy of the atom–cavity system depends upon its spatial configuration. The derivatives of these shifts with respect to the atomic position correspond to forces acting on the atom in the cavity. The effect of these forces on Rydberg atoms pushed towards metallic surfaces has been observed (Sandoghdar *et al.* 1992; Sukenik *et al.* 1993). Similar forces exist in the strong coupling regime and are predominantly important in the optical domain, as we will see below.

---

[3]See for instance Hinds (1990); Hinds and Sandoghdar (1991); Walther (1993); Sandoghdar *et al.* (1996); and Marrocco *et al.* (1998).

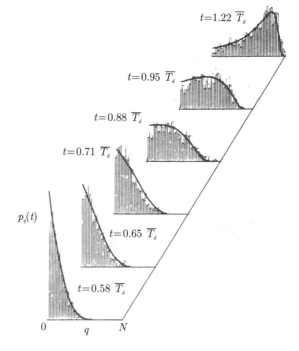

Fig. 5.4 Superradiance in a cavity. Statistical distribution of the number of de-excited atoms $p_q(t)$ for various atom–cavity interaction times, in units of the average time of half de-excitation $\overline{T_d}$. The solid lines correspond to the theory described in Section 4.7.3. Reprinted with permission from Raimond *et al.* (1982). © American Physical Society.

5.1.3   Superradiance in a cavity

Single-atom cavity QED effects are conceptually simple, but relatively difficult to observe because they require the manipulation and detection of very small numbers of atoms in order to avoid collective behaviours. The first experiments in cavity QED have in fact been performed with sample containing a large number of atoms. The first evidence of spontaneous emission enhancement by a resonant circuit was obtained in the NMR context (Feher *et al.* 1958), the sample containing a very large number of spins. The collective emission was then sped-up by the presence of the resonant coils around the sample.

A more quantitative test of superradiance in a cavity has been performed with Rydberg atoms in the microwave domain (Raimond *et al.* 1982). An ensemble of $\mathcal{N} \approx 3200$ atoms were prepared in the Rydberg state $|29S\rangle$ in a cavity resonant with the transition $|29S\rangle \rightarrow |28P\rangle$ at 162 GHz. The atomic sample being small compared to $\lambda \approx 2$ mm, all the atoms were symmetrically coupled to the cavity mode, realizing the superradiance conditions discussed in Section 4.7. The free-space spontaneous emission rate on the atomic transition was $\Gamma_0 = 43$ s^{-1} and the Purcell factor $\eta = 70$, leading to a superradiance time $T_r = 1/(\mathcal{N}\eta\Gamma_0) = 104$ ns. The cavity temperature was $T = 300$ K, corresponding to an average number of thermal photons $n_{\text{th}} = 38$, much smaller than the atom number. According to the discussion of Section 4.7, the delay of the superradiant emission was then expected to be $\overline{T_d} = T_r \log[\mathcal{N}/(1+n_{\text{th}})] = 0.45$ μs.

In each experimental sequence, the atoms were left to interact with the field for a time t, then suddenly decoupled from it by applying an electric field which shifted

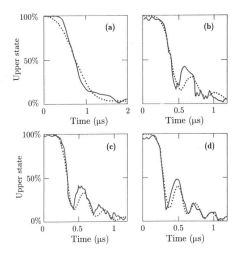

Fig. 5.5 Collective Rabi oscillation. Fraction of atoms in the upper state as a function of time for $\mathcal{N} = 2\,000$, $19\,000$, $27\,000$ and $40\,000$ [frames (a)–(d) respectively]. The collective Rabi oscillation is observed when the atom number becomes large enough [traces (b), (c) and (d)]. The solid line is the experimental signal and the dotted line is a theoretical fit. Reprinted with permission from Kaluzny *et al.* (1983). © American Physical Society.

by the Stark effect the atomic transition away from the mode frequency. This Stark-switching technique will be described in more detail below (Section 5.2). The system's evolution was then frozen until atomic detection. Many experimental runs were performed for each duration t and the number of atoms in the initial and final states recorded. Figure 5.4 presents the histograms of the number of de-excited atoms, $p_q(t)$, obtained from a statistical analysis of these runs, for various atom–cavity interaction times expressed in units of the average time of half de-excitation $\overline{T_d}$. There is excellent agreement with the theoretical model discussed in Section 4.7, whose predictions are plotted as solid lines. The characteristic features of the superradiant emission – the fast decay of the atomic energy and its large fluctuations at mid-emission point – are clearly displayed in these histograms.

5.1.4   The route towards strong coupling: collective oscillation in CQED

The strong coupling regime is reached when the atom–field interaction overwhelms dissipative processes. The first evidence of strong coupling was obtained with multi-atom samples, in which all the atoms were symmetrically coupled to the same mode. The characteristic atom–field coupling is then $\Omega_0\sqrt{\mathcal{N}}$ instead of Ω_0 (Section 4.7). The collective strong coupling regime condition thus becomes $\Omega_0\sqrt{\mathcal{N}} \gg T_a^{-1}$, T_c^{-1}, t_i^{-1}, which is easily met when $\mathcal{N}$ is large enough, even if the cavity Q factor is moderately high. Collective quantum Rabi oscillations have been observed on atoms in the microwave (Kaluzny *et al.* 1983) and optical (Brecha *et al.* 1995) domains, using Fabry–Perot cavities.

Figure 5.5 presents the collective quantum Rabi oscillation signal observed on $\mathcal{N}$ Rydberg atoms in a microwave cavity with a moderate Q. Sodium atoms were suddenly prepared in the upper state $|36S\rangle$ of the cavity-resonant transition towards $|35P\rangle$ at 82 GHz. As in the superradiance experiments, the atoms were symmetrically coupled to the cavity. The fraction of atoms in the upper state was recorded as a function of the atom–cavity interaction time. The Stark-switching technique already used in the superradiance experiment was employed to fix the atom–cavity coupling time.

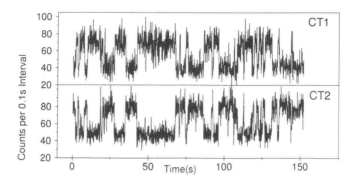

Fig. 5.6 Quantum jumps of the micromaser. The two traces present the atomic counting rates in the upper (CT1) and lower (CT2) states as a function of time. The random switching between a low and a high transfer rate corresponds to the micromaser jumps between two equally stable operating points. Reprinted with permission from Benson *et al.* (1994). © American Physical Society.

The cavity damping rate was $\kappa = T_c^{-1} = 5 \cdot 10^6$ s^{-1} and the single-atom vacuum Rabi frequency, $\Omega_0/2\pi$, was 20 kHz. For large enough atomic numbers, the collective Rabi oscillation frequency $\Omega_0\sqrt{\mathcal{N}}$ exceeded the cavity damping rate, which was the dominant cause of relaxation. Rabi oscillations, corresponding to a periodic energy exchange between the atomic sample and the cavity, were then clearly observed. The frequency of these oscillations increased with $\mathcal{N}$ as expected and excellent agreement was found between experiment and the theoretical predictions (Section 4.7).

Instead of a time-resolved experiment, one can also continuously probe the atom–cavity system by laser absorption. Its spectrum is then directly linked to the Fourier transform of the time-dependent Rabi oscillation signal. The collective atom–field coupling splits the spectrum into two components, symmetrical around the common atomic and cavity frequency, and separated by the collective Rabi frequency $\Omega_0\sqrt{\mathcal{N}}$. This splitting is observed provided it exceeds the width of the peaks, determined by the largest of the T_a^{-1}, T_c^{-1}, t_i^{-1} rates. This corresponds again to the strong coupling condition. These collective splitting has been observed on atoms in the microwave (Bernardot *et al.* 1992) and optical regimes [Section 5.1.6 and Raizen *et al.* (1989)]. Similar effects have also been observed in solid state devices (Section 5.1.7).

5.1.5 Strong coupling in microwave CQED: from micromasers to entanglement engineering

The trapping of microwave photons in very high-Q niobium superconducting cavities has led to a decisive breakthrough in CQED. These cavities can store photons over times in the millisecond to the tenth of a second range, providing a three to five orders of magnitude improvement over the Rydberg atom experiments described above. During its cavity crossing time lasting a few tens of microseconds, an atom can then undergo several Rabi oscillations in the vacuum field. Sending a stream of excited atoms one at a time across such a cavity, the group of H. Walther in Munich has realized, in 1985, a micromaser operating according to the dynamics described in Section 4.6 (Meschede *et al.* 1985).

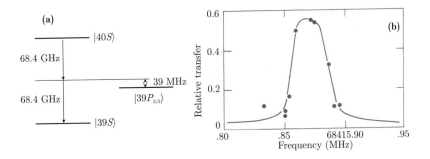

Fig. 5.7 Two-photon micromaser. (a) Relevant levels and transitions. The micromaser operates on the two-photon transition at 68.415 GHz between the $|40S\rangle$ and $|39S\rangle$ levels of rubidium. The relay level $|39P_{3/2}\rangle$ is 39 MHz away from the mid-point of the transition. (b) Transfer rate from $|40S\rangle$ to $|39S\rangle$ versus the cavity frequency. Reprinted with permission from Brune *et al.* (1987). © American Physical Society.

Rubidium atoms were velocity-selected and laser-excited in the Rydberg state $|63P_{3/2}\rangle$ before crossing the cavity, resonant on the transition towards the $|61D_{3/2}\rangle$ level at 21.5 GHz. A steady-state field of a few photons resulted from the competition between the cumulative atomic emissions and the losses. The dynamics and statistical properties of the field were followed by recording the state of the atoms emerging from the cavity. A rich variety of effects were observed, which have been theoretically analysed in Section 4.6: Rabi oscillations in a small quantum field (Rempe *et al.* 1987), sub-Poissonian (Rempe *et al.* 1990) and trapping field states (Weidinger *et al.* 1999), bistable behaviour (Benson *et al.* 1994), and Fock state generation (Varcoe *et al.* 2000; Brattke *et al.* 2001). As an illustration, Fig. 5.6 shows the switching of the micromaser between two stable operating states, corresponding to different transfer rates from initial to final levels. The numbers of atoms counted per unit time in these levels undergoes jumps revealing the system's quantum dynamics.

The extremely long photon storage time of these superconducting cavities makes it possible to achieve single-atom maser operation under exotic conditions. A micromaser operating on a two-photon transition (Brune *et al.* 1987) has been realized in a device which was the first maser/laser emitting light quanta by pairs. The rubidium atoms entered the cavity one at a time in the excited $|40S\rangle$ level. The cavity was tuned at 68.415 GHz, half the $|40S\rangle - |39S\rangle$ frequency interval, inducing each atom to emit a pair of photons. Figure 5.7(a) shows the relevant energy levels and transitions and Fig. 5.7(b) the resonant $|40S\rangle \rightarrow |39S\rangle$ transfer rate when the cavity frequency was scanned across the two-photon resonance. The effective two-photon coupling was enhanced by the presence of the relay level $|39P_{3/2}\rangle$, 39 MHz away only from the two-photon transition mid-point (Fig. 5.7a). The cavity bandwidth ($\omega_c/Q = 4.35$ kHz) was small enough to prevent the atom from emitting a single photon in a transition towards $|39P_{3/2}\rangle$. Such one-photon processes had been the main limiting factor in previous attempts to realize two-photon oscillators. CQED concepts thus played a double role here. The cavity favoured the weakly allowed two-photon process, while inhibiting the adverse one-photon emission. These two-photon oscillators have peculiar dynamical

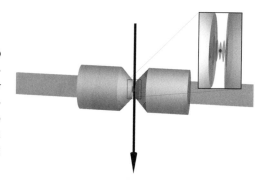

Fig. 5.8 Scheme of an optical cavity QED experiment. Atoms cross a high-finesse optical cavity (downwards arrow). The cavity transmission is probed by a weak laser (horizontal beam). The inset presents a close-up view of the space between mirrors, with an atom in the cavity mode Gaussian standing wave pattern.

properties (Davidovich *et al.* 1987; 1988; Maia-Neto *et al.* 1991).

In these experiments, the interactions of the successive atoms with the cavity were not individually controlled. This control was achieved in a series of Rydberg atom–cavity experiments starting in the mid-1990s. In these experiments, atoms are injected in the cavity at precise times, with well-defined velocities. Each atom's interaction time with the field can be varied at will, as well as the atom–cavity detuning. Each experimental sequence is much shorter than the cavity or atomic relaxation times. This leads to the engineering of complex entanglement involving several atoms and the cavity field, under well-controlled conditions. These experiments are described in detail in Sections 5.4 and 6.3.2.

5.1.6 Strong coupling regime in optical CQED

In microwave CQED experiments, atoms are detected and used to reveal the evolution of the field in the cavity. Optical CQED follows a complementary approach. It is the field escaping from the cavity which provides information about the atomic behaviour. Optical CQED has progressed in parallel with microwave experiments, reaching the single-atom strong coupling regime in 1992 [for a review, see Mabuchi and Doherty (2002); Vahala (2003)]. Here too, the challenge was the realization of a small-volume, high-Q cavity. The breakthrough came with the development of extremely high-finesse multi-dielectric mirrors, mounted in a Fabry–Perot configuration. The typical cavity length L is 100 μm and the mirror diameter is in the millimetre range. The Gaussian mode waist is a few tens of microns, resulting in a high field confinement. The photon storage time reaches the microsecond range, while the Rabi period can be as short as a few tens of nanoseconds, largely fulfilling the strong coupling condition.

In a typical set-up (Fig. 5.8), alkali atoms in their ground state $|g\rangle$ cross the cavity along a direction normal to its axis. The cavity is tuned close to resonance with the transition from $|g\rangle$ to the first excited state $|e\rangle$. The atom–cavity system is probed by a laser beam coupled to the cavity mode through the mirrors. This laser beam is so weak that it injects in the cavity less than one photon in average. The transmission of the probe beam can be monitored by sensitive homodyne detection. In early experiments, the atoms had thermal velocities, with atom–cavity interaction times of the order of the vacuum Rabi period. Later on, laser-cooled atoms were dropped across the cavity, increasing by several orders of magnitude the atom–cavity interaction time.

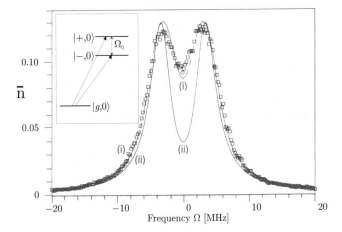

Fig. 5.9 Vacuum Rabi splitting. Experimental atom–cavity spectrum for a cavity containing one atom on the average. The intracavity photon number $\bar{n}$ is plotted as a function of the cavity–probe-laser detuning. The points are experimental. Curve (ii) presents the theoretical predictions for a single atom at the cavity centre. Curve (i) results from a numerical simulation taking atom number fluctuations and position dispersion into account. The inset depicts the dressed level structure. Reprinted with permission from Thompson *et al.* (1992). © American Physical Society.

Vacuum Rabi splitting and single-atom detection

The first spectacular manifestation of the strong coupling regime was the observation of the so-called 'vacuum Rabi splitting' with thermal atoms (Thompson *et al.* 1992) (Fig. 5.9). The weak tuneable laser source was exciting the coupled atom–cavity system and the transmitted signal was recording its spectrum. When the cavity was empty, the laser was exciting the $|0\rangle$ to $|1\rangle$ transition of the 'spring' mode, with a Lorentz line shape centred at the cavity frequency. When one atom was present in the mode, the laser probed the transition from the $|g, 0\rangle$ ground state to the first excited dressed states, $|\pm, 0\rangle = (|e, 0\rangle \pm i|g, 1\rangle)/\sqrt{2}$ (inset in Fig. 5.9 – see also Section 3.4). These levels are separated by the vacuum Rabi frequency Ω_0. Instead of a single Lorentzian resonance, a double peak was observed, since Ω_0 was larger than the average of the atom and cavity damping times. This simple analysis was complicated by various experimental features. Each atom probed a spatially varying standing wave pattern in the mode, resulting in a dispersion of vacuum Rabi frequencies. Moreover, the number of atoms coupled to the mode at a given time was fluctuating around 1, leading to a partial blurring of the splitting. These effects were well-understood as shown by the good agreement between the experimental points and the theoretical line labelled (i) in Fig. 5.9. More recent experiments (Boca *et al.* 2004; Maunz *et al.* 2005) provided even clearer evidence of the vacuum Rabi splitting in optical cavities.

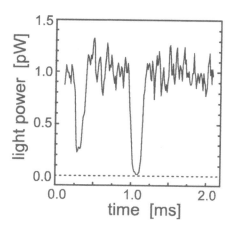

Fig. 5.10 Detecting single atoms. The probe laser is tuned at resonance with the empty cavity. The transmission is high when the cavity is empty, low when an atom is present. The two sudden drops in the output field reveal the passage of two slow atoms through the cavity mode waist. Reprinted with permission from Münstermann *et al.* (1999). © American Physical Society.

Single-atom detection

A well-resolved vacuum Rabi splitting effect provides an elegant and sensitive way to count one by one atoms crossing the mode (Hood *et al.* 1998; Münstermann *et al.* 1999; Ye *et al.* 1999). The probe laser beam, tuned at cavity resonance, is fully transmitted when the cavity is empty. When an atom crosses the mode, the cavity response is dramatically modified. The laser frequency now falls midway between the two Rabi peaks and the light transmission drops sharply. The laser beam is then reflected back on the entrance cavity mirror instead of being transmitted. Figure 5.10 shows a typical transmission signal exhibiting the passage of two atoms (Münstermann *et al.* 1999). The first is less well-coupled to the mode than the second, due to different trajectories across the cavity. This explains the difference in the depth of the two dips. This experiment was performed with cold atoms dropped through the cavity and spending a few tens of microseconds in the mode. Each atom stays in the cavity for a time much longer than T_c and can thus prevent many photons from reaching the detector. This amplification process is the trick which makes this single-atom detection scheme possible. We will see in Chapter 7 that similar effects are exploited in microwave experiments to detect and control mesoscopic fields made of many photons.

This atom counting scheme can also be used to trap deterministically atoms near cavity centre. When a dip in the cavity transmission indicates the presence of an atom, a more intense laser beam is switched on. It is tuned far on the red of the atomic transition. This trapping laser is resonant on a Gaussian cavity mode, with a moderate finesse. The position-dependent light shifts produced by the trapping beam result in a 'dipole' force (Cohen-Tannoudji 1992) attracting the atom towards the cavity centre. Atoms have been trapped in these FORT (Far Off-Resonance Traps) for times in the second range (Ye *et al.* 1999; McKeever *et al.* 2003b).

Single-atom microlaser

While trapped in the FORT beams, the atom is still interacting with the quantum cavity mode (a proper choice of the dipole trap wavelength makes the frequency of the $|g\rangle \rightarrow |e\rangle$ transition insensitive to the trapping potential). The trapped atom repeatedly emits photons in the cavity mode, realizing a one-atom microlaser (McKeever *et al.* 2003a). This device looks superficially similar to the micromaser. There are,

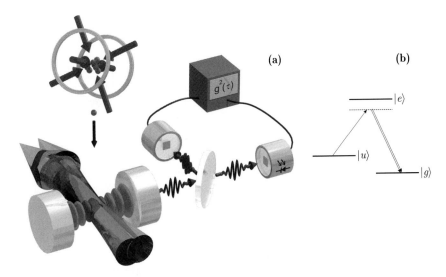

Fig. 5.11 Generation of single photons on demand: (a) Experimental set-up. Atoms from a magneto-optical trap are released in a high-finesse cavity. The triggering and repumping lasers are perpendicular to the cavity axis. The beam-splitter and photon counters are used to measure the output intensity autocorrelation function. Reprinted with permission from Hennrich *et al.* (2005). © American Physical Society. (b) Relevant atomic level scheme: the atom is transferred between hyperfine states $|u\rangle$ and $|g\rangle$ by a Raman process involving the transient virtual excitation into state $|e\rangle$ (simple arrow) followed by the downward transition corresponding to the emission of a photon in the cavity mode (double arrow). The atom is then rapidly pumped back in state $|u\rangle$.

however, marked differences between the two systems.

In the micromaser, a stream of atoms continuously enters the cavity in the upper state of the transition, sustaining a mesoscopic field in the cavity. In the one-atom microlaser, the same atom emits once and again.[4] The hyperfine sublevels of the ground and excited states realize a four-level laser scheme. Pumped by an auxiliary laser beam in the upper level of the laser transition, the atom emits a photon in the cavity. The lower level of the transition is emptied by an optical pumping cycle induced by an auxiliary laser. The cavity damping time being much shorter than the optical pumping process, the emitted photon escapes out of the mode before the atom is recycled in the upper state. The average photon number remains much smaller than unity. The complex micromaser behaviour related to the build-up of a many-photon field in the cavity is not observable in this regime.

The emitted field presents nevertheless interesting quantum characteristics. The microlaser cannot emit two photons in a short time interval, since a finite time is required to recycle the atom into the upper state. The emitted light is thus 'anti-bunched': the probability distribution of the time interval between successive photons

---

[4]We describe here the microlaser of the Caltech group. An optical oscillator operating, as the micromaser, with a stream of atoms crossing one by one the cavity has also been realized (An *et al.* 1994).

Fig. 5.12 Intensity autocorrelation function $g^{(2)}(\Delta t)$ of the CQED 'photon pistol'. The hatched area corresponds to spurious events due to a fortuitous coincidence between a photon emitted by the device and a detector's dark count. Reprinted with permission from Kuhn *et al.* (2002). © American Physical Society.

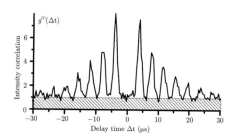

is nearly zero at the origin. This behaviour, analysed in Chapter 4 in a different context, is characteristic of single-emitter sources (Teich and Saleh 1989). It is very different from that of classical sources, for which the probability distribution of the time interval between photons is peaked at the origin.

Single-photon on demand

The photons emitted by the single-atom microlaser escape from the cavity at random times. For many quantum information applications it is important to produce single photons 'on demand', at preset times. This feat can be achieved with optical CQED technique, as has been demonstrated by an elegant 'photon pistol' experiment (Hennrich *et al.* 2000; Kuhn *et al.* 2002).

The set-up is sketched in Fig. 5.11(a). A triggering laser beam, transverse to the cavity axis, excites cold rubidium atoms falling one by one through the mode. Two hyperfine sublevels of the atomic ground state, $|g\rangle$ and $|u\rangle$ are involved in the process. The $|e\rangle \to |g\rangle$ transition is set off-resonance with the cavity mode. The atom is initially in $|u\rangle$ in an empty cavity. A Raman process, combining the absorption of a photon in the triggering laser pulse and the emission of a light quantum in the cavity, resonantly couples $|u, 0\rangle$ to $|g, 1\rangle$ (Fig. 5.11b shows the lambda-shaped atomic transitions). Performed in an adiabatic rapid passage sequence, this process realizes the preparation of $|g, 1\rangle$ with nearly unit efficiency (the excited level $|e, 0\rangle$ is only virtually reached and spurious spontaneous emission processes are negligible). One of the cavity mirrors has a weak transmission. The photon escapes through it in a well-defined Gaussian mode. The photon emission time is precisely controlled by the timing of the adiabatic sequence. The atom is then recycled from $|g\rangle$ to $|u\rangle$ by optical pumping, leaving the system ready to emit a new photon. The pistol is operated with a 4 μs repetition period and a few photons are emitted while a single atom falls through the cavity.

The operation of this single-photon source is probed by measuring its intensity autocorrelation function $g^{(2)}(\Delta t)$ (Loudon 1983). Basically, this function is proportional to the probability of detecting two photons separated by a time interval Δt. It is measured by separating the light field in two with a beam-splitter and correlating the photon clicks in the two output beams (Fig. 5.11a). One of these beams is retarded by Δt with respect to the other with the help on an optical delay line and the autocorrelation function is reconstructed by recording the photon coincidences as a function of this delay. A single photon must be channelled as a whole in one arm or the other (see Chapter 3), so that a photon pistol cannot yield click coincidences for $\Delta t = 0$. High coincidence rates are expected, on the other hand, when Δt is a multiple

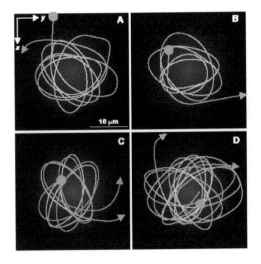

Fig. 5.13 Atom microscope. Four atomic trajectories through the cavity mode reconstructed from the cavity transmission signal. The dot represents the typical uncertainty on the atomic position. Reprinted with permission from Hood *et al.* (2000), © AAAS.

of the photon pistol repetition period. Figure 5.12 presents the observed autocorrelation signal. A comb of peaks reveals the emission of photons at regular time intervals corresponding to the 4 μs sequence period. More importantly, the peak at zero delay is, as expected, missing. The decay of the correlation function for large time intervals reflects the finite operation time of the device.

In this first experiment, an inherent randomness remained. The photon pistol was operated deterministically only within short periods starting at the random times when an atom happened to enter the mode. A variant using an atom trapped in the cavity has since been realized (Birnbaum *et al.* 2005), making the device fully deterministic over time intervals in the second range. Two photons emitted in two successive pulses are indistinguishable quantum particles, being emitted in the same field mode. They can thus interfere. The recent observation of this interference (Legero *et al.* 2004) demonstrates the coherence of the single-photon source.

Atom–cavity forces and the atom microscope

When the atom moves in the mode standing wave pattern, the energies of the dressed levels, $|\pm, 0\rangle$, depend upon its position. For a slow motion, the atom–cavity system follows adiabatically the dressed levels, whose energy acts as a potential, of the order of $\hbar\Omega_0$. This potential varies over the length scale of the transition wavelength λ. Hence, the atom feels a force, of the order of $\hbar\Omega_0/\lambda$, that can trap it near an anti-node of the field (Haroche *et al.* 1991; Englert *et al.* 1991). We have already evoked this mechanical effect of the atom–cavity coupling in Section 3.4.4.

In the microwave domain, these forces are very small due to the large wavelengths and the trapping potentials are overridden by gravity. In the optical domain, cavity forces are much larger. The trap depth may be significantly higher than the typical kinetic energy of a cold atom. There is no force, and hence no trapping in the ground state $|g, 0\rangle$. A weak laser beam exciting the cavity mode provides the photon required to exert a force on a ground-state atom.

After preliminary evidence of the cavity forces (Hood *et al.* 1998; Münstermann *et al.* 1999), the trapping of an atom in a single optical photon field has been realized

(Hood *et al.* 2000). In this experiment, cold atoms are sent through the mode. A weak probe laser, corresponding to much less than one photon on the average in the cavity, interrogates the mode continuously. A transmission change reveals the entry of an atom. The laser level is then suddenly raised, so that the cavity contains on the average about one photon. The atom is then trapped by the CQED force. Note that the trapping mechanism is very different from the FORT traps described above. There, the trapping field is a strong far-off resonant light beam, propagating in a mode different from the cavity one.

A remarkable feature of CQED trapping is that the position of the atom can be tracked in real time, through the transmission of the trapping field, which depends upon the atom–field coupling and hence upon the atomic position. The transverse motion of the atom in the mode is reconstructed using this information. One realizes in this way an 'atom–CQED microscope'. Figure 5.13 present four examples of reconstructed trajectories (Hood *et al.* 2000). A key feature is that the state and the position of the atom are encoded onto the very many probe photons that escape from the cavity on the time scale of the experiment. In this respect, a controlled, small amount of losses is an asset. This information can be used to monitor the atom and to react on its motion by way of tuning the trapping field amplitude (Fischer *et al.* 2002), providing a cooling mechanism for the atomic motion

Atom–cavity cooling

In fact, engineered feedback is not necessary to achieve atomic motion cooling. The complex interplay between the atomic motion, ruled by the intracavity intensity, and the cavity dynamics itself (the field inside the mode depends upon the atomic position) provides an efficient cooling mechanism (Gangl and Ritsch 2000; Vuletic and Chu 2000; Domokos and Ritsch 2002). Atomic motion cooling requires a dissipative process. For ordinary laser cooling, dissipation is provided by spontaneous emission. In atom–cavity cooling, dissipation corresponds to the photon leakage in the output beam. In some cases, this fast dissipation provides a cooling mechanism much stronger than standard light-cooling techniques relying on spontaneous emission. This effect has been recently demonstrated (Maunz *et al.* 2004; Nussmann *et al.* 2005).

When more than one atom are present, subtle motional effects due to the collective atomic coupling to the mode come into play (Domokos and Ritsch 2002). First experimental steps in this direction have demonstrated interatomic forces induced by the field (Münstermann *et al.* 2000) and cooling in a multiatom situation (Chan *et al.* 2003).

5.1.7   From fundamental physics to applications: solid-state CQED

The progresses of CQED in atomic physics have early on attracted the interest of the solid-state physics community (Yamamoto and Slusher 1993; Burstein and Weisbuch 1995; Ducloy and Bloch 1996). The fundamental concepts of CQED have progressively been married with the fast developing field of optronics, which extends the technology of semiconductor physics into optics. Optronics uses the powerful methods of microelectronics to realize novel types of heterostructures in which disparate semiconductors with remarkable optical properties are combined, at the micrometre or nanometre scales. The technology is largely based on molecular beam epitaxy which

(a)

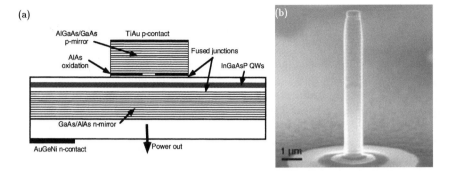

Fig. **5.14** Cavity QED in solid-state devices. (a) Structure of an electrically pumped VC-SEL laser. Quantum wells (InGaAsP QWs) are embedded in a distributed Bragg plane Fabry–Perot cavity made up of two n- and p-doped mirrors. The current is fed trough the mirrors (contacts at the top and the bottom of the structure). Reprinted with permission from Margalit *et al.* (1996). © American Institute of Physics. (b) Micro-photograph of a micro-pillar cavity, providing complete confinement of the optical field, coupled to a quantum dot emitter. Reprinted by permission from Macmillan Publishers Ltd: Nature, Reithmaier *et al.* (2004).

consists in depositing atomic monolayers in a ultra-high vacuum on a solid substrate, with masks to control the shape of the structures.

Basic optronics devices are either passive or active. Passive components include Bragg mirrors and photonic band-gaps (Burstein and Weisbuch 1995). The former are obtained by depositing multiple layers of semiconducting materials with alternately low and high indices, forming efficient multi-dielectric mirrors. Photonic band-gap materials are realized by modulating the medium index in three dimensions, either by alternating different materials or by embedding periodic gaps in an otherwise dense medium. The resulting index modulation prevents light from propagating when its frequency falls within a band. The effect is then similar to the existence of energy band-gaps for electrons in crystals.

Active optronics devices include light-emitting diodes of various kinds. Quantum wells (Bastard 1991) confining electrons along planes in two dimensions are made by alternating layers of different semiconductor atoms. Electrons confined in these wells have energy bands separated by optical transitions which can be excited by electrical or optical means (either feeding a current across the material layers or shining a laser onto it). Quantum dots are obtained with three-dimensional confining structures, restricting the electron motion within a very small region. Quantum dots behave, in a first approximation, like artificial atoms with a discrete energy spectrum. Their optical properties can be tailored by controlling the size and shape of the structure and its composition.

Active and passive optical elements are now combined in integrated CQED devices with a variety of possible applications, such as the efficient production of light by compact low-powered sources. It is rather easy to generate photons in semiconducting structures, with an excellent current-to-light conversion efficiency. The main problem is to couple this light out of the bulk. Due to the very high refraction index of these materials, most of the light is reflected back at the material–air interface and finally lost by internal absorption. The quantum efficiency of standard light-emitting diodes is thus limited to less than 20%. It is also essential, for practical applications, to couple light into a well-defined spatial mode, easy to couple into optical fibres. The techniques of CQED can considerably enhance the light collection efficiency. A common design uses a few layers of light-emitting quantum well structures, embedded in a Bragg-mirror Fabry–Perot cavity sustaining a low-order mode. The field propagation axis is parallel to the growth axis. The quality factor of these cavities is in the few hundreds to few thousand range. The upper mirror has a slightly lower reflectivity, so that the coupling to an outside mode is the dominant cause of photon losses. The light emitters are placed in the central antinode of the cavity. The enhanced spontaneous emission increases the fraction β of the photons coupled into the cavity mode, and hence the efficiency of the device. With state-of-the-art structures, β, which is of the order of 10^{-5} in a standard device without cavity, is raised up to a few percent. This design, directly based on CQED concepts, is at the heart of high-efficiency LEDs (Light Emitting Diodes) and VCSELs (Vertical Cavity Surface Emitting Laser), now in industrial use. Figure 5.14(a) presents, as an example, the structure of a VCSEL laser.

In spite of a dramatic modification of the spontaneous emission patterns, the emission rate in these planar structures is only weakly modified (Tanaka *et al.* 1995). Much stronger rate modifications are observed in structures providing three-dimensional field confinement (Gérard *et al.* 1998; Bayer *et al.* 2001). They are realized by etching the plane Fabry–Perot structure into micro-pillars, with a diameter in the micrometre range (Fig. 5.14b). The field is confined axially by the Bragg distributed reflectors and laterally by the semiconductor–air index gap. An alternative architecture uses a photonic band-gap material, providing a strong confinement of the optical field (Yoshie *et al.* 2004). Impressive progresses have been realized to increase the Q factors of these structures and decrease their volume. The strong coupling regime has been reached in devices whose dynamics, in the picosecond time range, is much faster than the one of atomic CQED systems (Reithmaier *et al.* 2004; Yoshie *et al.* 2004; Peter *et al.* 2005). Optronics components based on the coherent interaction of quantum dots with high quality photonic band-gap materials open promising perspectives for light sources and fast switching devices for optical communication (Fattal *et al.* 2004).

Moderate quality cavities can also be used to efficiently couple a single emitting quantum dot to the radiation field, providing a compact single-photon source with performances comparable to the atomic photon pistols (Santori *et al.* 2002). The dot is excited by a laser pulse much shorter than the radiative lifetime. At most one photon is emitted in each shot. Due to the CQED effect, the probability of coupling the photon in an useful light mode is high. This source is promising for quantum cryptography applications.

This brief journey into applied physics has shown how CQED ideas are likely to trigger fruitful developments into a multi-billion dollar market. Leaving this promising domain, let us now come back to more fundamental physics, where Rydberg atoms and microwave photons play together to unveil the strange features of the quantum.

5.2   Giant atom in a cavity: an ideal cavity QED situation

We have just seen that a rich variety of phenomena occur when atoms are introduced one by one in a box storing long-lived photons. We will focus in the following on one class of effect, the entanglement of atomic and field variables produced by the coherent interaction of single atoms with the cavity mode. This entanglement is, as discussed in Chapter 3, a direct manifestation of the spin–spring Hamiltonian dynamics. Ideally, this Hamiltonian describes a two-level atom and a field oscillator coupled coherently together and isolated from their environment. To get close to this ideal, we need very good cavities and very long-lived atomic states, interacting as strongly as possible with each other.

The longest cavity damping times, $T_c = Q/\omega_c = 1/\kappa$, are obtained in the microwave domain, in which very high quality factors Q are achievable at relatively low frequencies ω_c. Photon damping times ranging from the millisecond to the tenth of a second range can be obtained with superconducting cavities, leaving ample time for coherent atom–field manipulations. This, of course, requires an atom having a strong resonant transition in the microwave domain. Ordinary magnetic resonance (nuclear or electronic spin transitions between hyperfine or Zeeman levels) will not do. Their magnetic coupling to radiation is very small, with vacuum Rabi periods $2\pi/\Omega_0$ much larger than any realistic T_c value. Only microwave transitions between very excited atomic Rydberg states provide electric dipoles large enough to achieve the strong coupling regime, as demonstrated in the pioneering micromaser experiments.

In the energy spectrum of each atomic species, there is an enormous number of Rydberg states with a given energy, differing by their angular momentum. The levels with low angular momentum (used in the micromaser experiments) decay mostly towards low-lying levels via optical transitions, which are not affected by the microwave cavity. This decay process limits the time available for coherent manipulations and precludes the generation of complex entangled states. To overcome this difficulty, we must use Rydberg levels with the highest angular momentum. The 'circular Rydberg states' (Gallagher 1994) are not coupled to optical fields. They radiate only millimetre waves and survive much longer than low angular momentum states. Atoms in circular states behave as large antennas, exclusively for microwave radiation. As we shall see, strongly coupled atom–field systems with lifetimes reaching several tens of milliseconds are then available, with a vacuum Rabi period three orders of magnitude smaller.

The circular Rydberg atoms and the superconducting photon boxes are the main actors of the experiments described in the present and following two chapters. Their unique features made it possible to realize modern versions of thought experiments and to demonstrate various steps of quantum information processing. The principle of these experiments can be simply explained in terms of the spin and spring model of Chapter 3. Their actual realization, though, involves a variety of sophisticated techniques which are reviewed in this section.[5] These techniques involve insightful physical ideas, some of which illustrate the correspondence principle between the quantum and classical description of Nature. Circular Rydberg levels on the one hand and a field mode on the other hand are indeed two quantum systems whose main properties can be

---

[5]More details can be found in review papers like Haroche (1984); Haroche and Raimond (1985); Haroche (1992); Haroche and Raimond (1994); and Raimond *et al.* (2001).

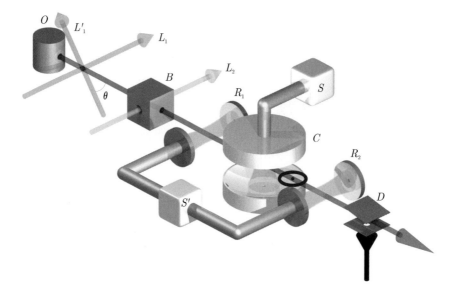

Fig. 5.15 General scheme of the CQED experiments with circular Rydberg atoms.

explained in simple classical terms. It is somewhat paradoxical that their coupling will lead us in an exploration of entanglement and non-locality, which are – arguably – the deepest expressions of the quantum.

A typical scheme of a cavity QED microwave experiment with circular atoms is shown in Fig. 5.15. Atoms effusing from oven O are velocity-selected (by lasers L_1 and L_1'). They are then prepared into a circular state in box B, by a combination of laser (L_2) and microwave excitation. They then interact with the superconducting cavity C before being counted in the detector D. The cavity C is coupled to a classical source S, which can be used to inject a small coherent field in the mode. Two low-quality cavities R_1 and R_2, driven by the classical source S', make it possible to manipulate the atomic state before or after the interaction with C. The set-up, from B to D, is enclosed in a cryogenic environment and shielded from room-temperature blackbody radiation.

We will now analyse the main components of this set-up. In Section 5.2.1 we discuss the properties of the circular Rydberg atoms and describe how they are prepared and detected. Section 5.2.2 is devoted to the description of the superconducting cavities and of the field stored in them. In Section 5.2.3, we analyse the atom–field coupling. We then describe (Section 5.2.4) how the cavities are tuned in resonance with the atoms, how their Q factors are measured and how they can be fed with coherent fields. We conclude by discussing the present and future limitations imposed by the finite cavity damping time.

5.2.1 Circular Rydberg atoms

Main properties of circular Rydberg states

A circular Rydberg state (Hulet and Kleppner 1983; Gallagher 1994) is a very high-lying alkali atom level (rubidium in our experiments), in which the single valence

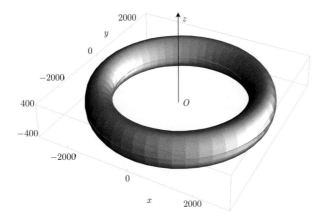

Fig. 5.16 Surface of equal value (50% of the maximum value) for the spatial probability distribution of the valence electron in $|50C\rangle$. The dimensions are in units of the Bohr radius, a_0.

electron has been excited close to the ionization limit, in a state with a large principal quantum number[6] n and a maximum value $\ell = |m| = n - 1$ for the orbital, ℓ, and magnetic, m, quantum numbers (ℓ and m are defined with respect to the quantization axis Oz). This state will be denoted $|nC\rangle$. Typically, in our experiments n is of the order of 50. The circular states correspond, in classical terms, to the circular orbits of the old theory of quanta, with a radius $a_0 n^2$ ($a_0 = 0.53$ Å is the Bohr radius).

This orbit is much larger than the singly charged ionic core of the alkali atom, which has a diameter of the order of a_0 and the electronic structure of a noble gas (krypton for rubidium). The ionic core is very compact and has an extremely low polarizability. To an excellent approximation, it behaves as a point charge, creating a nearly perfect $1/r$ potential for the valence electron. The energy of the circular states is thus given by the hydrogenic expression:

$$E_n = -\frac{R_{\mathrm{rb}}}{n^2} \, , \tag{5.1}$$

where R_{rb} is the Rydberg constant (slightly modified to take into account the rubidium core's mass effect).

The spherical harmonic of maximum ℓ and m values take a simple form, leading to the circular state wave function in spherical coordinates:

$$\Psi(r, \theta, \phi) = \frac{1}{(\pi a_0^3)^{1/2}} \frac{1}{n^n n!} \left(-\frac{r}{a_0} \sin\theta e^{i\phi} \right)^{n-1} e^{-r/n a_0} \, . \tag{5.2}$$

Figure 5.16 shows a surface of equal probability density (50% of the maximum value) for $n = 50$. This surface defines a torus centred on the Bohr orbit, in the plane perpendicular to the quantization Oz axis. The electronic orbital of the circular Rydberg

---

[6]For historical reasons, we use the same letter n for the photon number in a Fock state and the principal quantum number. There is no risk of confusion, since these two quantities are used in different contexts.

atom looks very much like a tyre and reminds us of the old Bohr model of atomic theory. The relative dispersion of r, $\Delta r/r$, and the fluctuation of θ, $\Delta \theta$, are equal:

$$\Delta r/r = \Delta \theta \approx 1/\sqrt{2n} \ . \tag{5.3}$$

The higher the level, the tighter is the confinement around the Bohr orbit. This wave function provides a representation of the electron state as close as it can get to a classical description. The phase of the orbital motion remains completely undetermined and the electron delocalized along the orbit. The azimuthal localization of the electron is precluded by Heisenberg uncertainty relations.

The analogy with the classical orbit goes beyond this simple geometrical property. Since all quantum numbers are large, most features of circular Rydberg atoms can be explained by classical arguments, in accordance with Bohr's correspondence principle. As a first example, let us consider the angular frequency $\omega_{nC,(n-1)C}$ of the transition between neighbouring circular states $|nC\rangle$ and $|(n-1)C\rangle$. It is, to the first non-vanishing order in $1/n$, obtained by differentiating the binding energy with respect to n (eqn. 5.1):

$$\omega_{nC,(n-1)C} \approx 2\frac{R_{\mathrm{rb}}}{\hbar}\frac{1}{n^3} \ . \tag{5.4}$$

The corresponding frequency is in the few tens of GHz range for n values of the order of 20 to 60 (millimetre wavelengths). In a classical picture, this is the frequency of the orbital motion of the electron in a Bohr orbit. The n^{-3} dependence of $\omega_{nC,(n-1)C}$ can be interpreted in classical terms, by invoking Kepler's third law which applies to all orbital motions in $1/r$ potentials. The period $2\pi/\omega_{nC,(n-1)C}$ of the electron must scale as the $3/2$ power of the orbit radius $a_0 n^2$. The relation between the Rydberg transition frequencies and the circular state sizes could thus have been derived by Kepler or Newton without any quantum consideration!

This simple analysis does not account for the fine or hyperfine structure contributions to the electron energy. These effects are very small in circular states. The fine structure, due to relativistic corrections including spin–orbit coupling, scales as $1/n^5$. It is only a few hundred hertz for n between 20 and 60 and will be neglected in the following. Hyperfine structures due to magnetic couplings between the atomic nucleus and the valence electron are three orders of magnitude smaller.

The $|nC\rangle \rightarrow |(n-1)C\rangle$ transition is σ^+-circularly polarized. Its dipole matrix element d is:

$$d = a_0|q|n^2/\sqrt{2} \ , \tag{5.5}$$

where $|q|$ is the absolute value of the electron charge. This dipole is extremely large, in the thousand atomic units range for n around 50.

In the absence of external fields, the circular state is degenerate with a large number of non-circular levels having the same n and smaller ℓ and m values. A small perturbing electric or magnetic field, transverse to the Oz axis, would efficiently couple the circular state to the levels with $\ell = n - 2$, which would in turn be coupled to other levels with lower angular momentum. The circular orbit would be rapidly lost, the atom evolving into an uncontrolled superposition of non-circular states (Gross and Liang 1986). A perfect cancellation of the stray fields is impossible in practice, but the circular orbit can be 'protected' by subjecting the atom to a directing electric field, aligned with Oz.

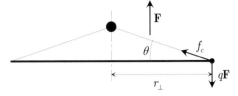

Fig. 5.17 Classical Bohr orbit in an electric field. The field is vertical and the orbit is seen from the side, appearing as a thick line. The large black dot represents the ionic core.

This field lifts the degeneracy between the circular level and the levels with $\ell = n - 2$. If it is much larger than the stray fields, the perturbing transitions are suppressed.

The circular state energy is shifted to second order by the directing field. This Stark shift is convenient to tune the atomic transition frequency, a procedure useful in many CQED experiments. A classical argument leads again to the electric polarisability of the circular state. Figure 5.17 shows a circular Bohr orbit – seen from the side as a straight line – normal to the applied electric field $\mathbf{F}$. The force produced on the electron, $q\mathbf{F}$, adds to the Coulomb force of the core, $\mathbf{f}_c$. The orbit remains circular (due to symmetry), but the core pops out of the orbit's plane, producing an induced electric dipole. Let us call θ the angle between $\mathbf{f}_c$ and the orbit plane and $r_\perp$ the radius of the perturbed orbit. When the external field is applied, the electron angular momentum, $m_0\omega r_\perp^2$ (m_0: electron's mass) remains constant, equal to $(n-1)\hbar \approx n\hbar$, since no torque is produced by $q\mathbf{F}$. The electron orbital angular frequency ω and orbit radius $r_\perp$, both affected by the electric field, remain linked by:

$$\omega \approx \frac{n\hbar}{m_0 r_\perp^2} \ . \tag{5.6}$$

The core-to-electron distance is larger than $r_\perp$ by the factor $1/\cos\theta$. The atomic Coulomb force $\mathbf{f}_c$ thus scales as $\cos^2\theta/(r_\perp)^2$ and its component along the vertical direction as $\cos^2\theta \sin\theta/(r_\perp)^2$. Let us project Newton's equation for the electron motion onto the electric field axis and express the balance between the force induced by the external field and the Coulomb force along this axis:

$$\cos^2\theta \sin\theta = \left(\frac{r_\perp}{a_0}\right)^2 \frac{F}{F_0} \ , \tag{5.7}$$

where:

$$F_0 = |q|/4\pi\varepsilon_0 a_0^2 = 5.14\,10^{11}\ \text{V/m} \ , \tag{5.8}$$

is the atomic electric field unit. In the orbit plane, the projection of the Coulomb force, proportional to $\cos^3\theta/(r_\perp)^2$, balances the centrifugal force, $m_0\omega^2 r_\perp$. Eliminating ω with the help of eqn. (5.6), we get:

$$\cos^3\theta = a_0 n^2/r_\perp \ . \tag{5.9}$$

In the weak field limit ($F \ll F_0$, $\theta \ll 1$), eqn. (5.9) shows that the variation of the orbit radius is negligible: $r_\perp \approx a_0 n^2$. Expanding (5.7) to first order in θ, we obtain the induced dipole d_i:

$$d_i = a_0 |q| n^6 F/F_0 \ , \tag{5.10}$$

proportional to the applied field. The atomic polarizability scales as the sixth power of the principal quantum number. The polarization energy, E_2, is computed by considering a process in which the dipole is built adiabatically in a field increasing from

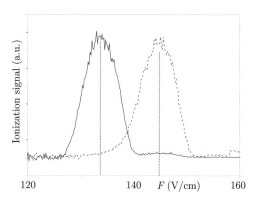

Fig. 5.18 Field-ionization signal. Electron flux versus the electric field applied in D. The atoms are initially either in $|50C\rangle$ (dotted line) or $|51C\rangle$ (solid line). The vertical lines define the ionization threshold.

zero to its final value F. Adding the elementary works done on the electric charges as the field is increased, we get $E_2 = -d_i F/2$. Taking into account eqns. (5.10) and (5.8), we finally express the second-order energy shift as:

$$E_2 = -R_\infty n^6 (F/F_0)^2 , \qquad (5.11)$$

where R_∞ is the Rydberg constant corresponding to an infinite core mass:

$$R_\infty = m_0 c^2 \alpha^2 / 2 , \qquad (5.12)$$

($\alpha = q^2/4\pi\varepsilon_0 \hbar c$ is the fine structure constant). This expression of the Stark shift agrees with the quantum calculation, in the asymptotic limit of a large n. For $n = 50$, the level shift is -1.8 MHz/(V/cm)^2. The differential shift of the $|51C\rangle \rightarrow |50C\rangle$ transition is ≈ -250 kHz/(V/cm)^2.

In the high-field limit, the variation of the orbit's radius cannot be neglected. Eliminating $r_\perp$ between eqns. (5.7) and (5.9), we get:

$$\cos^8 \theta \sin \theta = n^4 F/F_0 . \qquad (5.13)$$

The left-hand side term in this equation is bounded. Its maximum value, about 0.2, is reached for $\theta = \arcsin 1/3 \approx 19°$. There is thus a maximum value of the electric field compatible with a stable orbit, corresponding to an ionization threshold $F_i \approx 0.2 F_0/n^4$. The predicted values for $n = 50$ and $n = 51$ are, respectively 165 and 152 V/cm, to be compared to the measured values 145 and 134 V/cm. The differences are due to the ionization by tunnel effect in a field slightly smaller than the classical ionization threshold (Gallagher 1994). As far as the ionization is concerned, the classical limit is not yet fully reached for $n = 50$. The ionization thresholds correspond to relatively low fields, easily applied to the atoms.

The electron resulting from the ionization process can be detected with high efficiency, up to $90 \pm 10\%$ (Maioli et al. 2005), in a scheme sensitive to single atoms. The rapid variation of the ionizing field with n provides a state-selective detection. The electric field in the detector D is set to ionize a circular state with a selected n value in front of an electron multiplier. Figure 5.18 presents the average ionization signal collected as a function of the electric field applied in D. The atoms are initially prepared either in $|50C\rangle$ or $|51C\rangle$ (which ionizes in the lower field). The two ionization peaks are well-separated. The errors in the attribution of n (overlap of the two curves) are a few percent.

The spontaneous emission rate Γ_n and radiative lifetime $T_{a,n} = 1/\Gamma_n$ of a circular state $|nC\rangle$ are other essential parameters of a cavity QED experiment, which can also be derived by classical arguments. The electric dipole selection rule $\Delta\ell = \pm1$ allows a unique decay channel for this state: the microwave transition towards the lower circular state $|(n-1)C\rangle$. In a classical picture, the electron, accelerated on its circular orbit, radiates electromagnetic power proportional to the modulus square of its centripetal acceleration $\mathbf{a}$. Its mechanical energy accordingly decreases slowly, as it spirals down to the core, jumping between circular states of decreasing principal quantum numbers. The radiative lifetime of the initial state corresponds to the loss of an energy amount $\hbar\omega_{nC,(n-1)C} \approx 2R_\infty/n^3$ (we can neglect here the small difference between R_{rb} and R_∞).

The radiated power $\mathcal{P}_r$ is given by the Larmor formula (Jackson 1975):

$$\mathcal{P}_r = \frac{q^2 a^2}{6\pi\varepsilon_0 c^3} ,$$ (5.14)

and the spontaneous emission rate is thus:

$$\Gamma_n = \frac{n^3 \mathcal{P}_r}{2R_\infty} .$$ (5.15)

Writing the electron acceleration as:

$$a = |\mathbf{a}| = \frac{1}{m_0} \frac{q^2}{4\pi\varepsilon_0 (a_0 n^2)^2} = \frac{1}{m_0 n^4} \frac{2R_\infty}{a_0} ,$$ (5.16)

and replacing R_∞ by its expression $q^2/(8\pi\varepsilon_0 a_0)$, we combine eqns. (5.14)–(5.16) to obtain:

$$\Gamma_n = 2R_\infty \frac{q^2}{6\pi\varepsilon_0 c^3} \frac{1}{m_0^2 a_0^2} n^{-5} .$$ (5.17)

Writing a_0 as $4\pi\varepsilon_0\hbar^2/(m_0 q^2)$, we can factorize the cube of the fine structure constant α and get:

$$\Gamma_n = \frac{1}{T_{a,n}} = \frac{4}{3} \frac{R_\infty}{\hbar} \alpha^3 n^{-5} ,$$ (5.18)

which coincides, for large ns with the expression derived from a Fermi Golden rule argument in a quantum description of the spontaneous emission process (eqn. 4.73). Comparing eqn. (5.18) with (5.4), we can express Γ_n in terms of the electron frequency $\omega_{nC,(n-1)C}$ as:

$$\frac{\Gamma_n}{\omega_{nC,(n-1)C}} = \frac{2}{3}\alpha^3 n^{-2} = 1/Q_{a,n} .$$ (5.19)

The inverse of this very small dimensionless ratio defines the radiative quality factor $Q_{a,n} = 3n^2/2\alpha^3$ of the circular to circular state transition. The large $\alpha^{-3} = 137^3$ factor entails that usual excited atomic excited states (n small) decay slowly at the atomic time scale, with radiative damping times corresponding typically to $3 \cdot 10^6$

periods of the emitted field. This radiative quality factor is, in circular Rydberg states, increased by a factor of n^2. For $n \approx 50$, the decay takes $\sim 10^{10}$ periods of the emitted microwave. More precisely, $\Gamma_{51} = 28$ s^{-1}, corresponding to a lifetime $T_{a,51} = 36$ ms and to $Q_{a,51} = 1.14\,10^{10}$.

In spite of their extremely strong coupling to the millimetre-wave field, the circular Rydberg atoms are very stable. Among all possible bound orbits, the circular ones have the smallest average acceleration, the electron always remaining far from the core. The radiation loss is minimum, hence the advantage of using circular Rydberg atoms for CQED physics. Elliptical orbits (low ℓ quantum states) have a much shorter lifetime, proportional to n^{-3} instead of n^{-5}, due to the stronger acceleration of the electron near the core.

Circular atoms can travel over a few metres at thermal velocity within their lifetime. Spontaneous emission is thus negligible in an experimental set-up whose size is a few tens of centimetres. In the presence of a thermal field with n_{th} photons per mode on the average, the lifetime is reduced by a factor $1 + n_{\text{th}}$ (Chapter 4). It is thus essential to screen efficiently the room-temperature blackbody field, corresponding to tens to hundreds of photons per mode in the millimetre-wave domain.

van der Waals interactions and collisional entanglement of circular atoms

We have considered so far isolated circular atoms. Their mutual interactions also reveal unusual orders of magnitude and open interesting perspectives for atom–atom entanglement experiments. Let us consider as a simple example the collision between two circular atoms A_1 and A_2 flying past each other, the distance of their cores being a time-dependent function $R(t)$. The potential corresponding to their instantaneous electrostatic dipole–dipole interaction is given by the van der Waals expression (Cohen-Tannoudji *et al.* 1977):

$$V_{\text{vdW}}(R) = \frac{q^2}{4\pi\varepsilon_0} \frac{1}{R^3} \left[\mathbf{r}_1 \cdot \mathbf{r}_2 - 3(\mathbf{r}_1 \cdot \mathbf{u})(\mathbf{r}_2 \cdot \mathbf{u}) \right] , \qquad (5.20)$$

where $\mathbf{r}_1$ and $\mathbf{r}_2$ are the position operators of the valence electrons of each atom relative to their cores and $\mathbf{u}$ the unit vector from A_1 to A_2. This form of the interaction potential is valid when R satisfies the double condition:

$$a_0 n^2 \ll R \ll \lambda . \qquad (5.21)$$

If R becomes of the order of the atom's size, each atom's electric charge distribution cannot be described uniquely as a dipole and the expression (5.20) breaks down. For distances larger than the typical circular to circular transition wavelength λ, on the other hand, the interaction potential includes retardation effects corresponding to real photon exchange between the atoms. Since the typical wavelength of a $n \to n-1$ Rydberg transition is $n\alpha^{-2}$ times larger than the atomic size, there is a large range of R values fulfilling (5.21). Due to the fast $1/R^3$ dependence of V_{vdW}, the colliding atoms 'feel' the effect of their interaction mostly when they are near to their shortest distance b. We assume that b falls in the range of R values satisfying (5.21).

The van der Waals interaction is particularly effective when the colliding atoms are initially in two adjacent circular states $|n, C\rangle$ and $|n-1, C\rangle$, which we will call $|e\rangle$

and $|g\rangle$ to simplify the notation. The collision then couples strongly the initial state, denoted $|e_1, g_2\rangle$ to the degenerate state $|g_1, e_2\rangle$, in which the energies of the two atoms are exchanged. The electric dipole operators $q\mathbf{r}_1$ and $q\mathbf{r}_2$ have, in each atom, a very large matrix element scaling as n^2 between $|e\rangle$ and $|g\rangle$, so that the resulting van der Waals exchange collision has a probability amplitude scaling as n^4/R^3. This resonant interaction dominates the couplings to other states and the pair of circular atoms can, to an excellent approximation, be considered as a two-level system. It undergoes under the effect of the collision a Rabi oscillation between $|e_1, g_2\rangle$ and $|g_1, e_2\rangle$ and is finally found in:

$$|\Psi\rangle = \cos(\theta_c/2)|e_1, g_2\rangle + e^{i\Phi}\sin(\theta_c/2)|g_1, e_2\rangle , \qquad (5.22)$$

where the collision angle θ_c and the phase Φ are given by integrals depending upon the details of the atomic trajectories. For most values of θ_c, the two atoms are entangled. For $\theta_c = \pi/4$, they end up in a maximally entangled state of the EPR type.

Without entering into any detailed calculation, let us estimate the magnitude of θ_c. The evolution frequency at the minimum distance $R = b$ is of the order of:

$$\frac{V_{\mathrm{vdW}}(b)}{\hbar} \approx \frac{q^2 a_0^2 n^4}{4\pi\varepsilon_0\hbar} \frac{1}{b^3} . \qquad (5.23)$$

The interaction decreases rapidly with the distance. The order of magnitude of the effective collision time is thus b/v, where v is the relative velocity of the atoms, assumed to be constant. The mixing angle θ_c can finally be written as:

$$\theta_c \approx \frac{V_{\mathrm{vdW}}(b)}{\hbar}\frac{b}{v} \approx 2\alpha\frac{c}{v}\left(a_0 n^2/b\right)^2 , \qquad (5.24)$$

a simple expression involving, in addition to the fine structure constant α, the dimensionless ratios of the atomic size to the impact parameter b and of the speed of light c to the atomic velocity.

For a collision between thermal atoms, c/v is about 10^6. With $n \approx 50$, the collisional angle $\theta = \pi/2$ corresponding to maximal entanglement is reached for an impact parameter $b \approx 10~\mu$m. This value is in the range of our approximations. It is much smaller than the circular atom emission wavelength ($\lambda \approx 6$ mm) and much larger than the atomic size ($a_0 n^2 \approx 120$ nm). Under these conditions, the maximum interaction energy $[V_{\mathrm{vdW}}(b) \approx 10^{-8}$ eV$]$ corresponds to a 130 μK temperature, very small compared to the atomic kinetic energy. The collision thus has a negligible effect on the atomic trajectory, which justifies the assumption that v remains constant during the collision.

Thanks to their huge dipoles, two circular atoms can be entangled by a collision with a very large impact parameter. Controlling b precisely in a collision between free atoms would nevertheless be a tough experimental challenge. We will show below that atomic entanglement can be realized in a much simpler way, using the enhancement of the van der Waals interaction by a non-resonant cavity.

Which circular state to choose?

We have seen that the physical parameters of an $|nC\rangle$ state scale as positive or negative powers of n, with very large variations as one climbs the ladder of Rydberg states.

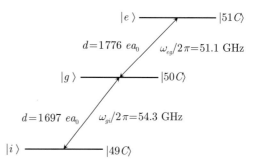

Fig. 5.19 The three relevant circular levels $|e\rangle$ ($|51C\rangle$), $|g\rangle$ ($|50C\rangle$) and $|i\rangle$ ($|49C\rangle$). The transition frequencies and the dipole matrix elements on these σ^+-polarized transitions are indicated.

Some of these parameters must be maximized. The atomic dipole and lifetime should be as large as possible, which favours large n values. On the other hand, the atomic Stark polarizability should not be too large in order to limit the sensitivity to stray fields. This pleads for relatively small n values. The sensitivity to the number of blackbody photons also favours a relatively high transition frequency, hence a low n. The states we have chosen, with n around 50, correspond to a compromise between these conflicting requirements.

The relevant levels in all the experiments described below are shown in Fig. 5.19. The cavity is resonant or nearly resonant with the transition between the states $|51C\rangle$ (noted $|e\rangle$) and $|50C\rangle$ ($|g\rangle$). The transition frequency is $\omega_{eg}/2\pi = 51.099$ GHz and it has a huge dipole matrix element d equal to 1776 $|q|a_0$. A third level $|i\rangle$ ($|49C\rangle$) is used as a reference in some experiments. The $|g\rangle \rightarrow |i\rangle$ transition, at 54.3 GHz, is far off-resonant with the cavity mode and $|i\rangle$ is thus unaffected by the atom–cavity coupling.

Circular state preparation

Rydberg states can be reached either by electron recombination with an ionic core in a plasma or by excitation from the atomic ground state (Fig. 5.20). The former process occurs naturally in the interstellar medium and accounts for the observation of Rydberg microwave transitions in radio-astronomy (Höglund and Metzger 1965). By recombination, circular (or nearly circular) Rydberg states are preferentially produced. The impinging electron has, on the average, a large impact parameter, corresponding to a high initial ion–electron angular momentum. The spontaneous emission of low-frequency photons puts the electron on a sequence of circular orbits with decreasing radii, corresponding classically to a trajectory spiralling towards the core (Fig. 5.20a). This process cannot be efficiently implemented in a laboratory experiment.

Optical excitation of Rydberg levels from a ground state atom, using a few high-energy laser photons only, prepares low angular momentum states, corresponding to highly elliptical orbits. Classically, the electron starts at a very small distance from the core and is suddenly boosted by the optical field. It gains only a few units of angular momentum and, periodically, has to come back close to the core (Fig. 5.20b).

The circular Rydberg states cannot be excited from the atomic ground state using only lasers. A two-step process needs to be implemented (Hulet and Kleppner 1983; Nussenzveig *et al.* 1993). A laser excitation first provides a large energy to the atom and brings it into a low angular momentum Rydberg level. The atom is then fed with a

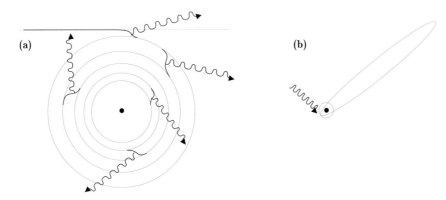

Fig. 5.20 Representations of the preparation of Rydberg states (a) by radiative recombination of an electron with an ionic core and (b) by optical excitation from the atomic ground state.

large number of radio-frequency photons, each adding one unit of angular momentum, with a very little amount of energy. The process is similar to a controlled change of orbit for a satellite, using successive rocket boosts.

The details of the method, implemented on rubidium atoms, are rather complex. We give here only its main outline. The first step is a pulsed (2 μs pulse width) laser excitation into a low angular momentum state with principal quantum number n around 50. Three laser diodes are used, at 780 nm ($|5S_{1/2}\rangle \rightarrow |5P_{3/2}\rangle$ transition), 776 nm ($|5P_{3/2}\rangle \rightarrow |5D_{5/2}\rangle$) and 1.26 μm for the final transition to the Rydberg state.

Figure 5.21 shows the Rydberg levels in the n-hydrogenic manifold when a small electric field (aligned along Oz) is applied (Gallagher 1994). The magnetic momentum m along Oz remains a good quantum number (due to cylindrical symmetry) while ℓ is no longer conserved. It is replaced by a 'parabolic' quantum number, which is required, in addition to m, to label the states (Bethe and Salpeter 1977). The levels with a given positive m value make up a ladder of energy-equidistant states, separated by an amount proportional to the electric field, about 200 MHz for 1 V/cm. The number of levels in each m ladder is $n - m$, equal to the number of ℓ values compatible with a given m in zero field. The length of the ladders thus decreases as m increases. The ladders with increasing m values make up a 'triangle' of levels, whose 'tip' corresponds to the circular state $m = n - 1$. Two levels with opposite m's are degenerate. The negative-m states thus form an identical ladder.

The laser excitation (dotted arrow in Fig. 5.21) is adjusted to prepare selectively the lowest Stark level in the $m = +2$ manifold. From this level, a series of $n - 3$ σ^+-polarized degenerate transitions between Stark levels (thick lines in Fig. 5.21) leads to the circular state $m = n-1$. The mere application of a resonant radio-frequency source would prepare a mixture of all the states involved in these transitions. A selective preparation of the circular state requires an adiabatic passage, similar to the process outlined in Section 3.3.2.

Let us recall that the transition between the $|0\rangle$ and $|1\rangle$ states of a spin-$1/2$ is efficiently produced by a slow rotation of the effective field, its angle with the OZ axis of the Bloch sphere evolving from zero to π. Here, the equidistant Stark states

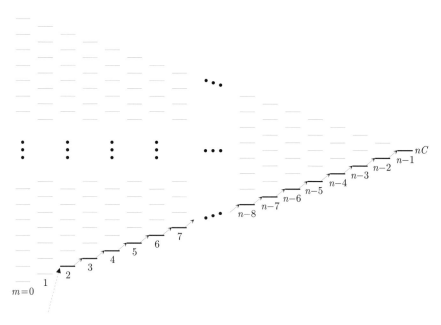

Fig. 5.21 Hydrogenic manifold (principal quantum number n) in a static electric field. The $n-m$ Stark levels with a given positive m value make up a ladder of states. The level spacing is the same for all ladders. Two consecutive ladders are separated by half this spacing. The Stark levels are thus arranged in a triangle, whose tip is the circular state $|nC\rangle$. The laser excitation (dotted arrow) prepares the lower state in the $m = 2$ manifold. The adiabatic circularization process uses radio-frequency transitions (solid arrows) between the levels represented as thick lines.

shown as thick lines in Fig. 5.21 can be viewed as making up the energy diagram of a large pseudo-angular momentum. The atomic system can thus be described by a generalized Bloch vector rotating around an effective field. The component along OZ of this field is, in the rotating frame, proportional to the atom/source detuning. Its transverse component corresponds to the constant radio-frequency amplitude.

Initially, the atomic pseudo-angular momentum points down ($m = 2$ state) and the source is tuned below the atomic transition frequency. The effective field thus points 'upwards', being anti-aligned with the atomic pseudo-angular momentum. The static electric field, and hence the atomic transition frequency, is then slowly reduced, so that the source finally ends up with a frequency above the atomic one. The effective field finally points 'down'. Provided the effective field rotation is much slower than the Rabi frequency in the radio-frequency field (a few megahertz), the atomic pseudo-angular momentum adiabatically follows this rotation, staying anti-aligned with the effective field. It finally points up, and the atom ends in the circular state $|nC\rangle$.

In the experiments described below, we use a slightly refined version of this general method (Raimond *et al.* 2001), achieving the preparation of $|50C\rangle$ or $|51C\rangle$ with purity better than 98%. A variant of this method using crossed slowly varying electric and magnetic fields can also be employed (Hare *et al.* 1988; Nussenzveig *et al.* 1991; Brecha *et al.* 1993).

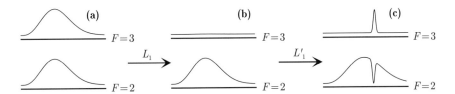

Fig. 5.22 Velocity-selection scheme. The horizontal lines represent the hyperfine sublevels of the rubidium ground state. The curves above them show the velocity distribution in the corresponding state at different stages of the process (the horizontal velocity axis and the vertical probability axis are omitted for clarity). (a) Initially, both levels are populated, with the same Maxwellian velocity distributions. (b) After optical pumping by laser L_1, level $F = 3$ is empty. (c) After interaction with L'_1, a narrow velocity class is optically pumped back from $F = 2$ into $F = 3$.

Velocity selection

Knowing the atomic positions during an experimental sequence is essential to be able to address them selectively as they fly one by one across the apparatus. This knowledge is achieved by controlling each atom's velocity v. The pulsed circular state generation determines when each atom leaves the preparation zone B, and hence, if v is known, where it is along the atomic beam at any subsequent time. Selecting the atomic velocity relies on simple laser techniques, based on optical pumping and making use of the hyperfine structure of the atomic ground state.

The $|5S_{1/2}\rangle$ ground state of rubidium has two hyperfine sublevels, $F = 2$ and $F = 3$, separated by a 3.5 GHz interval. When the atoms effuse from the oven O, both sublevels are populated in the ratio (7/5) of their degeneracy, with Maxwellian velocity distributions (Fig. 5.22a). The laser excitation of the Rydberg states starts from $F = 3$. The velocity selection relies on the preparation of a sample of atoms with a well-defined velocity in this level, before the circular state preparation occurs. First, all atoms in $F = 3$ are transferred to $F = 2$ by optical pumping using laser L_1 (Fig. 5.15). This laser crosses the atomic beam at right angle. It is resonant with all atoms in the beam on the $|5S, F = 3\rangle$ to $|5P, F = 3\rangle$ transition, irrespective of the atomic velocity (no first-order Doppler effect). During the crossing of the laser beam, the atoms undergo several optical pumping cycles: they are excited and decay spontaneously back onto the $F = 3$ or $F = 2$ ground states. In the latter case, they stay trapped in this state, which is non-resonant with L_1. After a few cycles, nearly all atoms, whatever their velocity, are in the $F = 2$ sublevel (Fig. 5.22b).

The atoms then interact with a second laser, L'_1, intersecting the atomic beam at an acute angle θ, so that the light frequency seen by the atoms is Doppler-shifted by an amount proportional to the atomic velocity. This laser is tuned on the $|5S, F = 2\rangle$ to $|5P, F = 3\rangle$ transition, achieving exact resonance for a well-defined velocity v. Only atoms within a narrow velocity window around v experience optical pumping cycles, bringing them eventually back into $F = 3$ (Fig. 5.22c). By scanning the frequency of L'_1, v can be tuned between 140 m/s and 600 m/s. Beyond these limits, the number of atoms in the thermal velocity distribution is too low for practical use. The width of the velocity window, ultimately determined by the linewidth of the optical transition, is about 15 m/s. Successive atoms in the same sequence can be prepared with different velocities.

To improve the precision of the velocity selection, we take advantage of the time-resolved circular state preparation. The laser L'_1 is pulsed and prepares a 2 μs bunch of atoms in $F = 3$, with a typical 1 mm length along the atomic beam axis. During the 30 cm travel from the velocity-selection zone to B, this bunch spreads due to its velocity dispersion. The Rydberg state excitation lasers L_2, at right angle with the atomic beam in B, are also pulsed (2 μs duration) and tightly focused. They excite atoms only in a narrow region of the spread atomic bunch, reducing velocity dispersion to 2 m/s. The circular state preparation time being known within 2 μs, the uncertainty on the atomic position during the 20 cm transit through the apparatus down from the preparation zone is less than 1 mm.

Single-atom preparation

In order to address individual atoms, one would ideally need an atomic version of the photon pistol described in Section 5.1.6, which would deliver on demand single circular atoms in the cavity. Although such a perfect device could be developed in the future (Maioli *et al.* 2005), it is not yet available. The experiments described below rely on a less efficient method to perform single-atom experiments. A weak excitation of the atomic beam is combined with a post-selection process in which are retained only the data corresponding to the desired sequence of events.

The weak laser excitation of the atomic beam produces Poissonian statistics for the atoms in circular states. Due to the finite detection efficiency, it is not possible to rely on exact atom counting at detection time. We choose instead to prepare in a sample much less than one atom on the average (typically 0.1). When an atom is counted, the probability that an undetected second atom is present is small ($< 10\%$).

The post-selection of single-atom events is obtained at the expense of a considerable lengthening of the data acquisition time. An atom is detected in less than 10% of the one-sample sequences. For experiments involving three atoms crossing in succession the apparatus at well-defined times, at most 1 out of 1 000 sequences contain useful data. The duration of one sequence is a few milliseconds, corresponding to the complete flight time from the velocity selection to the detection zone. A few seconds are thus required to record a single three-atom event, and many hours to accumulate statistically significant data.

5.2.2 Superconducting cavities

Closed cavities

The photon box storing the microwave field which interacts with the circular Rydberg atoms must have very highly reflecting walls. In the millimetre-wave domain, the best reflectors are superconducting metals. They are perfect conductors for direct currents. For oscillating fields, they present residual losses due to the weak coupling of the field with the unpaired charges. According to the BCS model of superconductivity (Poole *et al.* 1995), these losses decrease exponentially when the temperature T is reduced, varying as $\exp(-\Delta_s/k_bT)$, where Δ_s is the superconducting gap. Due to residual material imperfections, they eventually saturate at some finite value at very low temperature. They are then orders of magnitude smaller than the losses of an

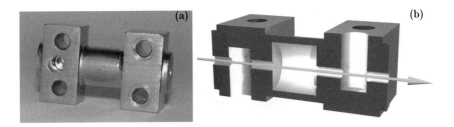

Fig. 5.23 Closed superconducting cavity used in a two-photon micromaser experiment (Brune *et al.* 1987). (a) Photograph (by the authors) of the niobium cavity. (b) Cut along the atomic beam axis. The inner cylindrical cavity is weakly coupled to the outside by two waveguides below cut-off. Elastic deformation of the thin cavity walls allow for a small amplitude tuning of the resonant frequency.

ordinary metal cavity.

Niobium is the choice superconducting material for high-quality resonators in the millimetre-wave domain. It has a large gap value (superconducting transition temperature $T_s = \Delta_s/2k_b = 9.2$ K), it can be obtained with a high purity and is easier to machine than other superconductors. It is widely used, in particular to realize the superconducting cavities providing the accelerating field in large particle colliders.

In CQED experiments, the photon box must not only have the highest possible Q, but also the largest possible field per light quantum, i.e. the smallest possible volume. For such cavities, of size of the order of the wavelength, the simplest geometry is a nearly closed tin-can-shaped box (Fig. 5.23). The cavity mode is coupled to the outside via two waveguides, below cut-off for the resonant frequency. These guides are also used to let the atomic beam pass through. The evanescent field coupling in these guides affects the mode only weakly. This coupling allows a direct measurement of the cavity damping time, $T_c = Q/\omega_c$, by observing the decay of the stored radiation as it leaks out of the box. Very long T_c's, up to a few tenths of a second, have been reached at 20 GHz (Brattke *et al.* 2001). These cavities have been used in the micromaser experiments described in Section 5.1. The resonant frequency of these closed resonators is determined by their size and geometry and cannot coincide *a priori* exactly with the atomic transition. The fine tuning required to bring the cavity on resonance is achieved by an elastic deformation of the cavity walls produced by a mechanical or piezoelectric pusher.

Open Fabry–Perot cavity

Such closed cavities, made of an ideal d.c. conductor, are equipotential volumes, incompatible with a non-vanishing static electric field inside. As discussed above, such a field is required to preserve and to manipulate circular Rydberg atoms (the micromaser experiments made use of low-angular momentum Rydberg states which do not need static fields). The CQED experiments with circular states thus use instead, as a photon box, an open Fabry–Perot cavity with two highly polished spherical mirrors facing each other (Raimond *et al.* 2001). A voltage applied across the mirrors produces a fairly homogeneous field inside the cavity.

The cavity is represented in Fig. 5.24, together with a photograph of the mirrors.

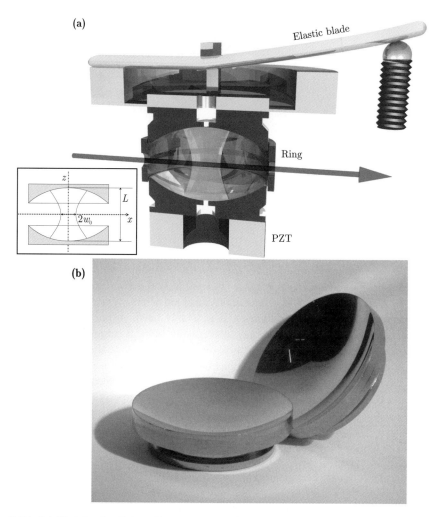

Fig. 5.24 (a) Sketch of a Fabry–Perot superconducting cavity. The inset defines the cavity geometric parameters. (b) Photograph (by the authors) of the mirrors.

They have a 50 mm diameter and an $R = 40$ mm radius of curvature. The cavity sustains a TEM_{900} Gaussian mode, with $k = 9$ anti-nodes along the Oz cavity axis. The distance between the mirrors along Oz is $L = 27$ mm$\approx k\lambda/2$, where λ is here the wavelength associated to the cavity mode. The cavity resonant frequency, $\omega_c/2\pi$, is close to the $|51C\rangle \rightarrow |50C\rangle$ transition at 51 GHz ($\lambda_{eg} = 2\pi c/\omega_{eg} \approx 5.9$ mm).

The relative mode amplitude f (Section 3.1) is real, with a proper choice of phase reference. Due to the cylindrical symmetry, f depends only on z and on the distance r to the z axis. We choose the origin on the z axis, halfway between the mirrors. In a paraxial approximation assuming that the mode angular spread is small, f can be written:

$$f(r, z) = \frac{w_0}{w(z)} e^{-r^2/w(z)^2} \cos \left[2\pi \frac{z}{\lambda} - \arctan \frac{\lambda z}{\pi w_0^2} + \frac{\pi}{r^2} R(z)\lambda + (k-1)\frac{\pi}{2} \right] , \quad (5.25)$$

with:

$$w(z) = w_0 \sqrt{1 + \left(\frac{\lambda z}{\pi w_0^2} \right)^2} \quad \text{and} \quad R(z) = z \left[1 + \left(\frac{\pi w_0^2}{\lambda z} \right)^2 \right] . \quad (5.26)$$

The cosine factor in f describes the longitudinal standing wave pattern of the mode and $R(z)$ is the wave-front radius of curvature at position z, which matches the mirrors curvature R on the cavity walls ($z = \pm L/2$). The transverse pattern is Gaussian, with a width $w(z)$. The minimum width (mode waist), w_0, at $z = 0$, is obtained by expressing the condition $R(L/2) = R$:

$$w_0 = \left[\frac{\lambda L}{2\pi} \sqrt{\frac{2R}{L} - 1} \right]^{1/2} \sim \lambda . \quad (5.27)$$

The mirror's curvature has been chosen to make w_0 nearly equal to λ.

The polarization of the cavity mode, ϵ_c, lies in the symmetry plane xOy. The cavity sustains two modes, M_a and M_b, with the same geometry and orthogonal linear polarizations. With perfect spherical mirrors, the cylindrical symmetry of the cavity geometry would ensure that these two modes are strictly degenerate. In practice, mirror shape imperfections lift this degeneracy by an amount $\delta \ll \omega_c$. Typically, $\delta/2\pi = 128$ kHz. We will call M_a the highest frequency mode and assume, for the sake of simplicity, that M_a is polarized along x and M_b along y.

Fluctuations of fields and currents in cavity mirrors

The mode volume $\mathcal{V}$ is obtained by integrating f^2 over space. The square of the cosine in eqn. (5.25) averages to $1/2$ and, after Gaussian integrations, $\mathcal{V}$ is given by the simple formula:

$$\mathcal{V} = \frac{\pi}{4} w_0^2 L \approx \frac{k\pi}{8} \lambda^3 \approx 700 \text{ mm}^3 . \quad (5.28)$$

According to eqn. (3.24), the r.m.s. amplitude of the vacuum field at the cavity centre is:

$$\mathcal{E}_0 = \sqrt{\frac{2\hbar\omega_c}{\pi\varepsilon_0 w_0^2 L}} \approx \frac{2}{\lambda^2} \sqrt{\frac{2\hbar c}{k\varepsilon_0}} \approx 1.5 \, 10^{-3} \text{ V/m} . \quad (5.29)$$

This inherently quantum parameter is of the order of a macroscopic unit, the millivolt per metre! Another instructive order of magnitude is provided by the current flowing along the mirror surface when the cavity operates at a quantum level. Maxwell's boundary conditions impose that the tangential electric field cancels at the cavity walls, which coincide with nodes of the standing wave $f(\mathbf{r}, t)$. The tangential magnetic field, on the other hand, has anti-nodes at the surface, with an r.m.s. vacuum field fluctuation on cavity axis equal to $(w_0/w_m)\mathcal{E}_0/c$ outside the metal and zero inside the metal $[w_m = w(L/2) \approx 1.25 \, w_0 = 7.5$ mm is the field waist at the mirrors]. The magnetic field jump across the surface divided by the vacuum permittivity $\mu_0 = 1/\epsilon_0 c^2$ is equal to the surface current density $\mathbf{j}_0$. Hence:

$$j_0 = \frac{w_0}{w_m} \frac{\mathcal{E}_0}{\mu_0 c} = \frac{w_0}{w_m} \epsilon_0 c \mathcal{E}_0 \approx 3 \cdot 10^{-6} \text{ A/m} . \tag{5.30}$$

The current flowing across the surface when the cavity stores a single photon is of the order of $j_0 w_m \approx 20$ nA. This is quite a large current, of the order of that circulating in a superconducting SQUID device. Replacing in (5.30) $\mathcal{E}_0$ by (5.29) and noting that $4\pi\varepsilon_0\hbar c = q^2/\alpha$, we find:

$$j_0 w_m \approx \frac{\sqrt{2}}{\sqrt{k\pi\alpha}} \frac{|q|c}{\lambda} \approx \frac{\sqrt{2}}{\pi} |q|\omega_c . \tag{5.31}$$

Noting that $|q|\omega_c/2\pi$ is, at resonance, the current circulating around the orbit of the circular atom, we conclude that the current quantum fluctuations of both CQED partners, the atom and the cavity, have the same order of magnitude. When an atom enters in the cavity, its electric image in the mirrors thus modifies the currents by an amount of the order of their initial quantum fluctuations. This is a big perturbation and we understand intuitively that the two systems are strongly coupled. We give in the next section a more quantitative analysis of this coupling.

5.2.3 Atom–cavity coupling

Vacuum Rabi frequency compared to other relevant parameters

The atoms cross the central anti-node of the cavity mode standing wave structure, where they experience the maximum field amplitude. The circular atomic transition polarization is $\boldsymbol{\epsilon}_a = (\mathbf{u}_x + i\mathbf{u}_y)/\sqrt{2}$ while the cavity polarization is $\mathbf{u}_x$ or $\mathbf{u}_y$. The vacuum Rabi frequency at the cavity centre is given by eqn. (3.168). Its value for the $|51C\rangle \rightarrow |50C\rangle$ circular state transition ($d = 1\,776\,|q|a_0$) coupled to the TEM$_{900}$ mode ($k = 9$; $\mathcal{E}_0 = 1.5\,10^{-3}$ V/m) is $\Omega_0/2\pi = 50$ kHz. It corresponds to a Rabi period $2\pi/\Omega_0 = 20\ \mu$s. This coupling frequency depends only upon the geometrical parameters of the atom and of the cavity. It can quite generally be expressed in a simple formula involving the two fundamental constants, α and R_∞, and the dimensionless quantum numbers associated to the atom and the cavity, n and k. Replacing in eqn. (3.168) d and $\mathcal{E}_0$ by their expressions (5.5) and (5.29), we get:

$$\Omega_0 = (2\alpha/\pi)^{3/2} \frac{1}{\sqrt{k}} \frac{1}{n^4} \frac{R_\infty}{\hbar} . \tag{5.32}$$

This expression has to be compared with the spontaneous emission rate Γ_n describing the strength of atom's coupling to the free space modes (in particular to the modes transverse to the cavity axis in which the circular atom can emit photons). Comparing with (5.18), we find:

$$\frac{\Omega_0}{\Gamma_n} = \frac{3n}{4\sqrt{k}} (2/\pi\alpha)^{3/2} . \tag{5.33}$$

This ratio is $\sim 10^4$ for the atom–cavity situation considered here ($n = 50, k = 9$). Its large value means that one of the conditions for the strong coupling regime ($\Omega_0 \gg$

$1/T_{a,n}$) is largely satisfied. We also remark that Ω_0/Γ_n increases linearly with n, which is another factor favouring large atoms in CQED experiments. This dependence is not very fast, though, so that increasing n too much becomes detrimental because of the much faster increase of the adverse effects mentioned above.

It is also instructive to compare the cavity frequency $\omega_c = \omega_{nC,(n-1)C}$ given by eqn. (5.4) with Ω_0. The ratio of these frequencies:

$$Q_{\min} = \frac{\omega_c}{\Omega_0} = n\sqrt{\frac{k}{2}} \, (\pi/\alpha)^{3/2} \, , \tag{5.34}$$

is again a very large number. For the situation of interest here ($n = 50$ and $k = 9$), it is $\sim 10^6$. The Rabi frequency is thus very small compared to the field frequency. This is an important feature of CQED, essential for justifying the rotating wave approximation leading to the Jaynes–Cummings Hamiltonian (Section 3.4). More practically, the large value of ω_c/Ω_0 implies that the cavity Q must be very high to achieve the strong coupling regime. The photon storage time, Q/ω_c, must exceed $1/\Omega_0$, which implies the threshold condition $Q > Q_{\min}$. In practice, for $n = 50$ and $k = 9$, Q must be much larger than 10^6 for CQED effects to be clearly observable. As we will see below, this condition is largely satisfied.

Comparing eqns. (5.33) and (5.34), we conclude by noting that Ω_0 satisfies the double inequality $\Gamma_n \ll \Omega_0 \ll \omega_c$. On a logarithmic scale, the vacuum Rabi frequency is close to the middle of the interval between Γ_n and ω_c. It is 4 orders of magnitude larger than the former and 6 orders of magnitude smaller than the latter. This strong hierarchy between the three frequencies is made possible by the huge quality factor of the circular to circular state transition.

Note finally that, in the microwave domain, the characteristic energy of the atom–field coupling, $\hbar\Omega_0$, is very small compared to the kinetic energy of the thermal atoms (≈ 25 meV, corresponding to a frequency $6 \cdot 10^{12}$ Hz). The atom–cavity coupling has then no influence on the atomic velocity v, which remains nearly constant during the atom–cavity interaction.

Effective atom–cavity interaction times

As the atom crosses the cavity and explores the Gaussian structure of the mode, it experiences a time-varying field amplitude. This, in general, complicates the analysis of the atom–field evolution during an experiment. Simple results can, however, be obtained in two limiting cases: either exact atom–cavity resonance or large atom–cavity detuning (dispersive limit). It is then possible to define an effective interaction time allowing us to describe experiments as if the coupling were constant throughout the atom–field interaction.

The effect of the interaction during the full atom–cavity crossing is described by the evolution operator U_∞, from $t \to -\infty$ to $t \to +\infty$ (we choose the time origin, $t = 0$, when the atom crosses the cavity axis). In the general case, $\Delta_c \neq 0$, there is no simple analytic expression for U_∞.

In the resonant case ($\Delta_c = 0$), however, the atom–field Hamiltonians $H(t)$ and $H(t')$ commute and U_∞ is:

$$U_\infty^r = \exp\left[-(i/\hbar) \int H(t)\, dt\right] = \exp\left[(i/\hbar)H(0)t_i^r\right] \, , \tag{5.35}$$

where:

$$t_i^r = \int f(vt)\, dt = \sqrt{\pi}\frac{w_0}{v} \ . \tag{5.36}$$

When the atom has crossed the mode, the system has evolved as if the maximum atom–field coupling (Ω_0) had been applied during the effective interaction time t_i^r. All atom–field state transformations described by eqns. (3.179) and (3.180) remain valid, with t being replaced by t_i^r. For instance, a 2π-quantum Rabi pulse is obtained for $t_i^r = 20$ μs, corresponding to a velocity $v = 503$ m/s. Up to a 7π-Rabi pulse can be realized in the accessible velocity range.

An analytical solution also exists in the dispersive limit $\Delta_c \gg \Omega$. We then use the effective Hamiltonian H_{eff} defined by eqn. (3.191), with a parameter s_0 proportional to the square of the atom–field coupling $f^2(vt)$. The Hamiltonians $H_{\text{eff}}(t)$ and $H_{\text{eff}}(t')$ for $t' \neq t$ still commute in this case. The total evolution operator U_∞^d can thus be written:

$$U_\infty^d = \exp\left[-(i/\hbar)H_{\text{eff}}(0)t_i^d\right] \ , \tag{5.37}$$

where the effective interaction time t_i^d is now defined by:

$$t_i^d = \int f^2(vt)\, dt = \sqrt{\frac{\pi}{2}}\frac{w_0}{v} \ . \tag{5.38}$$

The interaction with the mode now corresponds to the application of the maximum coupling for an effective dispersive interaction time t_i^d, smaller than t_i^r by a factor of $\sqrt{2}$.

5.2.4   Tuning the cavity and optimizing its quality

Controlling with precision the frequency of the cavity and storing photons in it for very long times are critical conditions for the success of CQED experiments. We conclude this section by discussing briefly how the cavity can be precisely tuned around the atomic resonance and how its Q factor is measured. We also describe the successive improvements which have led to a regular increase of the cavity Q factor, making it possible to reach deeper and deeper into the strong coupling regime.

The mode frequencies can be easily tuned by slightly changing the cavity length. The upper mirror is mounted on a deformable mount (Fig. 5.24). The motion of a micrometre screw is transmitted by an elastic bronze blade, with a very high tuning sensitivity (in the 100 kHz range) and a tuning range (20 MHz) wide enough to compensate the initial uncertainty on the mirrors distance. Fine tuning is achieved by a piezoelectric stack placed under the lower mirror, with a 1 MHz range and a few hertz sensitivity.

We have successively used three variants of our basic cavity design. In the first one, the mirrors were made of bulk niobium, with polished surfaces. The gap between the mirrors was left wide open. The cavity mode was coupled to the outside by small irises pierced at the apex of both mirrors. The Q value was obtained, as for the tin-shaped closed cavities, from recording of the field energy decay. The damping time T_c strongly depended on the quality of the mirror surface. From initial values in the few μs range obtained in early tries, T_c was progressively increased into the 100 to 200 μs range, which appeared to be the limit set by large-scale surface defects arising

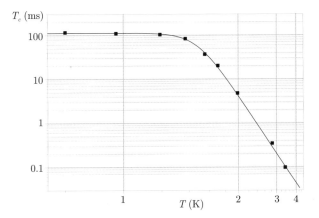

Fig. 5.25 Cavity damping time T_c versus mirror temperature T. The horizontal scale is reciprocal and the vertical one logarithmic. The BCS-limited damping time, observed above 1.6 K, plots as a straight line. The dots are experimental and the solid line is a fit, providing a superconducting gap value $\Delta_s/k_b = 16.6$ K. Below 1.2 K, the photon lifetime saturates at an extremely high value, 115 ms.

during the mirrors polishing process. These defects scatter photons outside the mode into free space through the centimetre gap between the mirror edges. In spite of these imperfections, Q factors $\omega_c T_c \approx 3 \cdot 10^7$ were obtained, largely above the $Q_{\min}$ limit given by eqn. (5.34).

A much longer damping time, $T_c = 1$ ms, corresponding to $Q = \omega_c T_c = 3 \cdot 10^8$, has then been obtained in a slightly modified cavity design, by plugging the gap between the mirror edges with an aluminium-alloy ring (shown in Fig. 5.24). This ring efficiently reflects back the photons scattered by the surface defects into the cavity mode. Surprisingly enough, this poorly conducting ring increases Q by an order of magnitude. The atomic beam, propagating along the x axis, is fed inside the cavity–ring structure through 3 mm diameter holes.

In both versions of the cavity (with or without recirculation ring), the microwave coupling irises may be used to inject a coherent field in the mode. In order to control fields at the single-photon level we employ an X-band source and produce its fourth harmonic in a non-linear Schottky diode. Reducing the X-band power provides a strong attenuation of the harmonic signal and allows us to obtain the very high attenuation required. Note that the X-band source, phase-locked to a reference crystal, has a linewidth much smaller than a hertz.

These two versions of the Fabry–Perot cavity have been used in most of the experiments described below. These cavities achieve a compromise between conflicting requirements: long photon storage time and existence of a non-vanishing static electric field applied across the mirrors and preserving the fragile circular Rydberg atoms. This compromise has, however, resulted in a cavity damping time at least two orders of magnitude smaller than for a closed cavity. Open cavities are obviously very sensitive to geometrical imperfections of their walls, which is not the case of closed structures.

Increasing the Q factor of open cavities requires a drastic improvement of the surface quality, which is not compatible with the use of bulk niobium. At the time we completed this book, we were testing a third version of the cavity with mirrors made of niobium-coated copper. The copper substrate is machined to an optical quality using techniques employed in the fabrication of infrared laser mirrors. We thus combine the geometrical quality of copper with the conducting quality of niobium.

We have built a cavity with the same mirror geometry as above, but with no photon recirculation ring. The mirrors have no coupling irises, which would spoil the surface quality around them. The field is injected in the mode by irradiating the cavity on the side and taking advantage of the mirror-edges residual diffraction. The decay of the stored field is probed by Rydberg atoms crossing the cavity.

We have measured in this way a 115 ms damping time at a 0.8 K temperature, corresponding to a quality factor $Q = 3 \cdot 10^{10}$. A single photon bounces $1.2\,10^9$ times on the mirrors, travelling over 34 000 kilometres before the amplitude decreases by $1/e$! This damping time is three orders of magnitude higher than for a cavity made up of massive niobium mirrors without the ring. It is of the same order as the expected T_c for closed cavities in this frequency range. It is even higher than the circular Rydberg atoms lifetime (36 ms)!

Figure 5.25 shows T_c as a function of the mirrors temperature T. It exhibits the exponential increase, predicted by the BCS theory, when T decreases down to 1.6 K. It then saturates to a value determined by the residual losses. The solid line in Fig. 5.25 is a fit combining BCS theory with an adjustable residual loss contribution. It yields a superconducting gap $\Delta_s/k_b = 16.6$ K and a transition temperature 8.3 K, slightly lower than that of the bulk niobium, as expected for a thin film. This new design, not yet used in actual CQED experiments, opens fascinating perspectives for future developments, which we will evoke at the end of Chapter 7.

5.3   Two experiments unveiling the quantum in a cavity

Setting up the one-atom–one-field-mode show has not been a simple affair. As we have seen, it has involved many tricks to train the spin and spring actors of this play and to put them on a very special stage, a box shielding them from unwanted influences from the outside world. Once the tricks have been learned and the stage prepared, the curtain can be raised on the opening scenes, which strikingly illustrate the quantum nature of light–matter interactions.

Quantum physics was born in the blackbody box which Planck envisioned to solve the thermal radiation puzzle. Many atoms and field modes were exchanging heat in the disorder of this box. It has been a matter of clever book-keeping to make sense of these random processes. The discrete nature of light was inferred from a brilliant intuition of the book-keeper, but the field graininess remained elusive – a mere mathematical requirement to avoid bothersome divergences in a theorist's computation. Since then, the box has remained one of the favourite stages of theorists, who keep using it as a thought experiment tool, to count discrete field modes throughout their analyses. They get rid of it in the end, like magicians making the walls vanish and letting the beauty of the physics stand alone.

Most of the quantum manifestations of light, from the simplest (photon clicks in detectors) to the most sophisticated ones (photon anti-bunching, Lamb shift) are free-space effects, observed on propagating fields or in a vacuum continuum, after the box scaffolding has been 'dismantled'. On the contrary, in CQED physics, we see for the first time the quantumness of light inside the box, in the discrete setting where theorists have unveiled it in the first place. We present in this section two experiments performed in this setting. One is the observation of the Rabi oscillation of an atom in a small field, which viscerally exhibits the field graininess in a cavity. The other is the measurement of an atomic Lamb shift, which demonstrates in its simplest form the perturbing effect of a cavity vacuum on an atom. While watching these experiments unfold, we will learn more about the CQED tools described in the previous sections and start to appraise their power for manipulating atoms and photons in well-controlled situations.

5.3.1   Field graininess unveiled by Rabi oscillation

The Rabi oscillation is the most fundamental manifestation of the interaction between an atom and a field mode at resonance. This phenomenon, analysed in Section 3.4.1, displays its simplest features when both partners are initially in energy eigenstates, the atom for instance in $|e\rangle$ and the field in a photon number state $|n\rangle$. The atom then periodically exchanges its energy with the cavity mode at a frequency proportional to $\sqrt{n+1}$.

When, as is generally the case, the initial field does not have a well-defined photon number, but is a superposition or a statistical mixture of Fock states, the Rabi oscillation signal becomes a superposition of contributions oscillating at incommensurate frequencies $\Omega_0\sqrt{n+1}$. The probability $P_e(t_i^r)$ of finding the atom in state $|e\rangle$ after an effective interaction time t_i^r is:

$$P_e(t_i^r) = \sum_n p(n)\frac{1 + \cos\Omega_0\sqrt{n+1}\,t_i^r}{2} \ , \tag{5.39}$$

where $p(n)$ is the photon number distribution (see also eqn. 3.182). We have analysed it theoretically in Chapter 3, for a coherent field having an average photon number $\overline{n}$ much larger than 1. The Rabi oscillation then exhibits collapses and revivals, a phenomenon which we will investigate experimentally in Chapter 7. We focus here on the observation – with the CQED set-up described in Section 5.2 – of the Rabi oscillation in vacuum and in a very small coherent field containing at most a couple of photons. We will see that this signal provides a direct insight into the quantum structure of radiation.

Vacuum Rabi oscillation

When the field is initially in vacuum, the oscillation occurs only if the atom is prepared in $|e\rangle$. The system then oscillates between the combined states $|e,0\rangle$ and $|g,1\rangle$. This phenomenon is known as the vacuum Rabi oscillation and can be viewed as a reversible spontaneous emission process in the high-Q cavity. In order to observe it, one of the cavity modes (M_a) is tuned at exact resonance with the $|e\rangle \rightarrow |g\rangle$ transition. The second mode (M_b), whose detuning with the atomic transition is larger than Ω_0 remains empty and plays a negligible role. Both cavity modes are cooled down to 0.6 K and contain a very small residual thermal field (average photon number $n_{\rm th} = 0.06$).

Figure 5.26(A) presents the probability $P_e(t_i^r)$ of finding the atom in $|e\rangle$ after interaction with the cavity, as a function of the effective interaction time t_i^r. This probability

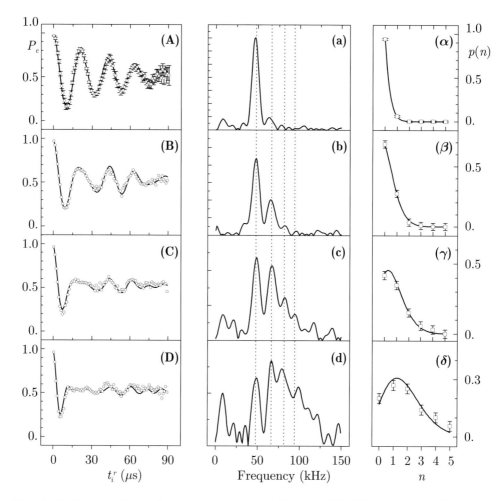

Fig. 5.26 Rabi oscillation in a small coherent field. (A), (B), (C) and (D): Rabi signal representing $P_e(t_i^r)$, for fields with increasing amplitudes. (A): no injected field (B), (C) and (D): coherent fields with 0.40 ($\pm$0.02), 0.85 ($\pm$0.04) and 1.77 ($\pm$0.15) photons on average. The points are experimental [errors bars in (A) only for clarity]; the solid lines correspond to a theoretical fit. (a), (b), (c), (d): corresponding Fourier transforms. Frequencies $\Omega_0/2\pi = 47$ kHz, $\sqrt{2}\Omega_0/2\pi$, $\sqrt{3}\Omega_0/2\pi$ and $2\Omega_0/2\pi$ are indicated by vertical dotted lines. Vertical scales are proportional to 4, 3, 1.5 and 1 from (a) to (d). (α), (β), (γ), (δ): Photon number distributions $p(n)$ inferred from the experimental signals (points). Solid lines show the theoretical thermal (α) or coherent [(β), (γ), (δ)] distributions which best fit the data. The inferred residual thermal photon number is $n_{\text{th}} = 0.06$ for (A). Reprinted with permission from Brune *et al.* (1996b). © American Physical Society.

is obtained by accumulating statistics on many runs in which a single atom is sent across the apparatus under the same conditions. There is here no velocity selection on the atomic beam. A sample with at most one circular atom and a large velocity dispersion is prepared at a given time. The atomic detection time measures the velocity and, hence, the effective interaction time t_i^r. For long t_i^r values, corresponding to slow atoms, the number of detected events is low and the statistical noise accordingly large. Due to the lack of fast atoms in the beam, short interaction times are obtained with the help of an electric field applied across the mirrors. It tunes the atomic transition out of the cavity resonance during a large part of the atomic transit through the mode, freezing the atomic evolution during the time it is applied.

We observe on Fig. 5.26(A) four complete oscillations. The coherent atom–field coupling dominates the dissipative processes, a clear signature of the strong coupling regime of CQED. In other words, the atomic transition is saturated by a single-photon field, an unusual situation in quantum optics. The Rabi frequency is obtained from the Fourier transform of the $P_e(t_i^r)$ signal (Fig. 5.26a). The large peak in the Fourier spectrum is centred at 47 kHz, a value in good agreement with the theoretical coupling (50 kHz). The small difference is accounted for by a slight misalignment of the atomic beam with the central anti-node in the cavity mode. The presence of the very small peak at frequency $\Omega_0\sqrt{2}$ in Fig. 5.26(a) is due to the finite temperature of the cavity, which has a small probability of containing one thermal photon.

Rabi oscillation in a small coherent field

Figure 5.26(B), (C) and (D) presents the Rabi oscillation signals obtained in the same way, when a small initial coherent field is prepared in C by the source S. The average photon numbers are $\bar{n} = 0.40, 0.85$ and 1.77 from (B) to (D). The signals clearly exhibit a beating between several frequencies, as is revealed by their Fourier transforms (Fig. 5.26b, c and d). The progressive appearance of discrete components with increasing frequencies is conspicuous. As expected, these frequencies scale as the square roots of successive integers, marked by the vertical dashed lines.

The weights of the Rabi frequency components give the probability $p(n)$ of the corresponding photon number in the initial coherent field. Figure 5.26(α)–(δ) present the inferred photon number distributions. The (α) curve is fitted with a Planck law at $T = 0.6$ K. The (β)–(δ) curves are fitted with the Poisson law which characterizes coherent states (solid lines). The average photon numbers are deduced from these fits.

The frequency distribution of the Rabi signal can thus be mapped directly into the photon number statistics. Remarkably, this mapping is not very sensitive to experimental imperfections, which damp the oscillations and broaden the Fourier peaks without affecting, to first order, their respective amplitudes. We do not enter here in a detailed discussion of these effects (Brune *et al.* 1996b), which we have simulated in the fits of Figure 5.26(A)–(D) by damping all the oscillating terms in eqn. (5.39) with a single time constant. These damping processes are not dominated by cavity or atomic relaxation but certainly by other imperfections (dispersions of the couplings due to atomic beam misalignment and size, collisions on background gas, inhomogeneous stray electric and magnetic fields shifting the atomic line out of cavity resonance, perturbing effect of the off-resonant M_b mode).

The presence of discrete peaks in these Fourier spectra is, as already noted in

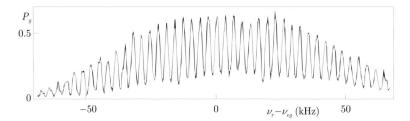

Fig. 5.27 Ramsey fringes on the $|e\rangle \rightarrow |g\rangle$ transition. Probability P_g of detecting the atom in the final state $|g\rangle$ as a function of the detuning between the Ramsey source frequency $\nu_r = \omega_r/2\pi$ and the $|e\rangle \rightarrow |g\rangle$ transition frequency $\nu_{eg} = \omega_{eg}/2\pi$. Reprinted with permission from Brune *et al.* (1994). © American Physical Society.

Chapter 1, a striking illustration of field quantization in a box. The atom acts as a very sensitive probe measuring, in frequency units, the electric field amplitude in the cavity. The quantization of this amplitude directly reveals the graininess of the field energy and hence the existence of photons. This is of course not the first experimental evidence of light quanta. It is nevertheless one of their most direct and dramatic manifestations.

5.3.2   Cavity Lamb shift unveiled by Ramsey interferometry

Atom and cavity influence each other when they are out of resonance too. The atomic transition frequency is modified when the atom crosses the cavity and, conversely, the field mode experiences a shift akin to a refractive index effect induced by a single atom. We have seen in Chapter 3 that the transition frequency of an atom at the centre of a cavity containing n photons is displaced by an amount $(1 + 2n)\Omega_0^2/4\Delta_c$ where Δ_c is the atom–cavity detuning. This expression is valid when Δ_c is much larger than Ω_0. A residual shift, $\delta_{LS} = \Omega_0^2/4\Delta_c$, subsists for $n = 0$. This simple expression assumes that only one mode appreciably contributes to the effect. In our experiments, we must take into account the contributions of the two closely spaced modes M_a and M_b, whose couplings to the atomic dipole are nearly identical. If Δ_c is the M_a–atom detuning, the total vacuum shift is $\delta'_{LS} = \Omega_0^2[1/4\Delta_c + 1/4(\Delta_c + \delta)]$. It results from the radiative correction produced on the atomic energies by the vacuum fluctuations of the two modes. In this way, it is a genuine quantum effect, akin to the famous Lamb shift, due to the influence of the vacuum fluctuations in the continuum of free-space modes surrounding the atom. The usual Lamb shift is still present, hidden in the renormalized atomic energies since the atom keeps interacting with all the modes (of short wavelength in particular) which are not affected by the presence of the cavity mirrors. The perturbing effect of the cavity only adds a specific contribution from the cavity modes lying very close to the atomic transition.

Observing this effect, beyond its textbook appeal, is important in the context of CQED experiments because it demonstrates the ability to control the atom–field system at the quantum level in a dispersive situation, a feat which is essential in many experiments described later on. It also illustrates the power of Ramsey interferometry in CQED experiments. To observe the shift, we must indeed probe the evolution of an atomic coherence as the atom crosses the cavity. The method of choice is to prepare an

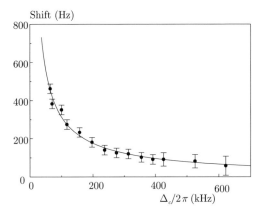

Shift (Hz)

Fig. 5.28 Cavity-induced Lamb shift $\delta'_{LS}/2\pi$ of the $|e\rangle \rightarrow |g\rangle$ transition, at the cavity centre, as a function of the detuning Δ_c. The points with error bars are measured by Ramsey interferometry. The vacuum effect is obtained after correction for the known residual thermal background. The line presents the theoretical predictions without adjustable parameters. Reprinted with permission from Brune *et al.* (1994). © American Physical Society.

atomic state superposition before the atom enters in the cavity, to let this coherence evolve as the atom drifts across the mode and then to detect it downstream. We recognize in this double atomic manipulation the ingredients of a Ramsey procedure.

In the Bloch vector picture, the atomic internal states are equivalent to those of a spin whose tip evolves on a sphere. The states $|e\rangle$ and $|g\rangle$ correspond to vectors pointing along the poles of this sphere and a state superposition with equal weights for $|e\rangle$ and $|g\rangle$ is represented by a vector pointing sidewise. Preparing such a state, which amounts to a $\pi/2$-rotation on the Bloch sphere, can be achieved by applying to the atom a pulse of classical resonant radiation produced in R_1, before the atom interacts with C, by the source S' (Fig. 5.15). The parameters of the spin rotation are determined by the amplitude, phase and duration of this classical pulse. In quantum information language, this pulse realizes the most general one-qubit quantum gate.

The detection of the atomic superposition along an arbitrary direction defined by the spin states $|0/1_{\theta,\varphi}\rangle$ follows similar steps. Since the field-ionization detector D distinguishes energy eigenstates ($\theta = 0$), the atomic spin must be rotated by another classical $\pi/2$-pulse applied in R_2 after the atom has left C (Fig. 5.15). This pulse, also produced by S', maps the $|0/1_{\theta,\varphi}\rangle$ states onto the poles of the Bloch sphere. The association of R_2 and D provides a general spin state detection, equivalent to a Stern–Gerlach analyser oriented in an arbitrary direction.

As described in Section 3.3.3, a Ramsey interferometer is obtained by combining two $\pi/2$-rotations of the atomic pseudo-spin in R_1 and R_2. The probability of detecting the atom in $|e\rangle$ or $|g\rangle$ after the two pulses is reconstructed by running the same sequence many times and the variations of this probability are measured as the phase of the interferometer is continuously swept. Modulations of the detected signal (Ramsey fringes) are observed versus this phase and reveal the atomic coherence.

The phase φ of the interferometer is tuned by changing the frequency offset between the source S' and the atomic frequency. Figure 5.27 shows the Ramsey fringes observed with a set-up using our very first open cavity, which had a Q factor $\sim 10^6$ and $T_c = 3\ \mu$s. The limited contrast of about 50% was due, in part, to imperfections of the microwave pulses and, in part, to stray fields experienced by the atoms during their flight between the Ramsey zones.[7] We then recorded similar fringes for various detunings Δ_c which were adjusted by translation of the cavity mirrors. We subtracted the phase of the fringes for a given detuning from the phase of the Ramsey pattern observed for a large Δ_c, corresponding to a negligible Lamb shift. Dividing the differential phase shift $\Phi(\Delta_c)$ by $2\pi t_i^d$, deduced from the known atomic velocity, we obtained the atomic frequency shift at cavity centre. The raw data were corrected to take into account a residual thermal field, corresponding to an independently measured average photon number $n_{th} = 0.3$ (slightly larger than in the Rabi oscillation experiment described above). The presence of these blackbody photons multiplied the vacuum effect by the factor $1 + 2n_{th} = 1.6$. We finally got the Lamb shift at the cavity centre $\delta'_{LS}/2\pi = \Phi(\Delta_c)/2\pi(1 + 2n_{th})t_i^d$, whose variations are shown versus Δ_c in Fig. 5.28. The solid line is theoretical, without adjustable parameter. Vacuum shifts varying in the range 100–600 Hz have thus been measured. The good fit with the experimental data showed that the dispersive atom–cavity interaction was well-understood.

In this early experiment, Q was too small to satisfy the strong coupling regime condition. The cavity damping time $T_c = 3$ μs, was smaller than the vacuum Rabi period (20 μs) and than the effective atomic transit time across the cavity ($t_i^d \sim 10$ μs). The fast damping of the field was not a problem, though, since the cavity was essentially in vacuum. The number of thermal photons was quickly fluctuating while each atom crossed the cavity and its effect was merely to multiply, as indicated above, the vacuum effect by the $1 + 2n_{th}$ factor. In all the CQED experiments we will be describing in the following, much better cavities have been used, satisfying the inequalities $t_i^d > T_c$ or $t_i^r > T_c$. As soon as photons are present in the cavity, these conditions are essential for preserving the quantum coherence during an experimental sequence.

5.4 An atom–photon entangling machine

The dynamics of the Rabi oscillation in vacuum corresponds to a unitary evolution in the Hilbert space spanned by the states $\{|e, 0\rangle, |g, 1\rangle\}$. This evolution is equivalent to a rotation of an atom–cavity pseudo-spin,[8] which generally produces entanglement between the atom and field partners (Section 3.4.1). By combining such rotations on composite systems involving the cavity field and successive atoms crossing the cavity, it is possible to engineer complex entangled states. These experiments provide an interesting testing ground for quantum information physics. Our present goal is to analyse vacuum Rabi pulses of various durations as operations of simple quantum circuits and to show how combinations of such circuits realize atom–field and field–field entangled states.

We start in Section 5.4.1 by defining the conditions under which the Rabi oscillation can be exactly described by a combination of gates involving an atom and a field qubit. We consider then (Section 5.4.2) three specific rotations, or Rabi pulses, whose angles are $\pi/2$, π and 2π. They correspond to the effective interaction times marked by the three arrows in Fig. 5.29 which represents the first period of the Rabi signal observed

---

[7]Since this early work, the fringe contrast has been increased to 80%.

[8]This rotation of a composite system must not be confused with the ordinary 'single-qubit' evolution in the $\{|e\rangle, |g\rangle\}$ subspace occurring during the Ramsey pulses.

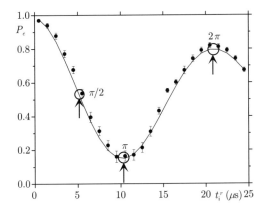

Fig. 5.29 Important atom–cavity interaction times in the vacuum Rabi oscillation. First period of curve (A) in Fig. 5.26 with $\pi/2$-, π- and 2π-quantum Rabi pulses highlighted. Reprinted with permission from Brune *et al.* (1996b). © American Physical Society.

with our CQED set-up. After a quarter of a period ($\pi/2$-quantum Rabi pulse), a maximally entangled atom–cavity state is prepared. Half a period pulse (π-pulse) corresponds to an energy swap between the atom and the field. Finally, a full period (2π-pulse) brings the system back to its initial state. It corresponds to a 2π-rotation of the atom–cavity pseudo-spin and, hence, to a reversal of the sign of the initial state.

Complex atom–field manipulations can be performed by combining these three pulses which constitute the elementary 'stitches' implemented to 'knit' entanglement. We illustrate this knitting process by describing three experiments combining various π- and $\pi/2$-stitches: realization of a quantum memory (Section 5.4.3), of an atomic EPR pair (5.4.4) and of a two-mode entangled state (5.4.5).

This step-by-step entangling method requires a very large cavity quality factor, conserving the atom–field coherence throughout the successive knitting stitches. It might be advantageous, in some cases, to replace this progressive procedure by an alternative method, achieving entanglement in a single non-resonant operation and thus less demanding in terms of cavity damping times. We describe this alternative procedure in Section 5.4.6. Other applications of the quantum Rabi oscillation to information processing in CQED experiments are described in Chapter 6.

5.4.1 Quantum circuit equivalent to vacuum Rabi oscillation

It is natural to describe the vacuum Rabi oscillation as resulting from the coupling of an atomic qubit (with $|e\rangle$ and $|g\rangle$ standing for $|1\rangle$ and $|0\rangle$) and a field qubit (with $|0\rangle$ and $|1\rangle$ represented by the vacuum and the one-photon states).[9] This immediately raises a difficulty. If the atom can be considered a *bona fide* two-level system, the field has an infinite numbers of levels besides $|0\rangle$ and $|1\rangle$. Even if the mode is initially prepared within the $\{|0\rangle, |1\rangle\}$ subspace, the evolution induced by the Jaynes–Cummings Hamiltonian does not leave invariant the tensor product of the atom and field subspaces defined above. Calling $U(t_i^r)$ the unitary operator describing this evolution, we see immediately that $U(t_i^r)|e, 1\rangle$ is generally a linear combination of $|e, 1\rangle$ and $|g, 2\rangle$. In more physical terms, bringing the atom in $|e\rangle$ in a cavity containing one photon

---

[9]We choose in this section an atomic qubit assignment different from that of Chapter 3. The lower state $|g\rangle$ is now $|0\rangle$. With this convention, the ground state of the atom–field system is $|0, 0\rangle$.

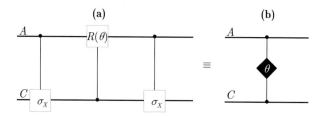

Fig. 5.30 (a) Quantum circuit equivalent to the action of the Jaynes–Cummings coupling on the $|e, 0\rangle$, $|g, 1\rangle$ and $|g, 0\rangle$ states of the composite atom–field system. (b) Simplified representation of this circuit.

leads to the stimulated emission of a second photon and the description of the field as a qubit breaks down.

If we prepare, however, the CQED system in one of the three states $|e, 0\rangle$, $|g, 1\rangle$ or $|g, 0\rangle$, or in any combination of these, the Jaynes–Cummings evolution leaves this subspace invariant. The first two states are merely admixed by the coupling, without any contamination from any other one, while $|g, 0\rangle$, the ground state, remains unchanged. A simple qubit description of the CQED physics remains thus possible if we restrict it to experiments in which the field mode contains at most one photon, with the additional condition that the atoms are never sent in $|e\rangle$ when the field is not in vacuum. The situations we will be considering in this section satisfy these conditions.

In order to make then the qubit analogy quantitative, let us consider the transformation $U_q(t_i^r)$ of a two-qubit system which coincides with the Jaynes–Cummings evolution operator when acting on $|e, 0\rangle$, $|g, 1\rangle$ or $|g, 0\rangle$ and which, in addition, leaves $|e, 1\rangle$, as $|g, 0\rangle$, invariant. Calling P_q the projector on the four-dimension two-qubit space and $P_{e,1} = |e, 1\rangle \langle e, 1|$ the projector on $|e, 1\rangle$, we easily find that $U_q(t_i^r)$ is:

$$U_q(t_i^r) = U(t_i^r)(P_q - P_{e1}) + P_{e,1} . \tag{5.40}$$

The unitarity of $U_q(t_i^r)$ in the two-qubit space follows directly from that of $U(t_i^r)$ and from simple properties of the P_q and $P_{e,1}$ projectors $[P_q^2 = P_q$ and $(P_q - P_{e1})P_{e1} = 0]$.

As a unitary transformation, $U_q(t_i^r)$ can be cast as a combination of one and two-qubit gates operations. This is a fundamental property of qubits, essential in quantum information physics (see Section 2.6.4). The quantum circuit representing this gate combination is shown in Fig. 5.30(a), where the top line represents the atom (active in state $|e\rangle$) and the bottom one the field (active in $|1\rangle$). Two control-not gates (conditional-σ_X operations) in which the atom is the control and the field the target are sandwiching a controlled-$R(\theta)$ gate of the atom-qubit, conditioned to the state of the field. The angle of the $R(\theta)$ rotation is $\theta = \Omega_0 t_i^r/2$ and its axis, in the Bloch sphere representation, is OY, so that:

$$R(\theta) = e^{-i\theta\sigma_Y/2} = \cos\left(\Omega_0 t_i^r/2\right) \mathbb{1} - i \sin\left(\Omega_0 t_i^r/2\right) \sigma_Y . \tag{5.41}$$

It is a simple exercise to show that this symmetrical circuit which contains three two-qubit gates, realizes the transformation defined by eqn. (5.40). Expressed explicitly in the computational basis of the two qubits, this transformation is:

$$U_q(t_i^r) |e, 0\rangle = \cos\left(\theta/2\right) |e, 0\rangle + \sin\left(\theta/2\right) |g, 1\rangle ;$$

$$\begin{aligned}
U_q(t_i^r)\,|g,1\rangle &= -\sin\left(\theta/2\right)|e,0\rangle + \cos\left(\theta/2\right)|g,1\rangle \ ;\\
U_q(t_i^r)\,|g,0\rangle &= |g,0\rangle \ ;\\
U_q(t_i^r)\,|e,1\rangle &= |e,1\rangle \ .
\end{aligned}\tag{5.42}$$

It is, by definition, the same transformation as the actual Jaynes–Cummings evolution operator for the initial states $|e,0\rangle$, $|g,1\rangle$ and $|g,0\rangle$ and it freezes the system when the true evolution operator would stimulate the emission of a second photon. As long as the system never evolves into the only state, $|e,1\rangle$, for which this stimulated process is possible, the quantum circuit shown in Fig. 5.30(a) faithfully describes the real atom–field evolution and all the features of the vacuum Rabi oscillation.

In this circuit, the atomic qubit is successively the control, the target and then the control again. The state of each qubit conditions the evolution of the other, in an intertwined symmetrical process. We will, in the following, represent systematically the evolution of the atom–field system by logical circuits emphasizing the quantum information content of the CQED experiments. Picturing a Rabi oscillation by three gates would rapidly become very cumbersome. We have chosen instead to symbolize it by a more compact pictogram, a mere vertical line connecting the two qubits with an indication in a lozenge box of the angle of the Rabi rotation (Fig. 5.30b). This representation stresses that the two qubits play a symmetrical role in the vacuum Rabi oscillation, both being in turn control and target. Before describing simple experiments in which combinations of vacuum Rabi operations are performed, we analyse in detail the properties of the three specific Rabi pulses defined by the values of t_i^r indicated by the arrows in Fig. 5.29.

5.4.2 Basic quantum Rabi stitches

π/2-vacuum Rabi pulse

Consider first the '$\theta = \pi/2$-Rabi pulse' corresponding to an effective interaction time $t_i^r = \pi/2\Omega_0$. Starting from $|e,0\rangle$, the atom–cavity state becomes:

$$U(\pi/2\Omega_0)\,|e,0\rangle = \left(|e,0\rangle + |g,1\rangle\right)/\sqrt{2}\ .\tag{5.43}$$

This is a maximally entangled atom–field state which remains frozen after the atom has left the cavity, as long as field relaxation, the dominant cause of decoherence, can be neglected. In practice, with a cavity having $T_c \sim 1$ ms, the atom flies away from the cavity centre over several tens of centimetres during the field damping time. The system thus exhibits strong non-local features which are worth revisiting in this CQED context (see also Section 2.4.5).

Assume that Alice and Bob are two experimentalists observing respectively the cavity field and the atom after their spatial separation. In a first experiment, Alice and Bob make independent measurements on their subsystems. The outcomes of these measurements are described by two identical partial density matrices, obtained by tracing the global state over the other system. These density matrices correspond to statistical mixtures of $|e\rangle$ and $|g\rangle$ or $|0\rangle$ and $|1\rangle$ with equal weights. Alice will find zero or one photon with equal probabilities if she measures the field energy. If she decides to measure the electric field instead, she will find a random value with zero average. Similarly, Bob will find $|e\rangle$ or $|g\rangle$ with equal probabilities in an energy measurement and a zero average transverse component of the atomic pseudo-spin. In independent measurements, Alice and Bob have no way to tell the difference between the actual entangled state and a mere juxtaposition of two independent systems in statistical mixtures.

In order to reveal the entanglement, Alice and Bob have to perform joint experiments of the EPR type. If Bob measures the atom in $|e\rangle$ or $|g\rangle$ and tells the result to Alice, she knows for sure that she has accordingly a $|0\rangle$ or $|1\rangle$ Fock state in the cavity. More generally, if Bob detects the atom in a given transverse pseudo-spin state and communicates this information to Alice, she will know that her field is in a corresponding coherent superposition of $|0\rangle$ and $|1\rangle$, with an average electric field having a well-defined phase. Measurements by Bob thus 'post-select' at a distance the state owned by Alice (see Section 5.4.4).

π-vacuum Rabi pulse

The effective interaction time $t_i^r = \pi/\Omega_0$ defines a 'π-quantum Rabi pulse' corresponding to the transformations:

$$|e, 0\rangle \longrightarrow |g, 1\rangle \quad \text{and} \quad |g, 1\rangle \longrightarrow -|e, 0\rangle \ . \tag{5.44}$$

The excitation is swapped between the atom and the cavity. More generally, if the atom enters the empty cavity in an arbitrary superposition of $|e\rangle$ and $|g\rangle$, $c_e |e\rangle + c_g |g\rangle$, the state transformation reads:

$$(c_e |e\rangle + c_g |g\rangle) |0\rangle \longrightarrow |g\rangle (c_e |1\rangle + c_g |0\rangle) \ . \tag{5.45}$$

The atom always ends up in $|g\rangle$ and the cavity is prepared in a quantum superposition of $|0\rangle$ and $|1\rangle$ which copies into the field the initial atomic superposition. In quantum information terms, the qubit initially carried by the atom is deposited into the cavity. An arbitrary superposition of $|e\rangle$ and $|g\rangle$ can be prepared on an atom by applying on it a classical pulse in R_1 before it enters in the cavity. Completing this procedure with a π-Rabi pulse provides a way to prepare a superposition of $|0\rangle$ and $|1\rangle$ photon states. Phase information initially contained in the classical R_1 pulse is copied onto the atom and then, with the help of the π-Rabi pulse, in the quantum field. This information exchange is reversible. Sending in the cavity an atom in $|g\rangle$ undergoing a π-Rabi pulse induces the transformation:

$$|g\rangle (c_e |1\rangle + c_g |0\rangle) \longrightarrow (-c_e |e\rangle + c_g |g\rangle) |0\rangle \ , \tag{5.46}$$

in which the field qubit is copied into the atom (with a phase change).

In atom–field manipulations in which the field is restricted to the $\{|0\rangle, |1\rangle\}$ Hilbert subspace, atom and field play, in principle, symmetrical roles. There is, however, a considerable practical difference between the atom and field qubits, since only the former is directly observable. The field remains 'hidden' in the cavity. Information we can extract from it must be carried away by a probe atom, before being read out on this atom by the 'Stern–Gerlach' method outlined in Section 5.3. The π-vacuum Rabi pulse, which realizes this field-to-probe-atom mapping, is hence an essential ingredient for the measurement of the field qubit. We describe in the next subsection an experiment which demonstrates a combination of qubit mapping operations.

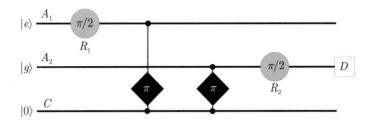

Fig. 5.31 Quantum circuit representing the quantum memory experiment.

2π-vacuum Rabi pulse

Let us finally consider the 2π-Rabi pulse corresponding to $t_i^r = 2\pi/\Omega_0$. After it, the atom–field system returns to the initial state so that one could think that such a trivial pulse is useless. In fact, it produces a global phase shift of the atom–field state, which turns out to be very useful for quantum information processing in CQED experiments. This pulse realizes the transformations:

$$|e, 0\rangle \longrightarrow -|e, 0\rangle \quad \text{and} \quad |g, 1\rangle \longrightarrow -|g, 1\rangle \;, \qquad (5.47)$$

which are reminiscent of the π-phase shift of a spin-$1/2$ undergoing a 2π-rotation in real space, demonstrated, for instance, in neutron interference experiments (Rauch *et al.* 1975; Werner *et al.* 1975; Rauch and Werner 2000). Most importantly for quantum logic applications, the atom–field phase shift is conditioned to the atom and cavity states. It occurs in $|g, 1\rangle$ (eqn. 5.47), while $|g, 0\rangle$ remains invariant. This conditional property can be used to realize a quantum logic gate, essential for the generation of complex entanglement. These experiments are described in detail in the next chapter. For the time being, we focus on simple quantum information manipulations achieved with the π- and $\pi/2$-Rabi pulses only.

5.4.3   State copying and quantum memory

The first example of such a manipulation combines the use of two π-Rabi pulses. It is a demonstration with three qubits (two atom and a cavity) of a quantum memory. A first atom A_1 writes a qubit state onto the cavity field, later read out by a second atom A_2. The write–read procedure is checked by comparing, via Ramsey interferometry, the final phase of the A_2 qubit state to that of A_1. The quantum circuit picture introduced above provides an elegant schematic description of this experiment (Fig. 5.31). The A_1, A_2 and field C qubits, represented from top to bottom by three horizontal lines, are initialized in states $|e\rangle$, $|g\rangle$ and $|0\rangle$ respectively.

Atom A_1 undergoes first a classical $\pi/2$-Ramsey pulse in R_1, equivalent to the operation of a one-qubit gate represented by the grey circle on the upper line in Fig. 5.31. A qubit state whose Bloch vector lies in the equatorial plane of the Bloch sphere is prepared in this way. The transverse orientation of the equivalent spin is determined by the phase of the R_1 classical pulse. This atom then crosses the cavity, undergoing a vacuum π-Rabi pulse represented by the vertical line joining the A_1 and C qubit lines. As discussed in Section 5.4.2, this operation swaps the states of these qubits, copying the A_1 state onto C and leaving A_1 finally in $|g\rangle$. Atom A_2 then crosses C and

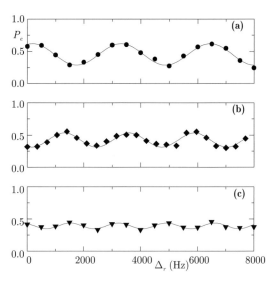

Fig. 5.32 Ramsey fringes observed on atom A_2 for three different time delays between the Ramsey pulses R_1 and R_2 [301, 436 and 581 μs for (a), (b) and (c) respectively]. The observation of these fringes proves the coherent operation of the quantum memory. Reprinted with permission from Maître *et al.* (1997). © American Physical Society.

undergoes also a π-Rabi pulse, represented by the vertical line joining the A_2 and C qubit lines. The process leaves C in vacuum and copies its state onto A_2. Finally, A_2 undergoes a $\pi/2$-classical pulse in R_2 and is detected by state-selective field-ionization in D. The R_2 pulse, equivalent to a one-qubit gate, is represented by the grey circle on the A_2 qubit line.

The atomic energy detection following the R_2 pulse amounts, as discussed in Section 5.3.2, to a final measurement of a transverse component of the A_2 pseudo-spin. The phase of the detected component is determined by that of the R_2 pulse. Since the qubit coherence, initially determined by the R_1 pulse, has been transferred from A_1 to C and then from C to A_2, the signal detected on A_2 must be similar to the one which would be observed in a Ramsey interferometer in which a single atom is subjected to a $R_1 - R_2$ sequence of pulses.

The observed signal is shown in Fig. 5.32, which represents the probability P_e of detecting A_2 in $|e\rangle$ as a function of the detuning Δ_r between the frequency of the Ramsey pulses R_1 and R_2 and the atomic frequency. Each experimental point has been obtained by accumulating statistics over many realizations of the same experimental sequence, before varying the detuning Δ_r and resuming the procedure for the next point. The three recordings correspond to three different time delays T_r between R_1 and R_2. Ramsey type fringes are clearly observed.

These fringes are rather unusual, since the two Ramsey pulses have been applied on different atoms. The existence of these fringes demonstrates that the 'two atom + cavity' system behaves as a single quantum object. That the coherence has been injected on one atom and finally detected on another does not matter for the interferometric process, as long as this coherence has been faithfully conserved all along. The Ramsey interferometer phase being, as for a single atom, $\phi = T_r \Delta_r$, the fringes period decreases with the increasing delay between the two Ramsey pulses, as can be checked by a mere inspection of the figure.

The contrast of the fringes also decreases with increasing delays. This is due to

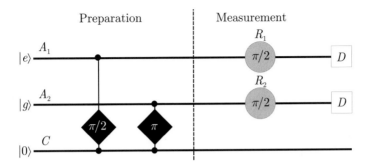

Fig. 5.33 Sequence of the EPR pair experiment.

cavity damping ($T_c = 160\ \mu$s in this early experiment). When the photon decays before A_2 enters C, A_2 exits C in $|g\rangle$ and is finally detected, after R_2, in $|e\rangle$ or $|g\rangle$ with equal probabilities, thus reducing the contrast of the fringes. The cavity state prepared by A_1 is a superposition of $|1\rangle$, which decays with a time constant T_c, and $|0\rangle$, which does not decay. The coherence lifetime is thus $2T_c$, in agreement with the observed fringes contrast damping time.

The observation of these fringes reveals the coherent operation of a CQED quantum memory, with its limitations due to the finite decay time of the cavity.[10] The goal of this experiment was not to test a long term memory, but rather to demonstrate a combination of coherent operations on multi-qubit systems. In this experiment, there was no qubit entanglement. We now turn to situations in which non-locality is involved.

5.4.4 An atomic EPR pair

To entangle an atom and a field qubit, we must use the $\pi/2$-quantum Rabi pulse. Let us illustrate it with an experiment in which we create first an atom–cavity entangled pair, which is then transformed into a detectable atom–atom EPR state via a second π-quantum Rabi pulse (Hagley *et al.* 1997). The complete procedure constitutes a first example of entanglement knitting in CQED experiments, using successively the $\pi/2$- and π-quantum stitches.

The sequence for this two-atom experiment is sketched as a quantum circuit in Fig. 5.33. The vertical dashed line separates the preparation of the entangled state from its subsequent measurement. Let us analyse, in turn, these two stages. A first atom, A_1, initially in $|e\rangle$, interacts with the empty cavity for a $\pi/2$-quantum Rabi pulse. This interaction prepares the entangled state given by eqn. (5.43). We then copy the cavity state onto the second atom, A_2. Prepared in $|g\rangle$, it undergoes a π-quantum Rabi pulse in C. The cavity returns to vacuum and the two atoms evolve into the entangled state:

$$|\Psi_{\mathrm{EPR}}\rangle = (|e_1, g_2\rangle - |g_1, e_2\rangle)/\sqrt{2}\ , \tag{5.48}$$

completing the preparation stage. Note that the atoms are entangled without having directly interacted. The entanglement is 'catalysed' by the cavity, which ends up unchanged and unentangled.

---

[10]With the cavities available at the time this book is completed ($T_c = 115$ ms), storage over time intervals thousand times longer than the write/read Rabi pulses duration (about 10 μs) becomes feasible.

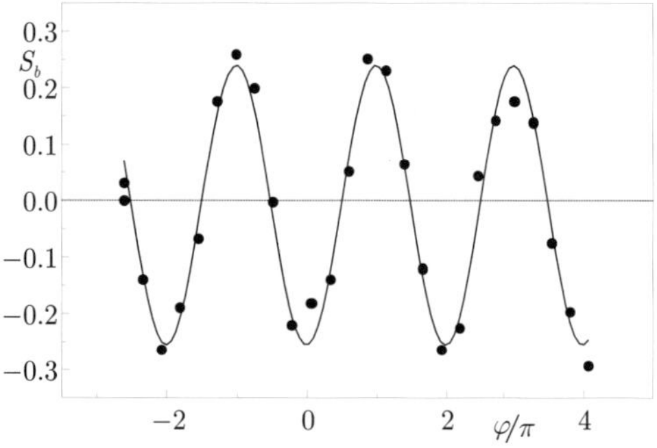

Fig. 5.34 'Bell signal' S_b plotted versus the relative phase φ (in units of π) of pulses R_1 and R_2. The line is a sine fit. Reprinted with permission from Raimond *et al.* (2001). © American Physical Society.

We now turn to the detection of this entanglement. The state $|\Psi_{\mathrm{EPR}}\rangle$ is, in pseudo-spin language, a rotation-invariant spin singlet. The two spins should always be found pointing in opposite directions along any measurement axis. In order to reveal this correlation, we analyse the atomic states along two directions in the horizontal plane of the Bloch sphere. These directions are determined by the phases of the two $\pi/2$-Ramsey pulses R_1 and R_2 applied to A_1 and A_2 before detection (Fig. 5.33). The phase of R_1 is taken as the reference. This pulse maps the pseudo-spin eigenstates $|0/1_X\rangle$ onto $|0/1\rangle$, i.e. $|g\rangle$ and $|e\rangle$. The atomic spin of A_1 is thus detected along OX by the combination of R_1 and D. The pulse R_2 is produced by the same source S' as R_1. The delay between R_1 and R_2 being T, the phase φ of R_2 is $\varphi = (\omega_r - \omega_{eg})T$. The combination of R_2 and D thus maps $|0/1_{\pi/2,\varphi}\rangle$ onto $|0/1\rangle$ and detects the atomic spin along an axis at an angle φ with OX in the equatorial plane.

The atomic detection events, averaged over many realizations of the experiment, yield the correlation signal:

$$S_b(\varphi) = \langle \sigma_{X,1} \sigma_{\varphi,2} \rangle \ , \tag{5.49}$$

where $\sigma_{X,1}$ and $\sigma_{\varphi,2}$ are the Pauli matrices corresponding to the detection axes. Alternatively, we can write $S_b(\varphi)$ in terms of the detection probabilities for the four possible channels:

$$S_b(\varphi) = P_{g_1,g_2}(\varphi) + P_{e_1,e_2}(\varphi) - P_{g_1,e_2}(\varphi) - P_{e_1,g_2}(\varphi) \ . \tag{5.50}$$

The Bell's inequalities (Section 2.4.5) involve a combination of four such signals for different φ settings. We thus call $S_b(\varphi)$ a 'Bell signal'.

In an ideal experiment, $S_b = -\cos\varphi$. This modulation can be explained by qualitative arguments. When φ is zero, both detections are performed along OX. The two atoms must be found in opposite states and $S_b = -1$. When $\varphi = \pi$, the two detections are performed along opposite directions. The two atoms are found in the same state and $S_b = +1$. In the intermediate situation, $\varphi = \pi/2$, the detection axes are orthogonal and there is no correlation: $S_b = 0$.

The experimental Bell signal, plotted in Fig. 5.34, exhibits the expected oscillations. Their contrast in this early experiment (Hagley *et al.* 1997) was 25%, limited,

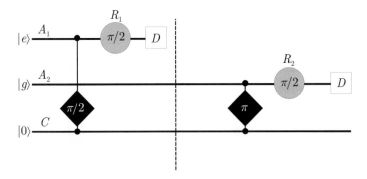

Fig. 5.35 Alternative timing for the EPR correlation experiment.

in part, by the imperfections of the EPR pair preparation.[11] Another limitation orig-
inated in the imperfections of the 'analysing' pulses R_1 and R_2.

These oscillations result from a Ramsey-type interference process. As in the quantum memory experiment, the two pulses are applied onto two different atoms. Phase
information is transferred between them by the 'spooky action at a distance' resulting
from the quantum correlations in the EPR pair.

It is also possible, with the same set-up, to record the correlations between A_1
and A_2 in the energy basis (Z axis), by merely removing the Ramsey pulses R_1 and
R_2 and directly detecting the atomic energies. The anti-correlations expressed by eqn.
(5.48) are then directly revealed (Hagley *et al.* 1997).

In the transverse spin-correlation experiment described above, the preparation and
detection stages are clearly separated in time by the dashed vertical line in Fig. 5.33.
The same fringes in the S_b signal are, however, expected if A_1 is detected in D before
A_2 is sent across C (Fig. 5.35). In this version, the experiment would be an illustration
of the post-selection procedure described in Section 5.4.2, consisting of the preparation
and subsequent read-out of a coherent superposition of $|0\rangle$ and $|1\rangle$ states.

The detection of the atom A_1 pseudo-spin along OX projects the cavity field onto
one of the superpositions $(|0\rangle \pm |1\rangle)/\sqrt{2}$, a state with a non-vanishing average electric
field. The second atom reads out this field, exiting C in a superposition state whose
phase is analysed by R_2 and D. In this case, no atom–atom entanglement is ever
created. Nevertheless, the correlation signals are the same as with the first timing.
They now reveal the atom–field entanglement created by the $A_1 - C$ interaction.
These two correlation experiments, with their different physical interpretations, should
yield the same signal. This reflects that quantum correlations do not depend on the
timing of the detection events, a property which has been checked in many EPR-type
experiments (Chapter 2).

5.4.5 Entangling two field modes

The EPR atom-pair experiment uses a cavity field as a catalyst to entangle two atoms.
Inversely, an atom can be employed in the same set-up to entangle two field modes

---

[11]The signal contrast in this experiment was not high enough to observe the violation of Bell's
inequalities. We describe in Chapter 8 more recent two-atom Bell inequality experiments performed
with ions in traps in which non-locality has been demonstrated (Rowe *et al.* 2001; Roos *et al.* 2004a).

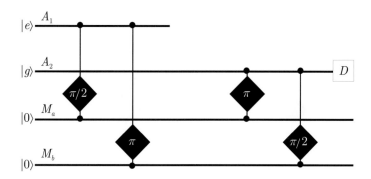

Fig. 5.36 Temporal sequence for the two field-mode entanglement generation and detection.

(Rauschenbeutel *et al.* 2001). It crosses C and deposits, with equal probability amplitudes, a photon in one of the two polarization modes M_a and M_b. The resulting state is an entangled superposition of $|1_a, 0_b\rangle$ and $|0_a, 1_b\rangle$, two field states in which a single photon is present in either one of two modes with slightly different frequencies. A second atom is used later to probe the entangled photonic state.

The experimental sequence is detailed in Fig. 5.36. The quantum circuit now involves four qubits, two atoms A_1 and A_2 and two field modes, M_a and M_b. The first atom A_1 is initially in $|e\rangle$. It is tuned successively in resonance with the initially empty modes M_a and M_b by applying a Stark electric field across the mirrors. First, the atom–mode M_a detuning, Δ_c, is set to zero. Atom A_1 undergoes a $\pi/2$-quantum Rabi pulse in M_a, preparing an $A_1 - M_a$ EPR pair, while M_b remains in vacuum:

$$|\Psi_1\rangle = \frac{1}{\sqrt{2}} \left(|e_1, 0_a\rangle + |g_1, 1_a\rangle\right) |0_b\rangle \ . \tag{5.51}$$

Atom A_1 is then rapidly tuned to resonance with M_b ($\Delta_c = -\delta$) for a time duration corresponding to a π-quantum Rabi pulse. Its state is thus copied onto M_b. Due to the large frequency difference $\delta \gg \Omega_0$ between the modes, the evolution of M_a is frozen during this time. The Stark field is then switched to a high value, freezing any further evolution of the atom–cavity system. Atom A_1 exits in $|g_1\rangle$ and the two modes are left in state:

$$|\Psi_2(0)\rangle = \left(|0_a, 1_b\rangle + |1_a, 0_b\rangle\right) / \sqrt{2} \ , \tag{5.52}$$

using a proper phase choice for $|0_a, 1_b\rangle$ and $|1_a, 0_b\rangle$. There is a striking analogy between $|\Psi_2(0)\rangle$ and the output state of a balanced beam-splitter fed with a single photon (Section 3.2.2). In both cases, there is a quantum ambiguity about the mode which is occupied by the photon. The atom plays here a double role. It is the source of the photon and, at the same time, a 'beam-splitter' channelling it into two different modes.

The two-mode entanglement is read-out by the probe atom A_2 entering C at a time T after the preparation of $|\Psi_2(0)\rangle$. During the time interval between preparation and detection, the relative phase of the components of the two-mode state has evolved, due to the $\hbar\delta$ energy difference between them. The probe atom starts to interact with modes in state $\left(e^{i\delta T}|0_a, 1_b\rangle + |1_a, 0_b\rangle\right) / \sqrt{2}$. Initially in $|g\rangle$, it is first resonantly

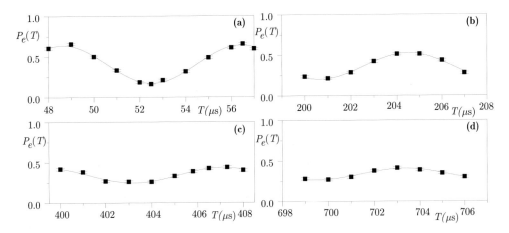

Fig. 5.37 Two-mode entanglement. Probability $P_e(T)$ of detecting A_2 in state $|e\rangle$ as a function of the time interval T. Figures (a), (b), (c) and (d) correspond to four different time windows in the $0 \rightarrow 710$ μs time interval. The dots are experimental and the curves are sine fits. Reprinted with permission from Rauschenbeutel *et al.* (2001). © American Physical Society.

coupled to M_a, undergoing a π-quantum Rabi pulse, which copies the state of M_a onto A_2. Cavity mode M_a ends up in $|0\rangle$ and factorizes out. Atom A_2 is then tuned to resonance with M_b, for a $\pi/2$-quantum Rabi pulse. The final $A_2 - M_b$ state is:

$$|\Psi_4\rangle = \tfrac{1}{2}\left[|g_2, 1_b\rangle(e^{i\delta T}e^{i\Phi} - 1) - |e_2, 0_b\rangle(1 + e^{i\delta T}e^{i\Phi})\right] , \qquad (5.53)$$

where Φ is the additional phase accumulated during the read-out sequence.

The probability $P_e(T)$ of finding A_2 in $|e\rangle$ is then:

$$P_e(T) = [1 + \cos(\delta T + \Phi)]/2 . \qquad (5.54)$$

It oscillates between zero and one, as a function of T, at the frequency $\delta/2\pi$. This oscillation reveals the coherence of the mode states superposition. Note the analogy with the physics of a Mach–Zehnder interferometer (see eqn. 3.128). The two atoms here play the roles of the two beam-splitters of the optical interferometer. In addition, the first is the source of the photon and the second an essential ingredient in its detection.

The experimental signals exhibiting $P_e(T)$ are presented in Fig. 5.37. The data sets (a), (b), (c) and (d) correspond to four different windows in the $0 \rightarrow 710$ μs time interval. The dots are experimental, with error bars reflecting the binomial detection statistics variance. The curves are sine fits oscillating at the beat note frequency $\delta/2\pi$ between M_a and M_b. The oscillation contrast decreases from (a) to (d) due to cavity damping.

5.4.6 Entanglement by non-resonant atom–cavity interaction

We have analysed so far entanglement procedures relying on two successive resonant atom–field interactions. A first vacuum Rabi oscillation creates an entangled atom–photon pair, which is subsequently transformed, via a second Rabi pulse, into either an atom–atom (5.4.4) or a field–field (5.4.5) EPR pair. This resonant method implies the conservation of the atom–field coherence during the interval between the two steps and thus requires very high cavity quality factors. We conclude this chapter by describing an alternative scheme in which the two successive resonant steps are replaced by a single non-resonant atom–field interaction. This procedure can be implemented with cavities having quality factors smaller than those required by the resonant methods, rendering it very promising for applications in quantum information.

A cavity-assisted collision between two circular atoms

The non-resonant atom–atom entanglement procedure presents a strong analogy with the collisional entanglement between two circular atoms qualitatively described in Section 5.2.1. The generation of an EPR atom–atom pair can be viewed as a collision between the two atoms assisted by a virtual photon exchange with the cavity mode (Kobayashi *et al.* 1995; Goldstein and Meystre 1997; Agarwal and Gupta 1998; Yang *et al.* 1999). Instead of colliding in free space, the two atoms, A_1 and A_2, initially in $|e_1, g_2\rangle$, simultaneously cross the cavity C and interact together with its field.

The detuning, Δ_c, between the $|e\rangle \to |g\rangle$ transition and the cavity mode is set to be larger than the vacuum Rabi frequency Ω_0.[12] No real energy exchange can take place between the atoms and the cavity in this dispersive limit. A second-order process is however possible, in which A_1 virtually emits a photon in the empty cavity, making a transition from $|e_1, g_2, 0\rangle$ to $|g_1, g_2, 1\rangle$ (the third symbol in the ket refers to the cavity state). This transition, which does not conserve energy, is immediately followed by the transition from $|g_1, g_2, 1\rangle$ to the final state $|g_1, e_2, 0\rangle$.

The initial and final states are degenerate and thus efficiently coupled. The free-space van der Waals interaction described in Section 5.2.1 results from similar transitions, in which the photon is virtually emitted in all the free space modes. Here, we are instead considering the interaction with one (or two) cavity modes. As we will see, the field confinement strongly enhances the exchange amplitude between the two atoms. More precisely, the cavity-assisted van der Waals interaction can be represented in the dispersive regime $\Delta_c \gg \Omega_0$ by an effective Hamiltonian (Zheng and Guo 2000):

$$H_{\text{eff}} = \frac{\Omega_0^2 f_1 f_2}{4\Delta_c} \left\{ \sum_{j=1}^{2} \left[|e_j\rangle \langle e_j| \left(1 + a^\dagger a\right) - |g_j\rangle \langle g_j| \, a^\dagger a \right] + \left(\sigma_+^1 \sigma_-^2 + \sigma_-^1 \sigma_+^2 \right) \right\} , \quad (5.55)$$

where f_1 and f_2 are the time-dependent mode amplitudes at the locations of A_1 and A_2 respectively and $\sigma_\pm^{1,2}$ are the atomic raising and lowering operators for A_1 and A_2. The first two terms on the right-hand side, involving the photon number operator $a^\dagger a$, describe atomic light and Lamb shifts produced by the cavity field (Section 3.4.4 and 5.3.2). The last two terms describe the atomic excitation transfer. When the cavity is initially in $|0\rangle$, the effective Hamiltonian is, in the $\{|e_1, g_2, 0\rangle ; |g_1, e_2, 0\rangle\}$ basis:

$$H_{\text{eff}} = \frac{\Omega_0^2 f_1 f_2}{4\Delta_c} \begin{pmatrix} 1 & 1 \\ 1 & 1 \end{pmatrix} . \quad (5.56)$$

---

[12]For the sake of simplicity, we consider here that the cavity sustains only one mode M_a and we will take into account the effect of M_b at a later stage.

The diagonal terms describe the common cavity-induced shift of the initial and final levels. They can be cancelled by a redefinition of the energy origin. The effective coupling between $|e_1, g_2\rangle$ and $|g_1, e_2\rangle$ is $\Omega_0^2 f_1 f_2 / 4\Delta_c$. We assume that A_1 and A_2 have different velocities v_1 and v_2, but reach the cavity axis simultaneously at time $t = 0$. Integrating the Gaussian mode amplitudes along the atomic trajectories, we get the collisional mixing angle:

$$\theta_{\text{cav}} = \frac{\Omega_0^2}{2} \frac{1}{\Delta_c} t_i^d \,, \tag{5.57}$$

where $t_i^d = \sqrt{\pi} w_0 / \sqrt{2} v_0$ is the effective dispersive interaction time for an atom crossing the mode waist w at the average velocity $v_0 = \sqrt{(v_1^2 + v_2^2)/2}$. Note that this expression of the mixing angle is exact in the dispersive limit, whereas eqn. (5.24) gave only an order of magnitude estimate. Combining eqns. (5.57) and (5.34) in which we express the fine structure constant α as $2\pi a_0 n^3 / \lambda$ or $a_0 n^3 \omega_c / c$ (λ is the wavelength of the cavity mode), we can cast θ_{cav} in a form reminding us of the free-space expression (5.24):

$$\theta_{\text{cav}} = 2\alpha \frac{\omega_c}{\Delta_c} \frac{c}{v_0} \left(\frac{a_0 n^2}{b_{\text{cav}}} \right)^2 \,, \tag{5.58}$$

where the effective 'impact parameter' b_{cav} is a simple function of the cavity mode geometry:

$$b_{\text{cav}} = \lambda \sqrt{\frac{k\lambda}{2 w_0 \sqrt{2\pi}}} \,. \tag{5.59}$$

This impact parameter, $b_{\text{cav}} = 0.81$ cm for our cavity ($k = 9$ and $\lambda \approx w$), is of the order of the field wavelength, much larger than the minimal distance $b \approx 0.7$ mm between the two 'colliding' atoms. The atomic beam is wide enough so that the probability of a close encounter adding a direct dipole–dipole interaction to the cavity mode effect is negligible.

The cavity-assisted mixing angle θ_{cav} is thus equal to the free-space mixing angle θ_c (eqn. 5.24) for two atoms colliding with velocity v_0 and an effective impact parameter b_{cav}, multiplied by the enhancement factor:

$$\eta = \omega_c / \Delta_c \,. \tag{5.60}$$

The detuning Δ_c, while being much larger than Ω_0, is extremely small compared to the mode frequency ω_c. The enhancement factor η can thus reach large values, in the 10^6 range. It more than compensates for the large value of the effective impact parameter. Under these conditions, a maximally entangled state can be produced, even though the two atoms never approach at a small mutual distance. The $A_1 - A_2$ entangled state produced by a cavity-assisted collision corresponding to the initial state $|e_1, g_2\rangle$ is similar to (5.22), in which θ_c is replaced by θ_{cav}. The probabilities P_{e_1, g_2} and P_{g_1, e_2} of finally finding the two atoms in $|e_1, g_2\rangle$ or $|g_1, e_2\rangle$ are, in the dispersive regime ($\Delta_c \gg \Omega_0$):

$$P_{e_1, g_2} = \cos^2(\theta_{\text{cav}}/2) \,; \qquad P_{g_1, e_2} = \sin^2(\theta_{\text{cav}}/2) \,. \tag{5.61}$$

Cavity damping perturbs marginally the system's evolution, since states in which there is one photon in the cavity are generated only virtually. More precisely, field

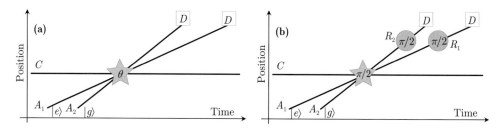

Fig. 5.38 Timing of the controlled collision experiment. (a) Test of population transfer. (b) Test of transverse correlations of the atomic EPR pair.

decay affects the intermediate state $|g_1, g_2, 1\rangle$. The (small) probability p_1 of finding the system in this state during the entanglement procedure is of the order of:

$$p_1 \approx \Omega_0^2/\Delta_c^2 . \tag{5.62}$$

The collision duration for the generation of a maximally entangled state ($\theta_{\text{cav}} = \pi/2$) is $t_i^d = \pi\Delta_c/\Omega_0^2$. The probability p_ℓ of a photon loss destroying the coherence is thus:

$$p_\ell = \kappa p_1 t_i^d \approx \pi\kappa/\Delta_c , \tag{5.63}$$

where κ is the cavity loss rate.

For a resonant atom–atom entanglement, a photon is really present in the cavity during the time interval required to perform a resonant atom–cavity state transfer, $t_i^r = \pi/\Omega_0$. The loss probability in the resonant regime, p_ℓ^r, is thus:

$$p_\ell^r \approx \pi\kappa/\Omega_0 . \tag{5.64}$$

The resonant loss rate thus exceeds the non-resonant one by a factor Δ_c/Ω_0, much larger than 1 when the dispersive regime condition is satisfied.

Remarkably, the cavity-induced collision is, to first order, insensitive to an initial field present in the cavity. A mere inspection of H_{eff} shows that it retains the matrix form given by eqn. (5.56) in the basis $\{|e_1, g_2, n\rangle ; |g_1, e_2, n\rangle\}$, corresponding to a situation in which the cavity initially contains a n-photon Fock state. The diagonal terms due to the cavity light shift effects are independent of n and can be removed by a redefinition of the atomic eigenenergies. The coupling between the levels is also independent of n. Thus all happens as if the cavity was empty. This result can be understood as a quantum interference effect. There are two paths connecting the initial state $|e_1, g_2, n\rangle$ to $|g_1, e_2, n\rangle$. The first corresponds, as in the $n = 0$ case, to a virtual photon emission towards $|g_1, g_2, n + 1\rangle$ followed by an immediate photon re-absorption. The effective coupling is $\Omega_0^2(n + 1)/4\Delta_c$, the $n + 1$ factor corresponding to the enhanced transition rate in an n-photon field (we assume here, for the sake of simplicity, that both atoms are motionless on the cavity axis, the generalization to moving atoms being straightforward). The other path (missing in the zero-photon case) involves a first virtual absorption by the second atom (transition to $|e_1, e_2, n - 1\rangle$), immediately followed by the emission of this photon. The transition amplitude is then $-\Omega_0^2 n/4\Delta_c$ (the detuning of the intermediate state is now $-\Delta_c$, hence the minus sign). The two

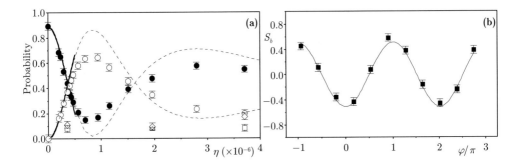

Fig. 5.39 Cavity-enhanced collision. (a) Joint detection probabilities versus the detuning parameter η. $P(e_1, g_2)$ and $P(g_1, e_2)$ (solid and open circles) oscillate in a symmetrical way, reflecting the atom–atom energy exchange enhanced by the cavity. The spurious channel probabilities, $P(e_1, e_2)$ and $P(g_1, g_2)$ (open squares and diamonds respectively), stay below 10%. The solid line represents the predictions of the simple model, in the $\eta < 5 \cdot 10^5$ range in which it applies. The dashed lines present the result of a numerical integration of the evolution equations. (b) 'Transverse' correlations: Bell signal $S_b = \langle \sigma_{X,1} \, \sigma_{\varphi,2} \rangle$ versus the relative phase φ in units of π. Reprinted with permission from Osnaghi *et al.* (2001). © American Physical Society.

transition amplitudes sum up to $\Omega_0^2/4\Delta_c$, a result independent of n.[13] The constraints on the cavity quality and temperature can thus be considerably relaxed when we use this cavity-induced collision scheme to generate atomic entangled states. This is very encouraging for complex entanglement manipulations.

Experimental realization

We have demonstrated this entanglement procedure with the circular atom CQED set-up (Osnaghi *et al.* 2001). The precise tuning of the atomic velocities is essential in this experiment, since both atoms must get close to each other inside the cavity and, yet, be separated when they reach the detector D by a distance large enough to allow for their independent detection. In order to achieve this, atom A_1 and A_2 are prepared, in this order, respectively in $|e\rangle$ and $|g\rangle$, inside box B (Fig. 5.15, page 252). Their velocities, $v_1 = 243$ m/s and $v_2 = 300$ m/s, are tuned so that A_2 catches up A_1 on cavity axis, overtakes it and is finally detected first in D. The timing of this experiment is sketched in Fig. 5.38, where we use a time-position representation instead of the quantum circuit picture. The two atomic space–time lines with different slopes cross with the horizontal cavity line at the point which defines the cavity-assisted collision event. The two atomic detections occur afterwards at the same point in space, but at two different times.

In a first experiment, we have observed the Rabi evolution between $|e_1, g_2\rangle$ and $|g_1, e_2\rangle$. We have measured, as a function of Δ_c and, hence, as a function of the

---

[13]This cancellation, making the non-resonant entanglement procedure insensitive to thermal photons in the cavity, is reminiscent of a similar effect occurring in a temperature-independent entangling scheme for ions in a trap (Mølmer and Sörensen 1999; Solano *et al.* 1999; Sackett *et al.* 2000), see Section 8.5.

enhancement factor η, the probabilities $P(e_1, g_2)$ and $P(g_1, e_2)$ of finding the two atoms in the expected detection channels and the probabilities $P(e_1, e_2)$ and $P(g_1, g_2)$ of finding them in spurious configurations. In this experiment, two detuned cavity modes M_a and M_b contribute to the cavity-assisted collision. These modes, with orthogonal linear polarizations, are split in frequency by $\delta/2\pi = 128$ kHz. Their effects add in the dispersive limit, and the enhancement factor becomes:

$$\eta = \frac{\omega_c}{\Delta_c} + \frac{\omega_c}{\Delta_c + \delta} \ . \tag{5.65}$$

The results are presented in Fig. 5.39(a). The initial and final state populations undergo a Rabi oscillation, as expected, while the spurious channel populations, produced by imperfections, remain below 10%. The observed oscillation agrees well with the simple model presented above (and plotted as solid lines in Fig. 5.39a) for $\Delta_c > 3\Omega_0$ i.e. $\eta < 5 \cdot 10^5$. The cavity-enhanced collision model is thus valid even when the dispersive regime condition $\Delta_c \gg \Omega_0$ is not strictly fulfilled. The maximally entangled state ($\theta_{\text{cav}} = \pi/2$) is reached in this regime, for an enhancement parameter $\eta \approx 4 \cdot 10^5$.

For smaller detunings, real photon emission and absorption become significant, reducing the contrast of the Rabi oscillations. The dashed lines in Fig. 5.39(a) present the results of a numerical simulation of the collision process. It takes into account the exact atom–field interaction and the residual thermal field. It is in fair agreement with the observed values. The fidelity of the atomic entanglement is noticeably reduced in this regime.

In a second experiment, we have checked the entanglement of the atomic state prepared at $\theta_{\text{cav}} = \pi/2$ by measuring, as in the EPR pair case (see Section 5.4.4), the Bell signal $S_b = \langle \sigma_{X,1}\sigma_{\varphi,2} \rangle$ versus the relative phase φ of the two detection axes in the horizontal plane of the Bloch sphere. This signal oscillates, in an ideal experiment, between -1 ($\varphi = 0$) and $+1$ ($\varphi = \pi$).

The experimental sequence is depicted in Fig. 5.38(b). The two atoms undergo, after their mutual interaction with the cavity, two independent $\pi/2$-Ramsey pulses with a relative phase φ between them. The final detection probabilities are recorded as a function of φ and the Bell signal computed from these data. The results are presented in Fig. 5.39(b). The Bell signal oscillations are clearly observed, albeit with a reduced contrast, of the order of 50%. This contrast reduction is mainly due to the imperfections of the analysing Ramsey pulses (the two atoms being separated by a short distance at the time of the pulses, it is difficult to avoid cross-talk effects). In spite of these imperfections, the existence of these oscillations reveals the coherent nature of the cavity-assisted collision.

In this experiment, quite remarkably, two atoms strongly interact at a distance of the order of a millimetre, about four orders of magnitude larger than the radius of their electron orbits. This distance could be increased still much further, the atoms being, for instance, sent across two distinct antinodes of the cavity.

Not only does the cavity enhance the coupling between the atoms, but it makes it easily tuneable, by ensuring a complete control of the collisional mixing angle θ_{cav}. This angle can be adjusted by properly setting the atomic velocities and the atom–cavity detuning. In free space, a precise tuning of θ_c would imply a complete control of atomic trajectories with submicrometre precision, which is in practice impossible to achieve. Without the help of a cavity, Rydberg–Rydberg collisions in an atomic beam would produce an incoherent mixture of the $|e_1, g_2\rangle$ and $|g_1, e_2\rangle$ states.[14] It is the precise control of the interaction parameters through CQED effects which makes the collision between the circular atoms fully coherent, an essential feature for possible quantum information applications. In this respect, cavity-assisted collisions are similar to the coherent atomic interactions tailored in cold atom or in ion traps, described in Chapters 8 and 9.

Through the diverse experiments described in this chapter, we have seen that atoms and fields can be manipulated in a photon box, realizing qubits and engineering entanglement between them via resonant or dispersive interactions. In these experiments, circular atoms play several roles. They act either as a source, a beam-splitter or a detector for the quantum field. Now that we have become acquainted with all the tricks the atoms and photons can perform, the stage is ready for more complex and subtler action.

---

[14]Note however that Rydberg–Rydberg collisions have been proposed to realize fast quantum gates between atoms localized at well-defined positions (Jaksh *et al.* 2000; Protsenko *et al.* 2002). Two laser-cooled atoms, trapped in their ground state in nearby sites, would carry two qubits coded in their hyperfine structure. A transient excitation to the Rydberg levels would produce, through the van der Waals interaction, an entangled two-atom state. The entanglement would then be transferred via photon emission back to the ground hyperfine states.

6
Seeing light in subtle ways

In the beginning, there was noting. God said: "Let there be light". And there was light. There was still nothing, but you could see it a whole lot better.

Ellen DeGeneres.

The photon box which Einstein and Bohr imagined at the Solvay meetings was supposed to store a photon and to release it on demand. Its realization in wood by Gamow shown in Fig. 1.7, on page 14, was attached to a spring used as a weighing scale and involved a mechanical shutter triggered by a clock to let the photon go at a preset time. Gamow jokingly stuck on it a patent stamp, signed A. E. and N. B., to acknowledge Einstein's and Bohr's intellectual property. The photon box described in the previous chapter is not made of wood, but of superconducting niobium. Microwave quanta stored in it are detected by their action not on a spring, but on single atoms crossing the box. In spite of these 'technical' differences, thought and real experiments pursue a similar goal, which is to catch quantum concepts 'in action' and hence understand quantum physics in a deeper way. In this way, the physicists working on present-day CQED set-ups are indebted to Einstein and Bohr, even if they do not have to pay royalties to their famous precursors.

In this chapter, we acknowledge this debt by describing, among others, two CQED studies guided by the spirit of thought experiments. The first one is a direct implementation of the recoiling slit interferometer imagined by Bohr. We have alluded to this device in Chapter 1. It was designed to illustrate the complementarity between the contrast of the fringes in an interferometer and the amount of information available about the path followed by the interfering particle. Bohr proposed to use a Young interferometer with a light moving slit (Fig. 1.6, on page 13), whose recoil was storing 'which-path information' about the particle. We discuss in Section 6.1 a closely related experiment (Bertet *et al.* 2001), in which a Ramsey interferometer is used instead of the unrealistic Young's one. The microscopic recoiling slit is replaced by a weak quantum field made of a few photons, which realizes one of the Ramsey pulses. This experiment provides a direct illustration of complementarity and emphasizes its relation to entanglement and decoherence.

The second experiment, described in Section 6.2, realizes another dream of early quantum mechanics by achieving a non-destructive detection of a single photon. Is it possible to 'see' a single photon without destroying it? Can one see again and again the same light quantum? In usual life, the answer to these questions is a definite 'no'. In standard photodetection schemes, the light is irreversibly absorbed by the detector, whether it is a photocathode or our retina. This destructive process, transforming photons into electric currents or chemical signals, seems so natural that it is often assumed to be a necessary condition of vision, illustrating what could be the proverb: '*you cannot have a photon and see it too*'. The demolition of light quanta, though, is in no way required by quantum laws and nothing prevents a photodetector from being fully transparent. This has been implemented in various quantum optics situations for fields involving macroscopic photon numbers, under the name of Quantum Non-Demolition (QND) light detection (Grangier *et al.* 1998). We show in Section 6.2 that a circular Rydberg atom is such a sensitive probe that it makes it possible to push these schemes to the quantum level and to detect the presence of a single photon in a cavity, while leaving it to be 'seen' again by another atom at a later time (Brune *et al.* 1990; Nogues *et al.* 1999).

We then remark, in Section 6.3, that the QND measurement of a single photon by a Rydberg atom corresponds to the operation of a control-not quantum gate in which the photon is the control bit and the atom the target. This universal gate can be used for general manipulations of atom–cavity entanglement. We show that multiple qubits can be entangled in a step-by-step process and we describe an experiment preparing three qubits in a maximally entangled state (Rauschenbeutel *et al.* 2000).

The atom–field manipulations described up to this point involve a resonant interaction and are restricted to single photons. We discuss in Section 6.4 their extension to larger fields, based on the dispersive atom–cavity interaction. A series of atoms crossing the cavity successively can be used to read-out the field intensity in a non-destructive way, realizing an optimal analogue/digital converter operating as a non-destructive photon counter (Haroche *et al.* 1992b). The first atom measures the photon number parity (last bit in the photon number). The second atom determines the next-to-last bit and so on. By repeating the measurement sequence many times on a large number of realizations of the same field, one reconstructs statistically the initial photon number distribution. We compare this optimal method, which requires the precise control of the successive atoms velocities, with a less demanding scheme which does not rely on atomic velocity selection, but requires a somewhat larger number of atoms.

The photon number distribution contains important statistical information about a quantum field, but this information is not complete, since it leaves undetermined the coherence between Fock states. Remarkably, there is an alternative statistical method, also based on the atom–field dispersive interaction, which allows us to extract all quantum information contained in the cavity field. The method, described in the final section (6.5), relies on the preliminary translation of the field in phase space by controlled complex amplitudes. The cavity field's Wigner function, whose knowledge is equivalent to that of the field density operator, is then obtained from an average of measurements of the photon number parity, performed by a single atom on the translated field (Bertet *et al.* 2002). We conclude the chapter by describing an experiment in which this method has been applied to the vacuum and to a single-photon Fock state.

(a)　　　　　　　　　　　　　(b)

(c)　　　　　　　　　　　　　(d)

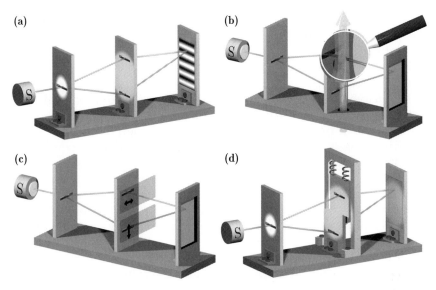

Fig. 6.1 Four *gedanken* experiments with a Young double-slit device. (a) A standard Young device, paradigm of a quantum interference experiment. (b), (c) and (d) Three different which-path arrangements. (b) A light beam illuminates the particle, which can be observed with a magnifying glass when crossing the upper slit. (c) Internal state labelling: the polarization of the particle is set to two orthogonal states in the two interfering paths. (d) A moving slit is set in motion when the particle takes the upper path.

6.1 Complementarity at quantum–classical boundary

6.1.1 A review of which-path experiments

A particle crossing a Young double-slit apparatus exhibits contradictory classical features. It behaves in some respect as a discrete entity, registering on a detector screen as a sharp spot. It has, at the same time, the property of a wave giving rise to constructive or destructive interference. Quantum physics reconciles these two aspects by postulating that the wave represents an amplitude whose squared modulus determines the probability that the particle will materialize at a given location if a measurement of its position is performed.

This quantum answer to the 'wave–particle riddle' refuses to give a meaning – what Einstein called an element of reality – to the position of a particle before it is actually measured or at least measurable. To a classical physicist asking the question '*through which slit did the particle pass?*', Bohr and his followers of the Copenhagen school typically answer '*your question has no meaning until you actually perform the measurement*'. And they add, '*if you do perform it, you will have to modify the set-up in such a way that the interference will disappear*'. The wave nature of the particle and the knowledge of its trajectory are two complementary properties, which cannot be determined together because they require different, incompatible experimental arrangements.

Figure 6.1 recalls four possible versions of the Young double-slit experiment, already mentioned in Chapter 2, in which the particle crossing the set-up exhibits either a wave nature (Fig. 6.1a) or a particle-like behaviour (Fig. 6.1b,c,d). In the first case, the set-up is a genuine interferometer leaving the path of the particle undetermined. Interference fringes with a full contrast are observed. In the three other situations, an element in the set-up allows the experimenter to gather 'which-path' information and the interference vanishes.

This information can be obtained by coupling the particle to an external probe such as a light beam off which the particle scatters a photon (Fig. 6.1b). Path information can also be acquired from an internal probe such as the polarization state of the particle, by inserting orthogonal polarizers in the two arms of the interferometer (Fig. 6.1c). Finally, as shown in Fig. 6.1(d), the interferometer itself may conspire to give away the particle's trajectory. In this situation, which Bohr explicitly considered, one of the slits is mounted on a movable stage and recoils when the particle is scattered off its edges.

As has been discussed in Chapter 2, complementarity and entanglement are intimately related. A which-path detector, whatever its nature, is a probe which stores information on the particle trajectory as it gets entangled with the particle state. When this information is unambiguous, the entanglement is maximum and the particle wave function is replaced by an incoherent mixture of states, resulting in a total suppression of interference. Partial information corresponds to a lesser degree of entanglement between the particle and the probe and to a partial coherence, with interference fringes exhibiting a limited contrast.

An interferometer with a which-path probe is, according to this analysis, similar to an EPR set-up in which the partner subsystems are the particle and the path detector. We have analysed in Section 2.4.7 this EPR game, in which Alice is trying to record fringes on the particle while Bob is performing measurements on the which-path detector. We have noticed that if Bob decides to detect an observable whose eigenstates are superpositions of the which-path states correlated to the particle's position (a 'transverse observable' with respect to these states), Alice will find her particle in a superposition of trajectories going through the two arms of the interferometer. Interference will then be restored, with a phase depending upon the outcome of Bob's measurement on the which-path probe. By deciding to measure an observable which does not commute with the one providing information on the particle's trajectory, Bob has voluntarily erased this information and re-established the fringes seen by Alice. This quantum eraser procedure works only if the measurement of the which-path detector in the transverse basis is actually performed. Alice observes fringes provided she *correlates* her results to a specific outcome of Bob's measurement. The quantum erasure, based on the properties of entanglement, is thus a conditional procedure, which requires a definite act of measurement by Bob.

Note that unconditional quantum eraser situations may also be imagined, in which a second interaction of the interfering particle with the which-path detector suppresses information stored during the first interaction. Such situations basically amount to cancelling the effective coupling of the particle with the detector. We describe below quantum eraser experiments of both conditional and unconditional kinds.

As analysed in Chapter 4, the which-path detector can also be a large environment, a thermodynamic bath, coupled to the particle during the crossing of the interferometer and inducing decoherence. Consider for instance an atomic Young double-slit experiment, performed with laser-cooled atoms (Shimizu *et al.* 1992). The environment is made up of the background gas in the vacuum chamber. A single collision between a cold atom and a background particle changes considerably their motional states. The environment then evolves into orthogonal states 'recording' which-path information. This information will never be accessible to any observer, being deeply buried in the thermodynamic complexity of the gas. The mere fact that it has been recorded is sufficient, though, to suppress the interference. There is then no way to play the quantum eraser game.

The connexions between complementarity, entanglement and decoherence have remained, for a long time after the advent of quantum mechanics, a subject for theoretical discussions and propositions of *gedanken* experiments (Schwinger *et al.* 1988; Scully and Walther 1989; Scully *et al.* 1991; Englert *et al.* 1992; 1994; 1995; Storey *et al.* 1995). This state of affairs has considerably evolved in the 1990s, due to technical progresses making these experiments feasible. Many demonstrations of complementarity have been realized, as well as various versions of the quantum eraser. Let us review them briefly while following loosely the classification suggested by Fig. 6.1.

A first kind of experiments realizes situations analogous to that depicted in Fig. 6.1(b). They involve either an interferometer with material gratings separating and recombining atomic paths (Chapman *et al.* 1995; Kokorowski *et al.* 2001), or an an optical standing wave which diffracts coherently an atomic beam (Pfau *et al.* 1994). In both cases, which-path information on the atomic position is provided by spontaneous emission. Each scattered photon localizes the emitting atom with a precision of the order of its wavelength and interference is suppressed when this wavelength is small enough. Variants of this experiment have also been realized, by correlating the scattered photon with the detection of the particle. As predicted by the discussion of Section 2.2, fringes are recovered in these conditional measurements (Chapman *et al.* 1995). When which-path information cannot be provided by a single photon, it can still be obtained through the scattering of several photons, each of them carrying partial information (Kokorowski *et al.* 2001; Hackermüller *et al.* 2003; 2004).

Complementarity and the effect of which-path detection have also been explored on photon interference. Let us mention ingenious experiments performed on photon pairs (Herzog *et al.* 1995; Kim *et al.* 2000), which directly play on the concept of entanglement. A measurement on one member of the pair leads to which-path information about the other. The entanglement between the two photons, one playing the role of which-path detector for the other, spoils the interference visibility. This situation is particularly well-adapted to the realization of quantum eraser situations. Several demonstrations of the erasure procedure have been made with this EPR-like system (Kwiat *et al.* 1992; Kim *et al.* 2000).

Another clever illustration of complementarity with an external which-path detector has been made in a solid-state electron interferometer (Buks *et al.* 1998). An electron gas is trapped in a two-dimensional layer inside a semiconducting structure (Imry 1997). The very high purity of the sample and its low temperature make scattering on impurities or phonons negligible. The electrons thus propagate as free quantum particles. The electron gas is confined on the sides by repelling electric fields produced by tailored electrodes. Two paths leading from a source contact to a drain, are left opened, giving rise to an interference effect in the device conductance. The relative phase of the two paths is tuned with the help of an external magnetic field, through the Aharonov–Bohm effect (Aharonov and Bohm 1959). An oscillation of the system's conductance is observed when this phase is swept. Close to one of the paths, a 'single-electron transistor' is used as a sensitive electrometer that detects the transit of the interfering particle. When this detector is switched on, the fringes contrast is reduced, albeit not cancelled due to a limited sensitivity of the which-path detector.

The situation depicted in Fig. 6.1(c), is easily realized. It is performed routinely with ordinary light beams, interferometers and polarizer sheets in undergraduate laboratory classes around the world. More refined versions are nevertheless worth mentioning. The labelling of the paths can be performed by acting on the spin of the particle in a neutron interferometer (Rauch *et al.* 1975; Badurek *et al.* 1986; Rauch and Werner 2000) or on the internal hyperfine state of an atom (Dürr *et al.* 1998a;b). In the latter case (Dürr *et al.* 1998a), the quantitative relations between the visibility of the fringes and the distinguishability of the paths (Jaeger *et al.* 1993; Englert 1996) have been checked.

A complementarity experiment in which the which-path element is part of the interferometer has also been performed (Eichmann *et al.* 1993), illustrating the situation shown in Fig. 6.1(d). The interferometer is made up of two ions held in a Paul trap and illuminated by a laser beam. The light scattered by these two microscopic 'slits' displays an interference pattern. Depending upon the laser beam polarization, the scattering of a photon may or may not change the internal state of the scattering atom. If this is not the case, no which-path information is available and fringes are observed, with a maximal contrast. When the atomic state change does occur, which-path information is stored in the scattering atom, and the fringes disappear.

This experiment is spectacular because it involves a truly microscopic system scattering light *à la* Young from two closely spaced locations. Its principle, though, has been known for a long time in neutron or photon scattering experiments. For instance, neutrons scattered from a crystal give rise to a coherent Bragg interference pattern if the nuclei in the crystal remain unperturbed in the process. If, on the contrary, the scattering flips a nuclear spin, the Bragg pattern vanishes and the scattering becomes 'incoherent'. This difference between elastic and inelastic scattering is again a manifestation of complementarity. We will analyse a similar situation in Chapter 9.

An important element in the *gedanken* set-up of Fig. 6.1(d) is still missing in the experiments described so far. With a moving slit of adjustable mass it is possible, in principle, to go continuously from a 'good' to a 'bad' which-path detector. By increasing the slit mass, the set-up can be made progressively less and less sensitive to the momentum transfer and the interferometer can continuously evolve from being quantum to genuinely classical. This aspect was particularly important to Bohr, who stressed that a good apparatus – here a *bona fide* interferometer – had to be classical. We describe below a CQED experiment which implements the *gedanken* situation that Bohr had in mind, with an interferometer gradually crossing the quantum–classical boundary.

6.1.2 A quantum optics version of Bohr's experiment

Before coming to the actual experiment, let us analyse in more depth the moving-slit device. Instead of the original Bohr arrangement, we consider a variant, based on a

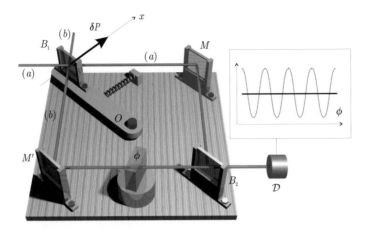

Fig. 6.2 A 'modern' version of Bohr's proposal. A Mach–Zehnder interferometer with a moving beam-splitter.

Mach–Zehnder interferometer. This change makes the geometry and the discussion somewhat simpler. In addition, the Mach–Zehnder configuration presents a strong analogy with the actual CQED experiment based on Ramsey interferometry. In this new *gedanken* experiment, sketched in Fig. 6.2, two beam-splitters B_1 and B_2 and the folding mirrors M and M' define two interfering paths (a) and (b) enclosing a square. A dephasing element on path (b) is used to scan the interference fringes, recorded by a detector $\mathcal{D}$. The role of the moving slit is played by the first beam-splitter B_1. It is bolted on a frame rotating around the geometrical centre O of the interferometer. A spring attached to this frame provides a restoring force when B_1 is moved away from equilibrium. For small displacements around this equilibrium, the moving beam-splitter behaves as a one-dimensional harmonic oscillator along the axis x (position X, momentum P), with an effective mass m and a resonant angular frequency ω. We assume that it is initially in its ground state $|0\rangle$. When the interfering particle, with momentum p, is reflected by B_1, the beam-splitter receives a momentum kick $\delta P = \sqrt{2}p$ along x. On the contrary, the state of the beam-splitter is unchanged when the particle is transmitted. The final motion of B_1 thus contains which-path information, making obvious the connexion with the Young design's moving slit.

Let us first analyse qualitatively the fringes visibility in terms of Heisenberg uncertainty relations. The beam-splitter is a *bona fide* which-path detector when the initial fluctuation of its momentum, ΔP, is much smaller than the momentum change imparted by the particle, δP:

$$\Delta P = \sqrt{m\omega\hbar/2} \ll \delta P = \sqrt{2}p . \tag{6.1}$$

Since B_1 is initially in its ground state, the position and momentum uncertainties, ΔX and ΔP, satisfy the Heisenberg equality $\Delta X \Delta P = \hbar/2$. The condition to get unambiguous which-path information is thus:

$$\Delta X \gg \frac{\hbar}{2\delta P} = \frac{\lambda}{4\pi\sqrt{2}} \approx \frac{\lambda}{20} , \tag{6.2}$$

where $\lambda = 2\pi\hbar/p$ is the de Broglie wavelength of the interfering particle. Typically, (6.2) is satisfied when ΔX is of the order of λ. This position uncertainty of the beam-splitter results in a similar fuzziness for the path difference, corresponding to an interferometer phase fluctuation of the order of 2π. As expected, the fringes are blurred as soon as B_1 is able to keep a record of the particle's path.

Note that (6.1) can also be written as $m \ll 4p^2/\hbar\omega$, which means that the mass of the beam-splitter has to be very small. Equivalently:

$$\frac{p^2}{2m} \gg \tfrac{1}{8}\hbar\omega \, , \tag{6.3}$$

which implies, not surprisingly, that the final beam-splitter kinetic energy should contain at least a few vibration quanta for its final state to be orthogonal to the initial ground state. We will now retrieve this condition by analysing this experiment in terms of entanglement.

The particle, when reflected into path (b), imparts to B_1 a momentum kick equivalent to a translation in phase space. The beam-splitter's state is a translated ground state, i.e. a coherent state $|\alpha\rangle$ with an imaginary amplitude:

$$\alpha = ip/\sqrt{m\omega\hbar} \, . \tag{6.4}$$

The interaction of the particle with the beam-splitter produces a combined state, a superposition of two contributions corresponding respectively to the transmission and the reflection of the particle by B_1:

$$|\Psi\rangle = \left(|\Psi_a\rangle |0\rangle + e^{i\phi} |\Psi_b\rangle |\alpha\rangle \right)/\sqrt{2} \, , \tag{6.5}$$

where $|\Psi_a\rangle$ and $|\Psi_b\rangle$ are the particle's states in the two paths. We have included the effect of the phase-shifter on the partial wave in (b). We recognize in eqn. (6.5) a particle–beam-splitter state of the EPR type, with a maximum entanglement when $\langle 0| \alpha\rangle \sim 0$, i.e. as soon as the kicked beam-splitter stores a few quanta. The probability of detecting the particle in the detector $\mathcal{D}$ located at $\mathbf{r}$ is:

$$|\langle\mathbf{r}|\Psi\rangle|^2 = \tfrac{1}{2}\left[|\langle\mathbf{r}|\Psi_a\rangle|^2 + |\langle\mathbf{r}|\Psi_b\rangle|^2 + 2\mathrm{Re}\left(e^{i\phi}\langle\Psi_a|\mathbf{r}\rangle\langle\mathbf{r}|\Psi_b\rangle\langle 0|\alpha\rangle\right)\right] \, . \tag{6.6}$$

The last term in the right-hand side of this equation is the modulated Mach–Zehnder interference signal. Its contrast, $\mathcal{C}$, is equal to the modulus of the overlap of the two final states of B_1:

$$\mathcal{C} = |\langle 0 | \alpha\rangle| = e^{-|\alpha|^2/2} = e^{-p^2/2m\hbar\omega} \, . \tag{6.7}$$

We find on this simple model the general features of an ideal complementarity experiment in which the interferometer is evolving from quantum to classical. By changing the oscillator's effective mass m while keeping ω constant, we can span a continuous transition from the microscopic to the macroscopic regime. For small masses, $|\langle 0 | \alpha\rangle|$ is exponentially small. We get complete which-path information, a fully entangled particle–beam-splitter state and no fringes. For large masses, α is negligibly small and $|\langle 0 | \alpha\rangle| \approx 1$. There is no entanglement; we get perfect fringe visibility but, obviously, no which-path information.

Standard interferometers, built with sturdy mirror mounts, are large-mass devices. It is certainly not quantum entanglement that is the limiting factor for the fringe visibility! These interferometers fulfil their task perfectly, which is to measure the relative phase of paths (*a*) and (*b*). If they were made of microscopic objects, they would fail in this task, getting entangled with the interfering particle. This is a clear illustration of the basic principle of the Copenhagen interpretation which we have recalled above. As has been often stated by Bohr, a measuring device should be a classical object, made of macroscopic parts.

The Mach–Zehnder with a moving beam-splitter is simpler to analyse than Bohr's moving slit set-up. It is however still a theorist's dream. It relies on the coupling between a microscopic object and an optomechanical device which should be sensitive to the momentum transfer from a single quantum particle. Even if great progress is being made in this direction (Cohadon *et al.* 1999), for instance by developing tiny mirrors mounted on cantilevers (Kleckner *et al.* 2006), the technology for a Mach–Zehnder which-path experiment of the kind analysed above is not yet ready. There is however an equivalent interferometer, realizable with present microwave CQED technology, which turns out to be an ideal tool to perform quite a similar experiment.

6.1.3 CQED implementation of Bohr's thought experiment

As pointed out in Section 3.3.3, the Ramsey interferometer is a close cousin of the Mach–Zehnder device. Instead of splitting the particle trajectory in real space, the Ramsey set-up performs a separation of an atom's path in the Hilbert space of its internal energy states (see in particular Fig. 3.16, on page 149, which emphasizes the topological analogy between the two types of interferometers). An atom, initially in $|e\rangle$, undergoes two $\pi/2$-Rabi pulses, R_1 and R_2, induced by a classical source nearly resonant with the $|e\rangle \rightarrow |g\rangle$ transition. The final probability amplitude of detecting the atom in $|g\rangle$ is a sum of two contributions, corresponding to interfering paths which differ by the pulse in which the $|e\rangle \rightarrow |g\rangle$ transition has occurred. The pulses R_1 and R_2 simply play the role of the beam-splitters in the Mach–Zehnder arrangement. The which-path information issue arises naturally in this context. The atomic transition from $|e\rangle$ to $|g\rangle$ stimulates the emission of one additional photon in the field which induces it. This photon plays the role of the momentum transfer between the particle and the beam-splitter in the Mach–Zehnder design. The question is whether this additional photon can provide information about the path in the interferometer.

In all standard Ramsey experiments, the classical pulses are induced by coherent states with a very large photon number. Adding a single photon to this huge number does not change the state appreciably. The modified field state has a near-unity overlap with the initial one. There is then negligible which-path information transferred to the field, and fringes can be observed with a good contrast. This is certainly good news for atomic clocks!

A more detailed analysis is required in the context of CQED experiments. The Ramsey pulses are then induced in a mode whose spatial extension is comparable with that of the cavity Gaussian mode. The pulse duration, a few microseconds, is of the order of the vacuum Rabi period. The number of photons in the Ramsey pulse mode is thus of the order of unity. Is it then legitimate to consider such a field pulse as classical?

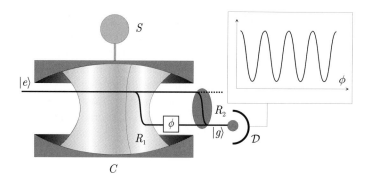

Fig. 6.3 Scheme of the CQED complementarity experiment. The 'atomic path' in the Ramsey interferometer is illustrated in a mixed 'position–quantum-state' representation. Time evolves along the horizontal direction, from left to right. The upper and lower lines symbolize respectively $|e\rangle$ and $|g\rangle$. The first Ramsey pulse is produced by a mesoscopic coherent field injected by the source S in the cavity C, just before the atom exits the mode. The timing of this pulse is determined by Stark switching. The region of space where the cavity mode is resonant with the atom is highlighted. The second pulse R_2 is classical and applied to the atom after it has left C. The atomic populations in $|e\rangle$ and $|g\rangle$ are finally probed by the detector $\mathcal{D}$. An electric field pulse is applied to the atom between C and R_2, producing by the Stark effect a tuneable differential phase shift ϕ between the two paths amplitudes. The probability $P_e(\phi)$ of finally detecting the atom in $|e\rangle$ exhibits fringes as a function of ϕ (inset). Reprinted by permission from Macmillan Publishers Ltd: Nature, Bertet *et al.* (2001).

The answer is fortunately yes (had not it been the case, Ramsey fringes would have never been observed in these experiments). The lifetime of a photon in the Ramsey mode is extremely short, in the subnanosecond range, since this mode is not sustained by a high-Q cavity. The 'single photon' interacting with the atom is frequently renewed by the classical source counteracting the large losses. Altogether, the Ramsey pulse involves a stream of many photons. That one photon is added by the atom in this huge flux cannot be detected. A more detailed and quantitative discussion can be found in Kim *et al.* (1999).

This analysis tells us that a Ramsey field can carry which-path information if it contains a small photon number and if these photons are stored in a high-Q cavity. Only then will path information be unambiguous and long-lived enough to make the initial and final states of the field distinguishable. Such a 'mesoscopic' field will be able to play the role of the moving beam-splitter in the Mach–Zehnder device. This remark immediately suggests the scheme represented in Fig. 6.3, a modified version of the CQED Ramsey set-up described in Chapter 5.

In this new design, a circular Rydberg atom, initially in $|e\rangle$, interacts first with a coherent field $|\alpha\rangle$ stored in the cavity mode. This interaction realizes the first Ramsey pulse R_1.[1] The coherent field is prepared in C before the beginning of the experimental sequence with the help of the classical source S. The average photon number $\bar{n} = |\alpha|^2$ can be tuned from zero to a few tens by means of calibrated attenuators. When the

---

[1] The low-Q microwave zone upstream the cavity is not used in this experiment.

atom enters C, it is tuned far off-resonance by a static field applied across the mirrors. The atom is set at resonance at a time such that a $\pi/2$-Ramsey pulse is exactly realized when the atom exits the mode. This condition can be fulfilled even when $\alpha = 0$ (the pulse then corresponds to the vacuum $\pi/2$-Rabi oscillation described in Section 5.4.2).

The atomic coherence created by the mesoscopic field in C is probed by a second, ordinary Ramsey pulse produced in R_2, after the atom has left C. Between C and R_2, a short pulse of electric field transiently shifts the frequency of the $|e\rangle \rightarrow |g\rangle$ transition, producing a phase shift of the atomic coherence and, hence, of the phase ϕ of the Ramsey interferometer. By averaging many realizations of the experiment, the probability P_g of finding finally the atom in $|g\rangle$ is reconstructed as a function of ϕ, for various value of the coherent state amplitude α.

Qualitatively, we expect high-contrast interference when α is large, the cavity field being then classical. On the contrary, when α is very small (ultimately zero), the atomic emission considerably modifies the field state and leaves which-path information, thus cancelling the fringes. For a more quantitative analysis, let us write the atom–cavity state just after the R_1 pulse. Calling t_i the effective atom–field interaction time corresponding to this pulse[2] and c_n the coefficients of the expansion of $|\alpha\rangle$ onto the Fock states basis, we get (eqn. 3.181, on page 156):

$$|\Psi(t_i)\rangle = \sum_n c_n \cos\frac{\Omega_0\sqrt{n+1}t_i}{2}|e,n\rangle + c_n \sin\frac{\Omega_0\sqrt{n+1}t_i}{2}|g,n+1\rangle . \quad (6.8)$$

The interaction time t_i is adjusted so that the probability P_g of finding the atom in $|g\rangle$ immediately after C is $1/2$:

$$P_g(t_i) = \sum_n |c_n|^2 \frac{1-\cos\Omega_0\sqrt{n+1}t_i}{2} = \frac{1}{2} . \quad (6.9)$$

The smallest time t_i fulfilling this condition ensures the $\pi/2$-pulse condition. After a simple rewriting of eqn. (6.8), the atom–cavity state at time t_i can be developed as:

$$\Psi(t_i) = (|e\rangle|\alpha_e\rangle + |g\rangle|\alpha_g\rangle)/\sqrt{2} , \quad (6.10)$$

where:

$$|\alpha_e\rangle = \sqrt{2}\sum_n c_n \cos\frac{\Omega_0\sqrt{n+1}t_i}{2}|n\rangle ; \quad (6.11)$$

$$|\alpha_g\rangle = \sqrt{2}\sum_n c_n \sin\frac{\Omega_0\sqrt{n+1}t_i}{2}|n+1\rangle . \quad (6.12)$$

The field states correlated to $|e\rangle$ and $|g\rangle$ are no longer coherent. Their photon number distribution has been modified by the atom–cavity interaction. That $|\alpha_g\rangle$ is not equal to $|\alpha\rangle$ is not surprising since the atom has emitted an additional photon in the

[2]The time t_i is defined as the effective time which would provide the same Rabi rotation angle if the atom–field coupling were constant, equal to Ω_0. It is different from t_i^r (eqn. 5.36), since the atom interacts resonantly with the mode for only a fraction of the atom–cavity crossing time.

cavity [c_n is now associated to $|n+1\rangle$ in (6.12)]. The modification of $|\alpha\rangle$ when the atom remains in $|e\rangle$ is less intuitive, since no energy has been added to the field. We have here another example of a state modification resulting from a 'negative measurement' (see Section 4.4.4 for a discussion of a similar situation in a different context). Measuring the atom's state and finding it in $|e\rangle$ at t_i has in general an effect on the field state, which can be understood by a classical argument involving implicitly Bayes law on conditional probabilities. We start by noticing that, if the cavity contained any number n of photons such that $\Omega_0\sqrt{n+1}t_i \approx \pi$, the atom would undergo a near π-Rabi pulse making it leave $|e\rangle$ and end up with a probability close to 1 in $|g\rangle$. *A contrario*, finding the atom in $|e\rangle$ is information which reduces the probability that the field contains a number n of photons which makes this measurement outcome unlikely. We understand in this way the cosine modulation in eqn. (6.11), which suppresses in the photon number distribution of α_e the photon numbers corresponding to a π-Rabi pulse condition.[3]

Equation (6.10) describes in general a partially entangled atom–cavity state. From the analogy with the Mach–Zehnder device (compare eqns. 6.5 and 6.10), we infer that the Ramsey fringe contrast $\mathcal{C}$ is equal to the overlap between the two final field states in C:

$$\mathcal{C} = |\langle \alpha_e | \alpha_g \rangle| \ . \tag{6.13}$$

Let us examine the two limiting cases of a large and small initial field in C. For large α values, the photon number distribution is peaked around the mean photon number $\bar{n}$. The $\pi/2$-pulse condition (6.9) is thus simply:

$$\Omega_0 \sqrt{\bar{n}}\, t_i = \pi/2 \ . \tag{6.14}$$

In the expansion of $|\alpha_e\rangle$ and $|\alpha_g\rangle$, the variation of n around $\bar{n}$ can be neglected in a zero-order approximation. Moreover, it is justified to make the approximation $c_n \approx c_{n+1}$ since the width of the photon number distribution, of the order of $\sqrt{\bar{n}}$, is much larger than 1. We then get:

$$|\alpha_e\rangle \approx |\alpha_g\rangle \approx |\alpha\rangle \ . \tag{6.15}$$

This means that the cavity field is not appreciably modified by the atomic emission and that the interference contrast $\mathcal{C}$ is close to one.

The other limit corresponds to an initial vacuum state in the cavity. The $\pi/2$-pulse condition is $\Omega_0 t_i = \pi/2$. We have then:

$$|\alpha_e\rangle = |0\rangle \qquad \text{and} \qquad |\alpha_g\rangle = |1\rangle \ , \tag{6.16}$$

an obvious result. The two final cavity states are now orthogonal. If the $|e\rangle \to |g\rangle$ transition has occurred in C, its photon number has changed from 0 to 1 revealing with certainty the atomic path. The Ramsey fringes contrast vanishes altogether. In between these two simple limits, it is easy to compute numerically the contrast from eqns. (6.11)–(6.13).

---

[3]We will show in the next section that it is possible to pin down exactly the photon number by gathering similar information from several atoms crossing the cavity successively.

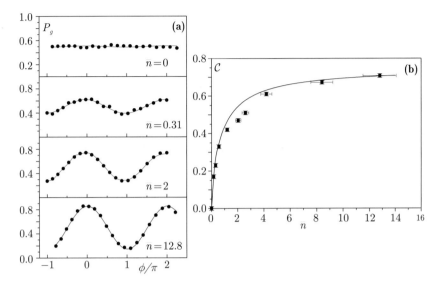

Fig. 6.4 Complementarity experiment. (a) Ramsey fringes observed for average photon numbers $\bar{n} = 0, 0.31, 2$ and 12.8. The lines are sine fits used for contrast determination. (b) Ramsey fringes contrast $\mathcal{C}$ as a function of the average photon number $\bar{n}$. The dots are experimental, with statistical error bars. The line represents the theoretical prediction, scaled to account for the finite contrast of the interferometer at large fields. Reprinted by permission from Macmillan Publishers Ltd: Nature, Bertet *et al.* (2001).

Figure 6.4(a) presents the Ramsey fringes obtained for different $\bar{n}$ values, ranging from zero to 12.8. The lines are sine fits, which provide a visual aid to determine $\mathcal{C}$ precisely. Small photon numbers correspond, as expected, to a low contrast. The fringes have good visibility for the highest photon number. Note, however, that their contrast does not reach the theoretical value, $\mathcal{C} = 1$. This is due to various imperfections of the Ramsey interferometer. Figure 6.4(b) presents (experimental dots) the observed contrast as a function of the initial average photon number $\bar{n}$. The line presents the theoretical predictions. In order to account for the intrinsic imperfections of the interferometer limiting the visibility of the fringes at large field, the predictions of eqn. (6.13) have been multiplied by a 0.75 factor. The agreement with the experimental data is then quite convincing.

This experiment is a faithful transposition of the original Bohr's proposal with a moving slit. The mesoscopic cavity field recording which-path information plays the role of the slit, its average photon number being the analogue of the slit's mass. We have spanned continuously, by changing the photon number, the boundary between the quantum and the classical worlds. At one limit (vacuum field), the interferometer contain a quantum component. It gets entangled with the interfering particle, and does not behave as a good phase-measuring device. At the other limit, the interferometer is genuinely classical and produces good fringes.

As its Mach–Zehnder counterpart, this interference experiment at the quantum–classical boundary can be qualitatively interpreted by a Heisenberg uncertainty rela-

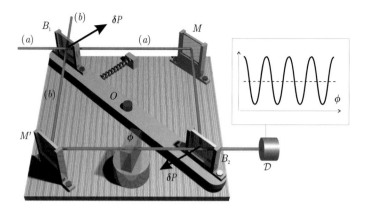

Fig. 6.5 An unconditional quantum eraser *gedanken* experiment.

tion. In a Ramsey set-up, the first pulse imprints its phase onto the atomic coherence and the second one probes this phase. The fringe contrast is thus sensitive to the phase correlation between the two pulses. Here, the R_2 pulse, being classical, has a well-defined phase. On the contrary, the pulse produced in C, being mesoscopic, exhibits a phase dispersion $\Delta\varphi$ limited by the 'Heisenberg photon-number–phase uncertainty relation' $\Delta N\Delta\varphi \approx 1$. When the field is large, the phase dispersion $\Delta\varphi \approx 1/\sqrt{\bar{n}}$ is small and the fringes have good contrast. For small initial fields, $\Delta\varphi$ is large and the phase correlation with the second pulse is lost, which explains why the interference contrast vanishes.

This qualitative argument should not lead to the misconception that the fringe blurring effect, observed when the field in C is small, is a trivial consequence of an irreversible random noise. The fringes are washed out by a quantum entanglement between the atom and the cavity, which can be reversed or manipulated in conditional experiments, leading to a restoration of the interference. Two CQED quantum eraser experiments illustrate this point.

6.1.4   Unconditional quantum eraser

A first example of fringe restoration is achieved by letting the atom interact twice with the cavity field, the second interaction suppressing the entanglement produced by the first. This is an unconditional eraser procedure, which does not require any correlation between the atom and the which-path field measurements.

It is again instructive to discuss the principle of this experiment in its Mach–Zehnder version, represented in Fig. 6.5. The two beam-splitters, B_1 and B_2, are now bolted to the same rotating frame. They behave thus as a single quantum oscillator. If the particle follows path (a), it sets the frame in motion when reflected by B_2. Path (b) is similarly revealed by a kick on B_1. In both cases, the beam-splitter assembly is finally moving in the same direction and no which-path information is available. Interference fringes should show-up. In fact, there is for a short time which-path information in the beam-splitter motion, while the particle is flying inside the interferometer after it has interacted with B_1. The second interaction with the rotating assembly erases unconditionally this entanglement and restores the fringes.

The CQED implementation of this eraser scheme is sketched in Fig. 6.6(a). The

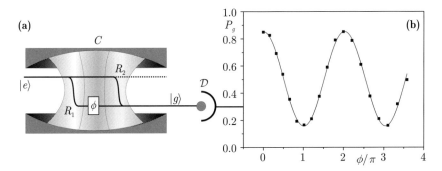

Fig. 6.6 (a) Principle of an unconditional quantum eraser experiment in cavity QED. (b) Observed fringes. The line is a sine fit. Reprinted by permission from Macmillan Publishers Ltd: Nature, Bertet *et al.* (2001).

atom, initially in $|e\rangle$, interacts first with the empty cavity mode, undergoing a $\pi/2$-quantum Rabi pulse which prepares the entangled state $(|e, 0\rangle + |g, 1\rangle)/\sqrt{2}$. Probed by a classical second Ramsey pulse, this state would reveal no interference. After this pulse, the atom is set far off-resonance with the cavity mode by an electric field, also used to tune the relative phase of $|e\rangle$ and $|g\rangle$. The atom–cavity state becomes $[|e, 0\rangle + \exp(i\phi)|g, 1\rangle]/\sqrt{2}$. Before the atom exits the cavity mode, the resonance condition is restored for another $\pi/2$-quantum Rabi pulse. The atom–field system then evolves into $[|e, 0\rangle + |g, 1\rangle + (|g, 1\rangle - |e, 0\rangle)\exp(i\phi)]/2$. The final probability of getting the atom in $|g\rangle$ is thus $P_g = \cos^2(\phi/2)$, exhibiting interference with unit contrast. The two interfering paths correspond to a transition from $|e\rangle$ to $|g\rangle$ either during the first interaction or during the second. Which-path information stored in a 0/1 photon signal after the first interaction is erased during the second quantum Rabi pulse. Whatever the path followed, the cavity contains always 1 photon when the atom is detected in $|g\rangle$.

The fringes observed under these conditions, shown in Fig. 6.6(b), have, as expected, good contrast. Note that we observe here Ramsey fringes without having to apply any external field to the atom. The only photon involved is generated by the atom itself in the vacuum Rabi oscillation process. We obtain Ramsey fringes without imposing external phase information. The atom interacts twice with the vacuum field fluctuations in the cavity. These fluctuations have no absolute phase, but keep a well-defined relative phase during the cavity damping time T_c. Since the atom crosses the cavity in a time smaller than T_c, it experiences two phase-correlated perturbations producing in principle maximum contrast interference.

6.1.5 Conditional quantum eraser

We now come back to the initial interferometer configuration with a $\pi/2$-vacuum Rabi pulse applied in C, followed by a classical $\pi/2$-pulse in R_2. No fringes are observed then but, as we will show, interference is restored by performing a measurement of the cavity field in a proper basis. The procedure is then conditional, the fringes being observed only in a correlation between the outcomes of the atom and the which-path field measurements. As already noted, microwave CQED experiments do not detect directly this field, but its state can be copied onto an atomic probe which can in turn

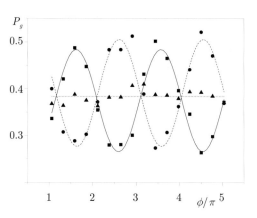

Fig. 6.7 Ramsey fringes observed on atom A_1 as a function of the interferometer phase ϕ, corresponding to a detection of the eraser atom A_2 in $|e\rangle$ (squares) or $|g\rangle$ (circles). Two π-out-of-phase fringe patterns are obtained. The sum of these two signals, corresponding to an unread measurement of A_2, exhibits no fringes (triangles). The lines are sine and constant function fits.

be manipulated and detected. This amounts to an indirect measurement of the field in C. The experiment involves now two atoms interacting successively with the cavity.

The first atom, A_1 and the field in C evolve, after the $\pi/2$-quantum Rabi pulse, into the entangled state $(|e,0\rangle + |g,1\rangle)/\sqrt{2}$. In order to erase which-path information, we send through the set-up a probe atom A_2, initially in $|g\rangle$. Its interaction time with C is tuned for a π-quantum Rabi pulse in the field of one photon. The cavity field thus ends up being in vacuum and the two atoms are left in the entangled state:

$$|\Psi_{\mathrm{EPR}}\rangle = (|e_1,g_2\rangle - |g_1,e_2\rangle)/\sqrt{2} , \qquad (6.17)$$

where the indices correspond to atoms 1 and 2. We recognize here the atomic EPR pair described in Section 5.4.4.

At this time, which-path information is not yet erased. It is merely transferred from C to A_2. In order to perform a conditional eraser, we need to measure A_2 in the $|0/1_X\rangle = (|e_2\rangle \pm |g_2\rangle)/\sqrt{2}$ basis, for instance. We thus apply a $\pi/2$-rotation on A_2 with a resonant classical field pulse that maps $|0/1_X\rangle$ onto $|e_2\rangle$ and $|g_2\rangle$, the eigenstates of the energy measurement in the field-ionization detector. In practice, this pulse is applied in the microwave zone R_2 which A_2 crosses, as does A_1, after it leaves C. When A_2 is finally detected in $|e_2\rangle$ ($|g_2\rangle$), A_1 is projected onto the state $(|e_1\rangle - |g_1\rangle)/\sqrt{2}$ $[(|e_1\rangle + |g_1\rangle)/\sqrt{2}]$, presenting a maximal e/g coherence. The interaction of A_1 with the Ramsey pulse R_2 then leads to fringes with unity contrast. The detection of A_2 has restored in a conditional way the fringes observed on A_1.

For the sake of clarity, we have presented the experiment as if A_2 was detected before A_1. In fact the opposite actually happens, since A_2 follows A_1 in the set-up. The $A_1 - A_2$ correlations are however, as already noticed, independent upon the order of the measurements. The two interference patterns observed on A_1 *in correlation* with a detection of A_2 in $|e\rangle$ or $|g\rangle$ are in phase opposition. An unread measurement of A_2 corresponds to a statistical mixture of these patterns, with no observable modulation.

At this stage, we realize that the procedure described here is precisely the scheme used to generate an atomic EPR pair and analyse its 'transverse' coherences (Section 5.4.4). The timing of the experiment has in fact been already depicted in Fig. 5.33, on page 285 (with a simple change in the labelling of the Ramsey pulses). From the experimental data used to construct the Bell signal (Fig. 5.34), we can extract the detection probability for A_1 as a function of the Ramsey phase, conditioned to the detected state of A_2. We obtain in this way the signals displayed in Fig. 6.7. The two patterns corresponding to A_2 being detected in $|e\rangle$ or $|g\rangle$ are, as expected, π-phase shifted with respect to one another. The signal corresponding to an unread measurement of A_2 is simply obtained by summing these two fringes patterns. The interference has completely disappeared in this sum (triangles in Fig. 6.7). This analysis shows clearly that a conditional quantum eraser experiment relies on a book-keeping procedure of experimental correlations which merely amounts to presenting in a different way the data of an EPR experiment.

6.2   Non-destructive photon number measurement

We have just shown that the addition of a single photon in a high-Q cavity can be detected by the large effect it produces on the visibility of interference fringes observed on atoms crossing the cavity. Besides exposing in a striking way the principle of complementarity, this experiment illustrates the extreme sensitivity of the CQED set-up for measuring fields at the single-photon level. We describe in this section a CQED experiment which demonstrates that, more than simply detecting a single photon, an atom crossing the cavity can do it in a special way, without destroying the light quantum while counting it. This experiment implements a procedure qualified, in the jargon of measurement theory as being 'quantum non-demolition'. Let us start by defining the meaning of this qualifier and present a brief review of the active field of quantum non-demolition measurement physics.

6.2.1   A review of quantum non-demolition measurements

We must first remark that most practical measurements are far from obeying the text-book projection postulate. Ideally, any quantum system should, upon measurement of an observable, end up in the eigenstate (or eigenspace) corresponding to the detected eigenvalue. This is not in general what happens. A case in point is the photo-detection process applied to a field mode. Here, the measured quantity is the photon number operator $N = a^\dagger a$. The counting is usually achieved with detectors (photomultipli-ers, photodiodes) whose efficiency is, in principle, not limited. The operation of all usual detectors relies however on the conversion of the incoming photons into electric charges through the photo-electric effect. The detected photons are certainly destroyed in the process. The quantum state of the light beam, instead of being projected onto a random photon number state as required by the quantum measurement postulate, is always cast onto the vacuum. The system's state is not 'projected', it is altogether 'demolished'.[4]

Quantum non-demolition measurements of light

This drastic destruction is not required by any fundamental law. A Quantum Non-Demolition (QND) measurement is precisely a detection process which, instead of erasing entirely all information in the state of the system it has detected, projects it into the eigenstate corresponding to the measurement result and thus preserves useful

---

[4]With the definition of Section 4.1.4, an usual photon counting process with a 100% efficient detector is a *generalized measurement* defined by the non-hermitian dyadic operators $M_0 = |0\rangle\langle 0|$, $M_1 = |0\rangle\langle 1|, \ldots, M_n = |0\rangle\langle n|, \ldots$ associated to the counting of $0, 1, \ldots, n, \ldots$ photons.

information for subsequent processing. A second essential element in the definition of a QND process is its repeatability. According to the projection postulate, after a first measurement has yielded a result, another projective measurement of the same observable performed *immediately* afterwards will, with certainty, yield the same result. This is not in general the case, though, if the two measurements are separated in time, since the eigenstate produced in the first measurement may evolve and expand along different eigenstates as time goes on. For the process to be QND, this evolution should not occur and successive measurements should yield the same result as the first, provided the system has not been perturbed in between. This puts severe limitations – to be analysed below – on the kinds of systems and observables on which QND measurements can be performed.

QND measurements fulfilling the above definition can be used as very sensitive probes of small perturbations acting on a system. Any deviation from the perfect correlation between two successive measurements can be traced to an external action. The first proposals in this field were made in the context of gravitational wave detection by mechanical resonators (Braginsky and Vorontosov 1974; Braginsky *et al.* 1977; Thorne *et al.* 1978; Unruh 1978). Huge aluminium bars at cryogenic temperatures are used as high-quality low-frequency mechanical oscillators. A gravitational wave, slow periodic deformation of space, induces tidal forces which can excite the bar oscillator. The detection of a gravity wave would thus amount to recording small changes in the oscillator's phonon number (or in one of its quadrature operators). Since these infrequent gravitational events may happen at any time, it is mandatory to observe continuously the system. QND measurements are ideally suited, since their repetition does not, in principle, add any noise or spurious evolution to the dynamics of the measured quantity.

Beyond this fundamental interest, quantum non-demolition measurements in optics could be used for unlimited distribution of optical signals. A QND device acts as an 'optical tap' making it possible to access information carried by an optical fibre without altering it. Practical information distribution is still far from being limited by such considerations, but further reductions of the number of photons used to code for a single bit could make this approach useful in the future.

Complete measurement repeatability makes QND measuring devices quite different from quantum state preparation set-ups which implement projections on an eigenstate (or an eigensubspace) of the measured system, while discarding it if it fails to project on the right state. A Stern–Gerlach apparatus, for instance, can filter out state $|0\rangle$ of a spin-$1/2$ system, by blocking the path corresponding to $|1\rangle$. Successive filters, all aligned for detecting $|0\rangle$, will let a spin go through when initially in $|0\rangle$. This is not, however, a QND situation. An external perturbation flipping the spin will result in it being stopped by the next filter and we will thus lose the system for further readings. A QND measurement, on the contrary, will register the change of the system's eigenvalue and will be able to carry on the continuous observation.

General conditions for QND measurements

A QND measurement relies on the coupling of the measured system, $\mathcal{S}$, with a meter $\mathcal{M}$ (or with a series of identical meters in the case of repeated measurements). As already described in Chapter 2, the meter evolves among an ensemble of pointer states, each 'pointing towards' an eigenvalue of the observable O_S measured on S. The ensemble of pointer states can be considered as an eigenbasis of a meter observable O_M. The Hamiltonian $H = H_S + H_M + H_{SM}$ of the $S + M$ system is made of three terms describing respectively the system and meter (H_S and H_M) and their mutual coupling (H_{SM}). For the measurement to be QND, these Hamiltonian terms should satisfy a set of conditions which have been defined in the early days of QND history (Caves *et al.* 1980). We briefly outline these conditions here without going into any detailed discussion.

First of all, there should be some information about the measured observable O_S (quadrature or photon number in the quantum oscillator case) encoded in the meter's pointer states after the interaction. The system–meter interaction Hamiltonian must thus not commute with the meter's observable O_M:

$$[H_{SM}, O_M] \neq 0 , \tag{6.18}$$

this necessary condition ensuring that the system induces an evolution of the meter. In addition, the essential QND requirement is that the measurement should not affect the eigenstates of O_S. This is the case if H_{SM} commutes with O_S:

$$[H_{SM}, O_S] = 0 . \tag{6.19}$$

Let us stress that this condition is sufficient, but not necessary. We will see later that some measurements, provided they are performed within a conveniently adjusted time, might satisfy the criterion of state non-demolition even if the measured observable is not a constant of motion of the interaction Hamiltonian.

A final important condition is related to measurement repeatability. If we do not restrict ourselves to extremely fast processes, the eigenstates of O_S should not evolve under the action of the free system's Hamiltonian H_S between two interactions with the meter. This condition implies that O_S must be a constant of motion for H_S:

$$[H_S, O_S] = 0 . \tag{6.20}$$

Not all observables can be measured in a QND way. For instance, in the case of the one-dimensional motion of a free particle, the position X cannot be a QND observable. Even though a proper interaction Hamiltonian with a meter could be designed, X does not commute with the free Hamiltonian, proportional to P^2, which violates the necessary condition (6.20) and breaks the correlations between repeated measurements. A first measurement of X projects the system's state onto a position eigenstate. According to the Heisenberg uncertainty relation, this state has infinite momentum dispersion. During the finite time before the next X measurement, this velocity indeterminacy leads to a complete dispersion of the results obtained for X in the subsequent measurement. This 'back-action' onto O_S of an observable which is conjugated with O_S should absolutely be avoided for a QND measurement.[5]

There is no such a 'back-action obstacle' for the counting of photons in a cavity since $N = a^\dagger a$ is obviously a constant of motion for the free-field Hamiltonian. It is

---

[5]Note that, on the contrary, the momentum P, a constant of motion, is a good QND observable for a free particle.

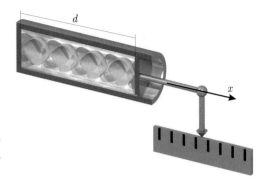

Fig. 6.8 A *gedanken* experiment measuring the photon number in a cavity through radiation pressure.

also possible to design for this operator a proper meter and a field–meter interaction Hamiltonian satisfying conditions (6.18)–(6.20). The photon number is thus a genuine QND observable and the demolition of photons upon measurement is not a fatality. This point had been first emphasized in proposals of opto-mechanical devices aiming at QND measurement of light.

A mechanical thought experiment for QND measurement of photons

As a simple example, let us consider the *gedanken* situation depicted in Fig. 6.8 (Caves 1979; Braginsky and Khalili 1996). A cylindrical cavity is closed by a 'piston', which can move freely along the x axis. The piston and the opposite fixed wall are perfect mirrors forming a Fabry–Perot optical resonator of length d, which sustains a standing wave field with angular frequency ω. The system S is the cavity mode, the observable O_S is the photon number operator N and the meter $\mathcal{M}$ is the moving piston, described quantum mechanically by its position X and momentum P.

Let us first analyse the situation in semi-classical terms. The radiation pressure on the wall, proportional to the energy density inside the cavity, produces on the piston a force $f = N\hbar\omega/d$ along the x direction. This force is given by a simple momentum book-keeping argument. Each photon bounces at a rate $c/2d$ with the cavity wall, yielding at each collision a momentum $2\hbar\omega/c$. The Heisenberg equation of motion for P:

$$\frac{dP}{dt} = N\frac{\hbar\omega}{d} \; , \qquad (6.21)$$

derives from a system–meter interaction Hamiltonian:

$$H_{SM} = -\frac{\hbar\omega}{d}NX \; , \qquad (6.22)$$

which fulfils the criteria for a QND measurement. Being proportional to the photon number, it preserves the Fock states and satisfies (6.19). Its X-dependency provides the meter with information about the measured observable. This information can be read by measuring the momentum P of the piston playing the role of $O_\mathcal{M}$ and satisfying (6.18). Note that the mode frequency ω depends also upon the piston's position since a change of the cavity length entails a drift of its frequency imposed by the field changing boundary conditions. This dependency, which we will describe more precisely below, does not modify the conclusions of this qualitative discussion.

The photon number is inferred in a QND way from the momentum acquired by the piston, which we assume to be initially in a minimum uncertainty state, with position and momentum fluctuations obeying $\Delta X_0 \Delta P_0 = \hbar/2$. The momentum acquired by the piston during the measurement time τ is $N\hbar\omega\tau/d$. The best precision possible on this quantity is equal to the initial momentum quantum fluctuation, ΔP_0, so that the minimum uncertainty on N obtained from this procedure is:

$$\Delta N = \frac{d}{\hbar\omega}\Delta P_0 \frac{1}{\tau} \ . \tag{6.23}$$

It can, in principle, be made as small as one wishes by choosing a long enough measurement time, the final field being arbitrarily close to a Fock state.

A precise determination of N spoils the initial phase of the field, since Fock states have a completely blurred phase. This effect can be understood on this simple model by estimating the field frequency change induced by a small displacement of the piston. Moving adiabatically the wall by an amount δx changes the cavity energy $\mathcal{E}$ by $\delta\mathcal{E} = -\mathcal{E}\delta x/d$, which is the work done by the piston against the radiation pressure force f. The photon number being invariant in this adiabatic wall motion, the energy change is directly related to a cavity frequency modification $\delta\omega$:[6]

$$\delta\omega = -\omega\delta x/d \ . \tag{6.24}$$

The quantum uncertainty ΔX_0 on the wall's initial position thus results in an uncertainty on the cavity frequency, $\Delta\omega = \omega\Delta X_0/d$. During the measurement time τ this frequency indeterminacy leads to a phase uncertainty:

$$\Delta\phi = (\Delta\omega)\tau = (\omega/d)\Delta X_0\tau \ , \tag{6.25}$$

which varies linearly with τ and is thus inversely proportional to the uncertainty on the photon number ΔN. More precisely, combining eqns. (6.23) and (6.25) yields:

$$\Delta\phi\Delta N = \left[\frac{\omega\tau}{d}\Delta X_0\right]\left[\frac{d}{\hbar\omega\tau}\Delta P_0\right] = \frac{1}{2} \ , \tag{6.26}$$

and we retrieve the well-known conjugation relation between phase and photon number fluctuations.

Recording the photon number by a macroscopic moving wall is certainly not realistic yet, even if experiments on the optomechanical coupling between a field mode and a high-quality quartz mechanical oscillator have made impressive progress recently (Pinard *et al.* 1995; Cohadon *et al.* 1999; Kleckner *et al.* 2006).

QND measurements of macroscopic fields based on the Kerr effect

Quantum optics provides more fertile ground for QND measurements and the first actual demonstrations have been realized in this domain. A detailed review can be found in Grangier *et al.* (1998). The fundamental principle of a large class of optical QND measurements is represented in Fig. 6.9. It exploits the non-linear optical properties

---

[6]This relation can also be derived by remarking that, in order to obey the changing boundary conditions, the field wavelength should vary proportionally to the cavity length d.

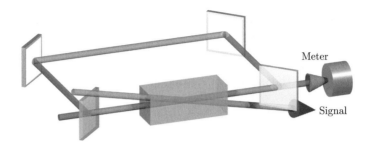

Fig. 6.9 General principle of a QND measurement through a Kerr non-linearity in a crystal. The signal and meter beams overlap in the crystal. The phase of the meter is read out with a Mach–Zehnder interferometer and an intensity detector.

of a transparent crystal. The absorption of this crystal is negligible and the real part of its refractive index varies linearly with the total light intensity it transmits. This is the Kerr effect. The beam to be measured ('signal') is sent through the crystal, superimposed with a weaker probe ('meter') beam. Both the signal and the meter contain large numbers of photons associated to 'macroscopic' field intensities. In these QND experiments, we are interested in analysing the quantum fluctuations around the large mean value of these fields.

The refractive index experienced by the meter during its propagation in the crystal contains a contribution proportional to the signal's intensity. The final phase of the meter thus records information about the signal's photon number. This phase is read-out in an interferometric arrangement represented in Fig. 6.9. The non-linear crystal is inserted in one arm of a Mach–Zehnder interferometer acting on the meter beam. The phase shift of the meter (measuring the change in the signal beam intensity) is detected as a variation of the intensity at one of the output ports of the Mach–Zehnder device.

A detailed analysis of quantum fluctuations shows that the interferometer output fluctuations are directly related to the signal intensity variations. The crystal being perfectly transparent, the incoming fluctuations of the signal beam are not affected. The system performs a QND measurement of the signal (Millburn and Walls 1983; Imoto *et al.* 1985). Phase information in the signal is spoiled, being blurred by the non-linear refractive index induced by the fluctuations of the meter's intensity.

Quantitative criteria based on quantities easily measurable in a practical experiment assess the quality of such a QND procedure (Grangier *et al.* 1992; Poizat *et al.* 1994). First, the meter should get information about the signal. The interferometer output intensity should be correlated with the signal intensity variations around the mean, either a weak classical modulation or quantum fluctuations. The signal fluctuations, on the other hand, should not be changed by the measurement when meter information is not read. Finally, the signal fluctuations should be reduced when they are conditioned to the detected meter level (projection of the signal state onto a Fock state). Two simple inequalities suffice to distinguish between trivial intensity measurements, quantum state preparation devices and genuine QND procedures.

The first attempts of QND measurements based on the Kerr effect have been realized in optical fibres, with c.w. beams (Levenson *et al.* 1986; Bachor *et al.* 1988) or with optical solitons (Friberg *et al.* 1992). The Kerr non-linearity of pure silica is notoriously small, but the signal and meter beams can propagate together over many kilometres. Additional noise sources, not foreseen in the initial stage, made it impossible, though, to achieve the QND criteria.

An atomic medium presents huge non-linearities close to resonance, which can be further increased by setting up an optical cavity around the sample. Adverse effects, such as spontaneous emission, can be efficiently counter-acted by quantum interference exploiting the multi-level energy diagram of the atomic system (Sinatra *et al.* 1998). The first actual QND measurements based on Kerr non-linearities have been realized in this context, either with atomic beams (Grangier *et al.* 1991; Poizat and Grangier 1993) or, more recently, with laser-cooled atomic samples (Roch *et al.* 1997). This latter experiment has achieved a record quality, largely fulfilling all QND criteria.

QND measurements based on the noiseless duplication of light

Another (less intuitive) approach to QND measurements, that will not be described in detail here, is based on Optical Parametric Amplifiers (OPAs). These phase-sensitive devices are able to amplify or attenuate a field quadrature without adding any noise (all quantum noise is rejected on the other quadrature, for the fulfilment of Heisenberg uncertainty relations). An arrangement of OPAs and beam-splitters makes it possible to amplify noiselessly the signal to be measured and to split away a part of it, restoring the initial intensity in one output beam and providing meter information in another. The whole process can be analysed with the same QND criteria as Kerr-effect experiments.

An early attempt to realize a QND situation was performed along these lines (LaPorta *et al.* 1989). Later experiments fulfilled the QND criteria (Levenson *et al.* 1993; Pereira *et al.* 1994; Bruckmeier *et al.* 1997a), but did not exceed the performance of cold atomic media. The relatively easy realization of OPA's, however, makes it possible to duplicate the measurement device. The repetition of QND measurements, implementing the 'optical tap' concept, has been tested in this context (Bencheikh *et al.* 1995; Bruckmeier *et al.* 1997b; Bencheikh *et al.* 1997).

Meeting the criteria for QND measurements does not necessarily require sophisticated quantum optics techniques. Two experiments have realized an authentic QND situation with simple off-the-shelf electronic devices (Goobar *et al.* 1993a;b; Roch *et al.* 1993). The signal beam is detected in a standard photodiode and is accordingly demolished. The photocurrent nevertheless faithfully represents the incoming quantum fluctuations. This current can be amplified, with a negligible added noise. Part of the amplified current is used as the meter. The other part is fed into a high-efficiency light emitter (LED), producing the output signal beam. With a proper tuning of the electronic gains, the input and output beams have identical quantum fluctuations (at least strong quantum correlations), and these fluctuations are reflected by the meter current.

Obviously, the inner workings of this device do not fulfil the sufficient QND criteria discussed above. The field–system Hamiltonian involves photon annihilation and creation and does not commute with the measured observable. The global process is though QND. Note finally that this 'cheap' QND device clearly shows that phase is spoiled: the input and output beams may have different colours!

The schemes mentioned so far deal with small quantum fluctuations superimposed on a large average amplitude. Relatively high intensities are required to provide enough optical non-linearities. These transparent detectors are not able to operate in a 'photon counting' mode, giving a click for each photon crossing them. We show below that the extremely large coupling of Rydberg atoms with millimetre-wave fields in high-Q cavities makes it possible to realize a genuine QND photon counting.

QND measurement of an electron in a trap

Before coming to this CQED experiment, it is worth mentioning another beautiful study in which a QND detection at the quantum level has been achieved, in a quite different context (Peil and Gabrielse 1999). The system is a single electron oscillating in a Penning trap (Ghosh 1995), combining a large magnetic field and a quadrupolar static electric field. The quantized cyclotron motion of the electron in the plane perpendicular to the magnetic field is measurably, if weakly, coupled to the axial oscillatory motion along the magnetic field. This axial motion is, in turn, detected through the current induced in an external circuit by capacitive coupling of the electron to the trap electrodes. This complex coupling scheme provides a direct, continuous, QND measurement of the quantum numbers associated to the cyclotron motion. Quantum jumps between cyclotron states due to the coupling to the environment have been directly observed.

6.2.2   Single-photon QND measurement in CQED

The QND detection of a single photon in a cavity uses the CQED set-up shown in Fig. 5.15, on page 252. The QND meter is a single Rydberg atom. Three circular states with principal quantum numbers $n = 51$ ($|e\rangle$), $n = 50$ ($|g\rangle$) and $n = 49$ ($|i\rangle$) (Fig. 5.19, on page 260) are involved in the QND detection scheme. The signal–meter coupling is the resonant vacuum Rabi oscillation on the $|e\rangle \rightarrow |g\rangle$ transition. The effective atom–cavity interaction time t_i^r is set for an exact 2π-vacuum Rabi pulse on this transition.

If the atom is initially in $|g\rangle$ with one photon in C, the atom–field system undergoes a complete oscillation passing transiently through $|e\rangle$ and comes back – up to a global phase shift – to the initial state when the atom leaves the cavity. Depending upon whether the cavity contains 1 or 0 photons, the state transformations are thus, according to eqn. (5.47):

$$|g, 1\rangle \rightarrow -|g, 1\rangle \; ; \qquad |g, 0\rangle \rightarrow |g, 0\rangle \; . \tag{6.27}$$

Remarkably, the presence of a photon simply produces a π-phase shift of the atomic state. This information is imprinted onto the atom without destroying the photon, which makes the process QND.

It is to detect this phase information that the third level $|i\rangle$ comes to the rescue. The transitions from this level are far-detuned from the cavity mode, so that the atom–field system stays invariant if it is initially in $|i, 0\rangle$ or $|i, 1\rangle$. The circular state $|i\rangle$ can thus be used as a phase reference. The phase change experienced by the atom in $|g\rangle$ is probed by a Ramsey interferometer preparing and detecting an atomic coherence between $|g\rangle$ and $|i\rangle$.

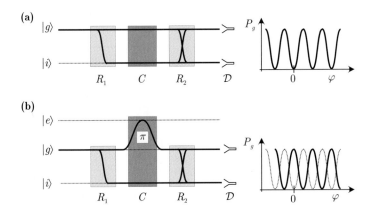

Fig. 6.10 Interfering paths in a single-photon QND experiment. The 'atomic path' convention of Fig. 6.3 is used again. The regions where the Ramsey pulses are applied are shown
by the grey rectangles and the cavity mode in between by a darker area. (a) C is empty.
The atom experiences two Ramsey pulses mixing the $|g\rangle$ and $|i\rangle$ states represented by the
two horizontal lines. The $|g\rangle$ and $|i\rangle$ paths interfere in the probability P_g of finding finally
the atom in $|g\rangle$, resulting in fringes in P_g when φ is scanned. (b) C contains a single pho-
ton. In the $|g\rangle$ path, the atom goes transiently through $|e\rangle$ (upper horizontal line) and the
corresponding amplitude accumulates an extra π-phase shift. As a result, the P_g fringes are
inverted. Reprinted by permission from Macmillan Publishers Ltd: Nature, Nogues *et al.*
(1999).

The atom, initially in $|g\rangle$, is put in a coherent superposition of $|g\rangle$ and $|i\rangle$ by a first
classical microwave pulse applied in zone R_1:

$$|g\rangle \longrightarrow (|g\rangle + |i\rangle)/\sqrt{2} \ . \tag{6.28}$$

The atomic state superposition is then probed at the exit of C by a second pulse R_2,
completing a Ramsey interferometer arrangement by realizing the transformations:

$$|g\rangle \longrightarrow (|g\rangle + e^{i\varphi}|i\rangle)/\sqrt{2} \ ; \qquad |i\rangle \longrightarrow (-e^{-i\varphi}|g\rangle + |i\rangle)/\sqrt{2} \ . \tag{6.29}$$

The relative phase φ of the two Ramsey pulses is tuned by sweeping around the $|i\rangle \rightarrow$
$|g\rangle$ transition the frequency of the classical source S' producing them.[7] The probability
of finding the atom in $|g\rangle$ exhibits a modulation when φ is scanned. This Ramsey
fringe pattern results from interference between two quantum paths symbolized in
Fig. 6.10(a) and (b), corresponding respectively to situations where there is 0 and 1
photon in C. In the first case, the probability of detecting finally the atom in $|g\rangle$:

$$P_{g|0}(\varphi) = (1 - \cos\varphi)/2 \ , \tag{6.30}$$

is equal to 0 when the interferometer phase is set to $\varphi = 0$. The atom is then detected
in $|i\rangle$ with unit probability.

---

[7]For the description of the ways to scan Ramsey fringes, see Section 3.3.3.

Fig. 6.11 Interferometric response of the meter atom to a 0/1 photon field. The $P_{g|0}(\varphi)$ and $P_{g|1}(\varphi)$ probabilities corresponding to zero (open diamonds) and one (solid squares) photon are π-out-of-phase. The error bars reflect the statistical fluctuations. The lines are sine fits. The interferometer setting $\varphi = 0$ is marked by a vertical line. Reprinted by permission from Macmillan Publishers Ltd: Nature, Nogues *et al.* (1999).

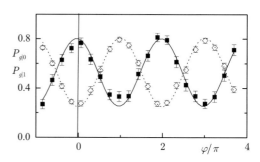

When the cavity contains initially one photon, the atom, if it enters C in $|g\rangle$, absorbs the radiation, goes transiently through $|e\rangle$ and finally returns to $|g\rangle$, releasing the photon. In the process, the corresponding probability amplitude undergoes a sign reversal with respect to the 0 photon case. In the other arm of the interferometer, corresponding to the atom crossing C in $|i\rangle$, nothing is changed with respect to the zero-photon case. As a result, the i/g coherence experiences a π-phase shift when the photon number goes from 0 to 1 and the Ramsey interference pattern in reversed (compare Fig. 6.10a and b). The probability of finding the atom in $|g\rangle$ if C contains one photon:

$$P_{g|1}(\varphi) = (1 + \cos\varphi)/2 , \tag{6.31}$$

is equal to 1 for $\varphi = 0$. The atom is found with certainty in $|g\rangle$ whereas it was in $|i\rangle$ when C was in vacuum. For this phase setting of the interferometer, the outgoing atomic state reveals unambiguously, and without destroying it, the presence of a photon.

We should remark at this stage that this procedure, based on an exact 2π-Rabi rotation, operates properly only if the cavity field is in the $\{|0\rangle, |1\rangle\}$ subspace. For other Fock states, the 2π-pulse conditions is not fulfilled (remember that the Rabi frequency in an n-photon Fock state is $\Omega_0\sqrt{n+1}$). There is thus, for $n > 1$, a finite probability of absorbing one of the photons. This restriction to zero- or one-photon fields can be removed, as will be shown in Section 6.4, by replacing the resonant 2π-Rabi pulse with a dispersive atom–field interaction. Let us however stick to this restrictive condition for the time being, in order to get a deeper understanding of the QND physics in a very simple two-state situation. To avoid all possible confusion, we will call Single-Photon-QND (or SP-QND) the method based on the resonant 2π-Rabi pulse.[8]

The experimental signals (Nogues *et al.* 1999) shown in Fig. 6.11 present a textbook illustration of this procedure. Before sending the meter atom through C, another atom is used as a source to prepare the signal field in the cavity. This source atom, initially

---

[8]The single-photon QND method satisfies the criterion of photon non-demolition, even though the atom–field Hamiltonian, a linear combination of a and $a^\dagger$, does not commute with $N = a^\dagger a$ (see eqn. 3.167). The commutation relation (6.19) is a strong sufficient condition, which ensures that the measuring QND process does not demolish the eigenstates of the measured observable, whatever the time it takes to complete the measurement. Here, we have chosen a special system–meter interaction time ($t_i^r = 2\pi/\Omega_0$), which preserves the system eigenstate $|1\rangle$, even though (6.19) is not satisfied.

in $|e\rangle$, undergoes a $\pi/2$-Rabi pulse in C and its final detection in $|e\rangle$ or in $|g\rangle$ signals the absence or presence of a single photon. The meter atom is then sent to read this field and its clicks in $|g\rangle$ or $|i\rangle$ are recorded and correlated with the outcome of the first atom measurement. The experimental sequence is repeated many times for various values of φ and the $P_{g|0}(\varphi)$ and $P_{g|1}(\varphi)$ signals are reconstructed from the correlated source and meter atom data.

Two Ramsey fringe patterns (open diamonds and solid squares in Fig. 6.11) are obtained, depending upon the reading of the first atom (0 or 1 photon in C). These fringes are π-out-of-phase with each other, demonstrating clearly the sign reversal produced on the meter atom by the 2π-Rabi pulse in a single-photon field. The experiment is not ideal, due to a limited contrast of the Ramsey interferometer. The data are fitted with sine curves obeying the equations:

$$P_{g|0}^{\mathrm{exp}}(\varphi) = (1 - \mathcal{C}\cos\varphi)/2 \; ; \qquad P_{g|1}^{\mathrm{exp}}(\varphi) = (1 + \mathcal{C}\cos\varphi)/2 \; , \qquad (6.32)$$

where $\mathcal{C} \approx 0.6$ is the experimental fringe contrast.[9] Setting the phase of the interferometer at $\varphi = 0$ (vertical line in Fig. 6.11), we see that the probability of detecting the atom in $|g\rangle$ when there is one photon in C is $\approx 80\%$ instead of the theoretical 1 value.

While this experiment demonstrates that a circular atom has the sensitivity to detect single photons in a non-destructive process, it is not sufficient to qualify the procedure as QND. On the one hand, we started with a known field in the cavity (0 or 1 photon), so that the experiment was verifying known information, and was not directly checking that the process was actually projecting an *a priori* unknown state onto a Fock state (or at least – due to the Ramsey pulses imperfections – onto an approximation of one). On the other hand, the repeatability of the procedure remained to be demonstrated. In other words, is it really possible to see the same photon twice?

To answer these questions, we have performed an experiment in which an unknown field in a mixture of $|0\rangle$ and $|1\rangle$ states was first measured by a QND meter atom A_1, then re-measured by a second atom A_2. The goal was to check that if a photon had been recorded by A_1, it was still there to be 'seen' by A_2, an unusual situation in photon counting. The A_2 atom was recording the photon by mere absorption. Prepared in $|g\rangle$, it was set on resonance with the cavity on the $|g\rangle \rightarrow |e\rangle$ transition, for the time required to perform a π-Rabi pulse in the field of a single photon.

The mode initially contained a small thermal field independently measured by atomic absorption, which was used as the 'unknown field'. The probabilities of 0 and 1 photon were $p_0 = 0.77$ and $p_1 = 0.18$, with a very small probability (5%) of having two photons or more, thus realizing a good approximation of the $0-1$ photon situation.

In each experimental sequence the meter atom A_1 was sent in C and read out by the $|g\rangle - |i\rangle$ Ramsey interferometer, then followed by A_2. For the second atom, the Ramsey pulses were switched off. The final states of both atoms ($|g\rangle$ or $|i\rangle$ for A_1, $|g\rangle$ or $|e\rangle$ for A_2) were recorded and the experimental sequence repeated many times for various values of the interferometer phase φ. The conditional probabilities $P_{e_2|g_1}(\varphi)$

---

[9]At the time this book is being completed, the fringe contrast has been improved to $\mathcal{C} \approx 0.9$, so that much higher fidelity QND single-photon signals are expected in future experiments.

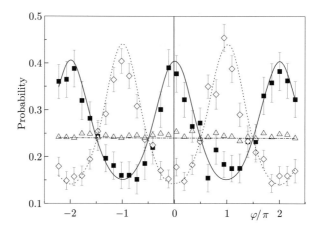

Fig. 6.12 Response of the field to a measurement by a QND atom and repeated measurement of a single photon. Conditional probabilities $P_{e_2|g_1}(\varphi)$ (solid squares) and $P_{e_2|i_1}(\varphi)$ (open diamonds) of detecting A_2 in $|e\rangle$, provided that the SP-QND meter A_1 has been detected either in $|g\rangle$ or $|i\rangle$. These probabilities are plotted versus the phase φ of the Ramsey interferometer acting on A_1. Ideally, they are identical to $p_{1|g}(\varphi)$ and $p_{1|i}(\varphi)$, the probabilities of finding one photon in C conditioned to a detection of A_1 in $|g\rangle$ or $|i\rangle$. The open triangles show, as a reference, the probability P_{e_2} of finding the probe atom in $|e\rangle$ (one photon in C), when A_1 is not sent. The vertical line indicates the $\varphi = 0$ setting discussed in text. The lines are theoretical fits taking into account experimental imperfections. Reprinted by permission from Macmillan Publishers Ltd: Nature, Nogues *et al.* (1999).

and $P_{e_2|i_1}(\varphi)$ of finding the second atom in $|e\rangle$ provided that the meter atom had been detected either in $|g\rangle$ or in $|i\rangle$ were reconstructed from these data as a function of φ.

The experimental signals, shown in Fig. 6.12, exhibit non-sinusoidal fringes which contain a lot of information about the SP-QND measurement process. They are plotted together with the reference signal obtained by recording the unconditional probability P_{e_2} of finding the probe atom A_2 in $|e\rangle$ when the QND atom A_1 is not sent. This reference, which obviously does not depend upon φ, corresponds to the probability p_1 that the initial field contained one photon.[10]

The analysis of the modulations in $P_{e_2|g_1}(\varphi)$ and $P_{e_2|i_1}(\varphi)$ is an interesting exercise on conditional probabilities. Let us first note that detecting A_2 in $|e\rangle$ means that there is one photon left in C after the measurement by the meter A_1. In other terms, $P_{e_2|g_1}(\varphi)$ and $P_{e_2|i_1}(\varphi)$ ideally represent the probabilities $p_{1|g}(\varphi)$ and $p_{1|i}(\varphi)$ of finding one photon in C *conditioned* to a 'click' $|g\rangle$ or $|i\rangle$ of the QND meter atom.[11] It is at first sight striking that these probabilities depend upon the Ramsey interferometer phase, which affects only A_1. It shows that the interaction with this atom modifies

[10]The difference between the measured value (0.24) of this reference background and the independently determined value $p_1 = 0.18$ is due to known experimental imperfections (errors in the reading of atomic levels, residual contribution of processes involving more than one photon in the thermal field).

[11]In this section, we use the capital letter P for probabilities of atomic states and the lower case letter p for probabilities of photon states.

the initial cavity field. Whereas the first experiment (Fig. 6.11) was studying the response of the meter atom to a known field (probability of detecting the meter in a given state *conditioned* to the number of photons in the field), this second experiment (Fig. 6.12) is studying the *reciprocal* response of the field to the detection of the meter (probability of having one or zero photon in C *conditioned* to a given reading of the meter atom). This reciprocal conditional probability is obviously the essential relevant quantity for a QND measuring device, since it determines the fidelity of the procedure.

In order to understand how the measurement by the meter atom modifies the knowledge about the field, let us first consider what happens for $\varphi = 0$ (vertical line in Fig. 6.12). The conditional probability $p_{1|g}(\varphi = 0)$ is then maximum, well above the reference level. This indicates that it is more likely to find one photon in C *after* a SP-QND atom has yielded the result $n = 1$ than *before* this measurement was performed (remember that, for $\varphi = 0$, A_1 in $|g\rangle$ records the QND detection of one photon in C). The detection of a photon by A_1 increases, immediately afterwards, the photon number. This, by itself, is a striking feature of a QND process. An ordinary absorptive photon counting process would instead reduce this probability since the photon would be annihilated. Ideally, the value of $p_{1|g}(\varphi = 0)$ should be 1, reflecting the projection of the field on the Fock state $|1\rangle$. The reduced fringe contrast, and hence reduced fidelity of the QND procedure, is mainly due to imperfections in the Ramsey pulses.

On the contrary, when the QND atom is found in $|i\rangle$, it indicates that C is empty. Accordingly, the probability of finding afterwards one photon is reduced below the reference level and the $p_{1|i}(\varphi = 0)$ signal is at its minimum. Ideally, it should be equal to 0. These conclusions are obviously reversed for $\varphi = \pi$. It is then $p_{1|g}(\varphi = \pi)$ which is at a minimum and $p_{1|i}(\varphi = \pi)$ at a maximum.

When the interferometer is set at $\varphi = \pi/2$, the two conditional probabilities cross each other and coincide with the unconditional probability of having one photon in C: $p_{1|g}(\pi/2) = p_{1|i}(\pi/2) = p_1$ (Fig. 6.12). The QND meter atom A_1 yields then no information about the cavity field, since it is found in $|g\rangle$ and $|i\rangle$ with equal probabilities (eqns. 6.30 and 6.31 for $\varphi = \pi/2$). Bringing no information on the field, the interaction with A_1 does not modify the photon number distribution.

The asymmetrical shape of the fringes observed in Fig. 6.12 can be understood by performing a simple inversion of indices on conditional probabilities, using Bayes law which states that $p_{1|g}$ and $p_{1|i}$ are given by:

$$p_{1|g} = \frac{P_{g|1}p_1}{P_{g|1}\,p_1 + P_{g|0}\,p_0} \quad ; \quad p_{1|i} = \frac{P_{i|1}p_1}{P_{i|1}\,p_1 + P_{i|0}\,p_0} \; . \tag{6.33}$$

These relations can be found by expressing in two different ways the *joint* probabilities $P_{g,1}$ and $P_{i,1}$ of detecting the meter atom in $|g\rangle$ or in $|i\rangle$ *and* finding one photon in C. We have for instance:

$$P_{g,1} = P_{g|1}p_1 = p_{1|g}P_g \; , \tag{6.34}$$

which yields $p_{1|g} = P_{g|1}p_1/P_g$. Combining this relation with $P_g = P_{g|1}p_1 + P_{g|0}p_0$, we readily deduce the first eqn. (6.33). The other relation giving $p_{1|i}(\varphi)$ is obtained in the same way. Replacing $P_{g|0}$, $P_{g|1}$ by their expressions (6.30) and (6.31) and noting that $P_{i|1} = 1 - P_{g|1}$ and $P_{i|0} = 1 - P_{g|0}$, we finally get:

$$p_{1|g}(\varphi) = \cfrac{1}{1 + \cfrac{1 - \cos\varphi}{1 + \cos\varphi}\cfrac{p_0}{p_1}} \; ; \qquad p_{1|i}(\varphi) = \cfrac{1}{1 + \cfrac{1 + \cos\varphi}{1 - \cos\varphi}\cfrac{p_0}{p_1}} \; . \qquad (6.35)$$

The variations of these conditional probabilities are periodic, but are more complex than the simple sine functions representing the direct Ramsey fringes. Note that the formulas (6.35) correspond to a perfect Ramsey set-up. To account for the limited fringe contrast of the actual interferometer, we must replace the $1 \pm \cos\varphi$ terms in these equations by $1 \pm C \cos\varphi$ where C is the experimental fringe contrast (≈ 0.6). Theoretical signals accounting for the set-up imperfections are represented by the lines in Fig. 6.12, in good agreement with the experimental points.[12]

When this experiment is presented in physics colloquia, a question often arises: '*Is it really the same photon that is inside the cavity at the end of the QND measurement, since it has been absorbed and re-emitted?*'. We should stress that such a question, natural in the macroscopic world, is meaningless in the quantum one (see Section 2.3 about identical particles). Photons, defined as elementary excitations of the cavity mode, are by essence indistinguishable bosons and any question about their identity has no pertinence. The right question to be asked is not '*is it the same photon?*', but rather '*is it the same quantum state that we have in the cavity after the measurement?*'. To this question, the answer is obviously '*yes*', since the state with a single excitation in C is unambiguously defined and preserved. Once the measurement has been performed, no thought experiment whatsoever could tell us whether the final photon in C is or is not '*the same*' as the one which was there at the beginning.

6.3 A quantum gate for multi-particle entanglement engineering

6.3.1 Single-photon QND measurement as a quantum gate operation

The SP-QND procedure extracts information about a photon in a cavity by imprinting it onto an atomic two-level system, while leaving the light quantum untouched. This is, by definition, the operation of a two-bit quantum gate in which the field (in $|0\rangle$ or $|1\rangle$) is the control bit while the atom (whose $|i\rangle$ and $|g\rangle$ states stand respectively for $|0\rangle$ and $|1\rangle$[13]) is the target. This analogy works as long as the number of photons does not exceed 1. The SP-QND procedure can hence be viewed as a textbook demonstration of the operation of a two-bit quantum gate.

As a simple exercise, let us rephrase more precisely the SP-QND method in the language of quantum information. We have seen in Section 2.6 that, in its simplest form, an ideal measurement of a qubit realizes a control-not, or control-σ_X gate i.e. a σ_X transformation of the target bit (the meter) conditioned to the state of the control. This gate leaves the meter in its initial state if the control is $|0\rangle$ and flips it to the other state if the control is $|1\rangle$. The final meter reading thus yields unambiguous information about the state of the signal, which is not altered in the process. This is the very definition of a QND process.

The same measurement can also be realized with other kinds of two-qubit gates, since any such gate, combined with one-qubit operations, can be used to construct a

---

[12]The fits in Fig. 6.12 take into account, in addition to the limited fringe contrast C, small effects due to Rabi pulse imperfections, to errors in reading the $|e\rangle$ and $|g\rangle$ atomic states and to the residual contribution of photon number states with $n > 1$.

[13]We now assign the state $|g\rangle$ to the qubit state $|1\rangle$, whereas $|g\rangle$ was $|0\rangle$ in the last chapter.

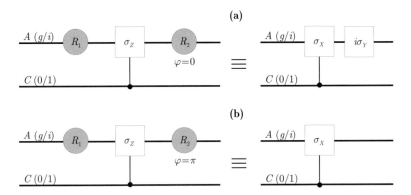

Fig. 6.13 Quantum circuit representing the SP-QND procedure. (a) Two equivalent circuits for $\varphi = 0$. (b) Two equivalent circuits for $\varphi = \pi$.

control-σ_X. In particular, a control-phase (or control-σ_Z) gate can be transformed into a control-σ_X by sandwiching it between two one-qubit Hadamard transformations U_H acting on the target bit (Fig. 2.20b).

This is precisely the kind of two-qubit gate transformations which we are using in the SP-QND procedure. The 2π-Rabi pulse implements on the atom–field system a control-σ_Z gate: when the control is in $|1\rangle$ it realizes on the atom the transformation $|\psi\rangle \to \sigma_Z |\psi\rangle$, i.e. it shifts by π the quantum phase of the $|g\rangle$ part of the target bit, and leaves the $|i\rangle$ part invariant. When the control is in $|0\rangle$ it leaves the atomic state untouched. This control-phase gate is sandwiched between two Ramsey $\pi/2$-pulses (with a phase difference φ between them) transforming the atomic qubit states into 'transverse states' whose pseudo-spins are in the equatorial plane of the Bloch sphere. These Ramsey pulses are not exactly Hadamard transformations, but they have similar properties. Sandwiching the control-σ_Z, they produce a control-σ_X, which might be combined with a one-qubit gate acting on the final state of the atom. This one-qubit gate depends upon the phase φ of the Ramsey interferometer.

More precisely, the first Ramsey pulse R_1, defined by eqn. (6.28), implements the one-qubit gate transformation:

$$U_{R_1} = e^{i\pi\sigma_Y/4} = (\mathbb{1} + i\sigma_Y)/\sqrt{2} . \tag{6.36}$$

When $\varphi = 0$, the same transformation is realized by R_2 $[U_{R_2}(\varphi = 0) = U_{R_1}]$. From:

$$U_{R_2}\sigma_Z U_{R_1} = e^{i\pi\sigma_Y/4}\sigma_Z e^{i\pi\sigma_Y/4} = e^{i\pi\sigma_Y/2}\left[e^{-i\pi\sigma_Y/4}\sigma_Z e^{i\pi\sigma_Y/4}\right] = (i\sigma_Y)\sigma_X , \tag{6.37}$$

we deduce immediately the two quantum circuits shown in Fig. 6.13(a), which describe equivalently the SP-QND procedure.[14] It can be seen either as a control-σ_Z phase gate sandwiched between the two Ramsey pulses (at left) or as a control-σ_X in which the atom is the target followed by the one-bit gate $i\sigma_Y$ acting on the atom (at right). The circuit at right transforms $|g\rangle$ into $|i\rangle$ if there is 0 photon in C (only the $i\sigma_Y$

---

[14]We must be careful, when drawing a quantum circuit to represent from left to right the transformations written from right to left in the mathematical expressions.

transformation is active in this case and it flips the atomic state). If there is one photon in the cavity, both σ_X and $i\sigma_Y$ act on the atom which flips twice and comes back to $|g\rangle$ (with a sign reversal). This is the assignment of the meter states we have found for the SP-QND process when the Ramsey interferometer is set to $\varphi = 0$.

When $\varphi = \pi$, the second Ramsey pulse becomes $U_{R_2}(\pi) = e^{-i\pi\sigma_Y/4}$ and the atomic transformation conditioned to the control state 1 is $U_{R_2}\sigma_Z U_{R_1} = e^{-i\pi\sigma_Y/4}\sigma_Z e^{i\pi\sigma_Y/4} = \sigma_X$. When the control is $|0\rangle$, the product of the two inverse transformations $U_{R_2}U_{R_1}$ reduces to the identity and the atomic state remains invariant. The quantum circuit representing the SP-QND procedure then reduces to a simple control-not gate. Figure 6.13(b) shows its two equivalent forms. The meter state assignment is then reversed with respect to $\varphi = 0$: a final $|g\rangle$ state corresponds to no photon and $|i\rangle$ to one photon.

At the heart of the SP-QND procedure we find a control-π phase gate, one of the fundamental two-qubit gates of quantum information. This gate is realized in CQED with a mere 2π-resonant Rabi pulse which brings transiently the atomic qubit state $|1\rangle$ outside the computational basis space, before returning it finally to this space.[15] This gate flips by π the phase of an atomic coherence when the field control is $|1\rangle$ and leaves it invariant when it is $|0\rangle$. This is exactly what is demonstrated in Fig. 6.11 where the phase of the atomic coherence is probed by the Ramsey interferometer. Likewise, the same gate shifts by π the phase of a 0/1 field coherence when the atom is in $|g\rangle$ and leaves it invariant when it is in $|i\rangle$, an effect which has also been checked experimentally (Rauschenbeutel *et al.* 1999).

6.3.2   QND detection and multi-bit entanglement

Entanglement knitting and generation of a GHZ state

The quantum-σ_Z gate produced by the 2π-resonant quantum Rabi pulse is a powerful tool for quantum information processing. Combined with the other basic quantum stitches – the $\pi/2$- and π-Rabi pulses described in Section 5.4.2 – it makes it possible to knit entanglement in ensemble of atoms crossing the cavity one by one. Let us describe an experiment performed on three atoms (Rauschenbeutel *et al.* 2000), the generation and analysis of a GHZ entangled state (Greenberger *et al.* 1990).

This state is a natural extension to triplets of particles of the EPR pair of atoms described in Section 5.4.4. The GHZ states are remarkable entangled states of three qubits, which allow simple tests of quantum non-locality (Mermin 1990b; Zeilinger 1999). We will not consider this aspect of the GHZ states here, but rather discuss their preparation and detection as a striking illustration of the complementarity principle and quantum eraser procedure.

Let us recall that an EPR entangled state of two atoms A_1 and A_2 can be generated in CQED by having atom A_1 emit a photon in the cavity with a 50% probability ($\pi/2$-Rabi pulse), and atom A_2 absorb at a later time the field with 100% probability (π-Rabi pulse). The Rabi pulses are produced, for both atoms, in a cavity exactly resonant with the $|g\rangle \rightarrow |e\rangle$ transition. The photon momentarily deposited in C 'catalyses' the entanglement between the atoms. As long as no information can be obtained about this transient state of the field and no answer given to the question '*was there a photon*

---

[15]Similar 2π-Rabi pulses are implemented in ion trap physics to realize phase gates. See Section 8.4.

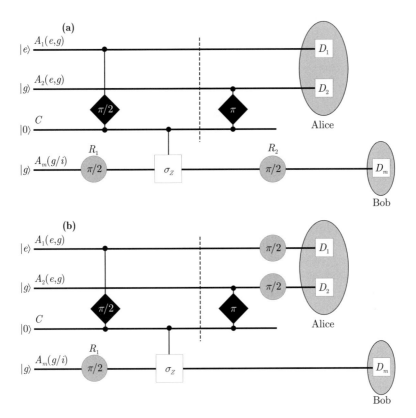

Fig. 6.14 Quantum circuits describing Alice's and Bob's three-particle entanglement experiment. (a) State preparation and test of qubit correlations in the energy basis. (b) Preparation and test of 'transverse' correlations: an illustration of the quantum eraser game.

in C between A_1 and A_2?', the system emerges in a linear superposition with equal weights of the two states $|e_1, g_2\rangle$ (corresponding to the 'no' answer to that question) and $|g_1, e_2\rangle$ (corresponding to the 'yes' answer). The A_1, A_2 system is then prepared in the EPR state whose expression was given by eqn. (5.48):

$$|\Psi(1,2)\rangle = (|e_1, g_2\rangle - |g_1, e_2\rangle)/\sqrt{2} . \tag{6.38}$$

A simple way to transform this experiment into a GHZ one is to add between A_1 and A_2 a third atom A_m which we use to measure the transient cavity field in a QND way. Reading this measurement would provide an answer to the question asked above and destroy the entanglement between A_1 and A_2. Before this reading is performed, though, an entangled triplet of particles is produced.

To make this experiment a little more dramatic, let us assume that it is performed by our friends Alice and Bob. She is making the EPR pair with A_1 and A_2 and the catalyst cavity C and she reads the final state of this pair with two detectors D_1 and D_2. Bob, on the other hand, is trying to find out the state of the field in C between A_1 and A_2. For this, he sneaks between A_1 and A_2 the QND meter atom A_m and measures the final state of this atom with a detector D_m.

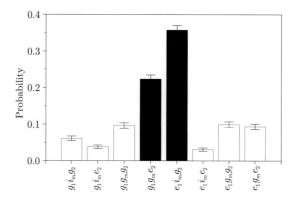

Fig. 6.15 Qubit measurements in the energy basis. Histograms of the detection probabilities of the eight possible detection channels. The two expected channels e_1, i_m, g_2 and g_1, g_m, e_2 (in black) clearly dominate the others (in white), populated by spurious processes. The error bars are statistical. Reprinted with permission from Rauschenbeutel *et al.* (2000), © AAAS.

The quantum circuit representing this joint experiment by Alice and Bob is shown in Fig. 6.14(a). Alice is performing the $\pi/2$- and π-Rabi pulses on her pair of atoms while Bob is applying the QND recipe on atom A_m, realizing a control-σ_Z phase gate with a 2π-Rabi pulse and sandwiching this gate between two Ramsey pulses. The Ramsey interferometer works on the $|g\rangle \rightarrow |i\rangle$ transition of A_m and its phase is set so that zero and one photons are correlated respectively to the states $|i\rangle$ and $|g\rangle$.

Note that Bob does need a QND meter for his attempt not to be a trivial tampering with Alice's procedure. If he used an ordinary detector instead – which would signal the presence of a photon by destroying it – the cavity would be left in vacuum after this measurement and A_2 would always be found by Alice in $|g_2\rangle$. Instead of an EPR pair, she would be left with an unentangled $A_1 - A_2$ system. By using a QND method, Bob can have access to information about the state of the field in C between A_1 and A_2, while leaving, until he performs a detection, a chance for A_2 to get entangled with A_1. In other words, the QND detection opens, as we will see, subtle possibilities for entanglement engineering and complementarity studies in a three-qubit system.

After Alice's and Bob's manipulations and before any detection is performed by the two partners, the system of the three atoms obviously emerges in the state:

$$|\Psi(1, m, 2)\rangle = (|e_1, i_m, g_2\rangle + |g_1, g_m, e_2\rangle)/\sqrt{2} , \qquad (6.39)$$

which is a coherent sum of two triplet states with equal weights. Moreover, in the first triplet component, each of the three qubits is in a state orthogonal to the one of the same particle in the second component. This is the definition of a GHZ state (Greenberger *et al.* 1990).[16]

---

[16]Let us remark that we do not have to wait until the time when the photon is absorbed by A_2 to get a GHZ state in this experiment. There is already, at the time marked by the vertical dashed line in Fig. 6.14(a), a GHZ state in the system. It involves then the two atoms A_1 and A_m (this one being in the middle of its measuring sequence) and the cavity field. We leave it as an exercise to find the expression of the 'two atom + field' GHZ state at this intermediate time.

Suppose that, in a first version of the experiment, Alice and Bob detect the energies of their qubits and correlate their results. Ideally, Alice must find with equal (50%) probabilities either the pair of states e_1, g_2, always correlated to Bob's finding i_m, or the pair g_1, e_2, always associated to Bob's g_m result. Out of the eight possible triplet combinations, only these two should be obtained. The actual result of this correlation experiment, performed with our CQED set-up, is displayed in Fig. 6.15 in a histogram giving the probabilities of the eight possible outcomes. The two expected correlations are clearly occurring with the highest probabilities.

The appearance with small probabilities of the six other state combinations and the imbalance between the two large peaks in the histogram result from imperfections of the set-up, among which the finite cavity damping time ($T_c = 1$ ms) plays a dominant role. Photon losses explain qualitatively why the population in the channel g_1, g_m, e_2 is lower than the one in e_1, i_m, g_2, which corresponds to an empty cavity between A_1 and A_2 and is thus not affected by field damping. A numerical simulation of the experiment taking into account the known sources of imperfections reproduces well the observed populations (Rauschenbeutel *et al.* 2000).

Seen from Alice's perspective, the meter atom A_m acts as an environment under Bob's control. The coherent superposition (6.39) expresses the entanglement between her system and this environment. If she does not know what Bob is measuring (or whether he measures anything at all), her knowledge about the $A_1 - A_2$ system is fully contained in the density operator obtained by tracing the three-qubit state over A_m. Alice's measurements alone are thus described by a statistical mixture, with equal probabilities, of $|e_1, g_2\rangle$ and $|g_1, e_2\rangle$. This expresses a well-known property of the GHZ states. When tracing over one qubit, they reduce to an incoherent mixture of tensor product states belonging to the two other qubits. As a result of the symmetrical form of this state, any of the three bits acts as a which-path detector for the two others.

Transverse correlations in the GHZ state: a quantum eraser again

To play a step further the which-path game, Bob can attempt to erase information he has got about the $|g_1, e_2\rangle$ and $|e_1, g_2\rangle$ quantum superposition. According to the recipe recalled in Section 6.1.5, he achieves this by mixing the meter states $|g_m\rangle$ and $|i_m\rangle$ before detecting the QND atom and communicating his result to Alice. She should then, in correlation with Bob's measurement, recover an entangled pairs of particles.

An obvious way for Bob to achieve this erasure scheme is merely to disconnect the R_2 Ramsey zone. This is equivalent to applying after this pulse a last transformation $U_{R_2}^{-1}$ undoing it and bringing A_m back to its state at cavity exit. The effect of $U_{R_2}^{-1}$ is to mix in equal proportions the two atomic states $|g_m\rangle$ and $|i_m\rangle$ so that a final detection of the atomic energy states in D_m amounts to a detection of one of the two $(|g_m\rangle \pm |i_m\rangle)/\sqrt{2}$ states in the first version of the experiment (i.e. when R_2 was active). When finding one of these states, Bob will be unable to tell whether there was a photon in C between A_1 and A_2 and the quantum ambiguity of Alice's entangled pair will be recovered.

Let us analyse in more detail this quantum eraser procedure. After the suppression of the U_{R_2} pulse (or equivalently the addition of the $U_{R_2}^{-1}$ one) and before any measurement, the triplet of qubits is prepared in the new state:

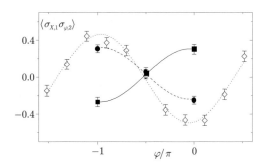

Fig. 6.16 Transverse correlations in the GHZ experiment. $A_1 - A_2$ atom Bell signal $\langle \sigma_{X,1}\sigma_{\varphi,2} \rangle_{\pm}$ versus φ. Open diamonds: no A_m atom sent. Solid circles: atom A_m detected in $|i\rangle$. Solid squares: atom A_m detected in $|g\rangle$. Error bars are statistical. Lines are sine fits. Reprinted with permission from Rauschenbeutel *et al.* (2000), © AAAS.

$$|\Psi^{\text{erase}}(1,m,2)\rangle = \frac{1}{\sqrt{2}}(|e_1,g_2\rangle + |g_1,e_2\rangle)\,|g_m\rangle + \frac{1}{\sqrt{2}}(|e_1,g_2\rangle - |g_1,e_2\rangle)\,|i_m\rangle \ . \quad (6.40)$$

The detection of A_m then projects Alice's state into one of the two entangled pairs $(|e_1,g_2\rangle \pm |g_1,e_2\rangle)/\sqrt{2}$ with the $+$ and $-$ sign correlated, respectively, to Bob's $|g_m\rangle$ and $|i_m\rangle$ results. In order to probe her entangled states, Alice can perform a transverse pseudo-spin measurement on A_1 and A_2, following the procedure described in Section 5.4.4.

She applies to both her atoms, before detection, a $\pi/2$-classical pulse. She tunes the common frequency of these two pulses around the $|e\rangle \to |g\rangle$ frequency, thus selecting for each value of this frequency an angle φ between the two transverse pseudo-spin components she is finally detecting. She then gets, upon detection, the product $\sigma_{X,1}\sigma_{\varphi,2}$. Correlating her results with those of Bob and averaging them over many realizations of the experimental sequence, she constructs two Bell signals $\langle \sigma_{X,1}\sigma_{\varphi,2}\rangle_{\pm}$, which correspond respectively to Bob's $|g_m\rangle$ and $|i_m\rangle$ readings.

The quantum circuit corresponding to this experiment performed jointly by Alice and Bob is shown in Fig. 6.14(b). It differs from the first experiment's circuit by the adjunction of two $\pi/2$-pulses on A_1 and A_2 and the suppression of the second pulse acting on A_m. A mere inspection of this circuit provides a simple interpretation of (6.40). The A_m atom undergoes a $\pi/2$-pulse before the cavity and crosses it in a superposition of $|g\rangle$ and $|i\rangle$. The effect of the π-phase gate is merely to shift by π, if A_m is in $|g\rangle$, the relative phase of the field $|0\rangle$ and $|1\rangle$ states.[17] After this field has been absorbed by A_2, this shift results in a change of the sign of the coherence between $|e_1,g_2\rangle$ and $|g_1,e_2\rangle$. Since no second Ramsey pulse is applied to A_m in this version of the experiment, the final state of the QND atom keeps information about this sign change and the final states $|g_m\rangle$ and $|i_m\rangle$ are correlated to two π-out-of-phase EPR states of A_1 and A_2. These states are detected with the transverse arrangement provided by the two $\pi/2$-pulses applied to these atoms before their detection.

The results of this correlation experiment, performed with our circular atom CQED set-up, are displayed in Fig. 6.16, where the two Bell signals are shown as a function of φ. They are compared with the ordinary transverse EPR Bell signal obtained by performing the same experiment without sending the QND atom A_m. These three signals have, as expected, a sinusoidal shape revealing the correlations observed between the

---

[17]A control-π phase gate is a symmetrical device. In Fig. 6.14, the field is the control and the atom the target. The same transformation is achieved by exchanging the qubits roles. The atom in $|g\rangle$ (active state) then shifts by π the relative phase of $|0\rangle$ and $|1\rangle$.

two partners in an entangled pair of pseudo-spins (Section 5.4.4). We observe moreover that the $\langle \sigma_{X,1}\sigma_{\varphi,2} \rangle_-$ signal, correlated to A_m detected in $|i\rangle$ is the same as the ordinary EPR signal, while the $\langle \sigma_{X,1}\sigma_{\varphi,2} \rangle_+$ signal, correlated to A_m found in $|g\rangle$ has the opposite phase. This phase shift is the one expected from a change of sign between the two components of an EPR pair. This is the result predicted by the above analysis and a mere inspection of eqn. (6.40).

Note that if Alice does not know Bob's results and collects all her results irrespective of what happens to A_m, these transverse correlations disappear. By summing the two $\langle \sigma_{X,1}\sigma_{\varphi,2} \rangle_\pm$ signals, she obtains a zero transverse $A_1 - A_2$ correlation, an evidence that she has got a statistical mixture instead of an EPR pair. We retrieve here an essential feature of the conditional quantum eraser. It is essential for the coherence to be restored that a proper reading is performed in the environment (here A_m) and the results of this reading must be correlated to the measurement performed on the system (here the $A_1 - A_2$ pair).

We have analysed, for the sake of simplicity, these correlations as if Alice and Bob could postpone their detections until after the preparation of their system has been completed and perform thereafter delayed-choice measurements. In the actual experiment, the precise timing of the three-atom interactions with the cavity field and the Ramsey pulses was very tight[18] and the meter atom was detected before A_2 had interacted with C. The entangled triplet actually prepared was made of A_1, A_m and the photon qubit in C (footnote 16, on page 330), A_2 being merely used to detect the field afterwards. We already noticed however, when studying EPR pairs in Chapter 5, that quantum correlations do not depend on the timing of the detection events in this kind of experiment. It is only important here that the QND meter atom interacts with C before the field produced by A_1 has appreciably decayed.

Extension to more particles: a possible route towards Schrödinger cats

The combination of the measurements performed in energy and transverse bases demonstrates that these CQED experiments prepare a three-particle entangled state involving non-classical correlations. The detailed analysis of measurement errors and the estimation of the fidelity of the procedure is rather technical, because we have to take into account many imperfections which introduce spurious detection channels in the first experiment and reduce the contrast of the Bell signal in the second (Rauschenbeutel *et al.* 2000). We will not enter in this discussion here. Let us only notice that experimental limitations precluded, at the time of this study, entanglement knitting between more than three particles.

With an improved version of the set-up, involving a better cavity and a higher contrast of Ramsey and Rabi pulses, these experiments could be generalized to larger atom numbers. The simplest extension will be to repeat the QND measurement of the field produced by atom A_1, using a sequence of meter atoms $A_{m_1}, A_{m_2}, \ldots, A_{m_q}$, before finally sending A_2 to absorb the field and bring it back to vacuum. The transient photon in the cavity will be used as a sort of optical tap, storing information which will repeatedly be transferred to successive meter atoms. The generalized GHZ or 'Mermin' state (Mermin 1990a):

---

[18]For a detailed analysis of this timing, see Raimond *et al.* (2001).

$$|\Psi(1, m_1, m_2, \ldots, m_q, 2)\rangle = \frac{1}{\sqrt{2}} \left[\left| e_1, i_{m_1}, \ldots, i_{m_q}, g_2 \right\rangle + \left| g_1, g_{m_1}, \ldots, g_{m_q}, e_2 \right\rangle \right] , \quad (6.41)$$

involving $q + 2$ entangled particles will be generated in this way. The first and last atoms A_1 and A_2 will be used to prepare a large number of meter atoms A_{m_i} in a superposition in which they are all in either one of two orthogonal states. This is a mesoscopic superposition, a kind of Schrödinger cat. The procedure suggested here is very challenging because it relies on successive interactions of many atoms with the same cavity field, which must be completed within a time very short compared to T_c. There are, as we will see in next chapter, less demanding methods to generate and study Schrödinger cat states in CQED experiments.

6.4   The quantum analogue/digital converter

6.4.1   A dispersive single-photon QND measurement

The SP-QND photon counter suffers from a severe limitation, since it can detect at most a single photon. We analyse now a general procedure which can count photons in much larger numbers. Its principle is still based on the acquisition of information by atoms crossing a cavity inserted between the Ramsey zones of an atomic interferometer. Information about the photon number is again imprinted onto the phase of an atomic coherence read by the interferometer.

The difference with the SP-QND scheme is that the interaction between the atom and the field is non-resonant. The effect of a dispersive coupling has been analysed in Section 3.4.4, in the perturbative limit of large atom–cavity detuning. Inside a cavity containing a n-photon Fock state, the energy of state $|g\rangle$ is shifted by an amount $-\hbar s_0 n$, with:

$$s_0 = \frac{\Omega_0^2}{4\Delta_c} . \quad (6.42)$$

This shift is proportional to the photon number and inversely proportional to the detuning Δ_c between the field and the $|e\rangle \rightarrow |g\rangle$ transition frequencies. Equation (6.42) is valid for relatively large detunings and moderate photon numbers such that $\Delta_c \gg \Omega_0 \sqrt{n}$. In practice, the detuning Δ_c must be not too large, typically of the order of a few Ω_0, for s_0 not to be negligibly small.

During the atom–cavity interaction, the state $|g\rangle$ accumulates a phase shift $n\Phi_0$, where:

$$\Phi_0 = s_0 t_i^d = \frac{\Omega_0^2 t_i^d}{4\Delta_c} , \quad (6.43)$$

is the phase-shift per photon, proportional to the effective atom–field interaction time t_i^d. Level $|i\rangle$, involved in transitions far-detuned from the cavity mode, remains practically un-shifted and can, as in the SP-QND scheme, serve as a phase reference.[19]

With slow atoms and moderate detuning values, it is in principle possible to fulfil the condition $\Phi_0 = \pi$. The g/i coherence is then shifted by π when the atom crosses

---

[19]In a dispersive experiment, the $|e\rangle \rightarrow |g\rangle$ transition should be preferred to the $|g\rangle \rightarrow |i\rangle$ one for the realization of the Ramsey interferometer. Its light shift is twice as large (levels $|e\rangle$ and $|g\rangle$ experience opposite shifts). For a more direct comparison with the resonant SP-QND procedure, we choose to discuss here a scheme using an interferometer based on the $|g\rangle \rightarrow |i\rangle$ transition.

a cavity containing one photon. This phase shift can be probed by the Ramsey interferometer described in Section 6.2.2. The dispersive version of the CQED photon counter is thus, for single photons, fully equivalent to the resonant SP-QND device.

The dispersive scheme has, however, the decisive advantage of working for larger photon numbers as well. The meter atom behaves as a qubit, its two states $|g\rangle$ and $|i\rangle$ standing, as in the previous section, for the logical states $|1\rangle$ and $|0\rangle$. It is convenient to introduce the Pauli matrices σ_Z, σ_X and σ_Y associated to this qubit and to express the interaction of the meter atom with the signal field as an effective Hamiltonian obtained by modifying eqn. (3.191) to take into account the fact that only one of the qubit states is now sensitive to the field:

$$H_{\mathcal{SM}} = \frac{\hbar s_0}{2}(\sigma_Z - \mathbb{1})a^\dagger a \ . \tag{6.44}$$

It is easy to check that this coupling produces the expected shift on an atom in $|g\rangle$, while leaving an atom in $|i\rangle$ unperturbed since $(\sigma_Z - \mathbb{1})|0\rangle = 0$. This meter–signal interaction satisfies all the QND criteria. The Ramsey interferometer is sensitive to a combination of the transverse components of the pseudo-spin σ_X and σ_Y (playing the role of $O_{\mathcal{M}}$) which does not commute with $H_{\mathcal{SM}}$, thus satisfying (6.18). In addition, this coupling commutes with $N = a^\dagger a$, satisfying the sufficient QND condition (6.19). Most importantly, the QND conditions are fulfilled for all the values of s_0 and t_i^d, and not, as was the case in the SP-QND scheme, for a very peculiar tuning of the atom–field coupling parameters only. This, as we will see, gives much more flexibility to the method, with the possibility of tuning the coupling parameters differently for successive atoms crossing the cavity.

The most important feature of this dispersive scheme is that the atomic coherence phase-shift is proportional to the photon number, for all photon numbers matching the dispersive regime condition. When $\Phi_0 = \pi$, any even photon number leads to the same fringe pattern as the vacuum and any odd photon number to the same pattern as a one-photon state. Tuning the interferometer phase at $\varphi = 0$, we get with unit probability an atom in $|g\rangle$ when the parity of the photon number is odd and an atom in $|i\rangle$ when the parity is even.

The generalized dispersive scheme realizes thus, for $\Phi_0 = \pi$, not just a non-destructive measurement of a single photon, but more generally, a QND measurement of the photon number parity. We will see in Section 6.5 that this measurement is an important ingredient for the direct determination of the cavity field Wigner function. We show here that this parity measurement is the first step in a full-fledged QND photon counting scheme using a string of atoms crossing the cavity, each being subjected to conveniently chosen dispersive phase Φ_0 and Ramsey interferometer phase φ (Brune *et al.* 1992; Haroche *et al.* 1992a).

6.4.2 Principle of a general QND measurement

Let us consider first an atom A_0 interacting with a n-photon Fock state. Using eqns. (6.28) and (6.29), we get the atom–field state transformation produced by the dispersive coupling and by the Ramsey interferometer:

$$|g, n\rangle \longrightarrow b_g(n)|g, n\rangle + b_i(n)|i, n\rangle \ , \tag{6.45}$$

with:

$$b_g(n) = \tfrac{1}{2}e^{-i\varphi}\left[e^{i(n\Phi_0+\varphi)} - 1\right] \; ; \qquad b_i(n) = \tfrac{1}{2}\left[e^{i(n\Phi_0+\varphi)} + 1\right] . \qquad (6.46)$$

The conditional probability $P_{g|n}^{(0)}$ of detecting atom A_0 in $|g\rangle$ knowing that the photon number is $|n\rangle$ is equal to $|b_g(n)|^2$, an oscillating function of the photon number n:

$$P_{g|n}^{(0)} = [1 - \cos(n\Phi_0 + \varphi)]/2 . \qquad (6.47)$$

If the cavity field is initially in a coherent sum of Fock states, $|\Psi_0\rangle = \sum_n c_n |n\rangle$, the dispersive interaction and the coupling with the Ramsey interferometer lead, by mere superposition, to the atom–cavity state:

$$|\Psi_0\rangle |g\rangle \longrightarrow \sum_n c_n \left[b_g(n) |g, n\rangle + b_i(n) |i, n\rangle\right] , \qquad (6.48)$$

which is generally entangled. The probability, $P_g^{(0)}$, of detecting the atom in $|g\rangle$ is entirely determined by the initial photon number distribution $p_0(n) = |c_n|^2$ and by the conditional probabilities $P_{g|n}^{(0)}$:

$$P_g^{(0)} = \sum_n p_0(n) P_{g|n}^{(0)} . \qquad (6.49)$$

Remarkably, this result, obtained by application of the quantum measurement postulate, can be retrieved by a simple classical argument. The probability of finding the atom in $|g\rangle$ must, by definition, be equal to the sum of the conditional probabilities of finding it in this state, provided that there are n photons in the field, weighted by the probabilities of occurrence of each photon number, as expressed by eqn. (6.49). That a classical argument applies here reflects that the non-resonant atom–field interaction, and thus the QND procedure, are insensitive to the coherence between Fock states. The system's evolution is the same as it would be for a classical mixture of Fock states with an identical initial photon number distribution.

The atomic detection results in a projection of the atom–cavity entangled state. If A_0 is detected in $|g\rangle$, for instance, the cavity state collapses into:

$$|\Psi_1\rangle = \frac{1}{\mathcal{N}_0} \sum_n c_n b_g(n) |n\rangle , \qquad (6.50)$$

where the normalization factor $\mathcal{N}_0$ is equal to $\sqrt{P_g^{(0)}}$. The new photon number distribution $p_1(n|g)$, conditioned to the detection of A_0 in $|g\rangle$, is thus:

$$p_1(n|g) = p_0(n) P_{g|n}^{(0)} / P_g^{(0)} . \qquad (6.51)$$

Similarly, an atomic detection in $|i\rangle$, occurring with the probability $P_i^{(0)} = 1 - P_g^{(0)}$, leads to the photon number distribution:

$$p_1(n|i) = p_0(n) P_{i|n}^{(0)} / P_i^{(0)} , \qquad (6.52)$$

where $P_{i|n}^{(0)} = 1 - P_{g|n}^{(0)}$.

The formulas (6.51) and (6.52) have been derived by a straightforward application of the quantum projection postulate but, here again, these results can be retrieved by a classical probabilistic argument. We recognize the expression of Bayes law which relates with each other reciprocal conditional probabilities (compare with eqn. 6.33).

The main result expressed by these formulas is that the first atom measurement has modified the initial photon number distribution, which has been multiplied by an oscillating function of the photon number $[P_{g|n}^{(0)}$ or $P_{i|n}^{(0)} = 1 - P_{g|n}^{(0)}]$. All photon numbers corresponding to $P_{g|n}^{(0)} = 0$ are eliminated altogether by a detection of A_0 in $|g\rangle$ and those close to these numbers end up with a reduced probability. The same happens to the photon numbers satisfying $P_{i|n}^{(0)} \approx 0$, if the atom is detected in $|i\rangle$. Through this first atomic detection, we get partial information on the cavity field, corresponding to a 'decimation' of some photon numbers in the initial $p_0(n)$ distribution. This result is quite general and occurs whatever the setting of the interferometer and the value of the phase shift per photon Φ_0.

We must stress that this modification does not result from photon absorption or emission (the atom–field interaction is non-resonant and forbids these exchanges) but follows from an acquisition of information about the field, which can be analysed classically and understood by simple logical arguments. If the atom has been detected in $|g\rangle$, for instance, it is clear that photon numbers which correspond to a vanishing probability of finding the atom in this level become *ipso facto* impossible. The way the photon number probability redistributes among the other n values depends upon the known $P_{g|n}^{(0)}$ distribution and is given by the Bayes relation (6.51), which follows directly from classical probability laws.

A classical probability game to illustrate the photon decimation process

The modification of a probability distribution induced by an acquisition of information might seem strange. This feeling has in fact nothing to do with quantum weirdness, but comes from a lack of intuition about classical conditional probabilities. It is easy to find situations implying 'probability decimations', in which wrong guesses are often made in a context quite similar to the QND process described here. An example from everyday life[20] will illustrate this point. Suppose that Alice and Bob are playing a simple game. Alice has hidden a gold coin in one out of three identical boxes labelled 1, 2 and 3 and she gives two guesses to Bob to find out where the bounty is. At the start of the game, Bob's knowledge about the position of the coin is described by a flat distribution: $p(1) = p(2) = p(3) = 1/3$ (Fig. 6.17a). He makes a first guess, say 1. Then Alice gives him a hint by opening one of the two other boxes, which is empty (let us assume it is box 2).

Should then Bob change his first guess? In other words, has his knowledge about the location of the gold coin been altered as a result of the negative measurement corresponding to Alice's hint? Many people picked at random (even statisticians) will – too rapidly – answer that the probabilities of finding the coin have become $p(1) = p(3) = 1/2$ and that Bob has no rationale to change his initial choice. This, in fact, is wrong. Bob had, a priori, a $2/3$ probability of having wrongly chosen box

[20]This example is borrowed from an actual TV game which has triggered long discussions about statistics and probabilities in the science page of the *New York Times* in the early 1990s.

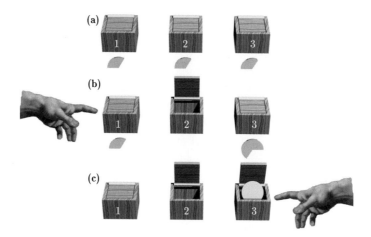

Fig. 6.17 A classical game based on a probability decimation process. (a) At the beginning of the game, the probability distribution of finding the coin in one of the three boxes is flat. (b) After Bob chooses box 1 and Alice's tells him that the coin is not in box 2, the distribution is modified, which suggests to Bob to change his guess and pick box 3 instead of 1. (c) Bob points to box 3, and is twice more likely to win than after his first choice.

1 and Alice's indication tells him that, had he been wrong, then the coin has to be *certainly* in box 3, whose probability thus jumps to $2/3$ (note the conditional argument made here). The probability distribution after Alice's *decimation* of box 2 has become $p(1) = 1/3$, $p(2) = 0$ and $p(3) = 2/3$ (Fig. 6.17b).

By picking box 1, Bob is asking Alice to tell him which of the two other boxes does not store the coin. By performing this 'experiment', he obtains precious information which alters his knowledge drastically. Based on this information, he is twice as likely to win if he changes his mind and picks box 3 instead of 1 (Fig. 6.17c)! This game exploits a probability decimation process which bears some similarity with the photons in the box game we are studying here. The analogy might give a better feeling about how information acquisition changes the statistical knowledge we have about the quantum field in the QND experiment.

Sending another atom A_1 through the cavity, with a different phase shift Φ_0 and/or a different interferometer setting φ, will decimate other photon numbers in the distribution. We may thus expect that, after a large enough sample of atoms has been detected, the photon distribution reduces to a single photon number, completing the QND measurement process. We show in the next paragraph that this complete decimation can be performed in an optimal way with a careful choice of the parameters for each detected atom.

'Inferred' versus actual initial photon number distribution

An important point at this stage is that the observer, performing the QND detection, does not have, in general, a complete knowledge about the initial cavity state and photon number distribution $p_0(n)$, which is 'hidden' from him. He can only 'infer' a photon number distribution $p^i(n)$ from information he might have got initially and from information he gathers from successive atomic detection events. If we assume no a *priori* knowledge whatsoever, the initial inferred distribution spans all photon numbers. As shown in the next paragraph, the decimation procedure then never converges.

We assume thus that the observer has been given an upper bound, n_M, on the photon number ($n < n_M$). This is, as we will see soon, a quite realistic hypothesis in most cases. The only reasonable guess the observer can make for the initial inferred distribution $p_0^i(n)$ is to take it as constant for all photon numbers between 0 and $n_M - 1$: $p_0^i(n < n_M) = 1/n_M$. Note that Bob made the same uniform guess at the beginning of his game with Alice!

The detection of A_0 provides the observer with information. The inferred distribution undergoes then the same decimation as the actual one. Provided A_0 is detected in $|g\rangle$, the new inferred distribution is:

$$p_1^i(n) = \frac{p_0^i(n) P_{g|n}^{(0)}}{\sum_n p_0^i(n) P_{g|n}^{(0)}} \, , \tag{6.53}$$

a result which follows again directly from Bayes law, merely replacing in eqn. (6.51) the hidden distribution $p_0(n)$ by $p_0^i(n)$. The decimation process goes on with subsequent atoms and finally reduces the inferred distribution to a single photon number, the final outcome of the QND process. The required number of atoms might, however, be larger than if the observer had initially complete information on $p_0(n)$. It is important to note that the detection probabilities of each atomic event are determined by the actual distribution (eqn. 6.49 for the first atom), and not by the inferred one. The observer, however, does not need information on the detection *probabilities* to perform his measurement. He only needs to list the atomic detection *results*.

6.4.3 An optimal convergence procedure

A proper tuning of the atomic probes sent through the cavity makes it possible to perform an 'optimal' (with respect to classical information theory) decimation of the photon number distribution (Haroche *et al.* 1992b). We first discuss the procedure in a very simple case, when the upper bound on photon number is $n_M = 4$. We then generalize it to arbitrary maximum photon numbers.

Initially the inferred probability distribution $p_0^i(n)$ is flat between 0 and 3, and there are *a priori* four possible measurement results. The photon number can be coded, in binary form, with two classical bits and we will see that two atoms are enough to find it out. We send in the cavity a first atom, A_0, whose interaction time is tuned for $\Phi_0 = \pi$ and we set the interferometer phase at $\varphi = 0$. The detection of A_0 in $|g\rangle$ indicates that the photon number has an odd parity, whereas this parity is even when the atom is detected in $|i\rangle$. Atom A_0 reveals the value of the least significant bit in the binary expansion of the photon number. Let us assume, for the sake of definiteness, that A_0 is detected in $|g\rangle$. The only two possible results left are $n = 1$ and 3. The inferred photon distribution, taking into account information provided by A_0, is $p_1^i(1) = p_1^i(3) = 1/2$.

A second atom A_1, providing an extra bit of classical information, is required to remove the remaining ambiguity. Considering eqn. (6.47), we choose the phases $\Phi_0 = \pi/2$ and $\varphi = -\pi/2$ so that $P_{g|1}^{(1)} = 0$ and $P_{g|3}^{(1)} = 1$. The detection of A_1 then unambiguously pins down the photon number by determining its most significant bit.

If it is found in $|g\rangle$ or $|i\rangle$, we obtain $n = 3$ or $n = 1$ respectively. Note that, when A_0 is detected in $|i\rangle$, the remaining photon numbers (0 and 2) can be distinguished by setting $\Phi_0 = \pi/2$ and $\varphi = 0$ for A_1. To conclude on this simple example, the four possible results $n = 0, 1, 2$ and 3 correspond to the atomic pair A_1, A_0 found, in that order, in the combined states $|i, i\rangle$, $|i, g\rangle$, $|g, i\rangle$ and $|g, g\rangle$. With our logical bit assignment, these readings correspond to the binary codings $[0, 0]$, $[0, 1]$, $[1, 0]$ and $[1, 1]$, i.e. to the expression of n in basis 2.

The $\Phi_0 = \pi/2$ condition can be satisfied by selecting for A_1 a velocity twice larger than for A_0, or by switching A_1 very far from cavity resonance through the Stark effect after a proper effective interaction time. Atom A_1 is twice less sensitive to the photon number than A_0. While A_0 measures the parity of n, A_1 determines the parity of the integer part of $(n/2)$, distinguishing 1 from 3 or 0 from 2.

The generalization of this 'photon decimation' procedure to larger fields, for which the initial bound on the photon number, n_M, is arbitrary, is straightforward. As suggested by this preliminary discussion, we use a binary representation for the photon number n:

$$n = [\epsilon_q, \ldots, \epsilon_k, \ldots, \epsilon_1, \epsilon_0] = \sum_k \epsilon_k 2^k \ , \tag{6.54}$$

where the $\epsilon_k = 0, 1$ are the successive bits of the photon number and q is the smallest integer such that $q \geq \log_2(n_M)$ ($\log_2$ being the base-two logarithm).

We first determine the least significant bit ϵ_0, i.e. the parity of the photon number, by sending a first atom A_0 with $\Phi_0 = \pi$ and $\varphi = 0$. The next bit is determined with the help of a second atom A_1 with $\Phi_0 = \pi/2$, the interferometer phase φ being set, as discussed above, according to the results of the detection of A_0. We assume, for an inductive argument, that all bits $\epsilon_0, \ldots, \epsilon_{k-1}$ have been measured with atoms $A_0, \ldots, A_{k-1}$. For the determination of ϵ_k, we send A_k, with the dispersive phase:

$$\Phi_0 = \pi/2^k \ . \tag{6.55}$$

From eqn. (6.47), we get then:

$$P_{g|n}^{(k)} = \frac{1 - \cos\left([\epsilon_q, \ldots, \epsilon_k]\pi + \varphi + [\epsilon_{k-1}, \ldots, \epsilon_0]\pi/2^k\right)}{2} \ . \tag{6.56}$$

Setting:

$$\varphi = -\frac{[\epsilon_{k-1}, \ldots, \epsilon_0]}{2^k}\pi \ , \tag{6.57}$$

we get finally:

$$P_{g|n}^{(k)} = \frac{1 - \cos\left([\epsilon_p, \ldots, \epsilon_k]\pi\right)}{2} = \frac{1 - \cos\left(\epsilon_k \pi\right)}{2} \ . \tag{6.58}$$

The detection of A_k in $|g\rangle$ indicates that the photon number's kth bit, ϵ_k, is one (a detection in $|i\rangle$ indicates on the contrary that this bit is zero). The inferred probability distribution at this stage, $p_k^i(n)$, is a comb of values, with a spacing 2^{k+1}, leaving $n_M/2^{k+1}$ possibilities open for n. There is a single possibility left, n_m, completing the photon number measurement, when $q \approx \log_2(n_M)$ atoms have been detected.

The convergence of the process requires an initial bound for n. This is, however, not a very restrictive condition. In most situations, it is quite easy to give a conservative upper bound for the field energy. At least, the photon number is certainly bounded by the cavity's mechanical resistance to the radiation pressure of the field it contains! The corresponding photon number might be quite large, but the logarithmic dependence of the atom count on n_M would still make the measurement possible (in principle only, since the dispersive regime condition is not likely to be met for such large fields!).

This 'photon decimation process' is optimal in terms of classical information theory since it uses the minimum number of bits required to code the number of photons. It bears a strong analogy with the internal workings of a standard Analogue/Digital (A/D) converter, which transform a bounded analogue input voltage V ($0 \leq V \leq V_M$) into the binary expansion of the ratio V/V_M. Most converters start by comparing the input voltage with $V_M/2$, providing the most significant bit. The voltage is then translated into the $[0, V_M/2]$ interval by a programmable voltage source, and the result is then compared to $V_M/4$, providing the next bit. The voltage being a continuous variable, the binary expansion of the V/V_M ratio is unbounded. The process stops when the required precision is obtained or when the incremental precision gain is lower than the electronic noise. Besides an obvious bit reversal (we determine here the least significant bit first), our QND decimation procedure is very similar to the A/D operation. We have thus coined the name 'quantum A/D converter' for the process.

What determines the probability of getting, at the end of the decimation, a photon number n_m? The answer is of course given by the quantum measurement postulate. We realize an ideal measurement of the photon number and the probabilities of the outcomes is given by the initial hidden distribution $p_0(n)$. By resuming very many realizations of the decimation process, starting each time afresh from the same initial field, we build the probability distribution of the results, $p_m(n_m)$, which must be identical to $p_0(n)$. The hidden probability distribution, and its evolution throughout the measurement process, determines the probabilities of the atomic detection outcomes, in such a way that the final measurement result obeys the proper statistics.

We have checked that the QND method does indeed work in this ideal way by performing simulations of the decimation procedure. Figure 6.18 presents the evolution of the photon number distribution in the simulation of the measurement of a coherent state with $\bar{n} = 10$ photons on the average (the corresponding Poissonian distribution of the photon number being the 'hidden' distribution). The initial photon number bound chosen by the observer is $n_M = 32$ (a good guess since the initial probability of having more than 31 photons is quite negligible). For each atom, we randomly choose the detection outcome using the hidden photon distribution (the procedure to make this random choice is the same as in the Monte Carlo simulations described in Chapter 4).

The simulation simultaneously follows the decimation of the inferred and actual photon number distributions. We plot in Fig. 6.18 the distribution $p_{k+1}^i(n)$ inferred by the observer after he sends atom A_k and the hidden distribution $p_{k+1}(n)$ at the same stage for $k = 0$ to 4. The initial distribution has a flat histogram over the 32 allowed photon number values. Atom A_0, detected in $|g\rangle$, indicates that the field parity is odd. After this detection, half of the photon numbers (the even ones) disappear from the distribution. Atom A_1 in turn removes half of the remaining photon numbers. The procedure continues until atom A_4, which leaves only one possible photon number ($n_m = 7$), completing the QND measurement. At this final stage, the inferred and

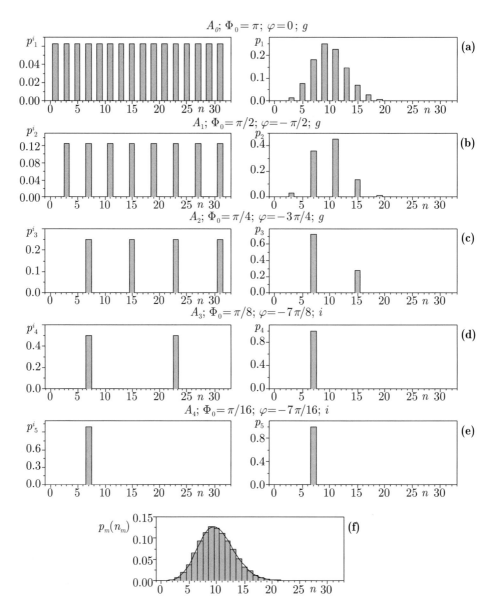

Fig. 6.18 Numerical simulation of a single realization of the optimal photon decimation process. The initial field contains less than 32 photons and can be expressed with five bits coded in five atoms. (a)–(e) Inferred (left) and hidden (right) photon number distributions $p^i_{k+1}(n)$ and $p_{k+1}(n)$ obtained after the interaction with atom A_k ($k = 0, \ldots, 4$). The values of Φ_0 and φ for atom A_k and the detection outcomes are indicated on top of each frame. (f) Measured photon number probability distribution, $p_m(n_m)$, averaged over 10 000 realizations of the process. The solid line corresponds to the initial hidden photon number distribution, $p_0(n)$ (coherent state with ten photons on the average).

hidden distributions, both reduced to a single photon number, of course coincide. Note that the hidden distribution is already pinned down to a single photon number by the detection of A_3, since the probability of having $n = 7 + 16 = 23$ photons in the initial state is negligible. By averaging very many realizations of the decimation process, we obtain the last histogram at the bottom of Fig. 6.18. Within statistical noise, it corresponds to the initial hidden photon number distribution for a 10-photons coherent state, recalled here as a solid line.

In the particular case in which the initial cavity state is a Fock state, with photon number n_0, there is no randomness left in the process. The detection state for each atom is unambiguously determined by the successive bits in the binary expansion of n_0 and the final outcome, revealed to the observer at the end of the procedure, is always n_0. This certainty of the result when the measurement is performed on an eigenstate of the measured observable is a characteristic feature of a QND procedure.

In a given run of the decimation process, the measured photon number n_m can be quite different from the initial average photon number. This is not in contradiction with the fact that the atoms, interacting dispersively with the cavity mode, do not modify its energy. A measurement of the photon number can give any result which is within the initial dispersion, without feeding or extracting energy. Of course, energy conservation is verified in an average over many realizations of this process.

The practical implementation of this optimal photon decimation process is very demanding. Setting the phase of the interferometer and the dispersive interaction time of the atoms according to the previous atomic detection is certainly within reach of modern computer control. The main obstacle is cavity field decoherence. Any photon loss during the measurement would make the process fail. This is easily understood by remarking that a single photon loss changes the parity of the photon number, the first bit to be determined. The whole process should thus be realized in a time short compared to the lifetime of the maximum photon number state, $T_c/(n_M - 1)$. The minimum time between successive atoms could be in the few hundreds microsecond range, corresponding to the time of flight between C and the field-ionization detector D (Fig. 5.15, on page 252). With a cavity lifetime in the 100 ms range, within reach of present technology, photon numbers up to $\sim 2^5 = 32$ could be measured. A train of five atoms – each crossing the apparatus within ≈ 100 μs – could perform a complete QND measurement within 0.5 ms, a time much shorter than $T_c/2^5 \approx 3.3$ ms.

6.4.4 The fate of the phase

By pinning down the photon number, a QND detection always randomizes phase information initially present in the field (see Section 6.2.1 for a discussion of this randomization process in an opto-mechanical detection scheme). The phase blurring occurring in the quantum A/D operation is an interesting phenomenon, which has a strong connexion with the generation of Schrödinger cat states of the field whose study is the main topics of next chapter. We focus here on the evolution of the phase of a coherent field, as successive atoms interact with it and determine bit by bit its photon number. To be specific, we choose an initial coherent field $|\Psi_0\rangle = |\alpha\rangle$ with $\bar{n} = |\alpha|^2 = 10$ photons on the average (corresponding to the numerical simulation presented above). For the sake of definiteness, we assume that α is a real amplitude (zero phase). The complete evolution of the field state can be followed throughout

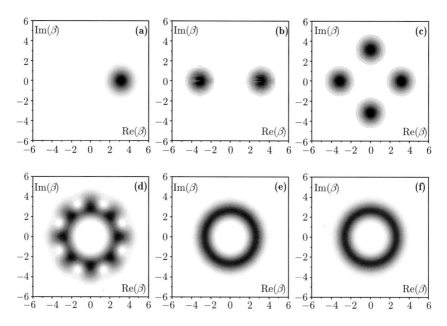

Fig. 6.19 Evolution of the field $Q(\beta)$ quasi-probability distribution in the optimal decimation process of a coherent state with $\bar{n} = 10$. Same simulation as in Fig. 6.18. (a) Initial state. (b)–(f) Density plots of $Q(\beta)$ in phase space after detection of atom A_k ($k = 0, \ldots, 4$).

the decimation process using eqn. (6.50) for each measuring atom A_k to obtain the new field state. We leave this as an exercise and we present here a more qualitative discussion.

Let us first examine the action of the first atom, A_0, which measures the field parity. The phase settings are $\Phi_0 = \pi$ and $\varphi = 0$. Atom A_0 enters the cavity in a superposition of $|g\rangle$ and $|i\rangle$. When in $|i\rangle$, it has no action on the non-resonant cavity field. On the contrary, when in $|g\rangle$, it transiently shifts the cavity resonance frequency and modifies the phase accumulation of the various $|n\rangle$ states over which the coherent state of the field is expanded. More precisely, the atom–field system evolves as:

$$|g\rangle\,|\alpha\rangle = e^{-|\alpha|^2/2}\sum_n \frac{\alpha^n}{\sqrt{n!}}\,|g\rangle\,|n\rangle \to e^{-|\alpha|^2/2}\sum_n \frac{\alpha^n e^{in\Phi_0}}{\sqrt{n!}}\,|g\rangle\,|n\rangle = |g\rangle\,\big|\alpha e^{i\Phi_0}\big\rangle \;. \quad (6.59)$$

The atom, if in $|g\rangle$, offsets by Φ_0 the phase of the coherent field. For $\Phi_0 = \pi$, this shift amounts to a reversal of the coherent field amplitude, whose state becomes $|-\alpha\rangle$. Immediately after its interaction with C and before R_2, A_0 is entangled with the field, its states $|i\rangle$ and $|g\rangle$ being respectively correlated to $|\alpha\rangle$ and $|-\alpha\rangle$. The atom then interacts with the second Ramsey pulse R_2 before being finally detected in $|g\rangle$ or $|i\rangle$. This detection, occurring after the state-mixing by R_2, does not yield information about the state of A_0 while it was inside C. It thus projects the field onto a quantum superposition $|\Psi_1\rangle$ of the two coherent states corresponding to the interaction with $|g\rangle$ and $|i\rangle$:

$$|\Psi_1\rangle = \frac{1}{\sqrt{\mathcal{N}}} (|\alpha\rangle \pm |-\alpha\rangle) \ . \tag{6.60}$$

The $+$ $(-)$ sign applies when A_0 is detected in $|i\rangle$ $(|g\rangle)$ and $\mathcal{N}$ is a normalization factor.

The field left in C is a quantum superposition of two coherent components with opposite phases, a non-classical mesoscopic state superposition. Most of the next chapter is devoted to the study of these states and to the analysis of their decoherence. We will not enter in a detailed discussion of their properties now. Note only that the superposition $|\alpha\rangle + |-\alpha\rangle$, when developed over the Fock state basis, involves even photon numbers, whereas $|\alpha\rangle - |-\alpha\rangle$ only expands on odd photon numbers (see Section 7.1). The detection of A_0 in $|g\rangle$ $(|i\rangle)$ thus corresponds, as expected, to a measurement of an odd (even) photon number. The phase of the mesoscopic superposition left by A_0 is less accurately defined than the one of the initial coherent state since it is now peaked around two values (0 and π) instead of one. This is the first step in the phase-blurring process.

A pictorial representation of the phase evolution is presented in Fig. 6.19 which shows the successive quasi-probability distributions $Q(\beta)$ of the field after each atomic detection. The Q function is introduced in Section 3.1.3 and in the appendix. It is represented here as a density plot in phase space.

The initial distribution (Fig. 6.19a) is centred on the real axis, at a distance from the origin equal to the field amplitude. The phase is well-defined, in the limit allowed by the photon-number–phase 'Heisenberg uncertainty relations'. The second panel, (b), corresponds to $|\Psi_1\rangle$. It presents two peaks, centred around α and $-\alpha$. The phase distribution has thus a large uncertainty.

The second measuring atom, A_1, is tuned for $\Phi_0 = \pi/2$. When crossing the cavity in $|g\rangle$, it imparts to the cavity field a $\pi/2$-phase kick, whereas it leaves the field invariant if it crosses the cavity in $|i\rangle$. A $\pi/2$-phase shift changes $|\alpha\rangle$ into $|i\alpha\rangle$ and $|-\alpha\rangle$ into $|-i\alpha\rangle$. Without going in too many details, we guess that the field left after the detection of A_1 is a quantum superposition of four coherent components, $|\pm\alpha\rangle$ and $|\pm i\alpha\rangle$. The corresponding $Q(\beta)$ distribution (Fig. 6.19c) exhibits four peaks.

The next atom, A_2, in turn, projects the field onto a superposition of eight coherent components. From then on, the components of the Q function overlap and cannot be distinguished (Fig. 6.19e and f) The final field state (Fig. 6.19f), after the detection of the last atom of the optimal sequence, A_4, is a superposition of 32 coherent components equally spaced over a circle. The relative quantum phases of these components depend upon the setting of the Ramsey interferometer and upon the atomic detection results. The corresponding Q function is very close to the Q function of the Fock state $|n_m\rangle$, where n_m is the measured photon number. It is a circular rim around the origin, whose radius is determined by the result of the decimation process. Initial phase information has been erased, as expected.

Note that a Fock state cannot be rigorously expressed as a finite sum of coherent components. The final cavity state after the detection of A_4 is thus not *exactly* the Fock state $|n_m\rangle$. In fact, all photon numbers $n_m + 32\,p$ (p: integer) are also possible (the photon number distribution is a comb with a period 32). The exact final Q function presents thus a set of rims with increasing radii. The relative amplitude of the outer rims is exponentially small, since the probability of having the corresponding photon

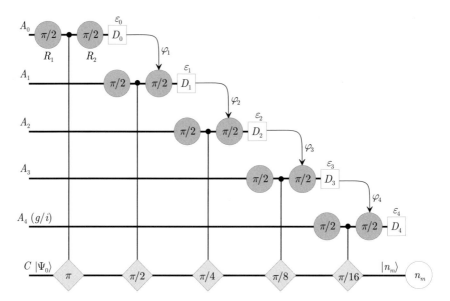

Fig. 6.20 Quantum circuit symbolizing the optimal quantum A/D converter in the five-bit case.

number is extremely small in the initial distribution. For all practical purposes, the final cavity state is indistinguishable from a Fock one.

This somewhat technical point arises only because we considered an initial coherent state, with no limit on the photon number. To fit with the requirement of an upper bound n_M on the photon number, we should have considered instead initially a truncated coherent state, expanding only on the first 32 Fock states. This state has a Q function slightly different from that of a coherent state. The final Fock state can be represented *exactly* as a weighted sum of 32 phase-shifted replica of this initial field.

6.4.5   The quantum A/D converter seen as a quantum circuit

To conclude this analysis of the optimal QND procedure, it is instructive to represent it as a quantum circuit. We will use – with some adaptations – the conventions which were introduced in the last chapter to describe atom–field interactions. To be specific, we show in Fig. 6.20 the circuit associated to the five-bit quantum A/D converter analysed above. Each atom and the field mode are represented by a horizontal line. The field can no longer be described as a qubit, since it involves up to 32 states. The action of each atom on it is nevertheless analogous to a two-bit gate conditional operation. The kth atom rotates the phase of the field by an angle $\pi/2^k$ if in state $|g\rangle$, while leaving the field unaffected if in state $|i\rangle$.

We describe this conditional transformation with the pictorial representation used for a quantum gate. A vertical line joins the 'control' bit (the atom in a superposition of $|g\rangle$ and $|i\rangle$) and the 'target' system, the field. A grey lozenge at the end of this line (with the value of the conditional phase shift inside) symbolizes the conditional phase rotation produced by the atom on the field ($\Phi_0 = \pi/2^k$ for atom A_k). Grey circles represent, as before, the single-qubit rotations associated to the R_1 and R_2 Ramsey

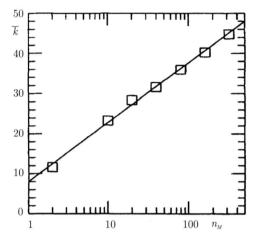

Fig. 6.21 Convergence of the random decimation process. Average number $\overline{k}$ of atoms required as a function of the initial upper bound n_M on the photon number. The dispersive phase at the average velocity v_0 is $\Phi_0(v_0) = \pi$. Reprinted with permission from Brune *et al.* (1992). © American Physical Society.

pulses acting on each atom.

The atomic detections are shown as squares at the end of each atomic line. The reading of detector D_k fixes the bit ϵ_k in the binary expression of the photon number, while determining also the phase setting φ_{k+1} of the Ramsey interferometer acting on the next atom. This 'feed-forward' operation is symbolized by the arrow joining the detector D_k of atom A_k to the interferometer acting on atom A_{k+1}.[21]

At the end of the procedure, the photon number $n_m = [\epsilon_4\epsilon_3\epsilon_2\epsilon_1\epsilon_0]$ is read on the dials of the atomic detectors. On the cavity line, the initial state $|\Psi_0\rangle$ (at left) has been gradually transformed into the Fock state $|n_m\rangle$ emerging at the right. If the initial state is a coherent one, each lozenge along the cavity line doubles the number of coherent components in the field, whose initially well-defined phase gets progressively blurred, as described above.

6.4.6 A random decimation process

The optimal decimation process relies on continuous adjustments of the interferometer phase, conditioned to the results of the successive atomic measurements. It thus requires that each atom should be detected before sending the next one. This puts a lower bound on the experiment duration and, hence, imposes severe requirements on the cavity damping time. An alternative procedure, using a much simpler scheme, also allows for a QND photon number measurement without the need for an optimal tuning of each atom's Ramsey phase.

Atoms are sent in the cavity–Ramsey interferometer arrangement, without velocity selection, corresponding to random Φ_0 values. The Ramsey phase φ remains constant during the process. Each atomic detection consists in a measurement of the atomic state *and* of the atomic velocity v (through the atomic preparation to detection time-of-flight measurement). The dispersive phase $\Phi_0(v)$, proportional to $1/v$, can thus be calculated for each detected atom, bringing information about the cavity state and multiplying the inferred probability distribution by a factor $P_{g|n}(v)$ when the atom

---

[21]The interferometer phase φ_{k+1} is adjusted by a proper tuning of the frequency of the source S' feeding the R_1 and R_2 pulses. For clarity, we have not shown S', but symbolized the feed-forward procedure by a line connecting D_k to R_2 in the A_{k+1} line.

is detected in $|g\rangle$ $[P_{i|n}(v) = 1 - P_{g|n}(v)$ when it is detected in $|i\rangle]$. The probability $P_{g|n}(v)$ is an oscillatory function of the photon number n, with a period depending upon Φ_0 and, hence, upon the atom's velocity.

Detecting one atom thus reinforces some photon numbers and decimates others in the probability distribution. The next atom, with a different velocity, performs a new decimation, with a different photon number period. After a few atoms have been detected, a single photon number is left, provided, once again, that the observer had initially an upper bound n_M on the photon number.

There is no simple analytical description of this random decimation process. Numerical simulations show that, in any case, the convergence to a Fock state finally occurs (Brune *et al.* 1992). We have determined the average number $\overline{k}$ of atoms required to achieve the convergence as a function of the initial bound on the photon number (Fig. 6.21). We assume that the atomic velocity distribution is Maxwellian, with an average velocity v_0. The phase Φ_0 is π for an atom at v_0. The number of atoms is significantly larger than for the optimal convergence procedure. Decimating a field with up to 10 photons, which can be performed with four atoms in an optimal way, requires about 25 atoms with the present procedure. However, $\overline{k}$ is still a logarithmic function of n_M, making the process efficient, if not optimal.

An interesting feature of this random convergence is that it does not require very large dispersive dephasing values. We have evaluated the length $\overline{k}$ of the converging sequence as a function of $\Phi_0(v_0)$. This length is independent of $\Phi_0(v_0)$, as soon as the condition $\Phi_0(v_0)n_M \geq 2\pi$ is fulfilled. This condition states that the 'average' decimation curve $P_{g|n}$ considered as a function of n should have at least one full oscillation over the extension of the initial photon number distribution.

This method puts much less severe constraints on the experiment than the optimal scheme. Larger velocities can be used, since the $\Phi_0 = \pi$ condition needs no longer to be fulfilled. Atoms can be sent in a shorter time interval since it is no more necessary to measure one before sending the next. Altogether, these two advantages more than compensate the increased size of the atomic sample and plead for the use of this random method in future experiments.

6.5   Photon number parity and Wigner function measurements

The first step in the optimal QND decimation process is the measurement of the photon number parity. The same operation naturally leads to a direct determination of the cavity field Wigner function, W, with many potential applications for the study of non-classical field states.

The Wigner (1932) function, $W(\alpha)$, described in detail in the appendix, is a quasi-probability distribution attached to any state of the field. It is defined as a distribution in phase space normalized to unity ($\int d^2\alpha \, W(\alpha) = 1$). It contains all possible information about the field state. The density operator of the field can be deduced from it and, hence, the expectation value of all field observables. It has also the important property that, integrated along the direction defining the quadrature $X_{\theta+\pi/2}$ in phase space, it provides directly the 'marginal' probability distribution of the conjugate quadrature X_θ. In simple terms, the distribution of X_θ is the 'shadow' of W projected along the $\theta + \pi/2$ direction. Note that W is the unique quasi-probability distribution exhibiting this simple property.

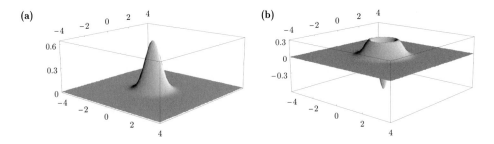

Fig. 6.22 Theoretical Wigner distributions for the vacuum (a) and for a single-photon Fock state (b).

The Wigner function is, however, not a classical probability distribution. It may be negative at some points in phase space. Figure 6.22 presents, for instance, the theoretical functions $W^{[|0\rangle\langle 0|]}$ and $W^{[|1\rangle\langle 1|]}$ for the vacuum, $|0\rangle$, and a one-photon Fock state, $|1\rangle$, respectively. For $|0\rangle$, as for any coherent, thermal or even squeezed state,[22] W is a positive Gaussian distribution. For most practical purposes, such a state can be understood as a fluctuating classical field, W being the probability density of the amplitude. On the contrary, W is negative close to the origin for the single-photon state $[W^{[|1\rangle\langle 1|]}(0) = -2/\pi]$. The properties of this state cannot be understood in terms of a classical fluctuating field. Quite generally, negativities in the Wigner function are a good indication of a quantum state's non-classical nature. The superpositions of coherent fields evoked in the previous section, for instance, present large negativities. We discuss this feature in more detail in the next chapter.

Measuring a field's Wigner function is a very powerful tool to investigate its non-classical properties. Several methods have been proposed or implemented to realize this measurement. The first one, historically, uses an homodyne scheme (Bertrand and Bertrand 1987). Beating the field to be measured with a reference coherent state having a tuneable phase provides, as shown in Section 3.2.4, a measurement of the probability density of any quadrature X_θ. Once many of these distributions have been determined for θ values sampling the $[0, 2\pi]$ interval, the W function can be inferred using the 'inverse Radon transformation', a complex mathematical operation.

Many experiments on quantum tomography have been realized on a variety of quantum states, including coherent states (Smithey *et al.* 1993) and squeezed states (Breitenbach *et al.* 1997). The Wigner function is then positive with a Gaussian shape. The same technique, applied to fields produced by entangled photon sources, has led to investigations of the Wigner function of a single-photon Fock state (Lvovsky *et al.* 2001) and of a coherent superposition of $|0\rangle$ and $|1\rangle$ (Lvovsky and Mlynek 2002), with, in both cases, a clear-cut observation of negativities. A similar method has led to the reconstruction of the Wigner function of Helium atoms in a Young's double-slit device (Kurtsiefer *et al.* 1997). In each of these achievements, the Radon inversion requires a large amount of experimental data and is quite sensitive to the quality of the signals. An excess of noise leads to bad reconstructions and to unphysical Wigner functions.

---

[22]Squeezed states have fluctuations along a quadrature smaller than the ones of a coherent state (Kimble 1992).

A more direct measurement of W can be envisioned by considering one of its expressions (Cahill and Glauber 1969):

$$W(\alpha) = \frac{2}{\pi}\mathrm{Tr}\left[D(-\alpha)\rho D(\alpha)\mathcal{P}\right] , \qquad (6.61)$$

where ρ is the density matrix describing the cavity field state and $D(\alpha)$ the field displacement operator (Section 3.1.3). A derivation of this expression is given in the appendix. The operator:

$$\mathcal{P} = \exp(i\pi a^\dagger a) , \qquad (6.62)$$

is the photon number parity observable. Its eigenstates are the Fock states $|n\rangle$:

$$\mathcal{P}|n\rangle = (-1)^n |n\rangle . \qquad (6.63)$$

Equation (6.61) shows that W is, within a normalization factor, the expectation value of $\mathcal{P}$ in a field described by the density matrix $\rho(\alpha) = D(-\alpha)\rho D(\alpha)$, obtained by performing on the cavity a coherent displacement with an amplitude $-\alpha$. This operation is easily realized with the classical source S connected for an adjusted time to C.

The Wigner function, being the expectation value of an observable, is a directly measurable quantity. The expectation value of $\mathcal{P}$ can be deduced from a determination of the complete photon number distribution (Banaszek and Wodkiewicz 1996). This reconstruction has been performed for a coherent state in a simple photon counting experiment (Banaszek *et al.* 1999b). The photon number distribution can also be deduced from the quantum Rabi oscillation signal. This principle has been applied in an ion trap experiment (see Section 8.2.2) and the Wigner function of a single-phonon state for the harmonic motion of an ion has been reconstructed in this way (Leibfried *et al.* 1996).

Lutterbach and Davidovich (1997) have proposed in the CQED context a very efficient way to measure the photon number parity. It is similar to the first stage of the optimal decimation procedure, being based on a $\Phi_0 = \pi$ dispersive phase shift experienced by the atom in a detuned cavity. In order to benefit from a large dispersive phase shift, we discuss the case of Ramsey fringes performed on the $|e\rangle \to |g\rangle$ transition, as it is the case in the experiment we describe below.

When the cavity is empty, the probability of finding finally the atom in $|e\rangle$, $P_{e|0}(\varphi)$, is an oscillating function of the interferometer phase φ. For a proper setting of the Ramsey pulses R_1 and R_2, $P_{e|0}(\varphi) = (1 + \cos\varphi)/2$. When the cavity contains a single photon, the e/g atomic coherence accumulates a $2\Phi_0 = \pi$ phase shift between the two Ramsey pulses.[23] The probability of detecting the atom in $|e\rangle$ is now $P_{e|1}(\varphi) = (1 - \cos\varphi)/2$. More generally, the phase shift being proportional to the photon number, the fringes pattern $P_{e|0}(\varphi)$ will be observed for all *even* photon numbers and $P_{e|1}(\varphi)$ for all *odd* photon numbers.

---

[23]The factor of 2 arises from the fact that the two levels $|e\rangle$ and $|g\rangle$ involved in the Ramsey interference now experience opposite phase shifts. In the experiments described up to now, based on $|g\rangle \to |i\rangle$ fringes, only one state, $|g\rangle$, was phase-shifted.

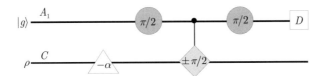

Fig. 6.23 Quantum circuit representing the procedure to determine the Wigner function of a cavity field. The upper and lower lines represent the measuring atom and the field. The atomic qubit ($|e\rangle,|g\rangle$) is initially in $|g\rangle$ and the field in the state described by the density operator ρ. The atomic interferometer is symbolized by the two grey circles, representing the two Ramsey pulses. The conditional dispersive action of the atom on the field (phase rotation by $\pm\pi/2$ depending upon the atomic state $|g\rangle$ or $|e\rangle$) is shown by a vertical line with a lozenge at the end, at the intersection with the cavity line. Prior to the interaction with the atom, the field injection realizing the field translation in phase space is symbolized by the white triangle with indication of the translation inside.

When the cavity contains an arbitrary quantum state, with a photon number distribution $p(n)$, the observed signal is:

$$P_e(\varphi) = \frac{1}{2}\left[1 + \sum_n p(n)(-1)^n \cos\varphi\right] . \tag{6.64}$$

The algebraic contrast $\mathcal{C}$ of the Ramsey fringes [positive when they have the phase of $P_{e|0}(\varphi)$, negative for $P_{e|1}(\varphi)$] is:

$$\mathcal{C} = \sum_n p(n)(-1)^n = \langle \mathcal{P} \rangle . \tag{6.65}$$

This contrast measures the parity operator, and it does it *in a direct way* since it does not require a full knowledge of the photon number distribution. The Wigner function measurement scheme proceeds thus according to a simple recipe (Lutterbach and Davidovich 1997):

- Prepare the cavity field in its initial state ρ.
- Perform a coherent displacement by an adjustable amplitude $-\alpha$ with the help of a classical source S coupled to the cavity mode.
- Send an atom, initially in $|g\rangle$, through the Ramsey interferometer, with a velocity such that $2\Phi_0 = \pi$ and record the output state.
- Repeat the whole sequence for many atoms and various interferometer phases φ and reconstruct the fringe signal $P_e(\varphi)$. From the contrast of the fringes, determine $\langle \mathcal{P} \rangle$ for this value of α and hence $W(\alpha)$.
- Repeat the previous scheme for a grid of amplitudes α over phase space and reconstruct the whole W distribution.

This scheme, summarized by the quantum circuit shown in Fig. 6.23, is very economical in terms of data acquisition. The Wigner function need not be measured on an equally spaced grid in phase space. Provided some preliminary knowledge on the field state is available, the measurement points can be adapted to the measured function

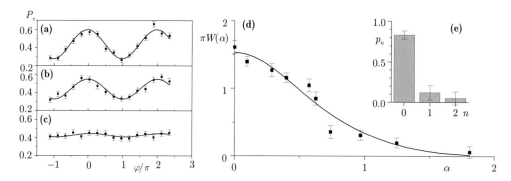

Fig. 6.24 Determination of the 'vacuum state' Wigner function. (a) Ramsey fringes for an injected amplitude $\alpha = 0$. Probability P_e of detecting the atom in state $|e\rangle$ as a function of the Ramsey interferometer phase φ/π. Dots are experimental with error bars reflecting the variance of the binomial detection statistics. The solid curve is a sine fit. (b) and (c) Ramsey fringes for $\alpha = 0.57$ and $\alpha = 1.25$ respectively. (d) Dots: resulting Wigner function (in units of $1/\pi$) versus α with error bars reflecting the uncertainty on the Ramsey fringes fit. The solid line is a theoretical fit. (e) Corresponding photon number distribution p_n. Reprinted with permission from Bertet *et al.* (2002). © American Physical Society.

(dense in regions where fast variations are important, more sparse in regions where W is close to zero). Such an adaptation is not possible in tomographic techniques. Moreover, the statistical noise or systematic errors on the atomic data reflect simply on the Wigner function noise but does not hinder the reconstruction process itself, as it is the case for Radon transformation-based techniques.

The Lutterbach–Davidovich method has been demonstrated with our generic set-up, for the vacuum and a one-photon Fock state (or rather, as we will see, for an 'approximation' of a zero- and one-photon field). We must be at this stage familiar with the experimental techniques used in these experiments. We will thus skip the details which can be found in Bertet *et al.* (2002) and present rapidly the results by commenting two figures.

In order to determine the Wigner function of the vacuum, residual thermal photons in the cavity are wiped out by absorbing them with resonant atoms crossing the cavity in state $|g\rangle$. This procedure is performed at the beginning of each data acquisition sequence, before the atom measuring $\mathcal{P}$ is sent across the set-up. This sequence is repeated many times, for various phases of the interferometer and different field translations, according to the recipe outlined above. Since there is no phase information in the cavity field, the phase of the displacements is not varied and assumed to be zero. Figure 6.24(a) to (c) show the Ramsey fringes obtained for the initial field, and after translation of this field by $\alpha = 0.57$ and $\alpha = 1.25$. Note that the phase of these fringes does not change, meaning that the Wigner function keeps the same sign as for $\alpha = 0$ and thus remains always positive.

By plotting the contrast of these fringes versus α we finally get, after proper normalization taking into account the system's imperfections (limited contrast of the interferometer), the Wigner function shown in Fig. 6.24(d). Since the field has then

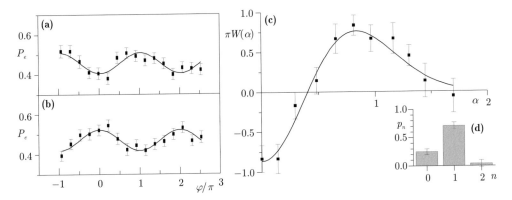

Fig. 6.25 Determination of the 'one-photon' Wigner function. (a) Ramsey fringes for an injected amplitude $\alpha = 0$. (b) Ramsey fringes for $\alpha = 0.81$ (c) Dots: experimental Wigner function in units of $1/\pi$. The solid line is a theoretical fit. (d) Inferred photon number distribution. Reprinted with permission from Bertet *et al.* (2002). © American Physical Society.

no privileged phase, we determine only the variations of W versus distance to phase space origin, i.e. a section of the Wigner distribution along a line of constant phase. The fit is Gaussian, as expected from the Wigner function of the vacuum state.

A closer inspection shows, however, that the width of this distribution is somewhat larger than predicted by theory for vacuum. It corresponds to the Wigner function of a small thermal field, whose photon number distribution is shown in Fig. 6.24(e) (the probability of finding 1 photon is ≈ 0.18). This experiment shows that the vacuum has been contaminated by a small thermal contribution which has built up in between the time the cavity was emptied from residual photons and the start of the Wigner function measurement sequence. It is important to emphasize here that this small thermal field – approximation of the vacuum – is 'classical', meaning that its Wigner function remains positive everywhere in phase space.

In a second experiment, we determine the Wigner function of the field after injecting one photon in the cavity. The same atom which determines the Wigner function is used to generate this photon first. Being initially in state $|e\rangle$, the atom starts by performing a π-Rabi pulse in the initially empty cavity and emits a photon in it. While still in the cavity, it is then rapidly detuned by the Stark effect from cavity resonance. The field translation is performed and the (now dispersive) interaction of the atom with the cavity field is used, as described above, to determine the parity of the translated field.[24]

---

[24]The timing of this experiment is subtle. The pulse R_1 required for the photon number parity measurement must be applied to the atom after it has undergone the π-Rabi pulse and while it is still in C. The Ramsey pulse is produced by a field generated by the source S' and leaking 'sidewise' inside the metallic ring surrounding the cavity. This field presents a standing wave structure and the R_1 pulse is fired while the atom is passing at an antinode. Since the atom is at this time detuned from C, S' has a frequency different from ω_c and does not perturb appreciably the cavity mode. More details can be found in Bertet *et al.* (2002).

Figure 6.25 presents the results of this experiment. We have plotted in frames (a) and (b) the Ramsey fringes obtained for the field translations $\alpha = 0$ and $\alpha = 0.81$. Contrary to what happened when the cavity was initially empty, these Ramsey patterns now exhibit a π-phase shift that reveals a sign change in the Wigner function. From these data, we get the normalized Wigner function displayed in Fig. 6.25(c). The negativity around the origin is clearly apparent, demonstrating the non-classical nature of the cavity field state.

Here again, as in the vacuum case, the field is not a pure Fock state but only an approximation of one. It has slightly decayed between the injection of the photon and the start of the Wigner function measurement. A small thermal field has also leaked in the cavity during this time interval. As a result, the actual field is a statistical mixture of zero-, one- and two-photon states. We fit the observed distribution on the Wigner function of such a mixture. The result of this fit, indicated as a solid line in Fig. 6.25(c) is in excellent agreement with the experimental data. It yields the photon number distribution plotted as an histogram in Fig. 6.25(d). The measured cavity field has a 70% probability of containing one photon, a 25% probability of being in vacuum and a residual probability ($\approx 5\%$) of storing two photons. The field's main component is the one-photon Fock state and the measured Wigner function has its characteristic shape. The zero- and two-photon contaminations are in fair agreement with estimations based on the known cavity damping time and independent measurements of the residual thermal field.

The Wigner function, more than a mere mathematical abstraction, is a directly measurable quantity. The method we have just described to determine it is very efficient in terms of data collection. It is moreover largely insensitive to technical imperfections, such as the reduced contrast of the Ramsey interferometer, and to the statistical noise on the data. This opens, as discussed in next chapter, very interesting perspectives for the study of the decoherence process of a mesoscopic state superposition.

We have seen that cavity quantum electrodynamics methods can be used to extract information from a quantum field in subtle ways. This information can be used to determine the path of an atom in an interferometer, providing textbook illustrations of complementarity and of quantum erasure procedures. Information can also be extracted from the field without destroying the photons. The field energy can be determined by counting the number of its light quanta, leaving these quanta in the cavity, ready to be detected again, or manipulated in various ways. Entanglement can be knitted between several atomic qubits by combining these QND photon detection and manipulation procedures. The parity of the number of photons can also be directly determined, without having to measure the photon distribution itself and, most importantly, without demolishing the field. Parity information, combined with translations in phase space, can be exploited to obtain directly the Wigner function of the field, from which all its statistical properties can be obtained, and its non-classical features directly deduced. Other applications of these non destructive methods will be discussed in the next chapter in the context of Schrödinger cat state studies.

7
Taming Schrödinger's cat

'All right,' said the cat; and this time it vanished quite slowly, beginning with the end of the tail, and ending with the grin, wihch remained some time after the rest of it had gone.

Lewis Carroll, *Alice's adventures in Wonderland*

In his famous paper about the meaning of the state vector in quantum mechanics, Schrödinger (1935) asks the question *'are the variables (described by the wave function) really blurred?'*. Attempting to answer, he notes that the indeterminacy of quantum physics starts to bother us only inasmuch as it invades our macroscopic world. To make this point clear, he adds that *'one can even set up ridiculous cases'* and exposes, as a provocative example already recalled in Chapter 2, the fate of the poor cat *'penned up in a steel chamber along with a diabolical device'*. A single atom trapped with the cat in the chamber has a 50% probability of decaying within an hour, thereby triggering a contraption which would kill the feline. As long as the blurring inherent to the wave function resides inside the atom, we are not really worried, but what when this fuzziness is amplified at the cat's scale? Can we accept that the wave function of the entire system (atom + cat) has in it *'the living and the dead cat (pardon the expression) mixed or smeared out in equal parts?'*.

Schrödinger did not really solve the mystery. He remarked that the cat's destiny is typical of a situation in which *'an indeterminacy, originally restricted to the atomic domain, becomes transformed into macroscopic indeterminacy which can then be resolved by direct observation. That prevents us'*, concluded Schrödinger, *'from so naively accepting as valid a blurred model for representing reality'*. In spite of this strong final statement, Schrödinger did not solve the quantum puzzle of measurement, since he did not explain what *'resolving the indeterminacy by direct observation'* really means. The cat's tale, initially intended as a pun in a serious debate, has since then raised innumerable discussions about the interpretation of quantum theory.

As discussed in Chapter 2, the most convincing 'for-all-practical-purposes' explanation for the absence of macroscopic superpositions is given by the concept of decoherence. Large systems are rapidly entangled with their environment. They become parts of a larger ensemble and, as such, lose their quantum identity. They cannot be described any longer by a quantum state, but only by a density operator, which is obtained by tracing over the environment variables. This introduces in their description a classical statistical indeterminacy instead of a quantum one. In short, decoherence makes the cat dead *or* alive, not dead *and* alive.

The transformation of quantum superpositions into statistical mixtures, typical of Schrödinger cat-state situations, plays an essential role in a quantum measurement when a micro-system is coupled to a macroscopic meter. In the absence of decoherence, this coupling would result in an entanglement between the two systems and in the generation of state superpositions involving different macroscopic states. These superpositions, and the interference they imply, are never observed, though, because the system instantaneously evolves under the effect of environment-induced decoherence into a statistical mixture. As recalled in Chapter 2, the coupling to the environment has the effect of leaving unaltered some specific states, the 'pointer states' of the meter, which are correlated to the eigenstates of the measured microscopic system, while the coherence between these states is rapidly destroyed. Elucidating the mechanisms by which the robust pointer states are selected out of the huge ensemble of states in the Hilbert space of the meter is an essential aspect of decoherence theory.

At a more practical level, macroscopic state superpositions, and the effect of decoherence on them, are central issues in quantum information physics. We have seen in Chapter 2 how ensembles of qubits evolving into state superpositions could be used to process information. These systems could be considered as 'computing cats', which would be maintained – to adopt Schrödinger's parlance – in a blurred state until a final measurement would extract desired information. Thus, monitoring and controlling decoherence, as well as correcting for its effects, are among the essential goals of quantum information physics.

For a long time, the decoherence process has been considered to be inaccessible to direct observation. Its formidable efficiency makes it so fast that we were supposed to see only its end result, the classical world around us. The technological progress of the quantum age, recalled in Chapter 1, has changed this situation drastically, by allowing us to manipulate increasing numbers of particles, in a well-controlled environment. It has then become possible to study systems at the boundary between the microscopic and the macroscopic worlds, which could be prepared in state superpositions before they would evolve, within an observable time scale, into a statistical mixture under the effect of decoherence.

The physics of Schrödinger cats has thus become accessible to laboratory investigations. We must acknowledge that the superpositions manipulated by present-day experiments remain remarkably simple compared to a real cat – or even a kitten. Calling the systems under study 'cats' implies some tongue-in-cheek-humour to which the editors of physics journals seem insensitive. When publishing their results, experimentalists must use a more technical wording, such as 'mesoscopic state superpositions'. The word *mesoscopic* indicates here that the systems under study are intermediate between the truly microscopic ones, in which state superpositions are usually observed, and the really macroscopic ones, which are expected to behave as classical mixtures. We are fortunately freer to share a bit of wit with our readers and will use throughout this chapter the more poetic name of 'Schrödinger cats', or simply 'cats', hoping that no animal rights activist will complain!

The Rydberg atom CQED set-up described in the last two chapters has proven to be especially well-adapted for the breeding of simple Schrödinger cats, undergoing decoherence under controlled conditions. These cats are not made of atoms, but of photons. A single Rydberg atom crosses a cavity storing a coherent field made of many light quanta and gets entangled with this field, which acquires two different phases at once. We recognize here the main ingredients of Schrödinger's tale: an atom with a blurred wave function imparts its quantum indeterminacy to a large system, made of many quanta. The process can be seen as the first stage of a quantum measurement, the field playing the role of a meter detecting the state of the atom.

The preparation of the cat state, which takes place during the atom–cavity crossing time, must occur faster than decoherence, due to the coupling of the cavity photons to the outside environment. This is achieved by storing the field in a cavity with very highly reflecting walls, ensuring that the cat state is created before a single photon has had time to leave the cavity. This condition becomes more and more difficult to fulfil as the number of photons is increased, which gives a visceral feeling about the fragility of these states when their size becomes large.

The coherence between the two parts of the system's wave function is revealed by quantum interference observed either on the atom which has prepared the cat state, or on another probe atom following the preparation atom after a delay. By studying how fast after the system's preparation this interference vanishes, one directly witnesses the decoherence process and catches in action the transformation of state superpositions into statistical mixtures. These experiments are textbook illustrations of the quantum theory of measurement and of the concept of complementarity.

This chapter is devoted to the physics of these CQED cats. We start in Section 7.1 with a general description of the superpositions of two coherent states and give a simple classification of possible cats. Different representations of these states in phase space are compared and the various kinds of interference effects they give rise to are reviewed.

Historically, cat states of radiation were first considered for optical fields. It has been suggested (Yurke and Stoler 1986) to exploit the non-linear coupling of a light wave with a transparent medium presenting a strong Kerr non-linearity in order to split the phase of the field and to prepare a field exhibiting several phases at once. These optical cats have turned out to be impossible to generate in the laboratory, but it is nevertheless interesting to discuss briefly how they could be produced, and to compare the procedure with the one successfully implemented in CQED. Strong analogies exist also between these optical cats and the mesoscopic state superpositions of matter waves in Bose–Einstein condensates that we will consider in Chapter 9. We briefly evoke these theoretical 'running wave optical cats' in Section 7.2.

We then turn to the description of specific experiments, differing by the process used to prepare the cat. The conceptually simplest method, described in Section 7.3, uses a non-resonant atom, interacting dispersively with the cavity field. The atom, prepared in a state superposition, acts as a microscopic piece of transparent dielectric material phase-shifting the field by two different angles at once. Several experiments performed in this way are described, which reveal various interesting features of these mesoscopic superpositions.

Cat states can also be prepared resonantly, by letting an atom undergo a Rabi oscillation in a mesoscopic cavity field. The physics of the cat states produced in this way is described in Section 7.4. It is intimately related to the features of the

Rabi oscillation collapse and revival phenomenon, already briefly studied in Chapters 3, 4 and 5. Analysing the intertwined features of the atom and field evolution, we discuss aspects of the cat-state physics related to the concepts of entanglement and complementarity.

The decoherence of the photonic cats is analysed in Section 7.5. We start by considering thought experiments, in which the evolution of the cat is monitored by collecting information in its environment. Simulating the field quantum jumps by the Monte Carlo method provides a striking illustration of the meter localization in a measurement process. We then describe a real experiment performed with two atoms, one preparing the cat state and the other playing the role of a 'quantum mouse', probing its coherence at a later time. This experiment catches the decoherence process 'in the act'. The comparison between the experimental observation and the theoretical prediction is particularly simple and direct in this system where the coupling to the environment is well-understood.

We then attempt, in Section 7.6, to foresee where these experiments are leading us to. Adding a second cavity along the atom's path, we dream about entangling two spatially separated cavity fields, each containing many photons. This would be a combination of the non-locality strangeness with the weirdness of Schrödinger's cat.

7.1   Representations of photonic cats

Let us start by defining a 'Schrödinger cat' (or more simply a 'cat') as a superposition with equal probabilities of two coherent states with different complex amplitudes β_1 and β_2:[1]

$$
\begin{aligned}
|\Psi_{\text{cat}}\rangle &= \frac{1}{\sqrt{\mathcal{N}}} \left(e^{i\phi_1} |\beta_1\rangle + e^{i\phi_2} |\beta_2\rangle \right) \\
&\approx \frac{1}{\sqrt{2}} \left(e^{i\phi_1} |\beta_1\rangle + e^{i\phi_2} |\beta_2\rangle \right) .
\end{aligned}
\tag{7.1}
$$

The denominator in the right-hand side of the first line ensures normalization and takes into account the overlap of the two coherent states $|\beta_1\rangle$ and $|\beta_2\rangle$. If, as we assume in the following, $|\beta_1 - \beta_2| \gg 1$, this overlap is negligible and the cat state is expressed by the simpler form given by the second line in eqn. (7.1).

The coherence between the two states distinguishes it from a statistical mixture. This is made clear by expressing the field density operator associated to the cat state:

$$
\rho_{\text{cat}} \approx \frac{1}{2} \left(|\beta_1\rangle \langle\beta_1| + |\beta_2\rangle \langle\beta_2| + e^{i(\phi_1 - \phi_2)} |\beta_1\rangle \langle\beta_2| + e^{i(\phi_2 - \phi_1)} |\beta_2\rangle \langle\beta_1| \right) .
\tag{7.2}
$$

The coherence is described by the off-diagonal part of this density operator [last two terms in the right-hand side of eqn. (7.2)].

7.1.1   Cat's Q function

Each component of the cat is represented in phase space by its Q function, a Gaussian centred at the amplitude β_1 or β_2 (see Section 3.1.3 and the appendix). The resulting Q function of the cat at point α in phase space:

---

[1]We will consider at the end of this section a generalization of this definition to coherent state superpositions with more than two components.

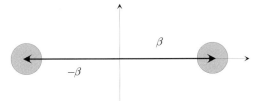

Fig. 7.1 Pictorial representation of an even or odd π-phase cat.

$$Q^{[\text{cat}]}(\alpha) = \frac{1}{2\pi} \left\{ |\langle \alpha | \beta_1 \rangle|^2 + |\langle \alpha | \beta_2 \rangle|^2 + 2\text{Re}\left[e^{i(\phi_1 - \phi_2)} \langle \alpha | \beta_1 \rangle \langle \beta_2 | \alpha \rangle \right] \right\} , \qquad (7.3)$$

is the sum of three terms. The first two are the Gaussian peaks associated to the cat components. The last term, contribution of the cat's coherence to $Q^{[\text{cat}]}$, is practically negligible since, for all α values, one at least of the scalar products $\langle \alpha | \beta_1 \rangle$ or $\langle \alpha | \beta_2 \rangle$ is vanishingly small. The Q-representation is not suited to reveal the coherence of the cat's state. We nevertheless use it as a convenient way to represent a state superposition, because it is simple and clearly exhibits the two parts of the state. Generalizing a picture introduced for coherent states (Section 3.1.3), we represent a cat by a plot in phase space showing as two circles the regions where its Q function is above $1/2\pi e$.

We focus mainly on 'phase-cats', in which the field amplitudes have the same modulus. As simple examples, let us consider two particular superpositions of coherent states with phases differing by π. These superpositions, which we call 'even' and 'odd' π-phase-cats, are:

$$\left| \Psi_{\text{cat}}^{\pm} \right\rangle = \frac{|\beta\rangle \pm |-\beta\rangle}{\sqrt{2 \left(1 \pm e^{-2|\beta|^2}\right)}} \approx \frac{1}{\sqrt{2}} \left(|\beta\rangle \pm |-\beta\rangle \right) , \qquad (7.4)$$

where β is the field amplitude (which we assume real without loss of generality by a proper choice of phase origin). These two states, which differ by the sign ($\pm$) between the two parts of the superposition, are represented by practically identical Q functions and have the same pictorial representation in phase space (Fig. 7.1). Their difference is displayed in interference signals which are vanishingly small terms in their Q functions.

7.1.2 Signatures of cat's coherence

The coherence of these cat states can be revealed by the distribution of a properly chosen field quadrature. Suppose, first, that we measure X_0. The probability amplitudes of finding a value x in state $|+\beta\rangle$ and $|-\beta\rangle$ do not appreciably overlap and thus cannot interfere. The resulting probability distribution $P_0^{(\pm)}(x)$ is, to a very good approximation, the sum of the distributions corresponding to the two state components:

$$P_0^{(\pm)}(x) \approx \frac{1}{\sqrt{2\pi}} \left[e^{-2(x-\beta)^2} + e^{-2(x+\beta)^2} \right] . \qquad (7.5)$$

If we measure instead $X_{\pi/2}$ along the orthogonal direction, the probability distribution is:

$$P_{\pi/2}^{(\pm)}(x) \approx \frac{1}{2} \left| \langle x_{\pi/2} | \beta \rangle + \langle x_{\pi/2} | -\beta \rangle \right|^2 , \qquad (7.6)$$

where the $\langle x_{\pi/2} |$ bra is an eigenstate of $X_{\pi/2}$, related to those of X_0 by a Fourier transform (eqn. 3.27). The probability $P_{\pi/2}^{\pm}(x)$ is the square of the sum of two amplitudes, which are now both non-vanishing. The scalar product of $|x_{\pi/2}\rangle$ with $|\beta\rangle$ is

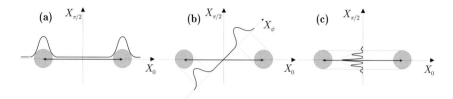

Fig. 7.2 Quadratures of a π-phase Schrödinger cat. (a) The X_0 quadrature exhibits well-separated Gaussian peaks corresponding to the cat's components. (b) For X_ϕ, with $\phi = \pi/4$, the Gaussian peak distance is reduced. (c) For $X_{\pi/2}$ the two peaks merge and fringes show up.

equal to the product of $|x_0\rangle$ (eigenstate of X_0 with eigenvalue x) with $|-i\beta\rangle$, as a rotation in phase space indicates. Using eqn. (3.68), we get:

$$\langle x_{\pi/2} \,|\beta\rangle = \langle x_0 \,|-i\beta\rangle = (2/\pi)^{1/4}\, e^{-2i\beta x}\, e^{-x^2}\ ,\tag{7.7}$$

and:

$$P_{\pi/2}^{(\pm)}(x) \approx (2/\pi)^{1/2}\, e^{-2x^2}\, [1 \pm \cos(4\beta x)]\ .\tag{7.8}$$

This probability distribution is a Gaussian function, centred at $x = 0$, modulated by an interference term with fringes having a period $1/(4\beta)$, inversely proportional to the 'cat size' β. This interference term is a conspicuous signature of the coherence of the state superposition. The fringes take opposite signs in the even and odd π-phase cats.

The distribution of any phase quadrature, X_ϕ, can be obtained in the same way. The interference term is non-vanishing only when ϕ is close to $\pi/2$. A graphical representation helps to understand this result (Fig. 7.2). For a coherent state, a field quadrature takes non-zero values in an interval corresponding to the projection of the state uncertainty circle on the direction of the quadrature. For a cat state, there are two such intervals, corresponding to the two state components. If $\beta \gg 1$ and $\phi = 0$ (Fig. 7.2a), the two intervals are non-overlapping and there is no interference. For a ϕ value between 0 and $\pi/2$ (Fig. 7.2b), the two intervals are getting closer, resulting in two still non-overlapping Gaussians without interference. It is only when ϕ is very close to $\pi/2$ that the two projected intervals overlap along the direction of the quadrature, leading to a large interference term (Fig. 7.2c).

Instead of rotating the quadrature for a given 'frozen' cat state, let us choose a given quadrature (e.g. X_0) and consider the time evolution of its probability distribution. The two coherent state components rotate while staying opposite to each other in phase space. Their projections along the (now fixed) direction of the field quadrature oscillate with opposite phases, and collide periodically at $x = 0$. When the two components merge, fringes with 100 % contrast appear under the Gaussian envelope. This periodic fringe pattern would not exist if the field were described by an incoherent superposition of coherent states (first two terms in eqn. 7.2). The fringes are getting narrower when the amplitude of the field is increased. We will see later that these fringes are fragile and efficiently washed out by decoherence.

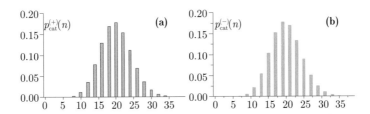

Fig. 7.3 Photon number distribution of an even (a) and odd (b) π-phase 20-photon cat ($|\beta|^2 = 20$).

An intuitive interpretation of this interference effect is given by the mechanical oscillator analogy developed at the beginning of Chapter 3. The X_0 and $X_{\pi/2}$ quadratures correspond to the position X and momentum P of a particle oscillating in a parabolic potential well. These observables periodically exchange their values. The π-phase cat of eqn. (7.4) describes a snapshot of the particle's wave function when it is in a superposition of its two turning points, at a time when its classical velocity cancels. Its momentum distribution then exhibits strong modulations, with an envelope centred at $p = 0$ (eqn. 7.8 in which x is replaced by p). A quarter of an oscillation period later, the position and momentum have exchanged their roles and the two initially separated wave packets interfere, with strong oscillations of the particle position distribution whose envelope is centred at $x = 0$. This exchange between position and momentum wave functions repeats itself periodically as the two cat components rotate in phase space.

The coherence of phase-cat states is also imprinted in their photon number distributions:

$$p_{\text{cat}}^{(\pm)}(n) = \frac{|\langle n \,|\beta\rangle|^2}{2}[1 \pm (-1)^n]^2 \,, \tag{7.9}$$

which are displayed in Fig. 7.3(a) and (b). The $\left|\Psi_{\text{cat}}^+\right\rangle$ and $\left|\Psi_{\text{cat}}^-\right\rangle$ states develop only along even and odd number states respectively. These modulations in the photon number distribution are lacking in statistical mixtures of $|\beta\rangle$ and $|-\beta\rangle$, which contain all photon numbers. The 'dark fringes' in the photon number distribution reveal the cats coherence, as does the existence of dark fringes in their $X_{\pi/2}$ quadrature.

The even and odd phase cats are eigenstates of the photon number parity operator $\mathcal{P}$ introduced in Chapter 6 (eqn. 6.62), with the eigenvalues $+1$ and -1 respectively:

$$\mathcal{P}\left|\Psi_{\text{cat}}^\pm\right\rangle = \pm\left|\Psi_{\text{cat}}^\pm\right\rangle \,. \tag{7.10}$$

A few properties of this operator are useful to remember. The action of $\mathcal{P}$ on a coherent state merely reverses its amplitude:

$$\mathcal{P}|\beta\rangle = |-\beta\rangle \,. \tag{7.11}$$

This result can be immediately derived from the expansion of the coherent states on the $|n\rangle$ states using eqn. (6.63). Another useful property of this operator is its action on an eigenstate of the quadrature operator X_ϕ:

$$\mathcal{P}|x_\phi\rangle = -|x_\phi\rangle \,. \tag{7.12}$$

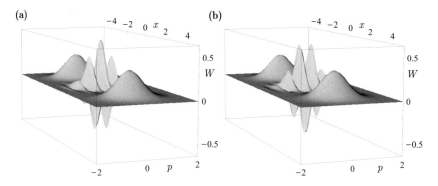

Fig. 7.4 Wigner functions of even (a) and odd (b) 10-photon π-phase cats

This relation, a direct consequence of the parity of the Hermite polynomials (see the appendix), can be considered as an alternative definition of $\mathcal{P}$, as the operator which 'reflects' any quadrature eigenstate around the origin in phase space.

Finally, we deduce from eqns. (3.51) and (7.4) that the action of the annihilation operator on an even (odd) phase cat results in the switching of the cat parity according to:

$$a\left|\Psi_{\text{cat}}^{+}\right\rangle = \beta\left|\Psi_{\text{cat}}^{-}\right\rangle \quad ; \quad a\left|\Psi_{\text{cat}}^{-}\right\rangle = \beta\left|\Psi_{\text{cat}}^{+}\right\rangle \ . \tag{7.13}$$

7.1.3   Cat's Wigner function

Measuring field quadratures or photon number distributions provide different ways to detect the interference between the two parts of a cat's state wave function. These various signatures of the cat's state coherence are coded in its density operator (eqn. 7.2) which contains all information about the state. Equivalently, this information is stored in the Wigner function W of the field, which is, as the Q function, a map of the field in phase space, whose knowledge is fully equivalent to that of the density operator (see the appendix). Whereas the cat's coherences are hidden in negligibly small terms of its Q function, they are conspicuous in W. Since W can be measured by various quantum optics methods, its shape provides an experimentally accessible and synthetic view of a cat's state main features.

As recalled in the appendix (eqn. A.31), the W function at the point in phase space whose C-number coordinate is $\alpha = x + ip$ is:

$$W(x,p) = \frac{1}{\pi} \int du\, e^{-2iup} \left\langle x + u/2\right| \rho \left|x - u/2\right\rangle \ . \tag{7.14}$$

It is the Fourier transform of a function built on off-diagonal matrix elements of the field density operator in a quadrature representation. This expression was derived first by Wigner (1932) in order to obtain a phase space distribution for a quantum particle resembling as closely as possible the probability distribution of classical statistical physics. The W function is, as Q, a real function, whose integral over phase space is equal to 1. Contrary to Q, however, it can take negative values in regions of phase space. These negative values are, as recalled below, a signature of non-classical behaviour for the corresponding states.

The W functions of coherent states can be computed from the definition (7.14) and from the expressions of the coherent states in a quadrature basis (eqn. 3.68). Like Q, they are Gaussian functions centred at the complex amplitude of the state, but their width is $\sqrt{2}$ times smaller (see Fig. A.1b). The W functions of even and odd π-phase cats are also easily derived (see eqn. A.61 in the appendix), since all their terms involve only Gaussian integrals:

$$W_{\mathrm{cat}}^{\pm}(\alpha) = \frac{2}{\pi(1 \pm e^{-2|\beta|^2})} \left[e^{-2|\alpha-\beta|^2} + e^{-2|\alpha+\beta|^2} \pm 2e^{-2|\alpha|^2} \cos(4\mathrm{Im}\,\alpha\beta^*) \right] . \quad (7.15)$$

In addition to the contributions from the two state components, the last term in eqn. (7.15) describes the coherence of the cat states. These W functions are plotted in Fig. 7.4 for even and odd π-phase cats containing on average 10 photons. The large interference pattern between the two Gaussian peaks presents an alternation of positive and negative ridges. This pattern, which we might call the cat's 'whiskers', is a signature of the coherence, lacking in the W function of a statistical mixture of coherent states. These whiskers take opposite signs in the even and odd cat states, which makes them easily distinguishable. The W representation is thus much better adapted than Q for the study of a mesoscopic state coherence.

Let us recall that the Wigner function is directly related to the marginal distributions of field quadratures (see Section 6.5 and the appendix). By integrating W over one quadrature, one gets the probability distribution of the orthogonal quadrature (eqn. A.46). We have seen above that Schrödinger cats have some non-occurring quadratures values ('dark fringes'). This destructive interference is fundamentally quantum. It implies that the integral of W along the orthogonal quadrature vanishes, which is possible only if W presents alternations of positive and negative values. We thus understand that negative values of W are directly related to non-classicality. Negative values are also encountered in the W function of Fock states. They have quadrature distributions described by Hermite polynomials, which vanish for some x and p values. This requires that their W functions should take negative values in some regions of phase space, a property already discussed in Chapter 6.

7.1.4 Multiple component cats

The zoo of cat states can be extended to superpositions with more than two components. Let us consider as a simple generalization the q states $\left| \Psi_q^k \right\rangle$:

$$\left| \Psi_q^k \right\rangle = \frac{1}{\sqrt{q}} \sum_{p=0}^{q-1} e^{-2ipk\pi/q} \left| \beta e^{2ip\pi/q} \right\rangle \qquad (k = 0, 1, 2, \ldots, q-1) , \quad (7.16)$$

superpositions of coherent states with equal amplitudes and phases regularly distributed between 0 and 2π. These states are, for $q < 2|\beta|$, nearly orthogonal to each other since:

$$\left\langle \Psi_q^{k'} \middle| \Psi_q^k \right\rangle \approx \frac{1}{q} \sum_{p=0}^{q-1} e^{-2ip(k-k')\pi/q} = \delta_{k,k'} . \quad (7.17)$$

The probability amplitude of the nth Fock state in $\left| \Psi_q^k \right\rangle$ is:

$$\langle n \left| \Psi_q^k \right\rangle = e^{-|\beta|^2/2} \frac{\beta^n}{\sqrt{q\,n!}} \sum_{p=0}^{q-1} e^{2ip\pi(n-k)/q} \ , \tag{7.18}$$

an expression which is non-zero only if $n = k + sq$ with s integer. In other words, $\left| \Psi_q^k \right\rangle$ develops only on the Fock states for which n is equal to k modulo q. We retrieve of course, for $q = 2$, the definition of the two-component even and odd phase cats, for which the n value is defined modulo 2. While the two-component phase cats are eigenstates of the photon number parity operator $\mathcal{P} = e^{i\pi a^\dagger a}$, their q-component version are eigenstates of the non-hermitian operator $\mathcal{P}_q = e^{2i\pi a^\dagger a/q}$:

$$\mathcal{P}_q \left| \Psi_q^k \right\rangle = e^{2ik\pi/q} \left| \Psi_q^k \right\rangle \ . \tag{7.19}$$

Finally, it is interesting to remark that the coherent state $|\beta\rangle$ develops along the $\left| \Psi_q^k \right\rangle$ according to:

$$|\beta\rangle = \frac{1}{\sqrt{q}} \sum_k \left| \Psi_q^k \right\rangle \ . \tag{7.20}$$

Similar multiple-component cat states have already been encountered in Chapter 6, where we have noticed that, during a photon optimal QND measurement process, an initial coherent field evolved successively into a state presenting first two, then four, then eight components and so on, regularly distributed around the origin in phase space (Fig. 6.19, on page 344).

7.2   A thought experiment to generate optical cats

We are now familiar with the main features of photonic cats and we have to learn the tricks to prepare and to probe them in their strange blurred state. Before describing the microwave cavity experiments which have effectively generated these states, let us briefly evoke an early method suggested for the preparation of radiation cat states (Yurke and Stoler, 1986). It involves the coupling of an optical field with a transparent non-linear medium and is different from the single-atom method which will be analysed in the remainder of this chapter. These optical cats present similarities with microwave cats as well as with mesoscopic superpositions of matter wave states discussed in Chapter 9. For these reasons, and even if the optical procedure has not been successfully implemented, its analysis will help us understand some features of the cats described in the following.

Let us consider a transparent slab of matter whose real refractive index varies with light intensity. This corresponds to the Kerr effect, the simplest phenomenon of non-linear optics. For a plane wave field described by its annihilation operator a, the medium index N_i is, up to first order in field intensity:

$$N_i = 1 + N_0 + N_1 a^\dagger a \ , \tag{7.21}$$

where N_0 and N_1 are constants. This index is expressed as an operator acting on the field states, which commutes with the free-field Hamiltonian $H_0 = \hbar\omega_0 a^\dagger a$. Its nth eigenvalue, $1 + N_0 + N_1 n$, is the index 'felt' by a field containing n photons. The modulus of the field wave vector, $k_0 = \omega_0/c$ in vacuum, becomes $N_i k_0$ and

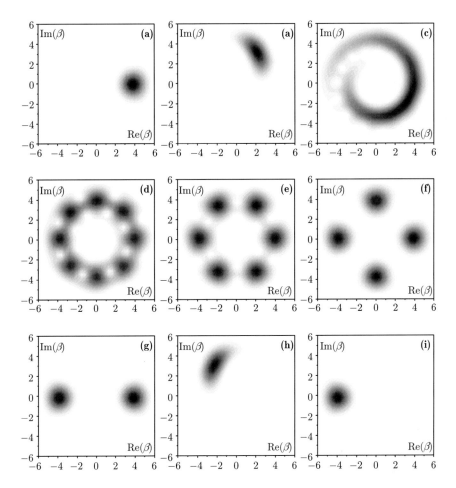

Fig. 7.5 Evolution of the Q function of an initially coherent field, with $\bar{n} = 15$ photons on average, propagating in a Kerr medium (interaction picture). Times τ for frames (a)–(i) correspond to $\gamma_k \tau / \pi = 0$, $1/100$, $1/4\sqrt{15}$, $1/8$, $1/6$, $1/4$, $1/2$, $99/100$, and 1. Note the formation of multiple component cats, the generation of a two-component phase cat for $\gamma_k \tau / \pi = 1/2$ and the revival of a coherent state at $\gamma_k \tau / \pi = 1$.

is also a field operator. The wave vector change, $(N_i - 1)k_0$, results in a photon-number-dependent phase shift, proportional to the propagation distance L. The field state after propagation over a length $L = c\tau$ in the Kerr medium, $|\Psi(\tau)\rangle$, is deduced from $|\Psi(0)\rangle$, corresponding to a propagation in vacuum over the same distance, by: $|\Psi(\tau)\rangle = U_k(\tau)|\Psi(0)\rangle$ where $U_k(\tau = L/c)$ is the unitary Kerr phase-shift operator:

$$U_k = e^{ik_0(N_i - 1)La^\dagger a} = e^{i[k_0 N_0 La^\dagger a + k_0 N_1 L(a^\dagger a)^2]} \ . \tag{7.22}$$

This corresponds to an evolution over propagation time L/c ruled by the Hamiltonian:

$$H_k = -\hbar\omega_0 \left[N_0 a^\dagger a + N_1 (a^\dagger a)^2 \right] \ . \tag{7.23}$$

The first term in the right-hand side accounts for a global field phase-shift produced by the linear index N_0, while the second describes the phase distortion produced by the Kerr effect. We will, from now on, disregard the global phase shift and define the Kerr frequency as $\gamma_k = \omega_0 N_1$, assumed to be positive.

Let us consider the action of the Kerr medium on an impinging coherent state $|\beta\rangle$. It follows immediately from the above definitions that this state, after propagating over a distance $L = c\tau$, is transformed into:

$$|\Psi(\tau)\rangle = e^{-|\beta|^2/2} \sum_n \frac{\beta^n}{\sqrt{n!}} e^{i\gamma_k n^2 \tau} |n\rangle , \qquad (7.24)$$

(we use here the interaction picture with respect to the linear index Hamiltonian). The non-linear phase distortions transform the field into a new state, in general no longer coherent. Most remarkably, this state turns, at well-defined times, into a q-component phase cat. Figure 7.5 shows the evolution of the field's Q function. The successive frames correspond to increasing field propagation distances. The Q function is numerically computed for an initial coherent field, represented in frame (a), with a real and positive amplitude and an average photon number $\overline{n} = 15$. The second frame (b) represents the field after a very short time $\tau = \pi/(100\gamma_k)$. Its initial circular shape, characteristic of a coherent field, has become elongated and 'banana-like'. This reflects the fast spreading of the field's phase, which evolves at a rate increasing with the distance from origin. This spreading of the phase is generally referred to as a 'phase collapse'. We estimate its characteristic time by computing how long it takes for the non-linear propagation to cancel the expectation value $\langle a \rangle$ of the field operator. A simple calculation yields at time τ:

$$\langle a(\tau) \rangle = \beta e^{-\overline{n}} \sum_n \frac{\overline{n}^{n-1}}{(n-1)!} e^{i\gamma_k[n^2-(n-1)^2]\tau} = \beta e^{-\overline{n}} \sum_n \frac{\overline{n}^{n-1}}{(n-1)!} e^{i\gamma_k(2n-1)\tau} . \qquad (7.25)$$

The dispersion $2n\gamma_k\tau$ of the Kerr phase over the width $2\sqrt{\overline{n}}$ of the coherent state photon number distribution is $\approx 4\sqrt{\overline{n}}\gamma_k\tau$. The phase collapse time T_{collapse} is reached when this dispersion corresponds to $\approx \pi$, hence:

$$T_{\text{collapse}} = \frac{\pi}{4\sqrt{\overline{n}}\gamma_k} . \qquad (7.26)$$

It corresponds to the third frame (c) in Fig. 7.5. After it has occurred, the Q function becomes smeared over a circle in phase space and its shape evolves in a complicated way. Most remarkably, from this apparent 'chaos' in the wide phase distribution, there emerges at precise times a transient Q distribution presenting evenly spaced Gaussian peaks centred on a circle around origin. The frames (d) to (g) represent the field Q function at $\tau = \pi/8\gamma_k, \pi/6\gamma_k, \pi/4\gamma_k$ and $\pi/2\gamma_k$. States with eight, six, four and two components appear at these successive times.

Frames (h) and (i), at times $\tau = 99\pi/100\gamma_k$ and $\tau = \pi/\gamma_k$, show the field re-focused into a single component whose phase is opposite to that of the initial field. Other q-component states (not represented in the figure) with $q = 7, 5$ and 3 are transiently generated in between the eight, six, four and two-component ones.[2] When propagating over longer distances $c\tau \geq c\pi/\gamma_k$, the field reaches again, at times $(q+1)\pi/q\gamma_k$, a succession of multiple superposition states with decreasing numbers of components (eight for $\tau = 9\pi/q\gamma_k$,... two for $\tau = 3\pi/2\gamma_k$). The initial field is reconstructed at time $\tau = 2\pi/\gamma_k$. The whole sequence then repeats itself periodically for still longer propagation times.

The re-focussing at regular times of the Q distribution into a single Gaussian alternatively symmetrical of or identical to the initial one is simple to understand. For $\gamma_k\tau = (2m+1)\pi$, (m integer), the exponentials in eqn. (7.24) become $e^{i\pi n^2} = (-1)^n$ and the field state evolves into $|\Psi\rangle = e^{-|\beta|^2/2}\sum_n (-1)^n \beta^n/\sqrt{n!}\, |n\rangle = |-\beta\rangle$. It is then a coherent field with an amplitude opposite to the initial one (Fig. 7.5i). For $\gamma_k\tau = 2m\pi$, all the phase terms in eqn. (7.24) come back to one and the initial field is simply retrieved. The evolution produced by the Kerr Hamiltonian results in periodic 'revivals' of a coherent state with a classical phase alternatively equal to π and 0.

Let us now focus on the q-component distributions appearing at times such that $\gamma_k\tau = \pi/q$. As suggested by their shape, they correspond to the generation in the Kerr medium of Schrödinger cat states, superpositions of coherent states with regularly distributed phases. At these specific times, the exponential factors in eqn. (7.24) become $e^{i\pi n^2/q}$. Expressing the photon numbers modulo q, we write them as $n = k + sq$ (s integer):

$$e^{i\pi n^2/q} = e^{ik^2\pi/q}e^{2iskπ}e^{is^2q\pi} = (-1)^{s^2q}e^{ik^2\pi/q} \ . \tag{7.27}$$

For q even, $(-1)^{s^2q} = 1$ and the quantum amplitudes of all Fock states with $n = k$ (modulo q) experience at time $\tau = \pi/q\gamma_k$ the same phase shift $k^2\pi/q$. The initial coherent field $|\beta\rangle$ can be expressed as a superposition of q-component phase-cats according to eqn. (7.20). Each state of this superposition, $\left|\Psi_q^k\right\rangle$, defined by (7.16), contains, according to eqn. (7.18), only Fock states with n equal to k modulo q. It it thus merely multiplied at time $\pi/q\gamma_k$ by the phase factor $e^{ik^2\pi/q}$. Hence, $|\beta\rangle$ is transformed according to:

$$U_k(\pi/q\gamma_k)\,|\beta\rangle = \frac{1}{\sqrt{q}}\sum_k e^{ik^2\pi/q}\left|\Psi_q^k\right\rangle \ , \tag{7.28}$$

or, replacing in this equation the $\left|\Psi_q^k\right\rangle$'s by their definition (7.16):

$$U_k(\pi/q\gamma_k)\,|\beta\rangle = \frac{1}{q}\sum_{p=0}^{q-1}\sum_{k=0}^{q-1} e^{ik(k-2p)\pi/q}\left|\beta e^{2ip\pi/q}\right\rangle \ . \tag{7.29}$$

It is easy to compute the probability amplitudes of the various components of these q-terms superpositions and to show that they have all the same modulus $1/\sqrt{q}$. The states in eqn. (7.29) are q-component Schrödinger cats analogous to the $\left|\Psi_q^k\right\rangle$, but

---

[2]State superpositions with more than eight components do not appear in the Kerr evolution of a coherent field with $\bar{n} = 15$ photons on average. The Gaussian peaks of the components, which are centred in phase space on a circle of radius $|\beta| = \sqrt{15}$, overlap for q values larger than eight, resulting in a blurred ring-shaped Q function at times $\tau = \pi/q\gamma_k$. Well-separated multiple peak distributions with more than eight components can be obtained with larger fields.

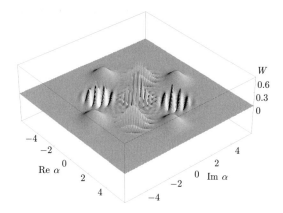

Fig. 7.6 Representation of the Wigner function of the four-component cat state defined by eqn. (7.31).

with different phases of the probability amplitudes associated to each component. For $q = 2$ and 4:

$$U_k(\pi/2\gamma_k)\,|\beta\rangle = \left(e^{i\pi/4}\,|\beta\rangle + e^{-i\pi/4}\,|-\beta\rangle\right)/\sqrt{2}\,, \tag{7.30}$$

and:

$$U_k(\pi/4\gamma_k)\,|\beta\rangle = \tfrac{1}{2}\left(e^{i\pi/4}\,|\beta\rangle + |i\beta\rangle - e^{i\pi/4}\,|-\beta\rangle + |-i\beta\rangle\right)\,. \tag{7.31}$$

The two-component cat (7.30) is a superposition of coherent states with opposite amplitudes, similar to the π-phase cats described in the previous section. Note however that, contrary to those cats, it does not have a well-defined photon number parity since the complex amplitudes of the superpositions are not real, but $\pi/2$-out-of-phase with respect to each other. The four-component cat (7.31) is an equal probability superposition of coherent state components with phases equal to 0, $\pm\pi/2$ and π. The Wigner function of this state is represented in Fig. 7.6. It exhibits four Gaussian peaks centred at the summits of a square, with wiggly structures presenting negativities and describing the cat state coherence around phase space origin and between the Gaussian peaks on each side of the square. An expression similar to eqn. (7.29), which we will not derive here, describes the cats with odd q values.

The simple model of Kerr effect presented here predicts an exact periodicity for the field evolution. This is of course a simplification of a real situation in which higher non-linear contributions to the index result in further phase spreading effects. These higher order non-linearities blur, after a few collapse and revivals, the periodic features of the simple model. More importantly, our theory has neglected the unavoidable imaginary contributions to the index, responsible for losses. We will see below that the loss of a single photon in the environment is generally enough to 'kill' the coherence of the mesoscopic state superposition. This explains why this clever proposal, made in the 1980s at the time when the research on non-classical and squeezed states of light was blossoming, has never been implemented successfully in the laboratory.

Even if it is not practical, the process we have discussed here is interesting, because of its similarities and differences with the cat state generation procedures implemented in real experiments, either in CQED (this chapter) or in matter wave physics (Chapter 9). The optimal QND method discussed in Chapter 6 has already introduced us to multiple-component superpositions of coherent states generated by the successive dispersive interactions of two-level atoms with a coherent field (note the strong analogy between Fig. 6.19, on page 344, and Fig. 7.5). We will see below that in CQED experiments, as in the ideal Kerr proposal, the cat state generation is a dynamical non-linear process, involving periodic revivals of the initial state, the cat states appearing between two of these revivals. As we present the CQED and the Bose–Einstein condensate experiments, we will have ample opportunities to compare various cat preparation procedures with the optical Kerr effect generation described in this section.

7.3 Dispersive cats in cavity QED

In the optical Kerr proposal, the splitting of the cat components is produced by a non-linear index effect in a macroscopic medium. If decoherence can be avoided, the field, evolving as a quantum system interacting with a classical phase shifter, is always in a pure state. In spite of obvious analogies, the CQED experiments we describe now correspond to a very different situation. The field interacts with a single atom acting as a refractive index in a superposition of states producing different phase shifts. The quantumness of the medium is an essential feature of these experiments, which is lacking in the optical cat proposal. In general, the CQED Schrödinger cat is entangled with the atom which has generated it. To isolate it as a pure state requires a projection resulting from a measurement procedure. In addition to these quantum features, the single atom is, as the Kerr medium, a non-linear system whose response to the field depends upon the number of photons. The physics of CQED cats thus marries the features of quantum and non-linear optics. This marriage produces a wealth of subtle effects.

Two types of cat-generating interactions need to be distinguished. In the non-resonant or dispersive case, the atom and the field cannot exchange photons, but only influence each other through phase-shift effects. In the resonant case, on the contrary, they can exchange energy and the atom undergoes a Rabi oscillation. At the same time, it also produces phase shifts on the field and cat states are produced by the intertwined evolution of the atom and field partners. We will discuss in turn these two regimes. In this section we start by considering the non-resonant coupling case, which is conceptually the simpler.

7.3.1 Imprinting a phase on a field with a single atom

A single atom, slightly off-resonant with a field mode, interacts strongly with it, even though real atomic transitions are forbidden by energy conservation rules. This dispersive coupling leads to important and potentially useful effects. The field produces a light shift on the atomic transition frequency, a feature which we have already exploited in Chapter 6 to measure non-destructively the photon number. Conversely, the atom exerts a back-action on a field in a coherent state, shifting its phase by an angle $\pm\Phi$, proportional to the atom–cavity mode interaction time. This shift, akin to a single-atom index effect, takes opposite values for an atom in $|e\rangle$ or $|g\rangle$, the upper and lower states of the atomic transition nearly resonant with the field mode.

Remarkably, we will see that Φ values as large as a few radians can be reached with realistic parameters of Rydberg atom CQED experiments, for fields containing up to a few tens of photons. The ability of a single atom to shift by a macroscopic angle the

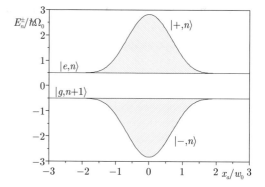

Fig. 7.7 Position of the dressed energies $E_n^\pm$, in units of $\hbar\Omega_0$, as a function of the atomic position x_a, in units of the cavity mode waist w_0, for $\Delta_c = \Omega_0$ and $n = 30$. The energy origin is taken as $(n + 1/2)\hbar\omega_c$.

phase of a field containing many quanta provides an ideal tool for cat experiments. If an atom is sent across the cavity in a superposition of $|e\rangle$ and $|g\rangle$, the field phase is blurred, to use Schrödinger's wording, and a mesoscopic superposition is produced in the cavity. Before describing this effect, we analyse the optimal conditions required to produce the largest possible phase shifts.

The reciprocal frequency shifts experienced by the $|e\rangle \rightarrow |g\rangle$ transition and the field when an atom sits at the cavity centre are proportional, in the perturbative limit of a large Δ_c detuning, to the 'shift per photon' s_0 defined by eqn. (3.187). The frequency shifts depend upon the position of the atom and the global phase shift induced on the field by an atom crossing the cavity simply results from the accumulation of the incremental shift experienced by the system along the atom's trajectory. Introducing once again the effective atom–cavity interaction time t_i^d, this perturbative shift is:

$$\Phi_0 = s_0 t_i^d = \frac{\Omega_0^2 t_i^d}{4\Delta_c} \,, \tag{7.32}$$

for an atom in $|e\rangle$ and takes the opposite value, $-\Phi_0$, for an atom in $|g\rangle$.[3] The dispersive $1/\Delta_c$ variation in eqn. (7.32) seems to imply that Φ_0 could be increased at will by tuning the atom arbitrarily close to resonance. This is deceptive because the perturbative expression is valid only for $\Delta_c \gg \Omega_0\sqrt{n+1}$ and thus breaks down for small detuning and large fields. As we will see, the shift saturates when Δ_c decreases, reaching an upper limit which depends upon the average number of photons. Since our goal is to imprint a macroscopic phase shift on a coherent field containing as many photons as possible, we need to understand this saturation process.

Let us start by following the tracks of a Rydberg atom in the CQED cavity set-up described in Section 5.2. This atom, propagating along the Ox axis, crosses a cavity containing a Fock state. The spatially dependent atom–field coupling $\Omega(x_a) = \Omega_0 \exp(-x_a^2/w_0^2)$ follows the Gaussian profile of the mode which has a waist w_0.[4] Starting from zero before the atom enters the cavity, it increases and reaches its

---

[3]Φ_0 is also the phase shift per photon of the Ramsey interferometer tuned on the $|g\rangle \rightarrow |i\rangle$ transition, when it is perturbed by a cavity field nearly resonant with the $|g\rangle \rightarrow |e\rangle$ transition (eqn. 6.43). It plays thus also an important role in the analysis of the QND photon detection scheme described in Section 6.4.

[4]x_a denotes here the atomic position, to avoid confusion with the field quadrature eigenvalues.

maximum value Ω_0 at the cavity centre, then decreases back to zero. The energies of the atom–cavity dressed states $|\pm, n\rangle$, linear combinations of the uncoupled $|e, n\rangle$ and $|g, n+1\rangle$ states, are given as a function of x_a by the simple formula derived from eqn. (3.174):

$$E_n^{\pm} = (n + 1/2)\,\hbar\omega_c \pm \frac{\hbar}{2}\sqrt{\Delta_c^2 + (n+1)\Omega^2(x_a)}\ . \tag{7.33}$$

The spatial dependence of these energies is shown in Fig. 7.7 for $\Delta_c = \Omega_0$ and $n = 30$. The $|+, n\rangle$ and $|-, n\rangle$ states evolve outside the cavity into the uncoupled states $|e, n\rangle$ and $|g, n+1\rangle$ respectively. Provided the atomic velocity v is small enough, the atom–cavity system, initially prepared in one of these two states, adiabatically follows the corresponding dressed state with negligible transition probability towards the other. The atom thus finally comes back to its initial state, the atom–cavity system merely undergoing a global phase shift, taking opposite signs for an atom in $|e\rangle$ or $|g\rangle$. This adiabatic phase shift, which has a magnitude proportional to the shaded areas in Fig. 7.7, is for $|e, n\rangle$:

$$\phi(n) = \int \left[\sqrt{\Delta_c^2 + (n+1)\Omega^2(x_a)} - \Delta_c\right]\frac{dx_a}{2v}\ . \tag{7.34}$$

The adiabatic evolution merely reflects the principle of energy conservation, amended by the Heisenberg time–energy uncertainty relation. Starting by a very qualitative argument, we remark that the perturbation experienced by the atom crossing the cavity possesses Fourier components up to a cut-off frequency ω_{cut} of the order of v/w_0, the reciprocal of its transit time across the mode. To avoid unwanted non-adiabatic transitions, ω_{cut} should remain smaller than Δ_c, the minimum frequency separation between the two dressed states. The atom velocity must thus obey at least the condition:

$$v/w_0 < \Delta_c\ . \tag{7.35}$$

The non-adiabatic transfer rate p_{na} between $|e\rangle$ and $|g\rangle$ can be evaluated more precisely (Messiah 1961). It is of the order of $1/\Delta_c^2$ multiplied by $|d\theta_n/dt|_{\text{max}}^2$, the square of the maximum variation rate of the dressed state mixing angle θ_n during the system evolution (see Section 3.4.1 for the definition of θ_n). For small detunings ($\Delta_c < \Omega_0\sqrt{n+1}$), the maximum of $d\theta_n/dt$ is reached when the atom enters and exits the mode, around the x_a values such that $\Delta_c \approx \Omega(x_a)\sqrt{n+1}$, i.e. $x_a \approx \pm w_0 \log^{1/2}(\Omega_0\sqrt{n+1}/\Delta_c)$. A simple calculation of $d\theta_n/dt$ at these points leads to:

$$p_{na} \approx \frac{v^2}{w_0^2 \Delta_c^2} \log\left(\Omega_0\sqrt{n+1}/\Delta_c\right)\ . \tag{7.36}$$

Imposing somewhat arbitrarily $p_{na} < 0.03$, we obtain an adiabaticity criterion:

$$\frac{v}{w_0} < \frac{\Delta_c}{6\log^{1/2}(\Omega_0\sqrt{n+1}/\Delta_c)}\ , \tag{7.37}$$

a bit more precise than but not so much different from the simple condition given by eqn. (7.35). Remarkably, p_{na} depends only logarithmically on n, making the adiabatic condition largely independent of the field intensity in the cavity.[5] Typically, this

---

[5]Equation (7.36) yields the linewidth Δ_{lw} of the resonance induced by the cavity field in an atomic beam of velocity v, defined by the implicit relation: $\Delta_{lw} \sim (v/w_0)\log^{1/2}(\Omega_0\sqrt{n+1}/\Delta_{lw})$.

condition is satisfied in our set-up for an atom having a thermal velocity (up to 300 m/s) and for photon numbers up to ~ 100 provided $\Delta_c > \Delta_c^{\min} \approx \Omega_0 = 2\pi \times 50$ kHz. We will take $\Delta_c^{\min}$ as the lower limit of the dispersive regime of CQED for thermal atomic beams. Smaller values of Δ_c, still satisfying the adiabatic condition, could be considered in experiments with cold atoms, having velocities down to the metre per second range.

Let us now consider an atom in $|e\rangle$ crossing under these adiabatic conditions a cavity containing a field in a coherent state $|\beta\rangle$ with $\overline{n} = |\beta|^2$ photons on average. By superposition of the various Fock states contributions, we express the final state $|\Psi_e\rangle$ of the atom–field system as:

$$|\Psi_e\rangle = e^{-|\beta|^2/2} \sum_n \frac{\beta^n}{\sqrt{n!}} e^{-i\phi(n)} |e, n\rangle , \tag{7.38}$$

where $\phi(n)$ is given by eqn. (7.34). We now proceed to show that this state describes, to a good approximation, an atom in $|e\rangle$ which has left in the cavity a coherent field whose phase has been shifted by an angle Φ. We start by expanding $\phi(n)$ around $n = \overline{n}$ as a Taylor series in powers of $n - \overline{n}$:

$$\phi(n) = \phi(\overline{n}) + (n - \overline{n})\phi'(\overline{n}) + O\left(1/\overline{n}\right) . \tag{7.39}$$

In a coherent state, the photon number fluctuation, $n - \overline{n}$, is of the order of $\sqrt{\overline{n}}$ and the ratio of two successive derivatives of $\phi(n)$ of the order of $1/\overline{n}$. It follows that the successive terms in eqn. (7.39) contribute to an expansion of the system's phase in powers of $1/\sqrt{\overline{n}}$. Assuming that the average photon number is at least of the order of a few tens (mesoscopic field approximation), we are justified to limit this expansion to first order. Replacing in eqn. (7.38) $\phi(n)$ by its expression (7.39), we break the phase $\phi(n)$ into two parts: an n-independent term contributing to a global phase shift of the system's state:

$$\psi_e(\overline{n}) = \phi(\overline{n}) - \overline{n}\phi'(\overline{n}) , \tag{7.40}$$

and a term proportional to n, whose coefficient $\phi'(\overline{n})$ corresponds to the phase shift $\Phi(\overline{n})$ of the field:

$$\Phi(\overline{n}) = \phi'(\overline{n}) = \int \frac{\Omega^2(x_a)}{\sqrt{\Delta_c^2 + (\overline{n} + 1)\Omega^2(x_a)}} \frac{dx_a}{4v} . \tag{7.41}$$

The system's state is finally re-expressed as:

$$|\Psi_e\rangle = e^{-i\psi_e(\overline{n})} \left|e, \beta e^{-i\Phi(\overline{n})}\right\rangle . \tag{7.42}$$

The interaction of an atom in $|g\rangle$ with the same field leads similarly to the final system state $|\Psi_g\rangle$, deduced from $|\Psi_e\rangle$ by exchanging e and g, reversing the sign of $\phi(n)$ and

---

The logarithmic dependence on n means that Δ_{lw} is largely insensitive to power broadening, a well-known property of atomic beam resonances (Ramsey 1985).

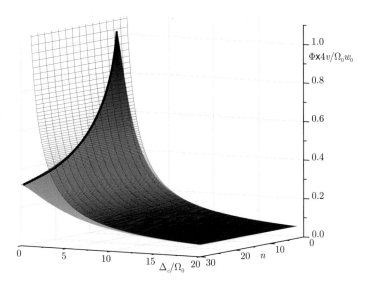

Fig. 7.8 Phase shift Φ (in units of $\Omega_0 w_0/4v$) produced by a single atom versus the average photon number $\bar{n}$ and the atom–cavity detuning Δ_c measured in units of Ω_0. The exact results are plotted as a solid surface. The asymptotic expression at large detunings, $\Phi_0 = \Omega_0^2 t_i^d/4\Delta_c$, is plotted as a wire frame. The variation of the saturating value $\Phi^{\max}(\bar{n})$ is shown as a thick solid line in the $\Delta_c = 0$ plane.

replacing in it n by $n-1$ (to account for the development of $|g, n\rangle$ along the dressed states $|\pm, n-1\rangle$ instead of $|\pm, n\rangle$). Defining the global phase of the system as:

$$\psi_g(\bar{n}) = -\phi(\bar{n}-1) + \bar{n}\phi'(\bar{n}-1) , \qquad (7.43)$$

and making the approximation $\Phi(\bar{n}-1) \approx \Phi(\bar{n})$ justified for large $\bar{n}$, we finally get:

$$|\Psi_g\rangle = e^{-i\psi_g(\bar{n})} \left|g, \beta e^{i\Phi(\bar{n})}\right\rangle . \qquad (7.44)$$

The phases ψ_e (ψ_g) and Φ in eqns. (7.42) and (7.44) play very different roles. The former describes a global quantum phase accumulated by the system's state as a whole. This overall phase change has no physical effect as long as we consider a coherent field interacting with an atom in one of the two states $|e\rangle$ or $|g\rangle$. This however becomes important, as we will see shortly, when the field is coupled to an atom in a superposition of $|e\rangle$ and $|g\rangle$. The difference $\psi_e(\bar{n}) - \psi_g(\bar{n})$ determines then the phase of the interference effects observed on the cat-state observables.

The phase Φ describes, on the other hand, the classical imprint of the atom onto the field complex amplitude. It takes opposite values, as expected, for atoms in $|e\rangle$ and $|g\rangle$. The shift magnitude, given by eqn. (7.41), depends upon $\bar{n}$ and Δ_c. Figure 7.8 shows the variations of Φ versus these two parameters in a three-dimensional plot. For large detunings, the shift, practically independent of the photon number, varies as $1/\Delta_c$. The asymptotic limit of eqn. (7.41) for $\Delta_c > \Omega_0\sqrt{\bar{n}}$ coincides of course with the perturbative solution. The final atom–field states for an atom initially in $|e\rangle$ or $|g\rangle$ are then:

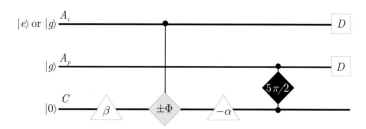

Fig. 7.9 Experimental sequence for the measurement of the Q function of a field phase shifted by a single atom.

$$|\Psi_e\rangle \approx e^{-i\Phi_0} \left|e, \beta e^{-i\Phi_0}\right\rangle \quad ; \quad |\Psi_g\rangle \approx \left|g, \beta e^{i\Phi_0}\right\rangle \,, \tag{7.45}$$

where Φ_0 is given by eqn. (7.32). The expressions of the global shifts of the atom–field system in eqn. (7.45) are easily deduced from eqns. (7.40) and (7.43), in which we replace ϕ and ϕ' by their asymptotic expressions for $\Delta_c/\Omega_0\sqrt{\bar{n}} \gg 1$: $\psi_e = \Phi_0$ and $\psi_g = 0$.

When Δ_c is of the order of or smaller than $\Omega_0\sqrt{\bar{n}}$, Φ falls below the perturbative solution and tends towards a maximum saturating value $\Phi^{\mathrm{max}}(\bar{n})$ as Δ_c approaches 0. The thick solid line in the $\Delta_c = 0$ plane of Fig. 7.8 describes the variation of this saturating value versus $\bar{n}$. It is obtained by expanding eqn. (7.41) to first order in $\Omega_0\sqrt{\bar{n}}/\Delta_c$:

$$\Phi^{\mathrm{max}}(\bar{n}) \approx \frac{\Omega_0 t_i^r}{4\sqrt{\bar{n}+1}} \approx \frac{\Omega_0 t_i^r}{4\sqrt{\bar{n}}} \,, \tag{7.46}$$

[the integration over x_a of the Gaussian function $\Omega(x_a)$ introduces in eqn. (7.46) the effective atom–cavity interaction time $t_i^r = \sqrt{\pi}w_0/v$; see eqn. (5.36)]. The condition of adiabatic following for the atom is satisfied down to $\Delta_c/\Omega_0 \sim 1$, so that the detuning can be decreased until obtaining a phase shift very close to $\Phi^{\mathrm{max}}(\bar{n})$, without observing unwanted atomic transitions between $|e\rangle$ and $|g\rangle$.

The maximum phase shift (7.46) is inversely proportional to the field amplitude. As could have been guessed intuitively, the phase-shift imprint of a single atom decreases when the field becomes more and more intense. There is only so much an atom can do to a large number of photons. Circular Rydberg atoms manage however surprisingly well and can have a macroscopic effect on a fairly large field. We can now justify the orders of magnitudes that we had announced at the beginning of this section. Typical parameters of our CQED experiment ($\Omega_0 = 3\cdot 10^5$ s^{-1}, $t_i^r = 100$ μs), yield $\Phi^{\mathrm{max}} \approx 1.5$ radian for a coherent field containing as much as $\bar{n} = 25$ photons!

The amazing ability of single atoms to affect large fields made of many quanta without being themselves perturbed is related to the adiabatic property. The slow branching on and off of the atom–field coupling is essential to ensure that the atom stays in its initial state $|e\rangle$ or $|g\rangle$, in spite of the very small value of the atom–field detuning.

7.3.2 Reading the phase imprint of a single atom

We must now explain how these phase shifts are observed. The usual method to measure the phase of a 'signal' field in quantum optics is homodyning (see Section

3.2.4): the signal is mixed with a reference of same frequency and variable phase φ. The phase distribution of the signal is obtained by measuring the intensity of the mixed signal + reference field as a function of φ.

We have adapted this general method to our CQED set-up and measured by homodyning the phase shift induced by a single atom on a coherent field (Maioli *et al.* 2005). The experimental sequence, sketched in Fig. 7.9, involves two successive field injections in the cavity and two atoms. First, a signal field with amplitude β is injected by connecting for a brief time interval (2 μs) the cavity to the classical source S. The modulus of β is calibrated in an independent experiment, measuring the dispersive shift produced by the field on the $|e\rangle \to |g\rangle$ transition (see Chapter 6). An off-resonant 'index' atom A_i, prepared with velocity v in $|e\rangle$ or $|g\rangle$ is then sent across the cavity and imprints its phase shift $\pm\Phi$ on the field whose amplitude becomes $\beta\exp(\pm i\Phi)$.

The Φ value is adjusted by independent control of the Δ_c and v parameters. The atom–cavity detuning is fixed by Stark shifting the atomic transition slightly away from the mode frequency. Detunings Δ_c of the order of a few tens to a few hundred kHz are realized in this way. A second reference field injection follows before the cavity field has had time to appreciably decay. This reference has a complex amplitude $-\alpha = -\beta\exp(i\varphi)$. It has the same mean photon number as the signal, but its phase φ can be continuously adjusted.

A direct measurement of the cavity field being impractical, we use a second atom, A_p, to probe the resulting field in the cavity. This atom is prepared in $|g\rangle$ and an electric field tunes A_p at exact resonance with the mode, so that it can absorb the homodyned field. Information provided by detecting the final state of A_p is used, in a repetitive experiment, to reconstruct the Q function of the phase-shifted signal field. The Q function of a field, defined by its density operator ρ, can be expressed in the following equivalent forms (see the appendix):

$$Q(\alpha) = \frac{1}{\pi}\langle\alpha|\,\rho\,|\alpha\rangle = \frac{1}{\pi}Tr\left[\rho\,|\alpha\rangle\,\langle\alpha|\right] = \frac{1}{\pi}\mathrm{Tr}\left[|0\rangle\langle0|D(-\alpha)\rho D(\alpha)\right]\;. \qquad (7.47)$$

The last expression means that $Q(\alpha)$ is equal to $1/\pi$ times the probability $p(0)$ of finding the cavity in the vacuum, after the field to be measured has been translated in phase space by the amplitude $-\alpha$. This translation is precisely realized by the injection of the reference field in C.

To perform the Q function reconstruction, we must be able to deduce from the statistics of the A_p detection signals the probability $p(0)$ of the cavity field being in vacuum. The probability of detecting A_p in $|e\rangle$ is in general:

$$P(e) = \sum_{n\geq1} P_e(n)p(n)\;, \qquad (7.48)$$

where $p(n)$ is the photon number distribution and $P_e(n)$ the probability of atomic excitation by an n photon field. The statistics of A_p detection events thus generally contains information about a combination of $p(n)$'s, not about $p(0)$ alone. What we need to realize is a situation in which the atom is equally sensitive to all photons numbers, i.e. $P_e(n) = P$ for $n \geq 1$, so that we would have $P(e) = P[1 - p(0)]$. Determining $P(e)$ would then directly yield $p(0)$. In a perfect set-up this is not the case, since the Rabi

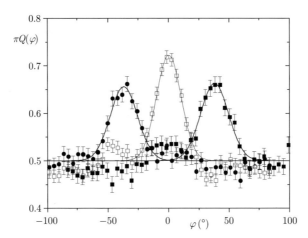

Fig. 7.10 Phase shift of a coherent field. Field phase distribution $\pi Q(\varphi)$ as a function of the displacement phase φ in degrees. Open squares: reference coherent field with $\bar{n} = 29$. Solid circles (squares): phase distribution after interaction with an atom in $|e\rangle$ ($|g\rangle$). The error bars represent statistical fluctuations. The solid lines are Gaussian fits. Reprinted with permission from Raimond *et al.* (2005). © Institute of Physics Publishing.

oscillation results in n-dependent excitation probabilities $P_e(n) = \sin^2(\Omega_0 t_i^r \sqrt{n}/2)$, which oscillate between 0 and 1 around an average of $1/2$. Fortunately for our present purpose our set-up is not ideal. Various experimental imperfections conspire to decrease the contrast of the Rabi oscillation. The main contrast reduction factor is the atomic beam finite transverse size which makes successive probe atoms experience different couplings, resulting in a fluctuation of the Rabi oscillation frequency. This fluctuation brings the experimental measured probability $P_e^{\exp}(n)$ closer to $1/2$ than the ideal value $P_e(n)$, the effect becoming more and more important when n increases. In practice, if we choose the optimal effective interaction time $t_i^r = t_{\rm opt} = 5\pi/2\Omega_0$, all the $P_e^{\exp}(n)$s with $n \geq 1$ satisfy the condition $|P_e^{\exp}(n) - 1/2| < 0.1$ and can be, within a good approximation, replaced by $p = 1/2$.[6] Hence, with this optimal interaction time, the probability $P_e^{\exp}$ of detecting the probe atom in $|e\rangle$ is:

$$P_e^{\exp} \approx [1 - p(0)]/2 , \qquad (7.49)$$

and, according to eqn. (7.47):

$$Q(\alpha) = [1 - 2P_e^{\exp}]/\pi . \qquad (7.50)$$

The A_p atom excitation probability $P_e^{\exp}$ is measured by repeating the same experimental sequence many times for each value of φ. The variation of $P_e^{\exp}$ when this angle is swept determines directly Q in phase space along a circle of radius $|\alpha|$, i.e. the phase distribution of the field left in the cavity by the index atom. The experimental signals are shown in Fig. 7.10. The coherent field contains in average 29 photons. The central

---

[6]The theoretical values of the $P_e(n)$ probabilities are then $P_e(1) = 0.5, P_e(2) = 0.44, P_e(3) = 0.25$.

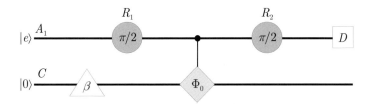

Fig. 7.11 Cat state generation sequence.

peak (open circles) represents Q for the signal field, when no index atom is injected. This peak serves as a reference for the signals shown as black circles and black squares, which correspond respectively to A_i crossing the cavity in $|e\rangle$ or $|g\rangle$. The atom–field detuning is $\Delta_c/2\pi = 50$ kHz and the velocity $v = 200$ m/s. The observed phase shift, $\pm\Phi = \pm39°$ is in excellent agreement with the theoretical predictions.

This experiment demonstrates directly that a single atom leaves a 'signature' on a large field. The dispersive single-atom effect is here used in a very effective way to amplify quantum information into a mesoscopic one. In a variant of this experiment, we have shown that a single atom can be detected non-destructively by a large photon count (Maioli *et al.* 2005). In short, we adjust the phase of the reference field so that it cancels the field in the cavity if no index atom has been sent. The passage of A_i modifies the relative phases of the signal and reference fields and results then in the appearance of a large field in C. This field is subsequently absorbed by a collection of resonant probe atoms sent across the cavity in $|g\rangle$.

This process is somewhat similar to the detection of a single atom in optical CQED (see Section 5.1.6). There, a single atom was controlling the flow of a large number of photons across the Fabry–Perot resonator while, here, a single atom is manipulating the phase of a large number of photons stored inside the cavity. Phase information is finally transformed into intensity information by the homodyne method so that, as in the optical CQED case, a single atom determines, without undergoing transitions, whether or not many photons are absorbed in a detector. Single atoms are, in both experiments, finally detected non-destructively by large photon counts.

7.3.3 Trapping a cat in an atomic interferometer

This single-atom amplification process can be exploited to generate cat states of the field. We merely have to inject the atom in a superposition of $|e\rangle$ and $|g\rangle$, prepared by a classical microwave pulse R_1 before it enters C. The field then acquires two distinct phases at once and gets entangled with the atom. This entanglement can be analysed by finally detecting the atomic state after the atom exits C. This analysis is carried out by combining a second microwave pulse R_2 with a state-selective detector D. The experimental sequence is sketched in Fig. 7.11, in which we recognize the basic features of a Ramsey interferometer with its two pulses sandwiching the atom–cavity interaction, the first being used for the cat state preparation and the second for its reading out. In short, the photonic cat, trapped inside an atomic interferometer, modifies the interferometric signal in ways which reveal its presence and its main features.

Let us follow the successive stages of this experiment. To keep the formulas as simple as possible, we assume that the detuning Δ_c is large and use the perturbative expressions (7.32) and (7.45) for phase shifts and states. The atom, initially in $|e\rangle$, is prepared by R_1 in $(|e\rangle + |g\rangle)/\sqrt{2}$. After it has crossed the cavity, the atom–field system evolves into the entangled state obtained by superposition of the $|e\rangle$ and $|g\rangle$ contributions:

$$|\Psi_1\rangle = \frac{e^{-i\Phi_0}}{\sqrt{2}} |e\rangle \otimes |\beta e^{-i\Phi_0}\rangle + \frac{1}{\sqrt{2}} |g\rangle \otimes |\beta e^{i\Phi_0}\rangle \ . \tag{7.51}$$

This is a typical quantum measurement situation in which a large 'meter' (the field) points simultaneously towards different directions corresponding to two orthogonal states of a microscopic system. If the atom were subsequently detected at the cavity exit in one of the two states $|e\rangle$ or $|g\rangle$, the field would be projected in the corresponding coherent state and the quantum ambiguity would be lost. The atom's measurement would tell us whether the dispersive index had taken one value or the other and this would yield precise information about the phase shift of the field.

The second Ramsey zone R_2 can be used to preserve the blurred state of the meter by erasing information about the value of the index experienced by the field. Let us apply in this zone a pulse realizing:

$$|e\rangle \rightarrow \left(|e\rangle + e^{i\varphi} |g\rangle\right)/\sqrt{2} \quad ; \quad |g\rangle \rightarrow \left(|g\rangle - e^{-i\varphi} |e\rangle\right)/\sqrt{2} \ , \tag{7.52}$$

where φ is adjusted freely by setting the relative phase of the R_1 and R_2 pulses. Immediately after R_2, the combined atom–field state is obtained by replacing, in eqn. (7.51), $|e\rangle$ and $|g\rangle$ by their transformed expressions (7.52). After rearranging the terms, the atom–field state becomes:

$$\begin{aligned}|\Psi_2\rangle \ = \ &\tfrac{1}{2} |e\rangle \otimes \left[e^{-i\Phi_0} |\beta e^{-i\Phi_0}\rangle - e^{-i\varphi} |\beta e^{i\Phi_0}\rangle\right] \\ &+ \tfrac{1}{2} |g\rangle \otimes \left[e^{i(\varphi-\Phi_0)} |\beta e^{-i\Phi_0}\rangle + |\beta e^{i\Phi_0}\rangle\right] \ . \end{aligned} \tag{7.53}$$

The two states $|e\rangle$ and $|g\rangle$ are now correlated to two field states of the cat type, mutually orthogonal when $\Phi_0 \gg 1/\sqrt{\bar{n}}$. The phase φ of R_2 can be adjusted to simplify the cat state expressions. Setting $\varphi = \Phi_0$, we get:

$$|\Psi_2\rangle = \frac{e^{-i\Phi_0}}{2} |e\rangle \otimes \left[|\beta e^{-i\Phi_0}\rangle - |\beta e^{i\Phi_0}\rangle\right] + \frac{1}{2} |g\rangle \otimes \left[|\beta e^{-i\Phi_0}\rangle + |\beta e^{i\Phi_0}\rangle\right] \ . \tag{7.54}$$

This equation shows that the final detection of the atom projects with equal probabilities the field into one of the two cat states $\left[|\beta e^{-i\Phi_0}\rangle \pm |\beta e^{i\Phi_0}\rangle\right]/\sqrt{2}$. When $\Phi_0 = \pi/2$, they are the even and odd π-phase cat states defined by eqn. (7.4):

$$|\Psi_2\rangle = -\frac{i}{2} |e\rangle \otimes \left[|\gamma\rangle - |-\gamma\rangle\right] + \frac{1}{2} |g\rangle \otimes \left[|\gamma\rangle + |-\gamma\rangle\right] \ , \tag{7.55}$$

where we have defined $\gamma = -i\beta$. It is impossible to predict the field state eventually obtained, which is revealed only by the outcome of the atomic measurement. The single-atom phase-shift $\Phi_0 = \pi/2$ could be reached, in the perturbative regime, with slow atoms spending a long time in the cavity. With atoms having a thermal velocity of a few hundred m/s, this large phase shift can also be produced using a smaller detuning Δ_c, for which the perturbative expressions (7.32) and (7.45) are no longer valid.

This cat experiment appears very similar to the QND Ramsey interferometry described in Chapter 6. It is in fact the same experiment, viewed from a complementary perspective. We are interested here in the effect produced by the atom on the field, seen as a meter measuring the atomic energy. At the same time, the field has a back-action on the atom, shifting its energy levels. This shift can be used, as discussed in Chapter 6, to measure the field's photon number. In this viewpoint, the roles of matter and radiation are exchanged: the atom becomes the meter measuring the field's energy. These two reciprocal measurements are simultaneously taking place in the apparatus and must be analysed together.

The complementary aspect of these two viewpoints is clearly apparent when $\Phi_0 = \pi/2$. This condition corresponds to a π-phase shift per photon in the QND Ramsey interferometer described in Chapter 6. We have seen that the atom undergoing such a phase shift realizes a measurement of the photon number parity $\mathcal{P}$. The field in a coherent state $|\beta\rangle$ can obviously be expressed as the sum $(|\beta\rangle+|-\beta\rangle)/2+(|\beta\rangle-|-\beta\rangle)/2$ of an even and an odd cat state. The atom measuring its parity thus projects the field in either the even or the odd cat state, which are eigenstates of $\mathcal{P}$. The cat state generation can, in this case, be reinterpreted as a projective measurement of the field parity operator.

We have seen in Chapter 6 that this measurement starts to pin down the photon number. Gaining information on this number results in a loss of information on the complementary quantity, the field phase. We have analysed this process in detail and noticed that the first step of the phase blurring process was the splitting of the field into two components with opposite phases (Fig. 6.19, on page 344). The generation of cat states is thus an unavoidable consequence of the field QND measurement. As a historical footnote, let us mention that we realized this point while analysing in detail the QND process, and this prompted us to write our first proposal about cat state generation in CQED (Brune *et al.* 1992).

To substantiate these qualitative remarks, let us analyse the Ramsey fringe signal versus the phase difference φ between the two Ramsey zones. Before feeding a field in the cavity, let us first assume that C is in vacuum. The probability of finding the atom in $|e\rangle$ or $|g\rangle$ as a function of φ is then given by the usual atomic Ramsey fringe signal. By setting $\beta = 0$ in eqn. (7.53), we obtain indeed for the final atom–field system:

$$\left|\Psi_2^{(\beta=0)}\right\rangle = \frac{e^{-i\Phi_0}}{2}\left[1-e^{-i(\varphi-\Phi_0)}\right]|e,0\rangle + \frac{1}{2}\left[1+e^{i(\varphi-\Phi_0)}\right]|g,0\rangle , \tag{7.56}$$

which yields the probabilities P_e^0 and $P_g^0 = 1 - P_e^0$ of finding the atom in $|e\rangle$ or $|g\rangle$:

$$P_e^0 = 1 - P_g^0 = \frac{1-\cos(\varphi - \Phi_0)}{2} . \tag{7.57}$$

Note that the fringes are phase-shifted by an angle Φ_0 with respect to their position if there were no cavity between R_1 and R_2. This effect, which reflects the Lamb shift of the atom in $|e\rangle$, has been analysed in Chapter 5.

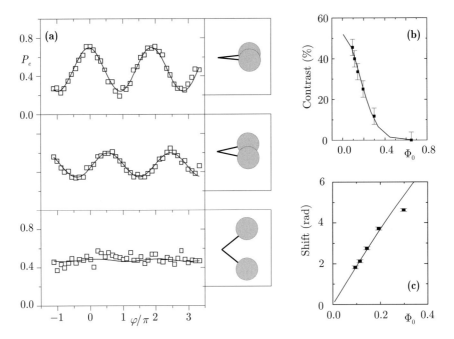

Fig. 7.12 Schrödinger cat and complementarity. (a) Ramsey fringes for $\beta = 3.1$ ($\bar{n} = 9.5$) and three Φ_0 values (0.1, 0.2 and 0.69 radians, corresponding to $\Delta_c/2\pi = 712$, 347 and 104 kHz respectively from top to bottom). The insets pictorially represent the two field phase components. (b) Ramsey fringes contrast versus Φ_0. The solid line corresponds to the theoretical predictions, scaled by the finite Ramsey interferometer contrast. (c) Fringes shift (in radians) versus Φ_0. The slope of the fitted line provides a calibration of the photon number. Reprinted with permission from Brune *et al.* (1996a). © American Physical Society.

When the cavity contains a coherent field with $\bar{n}$ photons on average, this field undergoes different phase shifts, depending on the state of the atom between R_1 and R_2. We recognize here the ingredients of a which-path experiment. The atom follows two paths at once in the interferometer and gets correlated to two final field states. If they are mutually orthogonal, which occurs for $\Phi_0 > 1/\sqrt{\bar{n}}$, they can, by the detection of their phase, tell away the path of the atom. This is enough to destroy the fringes. In more quantitative terms, the contrast of the fringes reveals the amount of entanglement between the atom and the field. When this entanglement is maximum, the fringes are fully suppressed. They have, on the other hand, a maximum contrast when the two systems are separable.

The fringe contrast reduction can alternatively and equivalently be understood as a phase-blurring effect of the atomic Ramsey fringes, due to a fluctuating light-shift effect. The atomic coherence phase shift per photon is equal to $2\Phi_0$. The $\Delta N = \sqrt{\bar{n}}$ dispersion of the photon number in the coherent field thus produces a fluctuation of the order of $2\Phi_0\sqrt{\bar{n}}$ of the phase of the atomic Ramsey fringes. They are fully blurred when this phase fluctuation is of the order of π, i.e. for $\Phi_0 \approx 1/\sqrt{\bar{n}}$. We retrieve the condition given above by the complementarity argument.

These qualitative features are quantitatively justified by an exact calculation of the Ramsey fringe pattern. The probabilities of finding the atom in $|e\rangle$ or $|g\rangle$ when the cavity contains a coherent field with $\bar{n}$ photons on average are obtained from (7.53):

$$
\begin{aligned}
P_e^{\bar{n}} &= 1 - P_g^{\bar{n}} = \frac{1}{2}\left\{1 - \mathrm{Re}\left[e^{i(\Phi_0-\varphi)}\langle \beta e^{-i\Phi_0}|\beta e^{i\Phi_0}\rangle\right]\right\} \\
&= \frac{1}{2}\left\{1 - \cos\left[\varphi - \Phi_0 - \bar{n}\sin(2\Phi_0)\right]e^{-\bar{n}[1-\cos(2\Phi_0)]}\right\}. \quad (7.58)
\end{aligned}
$$

We recognize in the first line the overlap between the two final field components, whose analytical expression is given in the last line. The argument $\bar{n}[1-\cos(2\Phi_0)]$ of the last exponential in eqn. (7.58) is equal to $D^2/2$, where D is the distance in phase space of the centres of the uncertainty circles associated to the two field components. The Ramsey fringes are phase-shifted by an angle proportional to $\bar{n}$ and their amplitude is suppressed by the factor $\exp(-D^2/2)$, which decreases exponentially with the separation of the field components.

We have observed these Ramsey fringes (Brune *et al.* 1996a) for different phase splittings Φ_0 obtained by varying Δ_c while keeping fixed the effective atomic transit time $t_i^d = 19$ μs and the average photon number $\bar{n} = 9.5$. For each φ phase setting, we repeat a large number of identical experimental sequences, preparing the field in the cavity, then sending an atom across the interferometer and recording its final state. We obtain in this way the average of the atomic transfer rate between $|e\rangle$ and $|g\rangle$ for a given value of φ. The Ramsey fringes are then reconstructed by sweeping φ.

The Ramsey signals for three different values of Φ_0 are shown in Fig. 7.12(a) with, in the insets, the corresponding final field state phase space representations. The collapse of the fringe amplitude when the field components separate is conspicuous. The fringe contrast versus Φ_0 is shown in Fig. 7.12(b) and the fringe phase shift in Fig. 7.12(c). In these plots, the points are experimental and the curves given by theory, with an overall contrast adjustment taking into account the imperfections of the Ramsey interferometer. Note that the theoretical formula (7.58), based on perturbative expressions is, for a field containing $\bar{n} = 9$ photons, valid for $\Delta_c > \Omega_0\sqrt{\bar{n}} \approx 3\Omega_0$. It does not apply for the smallest detuning $\Delta_c = 104$ kHz. To plot the theoretical curves, an exact expression of the phase shifts and atom–field states is used instead of the perturbative expressions valid for the largest values of Φ_0.

The experiment is found in excellent agreement with the theory. This is a clear indication that the field components are separated in phase space by the single atom crossing the cavity in a superposition of $|e\rangle$ and $|g\rangle$. The variation of the fringe phase with Φ_0 reveals the light shift experienced by the atom, whose transition frequency is modified by an amount proportional to the field photon number $\bar{n}$. This variation shown in Fig. 7.12(c) yields the calibration of the field in the cavity. This experiment measures the complex scalar product of the two final field components. The modulus of this product is measured by the fringe contrast and its phase by the positions of the fringe maxima. The combination of these two data provides a simultaneous measurement of the field amplitude and phase splitting.

We have analysed this Ramsey experiment as a complementarity test. It is instructive to compare it to the which-path experiment described in detail in Section 6.1. There, information about the path of the atom was coded in the number of photons in the first Ramsey zone, whereas now it is contained in the phase of the field stored in the cavity, between the Ramsey zones. Which-path information appears to be attached to complementary variables in the two experiments. This explains why the fringes were visible then when the field had a large photon number (and hence large photon fluctuation), whereas they have a full contrast now when the field is small (and hence has large phase fluctuation).

7.3.4 The quantum switch: an alternative route to cats

In the experiments considered so far, the cavity is first filled by a coherent field via its coupling to a resonant source, the index atom shifting the phase of the field only after the source has been disconnected. What happens if, without modifying anything else, the source is connected to the cavity during the time the atom crosses it? If the dispersive atom–field coupling is strong enough, the atom pushes the mode frequency towards higher or lower frequency (depending upon its state, $|e\rangle$ or $|g\rangle$) and makes the source off-resonant. The field is therefore reflected back instead of filling the mode. A single atom can thus, by the same dispersive effect responsible for the field phase shift, prevent the flow of a large number of photons inside the cavity.

We have encountered a similar effect in optical CQED experiments (Section 5.1.6). The field blockade is then used to signal the passage of single atoms in the cavity. In the context of microwave cat studies, this negative result does not seem at first sight very interesting, because the cavity remains empty after the exit of the atom, admittedly not a very exciting situation.

We can, however, give to the experiment a much more intriguing twist by adjusting the respective frequencies ω_s, ω_c and $\omega_{eg} = \omega_c + \Delta_c$ of the source S, the cavity C and the atom (Davidovich *et al.* 1993). Let us tune S to a frequency $\omega_c + \delta_s$ slightly higher than the cavity frequency. This mismatch obeys the condition $\delta_s \gg 1/T_c$ so that S, by itself, is unable to feed photons in C. We also adjust the atom–cavity mismatch Δ_c so that the detuning of the cavity mode when an atom in $|e\rangle$ sits at the cavity centre exactly compensates the cavity–source frequency mismatch:

$$\delta_s = \omega_s - \omega_c = \Omega_0^2/4\Delta_c . \tag{7.59}$$

We assume here that Δ_c is large enough so that it is legitimate to use the perturbative expressions of the cavity shifts (we discuss below in more detail the validity of this assumption and the limitation it imposes on the field amplitude).

The variations of the cavity field frequency as an atom crosses adiabatically the mode in $|e\rangle$ or $|g\rangle$ are represented in Fig. 7.13. The atomic index varies continuously and the field mode frequency is shifted symmetrically upwards or downwards, depending upon the state of the atom. If it is in $|e\rangle$, the atom-induced cavity shift tunes S in exact resonance with C when the atom reaches the cavity centre. By matching the source to the field mode, the atom opens, so to speak, a gate to the source photons, allowing the build-up of a coherent field with an amplitude α in the cavity. If, on the other hand, the atom is in $|g\rangle$, the source–cavity detuning increases, going from δ_s when the atom is outside C to $2\delta_s$ when it passes at the cavity centre. The gate remains always closed and no photon enters C.

The atom is a kind of switch with its 'on' and 'off' positions corresponding to $|e\rangle$ and $|g\rangle$. This switch has a very interesting quantum feature lacking in ordinary taps. It can be placed, with the help of R_1, in the superposition $(|e\rangle + |g\rangle)/\sqrt{2}$ of its on

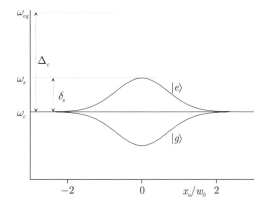

Fig. 7.13 Principle of the quantum switch: variation of the frequency of the cavity loaded with an atom in $|e\rangle$ or in $|g\rangle$, versus atomic position x_a in units of the cavity mode waist w_0. The relevant frequencies, ω_c, ω_s and ω_{eg} are indicated on the vertical axis.

and off states. As a result, the switch+field system ends up, when the atom leaves the cavity, in the entangled state:

$$|\Psi_1\rangle = (|e,\alpha\rangle + |g,0\rangle)/\sqrt{2} \ . \tag{7.60}$$

Detecting the atom at the cavity exit would yield two possible outcomes, each leaving the field in one of two mutually exclusive states. Using R_2 to mix again coherently the atomic states allows us to maintain the quantum ambiguity in the field. After the pulse R_2 (supposed here to be identical to R_1), the system has evolved into:

$$|\Psi_2\rangle = \tfrac{1}{2}\left(|e,\alpha\rangle + |g,\alpha\rangle + |g,0\rangle - |e,0\rangle\right) \ , \tag{7.61}$$

and a subsequent detection of the atom in $|e\rangle$ or $|g\rangle$ now leaves the field in one of the two coherent superpositions:

$$|\Psi_f\rangle = \frac{1}{\sqrt{2\left[1 \pm \exp(-|\alpha|^2/2)\right]}}\left(|\alpha\rangle \pm |0\rangle\right) \ . \tag{7.62}$$

After measurement of the final state of the 'switch' atom, the field is left in a quantum superposition of cavity 'filled' and empty states, which we will call an 'amplitude Schrödinger cat', to distinguish it from the phase cats studied above. Not surprisingly, the amplitude cat finally obtained (+ or − sign in eqn. 7.62) cannot be a *priori* predicted, but only found out after the atomic state has been recorded.

There is a direct connexion between amplitude and phase cats. It is very easy to turn one kind into the other. We simply have to connect the cavity to a resonant source in order to displace the state in phase space according to:

$$D(-\alpha/2)\left[(|\alpha\rangle \pm |0\rangle)/\sqrt{2}\right] = \left(|\alpha/2\rangle \pm |-\alpha/2\rangle\right)/\sqrt{2} \ , \tag{7.63}$$

where we have assumed $|\alpha| \gg 1$ to simplify the state normalization. Similarly, a phase cat can be turned into an amplitude one by the inverse translation.

The quantum switch and the dispersive phase shift methods appear at first sight to be two equivalent ways to get into the realm of mesoscopic superpositions. It seems just a matter of commuting the field injections and the atomic interaction sequences.

A more careful analysis of the system evolution will show us, however, that the quantum switch method is somewhat more demanding, requiring slower atoms and better cavities to prepare cats of the same size as the dispersive technique. This explains why, experimentally, the dispersive phase-shift method has been up to now preferred.

We have so far assumed that the atom behaves as a transparent medium having an index independent upon the field intensity. Clearly, this picture breaks down when the coupling between S and C becomes too strong. There is only so much that a single atom can do. It is only in a tale that a little boy can plug the big dike of the Netherlands polders with a single finger. What is the limit then? How many photons can a single atom superpose coherently with the vacuum in a cavity?

We must first realize that a proper operation of the quantum switch requires the atom–cavity detuning Δ_c to be large enough to satisfy the perturbative regime condition:

$$\Delta_c \gg \Omega_0 \sqrt{\bar{n}} \ . \tag{7.64}$$

It is only in this regime that the cavity mode frequency is independent of the field intensity. The cavity appears then to the source as a quantum oscillator, whose frequency has been merely displaced by the presence of the atom. The coupling between the source and the loaded cavity remains linear and the above analysis is valid. If Δ_c does not satisfy (7.64), the absorption spectrum of the dressed atom–cavity system becomes more complicated. The loaded cavity does not behave any more as a linear oscillator. It presents instead a complex energy ladder, offering many routes for the source to fill the mode, via single- or multi-photonic transitions. The atom is then unable to prevent the mode filling.

Another important condition is that the strength of the source–cavity coupling should be small enough so that the off position of the switch does not become leaky. The strength of this coupling can be loosely defined in frequency unit as the ratio $|\alpha|/t_s = \sqrt{\bar{n}}/t_s$, where t_s is the effective time during which the source-to-cavity connexion is open in the 'on' position of the switch. This time is of the order of the atom–cavity crossing time (we assume that S is switched off when the atom is not in C, which minimizes this leak). For the source not to sneak photons in when the atom is in $|g\rangle$, the strength of its coupling to C should be smaller that the loaded cavity detuning, which is then $2\delta_s$. We must thus satisfy the condition:

$$\sqrt{\bar{n}}/t_s < 2\delta_s \ . \tag{7.65}$$

This condition, combined with (7.59), can be expressed as another constraint on Δ_c:

$$\Delta_c < \frac{\Omega_0^2 t_s}{2\sqrt{\bar{n}}} \ . \tag{7.66}$$

Comparing it with eqn. (7.64), we conclude that the number of photons must satisfy:

$$\bar{n} \ll \Omega_0 t_s/2 \ . \tag{7.67}$$

The average number of photons in the amplitude cat must be much smaller than π times the number of vacuum Rabi oscillations that the atom would experience if it crossed at resonance the cavity. For an atom with $v = 100$ m/s (lower limit for a velocity-selected thermal atomic beam), we have $t_s \approx 100 \ \mu s$ and $\Omega_0 t_s/2 \approx 15$. Cat sizes are limited to a few photons at most. A CQED set-up employing laser-cooled atoms with $v \sim$ a few metres per second would put t_s into the millisecond range and raise the maximum $\bar{n}$ value to a few tens of photons.

These conditions are more stringent than those imposed on phase cats. This appears clearly if we rewrite eqn. (7.66) as $\Omega_0^2 t_s/2\Delta_c > \sqrt{\bar{n}}$, an inequality implying that the quantum switch atom, when used as a dispersive phase shifter operating in the perturbative regime, should be able to split the phase of a coherent field by an angle of $\sqrt{\bar{n}}$ radian. Phase cats require only a phase splitting of π radian, and even less if one is ready to settle for a cat whose components make a smaller angle. If condition (7.66) is not satisfied, the quantum switch is leaking in its off position, meaning that it prepares states of the form $(|\alpha_{\text{on}}\rangle \pm |\alpha_{\text{off}}\rangle)/\sqrt{2}$ with $|\alpha_{\text{on}}| > |\alpha_{\text{off}}|$. This is also an amplitude cat of the general form defined by eqn. (7.1).

Note finally that we have overlooked a last important limitation. The cat preparation time t_s should be much shorter than the loss of the mesoscopic coherence, which occurs within a time scale determined by the cavity damping time T_c. We will delay the discussions about the limitations imposed by the cavity Q factor until we have studied the decoherence of our photonic cat states (Section 7.5).

This digression about quantum gates simultaneously opened and closed contradicts the proverb made popular by the title of a famous play by the French nineteenth century author Alfred de Musset, '*Il faut qu'une porte soit ouverte ou fermée*'[7] (Musset, a famous member of the 'romantic school' opposed to the 'French classics' of the seventeenth century, has uttered here a 'classical' statement, at least as far as information theory goes). We will abandon now these strange devices and come back to the description of cats obtained by manipulating fields prepared beforehand in the cavity.

7.4 Resonant cats in cavity QED

In the dispersive regime, the phase splitting between the cat components increases when the atom–cavity detuning is reduced. This gives us a hint that the largest cats are obtained at resonance and leads us to study in detail what happens when $\Delta_c = 0$. The dynamics of the atom–field system is then quite different from the dispersive case. It is of course no longer possible to make the adiabatic assumption that the atom stays in $|e\rangle$ or $|g\rangle$ as it crosses the cavity. It undergoes instead a Rabi oscillation between these levels. If the field is mesoscopic, this oscillation presents collapses and revivals due to the photon graininess, which we have described in Chapter 3 and 4. As we will see, these peculiar features of the atomic evolution are intimately related to the creation of photonic phase cat states. The atom undergoing the complex Rabi oscillation reacts back on the field and the two systems pass through a succession of entanglement and separation stages, corresponding to transient apparitions and disappearances of cat states in the cavity. For a study of various aspects of this dynamics, see Eiselt and Risken (1989); Gea-Banacloche (1990); Averbukh and Perel'Man (1991); Gea-Banacloche (1991); Buzek *et al.* (1992); Miller *et al.* (1992); and Buzek and Knight (1995).

---

[7]'*A door ought to be opened or closed*', play in one act by Alfred de Musset (1845).

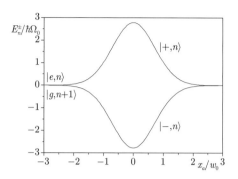

Fig. 7.14 Energies of the dressed levels (in units of $\hbar\Omega_0$) as a function of the atomic position x_a (in units of the cavity mode waist w_0) at exact resonance for $n = 30$. The energy origin is taken as $(n + 1/2)\hbar\omega_c$.

7.4.1 Photonic cats versus Rabi oscillation collapses and revivals

In order to analyse these effects, we will revisit the Rabi oscillation phenomenon, but instead of focusing as we did in Chapter 3 on the atomic dynamics alone, we now consider the joint evolution of the atom+field system. This evolution is, as in the dispersive case, simply described in the dressed state basis. At exact resonance these states are (eqn. 3.176):

$$|\pm, n\rangle = (|e, n\rangle \pm i\,|g, n + 1\rangle)/\sqrt{2} \,, \tag{7.68}$$

and their energies:

$$E_n^\pm = (n + 1/2)\hbar\omega_c \pm (\hbar/2)\Omega(x_a)\sqrt{n + 1} \,, \tag{7.69}$$

are shown, for $n = 30$, in Fig. 7.14 as a function of the atomic position x_a. The two dressed levels, degenerate outside the mode, repel each other as the atom enters the cavity, are maximally split at the cavity centre by $\hbar\Omega_0\sqrt{n + 1}$ and merge again when the atom leaves the cavity. The level degeneracy outside the cavity makes it now impossible to apply the adiabatic theorem to the atomic evolution starting from $|e\rangle$ or $|g\rangle$. These states are no longer eigenstates of the system's Hamiltonian outside the mode and the adiabatic approximation does not apply to them.

We can, however, invoke again the adiabatic theorem to analyse the evolution of the dressed states $|\pm, n\rangle$, which, for $\Delta_c = 0$, present the remarkable feature to be completely x_a-independent: the eigenstates of the system Hamiltonian do not evolve while the atom crosses the cavity. However small is the separation between the dressed energies in the Gaussian tail of the mode, these dressed states are unambiguously defined anywhere along the atomic trajectory and the time derivative of one state projected onto the other exactly vanishes, for any position x_a and atomic velocity v:

$$\langle -, n| \left(\frac{d}{dt} |+, n\rangle \right) = v\langle -, n| \left(\frac{d}{dx_a} |+, n\rangle \right) = 0 \,. \tag{7.70}$$

The system, prepared before the atom has entered C in one of the $|+, n\rangle$ or $|-, n\rangle$ states, remains thus stationary. When the atom has left the cavity, the global wave function has merely accumulated a quantum phase $\pm\phi_r(n)$ proportional to the area between the curves in Fig. 7.14:

$$\phi_r(n) = \frac{\Omega_0 t_i^r \sqrt{n+1}}{2} . \tag{7.71}$$

We have expressed, here again, the Gaussian integral in terms of the effective inter-action time t_i^r (eqn. 5.36). We recall that it represents the time it would take for the system to reach the same final state if the atom–field coupling were constant and equal to Ω_0. Varying continuously the atom velocity v provides a way to sweep t_i^r and to explore experimentally the dynamics of the atom–field system. Another method, sometimes more convenient, consists in keeping v constant and tuning, by the Stark effect, the atom in and out of resonance with the field between two positions x_a^i and x_a^f. The system evolution is frozen when the atom is outside this interval. The reso-nant phase accumulation in the $|\pm, n\rangle$ dressed states, suddenly switched on at x_a^i and interrupted at x_a^f becomes:

$$\phi_r(n) = \frac{1}{2v} \int_{x_a^i}^{x_a^f} \Omega(x_a) \sqrt{n+1} \, dx_a = \frac{\Omega_0 t_i \sqrt{n+1}}{2} , \tag{7.72}$$

where we have defined an effective interaction time for the frozen interaction:

$$t_i = \frac{1}{\Omega_0} \int_{x_a^i}^{x_a^f} \Omega(x_a) \frac{dx_a}{v} . \tag{7.73}$$

We systematically express in the following the system's evolution in terms of this effective time t_i, which allows us to interpret the results 'as if' the atom–field coupling were constant. This time, in general a complicated function of x_a^i, x_a^f and v, reduces to the simple expression $t_i^r = \sqrt{\pi} w / v$ when x_a^i and x_a^f are rejected to $-\infty$ and $+\infty$ respectively.

Combining the superposition principle with the adiabatic property of the dressed states, we compute easily the transformation of any initial atom–field state. In partic-ular, an atom in $|e\rangle$ injected in the cavity containing a coherent state $|\beta\rangle = \sum_n c_n |n\rangle$ corresponds to the initial state superposition:

$$|e\rangle \otimes |\beta\rangle = \frac{1}{\sqrt{2}} \sum_n c_n \left[|-, n\rangle + |+, n\rangle \right] , \tag{7.74}$$

which becomes, after the atom–field interaction:[8]

$$|\Psi(t_i)\rangle = \frac{1}{\sqrt{2}} \sum_n c_n \left[e^{i\phi_r(n)} |-, n\rangle + e^{-i\phi_r(n)} |+, n\rangle \right] . \tag{7.75}$$

It is then convenient to split this final state into two parts:

$$|\Psi(t_i)\rangle = (|\Psi_1\rangle + |\Psi_2\rangle) / \sqrt{2}, \tag{7.76}$$

where:

[8]We describe the evolution in the interaction representation which eliminates a fast phase rotation of the atom–field state at frequency ω_c. This amounts to disregarding the $(n + 1/2)\hbar\omega_c$ term in eqn. (7.69). To avoid heavy notation and since no confusion is possible, we will not mark the states with the tilde symbols which we should use in this case (see note 4, on page 108).

$$|\Psi_1\rangle = \frac{1}{\sqrt{2}}\left[\sum_n c_n e^{i\Omega_0\sqrt{n+1}t_i/2}\left(|e,n\rangle - i\,|g,n+1\rangle\right)\right],\tag{7.77}$$

and:

$$|\Psi_2\rangle = \frac{1}{\sqrt{2}}\left[\sum_n c_n e^{-i\Omega_0\sqrt{n+1}t_i/2}\left(|e,n\rangle + i\,|g,n+1\rangle\right)\right].\tag{7.78}$$

We next proceed to separate $|\Psi_1\rangle$ and $|\Psi_2\rangle$ into an atom and a field part. With a mere redefinition of the running index n, $|\Psi_1\rangle$ is rewritten as:

$$|\Psi_1\rangle = \frac{1}{\sqrt{2}}\left[\sum_n c_n e^{i\Omega_0\sqrt{n+1}t_i/2}\,|e,n\rangle - i\sum_n c_{n-1}e^{i\Omega_0\sqrt{n}t_i/2}\,|g,n\rangle\right].\tag{7.79}$$

This is still an exact expression. We now make, as in the dispersive case, the assumption that the average photon number and its standard deviation $\Delta N = \sqrt{\bar{n}}$ are much larger than unity. The c_n amplitudes are slowly varying functions of n and we can assume $c_n \approx c_{n-1}$. As usual in this kind of asymptotic expansion, we perform a zero-order approximation for the amplitudes of the terms contributing to $|\Psi_1\rangle$, but we treat with more care their phases. We use, in the exponentials, a first-order expansion of $\sqrt{n+1}$:

$$\frac{\Omega_0\sqrt{n+1}\,t_i}{2} \approx \frac{\Omega_0\sqrt{n}\,t_i}{2} + \frac{\Omega_0\,t_i}{4\sqrt{n}},\tag{7.80}$$

which leads to:

$$|\Psi_1\rangle = \frac{1}{\sqrt{2}}\left[\sum_n c_n e^{i\Omega_0\sqrt{n}t_i/2}\,|n\rangle\right]\left[e^{i\Omega_0 t_i/4\sqrt{n}}\,|e\rangle - i\,|g\rangle\right],\tag{7.81}$$

a product of independent atom and field states.

We then expand the $\sqrt{n}$ term around $n = \bar{n}$, up to the second order:

$$\sqrt{n} = \frac{\sqrt{\bar{n}}}{2} + \frac{n}{2\sqrt{\bar{n}}} - \frac{(n-\bar{n})^2}{8\bar{n}^{3/2}}.\tag{7.82}$$

This is legitimate since the photon number dispersion is much less than its average. We then get:

$$|\Psi_1\rangle \approx \frac{1}{\sqrt{2}}e^{i\Omega_0\sqrt{\bar{n}}t_i/4}\left[\sum_n c_n e^{in\Omega_0 t_i/4\sqrt{\bar{n}}}e^{-i\Omega_0(n-\bar{n})^2 t_i/16\bar{n}^{3/2}}\,|n\rangle\right]$$
$$\otimes\left[e^{i\Omega_0 t_i/4\sqrt{\bar{n}}}\,|e\rangle - i\,|g\rangle\right].\tag{7.83}$$

The second-order term in the phase of the Fock state amplitude (first line of eqn. 7.83) does not modify the photon number probability distribution, which remains the Poisson law of the initial coherent state. This term produces a field phase blurring. In more technical terms, its effect is to widen the Q or Wigner function distributions in a direction perpendicular to the field classical amplitude. We will discuss below in more detail the influence of this term at long times and show that its effect can be neglected for a qualitative analysis up to the first revival time $t_r = 4\pi\sqrt{\bar{n}}/\Omega_0$ of the Rabi oscillation (see Chapter 3). Keeping for the time being only the first-order terms in the phase expansion, the field state reduces to:

$$\sum_n c_n e^{in\Omega_0 t_i/4\sqrt{\bar{n}}}\left|n\right\rangle = \left|\beta e^{i\Omega_0 t_i/4\sqrt{\bar{n}}}\right\rangle \ , \tag{7.84}$$

a coherent state obtained from the initial one by a phase shift $\Omega_0 t_i/4\sqrt{\bar{n}}$.

A similar calculation is performed for $|\Psi_2\rangle$. Grouping all the terms, we finally obtain the atom–field wave function as:

$$|\Psi(t_i)\rangle \approx \frac{1}{\sqrt{2}}\left[\left|\Psi_a^+(t_i)\right\rangle\otimes\left|\Psi_c^+(t_i)\right\rangle + \left|\Psi_a^-(t_i)\right\rangle\otimes\left|\Psi_c^-(t_i)\right\rangle\right] \ , \tag{7.85}$$

where the atom wave functions are:

$$\left|\Psi_a^\pm(t_i)\right\rangle = \frac{1}{\sqrt{2}}e^{\pm i\Omega_0\sqrt{\bar{n}}t_i/2}\left[e^{\pm i\Omega_0 t_i/4\sqrt{\bar{n}}}\left|e\right\rangle \mp i\left|g\right\rangle\right] \ , \tag{7.86}$$

and the field wave functions:

$$\left|\Psi_c^\pm(t_i)\right\rangle = e^{\mp i\Omega_0\sqrt{\bar{n}}t_i/4}\left|\beta e^{\pm i\Omega_0 t_i/4\sqrt{\bar{n}}}\right\rangle \ . \tag{7.87}$$

We have arbitrarily split the global phase factor between the atomic and field states in a way which will prove convenient later on.

The final atom–field state (7.85) is generally entangled. The situation bears a strong similarity with the dispersive situation described in Section 7.3. The initial coherent field splits into two coherent components with opposite phases (eqn. 7.87), while getting correlated to two different atomic states. This leads, as in the dispersive case, to a cat-state situation. Note that the resonant phase splitting, $\pm\Omega_0 t_i/4\sqrt{\bar{n}}$, is precisely equal to the upper limit of the dispersive regime (eqn. 7.46). The resonant method is thus optimal for the preparation of photonic cats with many photons.

Generating a cat state in the dispersive regime requires to prepare by a classical microwave pulse the atom in a superposition state, whereas in the resonant experiment this preparation is automatically realized when the atom is injected in the energy eigenstate $|e\rangle$, linear superposition of $|\Psi_a^+(0)\rangle$ and $|\Psi_a^-(0)\rangle$. In this respect, the resonant cat experiment, which does not require a coherent preparation of the initial atomic state in the energy basis, appears to be simpler than its dispersive counterpart.

In the dispersive regime, the atomic states correlated to the photonic cat components are the energy eigenstates $|e\rangle$ and $|g\rangle$, which correspond to different refractive indices for the field. In the resonant regime, they are replaced by linear superpositions of $|e\rangle$ and $|g\rangle$ (eqn. 7.86). These states have non-zero expectation values of the atomic electric dipole operator. In the spin representation, they are associated to vectors lying in the equatorial plane of the Bloch sphere. In contrast to the dispersive case, these atomic states correlated to the cat components are not stationary. The corresponding Bloch vectors are rotating, their evolution being correlated to the phase rotation of the field components. We will analyse below this synchronous 'dance' of the atom and field parts of the global system's wave functions.

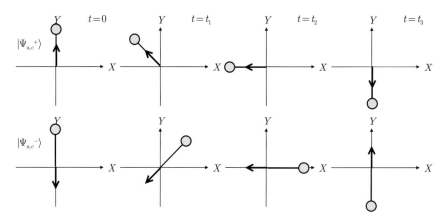

Fig. 7.15 Evolution of the atomic and field states represented by vectors in the Bloch equatorial plane and the Fresnel plane mapped onto each other. Top row: $\left|\Psi_a^+\right\rangle$ (thick arrow) and $\left|\Psi_c^+\right\rangle$ (arrow with uncertainty disk) at the initial time $t = 0$, at $t_1 = t_r/4 = \pi\sqrt{\bar{n}}/\Omega_0$, $t_2 = t_r/2 = 2\pi\sqrt{\bar{n}}/\Omega_0$ and finally $t_3 = t_r = 4\pi\sqrt{\bar{n}}/\Omega_0$. Bottom row: $\left|\Psi_a^-\right\rangle$ and $\left|\Psi_c^-\right\rangle$ at the same times.

We have so far overlooked the global phase factors appearing in the expressions of the $|\Psi_a^{\pm}\rangle$ and $|\Psi_c^{\pm}\rangle$ states (eqns. 7.86 and 7.87). These complex exponentials evolve at frequencies about $\bar{n}$ times larger than the drift frequency of the field and atomic dipole phases. If the system were prepared in the $|\Psi_1\rangle$ or $|\Psi_2\rangle$ part alone, these fast evolving global phases would have no physical effect. In the $(|\Psi_1\rangle + |\Psi_2\rangle)/\sqrt{2}$ superposition, though, their role is essential. They impart different quantum phases to the two parts of the wave function, giving rise to the interference effect responsible for the Rabi oscillation. The study of the photonic cat evolution is thus intimately related to the dynamics of the precession between $|e\rangle$ and $|g\rangle$.

The classical limit of this cat experiment is an instructive starting point to analyse the system's evolution. Consider a thought experiment in which we increase the average photon number while decreasing the atom–field coupling to keep the Rabi frequency constant. This could be realized by separating the mirrors by a larger and larger distance, thus increasing the cavity volume and diluting the vacuum field r.m.s. amplitude in the same ratio as the square root of the average photon number is augmented. At the limit $\Omega_0 \to 0$ and $\bar{n} \to \infty$, with $\Omega_0\sqrt{\bar{n}} = \Omega_c$=constant, we reach the classical situation. The phase drifts of the field and atom states become then infinitely slow. The $|\Psi_c^{\pm}\rangle$ field states coincide with the initial coherent field state. The $|\Psi_a^{\pm}\rangle$ atomic states become, within a global phase, the $|0/1_Y\rangle$ eigenstates of the atom–classical field interaction in the rotating frame.[9] The field is not affected by the coupling with the atom and, thus, no photonic cat is produced at this limit. The atom is unable to leave its imprint on the macroscopic field. Conversely, the field strongly influences the microscopic atom. The $|0/1_Y\rangle$ are the eigenstates of the atomic evolution. The phase of the corresponding atomic dipoles are aligned (or anti-aligned) with

[9]According to the convention of Chapter 5, we assign here $|g\rangle$ and $|e\rangle$ to the qubit states $|0\rangle$ and $|1\rangle$, corresponding respectively to the north and south poles of the Bloch sphere.

the direction of the field in the rotating frame and thus do not exchange any energy with it. The relative phase of the $|0_Y\rangle$ and $|1_Y\rangle$ states is, however, evolving at the fast Rabi frequency Ω_c and the probability of finding the atom in $|e\rangle$ or in $|g\rangle$ oscillates indefinitely at this frequency.

Let us come back to a finite average photon number $\bar{n}$. The fast Rabi oscillation is now accompanied by the slow drift of the field components and atomic state phases. This drift, absent at the classical limit, is a mesoscopic feature, directly related to the photon graininess. It means that the atom is now able to react on the field, slowly splitting and modifying its phase. At the same time, the atomic states continuously adjust their own phases to stay aligned or anti-aligned with the slowly rotating field components. This becomes clear in a simple pictorial representation of the atomic and field states. They can be represented as vectors rotating in two planes, the Fresnel plane for the coherent state and the Bloch equatorial plane for the atom. It is convenient to map these planes onto each other and to orient them so that the vectors representing $|\Psi_a^+\rangle$ and $|\Psi_c^+\rangle$ rotate in the same direction and are both initially aligned on the Y axis.[10]

Figure 7.15 presents the evolution of these vectors. The field states are represented with the usual unit radius uncertainty disk superposed to the classical amplitude. Initially the two atomic states, $|\Psi_a^+(0)\rangle = (|e\rangle - i|g\rangle)/\sqrt{2} = -i|0\rangle_Y$ and $|\Psi_a^-(0)\rangle = (|e\rangle + i|g\rangle)/\sqrt{2} = i|1\rangle_Y$, represented as thick arrows, are orthogonal and the representing vectors are opposite. Four different times are selected, expressed in units of the revival time $t_r = 4\pi\sqrt{\bar{n}}/\Omega_0$ (eqn. 3.184).

The $|\Psi_a^+\rangle$ and $|\Psi_c^+\rangle$ states rotate together in the anti-clockwise direction at the angular frequency $\Omega_0/4\sqrt{\bar{n}}$. They remain constantly parallel. The $|\Psi_a^-\rangle$ and $|\Psi_c^-\rangle$ states rotate at the same angular frequency in the opposite direction and remain anti-parallel. The two parts of the total wave function involve field states which, most of the time, have non-overlapping uncertainty disks and are thus orthogonal. The global state $|\Psi(t)\rangle$ is then generally entangled. This evolution occurs on a time scale $4\bar{n}$ times longer than the Rabi period and going to infinity at the classical limit. In practice, the situation will be considered as classical if $\sqrt{\bar{n}}/\Omega_0$ is longer than the duration of the atom–field interaction or than the shortest damping time in the system.

The quantum Rabi oscillation signal, $P_e(t)$, is retrieved from the reduced atomic density matrix, ρ_a, obtained by tracing the global density matrix, $|\Psi\rangle\langle\Psi|$, over the field:

$$\rho_a = \frac{1}{2}|\Psi_a^+\rangle\langle\Psi_a^+| + \frac{1}{2}|\Psi_a^-\rangle\langle\Psi_a^-| + \frac{1}{2}|\Psi_a^+\rangle\langle\Psi_a^-|\langle\Psi_c^-|\Psi_c^+\rangle + \text{h.c.}\,, \tag{7.88}$$

and:

$$P_e(t_i) = \langle e|\rho_a|e\rangle = \frac{1}{2} + \text{Re}\left[\langle e|\Psi_a^+\rangle\langle\Psi_a^-|e\rangle\langle\Psi_c^-|\Psi_c^+\rangle\right]\,. \tag{7.89}$$

The scalar product of the field states is easily obtained from eqn. (3.61):

$$\langle\Psi_c^-|\Psi_c^+\rangle = e^{-i\Omega_0\sqrt{\bar{n}}t_i/2}\exp\left[-\bar{n}\left(1 - e^{i\Omega_0 t_i/2\sqrt{\bar{n}}}\right)\right]\,. \tag{7.90}$$

We also have:

[10]This choice of initial condition implies that β is imaginary.

$$\langle e \left| \Psi_a^+ \right\rangle \langle \Psi_a^- \left| e \right\rangle = \frac{1}{2} e^{i\Omega_0 t_i \sqrt{\bar{n}}} e^{i\Omega_0 t_i / 2\sqrt{\bar{n}}} \;. \tag{7.91}$$

Equation (7.89) describes an interference effect between two amplitudes associated to the atomic states $\left| \Psi_a^+ \right\rangle$ and $\left| \Psi_a^- \right\rangle$. The contrast of this interference is proportional to the overlap of the field states $\left| \Psi_c^+ \right\rangle$ and $\left| \Psi_c^- \right\rangle$, given by eqn. (7.90). We now examine $P_e(t_i)$ at different time scales.

Short-time behaviour

For times t_i such that $\Omega_0 t_i / 2\sqrt{\bar{n}} \ll 1$, we can expand the argument of the second exponential in eqn. (7.90) in terms of this small quantity:

$$\left\langle \Psi_c^- \left| \Psi_c^+ \right\rangle \approx e^{-\Omega_0^2 t_i^2 / 8} \;. \tag{7.92}$$

In the same limit:

$$\langle e \left| \Psi_a^+ \right\rangle \langle \Psi_a^- \left| e \right\rangle \approx \frac{1}{2} e^{i\Omega_0 t_i \sqrt{\bar{n}}} \;, \tag{7.93}$$

and, finally:

$$P_e(t_i) = \frac{1}{2} \left[1 + \cos(\Omega_0 \sqrt{\bar{n}} t_i) e^{-\Omega_0^2 t_i^2 / 8} \right] \;. \tag{7.94}$$

These equations have a simple physical interpretation. Equation (7.92) shows that the overlap between the field states $\left| \Psi_c^+ \right\rangle$ and $\left| \Psi_c^- \right\rangle$ decays as a Gaussian function of time, over a time of the order of the vacuum Rabi period (1/e decay in time $t_c = 2\sqrt{2}/\Omega_0$). This is indeed the time it takes for the dephasing between the two components to reach a value $\sim 1/\sqrt{\bar{n}}$, of the order of the phase uncertainty of the initial coherent field. This field scalar product defines the contrast of the Rabi oscillations which thus exhibits a Gaussian envelope. This analysis confirms our qualitative discussion of the Rabi oscillation collapse developed in Chapter 3.

The collapse of the Rabi oscillation appears as a complementarity effect quite similar to the collapse of the Ramsey fringes in the dispersive experiment described in Section 7.3.3. The Rabi oscillation being an interference effect involving two quantum amplitudes associated to $\left| \Psi_a^+ \right\rangle$ and $\left| \Psi_a^- \right\rangle$, it is washed out when the field components (to which these dipole states are locked) carry information about these states. This happens as soon as these field components become distinguishable, i.e. quasi-orthogonal.

After the collapse: Schrödinger cat state generation.

As soon as $\Omega_0 t_i > 1$, $P_e(t_i) \approx 1/2$ and the atomic populations remain stationary. The system is however still evolving, even though it does not reflect on the evolution of the average atomic energy. The field and atomic states rotate in phase space, in opposite directions for the + and − components. An atom–field entanglement situation of the cat type is produced.

A remarkable situation occurs at $t_i = t_2 = 2\pi\sqrt{\bar{n}}/\Omega_0$. A mere inspection of Fig. 7.15 shows that the two atomic states $\left| \Psi_a^\pm \right\rangle$ then coincide, within an irrelevant global phase factor (their common Bloch vector is then anti-aligned with OX):

$$\left| \Psi_a^\pm(t_2) \right\rangle = \pm \frac{i}{\sqrt{2}} e^{\pm i\pi\bar{n}} (\left| e \right\rangle - \left| g \right\rangle) = \pm e^{\pm i\pi\bar{n}} \left| \Psi_a^0 \right\rangle \;. \tag{7.95}$$

The atom and field states thus factor out in the global state $|\Psi(t_i)\rangle$, which is:

$$|\Psi(t_i)\rangle = |\Psi_a^0\rangle \otimes |\Psi_c^0\rangle \ , \tag{7.96}$$

where:

$$|\Psi_c^0\rangle = \frac{1}{\sqrt{2}}\left(e^{i\pi\overline{n}/2}|i\beta\rangle - e^{-i\pi\overline{n}/2}|-i\beta\rangle\right) \ . \tag{7.97}$$

The field state is then a mesoscopic quantum superposition of two coherent fields with opposite classical phases.

It is interesting to notice that the atomic state $|\Psi_a^0\rangle = |1_X\rangle$ is reached at this particular time t_2, whatever the initial atomic condition. If the atom is initially in an arbitrary superposition of $|e\rangle$ and $|g\rangle$, its state can be expanded on the $\{|\Psi_a^\pm(0)\rangle\}$ basis. The wave function at any time is thus a linear superposition of $|\Psi_a^\pm(t_i)\rangle \otimes |\Psi_c^\pm(t_i)\rangle$, whose coefficients are determined by the initial condition. At time $t_i = t_2$, the atomic state $|\Psi_a^0\rangle$ factors out, and the initial condition only affects the expression of the cavity state. The initial atomic state superposition is mapped in this way onto the field superposition state.

Quantum revival

At the time $t_i = t_r = 4\pi\sqrt{\overline{n}}/\Omega_0$, the two field states have rotated by π in opposite directions. They thus overlap again, in phase opposition with the initial coherent state $|\beta\rangle$ (Fig. 7.15). The two atomic states are again orthogonal superpositions of $|e\rangle$ and $|g\rangle$. The field state factors out and the atom–field entanglement vanishes. Within our approximations, we return to a situation akin to the initial one: the Rabi oscillations revive with unit contrast. The revival time, t_r, coincides with the estimate made in Chapter 3. This revival effect appears here, as does the collapse, as a manifestation of complementarity. At the revival time, the field does not contain information about the dipole state in which the atom is and the interference between the corresponding atomic probability amplitudes reappears. Further identical revivals are predicted by this simple model for multiples of t_r.[11]

This analysis is only qualitative, since it relies on an approximate expression of the atom–field state. The exact Rabi oscillation signal exhibits revivals with a reduced contrast, progressively merging together (Fig. 3.20, on page 156). These features are explained by the influence of the quadratic phase terms neglected in eqn. (7.83). They widen the field phase distribution. The broadening effect at the revival time can be easily estimated. The field state is:

$$\left|\Psi_c'^\pm(t_i)\right\rangle = \sum_n c_n e^{\pm in\Omega_0 t_i/4\sqrt{\overline{n}}} e^{\mp i\delta\phi(n)t_i}|n\rangle \ , \tag{7.98}$$

where:

$$\delta\phi(n) = \Omega_0(n-\overline{n})^2/16\overline{n}^{3/2} \ . \tag{7.99}$$

This second-order term resembles the phase spreading of a coherent field propagating in a transparent medium with a non-linear index corresponding to a Kerr Hamiltonian

---

[11]The revival of atomic interference when the photonic cat components recombine after having made an integer number of half turns in phase space is also predicted in dispersive cat experiments. The Ramsey fringes studied in Section 7.3.3 are expected to revive when $\Phi = k\pi$ for the same reasons as discussed here in the resonant case. The large phase shifts required for this observation have not yet been reached.

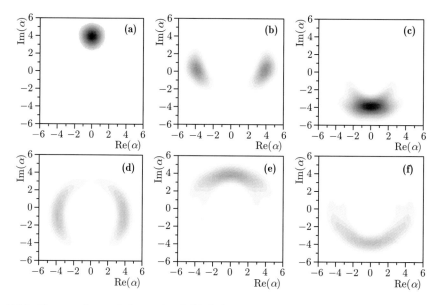

Fig. 7.16 Six snapshots of the cavity field Q function. The density of shade at the point α in phase space is proportional to $Q(\alpha)$. (a) Time $t_i = 0$. The field is in a coherent state with $\bar{n} = 15$ photons on the average (imaginary amplitude $\beta = i\sqrt{\bar{n}}$). (b) At $t_i = t_r/2 = 2\pi\sqrt{\bar{n}}/\Omega_0$, half the revival time. (c) At the revival time $t_i = t_r = 4\pi\sqrt{\bar{n}}/\Omega_0$. The progressive widening of the phase distribution is even more conspicuous at $t_i = 3t_r/2$, $2t_r$ and $3t_r$ [(d)–(f)].

$\gamma_k(a^\dagger a)^2$ (see Section 7.2). In order to estimate the Kerr-like phase diffusion during the Rabi oscillation, let us compute the expectation values of the field annihilation operator a in $\left|\Psi_c'^{\pm}(t_i)\right\rangle$:

$$\left\langle \Psi_c'^{\pm}(t_i)\left|\, a\,\right|\Psi_c'^{\pm}(t_i)\right\rangle = e^{\pm i\Omega_0 t_i/4\sqrt{\bar{n}}} \sum_n c_n^* c_{n+1}\sqrt{n+1}\, e^{\mp i[\delta\phi(n+1)-\delta\phi(n)]t_i} \,, \quad (7.100)$$

with:

$$\delta\phi(n+1) - \delta\phi(n) \approx \frac{\Omega_0}{8\bar{n}^{3/2}}(n - \bar{n}) \,. \quad (7.101)$$

This average value goes to zero around a time t_ϕ such that the phases corresponding to different n values spanning a $\sqrt{\bar{n}}$ interval around $\bar{n}$ are spread over 2π:

$$t_i > t_\phi \approx \frac{16\pi\bar{n}}{\Omega_0} \,. \quad (7.102)$$

At the revival time, $t_i = t_r = t_\phi/(4\sqrt{\bar{n}})$, the phase spreading is of the order of $\pi/2\sqrt{\bar{n}}$, i.e. is of the order of the initial field phase fluctuations. In simple terms, the field fluctuation in each component is approximately twice its initial value.

We now understand that the two field components overlap for a longer time interval around t_r than around $t_i = 0$. The build-up of the Rabi oscillation takes twice as long as the first collapse and then it takes again twice as long for the oscillation to vanish.

This spreading of the field phase results also in a lower degree of atomic coherence. The contrast of the reviving Rabi oscillations is expected to be of the order of 50%, in good agreement with a qualitative inspection of Fig. 3.20, on page 156.

These qualitative estimates are confirmed by a numerical investigation of the field state evolution based on an exact solution of the atom–field system Schrödinger equation. Figure 7.16 presents density plots of the cavity field's Q function at six different times for $\bar{n} = 15$. At $t_i = 0$ (Fig. 7.16a), the Q function is a Gaussian representing the initial coherent state. At $t_i = t_r/2$ (Fig. 7.16b), the Q function is split into two squeezed distributions with opposite phases. The non-linear phase terms have already a noticeable action and the peaks are distorted and widened. At $t_i = t_r$ (Fig. 7.16c), the phase distribution of the overlapping field states is about twice as wide as the initial one.

Beyond the revival time, the two field components continue their rotation in phase space (Fig. 7.16d for $t_i = 3t_r/2$) and separate again. The Rabi oscillation collapses once more, until the two field components merge again close to the initial phase, at the second revival time $t_i = 2t_r$ (Fig. 7.16e). At longer times, the quadratic phase terms and higher order contributions blur the phase distribution (Fig. 7.16f for $t_i = 3t_r$). At still later times, the two field states Q functions are generally spread over a full circle. Permanent Rabi oscillations are observed, with a complex envelope, as seen on Fig. 3.20.[12]

We can summarize this discussion by considering the various times which characterize the CQED system's dynamics. Four successive times, each being of an order $\sqrt{n}$ times larger than the previous, are involved. The shortest one, $t_R = 2\pi/(\Omega_0\sqrt{n})$ is the Rabi period. This is the only relevant time constant at the classical limit. Then comes, after about $\sqrt{n}$ Rabi oscillations, the collapse time $t_c = 2\sqrt{2}/\Omega_0$ followed by the first revival time $t_r = 4\pi\sqrt{n}/\Omega_0$ and finally the phase spreading time $t_\phi = 16\pi\bar{n}/\Omega_0$. Observing in practice the physical effects associated to these successive times becomes more and more difficult. It requires an experiment protecting the system against decoherence over longer and longer time intervals.

In some respects, the evolution of a coherent field coupled to a single resonant atom is reminiscent of the propagation of a coherent field in a Kerr medium described in Section 7.2. In both cases, the field experiences periodical phase splittings and refocusings. In the Kerr situation, the phase splittings are however more complex than in the CQED case. They are part of an intricate phase-spreading phenomenon which leads at specific times to the brief appearance of multi-component cat states. Two-state superpositions are the only ones to be generated in the CQED case (as long as the quadratic phase spreading can be neglected). They are only one of the possible avatars of the field in an optical Kerr experiment, among many other multi-state configurations. The phase spreading of the optical cats occurs in a very short time, of the order of $\pi/4\sqrt{n}\gamma_k$ and precedes the cats appearances. In the CQED cat generation, phase spreading is also present but it occurs at a much slower pace, so that it can be in first approximation neglected during the first collapse and revival

---

[12]At specific times much larger than t_r, the phase distribution presents transient interference patterns, corresponding to the appearance of multi-component cat states, reminiscent of the Kerr cats in quantum optics (Averbukh 1992).

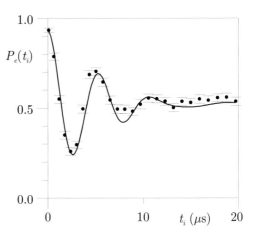

Fig. 7.17 Collapse of the Rabi oscillation signal. Probability $P_e(t_i)$ of detecting the atom in $|e\rangle$ as a function of the atom–cavity effective interaction time t_i. The atom is initially in $|e\rangle$ and the cavity contains a coherent field with $\bar{n} = 13$. Reprinted with permission from Meunier *et al.* (2005). © American Physical Society.

sequence (when it finally occurs, multi-component cat features can appear, see note 12).

In the Kerr as well as in the CQED cases, the first revival time is about $\sqrt{\bar{n}}$ times larger than the collapse time. The meanings of these phenomena are however quite different in the two situations. In the optical experiment, it is the average value of all the observables sensitive to the phase of the field (i.e. $\langle a \rangle$ or $\langle a^\dagger \rangle$) which vanish as soon as the phase spreading has occurred, in a time inversely proportional to the field amplitude. In the CQED case, it is the contrast of the Rabi oscillation observed on the atoms which cancels, in a time independent of the field photon number. Similarly, the term 'revival' refers to the reappearance of field coherences in the optical field experiment and to the rebirth of Rabi oscillations in the CQED one. In the first case, the revival time, π/γ_k, does not depend upon the field photon number, while in the second $(t_r = 4\pi\sqrt{\bar{n}}/\Omega_0)$, it is proportional to the field amplitude.

More importantly, the CQED phenomenon corresponds to an oscillation of entanglement between matter and radiation, a feature which is lacking in the optical experiment. The most striking property we should remember about the Rabi oscillation in a mesoscopic coherent field is this oscillation entanglement phenomenon. The atom–field system evolves quasi-periodically into a succession of entangled and separate atom–field states, this periodic exchange being revealed by an oscillation of the contrast of the 'Rabi interference' observed on the atom. We will encounter a similar situation in Chapter 9 when dealing with the entanglement of matter waves in Bose–Einstein condensates.

7.4.2 Observing the cat after the Rabi oscillation collapse

We now turn to the description of experiments demonstrating the existence of resonant cats in Cavity QED and probing the interplay between the cat's and the Rabi oscillations features. One might argue that the simplest smoking gun signalling the emergence of a cat state is the collapse of the Rabi signal. Figure 7.17 shows, as a function of t_i, the probability $P_e(t_i)$ of finding the atom, after cavity crossing, in the level $|e\rangle$ in which it has been initially prepared (Meunier *et al.* 2005). The field contains $\bar{n} = 13$ photons on average. The atomic velocity is $v = 154$ m/s corresponding to a

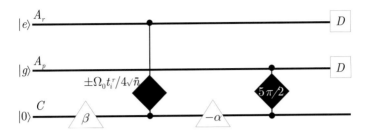

Fig. 7.18 Quantum circuit describing the homodyne measurement of the phase splitting induced on the field by its resonant interaction with an atom A_r. The procedure involves a second probe atom A_p. The lozenge at left describes the Rabi evolution of the $A_r + C$ system, wich lasts a variable time and imparts to the field the two phase shifts $\pm\Omega_0 t_i^r/4\sqrt{n}$. The second lozenge describes the $5\pi/2$ Rabi pulse of the probe atom, whose final probability of being found in $|e\rangle$ is directly related to the field's Q function. The triangles in the C line represent field injections.

total atom effective cavity crossing time $t_i^r = 70\ \mu s$. The effective interaction time t_i is adjusted to any value smaller than t_i^r by tuning suddenly the atom by the Stark effect out of resonance when it reaches a given position inside the cavity. The evolution is frozen from that position on and the atomic detector placed downstream counts the atoms in $|e\rangle$ and $|g\rangle$.

Repeating the sequence many times, the probability $P_e(t_i)$ at the freezing time is reconstructed. The Stark switching time is then swept and the variation of $P_e(t_i)$ versus t_i obtained, revealing a damped oscillating signal (Fig. 7.17) with an $1/e$ decay at $t_i \approx 6\ \mu s$, a little bit smaller than the ideal value $2\sqrt{2}/\Omega_0 \approx 9\ \mu s$ given by the theory. The extra damping is due to various experimental imperfections which we will not discuss here. The essential point is that the Rabi collapse has certainly occurred after a time of the order of $10\ \mu s$, while the atom is still inside the cavity. According to the complementarity argument, a cat state must exist in the cavity at this time. If the atom and cavity are left in resonant interaction without Stark switching, this cat state keeps evolving until the atom leaves the cavity at the effective time $t_i = t_i^r$.

Cat state phase distribution measured by homodyning

More direct evidence of resonant cat generation is given by homodyne detection of the phase distribution in the final field state (Auffeves *et al.* 2003). The experiment is similar to the homodyne detection of the dispersive phase shifts described in Section 7.3. It involves again two successive field injections and two atoms (Fig. 7.18). A first coherent state $|\beta\rangle$ is prepared with an average photon number ranging between 15 and 40. A first resonant atom A_r is then sent across C at a fixed velocity $v_1 = 335$ m/s or $v_2 = 200$ m/s (corresponding to the effective interaction times $32\ \mu s$ and $52\ \mu s$). No Stark switching is applied, letting the system evolve freely until the atom leaves the cavity ($t_i = t_i^r$). The atom imparts to the field the two phase shifts $\pm\Omega_0 t_i^r/4\sqrt{n}$ at the same time.

A second coherent field $|-\alpha\rangle = |-\beta e^{i\varphi}\rangle$ with the same photon number as the first field, but a variable phase φ, is then injected in C. A second probe atom, A_p, also resonant with the cavity field, immediately follows. Its absorption yields, as explained in

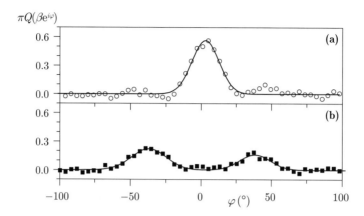

Fig. 7.19 Phase distribution $Q(\beta e^{i\varphi})$ (in units of $1/\pi$) for the resonant cat state. (a) Initial coherent field, with 29 photons on the average. (b) Phase distribution at $t_i^r = 52$ μs. The points are experimental and the solid curves are Gaussian fits. Reprinted with permission from Auffeves *et al.* (2003). © American Physical Society.

Section 7.3.2, the probability $p(0)$ of finding the resulting homodyned field in vacuum. This probability is directly related to the Q function of the field left in the cavity by A_r (eqn. 7.50).

The homodyned field develops along two coherent states whose amplitudes correspond to the cat components translated in phase space by $-\beta e^{i\varphi}$. More precisely, the atom–field system is after the second field injection:

$$|\Psi'\rangle = \frac{e^{-i\Omega_0\sqrt{\bar{n}}t_i^r/4}}{\sqrt{2}}\left|\Psi_a^+(t_i^r)\right\rangle \otimes \left|\beta(e^{i\Omega_0 t_i^r/4\sqrt{\bar{n}}} - e^{i\varphi})\right\rangle$$
$$+\frac{e^{i\Omega_0\sqrt{\bar{n}}t_i^r/4}}{\sqrt{2}}\left|\Psi_a^-(t_i^r)\right\rangle \otimes \left|\beta(e^{-i\Omega_0 t_i^r/4\sqrt{\bar{n}}} - e^{i\varphi})\right\rangle , \qquad (7.103)$$

and the probability $p(0)$ of finding the resulting field in the vacuum is the sum of two terms:

$$p(0) \approx \frac{1}{2}\left[\left|\langle 0\left|\beta(e^{i\Omega_0 t_i^r/4\sqrt{\bar{n}}} - e^{i\varphi})\right\rangle\right|^2 + \left|\langle 0\left|\beta(e^{-i\Omega_0 t_i^r/4\sqrt{\bar{n}}} - e^{i\varphi})\right\rangle\right|^2\right] . \qquad (7.104)$$

This sum, negligible unless φ coincides with the phase of one of the cat components, is expected to present two peaks equal to $1/2$, centred around $\varphi = \pm\Omega_0 t_i^r/4\sqrt{\bar{n}}$. Repeating many times the sequence of field and atom injections in C for different values of φ, we have reconstructed the variations of $p(0)$ and measured in this way the Q function of the field left by A_r. Figure 7.19(b) presents $Q(\beta e^{i\varphi})$ versus φ for $\bar{n} = 29$ and $t_i^r = 52$ μs. As expected, a double-peak distribution is obtained, with a phase splitting $2\phi = 74$ degrees, in fair agreement with the theoretical value $\Omega_0 t_i^r/2\sqrt{\bar{n}} = 86$ degrees. Figure 7.19(a) shows, as a reference, the measured Q function of the coherent field when A_r is not injected in C.

We have checked that the splitting is proportional to the interaction time and inversely proportional to the field amplitude. We have summarized the results by

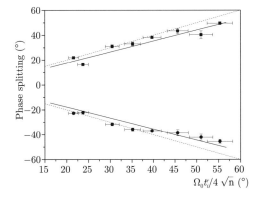

Fig. 7.20 Phase splitting of the cat components versus the dimensionless parameter $\Omega_0 t_i^r/4\sqrt{\bar{n}}$ (in degrees). The dotted and solid lines correspond to two theoretical models. Reprinted with permission from Auffeves *et al.* (2003). © American Physical Society.

plotting in Fig. 7.20 the phases of the observed peaks as a function of the dimensionless angle $\Omega_0 t_i^r/4\sqrt{\bar{n}}$. The dotted line corresponds to the phases predicted by the simple model presented here, which are in good agreement with the experimental points. This agreement is further improved by performing a numerical simulation solving the exact field equation of motion (including in particular the Kerr-like saturation effects) and taking into account cavity damping (solid lines in Fig. 7.20). Note that the maximum phase splitting obtained for $\bar{n} = 15$ and $t_i^r = 52 \ \mu$s is 90°.

It is interesting to remark that we have bred in this experiment an amplitude Schrödinger cat. When the phase of the reference field is adjusted to cancel one of the field components, we are left with a superposition of field states with different photon numbers. Setting $\varphi = \Omega_0 t_i^r/4\sqrt{\bar{n}}$ in eqn. (7.103), we find indeed that the atom–field state has become:

$$|\Psi'\rangle = \frac{e^{-i\Omega_0\sqrt{\bar{n}}t_i^r/4}}{\sqrt{2}}\left|\Psi_a^+(t_i^r)\right\rangle \otimes |0\rangle + \frac{e^{i\Omega_0\sqrt{\bar{n}}t_i^r/4}}{\sqrt{2}}\left|\Psi_a^-(t_i^r)\right\rangle \otimes \left|-2i\beta\sin(\Omega_0 t_i^r/4\sqrt{\bar{n}})\right\rangle .$$
(7.105)

Before its absorption by A_p, the field is now suspended between the vacuum and a large coherent field with amplitude $-2i\beta\sin(\Omega_0 t_i^r/4\sqrt{\bar{n}})$, containing on average $4\bar{n}\sin^2(\Omega_0 t_i^r/4\sqrt{\bar{n}})$ photons.

Correlation between the field and atom states

We have also checked the correlation between atomic and field states by selectively preparing the atom in one of the states $|\Psi_a^+\rangle$ or $|\Psi_a^-\rangle$ at the beginning of the atom–field interaction. This preparation requires a precise manipulation of the atomic Bloch vector, in a time short compared to the phase drift period. The evolution of this vector on the Bloch sphere is sketched in Fig. 7.21. To prepare $|\Psi_a^+\rangle$ (Fig.7.21a), an atom, initially in $|e\rangle$ (superposition of $|e\rangle \mp i |g\rangle$), is first coupled to the cavity field for a $\pi/2$-Rabi pulse. It undergoes the transformation:

$$|e\rangle \rightarrow \frac{1}{2}\left[e^{i\pi/4}(|e\rangle - i|g\rangle) + e^{-i\pi/4}(|e\rangle + i|g\rangle)\right] = \frac{1}{\sqrt{2}}(|e\rangle + |g\rangle) ,$$
(7.106)

which is deduced immediately from eqn. (7.86) in which we set $\Omega_0\sqrt{\bar{n}}t_i = \pi/2$. This pulse, during which the atom–field entanglement remains negligible, rotates the atomic

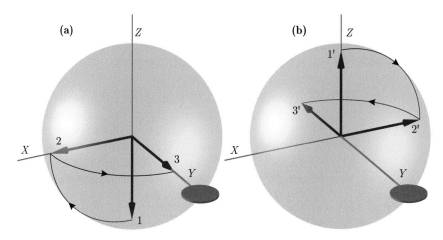

Fig. 7.21 Evolution of the atomic Bloch vector during the preparation of the $\left|\Psi_a^+\right\rangle$ and $\left|\Psi_a^-\right\rangle$ states. (a) The atom is initially in $|e\rangle$, corresponding to a Bloch vector pointing towards the south pole of the Bloch sphere (position 1). A $\pi/2$-Rabi pulse brings this vector in the equatorial plane along OX in position 2 and a subsequent $\pi/2$-Stark pulse in position 3 along OY, corresponding to state $\left|\Psi_a^+\right\rangle$. (b) Starting initially from $|g\rangle$ (north pole – position 1'), the Rabi and Stark pulses bring successively the Bloch vector in positions 2' and 3', corresponding finally to state $\left|\Psi_a^-\right\rangle$. The field coherent state during these transformations remains practically frozen, along OY (arrow with a black uncertainty circle).

Bloch vector from being anti-parallel to OZ to being aligned with OX, in the equatorial plane of the Bloch sphere. It then has a $\pi/2$-phase advance with respect to the field (aligned along OY). The atom is then detuned by the Stark effect, within a time much shorter than the Rabi period. A pulse of electric field is applied between the cavity mirrors, whose effect is to shift by $\pi/2$ the relative phases of $|e\rangle$ and $|g\rangle$:

$$\frac{1}{\sqrt{2}}(|e\rangle + |g\rangle) \rightarrow \frac{1}{\sqrt{2}}(|e\rangle - i\,|g\rangle) = \left|\Psi_a^+(0)\right\rangle \ . \tag{7.107}$$

This amounts to a rotation of the Bloch vector in the equatorial plane of the Bloch sphere, making it parallel to the field, along OY. The sequence of the Rabi and Stark pulses thus transforms the initial $|e\rangle$ state, a superposition of the interfering $\left|\Psi_a^+(0)\right\rangle$ and $\left|\Psi_a^-(0)\right\rangle$ states, into $\left|\Psi_a^+(0)\right\rangle$ alone.

In this way, the atom–field system is prepared in the slowly evolving quasi-stationary state $|\Psi_1\rangle$ (eqn. 7.83). After this fast preparation, the atom and the field subsequently drift in phase in one direction. The system being prepared in only one of the two states corresponding to the Rabi interference, we observe that the oscillation of the atomic populations is frozen from then on. Moreover, a homodyne measurement after the atom leaves the cavity reveals, as expected, only a single phase shifted component in the field (open circles in Fig. 7.22).

Similarly, the state $\left|\Psi_a^-(0)\right\rangle$ is prepared by applying the same Rabi and Stark switching pulse sequence starting from an atom in $|g\rangle$ (Fig. 7.21b). This state couples to the other component of the field, as revealed by the subsequent homodyne detec-

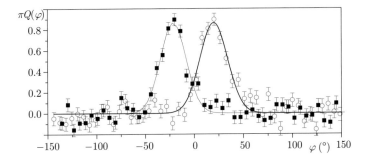

Fig. 7.22 Field phase distributions $\pi Q(\varphi)$ resulting from the interaction with an atom injected in C either in state $\left|\Psi_a^+\right\rangle$ (open circles) or in state $\left|\Psi_a^-\right\rangle$ (solid squares). The field has a mean photon number $\bar{n} = 27$ and $t_i = 32\ \mu$s. The error bars are statistical. Reprinted with permission from Auffeves *et al.* (2003). © American Physical Society.

tion (solid squares in Fig. 7.22). This experiment clearly demonstrates the correlation between the atom and the mesoscopic field states in the Rabi oscillation process.

7.4.3 Deconstructing the cat and reviving the Rabi oscillation

These resonant cat state experiments demonstrate the phase splitting of the field into two components, but do not yield information on the coherence of the field state superpositions. The simplest way to prove this coherence would be to observe the spontaneous revival of the Rabi oscillation. More precisely, the revival signal implies that the two recombining field states at time t_r have a well-defined relative quantum phase. This entails that the system had evolved coherently up to this time and, hence, that the cat produced at the intermediate times $0 < t_i < t_r$ was also in a coherent superposition of states.

Unfortunately we could not, in the experiments described here, wait until the occurrence of the revival. Even for a relatively small cat with $\bar{n} = 13$, the revival time $t_r = 4\pi\sqrt{\bar{n}}/\Omega_0 = 260\ \mu$s is prohibitively long. Observing the revival would require very slow atoms with v in the 30 to 40 m/s range and a cavity with a very long damping time, which was not available at the time these experiments were made. A numerical simulation of the Rabi signal including cavity damping confirms that the spontaneous revival contrast is negligibly small in the cavity we have employed, whose damping time was $T_c \approx 1$ ms. An increase of T_c by one or two orders of magnitude would be required. Such a cavity was just realized at the time this book was completed (see Section 5.2.2), but has not yet been used in an experiment.

The decay of the cat coherence raises the issue of mesoscopic state decoherence, which we address in detail in the next section. Let us just give here a qualitative explanation for the difficulty of observing the spontaneous Rabi revival. It requires to wait until the cat components have completed a long journey in phase space, getting maximally separated at time $t_r/2$, until they finally gather again with a phase opposite to the initial one at time t_r. As we will see in next section, the most critical time for the survival of the cat coherence is precisely when its components are separated by the largest distance. The revival signal, because it comes after this stage of fastest

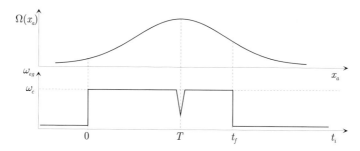

Fig. 7.23 Sketch of the timing of the echo experiment. Top: atom–cavity coupling as a function of atomic position. Bottom: Atomic transition frequency versus time. The atom is set at resonance with the cavity at $t_i = 0$. The Stark pulse is applied at time T, when the atom is at the cavity centre. The atom–cavity system evolution is frozen at time t_f. Reprinted with permission from Meunier *et al.* (2005). © American Physical Society.

decoherence, is thus intrinsically very fragile and extremely challenging to detect.[13]

It is however possible to bypass this difficulty and to force the system to undergo an early revival, occurring before decoherence has washed out the interference effects required for the reappearance of the signal. The trick is to let the field components separate in phase space until a time T shorter than $t_r/2$, then to force them to come back on their tracks and to recombine at the initial phase. The recombination time, $2T$, is then shorter that t_r and, more importantly, the cat components have avoided the maximum separation around time $t_r/2$.

Forcing a system to retrace its past trajectory in the opposite direction is a time-reversal operation. The way to achieve it is well-known in NMR, where it is called a 'spin echo' (Ernst *et al.* 1987; Levitt 2001). A collection of spins, prepared at time $t = 0$ in the equatorial plane of the Bloch sphere, is left free to evolve under the effect of its Hamiltonian, until a π-radiofrequency pulse is applied at time T. This pulse reverses the spins orientation in the equatorial plane. The evolution after this pulse is equivalent to a time-reversal operation, which deconstructs the unitary evolution of the spins between time 0 and T and brings the system back to its initial state at time $2T$. This refocusing technique applies only to the unitary evolution of the system. Irreversible relaxation processes are not compensated by the echo. Studying the echo decay as T is increased thus provides an elegant and widely used method to study spin relaxation processes in NMR physics.

The adaptation of the general spin echo method to CQED has been proposed by Morigi *et al.* (2002). The goal is to manipulate the system's evolution, ruled by the Jaynes–Cummings Hamiltonian, which at resonance and at the cavity centre is, in the interaction representation (eqn. 3.167):

$$H_{JC} = -i\hbar \frac{\Omega_0}{2}[a\sigma_+ - a^\dagger\sigma_-] \,. \tag{7.108}$$

---

[13]The revival signal in the presence of cavity damping has already been analysed in a different point of view (Monte Carlo approach – see Section 4.5.3). We have emphasized the fragility of this signal, which vanishes when a few photons have escaped from the cavity.

We will time the evolution with a 'clock' synchronized on the effective interaction time t_i. The experiment can then be described as if the atom–field coupling were fixed at the value Ω_0. In an echo sequence, the system first evolves from time 0 to T under the action of the unitary evolution operator $U_1 = \exp(-iH_{JC}T/\hbar)$. The atom then undergoes, at time T, a percussional controlled phase kick corresponding to the unitary operation σ_Z. The Jaynes–Cummings evolution then resumes for the remaining time $t_i - T$, with the evolution operator $U_2 = \exp[-iH_{JC}(t_i - T)/\hbar]$. The overall evolution operator is:

$$U = U_2 \sigma_Z U_1 = \sigma_Z^2 U_2 \sigma_Z U_1 = \sigma_Z e^{-iH_{JC}(2T-t_i)/\hbar} \ , \tag{7.109}$$

where we have used the identity $\sigma_Z^2 = \mathbb{1}$ and:

$$\sigma_Z H_{JC} \sigma_Z = -H_{JC} \ , \tag{7.110}$$

which directly follows from the identities $\sigma_Z \sigma_\pm \sigma_Z = -\sigma_\pm$. Equation (7.110) means that the evolution after the phase kick is the time-mirror image of the evolution between 0 and T. Equation (7.109) shows that, at time $t_i = 2T$, the evolution brings the system back to its initial state, up to a global π-phase shift between the amplitudes associated to $|e\rangle$ and $|g\rangle$ (contribution of the σ_Z term).

We have applied this time-reversal method to the study of the Rabi oscillation in our CQED set-up (Meunier *et al.* 2005). Figure 7.23 presents a time diagram showing the variations of the atom–field coupling $\Omega(x_a)$ and of the Stark-tuned atomic frequency as a function of the time t_i in an experimental sequence. The atomic velocity is set at $v = 156$ m/s and the average photon number in the field is $\bar{n} = 13.6$. The atom, when it enters the mode, is suddenly tuned into resonance with the cavity mode at time $t_i = 0$. The Rabi oscillation starts immediately and collapses while the atom is still near cavity entrance. In a preliminary experiment, this signal is recorded by freezing the evolution at increasing times in the ascending part of the field mode profile. The atomic state when the atom reaches the detector D is recorded and the Rabi collapse signal, already shown in Fig. 7.17, is obtained.

We then perform the echo experiment. In an experimental sequence, the phase-kicking pulse is applied to the atom when it reaches the cavity centre, long after the Rabi collapse is complete. This pulse is produced by a fast Stark-field-induced variation of the atomic frequency, whose amplitude and duration are set to perform an exact π-phase difference between $|e\rangle$ and $|g\rangle$. The atomic frequency then resumes its resonant value and the atomic evolution proceeds until a final time $t_i = t_f$, at which the atom is suddenly detuned. The frozen atomic state is subsequently analysed by the atomic detector.

The experiment is repeated many time for varying times t_f and the probability $P_e(t_i)$ of finding the atom in level $|e\rangle$ reconstructed from many such runs. To set the delay T of the echo pulse, we vary the position x_a^i at which the atom is put in resonance with the cavity. The echo is expected to occur at a point $x_a^f = -x_a^i$ inside the cavity, symmetrical of the starting point of the Rabi oscillation with respect to cavity centre.

The results of this experiment are plotted in Fig. 7.24. We have reproduced in trace 7.24(a), as a reference, the recorded Rabi collapse signal, when no phase kicking pulse is applied. Traces 7.24(b) and (c) show the signals obtained when the kicking pulse occurs at the effective time $T = 18$ μs and 22 μs respectively. The echo signals

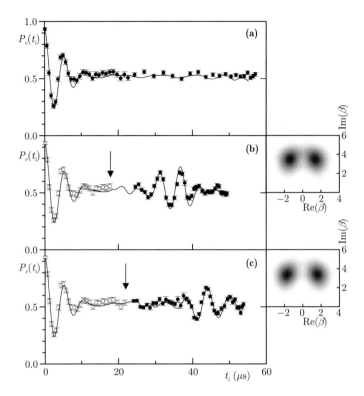

Fig. 7.24 Experimental echo signals. Probability $P_e(t_i)$ of detecting the atom in $|e\rangle$ as a function of the atom–cavity effective interaction time t_i. The initial field has $\bar{n} = 13.6$ photons on average. The points are experimental with statistical error bars and the solid lines are theoretical fits. (a) Rabi collapse without Stark pulse. (b) and (c) Stark pulse applied at 18 and 22 μs respectively (vertical arrows). The first part of the signal (open circles) is reproduced from (a) for visual convenience. The insets present the calculated $Q(\beta)$ function of the cavity field at the Stark pulse time. Reprinted with permission from Meunier *et al.* (2005). © American Physical Society.

around time $2T$ are clearly observed, with a contrast equal to 50% of the ideal value. Note that the kicking time $T = 22$ μs corresponds to 2.5 damping times of the initial Rabi signal. The maximum separation of the field components reaches 0.90 radians at time T, a value about three times the phase fluctuation of the initial coherent field ($1/\sqrt{\bar{n}} = 0.27$ radian). The components of the field are thus well-separated, before being recombined by the time reversal operation. We have represented in the inset of the figure the phase space plots of the calculated $Q(\beta)$ function of the cavity field at the kicking pulse time.

This echo signal reveals the existence of mesoscopic coherence in the atom–field system between the collapse and the induced revival time. The size of this mesoscopic superposition is naturally measured by the dimensionless distance D in phase space between its components:

$$D = 2\sqrt{\bar{n}}\sin(\Omega_0 T/4\sqrt{\bar{n}}) \, , \qquad (7.111)$$

which is about 2.6 for $T = 18$ μs. We note that this distance remains moderate. If we had looked for the spontaneous revival, the two field components would have had to go through a much larger maximum distance $D' = 2\sqrt{\bar{n}} \approx 7.3$. The comparison between D and D' explains why the induced revival signal is much easier to observe than the spontaneous one.

The induced revival signal could in principle be used to study quantitatively the cat state decoherence process. The method has been proposed by Morigi *et al.* (2002) precisely with this goal in mind. Ideally, the echo method should perfectly refocus the unitary evolution of the system and the reduced echo contrast should be directly related to irreversible decoherence integrated from $t_i = 0$ to $t_i = 2T$.

Such an analysis is difficult to carry out quantitatively because the echo is also affected by experimental imperfections, some of which have already been described earlier (limited contrast of Rabi pulses, residual thermal fields, etc). We have performed a numerical simulation of the system evolution (including the effect of decoherence due to field damping in the cavity and all known causes of imperfections). The predicted signals are shown as solid lines in Fig. 7.24. This simulation indicates that the contrast reduction is, in this experiment, largely dominated by mundane imperfections. We discuss in the next section more practical ways of observing the decoherence of these mesoscopic superpositions.

7.5   Decoherence of cavity cats

Macroscopic state superpositions are so fragile and so prone to be destroyed by their coupling to the environment that they cannot be practically observed. As soon as prepared, they turn in a flash into mundane statistical superpositions. We have given in Chapter 2 a qualitative description of this fundamental decoherence process, essential to understand the difference between the behaviour of Nature in the microscopic and macroscopic worlds.

At the fuzzy boundary between these two worlds, decoherence is however expected to be observable. In order to investigate it in 'real' time, we must perform experiments on systems which possess a physical parameter 'measuring' the size of the system and directly related to the rate of decoherence. This parameter should be continuously tuneable from a microscopic to a macroscopic value.

Mesoscopic cavity QED cats are ideal such systems. The tuneable parameter is their average photon number $\bar{n}$. The decoherence time, inversely proportional to $\bar{n}$, is comparable to the classical field energy damping time T_c for microscopic cats ($\bar{n} \approx 1$) and becomes much shorter when $\bar{n}$ increases to a value of the order of a few tens. These mesoscopic cat states are living in an interesting domain of Hilbert space where decoherence is fast enough to be distinguishable from classical field damping, yet slow enough to be observable on experimental signals. Moreover, the unitary system evolution and the irreversible coupling to environment are well-understood, making the comparison between experiments and theory particularly easy.

7.5.1   Exact solution of the decoherence master equation

The relaxation of a field in a cavity is a paradigmatic problem in quantum optics, discussed in detail in Chapter 4. The evolution of the field is described by a master equation (4.78), with rates depending in general on two parameters only, the inverse

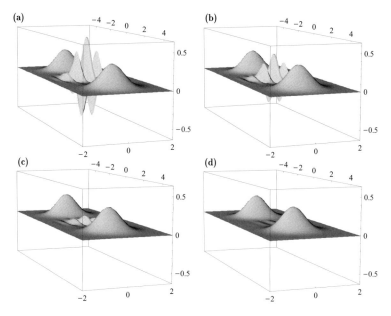

Fig. 7.25 Evolution of the Wigner function of an even cat state (10 photons on the average). (a)–(d): $t = 0$, $t = T_c/20$, $t = T_c/5$ and $t = T_c/2$ respectively.

of the cavity damping time $\kappa = 1/T_c$ and the temperature T of the environment. We assume here $T = 0$ K, which is a good approximation in our experiments. The environment is then in its ground state and the master equation for the field density operator ρ_A reduces (in interaction picture) to:

$$\frac{d\rho_A}{dt} = \frac{\kappa}{2} \left[2a\rho_A a^\dagger - a^\dagger a\rho_A - \rho_A a^\dagger a \right] \ , \tag{7.112}$$

a particularly simple form of the Lindblad equation. The field damping is fully characterized by a single jump operator $\sqrt{\kappa}a$ describing the photon escape from the cavity.

Solving this equation yields the density operator $\rho_A(t)$ at time t for a field initially prepared in an arbitrary state. The evolution of coherent states is particularly simple. They remain pure states, with an amplitude $\beta(t) = \beta e^{-\kappa t/2}$ decaying exponentially with time, their density operator reducing to a simple projector:

$$\rho_A(t) = |\beta(t)\rangle \langle \beta(t)| \ . \tag{7.113}$$

This equation means that coherent states are 'pointer states' of the field, impervious to entanglement with the environment. This fundamental property, a consequence of the fact that they are eigenstates of the jump operator a, is discussed in detail in Chapters 3 and 4.

Their status of pointer states gives to coherent states a privileged role in quantum optics. It is convenient to study the evolution of an arbitrary state by expanding it on a coherent state basis. This procedure is applied in the appendix to find the evolution of the density matrix of a field initially in an even or odd π-phase cat:

$$\rho_A^\pm(t) \quad = \quad \frac{1}{2}\Big[\,|\beta(t)\rangle\,\langle\beta(t)| + |-\beta(t)\rangle\,\langle-\beta(t)|$$

$$\pm e^{-2\bar{n}(1-e^{-\kappa t})}\,(|\beta(t)\rangle\,\langle-\beta(t)| + |-\beta(t)\rangle\,\langle\beta(t)|)\Big]\ . \qquad (7.114)$$

The fast decay of the coherence terms is described by the exponential coefficient of the off-diagonal terms in eqn. (7.114). The exponential term decays at short times as $\exp(-2\bar{n}\kappa t)$ and the coherent cat turns for $t > 1/\bar{n}\kappa$ into a mere mixture of states. This fast decoherence, whose rate increases proportionally to $\bar{n}$, is exhibited in a pictorial way on the analytical expression of the Wigner functions of the odd and even cats, easily derived from their density matrix (eqn. A.101 in the appendix). These distributions are represented at four different times in Fig. 7.25 for a cat state containing $\bar{n} = 10$ photons on average. The decoherence process appears clearly on these plots which exhibit two very different time constants. After a short time, the cat whiskers, which are a signature of its coherence, have been washed out, leaving only the two Gaussian peaks, which slowly relax down to vacuum.

Formal solutions of the master equation are fine, but they do not provide a clear explanation of the mechanism responsible for decoherence. To get a better feeling about this fundamental process and to understand how it is related to a loss of information about the field, we give now a heuristic derivation of eqn. (7.114) which does not make an explicit use of the master equation, but directly describes what happens to the environment during the field evolution. This calculation follows the line of reasoning already used in Section 3.2.5 to study the relaxation of a coherent field. Its results coincides of course with the exact solution of the Lindblad master equation (7.112).

We recall that the environment E responsible for decoherence can generally be described as an ensemble of oscillators, labelled by the index i, linearly coupled to the cavity mode. The spectral distribution of these oscillators spans a wide frequency domain around the frequency of the cavity. Starting from a coherent state $|\beta\rangle$ at time $t = 0$ and an empty environment, the global system evolves at time t into a separate non-entangled state which can be written (in the interaction picture) as:

$$\left|\Psi^{(AE)}(t)\right\rangle = |\beta(t)\rangle \otimes \prod_i |\epsilon_i(t)\rangle\ , \qquad (7.115)$$

where $\beta(t) = \beta e^{-\kappa t/2}$ and $\epsilon_i(t)$ is the very small complex amplitude at time t of the ith oscillator in E. It is important to note that, as a consequence of the linear coupling between the cavity mode and the environment oscillators, the phase of $\epsilon_i(t)$ is linearly related to the phase of β, whether the ith oscillator is exactly resonant with the cavity or not. Energy conservation requires moreover that the number of quanta in the environment oscillators equals at any time the number of photons lost by the field:

$$\sum_i |\epsilon_i(t)|^2 = \bar{n}\left(1 - e^{-\kappa t}\right)\ . \qquad (7.116)$$

Suppose now that we have prepared at $t = 0$ the field in a phase cat state:

$$\left|\Psi_{\text{cat}}^{(A)}(0)\right\rangle = \frac{e^{i\psi_1}}{\sqrt{2}}\,|\beta e^{i\Phi}\rangle + \frac{e^{i\psi_2}}{\sqrt{2}}\,|\beta e^{-i\Phi}\rangle\ . \qquad (7.117)$$

Fig. 7.26 Dissemination of Schrödinger kittens in the environment of the cavity.

The state of the field+environment system at time t is obtained by superposing the contributions of the two parts of the cat, each being given by an expression of the form (7.115). Note also that, when β is phase shifted by $\pm\Phi$, the same shift is experienced by the $\epsilon_i(t)$ amplitudes. This is a consequence of the linear coupling between the oscillators. We thus have:

$$\left|\Psi_{\text{cat}}^{(AE)}(t)\right\rangle = \frac{e^{i\psi_1}}{\sqrt{2}}\left|\beta(t)e^{i\Phi}\right\rangle \otimes \prod_i \left|\epsilon_i(t)e^{i\Phi}\right\rangle + \frac{e^{i\psi_2}}{\sqrt{2}}\left|\beta(t)e^{-i\Phi}\right\rangle \otimes \prod_i \left|\epsilon_i(t)e^{-i\Phi}\right\rangle .$$

(7.118)

The two parts of the cat are correlated to two states of the environment:

$$\left|E^+(t)\right\rangle = \prod_i \left|\epsilon_i(t)e^{i\Phi}\right\rangle ; \qquad \left|E^-(t)\right\rangle = \prod_i \left|\epsilon_i(t)e^{-i\Phi}\right\rangle ,$$

(7.119)

each of which is the product of minute copies of the cavity cat components disseminated in the environment (Fig. 7.26). Each environment oscillator is involved in a superposition of two states with very small amplitude ϵ_i and a large phase-splitting 2Φ, a 'Schrödinger kitten' so to speak. The product states $\left|E^+(t)\right\rangle$ and $\left|E^-(t)\right\rangle$ get very quickly mutually orthogonal. Using the expressions of coherent states scalar products (eqn. 3.61) and eqn. (7.116), we find:

$$\left\langle E^-(t)\left|E^+(t)\right.\right\rangle = \exp\left[-\sum_i |\epsilon_i(t)|^2 \left(1 - e^{2i\Phi}\right)\right] = \exp\left[-\overline{n}(1 - e^{-\kappa t})(1 - e^{2i\Phi})\right] ,$$

(7.120)

an expression which goes rapidly to zero as time increases. We understand now the fast loss of quantum coherence as a complementarity effect. It is fundamentally a matter of information about the cat state leaking into the environment. The final states of each oscillator in E have a near unity overlap (since $|\epsilon_i|^2 \ll 1$), but there are very many of them and the product of their scalar products quickly vanishes.

In other words, information about the phase $\pm\Phi$ of the cat state components could not be obtained from a single environment oscillator whose amplitude is small and hence phase fluctuations very large. It could however be quickly recovered, at least in principle, from a measurement of the environment as a whole by combining the small amounts of information disseminated among the elementary oscillators. This is enough to kill the cat state coherence and all the interference effects which are associated to it. When tracing over the environment, the cat coherence is multiplied by the overlap between the final states $|E^+(t)\rangle$ and $|E^-(t)\rangle$ and the field density operator finally is:

$$\rho_A(t) \;=\; \tfrac{1}{2}\Big\{ \big|\beta(t)e^{i\Phi}\big\rangle \big\langle\beta(t)e^{i\Phi}\big| + \big|\beta(t)e^{-i\Phi}\big\rangle \big\langle\beta(t)e^{-i\Phi}\big|$$
$$+\langle E^-(t)|E^+(t)\rangle\, e^{i(\psi_1-\psi_2)}\big|\beta(t)e^{i\Phi}\big\rangle \big\langle\beta(t)e^{-i\Phi}\big| + \text{h.c.}\Big\} \,. \quad (7.121)$$

Replacing in (7.121) $\langle E^-(t)|E^+(t)\rangle$ by its expression (7.120) and setting $\Phi = \pi/2$ and $\psi_1 - \psi_2 = 0$ or π, we recover, as expected, the exact formula (7.114) giving the evolution of an even or odd π-phase cat.

Photonic cat state decoherence provides a textbook illustration of the general analysis developed in Chapters 2 and 4 about pointer states and the measurement process. CQED cats constitute indeed an ideal example of pointer state superpositions and the study of their decoherence illustrates vividly this fundamental aspect of measurement theory. Let us now sharpen a bit more our analysis by imagining thought experiments in which we would try to recover explicitly information about the cat by performing measurements in its environment.

7.5.2 Monte Carlo scenarios of cat decoherence

Decoherence is the result of a dissemination of potential information about a system via its entanglement with an environment. When this information is not explicitly read in the environment, the system density operator ρ_A is obtained by adding up, with their respective probabilities of outcome, the projectors on the states associated to all possible environment measurement results. This is the meaning of the trace operation eliminating the irrelevant environment variables. A myriad of scenarios involving such virtual measurements can be imagined, resulting in as many equivalent expressions of the system density operator.

We have recalled in Chapter 4 how these virtual measurement games can be played with two actors, Alice and Bob, the first being the keeper of the system A and the second the observer of the environment E. Since Bob's measurements are virtual, they need not be realized in the real environment E, which is very complicated and spans in general a huge Hilbert space. The system A evolution can be exactly reproduced by considering an environment simulator B whose Hilbert space dimension is a cardinal exceeding only by one the number of Lindblad operators.

In many situations of interest, such as the spontaneous emission of an atom, or the decoherence of a field mode studied in this chapter, the effect of the coupling to the environment can be described by a single Lindblad operator. Hence, its decoherence can be understood with an environment simulator reducing to a single qubit with two levels $|0^{(B)}\rangle$ and $|1^{(B)}\rangle$. By performing measurements on this qubit in various bases, Bob picks up the equivalent final forms of the density operator of the field owned by Alice.

This approach to the decoherence problem involving Bob has many advantages. Assuming that Bob is effectively observing the environment and communicating the results of his measurements to Alice means that she can at all times describe her system by a pure state, which evolves randomly, according to the unpredictable outcomes of the measurements. This is the Monte Carlo point of view, which is reconciled with the density operator approach by averaging *in fine* all the possible system trajectories in Hilbert space. As discussed in Chapter 4, calculating a set of Monte Carlo trajectories can be much more economical than solving the master equation.

Moreover, this approach is more appropriate than the density operator point of view to describe a single realization of an experiment manipulating a unique quantum system. In many modern experiments, information is continuously extracted from the system's environment and its wave function follows a random trajectory exhibiting explicit quantum jumps (see for example Fig. 1.10, on page 17). The statistics of these observed jumps is reproduced by a Monte Carlo simulation involving an environment with as many states as the number of possible exclusive outcomes of the measurement. In most cases, this number remains small and an environment simulator B with a few quantum states is enough to simulate the experimental results.

We shall illustrate these general ideas by describing a few Monte Carlo scenarios of possible decoherence trajectories for a cat state in a cavity. One of these scenarios is directly related to a possible CQED experiment. The others are, at the present stage of the technology at least, of the *gedanken* type. They all provide interesting insights into fundamental aspects of the quantum measurement process.

According to the general discussion of Chapter 4, the evolution during an elementary time interval τ of a harmonic oscillator A in state $\left|\phi^{(A)}\right\rangle$ undergoing a decoherence process described by the jump operator $\sqrt{\kappa}a$ is obtained by coupling it to a qubit B in state $\left|0^{(B)}\right\rangle$, letting the $A + B$ system evolve according to the infinitesimal unitary transformation:

$$\left|\phi^{(A)}\right\rangle \otimes \left|0^{(B)}\right\rangle \rightarrow \left[1 - \frac{1}{2}\kappa\tau(a^{\dagger}a)\right]\left|\phi^{(A)}\right\rangle \otimes \left|0^{(B)}\right\rangle + \sqrt{\kappa\tau}a\left|\phi^{(A)}\right\rangle \otimes \left|1^{(B)}\right\rangle . \quad (7.122)$$

The probabilities p_1 and p_0 of finding at time τ the qubit B in state $\left|1^{(B)}\right\rangle$ or $\left|0^{(B)}\right\rangle$ are:

$$p_1 = 1 - p_0 = \kappa\tau\left\langle\phi^{(A)}\left|a^{\dagger}a\right|\phi^{(A)}\right\rangle . \quad (7.123)$$

Depending on the outcome of the measurement, A is projected in one of the normalized states defined respectively as:

$$\left|\phi_1^{(A)}\right\rangle = \frac{a\left|\phi^{(A)}\right\rangle}{\sqrt{p_1}} , \quad (7.124)$$

and:

$$\left|\phi_0^{(A)}\right\rangle = \frac{(1 - \kappa\tau a^{\dagger}a/2)\left|\phi^{(A)}\right\rangle}{\sqrt{p_0}} . \quad (7.125)$$

A Monte Carlo trajectory is obtained by iterating the process, making a random decision according to the probability law (7.123) to determine whether B is found in $\left|1^{(B)}\right\rangle$ or $\left|0^{(B)}\right\rangle$ at each stage of the procedure. The state of A at the beginning of the $(p + 1)$th step is determined by the outcome of the pth measurement and B is initialized to $\left|0^{(B)}\right\rangle$ at the beginning of each step.

The environment simulator B models here an ideal photon counter which clicks when it is found in state $\left|1^{(B)}\right\rangle$, signalling the annihilation of one quantum escaping from the cavity through a partially transmitting mirror. The corresponding thought

Fig. 7.27 An environment B detecting the photon lost by the cavity field.

experiment is sketched in Fig. 7.27. A Monte Carlo trajectory is defined by a series of random bit values 1 or 0 associated to the successive time bins of elementary duration τ. Depending upon whether the photon detector has clicked or not at each step, the wave function of the field undergoes the action of either the $(1 - i\kappa\tau a^\dagger a/2)$ or the a operator (with corresponding normalization). Note that the presence, as well as the absence, of clicks are information which, in general, change the system state. This procedure was applied in Chapter 4 to initial Fock or coherent states. The loss of photons results, for a Fock state, in a staircase evolution of the photon number n, which undergoes random jumps as the field energy gets dissipated. The usual cavity exponential decay is recovered by an average of a large number of individual staircase trajectories. We have also analysed in detail the non-intuitive evolution of a coherent state which stays unchanged when photons are lost and decays only during the time intervals when no photons are counted.

Parity jumps of a π-phase cat

Let us now use the *gedanken* set-up of Fig. 7.27 to follow the evolution of a cavity field initially prepared in the even π-phase cat state $\left|\Psi_{\mathrm{cat}}^+(0)\right\rangle = [|\beta\rangle + |-\beta\rangle]/\sqrt{2}$. The photon clicks are counted in the successive time bins of duration τ. Until the first click at time t_1, the non-unitary evolution simply shrinks the amplitude of the cat components exponentially, with the rate $\kappa/2$, turning β into $\beta(t_1) = \beta e^{-\kappa t_1/2}$. The cat state is continuously changed into $\left|\Psi_{\mathrm{cat}}^+(t_1)\right\rangle = [|\beta(t_1)\rangle + |-\beta(t_1)\rangle]/\sqrt{2}$.

According to eqn. (7.13), the first click corresponds to a sudden switch of the cat parity, the field state jumping from $\left|\Psi_{\mathrm{cat}}^+(t_1)\right\rangle$ to $\left|\Psi_{\mathrm{cat}}^-(t_1)\right\rangle = [|\beta(t_1)\rangle - |-\beta(t_1)\rangle]/\sqrt{2}$. The shrinking of the cat components amplitudes then resumes until a second jump restores the initial parity and so on. As long as the field energy has not appreciably decayed, the probability of occurrence of a click in a time bin is $p_1 = \kappa\bar{n}\tau$ and the average duration between clicks is $T_D = 1/(\kappa\bar{n})$. On each Monte Carlo trajectory, a fast and random telegraphic parity jumping is thus combined with a slow deterministic shrinking of the field amplitude components. The characteristic rates of these two very different evolutions are in the ratio $\bar{n}$.

It is quite remarkable that information provided by a photon click has an immediate effect on a field in a superposition of two coherent states, even if each of them, prepared separately in the cavity, would remain unchanged. This is a genuine quantum effect. Information provided by the click changes the relative quantum phase between the two parts of the field wave function and thus turns an even parity cat into an odd one and vice versa (eqn. 7.13). Paradoxically, this quantum effect seems to bring back some kind of intuitive behaviour in the system's evolution. It seems quite natural

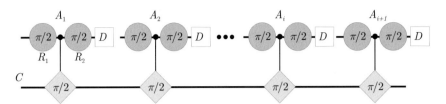

Fig. 7.28 Quantum circuit of a possible experiment to track the cat parity. A stream of atoms cross at regular time intervals the cavity C. A Ramsey interferometer (pulses R_1 and R_2) and a dispersive interaction with C provide a direct readout of the photon number parity.

that, by losing a photon, a field which contains only even photon numbers should be transformed into one containing only odd numbers.

This is indeed reasonable but one should still be cautious before trusting our intuition about photon numbers. A simple calculation left to the reader shows that the average photon number is practically unchanged when a photon disappears in a cat state [$\bar{n}$ changes by an exponentially small amount $\mp 2\bar{n}/\sinh(2\bar{n})$, totally negligible compared to 1 as soon as $\bar{n} > 3$]. As in the coherent state case (remember the discussion in Section 4.4.4), this results from a competition between the real loss of a photon and the acquisition of information at the click, which biases the photon number towards a larger value. An odd or even cat state is thus a strange system which switches its photon parity when it loses a quantum while keeping its average photon number nearly constant!

Assuming that the photon counter is perfect, the parity of the number of clicks recorded until time t would tell us without ambiguity whether the field is in the state $\left|\Psi_{\text{cat}}^{+}(t)\right\rangle$ or $\left|\Psi_{\text{cat}}^{-}(t)\right\rangle$. We could thus follow a Monte Carlo trajectory in which the field would be a pure, albeit random cat state. Observing the environment makes it thus possible, at least in principle, to keep the cat coherence alive for a time of the order of $T_c = 1/\kappa$. This continuous observation does not give us the power, though, to choose at a given time the parity of the cat. We can merely record it.

Detecting the photons leaking out of the cavity is not very realistic, especially in the microwave CQED experiments. We can, however, envision an equivalent detection scheme which would allow us to follow more practically the photon number parity $\mathcal{P}$ along single realizations of a cat experiment. We have shown that an atomic CQED Ramsey interferometer can be used to measure $\mathcal{P}$.

Sending a train of atoms across the Ramsey interferometer, we could thus fix the phases in such a way that the first atom would prepare, upon its detection, a $\left|\Psi_{\text{cat}}^{+}\right\rangle$ or $\left|\Psi_{\text{cat}}^{-}\right\rangle$ cat state, eigenstate of $\mathcal{P}$ with the $+1$ or -1 eigenvalue. The subsequent atoms, crossing the apparatus under the same conditions, would then re-measure $\mathcal{P}$. When a photon loss would induce a parity change, it would be recorded by a sudden change of the detected atomic state. We thus expect to find sequences of atoms now in $|e\rangle$, now in $|g\rangle$, the switching between these states signalling the loss of a photon and hence a jump of the cat parity. This experiment, whose principle is sketched in Fig. 7.28, is close to be feasible with a somewhat improved version of our CQED apparatus. We will describe below a variant in which, by detecting only a sequence of two atoms, we have been able to monitor the decoherence of the field.

As soon as we renounce counting the photon clicks or measuring the photon parity, or when these operations become imperfect due to finite detection efficiency, we have to describe the field by its density operator, obtained by averaging the Monte Carlo trajectories. After a time of the order of $T_D = 1/\bar{n}\kappa$ there are statistically as many trajectories in which the number of unrecorded clicks is even or odd, resulting in a field density operator which has become the sum with equal weights of projectors on even and odd cats:

$$\rho_A(t > T_D) \approx \tfrac{1}{2} \left| \Psi_{\text{cat}}^+(t) \right\rangle \left\langle \Psi_{\text{cat}}^+(t) \right| + \tfrac{1}{2} \left| \Psi_{\text{cat}}^-(t) \right\rangle \left\langle \Psi_{\text{cat}}^-(t) \right| . \tag{7.126}$$

This density operator can be equivalently expressed as an incoherent sum with equal weights of projectors on the $|\beta(t)\rangle$ and $|-\beta(t)\rangle$ states. We thus retrieve by the Monte Carlo approach the result predicted by an exact solution of the master equation [limit of eqn. (7.114) for $t > 1/\kappa\bar{n}$].

Pinning down the cat component by a homodyne measurement

Let us summarize the discussion up to this point and make it a bit more dramatic by introducing again our two characters, partners in classical and quantum communication, Alice and Bob. We have just considered a very simple scenario for the decoherence of a photonic π-phase cat by describing its environment as a single-qubit simulator. In this script, the photons lost by the field prepared by Alice are merely counted by a photon counter read by Bob. Information leaking about the field in the environment is dichotomic (field photon number parity $\mathcal{P}$ with ± 1 eigenvalues). If the photon clicks are not registered, or if Bob does not communicate his measurements to Alice, information she has about her system is very quickly scrambled, transforming for her the initial pure state into a mixture, naturally expressed in the basis of the photon parity eigenstates $\left| \Psi_{\text{cat}}^\pm \right\rangle$ (eqn. 7.126).

Let us consider now another scenario. Instead of merely counting the photons lost by Alice, Bob performs a homodyne measurement of the escaping field, which gives him information about the field quadrature aligned with one of the cat components and anti-aligned with the other. If he sends the result to Alice, this lifts for her the quantum ambiguity and collapses the field along the positive or negative quadrature component. The Monte Carlo trajectories of this scenario and the final states ($|\pm\beta\rangle$) are quite different from those ($\left| \Psi_{\text{cat}}^\pm \right\rangle$) of the parity-switch script discussed above.

If quadrature information is not available to Alice, though, the final field will be described by her as an incoherent mixture of the two cat components, which is fully equivalent to the mixture of the parity cat states obtained at the end of the 'environment-unread' version of the first scenario. Not only the final states of the two scenarios must be identical in the two unread versions, but the dynamics of the density operator evolution must be the same at all times. Not measuring or not reading the results of the measurement in the environment merely amounts to tracing on it and this operation is independent of any measurement not performed by Bob.

The description of her state by Alice could not even depend on any measurement Bob could have done without communicating with Alice, as can be shown by a simple causality argument. If, without classical communication with Bob, Alice described by different density operators her cat state in the two scenarios, this would mean that Bob could, by changing his experimental procedure (homodyning the photons he

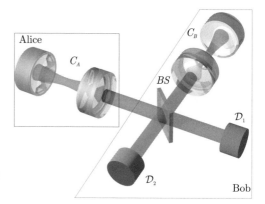

Fig. 7.29 Homodyne cat-measurement scenario.

receives versus counting them) communicate instantaneously information to Alice via a quantum correlation. This is of course impossible, as we noticed already in Chapter 2 when discussing the EPR paradox.

It is interesting to describe in more detail the homodyne measurement scenario as an exercise illustrating how the same density operator emerges from a perspective quite different from the cat parity measurements. This scenario is also instructive because it corresponds to a more complex, yet entirely calculable, measurement process. The collapse of the cat state in this case is reminiscent of the localization by the environment of a particle initially in a superposition of two different spatial positions (Section 4.4.5), a problem which has been considered in many theoretical papers (Zurek 2003) and which can be given an exact solution in this quantum optics context.

The homodyne method we describe here is similar to the description of the photon number QND counting analysed in Chapter 6. In both situations, the progressive acquisition of information about a quantum system results in a continuous change of its state. In the photon QND process, the measurement is carried out by a real observer and results in the final collapse of the field in a well-defined state. In the decoherence perspective considered here, this measurement is virtually performed by the system's environment and leads, *in fine*, to equivalent incoherent mixtures of states associated to the possible results of these virtual measurements.

The thought experiment now performed by Alice and Bob is sketched in Fig. 7.29. Alice prepares in her cavity C_A the π-phase cat $\left|\Psi_{\text{cat}}^{+(A)}\right\rangle = [|\beta\rangle+|-\beta\rangle]/\sqrt{2}$, which leaks at a rate κ through one of the cavity mirrors (β is here assumed to be real). Bob uses a symmetric beam-splitter BS to mix the leaking field with a reference homodyning field. The photons in the output modes are counted by the perfect detectors $\mathcal{D}_1$ and $\mathcal{D}_2$. Since Bob now collects information from two counters, the measurement process must be described by an environment simulator B possessing three states, $\left|0^{(B)}\right\rangle$, $\left|1^{(B)}\right\rangle$ and $\left|2^{(B)}\right\rangle$, coding respectively the 'no click', 'click in $\mathcal{D}_1$' and 'click in $\mathcal{D}_2$' information. In other words, B is no longer a qubit, but a 'qutrit'.

In order to calibrate the reference field, we assume that its source is a cavity C_B, identical to C_A. It contains a coherent state $\left|\alpha^{(B)}\right\rangle$ with $\alpha = \xi e^{-i\phi_r}\beta$, which leaks outside C_B with the same rate κ as C_A. The real and positive parameter ξ is thus equal to the ratio between the reference and the signal field amplitudes impinging on the beam-splitter. The phase ϕ_r is adjusted to yield a homodyne signal sensitive to a given quadrature of the field leaking from C_A. In order to probe the quadrature aligned or anti-aligned with Alice's cat components, we have to choose $\phi_r = \pi/2$ (see Section 3.2.4).

We choose to use a beam-splitter inducing the mode mixing described by eqn. (3.100) with $\varphi = 0$ and $\theta = \pi/2$. A simple inspection of the beam-splitter formulas will convince us that the field transmitted by BS towards $\mathcal{D}_1$ is identical to the one which would come directly from C_A, in the absence of beam-splitter, if the cavity leaking rate were $\kappa/2$ and if the field in C_A were translated in phase space by $i\alpha = i\xi e^{-i\phi_r}\beta$ (the factor i accounts for the phase shift experienced by the reference field reflected into $\mathcal{D}_1$). This translation is described by the operator $D(i\alpha)$ whose properties have been recalled in Section 3.1.3. When $\mathcal{D}_1$ clicks, the equivalent translated field undergoes a jump described by the Lindblad operator $\sqrt{\kappa/2}a$. The actual field stored in C_A thus undergoes the jump described by the operator L_1 obtained by undoing the translation $D(i\alpha)$:

$$L_1(\alpha) = \sqrt{\kappa/2}D^{-1}(i\alpha)aD(i\alpha) = \sqrt{\kappa/2}(a + i\alpha\mathbb{1}) . \tag{7.127}$$

We have used here the formula (3.48) describing the effect of a translation in phase space on the field operator a. Similarly, it is easy to show that a click in $\mathcal{D}_2$ corresponds to the jump operator:

$$L_2(\alpha) = i\sqrt{\kappa/2}D^{-1}(-i\alpha)aD(-i\alpha) = i\sqrt{\kappa/2}(a - i\alpha\mathbb{1}) . \tag{7.128}$$

Knowing these Lindblad operators, we can write the infinitesimal evolution of the $A + B$ system during an elementary time interval τ. Calling $\left|\phi^{(A)}\right\rangle$ the Monte Carlo wave function of the field at the beginning of this time interval and initializing the environment B into $\left|0^{(B)}\right\rangle$ we get:

$$\left|\phi^{(A)}\right\rangle \otimes \left|0^{(B)}\right\rangle \rightarrow \left\{1 - \frac{\tau}{2}\left[L_1^\dagger(\alpha)L_1(\alpha) + L_2^\dagger(\alpha)L_2(\alpha)\right]\right\}\left|\phi^{(A)}\right\rangle \otimes \left|0^{(B)}\right\rangle$$
$$+ \sqrt{\tau}\left[L_1(\alpha)\left|\phi^{(A)}\right\rangle \otimes \left|1^{(B)}\right\rangle + L_2(\alpha)\left|\phi^{(A)}\right\rangle \otimes \left|2^{(B)}\right\rangle\right] . \tag{7.129}$$

The second line in the right-hand side describes the quantum jumps experienced by the field if counters $\mathcal{D}_1$ or $\mathcal{D}_2$ click, which occur with the respective probabilities:

$$\begin{aligned} p_1 &= \tau\left\langle\phi^{(A)}\left|L_1^\dagger L_1\right|\phi^{(A)}\right\rangle = \frac{\tau\kappa}{2}\left\langle\phi^{(A)}\left|(a^\dagger - i\alpha^*\mathbb{1})(a + i\alpha\mathbb{1})\right|\phi^{(A)}\right\rangle ; \\ p_2 &= \tau\left\langle\phi^{(A)}\left|L_2^\dagger L_2\right|\phi^{(A)}\right\rangle = \frac{\tau\kappa}{2}\left\langle\phi^{(A)}\left|(a^\dagger + i\alpha^*\mathbb{1})(a - i\alpha\mathbb{1})\right|\phi^{(A)}\right\rangle , \end{aligned} \tag{7.130}$$

while the first line in eqn. (7.129) accounts for the continuous evolution of the system's wave function in the no-click case, which occurs with the probability $p_0 = 1 - p_1 - p_2$.

This infinitesimal unitary evolution corresponds to Monte Carlo trajectories which, when averaged, yield a density operator evolving under the master equation put in Lindblad form:

$$\frac{d\rho_A}{dt} = \frac{1}{2} \sum_{i=1,2} \left[2L_i(\alpha)\rho_A L_i^\dagger(\alpha) - L_i^\dagger(\alpha)L_i(\alpha)\rho_A - \rho_A L_i^\dagger(\alpha)L_i(\alpha) \right] . \qquad (7.131)$$

This equation looks different from (7.112), but it is indeed the same as can be verified easily by replacing the L_i operators by their expressions (7.127) and (7.128) and checking that all the terms depending upon α cancel. We have here a typical example of ambiguity of the Lindblad formulation. Different expressions, with even different numbers of jump operators, describe the same physical evolution for Alice, as long as she does not have information about the results of possible measurements performed by Bob in B.

Let us see now what happens if Alice does know the results of Bob's homodyne measurements. Choosing $\phi_r = \pi/2$, we set $\alpha = -i\xi\beta$. We can still choose freely the relative amplitude ξ of the reference field. Let us first fix $\xi = 1$, meaning that Bob adjusts the reference so that it has exactly the same amplitude as the field leaking from Alice's cavity. The jump operators then reduce to $L_1 = \sqrt{\kappa/2}(a + \beta\mathbb{1})$ and $L_2 = i\sqrt{\kappa/2}(a - \beta\mathbb{1})$. What are then the initial probabilities p_1 and p_2 to get a click in $\mathcal{D}_1$ or $\mathcal{D}_2$? Replacing in eqn. (7.130) $|\phi^{(A)}\rangle$ by the initial state $\left|\Psi_{\text{cat}}^{+(A)}\right\rangle$, we get $p_1 = p_2 = \kappa\bar{n}\tau$. The two detectors are thus as likely to click. If $\mathcal{D}_1$ registers the first photon, the cat state undergoes the jump:

$$\left|\Psi_{\text{cat}}^{+(A)}\right\rangle \rightarrow \frac{1}{2\sqrt{\bar{n}}}(a + \beta\mathbb{1})\left(\left|\beta^{(A)}\right\rangle + \left|-\beta^{(A)}\right\rangle\right) = \left|\beta^{(A)}\right\rangle , \qquad (7.132)$$

and if it is $\mathcal{D}_2$:

$$\left|\Psi_{\text{cat}}^{+(A)}\right\rangle \rightarrow \frac{1}{2\sqrt{\bar{n}}}(a - \beta\mathbb{1})\left(\left|\beta^{(A)}\right\rangle + \left|-\beta^{(A)}\right\rangle\right) = \left|-\beta^{(A)}\right\rangle . \qquad (7.133)$$

The first detection event results in the collapse of the cat state in one of its components $\left|\pm\beta^{(A)}\right\rangle$. Once this collapse has occurred, the p_1 and p_2 probabilities are completely unbalanced. $\mathcal{D}_2$ can no longer click if the field has collapsed in $\left|\beta^{(A)}\right\rangle$ since $(a - \beta\mathbb{1})\left|\beta^{(A)}\right\rangle = 0$. All the subsequent photons are then detected by $\mathcal{D}_1$. Similarly, if the first click is recorded by $\mathcal{D}_2$, so are all the following ones. The randomness of the Monte Carlo trajectories resides only in the first step, which completely determines all the others.

The first click occurs after a characteristic time $1/\bar{n}\kappa$. Since the system has equal probabilities of collapsing in $\left|\beta^{(A)}\right\rangle$ or $\left|-\beta^{(A)}\right\rangle$, the averaged density operator thus reduces, after a time of the order of T_D, to an incoherent superposition with equal weights of the projectors on these two states. We understand again in this scenario why the decoherence process is so much faster than the field energy damping, which occurs on a $1/\kappa$ time scale. Note that we are justified, during the early decoherence stage, to neglect the non-unitary evolution of the system state which slowly shrinks the field amplitude.

In this thought experiment, Bob has cleverly balanced the amplitude of the reference so that it cancels the field corresponding to the cat components in one output or the other of the beam splitter. A single classical bit of information (click in $\mathcal{D}_1$ or $\mathcal{D}_2$) then suffices to pin down the phase of the field. One might think however that Bob has cheated a bit, because he had to know in the first place the amplitude of the field coming from Alice's cavity. What could Bob do if he does not have preliminary information?

Bob will then decide to use a reference with a relative amplitude $\xi \gg 1$, much more intense than the very weak leaking field he wants to measure. This has the advantage to increase proportionally to ξ^2 the photon count rate at the detectors. The Lindblad operators now become $L_1 = \sqrt{\kappa/2}(a + \xi\beta\mathbb{1})$ and $L_2 = i\sqrt{\kappa/2}(a - \xi\beta\mathbb{1})$. We will see that the cat state collapse becomes a continuous process, as the successive clicks registered by the detectors progressively increase Bob's knowledge about the field.

Each click in $\mathcal{D}_1$ and $\mathcal{D}_2$ will result in a slight increase of the probability amplitude of $|\beta^{(A)}\rangle$ and $|-\beta^{(A)}\rangle$ respectively. On each Monte Carlo trajectory, a majority ruling will decide which of the two components eventually wins. Since the ballot is decided in a very short time compared to $1/\kappa$, we will neglect the slow shrinking of the field amplitude produced by the non-hermitian contribution to the Monte Carlo trajectories and consider only the effect of the quantum jumps.

The field state at any time can be written as a superposition of $|\beta^{(A)}\rangle$ and $|-\beta^{(A)}\rangle$:

$$\left|\Psi^{(A)}(t)\right\rangle = \cos[\theta(t)/2]\left|\beta^{(A)}\right\rangle + \sin[\theta(t)/2]\left|-\beta^{(A)}\right\rangle , \qquad (7.134)$$

where $\theta(t)$ is a random angle comprised between 0 and π, whose initial value is $\theta(0) = \pi/2$. Identifying $|\pm\beta^{(A)}\rangle$ with the states $|0/1^{(A)}\rangle$ of a 'field qubit', we can assimilate the field state to a pseudo-spin on the surface of the Bloch sphere, which starts in the equatorial plane and undergoes a diffusion process along a meridian towards either the north or the south pole. A straightforward calculation yields the probabilities of $\mathcal{D}_i$ clicks $(i = 1, 2)$ in an elementary time interval $[t, t + \tau]$, as a function of the $\theta(t)$ angle:

$$p_i = \frac{\kappa\tau}{2}\left\langle\Psi^{(A)}(t)\right|(a^\dagger + \epsilon_i\xi\beta\mathbb{1})(a + \epsilon_i\xi\beta\mathbb{1})\left|\Psi^{(A)}(t)\right\rangle = \frac{\kappa\bar{n}\tau}{2}[1 + 2\epsilon_i\xi\cos\theta + \xi^2], \quad (7.135)$$

with $\epsilon_1 = +1$ and $\epsilon_2 = -1$ corresponding respectively to clicks in $\mathcal{D}_1$ and $\mathcal{D}_2$.

Let us now compute the incremental change of the state vector after detection of a click in $\mathcal{D}_i$. Applying to $|\Psi^{(A)}\rangle$ the L_i operators, we find that, depending upon which detector has clicked, the field evolves into one of the two $(i = 1, 2)$ (non-normalized) states:

$$\left|\Psi_i'^{(A)}\right\rangle \approx (\epsilon_i x + 1)\cos(\theta/2)\left|\beta^{(A)}\right\rangle + (\epsilon_i\xi - 1)\sin(\theta/2)\left|-\beta^{(A)}\right\rangle$$

$$\approx \cos(\theta/2)\left|\beta^{(A)}\right\rangle + \left(1 - \frac{2\epsilon_i}{\xi} + \frac{2}{\xi^2}\right)\sin(\theta/2)\left|-\beta^{(A)}\right\rangle . \quad (7.136)$$

The two kinds of jumps $(\epsilon_i = \pm 1)$ correspond to rotations in opposite directions of the field pseudo-spin. A simple trigonometric calculation yields the corresponding incremental change of θ, up to second order in $1/\xi$:

$$\delta\theta_i = -\frac{2\epsilon_i \sin\theta}{\xi} + \frac{\sin 2\theta}{\xi^2} . \tag{7.137}$$

At the beginning of the cat homodyne measurement ($\theta = \pi/2$), the probabilities p_1 and p_2 are equal (eqn. 7.135). The evolution of θ starts as a symmetrical one-dimensional random walk with an elementary step $\pm 2/\xi$. After a while, the steps get smaller and the probabilities for θ to increase or to decrease become different. An inspection of eqn. (7.135) shows that, for $\theta > \pi/2$, p_2 becomes larger than p_1. There are thus statistically more steps towards larger θ values, which amplifies the imbalance between the amplitudes of the two cat components. A symmetrical effect occurs for the trajectories in which θ happens to be smaller than $\pi/2$, leading again to an amplification of the drift, this time towards $\theta = 0$. We infer from this qualitative analysis that the continuous measurement makes the cat state highly unstable. The first random steps push θ away from the $\pi/2$ value, with equal probabilities in both directions. Once the field state has departed from this unstable equilibrium position, the motion accelerates and the system gets carried away towards $\theta = 0$ ($\left|\beta^{(A)}\right\rangle$) or $\theta = \pi$ ($\left|-\beta^{(A)}\right\rangle$).

The characteristic time of this localization process can be estimated qualitatively. During a time interval t, there are about $\kappa t \bar{n} \xi^2$ detected photons. Assimilating roughly the θ angle diffusion to a usual random walk, we expect the departure from $\theta = \pi/2$ to reach at time t a value $\Delta\theta$ of the order of the square root of the number of steps multiplied by the average step length $1/\xi$, i.e. $\Delta\theta \approx (\kappa t \bar{n} \xi^2)^{1/2} \xi^{-1} = (\kappa t \bar{n})^{1/2}$. This value does not depend upon ξ. We thus predict that the field state will end up in $\left|\beta^{(A)}\right\rangle$ or $\left|-\beta^{(A)}\right\rangle$ in a time T_D such that $(\kappa \bar{n} T_D)^{1/2} \approx 1$, i.e. $T_D \approx 1/\kappa\bar{n}$. We find again the same decoherence time as in the previous scenarios.

This time does not depend on the intensity of the reference field, as could have been expected since this intensity, left to the free choice of Bob, cannot obviously influence the decoherence process as it is seen by Alice. When ξ is increased, more photons are detected per unit time, but each click has a smaller effect on the field state. More detection events are necessary to collect information required to pin down the cat state to one of its two components.

Let us now be more quantitative and describe the evolution of a statistical ensemble of Monte Carlo trajectories. We will write an equation for the distribution of the tipping angle θ, analogous to the Fokker–Planck equation (Louisell 1973) describing the statistical evolution of a particle undergoing a one-dimensional diffusive motion. Establishing this equation requires some care because the jump probabilities and the elementary steps $\delta\theta_i$ of this random process are variable, depending upon the actual position of the pseudo-particle representing the system.

Let us call $P(\theta, t)d\theta$ the probability of finding at time t the field state with an angle in the interval $[\theta - d\theta/2, \theta + d\theta/2]$. This probability evolves under the effects of two processes. On the one hand, it gets fed by the jumps bringing, from a different value, the tipping angle into the $d\theta$ interval around θ. On the other hand, it gets depleted by the jumps which make θ leave this interval.

To compute the probability of the first process, we must determine the mean starting angle $\theta - \delta'\theta_i$ of the jumps bringing the system into the final $d\theta$ interval. The incremental change $\delta\theta_i$ given by eqn. (7.137) is expressed in term of the initial angle of the jump. We need here the expression $\delta'\theta_i$ of this increment in terms of its final angle. It is simply obtained by replacing θ by $\theta - \delta\theta_i$ in eqn. (7.137). Keeping terms up to second order in $1/\xi$, we get:

$$\delta'\theta_i = -\frac{2\epsilon_i \sin\theta}{\xi} - \frac{\sin 2\theta}{\xi^2} . \tag{7.138}$$

We need also to determine the width $d\theta_i$ of the interval which is mapped into the final $d\theta$ one. By computing the initial angles corresponding to the final values $\theta - d\theta/2$ and $\theta + d\theta/2$, we find:

$$d\theta_i = \left[1 - \frac{d(\delta'\theta_i)}{d\theta}\right] d\theta = \left[1 + \frac{2\epsilon_i \cos\theta}{\xi} + \frac{2\cos 2\theta}{\xi^2}\right] d\theta . \tag{7.139}$$

We can now express the incremental change of $P(\theta, t)d\theta$ during an infinitesimal time τ as:

$$P(\theta, t + \tau)d\theta = \sum_{i=1,2} p_i(\theta - \delta'\theta_i, \tau)P(\theta - \delta'\theta_i, t)d\theta_i + \left[1 - \sum_{i=1,2} p_i(\theta, \tau)\right] P(\theta, t)d\theta . \tag{7.140}$$

This equation has a transparent interpretation. It expresses that with the probabilities $p_i(\theta - \delta'\theta_i, \tau)$ the distribution P is translated by $\delta'\theta_i$, while it remains unchanged with the probability $1 - p_1(\theta, \tau) - p_2(\theta, \tau)$. The p_i probabilities have to be computed for the values of the tipping angles *at the start* of the corresponding jumps. The change of the jumping step over $d\theta$ is accounted for by the multiplying factor $d\theta_i$, different from $d\theta$, in the first right-hand side term of eqn. (7.140).

In order to transform finally eqn. (7.140) into a partial differential equation, we expand P to second order in $\delta'\theta_i$:

$$P(\theta - \delta'\theta_i, t) = P(\theta, t) - \delta'\theta_i \frac{\partial P(\theta, t)}{\partial\theta} + \frac{1}{2}\delta'\theta_i^2 \frac{\partial^2 P(\theta, t)}{\partial\theta^2} + O(\delta'\theta_i^3) , \tag{7.141}$$

we plug this expression in (7.140) and replace $\delta'\theta_i$, the p_i's and $d\theta_i$ by their expressions (7.138), (7.135) and (7.139). We finally get for $P(\theta, t)$ an equation of the Fokker–Planck type, which does not depend upon ξ:

$$\frac{1}{\kappa\bar{n}}\frac{\partial}{\partial t}P(\theta, t) = \frac{\partial}{\partial\theta}[\sin(2\theta)P(\theta, t)] + 2\frac{\partial^2}{\partial\theta^2}[\sin^2(\theta)P(\theta, t)] . \tag{7.142}$$

The first-order θ-derivative in the right-hand side is a 'drift' term which pushes the distribution away from $\theta = \pi/2$, as soon as $\sin(2\theta)$ is different from 0, i.e. $\theta \neq \pi/2$. The second-order derivative is a diffusion term, which controls the rate at which the distribution is spreading. The $\sin^2\theta$ coefficient multiplying P in this term is maximum for $\theta = \pi/2$ and decreases when θ drifts towards 0 or π. This slowing-down of the diffusion process reflects the decrease of the θ-random-walk step length when the field state departs from its initial position.

Figure 7.30 shows the time evolution of $P(\theta)$ as it results from a numerical integration of the diffusion equation. The distribution peaked at $\theta = \pi/2$ for $t = 0$ very quickly widens and develops two side peaks which drift towards $\theta = 0$ and $\theta = \pi$. The

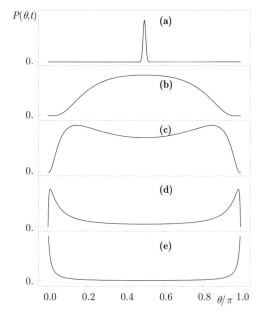

Fig. 7.30 Evolution of the $P(\theta, t)$ distribution illustrating the localization of the cat state. The initial distribution is a narrow Gaussian centred at $\theta = \pi/2$, standing for a δ distribution. Frames (a)–(e) correspond to $t = 0, 0.1, 0.2, 0.5$ and 0.75 in units of $1/(\kappa\bar{n})$.

final distribution, reached in a time of the order of $1/\kappa\bar{n}$, is made of two Dirac peaks of equal magnitude. This means that, as expected, half of the Monte Carlo trajectories end up in each of the two $\left|\pm\beta^{(A)}\right\rangle$ states. This figure gives a striking illustration of the cat localization due to being watched by the environment.

Observing the diffusion of the cat quantum phase

We now consider a last decoherence scenario. In the previous one, Bob was watching the quadrature of the field in phase with one of the cat's components. What would happen if he decides to observe the complementary quadrature? This is the observable whose probability distribution presents a characteristic interference pattern which we have considered as a signature of the cat's coherence (eqn. 7.8). In order to gain information on this quadrature, Bob must use a reference field with an amplitude $\alpha = \xi\beta$ in phase with β. The Lindblad operators associated to the photon counting process are now $L_1 = \sqrt{\kappa/2}(a + i\xi\beta\mathbb{1})$ and $L_2 = i\sqrt{\kappa/2}(a - i\xi\beta\mathbb{1})$. We will assume again that $\xi \gg 1$ (intense reference field). In order to analyse the effect of each click on the field state, it is now convenient to write it as:

$$\left|\Psi^{(A)}(t)\right\rangle = \left[\left|\beta^{(A)}\right\rangle + e^{i\psi(t)}\left|-\beta^{(A)}\right\rangle\right]/\sqrt{2}, \tag{7.143}$$

where $\psi(t)$ is a random quantum phase starting from 0 at $t = 0$. The continuous measurement process will now result in a diffusion of this quantum phase.

It is easy to check that the probabilities of getting a click during an elementary time interval τ are, in this case, the same in the two detectors, and independent of the random state of the field:

$$p_i = \frac{\kappa\tau}{2}\left\langle\Psi^{(A)}(t)\right|(a^\dagger - i\epsilon_i\xi\beta\mathbb{1})(a + i\epsilon_i\xi\beta\mathbb{1})\left|\Psi^{(A)}(t)\right\rangle = \frac{\kappa\bar{n}\tau}{2}(1 + \xi^2) \approx \frac{\kappa\bar{n}\tau\xi^2}{2}. \tag{7.144}$$

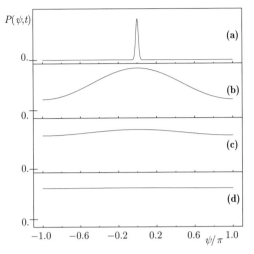

$P(\psi,t)$

(a)

(b)

(c)

(d)

−1.0   −0.6   −0.2   0.2   0.6   1.0 ψ/π

Fig. 7.31 Evolution of the $P(\psi,t)$ distribution illustrating the diffusion of the cat state quantum phase. The initial distribution is a narrow Gaussian centred at $\psi = 0$ (mimiking a δ-distribution). Frames (a)–(d) correspond to $t = 0,\ 0.15,\ 0.4$ and 0.75 in units of $1/(\kappa\bar{n})$.

Depending upon the detector which clicks ($\epsilon_i = +1$ or -1), the field state gets transformed according to:

$$
\begin{aligned}
\left|\Psi^{(A)}\right\rangle &\approx (i\epsilon_i\xi + 1)\left|\beta^{(A)}\right\rangle + (i\epsilon_i\xi - 1)e^{i\psi}\left|-\beta^{(A)}\right\rangle \\
&\approx \left|\beta^{(A)}\right\rangle + \left(1 + \frac{2i\epsilon_i}{\xi}\right)e^{i\psi}\left|-\beta^{(A)}\right\rangle ,
\end{aligned}
\tag{7.145}
$$

which amounts to an incremental change of the quantum phase by an angle:

$$
\delta\psi_i = 2\epsilon_i/\xi .
\tag{7.146}
$$

Detecting a photon in $\mathcal{D}_1$ or $\mathcal{D}_2$ produces a rotation in one way or the other of the quantum phase of the superposition, without changing the statistical weight of the two components. While the pseudo-qubit state was randomly evolving along a meridian of the Bloch sphere when the measurements were pinning down its components, it now rotates along the equator of the sphere.

The quantum diffusion corresponds now to the usual random walk problem, with a fixed elementary step and equal and time-invariant probabilities for steps in both directions. This situation has been analysed in detail in Milman *et al.* (2000). A derivation similar to that developed above leads to a standard Fokker–Planck equation for the distribution $P(\psi,t)$:

$$
\frac{1}{2\kappa\bar{n}}\frac{\partial P(\psi,t)}{\delta t} = \frac{\partial^2 P(\psi,t)}{\partial\psi^2} .
\tag{7.147}
$$

This equation is simpler than eqn. (7.142). It has no drift term (because the click probabilities are equal in both counters) and a ψ-independent diffusion coefficient (because the random walk steps are constant). The evolution of $P(\psi)$, starting from the initial condition $P(\psi,0) = \delta(\psi)$, is given by the usual one-dimensional random walk solution:

$$P(\psi, t) = \frac{e^{-\psi^2/8\kappa\bar{n}t}}{\sqrt{8\pi\kappa\bar{n}t}}\ . \tag{7.148}$$

A graphical representation of the quantum phase spreading is shown in Fig. 7.31. The final state, averaged over Monte Carlo trajectories, is reached after a time of the order of $T_D = 1/\kappa\bar{n}$. It is a sum of projectors on states $(\left|\beta^{(A)}\right\rangle + e^{i\psi}\left|-\beta^{(A)}\right\rangle)/\sqrt{2}$, with ψ equally distributed between 0 and 2π. This sum is equivalent to an incoherent mixture with equal weights of $\left|\beta^{(A)}\right\rangle$ and $\left|-\beta^{(A)}\right\rangle$. We retrieve of course the result given by the other scenarios.

Even though we loosely called it a quadrature measurement, it is important to note that the continuous data acquisition process described here is quite different from a von Neumann projective measurement of the quadrature operator X_ϕ of the field, as we had defined it in Chapter 3. Here, measurements are performed in the environment to which the field is entangled. It is a generalized measurement of the cavity field (in the meaning defined in Chapter 4), not a projective one. An ideal von Neumann measurement of the field quadrature would yield a random eigenvalue x for X_ϕ, corresponding to the collapse of the field in an $\left|x_\phi^{(A)}\right\rangle$ eigenstate of X_ϕ, a highly non-classical superposition of coherent states which is never reached in the Monte Carlo trajectories of the continuous simulations considered above.

7.5.3   A quantum mouse to monitor the cat's decoherence

Let us now leave the dreamland of thought experiments and describe how we have been able to monitor the decoherence of a photonic cat (Brune *et al.* 1996a). It is of course impossible to measure directly the field in the cavity, as well as to detect the microwave radiation escaping from it, an extremely weak photon flux scattered in a large ensemble of outside modes.

As in the other cavity QED experiments, we must rely on an atomic signal to get indirect information about the field. The principle of the experiment is to use a first atom, A_1, to prepare the field in a cat state, to let it evolve for a given time, then to probe this state with the help of a second atom, A_2. To go on with Schrödinger's metaphor, we can say that A_2 is a 'quantum mouse' sent in the cavity to probe the coherence of the cat state left by A_1.

The analysis of the experiment is straightforward if, using the dispersive method (see Section 7.3), we adjust the single-atom index so that $\Phi_0 = \pi/2$. The atoms crossing the Ramsey interferometer simply measure in this case the cat photon number parity $\mathcal{P}$. Suppose that A_1 is detected in $\left|g\right\rangle$, meaning that the field has collapsed in $\left|\Psi_{\mathrm{cat}}^+\right\rangle$. If no decoherence has had time to take place before A_2 crosses the apparatus at time t, the conditional probabilities $P_{g|g}$ and $P_{e|g}$ for the probe to be detected in the same state $\left|g\right\rangle$ or in the other state $\left|e\right\rangle$ are obviously equal to 1 and 0, a direct consequence of the measurement postulate applied to the $\mathcal{P}$ observable.

If, on the other hand, decoherence has fully taken place during the time interval t, the density operator of the field has become an incoherent mixture with equal weights of the two parity states $\left|\Psi_{\mathrm{cat}}^+\right\rangle$ and $\left|\Psi_{\mathrm{cat}}^-\right\rangle$ (we neglect here the slow amplitude damping of the field). The probabilities of finding A_p in $\left|g\right\rangle$ and $\left|e\right\rangle$ have hence become equal to $1/2$. This leads us naturally to measure the quantum coherence of the cat by the value of the atomic correlation signal η, difference of the two-atom conditional probabilities:

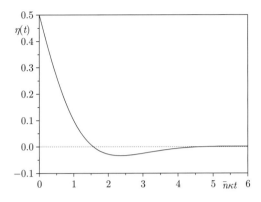

Fig. 7.32 Theoretical value of the correlation signal $\eta(t)$ for $2\Phi_0 = \pi/2$, as a function of $\overline{n}\kappa t$.

$$\eta = P_{g|g} - P_{e|g} \,, \tag{7.149}$$

which is equal to 1 in the absence of decoherence, to 0 if decoherence is complete. In order to find out how this correlation signal varies between these two values as decoherence proceeds, we remark that $\eta(t)$ is simply equal to the expectation value of $\mathcal{P}$ in the cavity field, at the time when the second atom measures it. We can thus express $\eta(t)$ as:

$$\eta(t) = \mathrm{Tr}[\rho_A(t)\mathcal{P}] \,, \tag{7.150}$$

where $\rho_A(t)$ is given by eqn. (7.114). Recalling the action of $\mathcal{P}$ on $|\pm\beta\rangle$ (eqn. 7.11) and assuming that $\langle\beta|-\beta\rangle$ is negligibly small, we get immediately:

$$\eta(t) = \exp\left[-2\overline{n}(1 - e^{-\kappa t})\right] \approx e^{-2\overline{n}\kappa t} \,. \tag{7.151}$$

The last expression in eqn. (7.151) is an approximation justified by the condition $t \ll 1/\kappa$, verified during the short time of the decoherence process. The atomic correlation signal $\eta(t)$, which decreases quasi-exponentially towards 0 with the time constant $1/2\overline{n}\kappa$, provides a direct signature of the coherence left in the cat state at the time the quantum mouse atom probes it.

The quantum mouse trick is not restricted to the study of π-phase cats decoherence. The two-atom correlation signal $\eta(t)$ is also a good indicator of the decoherence of a cat when the phase splitting $2\Phi_0$ between its components is different from π. We can no longer in this case invoke the photon number parity argument and we must return to the description of the interaction of A_2 with the field left in the cavity by A_1. Suppose that this latter atom has been detected in $|g\rangle$. Then, according to eqn. (7.54), the field has been prepared at $t = 0$ in the state:

$$\left|\Psi_{\Phi_0}^{+(A)}\right\rangle = \frac{1}{\sqrt{2}}\left[|\beta e^{-i\Phi_0}\rangle + |\beta e^{i\Phi_0}\rangle\right] \,, \tag{7.152}$$

where we assume $\beta \gg 1$. A time t later, when A_2 enters the apparatus, this field state has been partially entangled with its environment E, the $A + E$ system being described by the global state:

$$\left|\Psi^{(AE)}(t)\right\rangle = \frac{1}{\sqrt{2}}\left[|\beta e^{-i\Phi_0}\rangle \otimes |E^-(t)\rangle + |\beta e^{i\Phi_0}\rangle \otimes |E^+(t)\rangle\right] \,, \tag{7.153}$$

where $|E^-(t)\rangle$ and $|E^+(t)\rangle$ are the environment states defined by eqn. (7.119). The probe atom A_2 then interacts with the cavity field. As A_1, it crosses the field mode in a time short compared to the decoherence time. The environment remains unaffected during this process. Using the superposition principle, we can compute separately the effect of this interaction on the two parts of the cat state and add the resulting contributions. Since A_2 shifts again the phase of the field by $\pm\Phi_0$, the $|\beta e^{-i\Phi_0}\rangle$ state gives birth to two second-generation cat components $|\beta e^{-2i\Phi_0}\rangle$ and $|\beta\rangle$. Similarly, the $|\beta e^{i\Phi_0}\rangle$ state is split into the two states $|\beta e^{2i\Phi_0}\rangle$ and $|\beta\rangle$. From eqn. (7.153), we obtain the final state of the $A + E + A_2$ system:

$$
\begin{aligned}
\left|\Psi^{(AEA_2)}(t)\right\rangle \;=\; & \frac{1}{2\sqrt{2}}\Big[e^{-i\Phi_0}\,|e_2\rangle \otimes \big(|\beta e^{-2i\Phi_0}\rangle - |\beta\rangle\big) \\
& + |g_2\rangle \otimes \big(|\beta e^{-2i\Phi_0}\rangle + |\beta\rangle\big)\Big] \otimes \left|E^-(t)\right\rangle \\
& + \frac{1}{2\sqrt{2}}\Big[e^{-i\Phi_0}\,|e_2\rangle \otimes \big(|\beta\rangle - |\beta e^{2i\Phi_0}\rangle\big) \\
& + |g_2\rangle \otimes \big(|\beta\rangle + |\beta e^{2i\Phi_0}\rangle\big)\Big] \otimes \left|E^+(t)\right\rangle \;,
\end{aligned}
\tag{7.154}
$$

where $|e_2\rangle$ and $|g_2\rangle$ are the states of A_2. The first two and last two lines in the right-hand side describe the 'offsprings' of the $|\beta e^{-i\Phi_0}\rangle$ and $|\beta e^{+i\Phi_0}\rangle$ components of the cat left by A_1 in the cavity. Each one is tagged by a different environment state, carrying information about the phase of this field. From this expression, we get the probabilities $P_{g|g}$ and $P_{e|g}$ of detecting the second atom in $|g\rangle$ or $|e\rangle$ (after having found A_1 in $|g\rangle$):

$$
P_{g|g} = \tfrac{1}{2} + \tfrac{1}{4}\,\mathrm{Re}\,\langle E^-(t)\,|\,E^+(t)\rangle \;; \qquad P_{e|g} = \tfrac{1}{2} - \tfrac{1}{4}\,\mathrm{Re}\,\langle E^-(t)\,|\,E^+(t)\rangle \;.
\tag{7.155}
$$

Replacing finally $\langle E^-(t)\,|\,E^+(t)\rangle$ by its expression (7.120), we obtain the two-atom correlation, in the case $2\Phi_0 \neq \pi$:

$$
\begin{aligned}
\eta(t) \;=\; & P_{g|g} - P_{e|g} = \tfrac{1}{2}\,\mathrm{Re}\,\exp\left[-\bar{n}(1 - e^{-\kappa t})(1 - e^{2i\Phi_0})\right] \\
\;=\; & \tfrac{1}{2}\,e^{-2\bar{n}(1 - e^{-\kappa t})\sin^2\Phi_0}\cos\left[\bar{n}(1 - e^{-\kappa t})\sin(2\Phi_0)\right] \;.
\end{aligned}
\tag{7.156}
$$

For $\kappa t \ll 1$ this expression can be approximated by:

$$
\eta(t) \approx \tfrac{1}{2}\,e^{-2\bar{n}\kappa t \sin^2\Phi_0}\cos[\bar{n}\kappa t \sin(2\Phi_0)] \;.
\tag{7.157}
$$

The two-atom correlation signal varies now from $1/2$ to 0 as decoherence proceeds. Its evolution is described by the product of a decaying exponential by a cosine function of time. The decay rate of the exponential, $1/T_D = 2\kappa\bar{n}\sin^2\Phi_0$, is equal to the cavity damping rate κ multiplied by half the square of the distance in phase space $D^2/2 = 2\bar{n}\sin^2\Phi_0$ of the two cat components. The cosine factor is very close to 1 at the beginning of the evolution. When $\eta(t)$ is already strongly reduced by decoherence, this factor can change sign, adding a modulation in the tail of the $\eta(t)$ function. For $2\Phi_0 \approx \pi/2$ (cat state with orthogonal components in phase space) the inversion of the sign of η occurs for $t = \pi/2\kappa\bar{n}$. Figure 7.32 represents the theoretical variation of $\eta(t)$ in this case.

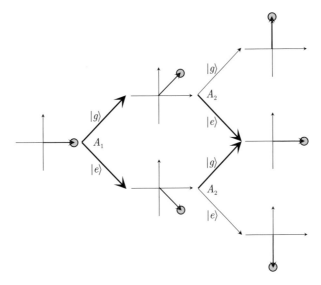

Fig. 7.33 The interfering pathes contributing to the two-atom correlation η (for $\Phi_0 = \pi/4$) are highlighted by the thick arrow lines. In the upper path, the first atom crosses C in $|g\rangle$ and the second in $|e\rangle$. In the lower path, these state-assignments are reversed.

The correlation signal $\eta(t)$ can be understood as a two-atom quantum interference effect. The conditional probabilities $P_{g|g}$ and $P_{e|g}$ are the sum of a $1/2$ background plus a contribution corresponding to two atom–field paths in which the field ends up in the final state $|\beta\rangle$.

These two paths are pictured in Fig. 7.33. In the first, A_1 crosses the cavity in $|g\rangle$, producing the $|\beta e^{i\Phi_0}\rangle$ state in the cavity, and the quantum mouse atom crosses the cavity in $|e\rangle$, bringing back the phase of the field to its initial value. In the second path, the roles of the two atoms are exchanged, the field being phase-shifted first into the $|\beta e^{-i\Phi_0}\rangle$ state by A_1 crossing C in $|e\rangle$, before being transformed back into $|\beta\rangle$ by A_2, crossing the cavity in $|g\rangle$. Both paths contribute to the probabilities associated to the detection of A_2 in $|e\rangle$ and $|g\rangle$ because the second Ramsey zone R_2 mixes these two states before detection, erasing information about the state of the atom when it was inside the cavity. The amplitudes associated to these paths add in $P_{g|g}$ and subtract in $P_{e|g}$. The interference term is thus doubled in the difference $P_{g|g} - P_{e|g}$, while the constant background cancels.

Note that the maximum two-atom correlation signal is larger (equal to 1) when $\Phi_0 = \pi/2$, because then the two states $|\beta e^{-2i\Phi_0}\rangle$ and $|\beta e^{+2i\Phi_0}\rangle$ coincide. This gives rise to another interference effect in the $P_{g|g}$ and $P_{e|g}$ probabilities, which doubles the amplitude of the maximum $\eta(t)$ signal.

The two-atom interference effect exists as long as one cannot distinguish between the two paths leading to the final state $|\beta\rangle$ of the field by observing the environment, i.e. as long as the final environmental states $|E^-(t)\rangle$ and $|E^+(t)\rangle$ are overlapping (again a complementarity effect!). The two-atom interference signal thus follows the decay of the environment states scalar product, which reflects precisely the evolution of the coherence term in the expression of the density operator of the field left by A_1 in the

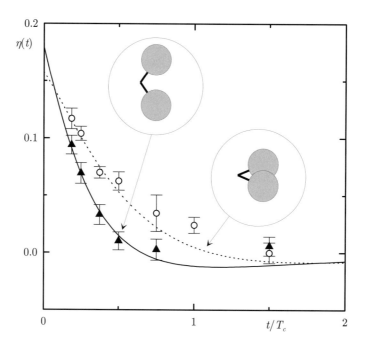

Fig. 7.34 Two-atom correlation signal η as a function of t/T_c for an atom–cavity detuning $\Delta_c/2\pi = 170$ kHz (open circles) and $\Delta_c/2\pi = 70$ kHz (solid triangles). The average photon number is $\bar{n} = 3.3$. The points are experimental, with statistical error bars. The curves are a theoretical fit. The cat phase components are pictorially depicted in the insets. Reprinted with permission from Brune *et al.* (1996a). © American Physical Society.

cavity.

Sending pairs of atoms across our CQED Ramsey interferometer in which a coherent field was beforehand injected, we have observed the evolution of η (Brune *et al.* 1996a). The cavity damping time in this early experiment was only 160 μs, making the decoherence process very fast and limiting the experiment to relatively small values of $\bar{n}$. The cat state prepared by the first atom contained $\bar{n} = 3.3$ or 5.1 photon on average (admittedly only a kitten!). The phase splitting $2\Phi_0$ was set at two values (100 and 50 degrees) by choosing the detuning Δ_c (70 kHz and 170 kHz respectively).

The separation between the two atoms was varied between 30 μs and 250 μs. The experiment consisted in repeating experimental sequences in which two atoms separated by a fixed delay were sent across the Ramsey interferometer. The final state of the atoms were detected and the η signal reconstructed from a large number of realizations. We then resumed the whole procedure for different delays t.

Figure 7.34 shows the $\eta(t)$ correlation for cat states with $\bar{n} = 3.3$ corresponding to the two splitting angles. The points are experimental and the lines theoretical. Note that the theory is not restricted to the simple perturbative analysis developed above for the sake of simplicity. It includes higher order terms in Ω_0/Δ_c correcting the expression of the cat states at $\Delta_c = 70$ kHz. We have also incorporated in the analysis the finite single-atom Ramsey fringe contrast, which explains why the maximum correlation is

0.18 instead of the theoretically expected value 0.5.

We clearly observe that the correlation signals decrease with the delay t, revealing directly the dynamics of decoherence. The agreement with the theoretical model is excellent. Most strikingly, we observe that decoherence proceeds at a faster rate when the distance between the two state components of the cat is increased. An effective decoherence time $\approx 0.24/\kappa$, much shorter than the photon decay time, is found for the larger cat state ($\Delta_c = 70$ kHz). A similar agreement with theory is obtained when comparing, for the same $\Delta_c = 70$ kHz, the correlation signals corresponding to different $\bar{n}$ values (5.1 and 3.3).

The CQED cat experiments are, as discussed above, models for an ideal quantum measurement. The coherent field can be viewed as a meter measuring the state of the first atom crossing the cavity and evolving under the effect of the atom–field interaction into a mesoscopic superposition. This decoherence experiment illustrates quite vividly the fast evolution in a measurement process of the atom–meter state towards a statistical mixture and the increasing difficulty to maintain quantum coherence when the distance between the components of the mesoscopic superposition is increased.

7.5.4 How big a cat?

Since this experiment, our photon boxes have been considerably improved. The longest cavity damping time obtained in 2006 is $T_c = 115$ ms (Section 5.2.2), a three-order of magnitude increase with respect to the 1996 experiment, opening the way to much fatter photonic cats. The limit to the cat size can be evaluated by comparing the time required to prepare or to probe it to its decoherence time. Assuming that the phase splitting is optimal (which implies resonant or quasi-resonant coupling, see Section 7.4), the effective interaction time t_i^r needs to be such that $\Omega_0 t_i^r / 2\sqrt{\bar{n}} \approx \pi$, i.e. $t_i^r \approx (2\pi/\Omega_0)\sqrt{\bar{n}}$. This time should be shorter than the decoherence time, $1/2\kappa\bar{n} = T_c/2\bar{n}$, of an $\bar{n}$-photon π-phase cat which implies that $\bar{n}$ must satisfy the condition:

$$\bar{n} < \left(\frac{\Omega_0 T_c}{4\pi}\right)^{2/3} = \left(\frac{\Omega_0 Q}{4\pi\omega}\right)^{2/3}. \tag{7.158}$$

Having succeeded in increasing T_c and Q by three orders of magnitude corresponds to a ≈ 100-fold improvement over the limit ($\bar{n} \approx 5$) of the early cat experiments. We can thus hope to prepare and study now cat states with $\bar{n}$ as large as 500 and we are planning several experiments to complete with these larger cats our preliminary exploration of decoherence.

We could first study the decay of the expectation value of $\mathcal{P}$ in these states. This would be extracting only very partial information from the field. Exploiting optimally the properties of $\mathcal{P}$, we could also measure its expectation value in a cat state translated by an amount α in phase space, according to the method described in Chapter 6. We would obtain in this way the complete Wigner function $W(\alpha)$ of this state, revealing the coherent whiskers around $|\alpha| = 0$, signature of the cat's coherence. This would be again a two-atom correlation experiment but it would require us, just before sending atom A_2, to inject a second reference coherent field in the cavity with a variable amplitude $-\alpha$. By repeating this experiment with increasing delays, we would reconstruct the evolution of W pictured in Fig. 7.25 and witness directly the progressive washing out of the cat interference features.

It will even become possible to go beyond two-atom correlation experiments. The long T_c time now available makes it possible to send several atoms across the cavity in a time short compared to the decoherence time. For instance, a cat with $\bar{n} = 20$ photons on average would undergo decoherence in ≈ 2.5 ms, while it will take $t_i^r = 2(\pi/\Omega_0)\sqrt{\bar{n}} \approx 90$ μs only for single atoms to prepare or probe it. About 25 atoms could be sent successively in the cavity within the decoherence characteristic time, making it possible to observe the continuous evolution of the cat state as its parity jumps when successive photons get lost (see section 7.5.2).

What is the limit to the cat size we can realize? The Q factor is still presently limited by slight mirror imperfections. With ideally smooth mirrors, and low enough temperature we will eventually reach the diffraction limit, when the residual losses occur by photon scattering at the edge of the mirrors. With our current Fabry–Perot geometry, this would correspond to $Q \approx 4 \cdot 10^{11}$ or $T_c = 1.4$ s, again a factor of 10 improvement over the present best value. We can thus expect another 4-fold increase in the cat size, which would now contain up to 2000 photons. This seems to be a limit difficult to break, but it leaves a lot of room for fascinating decoherence experiments, some of which will be considered in the next section.

7.5.5   Keeping the cat's coherence alive with quantum feedback

Decoherence is the big enemy of quantum information science (Haroche and Raimond 1996). Keeping the coherence of a large quantum system alive as long as possible is of primordial importance to be able to achieve some of the tasks we have briefly described in Chapter 2. The first approach to the decoherence problem is to try to isolate the system as well as possible. In the case of our photonic cats, we just have seen how far we can push in this direction. To go beyond requires methods more sophisticated than the passive protection of the system.

One of these methods is quantum feedback. It consists in observing the system or its environment in order to find out when a quantum jump occurs, and then to correct it by reacting back on the system. It is usually preferable to get the quantum jump signature from the system itself and not from the environment, because the former is supposed to be much better known and controlled.

The major conceptual difficulty of the feedback method is that, in general, the jump operators are non-unitary and describe irreversible processes. The correction, on the other hand, is performed by a unitary manipulation. To undo a non-unitary transformation by a unitary one is in general impossible. Tricks can be played, however, by coding the useful information in subspaces of Hilbert space in which the restriction of the jump operation is unitary, even if the full operator acting on the whole Hilbert space is not. The error correction protocols recently developed in quantum information are the most sophisticated forms of quantum feedback methods applied to ensemble of qubits (Nielsen and Chuang 2000).

The study of these methods is far beyond the scope of this book. We will conclude this section by presenting an especially simple example of quantum feedback applied to cat state situations in CQED. It has been proposed to maintain a π-phase cat coherent over a time much longer than its natural decoherence time (Fortunato *et al.* 1999; Zippilli *et al.* 2003). This experiment has not been performed yet, but it is within the reach of future cavity QED experiments.

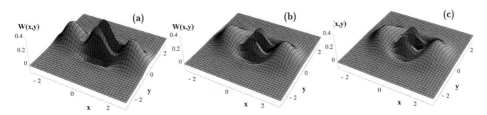

Fig. 7.35 Protection of a Schrödinger cat by quantum feedback. Wigner function of the cavity field as a function of time. Frames (a)–(c) correspond to $t = 0$, $t = 1/\kappa$ and $t = 2/\kappa$ (0, 13 and 25 feedback cycles). The initial state is $\left|\Psi_{\text{cat}}^{+}\right\rangle$ with $\bar{n} = |\beta|^{2} = 3.3$. Reprinted with permission from Fortunato *et al.* (1999). © American Physical Society.

Its principle is simple. It involves the Ramsey set-up we have just described and two different set of atoms A_i^p and A_j^c sent one by one across the cavity. The A_i^p atoms are non-resonant probes, sent in C at regular intervals, short compared to the decoherence time. They are used to check the cat state parity according to the method described above.

As soon as a parity jump is observed by detecting a change of quantum state in one of the A^p atoms, an A^c atom, prepared in the upper state $|e\rangle$, is sent across C. The Ramsey fields are switched off for a brief time interval and the A^c atom is made resonant with the field by the Stark effect. Its interaction time is set so that it undergoes, on average, a π-Rabi pulse in the field of the cavity ($\Omega_0 \sqrt{\bar{n}} t_i^r = \pi$). With a high probability, this atom adds a quantum of energy in the cavity field. It also restores, with a good approximation, the parity of the photon number to the value it had before the jump. This can be checked by expanding the field state just before the injection of the A^c atom over a Fock basis. Assuming, for example, that the cat has just jumped from an odd to an even parity state, the field+A^c atom system will evolve as:

$$
|e\rangle \otimes \left|\Psi_{\text{cat}}^{+}\right\rangle \;\rightarrow\; \sum_n c_n [1 + (-1)^n] \left[\cos\left(\Omega_0 \sqrt{n+1}\, t_i^r /2\right) |e, n\rangle \right.
$$
$$
\left. + \sin\left(\Omega_0 \sqrt{n+1}\, t_i^r /2\right) |g, n+1\rangle \right] . \tag{7.159}
$$

We then make the approximations consisting in replacing the arguments of the cosine and sine by their average value $\Omega_0 \sqrt{\bar{n}} t_i^r = \pi$ and neglecting the difference between the modulus of c_n and c_{n-1}:

$$
|e\rangle \left|\Psi_{\text{cat}}^{+}\right\rangle \rightarrow |g\rangle \otimes \sum_n c_n [1 + (-1)^n] |n+1\rangle \approx |g\rangle \otimes \sum_p c_p [1 - (-1)^p] |p\rangle . \tag{7.160}
$$

Within these approximations, the feedback atom has corrected the cat parity, bringing it back to its initial -1 value. Obviously, the approximation neglecting the variation of the Rabi phase over the width of the coherent state distribution is a bit naive.[14]

---

[14]That eqn. (7.160) is only an approximation can be guessed from the discussion of Section 7.5.2. We remarked then that a parity jump in a cat state does not change the average photon number in the field. *A contrario*, a transformation such as the π-Rabi pulse atomic feedback, which does change the average photon number by one unit, cannot be equivalent to an exact parity jump of the field.

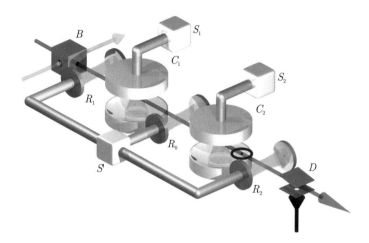

Fig. 7.36 The two-cavity set up.

We have seen in Section 7.4 that we must in general be more careful in dealing with these phases. Much better approximations are obtained by expanding them to first or second order in $n - \bar{n}$. We were then interested in the evolution of the system over long times, whereas we are dealing here with a very short interaction lasting only half a Rabi period. The naive approximation is then not too bad.

If we repeat the process to track the evolution of the field system as it keeps being fed-back by the atoms correcting the effects of the successive jumps, the errors produced by these approximations cumulate and can no longer be overlooked. An exact simulation of the system's evolution can be performed (Fortunato *et al.* 1999), which shows that the cat state gets slowly and irreversibly modified. Figure 7.35 shows the results of this simulation on the Wigner function of an even cat state, being watched and fed-back upon over a very long time. The artificial prolongation of the cat coherence is quite spectacular. At long times though, the cat components undergo a diffusion in phase space and the coherence is eventually lost. This diffusion reflects the fact that the irreversible effects of the jumps cannot be exactly corrected by the unitary Rabi rotation.

7.6 Non-local cats

The control of a mesoscopic field by a single atom can be extended to even stranger situations. The fields of spatially separated cavities can be entangled together via their interaction with the atom, leading to situations which would add the flavour of non-locality to the strangeness of cat states. These experiments are, at the time of the writing of this book, in planning stage in our laboratory (Milman *et al.* 2005). We give in this section a brief description of what we hope to achieve.

7.6.1 Two weirdnesses combined: toward mesoscopic non-locality

The principle of the experiment consists in combining together two (or more) set-ups of the kind described above, in order to knit entanglement between mesoscopic fields stored in distinct boxes. Restricting ourselves to the case of two cavities, we have in mind the set-up sketched in Fig. 7.36. An atom crosses successively two identical cavities C_1 and C_2, connected by waveguides to two classical sources S_1 and S_2. The separation between the two cavity centres is typically 10 cm and it takes a time of the order of $10t_i^r \sim 1$ ms for the atom to fly from C_1 to C_2. The atom can be submitted to microwave pulses R_1 and R_2 before and after its interaction with C_1 and C_2. An additional pulse between the two cavities could be optionally applied in R_0. Finally, the atom is counted downstream by the state-selective ionization detector D.

Using the state-dependent dispersive atom–field index effect, this set-up will allow us to entangle the atom with mesoscopic fields stored in the two cavities. After atomic detection, the field will be projected into an entangled state between the two cavities, which could contain up to several tens of photons. A large variety of states could be prepared, depending upon the sequence of operations applied to the atom (via the R_i pulses) and to the fields [using the sources S_1 and S_2 to produce phase space translations $D_1(\alpha)$ and $D_2(\alpha)$ of the fields in the two cavities]. The cat states could be prepared either by using an atom as a quantum phase-shifter for coherent fields injected beforehand in the cavities, or by implementing quantum switch methods, employing the atom as a tap in a superposition of its on and off positions (Davidovich *et al.* 1993). We restrict the present discussion to the phase-shifting methods.

We consider first state superpositions involving coherent fields with opposite complex amplitudes:

$$|\phi_{\rm cat}^{\pm}\rangle = (|\gamma,\gamma\rangle \pm |-\gamma,-\gamma\rangle)/\sqrt{2} \; ; \quad |\psi_{\rm cat}^{\pm}\rangle = (|\gamma,-\gamma\rangle \pm |-\gamma,\gamma\rangle)/\sqrt{2} \; , \qquad (7.161)$$

where the first and the second symbol in each ket is the field amplitude in C_1 and C_2 respectively. We have assumed a negligible overlap between $|\gamma\rangle$ and $|-\gamma\rangle$ ($|\gamma| \gg 1$), which simplifies the normalization in these equations. Making the notation change $|\gamma\rangle \rightarrow |0_f\rangle, |-\gamma\rangle \rightarrow |1_f\rangle$ which amounts to assimilating the coherent state in the cavity to a pseudo-qubit, we note that these entangled field states are analogous to the Bell states of two qubits defined by eqn. (2.54).

To prepare these states we need an atom which produces, depending upon its state, a $\pm\pi/2$ phase shift on a coherent field. We have seen above that this is achievable with atomic cavity crossing times of the order of 100 µs. To prepare $|\phi_{\rm cat}^{\pm}\rangle$, we can start from a field prepared in $|i\gamma, i\gamma\rangle$ and send across the set-up an atom prepared by R_1 in $(|e\rangle + |g\rangle)/\sqrt{2}$ (no pulse is applied in this case in R_0).

The part of the atomic wave function in $|e\rangle$ transforms $|i\gamma, i\gamma\rangle$ into $|\gamma, \gamma\rangle$, while the part in $|g\rangle$ turns the two-cavity state into $|-\gamma, -\gamma\rangle$. After a final mixing of the atomic levels in R_2, the atomic detection will finally project the field into one of the two $|\phi_{\rm cat}^{\pm}\rangle$ states, depending upon the random outcome of the measurement.

Generating $|\psi_{\rm cat}^{\pm}\rangle$ involves the same kind of procedure, with an added π-pulse applied to the atom in R_0. This pulse switches the atomic states when the atom flies from one cavity to the other, giving opposite phases to the cavity fields in the two parts of the final superposition.

The coherence of these non-local cats could be probed by a second atom crossing the cavities after a variable delay, following a procedure similar to the single-cavity decoherence experiment described above. Decoherence due to photon loss in the environment would again result in a reduction of the two-atom correlation signal $\eta(t)$ as the delay t between the two atoms increases (Davidovich *et al.* 1996).

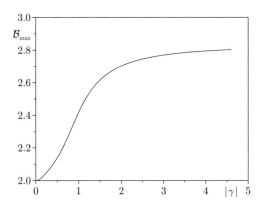

Fig. 7.37 The maximum Bell sum, $\mathcal{B}_{\mathrm{max}}$, as a function of modulus of the amplitude γ of the non-local cat state. Reprinted with permission from Milman *et al.* (2005).

Two-atom correlation experiments could be used to study the effect of decoherence on non-locality. As for two maximally entangled spins, it has been shown (Banaszek and Wodkiewicz 1999) that there is a 'Bell sum', $\mathcal{B}$, combination of correlations between observables pertaining to the two entangled fields, whose expectation is larger than the maximum value authorized by any local hidden variable theory. In the case of qubits, the correlations are products of spin operator components along well-chosen directions (eqn. 2.58). For our pseudo-qubit cat states, the correlation observables are two-mode Wigner functions $W(\alpha, \beta)$, taken for four couples of points in phase space (α, β), (α, β'), (α', β) and (α', β').

Generalizing the single-mode case, $W(\alpha, \beta)$ is equal to the expectation values of the global photon number parity operator $\mathcal{P}_{(1+2)} = \exp[i\pi(a_1^\dagger a_1 + a_2^\dagger a_2)]$ in the state obtained by translating the field by $-\alpha$ in C_1 and by $-\beta$ in C_2:

$$W(\alpha, \beta) = \frac{4}{\pi^2}\langle D_1(\alpha)D_2(\beta)\mathcal{P}_{(1+2)}D_1^{-1}(\alpha)D_2^{-1}(\beta)\rangle \; . \tag{7.162}$$

The Bell sum is obtained by combining the values of this function at four points in phase space:

$$\mathcal{B} = \frac{\pi^2}{4}\left|W(\alpha, \beta) - W(\alpha, \beta') - W(\alpha', \beta) - W(\alpha', \beta')\right| \; . \tag{7.163}$$

This expression is obviously very similar to the CHSH form of the two-spin Bell sum (eqn. 2.58). According to the argument developed in Chapter 2 about two spins, a believer of local realism would consider the $+1$ or -1 value of the photon number parity in each cavity as a random variable existing in the system before it is actually measured. He would then infer, following a line of reasoning very similar to the spin discussion, that the combination of expectation values making up $\mathcal{B}$ should always be smaller than 2. Quantum mechanics, in contradiction with this point of view, predicts $\mathcal{B} > 2$ for adequately chosen α, α', β, β' values in phase space (Banaszek and Wodkiewicz 1999).

A proposal to measure the Bell sum in the two-cavity experiment is analysed in Milman *et al.* (2005). The optimization of the (α, β) values to obtain the maximum violation of the Bell's inequality is discussed, as well as the dependence of $\mathcal{B}$ upon the delay t between the two atoms. In the absence of decoherence, the maximum $\mathcal{B}_{\mathrm{max}}$ of

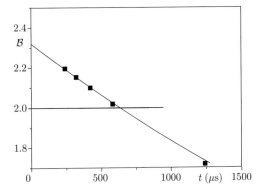

Fig. 7.38 Decoherence of the Bell signal: $\mathcal{B}$ as a function of time delay t. The cat amplitude is $\gamma = \sqrt{2}$. The cavity damping time is $T_c = 30$ ms. Reprinted with permission from Milman et al. (2005).

the Bell sum is shown as a function of the cat amplitude $|\gamma|$ in Fig. 7.37. It reaches the value $2\sqrt{2}$ for large cats ($|\gamma| \gg 1$). When the cat size decreases, $\mathcal{B}_{\max}$ diminishes and tends towards the classical limit, 2, when $\beta \to 0$. This is not unexpected, since the cavities are then empty, obviously a classical situation.

In a real experiment, $\mathcal{B}$ is expected to fall very quickly below the classical limit, revealing how decoherence washes out the non-classical features of the cat Wigner function and very efficiently destroys the mesoscopic non-locality. We have to choose a compromise between too big cats, which would decohere too fast, and too small ones, for which $\mathcal{B}_{\max}$ would be too close to 2. With the parameters of our CQED set-up, assuming a $T_c = 30$ ms cavity damping time, a Bell inequality violation should be observable at small delays for 'kitten' states made of a few photons.

Figure 7.38 shows how $\mathcal{B}$ would decay as a function of t when $\overline{n} = |\gamma|^2 = 2$. The points representing $\mathcal{B}(t)$ at various delays t are given by a simulation taking into account a realistic experimental situation. The Bell sum starts from an initial value $\mathcal{B} = 2.32$ and falls below the classical value 2 for $t = 500\mu s$. The experiment could be performed with larger cat states ($\overline{n} \sim 10$) with the cavities now available having a T_c of the order of a hundred of milliseconds.

The measurement of the two-mode Wigner function is straightforward in our CQED set-up, which is well-adapted to the detection of the photon number parity. Without entering in any detail, let us describe briefly how the experiment proceeds. First, we use the sources S_1 and S_2 and an atom A_1 to prepare one of the four pseudo-Bell states (7.161) in C_1 and C_2. Then, after a delay t, we translate the fields by $-\alpha$ in C_1 ($-\beta$ in C_2), again using S_1 and S_2. A probe atom, A_2, is then injected in the set-up[15] turned into a Ramsey interferometer to measure the global photon number parity $\mathcal{P}_{(1+2)}$ (the R_0 Ramsey zone is then de-activated). The sequence of events is repeated many times to obtain the expectation value of $\mathcal{P}_{(1+2)}$ yielding the $W(\alpha, \beta)$ value. The same procedure is reproduced for the three other couples of displacements in phase space and $\mathcal{B}$ is reconstructed by combination of all the results.

This experiment, technically very challenging, would be an extension of Bell's measurement towards the mesoscopic world, an attempt to test the limits of quantum

---

[15]In fact, decoherence is so fast that A_2 should be sent in C_1 immediately after A_1 has left it i.e. before the Bell state has been fully generated. According to quantum theory, the Bell sum, however, does not depend upon the timing of the detection events and is thus the same as if we waited for the detection of A_1 before sending A_2.

Fig. 7.39 A classical beam-splitter filling two cavities.

weirdness in a new direction. Since the two cavities are probed by the same atom, the experiment could not rule out hidden variables models and would thus not be a direct test of quantum non-locality. If we take however quantum theory for granted, this experiment would study the dynamical effect of decoherence on non-locality. The system considered here would be not only macroscopic since its two parts would be separated by a large distance at the atomic scale, but also would become 'big' since each of its parts would contain several quanta. Studying how a size-dependent decoherence process washes out the system's non-local features would be an interesting contribution to the exploration of the quantum–classical boundary.

7.6.2 A mesoscopic quantum beam-splitter

Once one of the pseudo-Bell cats has been produced, other non-classical states can be generated. The action of $D_1(\alpha)D_2(\alpha)$ on $\left|\psi_{\text{cat}}^{\pm}\right\rangle$, for instance, produces state superpositions in which the fields have different amplitudes in the two cavities. In the particular case $\alpha = \gamma$, we obtain:

$$\left|\chi_{\text{cat}}^{\pm}\right\rangle = D_1(\gamma)D_2(\gamma)\left|\psi_{\text{cat}}^{\pm}\right\rangle = (|2\gamma, 0\rangle \pm |0, 2\gamma\rangle)/\sqrt{2} , \qquad (7.164)$$

representing a field with many photons coherently suspended between two cavities, the mesoscopic analogue of a particle crossing 'at the same time' the two slits of a Young interferometer, or propagating simultaneously along the two output modes of a beam-splitter.

It is important to remark that this strange state could not be produced by usual quantum optics techniques with ordinary beam splitters acting on coherent fields. We have represented such a possible experiment in Fig. 7.39. A coherent field wave packet of amplitude α is impinging in one of the input modes of a symmetrical beam-splitter [$\theta = \pi/4$ with the notation of eqn. (3.99)], the other input mode being in vacuum. We couple the output modes to the two cavities. This set-up would realize the following transformation (see Chapter 3):

$$|\alpha, 0\rangle \rightarrow \left|\alpha/\sqrt{2}, i\alpha/\sqrt{2}\right\rangle . \qquad (7.165)$$

We would obtain approximately $\bar{n}/2 = |\alpha|^2/2$ photons in each cavity (with $\sqrt{\bar{n}}$ fluctuations around this average). This is a situation quite different from the bimodal

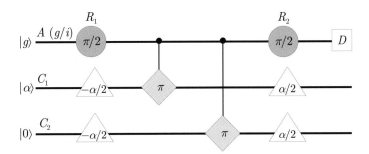

Fig. 7.40 The quantum circuit equivalent to the mesoscopic field beam-splitter.

photon distribution in the state described by eqn. (7.164), with a Dirac peak at $n = 0$ and a second peak around $\bar{n} = 4|\gamma|^2$. In addition, there are strong correlations between the fields in the two cavities. If one is empty, the other contains plenty photons and vice versa.

To understand clearly the analogy and differences with the usual beam-splitter situation, it is instructive to follow the sequence of operations leading in a CQED experiment from an initial state $|\alpha, 0\rangle$ to a final superposition with equal weights of $|\alpha, 0\rangle$ and $|0, \alpha\rangle$. Adopting the language of quantum circuits introduced in Section 5.4, we will analyse these operations by simple diagrams. While using this representation borrowed from quantum information, we should once again keep in mind that the fields in the cavity are not real qubits here, but harmonic oscillator modes whose phase is not a bivalued quantity, but a continuous variable controlled by the atomic qubit.

In usual gates, only one of the control states produces a change in the target. In the system we have considered so far, the two states $|e\rangle$ and $|g\rangle$ affect the phase of the field. This complication could be avoided by considering a slightly different experimental situation, in which only one of the two atomic levels induces a phase shift on the cavity fields. For this, we come back to the three-level configuration already considered in Chapter 6 when we discussed the photon counting QND experiment.

We still assume that the cavities are nearly resonant with the $|g\rangle \rightarrow |e\rangle$ transition (with a small detuning Δ_c). The Ramsey fields are, however, no longer resonant with this transition. They mix $|g\rangle$, the initially prepared atomic state, with a third state $|i\rangle$ (the circular state with principal quantum number 49, while the cavity is nearly resonant with the transition between the circular states 50 and 51).

If an atom crosses the cavities in $|i\rangle$, it has practically no action on the field phase because the transitions originating from this level are far-detuned from the cavity. If it is in $|g\rangle$ (coupled to $|e\rangle$ by a transition nearly resonant with the cavities), it induces on the other hand a phase shift Φ_0 on both fields. We now want this angle to be equal to π. The atom must thus cross each cavity in a time twice longer than was required in the earlier version of the experiment to produce a $\pm\pi/2$-phase shift. This puts stronger constraints to the experiment, but here we are discussing only principles. In this variant of the cat state preparation, $|e\rangle$ is eliminated from the game and $|i\rangle$ and $|g\rangle$ play the roles of the $|0\rangle$ and $|1\rangle$ control qubit states respectively.

The system, initialized in $|g; \alpha, 0\rangle$, is submitted to the sequence of operations schematized by the quantum circuit shown in Fig. 7.40. The horizontal lines represent from top to bottom the atom and the C_1 and C_2 fields. The single-qubit operation R_1 is, in quantum information language, a qubit $\pi/2$-rotation, similar to a Hadamard gate, transforming $|g\rangle$ into $(|g\rangle + |i\rangle)/\sqrt{2}$ while R_2 is also a $\pi/2$-rotation, with a different phase, defined by:

$$|g\rangle \to (|g\rangle + i\,|i\rangle)/\sqrt{2} \;; \qquad |i\rangle \to (|i\rangle + i\,|g\rangle)/\sqrt{2} \;. \tag{7.166}$$

The two-qubit gates, symbolized by the vertical lines joining the control atom and the target field, switch, conditionally to the state of the atom, the signs of the field amplitudes in both cavities. They are analogous to control-not gates acting on the field pseudo-qubits. These gates are sandwiched between two displacement operators $D(-\alpha/2)$ and $D(+\alpha/2)$ acting on each cavity field. Finally, the atomic state is read out. It is a simple exercise to show that the field has undergone the transformation:

$$|\alpha, 0\rangle \to (|\alpha, 0\rangle + i\,|0, \alpha\rangle)/\sqrt{2} \;, \tag{7.167}$$

if the atom is detected in $|i\rangle$ (the relative signs of the two quantum amplitudes being reversed for an atom detected in $|g\rangle$). Note that the factor i represents a quantum phase of no fundamental significance. It could be changed by slightly modifying the circuit. One could add a single-qubit gate on the atomic line, which would phase-shift by different amounts the qubit states $|g\rangle$ and $|i\rangle$. A proper choice of phase would result in a superposition of $|\alpha, 0\rangle$ and $|0, \alpha\rangle$ with real coefficients, as in eqn. (7.164).

Comparing with eqn. (7.165), we realize again that this quantum transformation is very different from an usual beam-splitter operation. Whereas a classical beam-splitter acting on coherent fields leaves them unentangled, this quantum circuit generates a maximum entanglement between the two cavity modes. We will call the sequence of operations described by this circuit a 'quantum beam-splitter'. This device has two possible final states, depending upon the random outcome of the atom measurement. We choose, by convention, to retain only the operations leading to the final atomic state $|i\rangle$. If we exchange the roles of the two input ports, we find symmetrically, again for a final atomic detection in state $|i\rangle$:

$$|0, \alpha\rangle \to (|0, \alpha\rangle + i\,|\alpha, 0\rangle)/\sqrt{2} \;. \tag{7.168}$$

The final states corresponding to the input states $|\alpha, 0\rangle$ and $|0, \alpha\rangle$ are orthogonal for $\alpha \gg 1$,[16] as can be checked by comparing eqns. (7.167) and (7.168). We can thus distinguish, by looking at the final field state, which input channel has been initially fed. This is a property of a classical beam-splitter and it is reasonable to have it satisfied by this quantum version of the device.

7.6.3   An interferometer for mesoscopic fields with ultra-narrow fringes

The coherent separation of a single-particle trajectory in two paths is demonstrated by a combination of two ordinary beam-splitters in a Mach–Zehnder interferometer, a device which we have analysed in detail in Chapter 3. What would be its mesoscopic

---

[16]This is not an obvious feature since the quantum circuit involves a non-unitary measurement. The phase of the second Ramsey gate (eqn. 7.166) has been chosen to ensure the orthogonality of the two final projections resulting from the detection of the atom in $|i\rangle$.

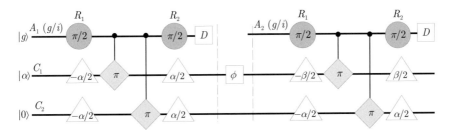

Fig. 7.41 The quantum circuit equivalent to the mesoscopic Mach–Zehnder interferometer $[\beta = \alpha \exp(i\phi)]$.

cousin? Starting again from a field prepared in $|\alpha, 0\rangle$, we must now use two beam-splitter atoms, successively interacting with the cavity field. Both are then detected and we keep only the experimental sequences resulting in a $(|i\rangle, |i\rangle)$ reading (25% of all the outcomes).

The mesoscopic field in one of the cavities (say C_1) must be phase-shifted, between the interactions with the two atoms, by a variable angle ϕ. This can, in principle at least, be achieved by shaking briefly one cavity mirror. The transient Doppler effect produced on the field bouncing on the moving cavity wall produces the desired phase shift. We finally measure the mean photon numbers $\langle a_i^\dagger a_i \rangle$, $(i = 1, 2)$ in C_1 and C_2 as a function of ϕ. The complete quantum circuit achieving these operations (not including the final field measurement) is represented in Fig. 7.41. Note that the variable phase shift experienced by C_1 must also be applied to the subsequent phase space displacements of the field in this cavity [α being replaced by $\beta = \alpha \exp(i\phi)$].

The final state produced by this quantum circuit is simply expressed with the help of the superposition principle. The $|\alpha, 0\rangle$ output channel of the first beam-splitter becomes $|\alpha e^{i\phi}, 0\rangle$ after the phase shift and $(|\alpha e^{i\phi}, 0\rangle + i |0, \alpha e^{i\phi}\rangle)/\sqrt{2}$ after the second beam-splitter. Similarly, the $|0, \alpha\rangle$ component after the first atom operation becomes $(|0, \alpha\rangle + i |\alpha, 0\rangle)/\sqrt{2}$ after the second. Combining these expressions, we finally get the complete transformation:

$$|\alpha, 0\rangle \rightarrow \frac{1}{2} \left[|\alpha e^{i\phi}, 0\rangle + i |0, \alpha e^{i\phi}\rangle + i |0, \alpha\rangle - |\alpha, 0\rangle \right] . \tag{7.169}$$

After a straightforward calculation involving once more the evaluation of the scalar product between two coherent states, we obtain the final photon number expectation values $\bar{n}_i$ in C_i $(i = 1, 2)$:

$$\bar{n}_i = \frac{\bar{n}}{2} \left[1 - \epsilon_i \mathrm{Re} \left(e^{i\phi} \langle \alpha | \alpha e^{i\phi} \rangle \right) \right] = \frac{\bar{n}}{2} \left[1 - \epsilon_i \cos(\phi + \bar{n} \sin \phi) e^{-\bar{n}(1 - \cos \phi)} \right] , \tag{7.170}$$

with $\epsilon_1 = +1$ and $\epsilon_2 = -1$. The average number of photons in C_1 when ϕ is scanned around 0 exhibits narrow fringes, displayed in Fig. 7.42 for $\bar{n} = 100$. The fringe spacing, $\Delta \phi \approx 2\pi/\bar{n}$, is inversely proportional to the average photon number. The interference is washed out when the phase difference between the cavities reaches a value of the order of $2\pi/\sqrt{\bar{n}}$ and only $\sim \sqrt{\bar{n}}$ fringes are visible.

To observe this signal, we would use absorbing atoms crossing the cavities as a probe. These atoms, sent in $|g\rangle$ after the quantum circuit has been operated, would

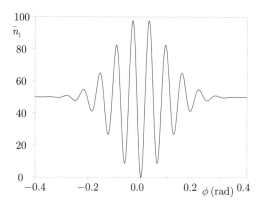

Fig. 7.42 The interference fringes of the quantum Mach–Zehnder circuit: average number $\bar{n}_1$ of photons in C_1 versus phase ϕ.

interact selectively and resonantly either with the field in C_1 or in C_2. An electric field would be used to put the probe atoms in resonance with the field of the chosen cavity while they cross it. The transfer rate of the atoms between $|g\rangle$ and $|e\rangle$ would provide a measurement of the average photon number $\bar{n}_1$ or $\bar{n}_2$.

The main features of this mesoscopic interference can be given a simple physical interpretation. It is instructive to consider first the operation of the quantum circuit for small phase differences $\phi \ll \sqrt{\bar{n}}$. We can then neglect the differences between the phases $n\phi$ of the Fock states in the expansion of the coherent state and approximate $|\alpha e^{i\phi}\rangle$ by $e^{i\bar{n}\phi} |\alpha\rangle$. The transformation (7.169) then simplifies as:

$$|\alpha, 0\rangle \rightarrow \frac{1}{2}\left[\left(e^{i\bar{n}\phi} - 1\right)|\alpha, 0\rangle + i\left(e^{i\bar{n}\phi} + 1\right)|0, \alpha\rangle\right] \ . \tag{7.171}$$

This formula looks like the usual Mach–Zehnder operation on a single photon (compare with eqn. 3.125), except that it is now the entire field, with $\bar{n}$ photons on average, which is channelled as a whole in a superposition of the two interfering paths, as shown in Fig. 7.43. When the quantum amplitude associated to one photon is phase-shifted by ϕ, the $\bar{n}$ photon field, considered as a single quantum object, has its wave function phase shifted by $\bar{n}\phi$. This explains the fast variation of the quantum amplitude with ϕ and the narrow fringes observed in the Mach–Zehnder output.

The collective narrowing of the interference pattern can be understood by a simple analogy with atomic or molecular interferometry. The interval between fringes observed in an atomic interferometer is proportional to the de Broglie wavelength λ of the atom or molecule, which is in turn inversely proportional to the particle mass. Hence, a molecule made of n identical atoms, such as C_{60}, exhibits an interference pattern with fringes n time narrower than those which would be displayed by the constituent atoms, sent with the same velocity through the interferometer (Hackermüller *et al.* 2003). One can say that the narrow pattern of mesoscopic interference reveals the 'collective de Broglie wavelength' of the photons when they are all forced to cross the interferometer as a single quantum object (Jacobson *et al.* 1995).

The same interferometer can be used for the single-atom or the n-atom molecule because the beam-splitters act on the system as a whole, their interaction with the particle being much too weak to affect the binding between the constituents. The $1/n$ narrowing effect of the fringe pattern is thus quite a natural effect. In the optical

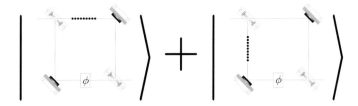

Fig. 7.43 Representation of the two paths in the collective Mach–Zehnder equivalent to the quantum circuit for $\phi \ll 1/\sqrt{n}$. All the photons are globally channelled in one arm or in the other. The resulting fringes have a $1/\bar{n}$ spacing when ϕ is scanned.

version of the experiment, the situation is more subtle because the photons are not bound together. We have seen in Chapter 3 that n photons sent through an ordinary Mach–Zehnder interferometer are independently dispatched in the two arms of the device, and the resulting signal is merely the sum of the patterns produced by photons sent one at a time, with the same n-independent fringe interval. In order for the whole field to behave as a collective entity, we need a very special kind of beam-splitter, an atom which is by itself a quantum object in a superposition of two states affecting the field in different ways.

The loss of fringe contrast when ϕ increases corresponds to the breakdown of the approximation made above. When ϕ becomes of the order of $\sqrt{n}$, the fluctuation of the photon number in the coherent field becomes significant. The situation is similar to classical interference fringes observed with a broadband light containing a continuum of wavelengths. The dispersion of the fringe spacing, proportional to the wavelength, then washes out all but a few fringes around the central one. In the quantum situation considered here, the field is monochromatic, but the experiment is sensitive to a different kind of interference, with a fringe spacing depending upon the particle number. Since this number fluctuates, only a few fringes remain visible. Note also that these fringes are extremely sensitive to decoherence. Their observation requires us to complete the whole quantum circuit operation before a single photon has been lost in the environment.

Let us stress that the observation of fringes in the output of our quantum circuit results from a post-selection process. The quantum beam-splitters are conditioned to a specific outcome of the atomic detection event. It is a simple exercise to show that the interference terms in the cavity average photon number have different phases, depending upon the recorded combination of final atomic states ($|i\rangle , |i\rangle$), ($|g\rangle , |g\rangle$), ($|g\rangle , |i\rangle$) or ($|i\rangle , |g\rangle$). If we were not recording this final state, we would have to trace over them, which would amount to add contributions with opposite interference terms, cancelling each other. We would observe no collective fringes. In a real experiment, it would be wise to keep all the data corresponding to the four possible outcomes and to combine them with the appropriate signs to maximize the interference signal.

Collective fringes with $1/n$ spacing have been given a lot of attention in modern quantum optics. We will briefly mention in the next section some of the experiments in which such fringes have been produced. Their narrow spacing make these signals good sensors of small phase fluctuations $\Delta\phi$. This might have interesting applications in high-resolution spectroscopy or metrology [see Leibfried *et al.* (2004) and Section 8.5.2]. We have shown in Chapter 3 that an usual single-particle Mach–Zehnder interferometer has a phase sensitivity of the order of $\sqrt{n}$, where n is the total number of photons participating in the signal. Our collective photonic interferometer would have much higher phase sensitivity, of the order of $\bar{n}$.

7.6.4 'High noon' states in a cavity

An atomic CQED quantum beam-splitter is well-suited to channel a coherent field as a whole in a superposition of its two output modes. Can we generalize this property to other kinds of input states? What would be for instance the action of the quantum circuit of Fig. 7.40 on the initial state $|n, 0\rangle$ representing C_1 with exactly n photons while C_2 is in vacuum? Could we realize the transformation:

$$|n, 0\rangle \rightarrow (|n, 0\rangle + i\,|0, n\rangle)/\sqrt{2}\,, \qquad (7.172)$$

generalizing for Fock states the transformation (7.167)?

Obviously the quantum circuit of Fig. 7.40 would not do the job. The action of the D translation operators on Fock states produces complex superpositions of Fock states whose amplitudes involve Laguerre polynomials of α (Barnett and Radmore 1997). The simplicity of the Mach–Zehnder operation is entirely lost.

Even if we knew how to build a direct beam-splitter realizing (7.172), we would still have the problem to prepare the input state $|n, 0\rangle$, which is, in cavity QED experiments, much more difficult to generate than $|\alpha, 0\rangle$. A realistic way to build a Fock state with n sensibly larger than 1 has been described in Chapter 6. It consists in performing a QND measurement of the photon number on a coherent state. The measurement pins down the photon number to a single value, within the distribution of the initial state. We do not know, when we start the procedure, which number we will find, roughly between $\bar{n} - \sqrt{\bar{n}}$ and $\bar{n} + \sqrt{\bar{n}}$ but this number is unambiguously determined at the end. This remark leads us to replace the above problem by a simpler one. Instead of trying to perform the operation (7.172), we will settle for a simpler transformation:

$$|\alpha, 0\rangle \rightarrow (|n, 0\rangle + i\,|0, n\rangle)/\sqrt{2}\,. \qquad (7.173)$$

To achieve it, we have to combine the tricks learned in this chapter and the last one. We first perform a quantum beam-splitter operation on the initial field $|\alpha, 0\rangle$, transforming it into $(|\alpha, 0\rangle + i\,|0, \alpha\rangle)/\sqrt{2}$, followed by a QND measurement of the global photon number $a_1^\dagger a_1 + a_2^\dagger a_2$ in the two cavities. The complete quantum circuit achieving the transformation (7.173) is shown in Fig. 7.44. All atoms behave as qubits on the $|g\rangle$ to $|i\rangle$ transition while the cavity field is nearly resonant with the transition from $|g\rangle$ to $|e\rangle$. The left part of the circuit, with atom A_0, realizes the mesoscopic beam splitting. Then a succession of atoms $A_1, A_2, \ldots$ perform the QND measurement, according to the optimal dichotomic procedure outlined in Chapter 6 (only the first two atoms are represented for clarity in Fig. 7.44).

The QND atoms are, as A_0, submitted to two Ramsey pulses with adjustable phases on the $|g\rangle$ to $|i\rangle$ transition. They control, in between, the phases of the cavity

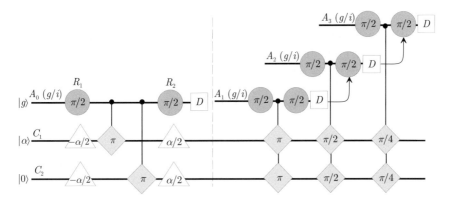

Fig. 7.44 The quantum circuit preparing a noon state.

fields, shifting them together, if in $|g\rangle$, by π (A_1 atom), $\pi/2$ (A_2), $\pi/4$ (A_3) and so on. They do nothing to the field if they are in $|i\rangle$. The atoms are then detected and each measurement outcome ($|g\rangle$ or $|i\rangle$) provides a common bit of information about the photon number. The result of each measurement determines also the setting of the phases of the Ramsey pulses for the next step, as discussed in Chapter 6.

We should note that nothing in this experiment would allow us to distinguish between the two cavities. At each stage of the decimation process, the two fields are projected into the same state. In the end, the field must thus have collapsed in the same Fock state in both cavities. The initial state is transformed into $(|n, 0\rangle + i |0, n\rangle)/\sqrt{2}$, where n is a value falling randomly within the initial photon number distribution of $|\alpha\rangle$ and depending upon the specific series of bit values obtained in the experiment. We have thus achieved the desired transformation (7.172), with the important qualification that we are not free to choose *a priori* which exact value of the photon number we will obtain in the end.

Delocalized states of the form (7.172) have been called 'high noon states', probably by a Gary Cooper fan, in reference to the way they are expressed in Dirac notation [writing them as $|0n\rangle + |n0\rangle$ (Oh no!) would be less glamorous]. The CQED method is a way to prepare them in cavities, whereas other methods have been proposed and implemented in propagating light beams.[17] These highly non-classical states are, as the other cat states studied in this chapter, very fragile. Decoherence limits n to ~ 10 at most in any conceivable CQED experiment.

With this visit to the zoo of cat states we conclude our study of CQED, the physics of photons and atoms interacting strongly in boxes. Using these simple ingredients, we have realized thought experiments illustrating the strange features of the quantum. More experiments are yet to come, once obvious improvements will have been made on our set-up. Remarkably, the tools we are using for these experiments are very

---

[17]We have described in Chapter 2 how a state of the form $|2, 0\rangle + |0, 2\rangle$ can be prepared with two photons impinging symmetrically on a beam-splitter (Ghosh and Mandel 1987). Interference signals involving these biphoton states with a fringe spacing twice smaller than for ordinary single-photon interference has been obtained (Edamatsu *et al.* 2002). These experiments have been extended recently to noon states with $n = 3$ (Mitchell *et al.* 2004) and 4 (Walther *et al.* 2004). A related experiment exhibiting a fringe spacing scaling as $1/n$ has been performed in an ion trap (Chapter 8).

simple and borrowed from a technology developed in nuclear magnetic resonance and atomic spectroscopy. Among these tools, Ramsey interferometry has turned out to be especially useful and versatile. While it is generally used as a method of high-resolution spectroscopy (most notably in atomic clocks), it has turned out in CQED as a very general and very flexible method to analyse the coherence of the atom–field system, to perform QND measurements and which-path complementarity experiments, to prepare and analyse mesoscopic states superpositions. We could indeed have called this book '*All we can do with Ramsey fringes*'.

This title will seem even more adequate after the two chapters yet to come, in which we explore two domains of physics which bear strong similarities with CQED. We describe in these concluding chapters how entanglement can be engineered in ion traps and in ultra-cold atom matter waves and we analyse the analogies and differences of this physics with CQED. Here again, Ramsey interferometry will come back, time and again, as the ideal method to investigate and manipulate quantum systems in a simple and controllable way.

8
Atoms in a box

Il faut qu'une porte soit ouverte ou fermée

Alfred de Musset

We now explore a class of experiments which features the spring–spin model on a different stage. The oscillator in the physics of this chapter is no longer a field whose photons are confined in a cavity made of reflecting walls, but a collective mode of motion of an ensemble of ions held in space by electric and magnetic forces forming an immaterial box. This oscillator is manipulated by lasers irradiating the ions. These lasers excite a transition between two ion's electronic levels playing the role of pseudo-spin or qubit states. By exploiting the dynamics of the collective oscillator-qubit state couplings, the ions spin and spring degrees of freedom are entangled in a controlled way, leading to the demonstration of various quantum information procedures.

The recipes to generate and manipulate entanglement in ion trap physics recall the methods of CQED. We are dealing in both cases with similar equations of motion. In spite of this analogy, there are also marked differences between the two situations. At a conceptual level, the CQED Hamiltonian is more fundamental than the ion trap one. It derives directly from the charge–field coupling of quantum electrodynamics and describes an exchange of excitation between the atom and the field, which is very simple to grasp intuitively. The spin–spring interaction in ion trap physics is more subtle. It results from an effective coupling mediated by laser light irradiating the ions in the trap. On the one hand, the interaction is less direct than the CQED one. On the other hand, it gives rise to a richer variety of effects since the details of the coupling can be fine-tuned by changing the parameters of the laser light, a possibility lacking in CQED physics.

Another difference between CQED and ion trap experiments is practical. It has to do with the lifetime of the systems involved. In microwave CQED, the qubits are coded either in a photon, which cannot survive longer than a fraction of a second in the cavity, or in an atom, which flies across the set-up before being destroyed upon detection. The quantum information procedure must thus be implemented in a short time, before the system has vanished. The ion 'box', on the contrary, keeps the particles confined basically forever, and the qubit states are long-lived, making the system more realistic than CQED set-ups for implementing practical quantum information. The two systems are thus complementary. CQED experiments are conceptually simpler and ion trap ones more promising for applications.

Before focusing on quantum information, we must recall that particle trap physics has been developed first and foremost as a very efficient tool for high-precision spectroscopy and metrology, a domain in which it is still making impressive advances. A particle held isolated in space at a large distance from other potentially perturbing matter is an ideal system for these studies. Long interrogation times make it possible to probe the internal structure of the trapped particle with an exquisite precision. A landmark in this field has been the spectroscopy of electron and positron spin states by the Seattle group of H. Dehmelt (1990). The anomalous gyromagnetic ratio of these particles, $g - 2$ (Chapter 1), has been determined at the 10^{-10} level (van Dyck *et al.* 1977; 1987).

As another impressive experimental achievement, let us mention the measurement by the same group of the relativistic corrections to the motion of an electron in the trap, at a kinetic energy of a few milli-electron-volts only (Gabrielse *et al.* 1985). In this experiment, the same electron was held in the trap for several months. The experimenters developed quite a personal relationship with this unique particle, which, unfortunately, ended up drowning in the Fermi sea of an electrode.

More complex than electrons, ions have a rich internal structure with optical transitions which can be probed by laser beams. Strong lines generate intense laser-induced fluorescence signals. The ion, under resonant excitation, scatters so many photons that it can be directly imaged, as shown in Fig. 1.10, on page 17, or even seen with the naked eye. Other lines, corresponding to weak transition probabilities, have an extremely narrow linewidth, with an excited state lifetime in the second to minute range. A very stable laser can be locked to a single-ion transition, providing a frequency reference in the optical domain. A radiofrequency signal can be phase-locked to the laser probing the ion by means of the recently developed frequency comb technique (Holzwarth *et al.* 2000; Cundiff and Ye 2003). The accuracy and repeatability of these prototype optical clocks outperform by far the standard caesium clocks (Diddams *et al.* 2001; Bauch and Telle 2002) and compete with the cold atom clock mentioned in Chapter 1 (Lemonde *et al.* 2000). The development of high accuracy ion clocks could have a major impact on metrology and, more practically on the global positioning system, with a possible dramatic improvement of its precision.

Let us also mention that ion traps are well-suited for precise measurements of ionic masses (Rainville *et al.* 2004), with potential applications for an atomic definition of the mass unit. They also open the way to the study of fragile or reactive species, a spectacular example being the production of anti-hydrogen atoms in an anti-proton/positron trap (Amoretti *et al.* 2002; Storry *et al.* 2004).

In ion trap experiments, laser beams are not only used to induce transitions and to detect the particles via their fluorescence. They can also be employed – as we will show in more detail below – to manipulate the internal states of the ions and to couple them to the motional degrees of freedom in the trap. The first application of this coupling is the laser cooling of the ionic motion. The 'Doppler cooling' method, used as a first step in most cold atom experiments (Cohen-Tannoudji 1992), has been first implemented in the ion trap context (Wineland *et al.* 1978). Improving on Doppler cooling, the 'sideband cooling' procedure reaches the ultimate limit, the ion being finally brought, with a probability very close to unity, in its motional ground state (Diedrich *et al.* 1989; Heinzen and Wineland 1990; Monroe *et al.* 1995b). It is then strongly localized, the ground state wave function extension in tight ion traps being only a few nanometres (Guthöhrlein *et al.* 2001).

Starting from this state, an ion can be manipulated by coupling its spin and spring degrees of freedom in a controlled way (Leibfried *et al.* 2003a). In the simplest situation, the effective interaction reproduces the Jaynes–Cummings model of CQED, a coherent excitation exchange between the spin and the spring. In this regime, ion trap and CQED experiments are formally identical. Any quantum state manipulation implemented in CQED can be immediately transposed to the ion trap context.

The converse is not always true. New kinds of spin–spring interactions, unknown in CQED, can be realized through a proper tailoring of the driving lasers. In the 'anti-Jaynes–Cummings' situation, for instance, the spin and the spring both gain or lose one excitation quantum in the elementary interaction process (the extra energy being provided by the driving laser). In the non-linear Jaynes–Cummings model, the spin undergoes a transition, while producing or absorbing $p > 1$ vibration quanta or 'phonons'. The experimenter, having control over the coupling laser light, can tailor and combine at will these processes. The ion trap system offers thus a remarkable flexibility, which makes it ideal for the preparation of complex quantum states.

Experiments with a few ions simultaneous held in the same trap open even more possibilities. At the very low temperatures reached by laser cooling, the ions arrange themselves in a regular crystal structure under the combined action of the trapping potential and the inter-ion Coulomb repulsion. In linear traps, this crystal is a one-dimensional chain, with ionic spacing in the few micrometres range. Each of the ions in this chain acts as a qubit, which can be addressed with laser beams. In quantum information terms, this line of ions behaves as a quantum register. Laser-induced fluorescence provides a nearly ideal qubit state detection. The qubit states can be entangled with a common vibration mode of the ionic crystal, which provides a relay, a 'quantum bus' for the indirect entanglement of two qubits. Two-qubit quantum gates can thus be operated between any pair of qubits in the chain (Cirac and Zoller 1995).

We recognize here the premises of a general-purpose quantum computer. In principle, a long chain of ions could run any quantum algorithm with a simple sequence of coupling laser pulses. Significant progresses have been made towards this dream in recent years: implementation of a variety of two-ion quantum gates (Leibfried *et al.* 2003b; Schmidt-Kaler *et al.* 2003), non-local entanglement of two ions (Rowe *et al.* 2001), teleportation using three ions in the same trap (Riebe *et al.* 2004; Barrett *et al.* 2004), implementation of quantum error correction codes (Chiaverini *et al.* 2004) or of simple quantum algorithms (Gulde *et al.* 2003). The extension to higher qubit numbers is in progress. Complex entanglement sequences implying a few, perhaps a few tens qubits, are a mid-term perspective (Kielpinski *et al.* 2002).

This chapter is devoted to a description of the ion traps essential features and to the quantum information science that these systems make possible. More details can be found in textbooks and review papers (Ghosh 1995; Leibfried *et al.* 2003a). In Section 8.1, we first analyse the Paul trap design generally used in these experiments and we describe how this trap operates with a single ion or with a set of ions making up a string of qubits (Schmidt-Kaler *et al.* 2003). We next define some important modes of vibration of this ion chain. We then describe the energy diagram of the two kinds of ions which are mainly used for quantum state manipulation experiments (Ca^+ and Be^+). We show how lasers exciting the ions can be used to realize various kinds of spin–spring couplings. We then analyse the tomographic procedure which, by detecting repeatedly each ion in various qubit states, makes it possible to reconstruct the density operator of an $\mathcal{N}$-qubit system. Many quantum information experiments require an initialization of the ions in the motional ground state of the ion's chain. We conclude the section by analysing the laser mechanisms which achieve this cooling.

In Section 8.2, we focus on the manipulation of the ion's motional states. We show how one can prepare a Fock state of motion of a single ion with n phonons, and how one can detect this state by a Rabi oscillation experiment. We also describe the preparation and detection of coherent states of an ion's vibration, obtained by implementing a translation in phase space of the ion's ground state. We show how this translation in phase space can be made dependent upon the ion's internal state, realizing a Schrödinger cat state of motion of a single particle.

In Section 8.3, we include in the spin–spring model the ionic relaxation due to spontaneous emission, using the quantum jump language introduced in Chapter 4. We show that spontaneous emission can be used in a clever way to implement a tailored environment for the ionic motion (de Matos Filho and Vogel 1996; Poyatos *et al.* 1996; Carvalho *et al.* 2001). The simplest of these environments is a phonon sink, at zero temperature. The coupling to this sink realizes a continuous version of the sideband cooling procedure. Other, more complex environments can be engineered, in which the ion motion relaxes towards non-trivial quantum states. We show that coherent states, and even Schrödinger cat states can be efficiently prepared and, to some extent, protected in this way.

The next two sections are devoted to a brief and necessarily incomplete survey of quantum information processing with ion traps. In Section 8.4, we focus on the conceptually simple scheme proposed by Cirac and Zoller (1995) which has been implemented in the laboratory on Ca^+ ions by the Innsbruck group (Roos *et al.* 2004b). In this scheme, the ions are addressed separately and sequentially, according to a procedure which is, at least in principle, scalable to a large number of ions and operations. We illustrate this scalability by describing an experiment which has produced and analysed a three-ion GHZ state, according to a procedure very similar to the one implemented in CQED (Section 6.3.2). We also briefly present the results of a teleportation experiment involving three ions. In an alternative scheme, the Boulder group working on Be^+ has realized gates in which the ions are addressed collectively. We describe these gates in Section 8.5 and show how collective ion addressing has made it possible to prepare and analyse GHZ states with up to six ions. We conclude the chapter (Section 8.6) by discussing some of the perspectives opened by ion trap physics in quantum information science.

8.1 Ion trap physics

8.1.1 The Paul trap

Let us recall how a charged particle can be trapped by electromagnetic forces. We assume that the ion has a positive unit charge, e_i. What is needed is a field configuration which provides a stable equilibrium position in free space. If we use an electrostatic

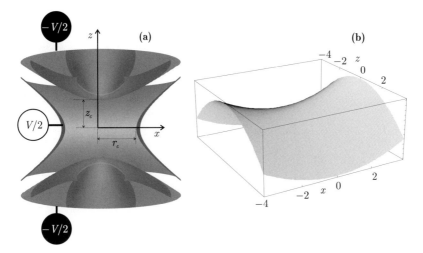

Fig. 8.1 (a) Electrode configuration creating a quadrupolar potential $U_q(\mathbf{r})$ near the origin. (b) Electrostatic potential $U_q(x, 0, z)$ in the xOz plane. The origin is a saddle stationary point. For positive V, the ion is trapped along the x axis and anti-trapped along the z one.

potential $U(\mathbf{r})$, such an equilibrium, which must be a minimum of the potential, does not exist. There is indeed – as a direct consequence of Gauss theorem – no extremum of U in empty space.

The only stationary points of the electrostatic potential in free space are 'saddle points', the simplest example being the origin for the quadrupolar potential:

$$U_q(\mathbf{r}) = \frac{V}{4z_c^2}(x^2 + y^2 - 2z^2) \,. \tag{8.1}$$

Such a potential – an obvious solution of Laplace equation $\Delta U_q(\mathbf{r}) = 0$ – is created by the electrode configuration represented in Fig. 8.1(a). The three electrodes are hyperboloids, with Oz as a cylindrical symmetry axis. The central ring, connected to a potential source at $V/2$, has an inner radius r_c. The two 'end caps' are connected to a source at $-V/2$. Their summits are at $\pm z_c$ on the Oz axis, with $z_c = r_c/\sqrt{2}$.

The function $U_q(x, 0, z)$ is represented as a three-dimensional plot in Fig. 8.1(b), in the case $V > 0$. The origin is a minimum along x, but a maximum along the z direction. The ionic motion, trapped along the x axis, is anti-trapped along z. The ion is thus accelerated towards one of the end caps. The situation is reversed when the sign of V is changed. This configuration produces only an unstable equilibrium at the origin.

Two different routes can be used to reach a stable trapping in spite of the constraints imposed by electrostatic laws. In the Penning trap (Ghosh 1995), a strong homogeneous magnetic field aligned with Oz is added to the quadrupole potential (set with $V < 0$). The electric field provides the stability in the z direction. The magnetic field, around which the ion orbits in a tight nearly circular cyclotron motion, provides an effective confinement in the transverse direction. This trap design has been used by the Seattle group for single electrons and positrons. The comparison of the spin up/spin down transition frequency in the trap's magnetic field with the frequency

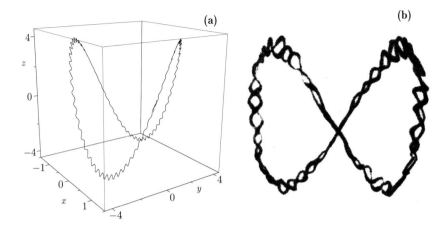

Fig. 8.2 (a) Numerical simulation of an ion motion in a Paul trap. The time range is 75 periods of the trapping frequency. The trap parameters are given in the text. The micro- and macro-motions are clearly visible. (b). Photograph of an aluminium dust particle trajectory in a Paul trap. Reprinted with permission from Wuerker *et al.* (1959). © American Institute of Physics.

of the cyclotron motion provides a direct measurement of the gyromagnetic ratio of these particles.

For quantum state engineering and quantum information manipulations, the Paul trap is generally preferred. There is no magnetic field, but the voltage V applied across the electrodes is now a rapidly oscillating function of time:

$$V(t) = V_0 \cos(\omega_{\rm rf} t) , \qquad (8.2)$$

the frequency $\omega_{\rm rf}/2\pi$ being in the range $10 - 100$ MHz. The potential in a given direction thus oscillates rapidly between a trapping and an anti-trapping configuration. This oscillation provides an effectively stable, nearly harmonic trap.

Let us examine briefly the classical equations of motion in the potential $U_q(\mathbf{r}, t)$. The evolutions along the three axes are decoupled, and we focus on the motion along Oz, described by:

$$\frac{d^2 z}{dt^2} - \frac{e_i V_0}{m z_c^2} \cos(\omega_{\rm rf} t) z = 0 , \qquad (8.3)$$

which can be cast under the form of a Mathieu equation (Ghosh 1995) using the dimensionless time $\tau = \omega_{\rm rf} t/2$:

$$\frac{d^2 z}{d\tau^2} - 2 q_z \cos(2\tau) z = 0 , \qquad (8.4)$$

with the dimensionless parameter q_z defined as:

$$q_z = \frac{2 e_i V_0}{m z_c^2 \omega_{\rm rf}^2} = \frac{4 e_i V_0}{m r_c^2 \omega_{\rm rf}^2} . \qquad (8.5)$$

The equations of motion along x or y are similar, with q_z replaced by $q_x = q_y = -q_z/2$. Equation (8.4) can be solved analytically using Floquet's theorem (Ghosh 1995). We shall not enter into the discussion of this solution here. Suffice it to say that the motion is bounded only when q_z is adjusted within well-defined values. Otherwise, the ion position diverges exponentially. The first stability domain, generally used in practical Paul traps, corresponds to $0 \leq q_z \leq 0.908$. The motion is then stable, for a fixed V_0, when the driving frequency $\omega_{\rm rf}$ satisfies:

$$\omega_{\rm rf} > \frac{\sqrt{2e_i V_0}}{z_c \sqrt{0.908m}} . \tag{8.6}$$

This threshold can be understood by noticing that the potential oscillation period, $\sim 1/\omega_{\rm rf}$, must be smaller than $\sim z_c/v$, where $v = \sqrt{2e_i V_0/m}$ is the velocity of the ion accelerated from rest by the potential V_0. In other words, the potential should change sign before the ion, accelerated in the anti-trapping direction, has had time to reach an electrode.

Figure 8.2(a) presents a numerical simulation of the motion of a calcium ion, ^{40}Ca$^+$, in a trap with $z_c = 1$ mm, $V_0 = 100$ V and $\omega_{\rm rf}/2\pi = 10$ MHz. These typical parameters, which we will keep in mind for forthcoming discussions, correspond to $q_z = 0.15$; $q_x = q_y = -0.075$, well inside the stability domain. The ion undergoes a fast harmonic 'micro-motion' at the trap drive frequency $\omega_{\rm rf}/2\pi$, with a very small amplitude ζ_0, which nearly cancels when the ion is near the trap centre. The much larger and slower 'macro-motion' corresponds to the evolution of the trapped ion in an effective anisotropic harmonic potential whose oscillation frequencies are in the two-to-one ratio along the Oz and the Ox or Oy directions. The slow motion trajectory has the shape of a closed Lissajous curve in which a full swing along Oz corresponds to half an oscillation in the xOy plane.

These features can be explained by a simple analysis based on an approximation separating the fast and the slow motions. The occurrence of two different length scales in the motion suggests that we express the ion's position as $z = Z + \zeta$, where ζ is the small, rapidly oscillating micro-motion component, with a zero average value. The ion's position averaged over a period of the micro-motion, Z, evolves over a much larger time scale. Making the assumption (to be verified a *posteriori*) that the macro-motion acceleration is much smaller than the micro-motion's ($d^2 Z/dt^2 \ll d^2\zeta/dt^2$), the equation of motion for ζ can be approximated as:

$$m\frac{d^2\zeta}{dt^2} \approx \frac{e_i V_0}{z_c^2} \cos(\omega_{\rm rf}t)\, Z , \tag{8.7}$$

with the obvious solution:

$$\zeta = -\frac{q_z}{2} \cos(\omega_{\rm rf}t)\, Z . \tag{8.8}$$

This equation shows that the trap's electric field induces a quiver motion of the ion whose amplitude is proportional to the mean distance Z to the trap centre. The time-averaged kinetic energy of this micro-motion is:

$$\overline{E_c}(Z) = \tfrac{1}{16}mq_z^2\omega_{\rm rf}^2 Z^2 = \tfrac{1}{2}m\omega_z^2\, Z^2 , \tag{8.9}$$

with:

$$\omega_z = \frac{1}{2\sqrt{2}} q_z \omega_{\mathrm{rf}} \,. \tag{8.10}$$

When the ion moves slowly in the trap, the quiver energy $\overline{E_c}(Z)$, which has to be borrowed from the kinetic energy of the macro-motion, plays thus the role of a harmonic potential energy for the Z-evolution.[1] The ion's trapping, due to the 'ponderomotive' force corresponding to the gradient of this potential energy, produces an oscillation along Oz at frequency $\omega_z \ll \omega_{\mathrm{rf}}$. The macro-motions along x and y are produced by smaller restoring ponderomotive forces, such that $\omega_x = \omega_y = \omega_z/2$. We thus understand the characteristics of the wobbling Lissajous trajectory displayed in Fig. 8.2(a).

The motion appears, in this simple model, as unconditionally stable. In fact, near the upper limit of the stability criterion, $q_z \approx 0.908$, the macro- and micro-motion frequencies get close (eqn. 8.10) and our initial approximation, which requires $\omega_z \ll \omega_{\mathrm{rf}}$, breaks down. Subtle effects, like parametric resonance between micro- and macromotion also come into play to reduce the stability domain. Note finally that the macro- and micro-motion accelerations are in a ratio of the order of $(\omega_z^2/\omega_{\mathrm{rf}}^2)(Z/\zeta) \sim q_z/4 \ll 1$, which satisfies the initial hypothesis made to derive eqn. (8.7).

This effective potential description suggests an analogy with the trapping of an atom in a detuned laser field (Cohen-Tannoudji 1992). The polarization energy of the atom, proportional to the field intensity, plays then the role of an effective potential. For a negative atomic polarizability (i.e. for a laser tuned below the atomic resonance), the atom is attracted towards an intensity maximum. This is the standard configuration for dipole traps. For positive polarizabilities (above resonance, 'blue detuning'), the atom is expelled from high-intensity regions. In the ion trap case, the laser is replaced by the radio-frequency field. We deal with a free charge, whose resonance frequency vanishes. The trap field is thus blue-detuned for all ω_{rf} values, and the ion is attracted to the minimum of the oscillating field amplitude, at trap centre. This analogy is fully quantitative, $\overline{E_c}$ being precisely the polarization energy of a free charge in the trap field.

Let us give a few orders of magnitude, based on the trap parameters used for Fig. 8.2, with $q_z = 0.15$. The macro-motion frequency is $\omega_z/2\pi = 530$ kHz, about twenty times smaller than ω_{rf}, in fair agreement with the numerical simulation. The trap depth is measured by the effective potential value on the end caps:

$$\overline{E_c}(z_c) = \tfrac{1}{8} q_z \, e_i V_0 \,. \tag{8.11}$$

It is about 2% of the static potential depth, i.e. of the order of one electron-volt. Ponderomotive forces are much smaller than electrostatic ones, but the trap is still deep enough to catch energetic ions with effective temperatures in the few thousand kelvin range. The effective trap is conservative. Capturing ions requires thus either a dissipative process (collisions on a background gas, for instance) or the production of the ions directly near the trap centre. Most experiments use the second solution, with the direct ionization of an atomic beam crossing the trap by electron impact or photo-ionization.

---

[1]Note the analogy between ion traps and molecular dynamics. In a molecule Hamiltonian, the energy associated to the fast electronic motion plays the role of a potential energy for the slow motion of the nuclei. Similarly, in an ion trap, the fast micro-motion energy appears as a potential energy for the slow macro-motion of the ion.

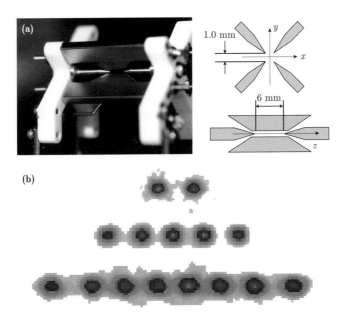

Fig. 8.3 Linear ion trap. (a) Photograph of the actual trap, together with a scheme of the electrode configuration. (b) Linear strings of two, five and eight ions in the trap. Courtesy of R. Blatt.

We should note that Paul traps need not be sophisticated devices in an ultra-high vacuum cryogenic environment. Building a Paul trap able to hold small charged dust particles is fairly simple. Figure 8.2(b) shows the trajectory of an aluminium dust particle suspended in a Paul trap operating with moderate voltages and a rough vacuum in an early experiment (Wuerker *et al.* 1959).

The quadrupolar trap geometry can be modified by changing the shape of the trap electrodes to confine ions in different spatial configurations. The linear trap shown in Fig. 8.3(a) is used in the quantum information experiments of the Innsbruck group (Gulde *et al.* 2003). Four elongated electrodes, symmetrically arranged around the z axis, create a quadrupolar potential in the xOy plane, almost invariant by translation along Oz. A strong radiofrequency voltage applied across these electrodes provides a tight transverse confinement. The typical transverse oscillation frequencies, $\omega_{x,y}/2\pi$ are in the two to four Megahertz range. The longitudinal confinement along the z axis is provided by a weak positive d.c. potential applied on the two needles closing the structure. The corresponding oscillation frequency, $\omega_z/2\pi$, is below 1 MHz. When this trap is loaded with a few cold ions (see cooling procedure below), they stay close to trap axis, with negligible contribution from the micro-motion in the transverse direction. The longitudinal motion along trap axis is ruled by electrostatic forces, without any micro-motion in that direction. The fast evolution at the rf frequency is thus completely negligible and the ions motion is fully described by the slow macromotion at frequencies $\omega_{x,y,z}/2\pi$.

8.1.2   One dimension $\mathcal{N}$-ion chain in linear Paul trap

When several ions are loaded in a trap, they experience the competing effects of the trapping potential, which attracts them to the trap centre and the Coulomb repulsion which tends to keep them apart. The total potential energy adds to the ion's kinetic energy, measured by the ion temperature T. When T is large, the ions move randomly, occupying the whole trap volume. At low temperatures, the particles organize themselves in a regular pattern, a pseudo-crystal, which minimizes the overall energy. For large ion numbers, the crystallization is a first-order phase transition, which has been studied in detail (Walther 1992).

We focus here on systems with a few ions, in the linear trap configuration of Fig. 8.3(a). Let us start by considering two ions at zero temperature. At equilibrium, the ions are located at $\pm z_C/2$, where z_C, characteristic of the Coulomb interaction, minimizes the overall energy $m\omega_z^2 z_C^2/4 + e_i^2/4\pi\varepsilon_0 z_C$:

$$z_C = \left(e_i^2/2\pi\varepsilon_0 m\omega_z^2\right)^{1/3} . \tag{8.12}$$

For a Ca^+ ion with $\omega_z/2\pi = 530$ kHz, we find $z_C = 8.8$ μm. When larger numbers of ions are trapped, z_C still gives the order of magnitude of the interionic distances at $T = 0$ K. In qualitative terms, we understand intuitively that a linear crystal forms when the residual ionic oscillations have an amplitude much smaller than z_C. This is clearly not the case when ions are captured in the trap at high temperature, since they then explore during their random motion the whole trapping region. After laser cooling (see below), the motion excursion becomes much smaller than z_C and the crystal is certainly formed.

Three linear ion chains, with $\mathcal{N} = 2$, 5 and 8 ions imaged by laser-induced fluorescence, are shown in Fig. 8.3(b). The ion spacing is larger on the sides of the crystal (the last ion feels the repulsion from the $\mathcal{N} - 1$ other ones). For the small $\mathcal{N}$ values considered here, the ions are fairly well-separated, and individual addressing by focused laser beams is possible (Nägerl *et al.* 1999).

The motion of the ions in the trap can be expanded over the crystal vibration eigenmodes (James 1998). Let us assume that the temperature is low enough so that the transverse modes along Ox and Oy are not excited. The motion of the ions is then unidimensional along Oz. For $\mathcal{N}$ ions with positions $z_1, z_2, ..., z_{\mathcal{N}}$, there are $\mathcal{N}$ independent modes. They can be easily computed analytically up to $\mathcal{N} = 3$. For two ions, the collective mode variables are obviously the centre of mass position $Z_1 = (z_1 + z_2)/2$ and the inter-ion distance $Z_2 = z_1 - z_2$. A simple expansion of the Coulomb interaction around $Z_2 = z_C$ leads to the following expression for the two-ion motional energy in the trap, expressed in terms of the collective positions and velocities:

$$E = m[\dot{Z}_1{}^2 + \omega_z^2 Z_1^2] + \frac{m}{4}\left[\dot{Z}_2{}^2 + 3\omega_z^2(Z_2 - z_C)^2\right] , \tag{8.13}$$

in which we have removed a constant offset equal to the crystal rest energy. The motional Hamiltonian thus separates into two independent terms describing a *centre of mass* mode (CM) with frequency $\omega_{z1} = \omega_z$ and a *stretch mode* with frequency $\omega_{z2} = \omega_z\sqrt{3}$. The former corresponds to an oscillation of the two ions as a whole, leaving their distance fixed. In the stretch mode, on the contrary, the ion centre of mass

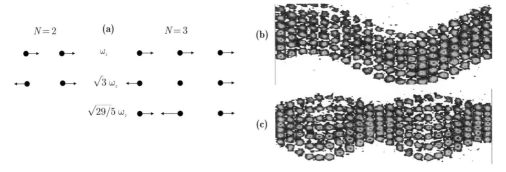

Fig. 8.4 Eigenmodes of the ion crystal. (a) Modes for $\mathcal{N} = 2$ and $\mathcal{N} = 3$ ions. The arrows indicate the phase and the amplitude of the displacement. (b) Time-resolved stroboscopic pictures of an $\mathcal{N} = 7$ ion crystal showing oscillations in the centre-of-mass mode. (c) Oscillations in the stretch mode. In (b) and (c), the ion crystal is vertical and successive snapshots are presented from left to right. Reprinted with permission from Nägerl *et al.* (1998b). © Optical Society of America.

remains motionless and only the inter-ion distance oscillates around its equilibrium value z_C. Note that the CM mode is an oscillator with mass $2m$ and a zero-point fluctuation $\Delta Z_1 = \sqrt{\hbar/4m\omega_z}$, which is $\sqrt{2}$ smaller than that of a single ion in the same trap, while the stretch mode has a relative mass $m/2$ and a zero-point fluctuation $\Delta(Z_2 - z_C) = (\hbar/m\omega_z\sqrt{3})^{1/2}$, $(4/3)^{1/4}$ larger than the one of a single ion in the same trap.

For three ions, a similar analysis leads us to define the collective coordinates as $Z_1 = (z_1 + z_2 + z_3)/3$, $Z_2 = (z_3 - z_1)/2$, $Z_3 = (z_1 - 2z_2 + z_3)/6$ and the ions motional energy, again developed for small amplitude oscillations around the ions equilibrium positions ($Z_1 = Z_3 = 0$, $Z_2 = 5^{1/3}z_C$), is, within a constant offset:

$$E = \frac{3m}{2}\left[\dot{Z_1}^2 + \omega_z^2 Z_1^2\right] + m\left[\dot{Z_2}^2 + 3\omega_z^2\left(Z_2 - 5^{1/3}z_C\right)^2\right] + 3m\left[\dot{Z_3}^2 + \frac{29}{5}\omega_z^2 Z_3^2\right].$$

$$(8.14)$$

The Hamiltonian now separates into three independent contributions, a CM and a stretch mode having the same frequencies as in the two-ion case, and a 'scissor mode' with frequency $\omega_{z3} = \omega_z\sqrt{29/5}$. In this highest frequency mode, the two extreme ions move together in one direction with the same amplitude while the centre ion oscillates with double amplitude in the opposite direction. Figure 8.4(a) presents the various modes for $\mathcal{N} = 2$ and $\mathcal{N} = 3$, with their frequencies and corresponding ion motion patterns.

The characteristics of the two lowest frequency modes for $\mathcal{N} = 2$ and 3 generalize to higher $\mathcal{N}$ values. The fundamental mode with the lowest frequency is always the CM one. The crystal then oscillates as a rigid body at the trap frequency ω_z. A time-resolved picture of the CM oscillation is shown in Fig. 8.4(b) for a seven-ion crystal. The next mode in order of increasing frequencies is always the 'stretch mode', displayed in Fig. 8.4(c), again for $\mathcal{N} = 7$. The displacement of each ion is then proportional to its distance from the trap centre at rest. The oscillation frequency of the stretch mode

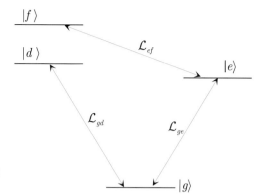

Fig. 8.5 Four-level scheme implementing an ionic qubit

is $\omega_z \sqrt{3}$ for all $\mathcal{N}$ values, a remarkably simple result. There is no such simple rule for higher frequency modes.

8.1.3   Implementing the qubits: the Ca$^+$ and Be$^+$ ions

In order to turn an ion string into a qubit register, the particles must first be initialized in an electronic quantum state coding for $|0\rangle$. Binary information is then stored in each ion by driving it in a transition linking $|0\rangle$ to an upper state coding for $|1\rangle$. After qubit manipulation, information must be read out by measuring efficiently and selectively the state of each ion. Coupling between the ionic qubits is, as we will see, achieved via their Coulomb interaction, by entangling with the help of lasers the ion's electronic states to the collective oscillation states of the ion chain. This procedure often requires the ionic vibrations to be initially in their ground state, which implies a cooling mechanism much more efficient than the one needed for the crystallisation of the ion chain. All these operations – qubit initialization, vibrational ground state cooling, qubit manipulation and detection – require a proper choice of ion species, with an energy diagram convenient for laser manipulations. A couple of long-lived qubit levels must be selected in the ion spectrum. These levels must be coupled between themselves, as well as to other short-lived levels, by sets of laser beams with controlled polarizations, frequencies, phases and amplitudes.

A simple energy diagram for quantum information requires four ionic levels, as shown in Fig. 8.5. Two long-lived states $|g\rangle$ and $|e\rangle$ code the $|0\rangle$ and $|1\rangle$ qubit logical states. A laser $\mathcal{L}_{ge}$ resonant or nearly resonant with the transition between these states achieves one-qubit operations by mixing coherently the qubit levels. It can also, as we will show below, realize sideband excitations of the qubit transition. When tuned at the qubit frequency plus or minus an integer multiple of one of the normal modes oscillation frequencies, $\mathcal{L}_{ge}$ excites the qubit and simultaneously feeds or absorbs phonons of the ionic chain.

Two auxiliary short-lived upper levels $|f\rangle$ and $|d\rangle$ are required for qubit state preparation at the beginning of a quantum information sequence and for its detection at the end. Laser beams $\mathcal{L}_{ef}$ and $\mathcal{L}_{gd}$, respectively resonant or nearly resonant with the $|e\rangle \rightarrow |f\rangle$ and $|g\rangle \rightarrow |d\rangle$ transitions, are used during these preparation and detection stages. In the initial step of the experiment, $|e\rangle$ and $|f\rangle$ are coupled by $\mathcal{L}_{ef}$. This excitation, combined with spontaneous emission from $|f\rangle$ to $|e\rangle$ and $|g\rangle$ realizes an

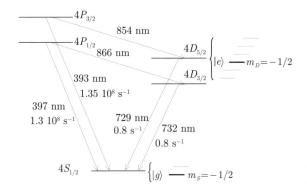

Fig. 8.6 Energy diagram of the Ca$^+$ ion.

optical pumping process which brings, after a few cycles, all the ions in the dark state $|g\rangle$, unable to absorb the $\mathcal{L}_{ef}$ light. Laser $\mathcal{L}_{gd}$, slightly detuned on the red side of the $|g\rangle \rightarrow |d\rangle$ atomic transition, is combined with the optical pumping to eliminate by Doppler cooling a large number of phonons from all the vibration modes of the ion chain. This pre-cooling procedure is completed by a more efficient sideband cooling scheme, which combines the use of laser $\mathcal{L}_{ge}$ tuned to the first red sideband of the $|g\rangle \rightarrow |e\rangle$ qubit transition and laser $\mathcal{L}_{ef}$ coupling the upper qubit state $|e\rangle$ to state $|f\rangle$ (these cooling procedures are described in more detail below). The ion chain is then ready for the quantum information procedure, which involves only the long-lived qubits states $|g\rangle$ and $|e\rangle$ coupled by $\mathcal{L}_{ge}$ and well-protected from the environment constituted by the radiation modes interacting with the short-lived $|f\rangle$ and $|d\rangle$ states.

At the end of the experimental sequence, laser $\mathcal{L}_{gd}$ is switched on to excite the $|g\rangle \rightarrow |d\rangle$ transition. The ions undergo a sequence of absorption–fluorescence cycles, at the fast rate determined by the short spontaneous emission time of $|d\rangle$. This fluorescence signal is used to detect the ion qubit states with near 100% efficiency. After this readout process, the ion chain is initialized again as described above and a new experimental sequence can start. For the scheme to discriminate unambiguously between the qubit states $|e\rangle$ and $|g\rangle$, the detection transition must be closed, with negligible spontaneous emission rate of its upper state towards $|e\rangle$. This precludes the use of $|f\rangle$ for detection and explains why we need different auxiliary states to realize the preparation and detection of the qubit.

This simple model is an idealization. Real ions have more levels, with state degeneracies. In some cases, the transition between the qubit states cannot be easily induced by a single resonant laser, but requires two lasers involved in a Raman process. The above description must be adapted to these more realistic situations. We focus here on the two species – Ca$^+$ and Be$^+$ – which have become the workhorses of most quantum information experiments with trapped ions (Leibfried *et al.* 2003a).

Implementing the qubits with Ca$^+$ ions

The ^{40}Ca$^+$ calcium ion, favourite species of the Innsbruck team, has a single electron outside the complete electronic shells and assumes the simple electronic configuration of the alkaline potassium atom. The main relevant levels, transition wavelengths and radiative decay rates of this ion are shown in Fig. 8.6 (James 1998).

The ground state is $4S_{1/2}$, with no hyperfine structure (there is no nuclear spin).

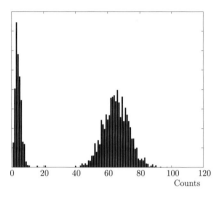

Fig. 8.7 Histogram of the probability distribution of the number of fluorescence photons, emitted by a calcium ion, detected during a 9 ms time slot. The two peaks corresponding to an ion in $|e\rangle$ (low photon number) or $|g\rangle$ (high photon number) are clearly separated, providing a 99.85% detection efficiency. Reprinted from Blatt (2006) with permission of Wiley.

The Zeeman degeneracy between the two $m_S = \pm^1\!/_2$ ground states is removed by a small magnetic field. The $m_S = -^1\!/_2$ substate plays the role of the ground qubit state $|g\rangle$. The first excited states of Ca$^+$ are the $4D_{3/2}$ and $4D_{5/2}$ levels, which can decay towards the ground state only through extremely low-probability quadrupole transitions. As a result, these states have very long radiative lifetimes. The $m_D = -^1\!/_2$ Zeeman substate of the $4D_{5/2}$ level (lifetime 1.2 s) plays the role of the qubit $|e\rangle$ upper state. The quadrupole transition between $|e\rangle$ and $|g\rangle$ at 729 nm is induced by a very stable and intense red laser beam.

The two fine-structure components of the $4P$ level, $4P_{3/2}$ and $4P_{1/2}$, lie above the $4D$ states. They are coupled to the $4S$ ground state by strong electric dipole near-ultraviolet lines at 393 and 397 nm and by near-infrared electric dipole transitions to the $4D$ states (among which the $4P_{1/2} \rightarrow 4D_{3/2}$ line at 866 nm and the $4P_{3/2} \rightarrow 4D_{5/2}$ line at 854 nm). The $4P$ states have very large radiative spontaneous emission rates of the order of $\Gamma_{ps} = 1.3\,10^8$ s^{-1}.

The preparation of $|g\rangle$ is achieved by optical pumping with σ^--circularly polarized laser light on the $4S_{1/2} \rightarrow 4P_{1/2}$ UV transition at 397 nm. A sequence of absorption–spontaneous emission cycles feeds angular momentum in the Ca$^+$ ion, until it falls and stays trapped into the $4S_{1/2}$, $m_S = -^1\!/_2$ state which can no longer absorb light. Another laser at 866 nm is simultaneously exciting the ions out of the $4D_{3/2}$ state in which they may end-up by spontaneous emission following their excitation in the $4P_{1/2}$ level by the optical pumping beam.

Once the $4S_{1/2}$, $m_S = -^1\!/_2$ state is prepared, the ions can be driven by a π-polarized laser pulse at 729 nm into a superposition of this state with the $4D_{5/2}, m_D = -^1\!/_2$ substate. The ions then evolve on a closed two-level transition and all the other ionic levels can be disregarded. The 729 nm laser pulse can either be tuned in exact resonance with the qubit line, or set to resonance on a sideband transition at frequency $\omega_{eg} \pm q\omega_{zi}$ (q: integer, ω_{zi} is one of the vibration frequencies of the ion chain). In the first case, the laser pulse manipulates the internal qubit state, without changing the external motion of the ions. In the latter one, the qubit transition is accompanied by the absorption or emission of q phonons, leading in general, as we will see below, to entanglement between the external and internal degrees of freedom of the ions.

The detection of an ion in $|g\rangle$ is based on resonant laser scattering (see Section 1.3). A Ca$^+$ ion driven by a laser resonant on the $4S_{1/2} \rightarrow 4P_{1/2}$ transition scatters at saturation $\Gamma_{ps}/2 = 7 \cdot 10^7$ photons per second. In order to avoid the interruption

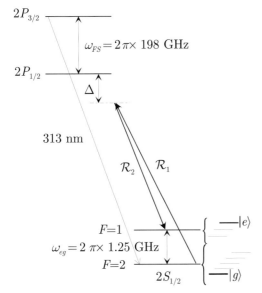

Fig. 8.8 Energy diagram of the Be^+ ion. The Zeeman sublevels of the $2S_{1/2}$ ground state in a small magnetic field are shown at right (opposite Landé factors in $F = 1$ and $F = 2$). The qubit states $|g\rangle$ and $|e\rangle$ (hyperfine sublevels $F = 2, M_F = -2$ and $F = 1, M_F = -1$) are coupled by a Raman transition involving lasers $\mathcal{R}_1$ and $\mathcal{R}_2$ detuned from the $2S_{1/2} \rightarrow 2P_{1/2}$ and $2S_{1/2} \rightarrow 2P_{3/2}$ transitions by Δ and $\Delta + \omega_{FS}$.

of the ion's fluorescence due to a quantum jump from the $4P_{1/2}$ state towards $4D_{3/2}$, the 866 nm repumping laser is used to drive the ion back into the cycling $4S_{1/2} \rightarrow 4P_{1/2}$ transition and restore the fluorescence light. Not all the fluorescence photons are detected. The solid angle of the detection optics in a typical experiment is 10^{-2} sr and the photon detectors quantum efficiency in the range 0.1 to 0.2. Altogether, a single ion produces at least 10^4 photon counts per second, a very large signal. A few milliseconds are enough to detect it.

When the ion is initially in $|e\rangle$, the detection laser is not resonant and there is no photon scattering. The fluorescence level is thus very different for the two qubit states. In a typical experiment, less than 1 background photon is detected per millisecond when the ion is in $|e\rangle$, while nearly ten are counted on the average when the ion is in $|g\rangle$ (Roos *et al.* 1999). Figure 8.7 shows the experimental histogram of the detected photon number probability distribution. The detection time slot is 9 milliseconds. The state of the ion can be inferred from the detected photon number with a 99.85% fidelity. This efficient and selective detection is a major asset of ion traps for quantum information.[2]

Once the ions have been detected, they are ready to be used again in a new experimental sequence. The upper qubit state $|e\rangle$ is rapidly recycled to the ground state by applying a pulse of laser light at 854 nm exciting the ions to $4P_{3/2}$ from which they decay rapidly to $4S_{1/2}$, before being optically pumped again into $|g\rangle$.

Implementing the qubits with Be^+ ions

The Boulder group prefers the $^9Be^+$ beryllium ion (Leibfried *et al.* 2003b), whose relevant levels are shown in Fig. 8.8. The $|e\rangle$ and $|g\rangle$ levels are then the two substates

---

[2]The same efficient detection by laser-induced fluorescence also applies to neutral atoms in optical traps.

with minimum angular momentum of the ion's hyperfine levels, $F = 1, m_F = -1$ and $F = 2, m_F = -2$, in the $2S_{1/2}$ ground state (the nuclear spin of ^{9}Be is $I = 3/2$). These substates are basically stable and separated by a hyperfine frequency interval in the microwave range ($\omega_{eg}/2\pi = 1.25$ GHz). They are selected among all the hyperfine levels by optical pumping, using σ^--circularly polarized light exciting the ions from the $F = 1$ and $F = 2$ ground states to the $2P_{1/2}$ and $2P_{3/2}$ states (UV transitions around 313 nm; the fine structure splitting between the $2P_{3/2}$ and $2P_{1/2}$ states is $\omega_{FS}/2\pi = 198$ GHz).

A circularly polarized laser tuned on the $2S_{1/2}, F = 2, m_F = -2 \rightarrow 2P_{3/2}, F = 3, m_F = -3$ closed transition is used for detection. It produces a large number of fluorescence photons at 313 nm whose detection signals the presence of the ion in $|g\rangle$. On the contrary, the ions do not scatter photons when they are in $|e\rangle$, which is not coupled to the detection laser. As in the Ca$^+$ case, a very good discrimination between the two qubit states is achieved within a time of a few milliseconds.

In between preparation and detection, the coupling between the Be$^+$ qubit states is achieved by a resonant radiative process. A direct excitation of the $|g\rangle \rightarrow |e\rangle$ transition by absorption of a single photon (as in the Ca$^+$ ion case) would not be convenient. The long wavelength of this microwave photon would make it difficult to address the ions of a chain individually. Even more importantly, the small momentum of the photon would result in an extremely weak coupling between the internal and external degrees of freedom of the ions.

Instead of a direct one-photon coupling, one resorts to a two-photon Raman process. The ions are excited with a combination of two lasers $\mathcal{R}_1$ and $\mathcal{R}_2$ whose wavelengths are close to 313 nm. Their frequency difference $\omega_{\ell 1} - \omega_{\ell 2}$ matches the frequency of the transition to be induced in the ion. This transition is accompanied by the absorption of one photon of beam $\mathcal{R}_1$ at $\omega_{\ell 1}$ and the stimulated emission of a photon in beam $\mathcal{R}_2$ at frequency $\omega_{\ell 2}$. The polarizations and detunings of the laser beams are chosen to ensure that $|g\rangle$ (angular momentum $m_F = -2$ along the quantization axis) is coupled only to $|e\rangle$ (angular momentum $m_F = -1$), the other $2S_{1/2}$, $m_F = 0, +1$ substates of the ion's $F = 1$ hyperfine ground state remaining unexcited.

The momentum transfer from the lasers to the ions during a Raman transition is $\Delta \mathbf{k} = \mathbf{k}_{\ell 1} - \mathbf{k}_{\ell 2}$ ($\mathbf{k}_{\ell 1}$ and $\mathbf{k}_{\ell 2}$ being the wave vectors of the two beams). Its absolute value depends upon the angle between the two laser beams. For counter-propagating Raman beams, it reaches the value $2\omega_{\ell 1}/c \approx 2\omega_{\ell 2}/c$, much larger than it would be if a single microwave photon were absorbed by the ions. We will soon see the importance of this large momentum transfer for ion manipulation.

The Raman lasers are detuned by Δ and $\Delta + \omega_{FS}$ – both quantities being of the order of a few hundred GHz – from resonance with the $2S \rightarrow 2P_{1/2}$ and $2S \rightarrow 2P_{3/2}$ transitions. This Raman detuning, while large enough to prevent the population of the $2P$ states during the qubit manipulation, is small enough to produce large Raman amplitudes with available UV laser powers.

The qubit manipulation in Be$^+$ seems more complex than in Ca$^+$, since it requires two laser beams instead of one. These two beams are however derived from a single master laser, by using an acousto-optic crystal which produces two beams whose frequency difference $\omega_{\ell 1} - \omega_{\ell 2}$ is tuned to the appropriate value. The spectral purity of the frequency difference is determined by that of the microwave signal driving the

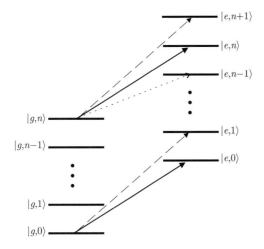

Fig. 8.9 Uncoupled states of the ion–spring system. The thick arrow represents a carrier frequency transition, the dotted and dashed arrows the first red and blue sidebands.

acousto-optic crystal, and not by the frequency stability of the UV laser feeding this crystal. Since only $\omega_{\ell 1} - \omega_{\ell 2}$ has to be precisely defined, the Raman lasers need not be individually as stable as the single laser used in the Ca^+ case. This is an asset for the Be^+ experiments.

8.1.4 Manipulating the qubits with laser light

Qubit manipulation is achieved by laser radiation resonantly coupling $|g\rangle$ and $|e\rangle$. This interaction can be tailored to excite not only the ions internal states, but also their vibrational motion in the trap. Let us examine in more detail this coupling, which is at the heart of all quantum information experiments with trapped ions. We start by describing a single ion oscillating in a linear trap at frequency ω_z and assume that the $|g\rangle \rightarrow |e\rangle$ transition falls in the optical domain (Ca^+ case). We will then extend the analysis to a microwave $|g\rangle \rightarrow |e\rangle$ transition (Be^+) and to an $\mathcal{N}$-ion crystal.

Let us first consider the uncoupled states, $|g, n\rangle$ and $|e, n\rangle$, representing the ion in $|g\rangle$ or $|e\rangle$ with n vibrational quanta. They are the eigenstates of the Hamiltonian:

$$H_0 = H_z + H_i = \hbar\omega_z a^\dagger a + \frac{\hbar\omega_{eg}}{2}\sigma_Z \,, \tag{8.15}$$

where $\sigma_Z = |e\rangle\langle e| - |g\rangle\langle g|$ and a and $a^\dagger$ are the phonon annihilation and creation operators. Figure 8.9 shows these uncoupled levels.

A laser at frequency ω_{eg} couples resonantly $|g, n\rangle$ with $|e, n\rangle$. We show now that a laser beam can also induce sideband transitions between $|g, n\rangle$ and $|e, n \pm p\rangle$ when it is tuned at the frequencies $\omega_{eg} \pm p\omega_z$.

Ion–laser interaction Hamiltonian

The laser beam (frequency ω_ℓ, wave vector $\mathbf{k}_\ell$), which contains a huge number of photons, is described classically. To study its interaction with the ion, we use the general model of Section 3.3.2, which views the laser–ion coupling as the interaction of a classical field with a spin. We must, however, complete this model in one important aspect. An essential point in the present discussion, which we did not consider then, is that the spin interacts with a field whose phase depends upon the ion's changing position Z. Taking this effect into account, the interaction Hamiltonian is:

$$H_{iz} = -i\hbar \frac{\Omega_\ell}{2} e^{-i\omega_\ell t} e^{ik_\ell Z \cos\theta} e^{-i\varphi} \sigma_+ + \text{h.c.} , \tag{8.16}$$

where $\sigma_+ = |e\rangle \langle g|$ is the spin-raising operator, Ω_ℓ the laser-induced Rabi frequency,[3] θ the angle between $\mathbf{k}_\ell$ and Oz and φ the laser phase at the origin, which is under experimenter's control. The factor $\exp(ik_\ell Z \cos\theta)$ in eqn. (8.16) describes the laser phase slip between the actual position of the ion and the origin. Let us stress that, in this quantum treatment, Z is the position operator of the ion. The interaction Hamiltonian H_{iz} is thus an operator both for the spin (σ_+ term) and for the spring.

Expressing Z as $z_0(a + a^\dagger)$, we write the position-dependent phase factor as:

$$e^{ik_\ell Z \cos\theta} = e^{i\eta(a+a^\dagger)} = e^{-\eta^2/2} e^{i\eta a^\dagger} e^{i\eta a} , \tag{8.17}$$

where we make use of Glauber's identity (eqn. 3.39) and define:

$$\eta = \eta_0 \cos\theta , \tag{8.18}$$

with:

$$\eta_0 = k_\ell z_0 . \tag{8.19}$$

The dimensionless quantity η_0 is called the Lamb–Dicke parameter. It measures (within a 2π-factor) the extension of the oscillator's ground state wave function, z_0, in units of the optical wavelength, $2\pi/k_\ell$. Typical ion traps are tightly confining and the Lamb–Dicke parameter is small (varying from 0.05 to 0.3 in most experiments). We are thus justified to expand the interaction Hamiltonian in power series of η.

Before doing this, let us note an equivalent expression of η_0:

$$\eta_0 = \sqrt{\frac{\hbar k_\ell^2}{2m\omega_z}} = \sqrt{\frac{E_r}{\hbar\omega_z}} , \tag{8.20}$$

where $E_r = \hbar^2 k_\ell^2 / 2m$ is the recoil of the ion under photon absorption. The Lamb–Dicke parameter appears also as the square root of the ratio between the ion recoil and the phonon energies. It measures the impact of a single photon emission or absorption on the ionic motion. When η_0 is large, the motion is appreciably modified by the absorption of a laser photon, since many phonons can be created in the recoil process, while it is insensitive to photon exchange when η_0 is vanishingly small. Note also that η can be tuned between 0 and η_0 by adjusting the incidence angle θ (eqn. 8.18).

Let us now expand the exponential operators in eqn. (8.17) in power series of η:

$$e^{ik_\ell Z \cos\theta} = e^{-\eta^2/2} \sum_{m=0;\, n=0}^{\infty} (i\eta)^{m+n} \frac{(a^\dagger)^m a^n}{m!n!} . \tag{8.21}$$

---

[3]In the Ca$^+$ experiment, this frequency is proportional to the matrix element of the electric quadrupole operator between $|e\rangle$ and $|g\rangle$. This operator replaces the electric dipole in the spin–spring model of Section 3.3.2, which remains valid with this change.

The double sum can be split into three parts, the sum over $m < n$, i.e. $n = m+q$, $q > 0$, the sum over $m > n$, i.e. $m = n + q$, $q > 0$ and finally the sum over $m = n$. After a term rearrangement, we get:

$$e^{ik_\ell Z \cos\theta} = e^{-\eta^2/2} \left[f_0(a^\dagger a) + \sum_{q>0} (i\eta)^q f_q(a^\dagger a) a^q + \sum_{q>0} (i\eta)^q \left(a^\dagger\right)^q f_q(a^\dagger a) \right] , \quad (8.22)$$

where:

$$f_q(a^\dagger a) = \sum_{m=0}^{\infty} (-1)^m \eta^{2m} \frac{\left(a^\dagger\right)^m a^m}{m!(m+q)!} . \quad (8.23)$$

Note that, for a very small Lamb–Dicke parameter:

$$f_q(a^\dagger a) \approx 1/q! . \quad (8.24)$$

Inserting these expressions into the interaction Hamiltonian H_{iz} and switching to an interaction representation with respect to the uncoupled Hamiltonian H_0, we finally get:

$$\begin{aligned}
\widetilde{H_{iz}} = {} & -i\hbar\frac{\Omega_\ell}{2}e^{-\eta^2/2}e^{-i\varphi}\Big[f_0(a^\dagger a)e^{i(\omega_{eg}-\omega_\ell)t}\,\sigma_+ \\
& + \sum_{q>0} (i\eta)^q e^{i(\omega_{eg}-\omega_\ell-q\omega_z)t}\, f_q(a^\dagger a)a^q\sigma_+ \\
& + \sum_{q>0} (i\eta)^q e^{i(\omega_{eg}-\omega_\ell+q\omega_z)t} \left(a^\dagger\right)^q f_q(a^\dagger a)\sigma_+ \Big] + \text{h.c.} \quad (8.25)
\end{aligned}$$

This equation will allow us to describe the combined manipulation of the qubit and oscillator degrees of freedom under various circumstances. It exhibits modulated terms corresponding to a *carrier* resonance for $\omega_\ell = \omega_{eg}$ and to *red* and *blue* sideband resonances below and above the carrier frequency for $\omega_\ell = \omega_{eg} \mp q\omega_z$. These resonances are easy to account for by a classical argument. The laser field is 'seen' in the ion's frame as a monochromatic carrier perturbation, whose frequency is modulated by the Doppler effect due to the ion's motion. This frequency modulation produces sidebands in the laser light spectrum seen by the ion. When the carrier frequency, or one of the sideband frequencies of this spectrum corresponds to the transition frequency ω_{eg}, the laser induces a resonant process between the qubit states $|e\rangle$ and $|g\rangle$.

The general expression (8.25) looks complicated, but it considerably simplifies with the help of the rotating wave approximation (RWA) (Chapter 3). When the laser frequency fulfils one of the resonance conditions $\omega_\ell = \omega_{eg} \pm q\omega_z$, a single term in eqn. (8.25) becomes time-independent. All others evolve at frequencies at least equal to ω_z. Assuming that ω_z^{-1} is much smaller than all time constants involved, we can invoke the RWA approximation and neglect in the Hamiltonian all terms but the resonant one. This 'resolved sidebands' approximation requires, in particular, $\Omega_\ell \ll \omega_z$ and breaks down for an ion driven by too intense a laser. From now on, we shall only consider this limit and examine successively the coupling Hamiltonian for a carrier excitation, a red or a blue sideband resonance.

Carrier frequency excitation

Let us consider the $q = 0$ term in eqn. (8.25) and assume first that η is so small that $f_0 \approx 1$ and $\exp(-\eta^2/2) \approx 1$. The interaction Hamiltonian then becomes:

$$\widetilde{H_{iz}} \to H_c = -i\hbar \frac{\Omega_\ell}{2} \left(e^{-i\varphi}\sigma_+ - e^{i\varphi}\sigma_- \right) . \tag{8.26}$$

We recover eqn. (3.152) describing the interaction of a spin with a classical field. The carrier frequency excitation addresses only the spin degree of freedom and leaves the motion unchanged: it couples $|g, n\rangle$ and $|e, n\rangle$. All single-qubit gates (arbitrary rotations of the spin on the Bloch sphere) can be achieved by properly tuning the field phase φ, the Rabi frequency Ω_ℓ and the atom–laser interaction time.

When the Lamb–Dicke parameter is not very small, the $\exp(-\eta^2/2)f_0$ factor in the first line of eqn. (8.25) must be taken into account. Since it is diagonal in the Fock state basis, $\widetilde{H_{iz}}$ still couples $|g, n\rangle$ and $|e, n\rangle$ only, albeit with an effective Rabi frequency, $\Omega_\ell \langle n| f_0 |n\rangle \exp(-\eta^2/2)$, which depends upon the phonon number. This dependency can create an entanglement between spin and motion if the system is initially in a superposition of motional states. However, this effect occurs only for very long ion–laser interaction times and it can be safely neglected for small η values.

Red sideband excitation

Assume now that the driving laser is tuned to the qth red sideband: $\omega_\ell = \omega_{eg} - q\omega_z$. We will focus, in this paragraph and in the following, to the small Lamb–Dicke parameter regime, in order to avoid unnecessary complications. We will also, for the sake of definiteness, assume that $\varphi = \pi/2$. With these conventions, the interaction Hamiltonian becomes:

$$\widetilde{H_{iz}} \to H_{r;q} = \hbar g_q \left(D_q \sigma_+ + D_q^\dagger \sigma_- \right) , \tag{8.27}$$

with:

$$D_q = (ia)^q , \tag{8.28}$$

and:

$$g_q = -\frac{\Omega_\ell \eta^q}{2q!} . \tag{8.29}$$

In the simple case $q = 1$, $H_{r;q}$ reduces to:

$$H_{r;1} = -i\hbar \frac{\Omega_\ell \eta}{2} \left(a\sigma_+ - a^\dagger \sigma_- \right) . \tag{8.30}$$

The analogy with the resonant CQED Jaynes–Cummings Hamiltonian is conspicuous (eqn. 3.167). The first red sideband couples $|g, n\rangle$ and $|e, n-1\rangle$ and induces a quantum Rabi oscillation between these states. The excitation of the spin comes together with the absorption of a phonon in the motional state. The effective 'vacuum Rabi frequency' is, here, $\Omega_\ell \eta$. It can be tuned to some extent. We should keep in mind, though, that the resolved sidebands approximation requires $\Omega_\ell \ll \omega_z$. Hence, the coupling cannot be made arbitrarily large. Note that, as in CQED, there is no evolution when the initial state is $|g, 0\rangle$. The red sideband excitation leads to an ion dynamics conditioned by the phonon number, an ideal situation to realize entanglement between internal and external degrees of freedom of the ions. Starting for instance from $|e, 0\rangle$, a $\pi/2$-pulse on the first red sideband prepares a superposition with equal weights of $|e, 0\rangle$ and $|g, 1\rangle$, with maximum entanglement between the qubit and ion motional states.

Blue sideband excitation

When the blue sideband resonance condition, $\omega_\ell = \omega_{eg} + q\omega_z$, is fulfilled the interaction Hamiltonian becomes:

$$\widetilde{H_{iz}} \rightarrow H_{b;q} = (-1)^q \hbar g_q \left(D_q^\dagger \sigma_+ + D_q \sigma_- \right) , \qquad (8.31)$$

reducing, for $q = 1$ (first blue sideband) to:

$$H_{b;1} = -i\hbar \frac{\Omega_\ell \eta}{2} (a^\dagger \sigma_+ - a\sigma_-) . \qquad (8.32)$$

The first blue sideband interaction implements an 'anti-Jaynes–Cummings model', in which the spin makes a transition from $|g\rangle$ to $|e\rangle$ while *emitting* a phonon in the vibration mode. Such a process does not conserve energy in CQED and is therefore negligible in this context. In ion traps, the driving laser provides the extra energy required for this counter-intuitive transition. The anti-Jaynes–Cummings coupling can be used, as well as the Jaynes–Cummings one, to achieve entanglement between the qubit and motional states of an ion.

Raman excitation

This analysis can be extended straightforwardly to the case of a qubit realized with two hyperfine ground states and excited via a Raman process involving two lasers with frequencies $\omega_{\ell 1}$ and $\omega_{\ell 2}$ and wave vectors $\mathbf{k}_{\ell 1}$ and $\mathbf{k}_{\ell 2}$ (Be$^+$ ion experiments). We merely have to replace in eqn. (8.25) Ω_ℓ by Ω_R, the Rabi frequency associated to the Raman process (which is proportional to the product of the two laser fields amplitudes), ω_ℓ by $\omega_{\ell 1} - \omega_{\ell 2}$ and $\eta = k_\ell z_0 \cos\theta$ by $(\mathbf{k}_{\ell 1} - \mathbf{k}_{\ell 2}) \cdot \mathbf{u}_z z_0$ [the same replacement of η has of course to be made in the expression of the f_q functions in eqn. (8.23)]. The conditions for carrier, red and blue sideband excitations become simply $\omega_{\ell 1} - \omega_{\ell 2} = \omega_{eg}$ and $\omega_{\ell 1} - \omega_{\ell 2} = \omega_{eg} \mp q\omega_z$ respectively.

Note that in the Be$^+$ case the Lamb–Dicke parameter can be tuned by merely changing the angle between the two Raman beams. From a value essentially equal to zero when these beams are parallel, it increases up to the value $\sqrt{2\hbar k_\ell^2 / m\omega_z}$ when they are counter-propagating along the direction of the ion's oscillation. Adjusting η by changing the relative directions of the Raman beams will prove useful in the following.

Laser manipulation of an $\mathcal{N}$-ion crystal

The above analysis also applies, with slight modifications, to describe the manipulation of several ions trapped in a one-dimensional potential. In the simplest situation, the ions are addressed individually, with laser beams whose waist is smaller than the ions separations. Assume that the ith ion in the linear chain is excited by a single laser (or a couple of Raman lasers). Its coordinate z_i can be expanded as a linear combination of the Z_j coordinates of the collective ion oscillators (with frequencies ω_{zj}). Classically, the ion motion has Fourier components at all the ω_{zj}'s, resulting in sidebands at frequencies $\omega_{eg} \pm q\omega_{zj}$ in its spectrum. Since these frequencies are well-separated and incommensurate with each other, a laser tuned at a given frequency ω_ℓ (or a couple of Raman lasers tuned at $\omega_{\ell 1} - \omega_{\ell 2}$) is able to excite only one of the carrier or sideband lines at a time (again a consequence of the RWA approximation).

By choosing properly the laser frequencies, one can either manipulate the ion's internal qubit state (by driving the carrier transition), or excite a well-defined vibration mode (by driving the corresponding red or blue sideband transition). Note that the Lamb–Dicke parameter generally depends upon the vibration mode which is being excited. If it is the CM mode, the mass of the collective oscillator is $\mathcal{N}m$ and its zero-point fluctuation, hence its Lamb–Dicke parameter, scale as $1/\sqrt{\mathcal{N}}$. This can be understood intuitively. When the number of ions increases, it becomes more and more difficult to set the whole crystal in motion by acting on a single ion, which is reflected by the decrease of the corresponding Lamb–Dicke parameter.

In some experiments, notably those performed with Be^+ (Leibfried *et al.* 2003b), all the ions are addressed collectively by the same laser excitation. The laser–ion Hamiltonian is then the symmetrical sum of $\mathcal{N}$ ion–laser couplings. Here again, by expanding the z_i's on the Z_j's, one obtains an effective Hamiltonian depending on the resonance condition. If the carrier transition is excited, this Hamiltonian acts symmetrically on the internal states of the $\mathcal{N}$-qubit system, while leaving the vibration modes unperturbed. If a sideband resonance condition is fulfilled, the collective ion–laser coupling results in a complex entangled state involving multiple excitations and de-excitations of the vibration mode. We will not consider such sideband excitations here.

8.1.5   Tomography of the qubit density operator

Detecting the ions in the qubit states $|g\rangle$ or $|e\rangle$ by laser-induced fluorescence can be combined with single-qubit operations to yield a general method for determining the complete density matrix of the qubit states of an ionic crystal. Let us first consider the case of a single ion repeatedly prepared in a quantum state defined by the density operator ρ. The fluorescence data collected on an ensemble of realizations of the experiment determine the probability of finding the ion in $|g\rangle$ and in $|e\rangle$, thus realizing a measurement of the average value $\mathrm{Tr}(\rho\sigma_Z)$ of the operator σ_Z.

One can similarly measure the average values $\mathrm{Tr}(\rho\sigma_X)$ and $\mathrm{Tr}(\rho\sigma_Y)$. For this, we simply have, before shining the detection laser, to rotate the qubit state by applying a $\pi/2$-laser pulse with a convenient phase φ on the $|g\rangle \to |e\rangle$ transition. The phase φ is set so that the pulse brings the OX or OY axis of the Bloch sphere along OZ. A repetitive detection of the $|e\rangle$ and $|g\rangle$ populations on a large number of experimental realizations then corresponds to a measurement of the average value of σ_X and σ_Y. The qubit rotation is produced by a single 729 nm laser pulse at frequency ω_{eg} in the case of Ca^+, and by a pair of Raman lasers such that $\omega_{\ell 1} - \omega_{\ell 2} = \omega_{eg}$ in the case of Be^+. Once the three measurements $P_i = \langle \sigma_i \rangle$ ($i = X, Y, Z$) of the qubit polarization components have been made, the qubit density operator is obtained as (eqn. 4.12):

$$\rho = \frac{1}{2}\left(\mathbb{1} + \sum_i P_i\sigma_i\right) . \tag{8.33}$$

The method generalizes straightforwardly to $\mathcal{N}$ ions, provided one can detect separately the fluorescence from each ion and rotate its qubit states independently from the others. This requires a separation between the ions larger than the laser waists. The density operator $\rho_{\mathcal{N}}$ of an $\mathcal{N}$ qubit system has $2^{2\mathcal{N}}$ matrix elements. Generalizing eqn. (8.33), one can expand $\rho_{\mathcal{N}}$ on the basis of the $4^{\mathcal{N}}$ operator products $\sigma_{j_1}^{(1)}\sigma_{j_2}^{(2)}\cdots\sigma_{j_{\mathcal{N}}}^{(\mathcal{N})}$ (with $j_1, j_2, \ldots, j_{\mathcal{N}} = 0, X, Y, Z$ and $\sigma_0 = I$):

$$\rho_{\mathcal{N}} = \frac{1}{2^{\mathcal{N}}}\sum \lambda_{j_1,j_2,\ldots,j_{\mathcal{N}}}\sigma_{j_1}^{(1)}\sigma_{j_2}^{(2)}\cdots\sigma_{j_{\mathcal{N}}}^{(\mathcal{N})} . \tag{8.34}$$

The $\lambda_{j_1,j_2,\ldots,j_{\mathcal{N}}}$ coefficients are real, between -1 and $+1$. For $\mathcal{N} = 1$, we retrieve the polarization components of a single qubit with the correspondence $\lambda_0 = 1, \lambda_i = P_i$. Noticing that the trace of a product of two σ_j operators belonging to the same ion with different j's is zero, it is easy to deduce that each λ coefficient is equal to the expectation value of the corresponding product of σ_j operators:

$$\lambda_{j_1,j_2,\ldots,j_{\mathcal{N}}} = \mathrm{Tr}(\rho_{\mathcal{N}}\sigma_{j_1}^{(1)}\sigma_{j_2}^{(2)}\cdots\sigma_{j_{\mathcal{N}}}^{(\mathcal{N})}) . \tag{8.35}$$

It is simply obtained from separate measurements of the $\mathcal{N}$ ion qubits, after a proper combination of rotations which map, for each ion for which $j \neq 0$, the eigenvectors of σ_i onto those of σ_Z. When an ion fluoresces, the corresponding σ_i operator assumes the value -1, while it takes the value $+1$ if the ion does not fluoresce. For ions with $j = 0$, the corresponding $\sigma_0 = \mathbb{1}$ operator always assumes the value $+1$ (which need not be measured). The result of a measurement sequence, being a product of $+1$ and -1 terms, is necessarily equal to ± 1. Each such measurement is repeated a large number of times, its average yielding the corresponding $\lambda_{j_1,j_2,\ldots,j_{\mathcal{N}}}$ coefficient, whose absolute value is obviously bounded by 1.

The procedure is resumed with a different rotation mapping, a new λ coefficient is obtained and so on. Altogether, $3^{\mathcal{N}}$ pulse settings are required, since there are three possible choices for each ion (do nothing on a given ion if σ_0 or σ_Z has to be measured on it, or apply a pulse mapping either OX or OY onto OZ). This method, which fully determines the state of the $\mathcal{N}$ qubit system, is called 'ion qubit tomography'. It requires a time increasing exponentially with the number of qubits, becoming impractical for large systems.

8.1.6 Doppler cooling

The initial kinetic energy of the ions after their capture following electron impact ionization or photo-ionization is of the order of the trap depth, a fraction of an electron-volt. It corresponds to a huge number of phonons ($\sim 10^9$) which must be suppressed before starting quantum state manipulation. This is achieved by a combination of two procedures, called Doppler and sideband cooling (Eschner *et al.* 2003). Let us start by analysing Doppler cooling. We consider first the case of a single ion, and then generalize the discussion to an ion ensemble.

For the sake of simplicity, we will discuss Doppler cooling for an ion having the simple state configuration of Figure 8.5. The ion, initially in $|g\rangle$, interacts with the laser $\mathcal{L}_{gd}$ tuned on the red side of the $|g\rangle \rightarrow |d\rangle$ transition, with a negative detuning Δ_0 equal to a fraction of the transition natural width Γ_{dg}. We can disregard the other states in the analysis of the cooling process. This simple model can easily be adapted to the Doppler cooling of the real Ca^+ and Be^+ ions, with slight complications arising from the combination of the cooling process with the optical pumping which initializes the qubit state. We assume that the $|d\rangle \rightarrow |g\rangle$ transition has a spontaneous emission rate $\Gamma_{dg} \sim 10^8$ s^{-1}, of the order of the spontaneous emission rates of the $4P_{1/2} \rightarrow 4S_{1/2}$ Ca^+ transition or of the $2P_{1/2} \rightarrow 2S_{1/2}$ transition in the Be^+ ion. This rate is in any case much larger than the ion oscillation frequency:

$$\Gamma_{dg} \gg \omega_z . \tag{8.36}$$

This condition implies that the sidebands of the ion oscillation are not resolved in the optical spectrum of the cooling $|g\rangle \rightarrow |d\rangle$ transition.

The ion oscillates along the z axis of the linear trap, with the 'cooling' laser beam $\mathcal{L}_{gd}$ propagating along the same direction. Under these conditions, the ion scatters more photons during half of its oscillation cycle, when it is moving in the direction opposite to the laser beam, since it is then tuned closer to resonance by the Doppler effect. The net effect, averaged over many oscillations is a 'viscous force', which damps the ion vibration. Hence the name of 'Doppler cooling' given to this process first proposed by D. J. Wineland and H. Dehmelt (1975).[4]

Average cooling force

The spontaneous emission rate on the cooling transition is so strong that the ion scatters tens of photons during an oscillation period (typically $2\pi/\omega_z = 2 \cdot 10^{-6}$ s ; we adopt here the trap parameters discussed in Section 8.1.1). The effect of the laser beam can then be described as a radiation pressure force acting on the ion at a well-defined point in phase space, as if it were a free particle. This force merely adds to the restoring force of the trap, and the ion motion can be described classically. The book-keeping of the momentum exchange between the ion and the laser beam is then very simple. Calling Ω_ℓ the classical Rabi frequency of the ion in the intense laser field, $\mathbf{k}_\ell$ the laser wave vector aligned with the motion axis Oz and v the ion's velocity, we compute first the probability $P_d(v)$ of finding the ion in the excited state $|d\rangle$, which depends through the Doppler effect on v:

$$P_d(v) = \frac{1}{2} \frac{\Omega_\ell^2/2}{\Omega_\ell^2/2 + \Gamma_{dg}^2/4 + (\Delta_0 - k_\ell v)^2} . \tag{8.37}$$

During each scattering cycle, the ion receives a net momentum $\hbar(\mathbf{k}_\ell - \mathbf{k}'_\ell)$ where $\mathbf{k}'_\ell$ is the wave vectors of the scattered photon. Since this photon has equal probabilities of being emitted in opposite directions, the average recoil of the ion per scattering event is $\hbar k_\ell$ along the laser beam direction. The average number of scattering events per unit time is $\Gamma_{dg} P_d(v)$ and the mean radiation force experienced by the ion is:

$$F(v) = \hbar k_\ell \Gamma_{dg} P_d(v) . \tag{8.38}$$

---

[4]A similar cooling mechanism was proposed by T. W. Hänsch and A. L. Schawlow (1975) for neutral atoms in a standing wave. Doppler cooling of bound particles was demonstrated for the first time on magnesium ions in a Penning trap (Wineland et al. 1978).

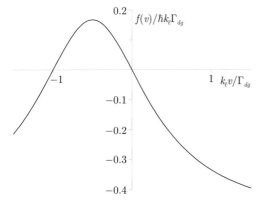

Fig. 8.10 Principle of Doppler cooling: variation versus v (in units of Γ_{dg}/k_ℓ) of the velocity-dependent part $f(v)$ of the radiation pressure force experienced by an ion moving along a red detuned laser beam (in units of $\hbar k_\ell \Gamma_{dg}$). The negative slope of $f(v)$ around $v = 0$ corresponds to a damping effect for slow atoms (such that $|k_\ell v| < \Gamma_{dg}/2$). The force is computed for the optimal values of Ω_ℓ and Δ_0 given in text.

It is convenient to express this force as the sum of a constant part, $F(0)$, plus a velocity-dependent contribution $f(v)$, which amounts to defining $f(v)$ as $F(v) - F(0)$. The work of the constant force $F(0)$ averages out over an oscillation period and does not contribute to the ion's damping. The $f(v)$ part, represented versus v in Fig. 8.10, reflects the variation of the photon scattering rate due to the Doppler effect and is responsible for the cooling.

We will not consider its general expression, but only its limit when the condition $|k_\ell v| \leq \Gamma_{dg}/2$ is satisfied. The excursion of the Doppler shift $k_\ell v$ experienced by the ion during an oscillation is then smaller than the transition linewidth, meaning that the absorption rate varies quasi-linearly with the ion velocity. The velocity-dependent force $f(v)$ is then, to a good approximation, equal to $vdF(v)/dv$. Computing this force for the optimal values of the laser parameters $\Omega_\ell = \Gamma_{dg}$ and $\Delta_0 = -\Gamma_{dg}/2$, we find:

$$f(v) = -\frac{\hbar k_\ell^2}{4}v \ . \tag{8.39}$$

The ion's motion is thus ruled by the equation:

$$\frac{d^2 z}{dt^2} = -\frac{\hbar k_\ell^2}{4m}\frac{dz}{dt} - \omega_z^2 z \ , \tag{8.40}$$

describing (for $\hbar k_\ell^2/4m \ll \omega_z$) an exponentially damped oscillation, whose average energy decreases with the rate:

$$\frac{1}{T_d} = \frac{\hbar k_\ell^2}{4m} \ , \tag{8.41}$$

equal to $1/2\hbar$ times the recoil energy induced by the absorption of a laser photon.

Fluctuation of the radiative force: the Doppler cooling limit

For the parameters of a calcium (or beryllium) ion, the energy damping time T_d, of the order of 10^{-5} s, is very short. The Doppler cooling process is fast and efficient. It extracts a lot of energy from the ion's motion, but it does not bring it to its ground state. In fact, the Doppler cooling eventually enters in competition with an unavoidable heating mechanism.[5] The viscous force we have considered so far is indeed an average

---

[5]The competition between an average dissipative force which tends to cool the system and a fluctuating heating mechanism is quite general in statistical physics (fluctuation–dissipation theorem).

effect. The ion experiences in addition a fluctuating force due, on the one hand, to the random direction of the spontaneously emitted photons and, on the other hand, to the fluctuation of the number of absorbed photons per unit time. Both effects contribute by terms of the same order of magnitude to the heating of the ion.

Let us consider first the heating term coming from the second effect. With the optimal laser parameters indicated above ($\Omega_\ell = \Gamma_{dg}$, $\Delta = -\Gamma_{dg}/2$), the mean number of absorption events per second is $\Gamma_{dg}/4$. If the distribution of this number is Poissonian, its variance is equal to its mean. Making this assumption, we find that the increase of the atomic momentum variance per unit time is equal to $(\hbar k_\ell)^2 \Gamma_{dg}/4$ and the corresponding increase in the ion's energy is $(\hbar k_\ell)^2 \Gamma_{dg}/8m$. Assuming that the contribution to the ion momentum diffusion coming from the random direction of the photon emission is of a similar magnitude, we get an overall heating rate $\sim (\hbar k_\ell)^2 \Gamma_{dg}/4m$. The net energy change of the ion per unit time, $(dE/dt)_{\text{net}}$, is the algebraic sum of the heating and Doppler cooling rates: $(dE/dt)_{\text{net}} = -(\hbar k_\ell^2 E)/4m + \hbar^2 k_\ell^2 \Gamma_{dg}/4m$. An equilibrium is reached when this sum cancels, which occurs for:[6]

$$E \sim \hbar \Gamma_{dg} \ . \tag{8.42}$$

This equilibrium corresponds, on average, to $n_{\text{th}} \sim \Gamma_{dg}/\omega_z$ phonons left in the ionic motional state. Typically, for the parameters of a calcium ion, n_{th} is of the order of 50 corresponding to a motional temperature in the millikelvin range, with a residual ionic macro-motion amplitude $\sqrt{n_{\text{th}}} z_0 \approx 100$ nm. In a typical Be$^+$ trap, ω_z is about an order of magnitude larger and at most a few phonons are left at equilibrium.

Doppler cooling generalizes easily to an $\mathcal{N}$-ion register. If the laser irradiates symmetrically all the ions, the viscous light force extracts efficiently phonons from the $\mathcal{N}$ vibration modes, cooling each of them to the limit defined by eqn. (8.42). It is not even required that all the ions interact with the cooling laser light. The coupling of $\mathcal{L}_{gd}$ with a single ion of the chain is enough to extract phonons from all the collective modes in which this ion is not motionless. The suppression of collective phonons corresponds to a decrease of the kinetic energy of all the other ions, even though they are not irradiated. This indirect cooling process mediated by the Coulomb interaction between the ions is called 'sympathetic cooling'.

8.1.7 Sideband cooling

In many experiments, Doppler cooling is not sufficient. It must be complemented by the more efficient sideband laser cooling process, which brings all the collective vibration modes of an ion register in their ground state, with a probability which can exceed 99%. The first demonstration of sideband cooling was performed by Diedrich *et al.* (1989) on a mercury ion. We give here a qualitative picture of the sideband cooling mechanism applied to the Ca$^+$ and Be$^+$ ions used in the quantum information

---

In a different context, the balance between fluctuation and dissipation explains, for instance, the steady-state Brownian motion.

[6]The argument presented here is qualitative. The number of photon scattering events per unit time is in fact a non-Poissonian variable, which introduces a correction in the expression of the corresponding heating term. The equipartition between the two heating terms is also an approximation. The cooling limit given by eqn. (8.42) must be taken as an order of magnitude [see Wineland and Itano (1979) and Cohen-Tannoudji (1992) for more complete treatments].

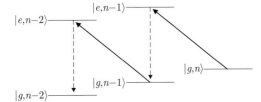

Fig. 8.11 Transitions implied in the sideband cooling mechanism. The laser-induced transitions are depicted by solid arrows, the spontaneous emission processes by dashed arrows.

experiments described below (see also Section 8.3.2). We start again with a single ion, and then generalize to an $\mathcal{N}$ ion system.

Instead of coupling laser light to a transition with a large natural width, as is the case for Doppler cooling, we now irradiate the Ca^+ or Be^+ ion with a laser (or a couple of Raman lasers) inducing the first red sideband transition between the two long-lived qubit states $|g\rangle$ and $|e\rangle$. The laser frequency is set to $\omega_\ell = \omega_{eg} - \omega_z$ in the Ca^+ case, while, for Be^+, the Raman lasers are tuned to a couple of frequencies obeying the condition $\omega_{\ell 1} - \omega_{\ell 2} = \omega_{eg} - \omega_z$. By absorbing laser light on this sideband transition, an ion initially in $|g, n\rangle$ undergoes a transition towards $|e, n-1\rangle$, simultaneously jumping from the lower to the upper qubit state and losing a vibration quantum. The Rabi frequency corresponding to this single-phonon absorption process, $\Omega_\ell \eta$ or $\Omega_R \eta$, can be made of the order of 10^5 s^{-1} with reasonable laser power and Lamb–Dicke parameter in the 0.03–0.2 range.

In order to lose a second phonon, the ion must be recycled in the lower qubit state $|g\rangle$. This is achieved by another laser-induced process coupling $|e\rangle$ to a short-lived state, from which the ion decays spontaneously towards $|g\rangle$. In the Ca^+ case, this coupling is induced by the infrared laser at 854 nm connecting the $4D_{5/2}$ to the excited $4P_{3/2}$ state. In a Be^+ experiment, this recycling stage is achieved by an UV laser at 313 nm inducing a spontaneous Raman process: a laser photon is absorbed while another photon is scattered in an empty mode, accompanied by a jump of the ion from $|e\rangle$ to $|g\rangle$.

In both cases, this recycling transition involves the spontaneous emission of a photon. This process has an amplitude proportional to the matrix element of the ion–radiation field coupling between the initial and final states involved. If the process is accompanied by the gain or loss of a vibration phonon, this amplitude is multiplied by the Lamb–Dicke parameter corresponding to the momentum of the emitted photon. Since this parameter is very small compared to 1, the ion is most likely to emit a photon without changing the phonon number.[7] The recycling transition thus brings the ion from $|e, n-1\rangle$ to $|g, n-1\rangle$. The ion is then subjected again to a red sideband transition $|g, n-1\rangle \rightarrow |e, n-2\rangle$ and so on. Phonons are thus dissipated one by one through a succession of sideband pulses sandwiching recycling pulses. The cascading transitions involved in this cooling mechanism are sketched in Fig. 8.11.

After n such cycles, the ion ends up in the final state $|g, 0\rangle$, a 'dark state' for the sideband excitation. The laser (or the couple of lasers) tuned at frequency $\omega_{eg} - \omega_z$ has no effect on this state since there is no available final state for the excitation process. The ion, trapped in $|g, 0\rangle$, has been cooled to its ground state of motion. This cooling

---

[7]The momentum conservation is then ensured by the recoil of the trap as a whole, a process which is reminiscent of the Mössbauer effect for gamma emission by nuclei embedded in a crystal.

Fig. 8.12 Absorption spectrum of a Ca$^+$
ion from $|g\rangle$ to $|e\rangle$ on the red (a) and blue
(b) first sidebands. Open circles: between
Doppler and sideband cooling. Solid circles:
after sideband cooling. The red sideband
has disappeared after sideband cooling, in-
dicating a very low motional temperature.
Reprinted with permission from Roos *et al.*
(1999). © American Physical Society.

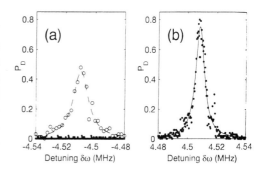

mechanism requires that the sideband transitions are resolved within the linewidth
Γ_{eg} of the qubit transition, which corresponds to the condition:

$$\Gamma_{eg} \ll \omega_z , \qquad (8.43)$$

opposite to the condition for Doppler cooling. If the inequality (8.43) is not satisfied,
the transitions $|g,n\rangle \rightarrow |e,n-1\rangle$, $|g,n\rangle \rightarrow |e,n\rangle$ and $|g,n\rangle \rightarrow |e,n+1\rangle$ overlap.
The processes which absorb phonons compete with those which conserve the phonon
number or increase it and the cooling mechanism loses its efficiency.

Sideband cooling is implemented in slightly different ways in Be$^+$ and Ca$^+$. For
Beryllium (Monroe *et al.* 1995b), the procedure alternates, as described above, the
phonon absorption and ion recycling stages. The cooling of the three vibration modes
of the ion has been achieved in an anisotropic trap with $\omega_x \neq \omega_y \neq \omega_z$. All the modes
have been cooled down, reaching a ground state occupation probability larger than
99%. In the case of Ca$^+$ (Roos *et al.* 1999), the sideband excitation was performed
continuously, while the upper $4D_{5/2}$ qubit level was interacting with a c.w. 854 nm laser
beam admixing in it a small fraction of the fast decaying $4P_{3/2}$ state natural width.
The upper qubit state then acquired a 'dressed' linewidth of the order of 10^5 s^{-1}, still
much smaller than the ion vibration frequency $\omega_z \sim 10^6$ s^{-1}.

The final stage of the cooling process is monitored by measuring the sideband
spectrum. Immediately after the cooling stage, a probe laser pulse (or a couple of
Raman lasers) resonant with either the first red or blue sideband is applied to the
ion. The probability that this probe laser (or couple of lasers) excites the ion in $|e\rangle$ is
obtained, after this sideband excitation, by measuring the ion's fluorescence induced
by a strong final detection laser irradiation (see Section 8.1.3). When the probe laser
(or couple of lasers) is applied on the red sideband, the $|g\rangle \rightarrow |e\rangle$ transfer occurs only
if the ion is not in the ground state of motion (since $|g,0\rangle$ is then not coupled to any
level). On the contrary, the excitation probability from $|g,0\rangle$ by the blue sideband
probe remains important at zero temperature (transition towards $|e,1\rangle$). The ratio of
the red to blue sideband excitation transfers to $|e\rangle$ gives thus a measurement of the
residual average phonon number, i.e. of the final motional temperature of the ion.

Figure 8.12 presents the results of the sideband interrogation thermometry on a
Ca$^+$ ion, after Doppler cooling and 6 ms of sideband cooling (Roos *et al.* 1999). The
final probability of finding the ion in $|e\rangle$ is plotted as a function of the probe laser
frequency, tuned around the first red sideband (left) or first blue sideband (right)
resonances. Before sideband cooling (open circles in the left frame), both excitation

probabilities are of the same order of magnitude. After sideband cooling (solid circles in both frame), the red sideband signal is completely negligible compared to the blue one. The inferred population of the ground motional state is 0.999. This corresponds to an extremely low ionic temperature, of the order of 30 μK.

The sideband cooling method can also be applied to an ionic crystal. Each vibration mode is cooled by addressing one of the ions of the string on the corresponding red sideband frequency. The scheme becomes rapidly complex when the ion number and hence the number of modes increases. Alternative methods can be used to reach low temperatures on many modes at the same time (Roos *et al.* 2000).

Sideband cooling is generally used as a second cooling mechanism, after Doppler cooling has removed the bulk of the phonons from the ion's motion. Both processes use light to extract energy from the ion. Doppler cooling uses an intense laser, tuned on a transition having a large natural linewidth. As we have seen above, the vibration quanta are not resolved in this width, so that the ion can be considered as a classical oscillator submitted to a velocity-dependent radiation pressure effect. The process is very similar to the Doppler cooling of atoms in free space. The radiation pressure presenting an unavoidable fluctuation, this cooling mechanism comes to a stop when the conditions for equilibrium between fluctuation and dissipation are fulfilled (eqn. 8.42). Sideband cooling operates on a weak transition, whose linewidth Γ – even though it might be artificially broadened – remains small compared to the ion oscillation frequency. As a result, the sideband spectrum is resolved and a combination of lasers can be employed to extract phonons one by one, in a process which is inherently quantum. In its final stage, the ion falls in a dark state which is not coupled any longer to radiation and which is thus immune to spontaneous emission fluctuations. The limitations of the fluctuation-dissipation theorem are thus circumvented and the ion motional temperature can be reduced to very low values, corresponding to extremely small residual average phonon numbers.

8.2 Engineering ionic states of motion

Once the ions are initialized in their ground vibrational state, laser-light-induced forces can be employed to set them in motion in a controlled way. We describe now various experiments, performed on a single trapped ion, in which Fock and coherent states of vibration are prepared and manipulated. Superpositions of coherent states realize ionic versions of Schrödinger cats which bear strong similarities with their CQED counterparts. The detection of these states is realized by various methods, involving the observation of Rabi oscillations and other atomic interference methods which are again strongly reminiscent of CQED.

8.2.1 Fock states of motion

Starting from the vibrational ground state, phonon-number states are generated by a combination of π-sideband Rabi pulses. In order to prepare the $n = 2$ Fock state of vibration, one realizes for instance the sequence of π-pulses $|g, 0\rangle \rightarrow |e, 1\rangle \rightarrow |g, 2\rangle$, which involves successively the first blue and red sidebands. Alternating in this way

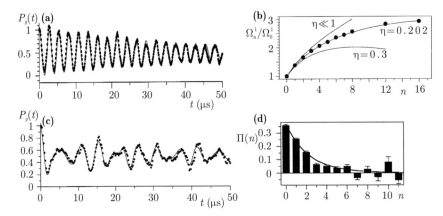

Fig. 8.13 Rabi oscillation of a Be^+ ion. (a) The first blue sideband oscillation corresponding to an $n = 0$ initial phonon number. (b) The Rabi frequency Ω_n^1 of the $|g, n\rangle \rightarrow |e, n+1\rangle$ oscillation plotted versus n in units of Ω_0^1. The theoretical solid lines correspond to the three indicated values of η. The experimental points show that $\eta = 0.202$ in this experiment. (c) The Rabi oscillation observed for a thermal state of motion (average phonon number $n_{th} = 1.5 \pm 0.1$). (d) Corresponding thermal phonon distribution histogram, deduced from a Fourier analysis of the time-dependent signal. Reprinted with permission from Meekhof *et al.* (1996). © American Physical Society.

excitations on the two sidebands, one can prepare any n-phonon state.[8]

The Fock state of the ion's motion is analysed by observing the subsequent Rabi oscillation. Immediately after the sequence of pulses preparing $|g, n\rangle$, a laser resonant with the first red or blue sideband is applied to the ion for a given time t. The qubit is then detected by the fluorescence method. The sequence is repeated many times, the statistics yielding the probability $P_g(t)$ of finding the ion in $|g\rangle$ at time t. The procedure is then resumed for a different time interval and the Rabi oscillation is reconstructed. This experiment has been carried out on both Be^+ (Meekhof *et al.* 1996) and Ca^+ (Roos *et al.* 1999) ions. Figure 8.13(a) shows the first blue sideband Rabi oscillation signal observed on Be^+ for the $n = 0$ Fock state. Many oscillations are observed, with a slow decay due to various experimental imperfections. The analogy with the CQED vacuum Rabi oscillation signal (Fig. 5.26A) is conspicuous. Similar oscillations have been observed for n up to 16.

The frequency of this oscillation is an increasing function of n. The exact expression of the Rabi frequency for the Fock state $|n\rangle$ on the first blue sideband is, according to eqn. (8.25):

$$\Omega_n^1 = \Omega_\ell e^{-\eta^2/2} \eta \langle n+1| a^\dagger f_1(a^\dagger a) |n\rangle . \tag{8.44}$$

Expanding it in powers of η with the help of eqn. (8.23), we get:

$$\Omega_n^1 = \Omega_\ell \eta \sqrt{n+1} \left[1 - \frac{\eta^2}{2}(n+1) + \ldots \right] . \tag{8.45}$$

---

[8] A direct excitation of the nth blue sideband can also be employed, but it turns out to be less practical because it requires a much larger laser power (due to the η^n term in the expression of the transition amplitude).

We note, in front of the bracket, a term which increases as the square root of the phonon number plus one. This CQED-like Rabi frequency is corrected by a saturation factor, function of the Lamb–Dicke parameter. The Rabi frequencies measured in the Be^+ experiment are plotted as a function of n in Fig. 8.13(b). The solid lines show the theoretical variations of Ω_n^1/Ω_0^1 obtained from eqn. (8.44) for $\eta \ll 1$ ('CQED–like' frequency), $\eta = 0.202$ and $\eta = 0.3$. The experimental points fit well the curve corresponding to the intermediate η value. The measurement of the Rabi oscillation thus provides a precise method for the determination of the Lamb–Dicke parameter.

When the ion's vibration is in a superposition or a mixture of Fock states, the Rabi signal reflects it, as in the CQED case, by exhibiting a beating between the corresponding frequencies. Figure 8.13(c) shows the oscillation observed when the ion's motion is probed after the Doppler cooling stage, without applying the sideband cooling procedure (Meekhof *et al.* 1996). The ion's motional temperature is then controlled by the detuning of the Doppler cooling laser. The complex oscillation pattern provides a signature of this temperature. The weights of the Fourier components of this oscillation, shown in Fig. 8.13(d), yield the probabilities $\Pi(n)$ of the successive phonon numbers in the motional state. The exponential variation of $\Pi(n)$ is characteristic of a thermal distribution. From these data, an average phonon number $n_{\text{th}} = 1.5 \pm 0.1$ is obtained, which is in good agreement with an independent determination based on the comparison between the first red and blue sideband components in the ion absorpion spectrum.

8.2.2 Coherent states of motion

Preparing and probing a coherent state

As in the CQED case, a rich variety of effects can be observed by exciting classically the ion oscillator, preparing it in a coherent state or in a superposition of such states. In CQED, coherent states are obtained by coupling the field mode, initially in vacuum, to a classical resonant current. The evolution operator describing this field-feeding process is equivalent to a translation of the field state in its phase space (see Section 3.1.3). In ion trap physics, the same evolution can be obtained by applying on the ion, initially motionless, an electric force oscillating at the ion's vibration frequency. For instance, an oscillating voltage applied across the end cap electrodes in a quadrupolar Paul trap produces near the trap centre a modulated electric field with amplitude E_0, which interacts with the ion through the potential:

$$V(t) = -e_i E_0 Z \sin(\omega_z t - \varphi) = i \frac{e_i z_0 E_0}{2}(a + a^\dagger)\left[e^{i(\omega_z t - \varphi)} - e^{-i(\omega_z t - \varphi)}\right] . \quad (8.46)$$

The RWA approximation in interaction representation keeps only two of the four terms in the development of this expression:

$$\widetilde{V}(t) = i\frac{e_i z_0 E_0}{2}(ae^{-i\varphi} - a^\dagger e^{i\varphi}) . \quad (8.47)$$

The evolution of the ion's motional state resulting from the application of this electric force during a time τ is described by the unitary operator:

$$\widetilde{U}(\tau) = \exp\left[\frac{e_i z_0 E_0 \tau}{2\hbar}(ae^{-i\varphi} - a^\dagger e^{i\varphi})\right] , \quad (8.48)$$

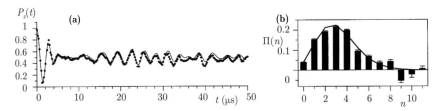

Fig. 8.14 (a) Rabi oscillation of a Be^+ ion prepared in a coherent state of motion with $\bar{n} = 3.1$ phonons on average. The collapse and revival is clearly apparent. (b) Phonon number distribution, obtained by Fourier analysis. Reprinted with permission from Meekhof *et al.* (1996). © American Physical Society.

and we recognize in this expression the Glauber displacement operator $D(\alpha)$ corresponding to the complex amplitude:

$$\alpha = -\frac{e_i z_0 E_0 \tau}{2\hbar} e^{i\varphi} . \tag{8.49}$$

Starting from the ground motional state, any coherent state $|\alpha\rangle$ can be prepared by applying to the ion a resonant force of proper phase and duration. The coherent state can be analysed by observing the subsequent Rabi oscillation of the ion. Figure 8.14(a) shows the time-dependent oscillation observed on the first blue sideband following the preparation of a coherent state of motion with $\bar{n} = 3.1 \pm 0.1$ phonon on average (Meekhof *et al.* 1996). The signal displays the collapse and revival phenomenon which has been analysed in detail in the CQED context (Chapters 3, 5 and 7). The Fourier analysis of this oscillation yields the probability distribution $\Pi(n)$ of the phonon number (Fig. 8.14b), which displays the Poissonian shape characteristic of a coherent state.

Tomography of the ion's motional state

Exciting the ion's motion with a resonant force of controlled amplitude, phase and duration realizes an arbitrary displacement in the ion's phase space. We have seen in Section 6.5 that such displacements can be used to determine the Wigner function of a quantum oscillator. The procedure, described then for a field in a cavity, works also for the vibration mode of an ion in a trap. The oscillator is first displaced by a controlled push by an amount $-\alpha$ in phase space. The expectation value of the parity of the phonon number in the displaced state then directly yields $W(\alpha)$. To obtain this value, the Rabi oscillation of the ion in the displaced mode is recorded and the phonon probability distribution $\Pi(n)$ reconstructed from a Fourier analysis of this oscillation. The parity signal is then obtained[9] as $\sum_n (-1)^n \Pi(n)$. By repeating the measurement for different α values, one reconstructs the whole Wigner function of the oscillator mode. The method has been applied by the Boulder team to reconstruct the Wigner function of an $n = 1$ Fock state (Leibfried *et al.* 1996).

---

[9]In the CQED case, the parity of the photon number can be measured in a direct way, without determining first the whole photon number distribution (see Section 6.5).

8.2.3   Schrödinger cat states of an ion's motion

Superpositions of coherent states of motion of a single Be^+ ion have been generated and analysed in an ensemble of elegant experiments by the Boulder group (Monroe *et al.* 1996; Myatt *et al.* 2000). As in CQED cat studies, the mesoscopic superposition is realized by exploiting the entanglement between the spring and spin systems. The trick is here to realize a displacement of the ion oscillator which is conditioned to its qubit state. By preparing this qubit in a superposition, the experimenters force the ion to follow – so to speak – two displacements at the same time, resulting in a non-classical cat state situation.

Pushing the ion with a state-dependent force

In these experiments, the ion is subjected to a force produced by the beating between two lasers beams $\mathcal{L}_1$ and $\mathcal{L}_2$ whose frequency difference $\omega_{\ell 1} - \omega_{\ell 2}$ precisely matches the ion's oscillation frequency ω_z. When the ion is in the qubit state $|e\rangle$, phonons are fed in the ion oscillator by a Raman process inducing the $|e, n\rangle \rightarrow |e, n+1\rangle$ transitions. No such excitation occurs when the ion is in state $|g\rangle$. This state selectivity is achieved by a proper setting of the lasers frequencies and polarizations. The lasers $\mathcal{L}_1$ and $\mathcal{L}_2$ are tuned close to the frequency of the $2S_{1/2} \rightarrow 2P_{1/2}$ transition, $\mathcal{L}_1$ having a σ^- circular polarization with respect to the direction of the magnetic field defining the ion's quantization axis. With this polarization choice, the $|e\rangle$ state (with $F = 1, m_F = -1$) is experiencing a strong Raman excitation via a virtual transition involving the $F = 2, m_F = -2$ relay level in the upper $2P_{1/2}$ state. The other qubit state, $|g\rangle$ $(F = 2, m_F = -2)$, is on the other hand insensitive to the $\mathcal{L}_1 - \mathcal{L}_2$ excitation since there is no accessible relay level with $m_F = -3$ in the $2P_{1/2}$ configuration[10].

In order to produce a large Raman drive, $\mathcal{L}_1$ and $\mathcal{L}_2$ are oriented at right angle of each other and are inclined at 45 degrees with respect to the ion oscillation. In this way, the Raman exchange of a photon between $\mathcal{L}_1$ and $\mathcal{L}_2$ transfers to the ion a momentum $\Delta k = k_\ell \sqrt{2}$ along the oscillator axis, with a Lamb–Dicke parameter $\eta = \sqrt{\hbar k_\ell^2 / m\omega_z}$. The effective Hamiltonian describing the displacement produced by this Raman process is, in interaction representation:

$$\widetilde{H}_d = -i\hbar \, |e\rangle \, \langle e| \, \frac{\Omega_d \eta}{2} (ae^{-i\varphi} - a^\dagger e^{i\varphi}) \, , \qquad (8.50)$$

where Ω_d is a Raman Rabi frequency proportional to the product of the laser electric fields E_1 and E_2 and φ the relative phase of the two lasers, under the experimenter's control. The evolution induced during a time τ by the $\widetilde{H}_d$ Hamiltonian on the motional state of the ion in $|e\rangle$ is described by a displacement operator $D(\alpha)$ with:

$$\alpha = \frac{\Omega_d \eta \tau}{2} e^{i\varphi} \, . \qquad (8.51)$$

This displacement can be understood as an effect of the modulated dipole-light force produced on the ion by the interfering amplitudes of $\mathcal{L}_1$ and $\mathcal{L}_2$. These light beams beat together to produce a 'walking standing wave' $E_1 E_2 \cos[(\omega_{\ell 1} - \omega_{\ell 2})t - k_\ell Z\sqrt{2} - \varphi]$, which

---

[10] An accessible relay state exists in the $2P_{3/2}$ level, but the detuning of the Raman transition with this state is much larger, making the corresponding process negligible.

generates a time-dependent optical potential acting on the ion's motional degree of freedom. The spatial gradient of this potential, which is time-modulated, corresponds to the oscillating force experienced by the ion. When the replacement $Z \to z_0(a + a^\dagger)$ is made, this gradient becomes proportional to $k_\ell z_0 \sqrt{2} = \eta$ and we retrieve the main features of the interaction expressed by eqn. (8.50).

Preparing and detecting a cat state of motion

Once this selective pushing force is available, the recipe to prepare a cat state of the ion's vibration proceeds in three simple stages. First, the ion is initialized in the ground state $|g, 0\rangle$ by Doppler and sideband cooling. Next, a $\pi/2$-Raman pulse is applied on the ion carrier transition, with a couple of lasers $\mathcal{R}_1$ and $\mathcal{R}_2$ having a frequency difference $\omega_{\ell 1} - \omega_{\ell 2} = \omega_{eg}$. These laser beams, different from $\mathcal{L}_1$ and $\mathcal{L}_2$, propagate along the same direction, with an $\eta = 0$ Lamb–Dicke parameter. They thus do not couple to phonons and affect only the internal state of the qubit, preparing the system in state $(|g, 0\rangle + |e, 0\rangle)/\sqrt{2}$. Finally, the pushing lasers $\mathcal{L}_1$ and $\mathcal{L}_2$ are applied for a finite time, to induce on the ion in $|e\rangle$ a displacement with amplitude α. The ion ends up in:

$$|\Psi_1\rangle = (|g, 0\rangle + |e, \alpha\rangle)/\sqrt{2} , \qquad (8.52)$$

which exhibits entanglement between the qubit and motional states of the ion. This situation, according to the terminology of Chapter 7, corresponds to an 'amplitude cat' of motion: the ion is in a superposition of two classically distinct states, one in which it is motionless, and the other in which it oscillates in the trap with a finite amplitude which may correspond to an average phonon number larger than one.

A simple manipulation can then transform this amplitude cat into a phase cat similar to the CQED photonic cats studied in Chapter 7. The qubit states are next exchanged by a π-carrier pulse and another pushing pulse is applied to the ion, for the same duration as the first one, but with an adjustable phase ϕ. The ion's state then becomes:

$$|\Psi_2\rangle = (|e, \alpha e^{i\phi}\rangle + |g, \alpha\rangle)/\sqrt{2} . \qquad (8.53)$$

The situation is now reminiscent of the photonic phase cats analysed in the previous chapter. If we finally mix by a $\pi/2$-carrier pulse the two qubit states and detect them, the ion's motional state will be projected into one of the two $|\alpha e^{i\phi}\rangle \pm |\alpha\rangle$ states, which for $\phi = \pi$ are even- and odd-phonon number states. For an arbitrary ϕ, the state obtained after the $\pi/2$-carrier pulse is:

$$|\Psi_3\rangle = \frac{1}{2} |e\rangle (|\alpha\rangle + |\alpha e^{i\phi}\rangle) + \frac{1}{2} |g\rangle (|\alpha\rangle - |\alpha e^{i\phi}\rangle) , \qquad (8.54)$$

and the probability of finding finally the ion in $|g\rangle$ is a function of ϕ exhibiting interference features:

$$P_g(\phi) = \frac{1}{2} \left[1 - \mathrm{Re} \, \langle \alpha | \alpha e^{i\phi} \rangle \right] = \frac{1}{2} \left[1 - e^{-|\alpha|^2 (1 - \cos\phi)} \cos(|\alpha|^2 \sin\phi) \right] . \qquad (8.55)$$

This quantum interference signal, observed by the Boulder team (Monroe *et al.* 1996), is shown in Fig. 8.15. The variations of $P_g(\phi)$ versus ϕ have been recorded, from (a) to (d), for increasing amplitudes $|\alpha|$. Each $P_g(\phi)$ point results from the averaging

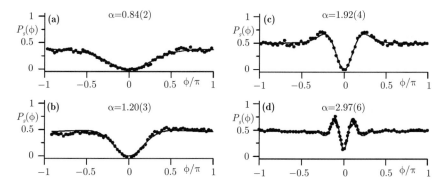

Fig. 8.15 Interference signal $P_g(\phi)$ observed after separation and recombination of the Schrödinger cat state of a Be^+ ion motion. Curves (a)–(d) correspond to increasing amplitudes $|\alpha|$. Reprinted with permission from Monroe *et al.* (1996), © AAAS.

of a large number of recordings realized under identical conditions. The experimental signals (points) are in excellent agreement with theory (solid lines). The interference is visible around $\phi = 0$, with a fringe interval decreasing when the coherent state amplitude increases.

These features can be understood by general complementarity arguments. This experiment, in which the ion has been subjected to two $\pi/2$-carrier pulses mixing the qubit states $|g\rangle$ and $|e\rangle$ (with a π-pulse in-between) is a variant of a Ramsey procedure. The fringes result from the quantum interference of two atomic paths, one in which the first Ramsey pulse has left the ion in $|g\rangle$ and the other in which it has been brought by this pulse in $|e\rangle$. These paths are tagged by different coherent states of the ion motion. If these coherent states are non-overlapping, i.e. if $|\alpha(1 - \cos\phi)|$ is large, they carry information about the path followed by the qubit in the interferometer and interference should not be visible. If on the other hand the two motional state overlap, which occurs around $\phi = 0$ over an interval inversely proportional to $|\alpha|$, the motional state does not provide information about the qubit path and interference is observable. We have encountered this argument time and again in previous chapters.

Where is actually the cat state in this experiment? We should note that there is no cat immediately before detection for the values of ϕ close to 0 for which an interference fringe is observed, since this is precisely a situation in which the coherent state components strongly overlap. There is a cat state, on the contrary, when ϕ is non-zero, but this cat state does not give rise to interference. We can also remark that the appearance of the interference signal near $\phi = 0$ is an evidence that an amplitude cat state has been prepared in the intermediate stage of the Ramsey experiment, before the second push on the ion. The coherence between the two components of $|\Psi_1\rangle$ in eqn. (8.52) is indeed revealed by the interference signal obtained after achieving the second translation in phase space and finally recombining the two motional state components. The largest cat state obtained in this experiment ($|\alpha| \sim 3$) corresponds to a superposition of two states separated by about $3z_0 \approx 21$ nm, three time the size of the ground state wave packet and hundreds of times the diameter of the ionic electronic cloud.

Decoherence studies have been performed on these cat states by the Boulder group

(Myatt *et al.* 2000). A controlled noise has been applied to the trap electrodes, realizing an artificial environment for the motional states. The sensitivity of the ion oscillator state superpositions to this noise has been studied by measuring the contrast of the interference fringes as a function of the delay between the generation of the two coherent fields. The decay rate of the fringe contrast has been found to be inversely proportional to the square of the separation of the coherent components in phase space. This feature, which underscores why these mesoscopic superpositions are difficult to generate and conserve, has been discussed in detail in the CQED context, as a striking example of the quantum–classical boundary.

8.3   Ion relaxation and engineered environments

As we have just shown, properly tailored laser excitation provides useful tools to engineer non-classical states of the ion motion such as Fock or Schrödinger cat states. To prepare and study them, the coherent laser–ion coupling must dominate the damping processes which, in a real experimental situation, also affect the ion internal and external degrees of freedom.

Two relaxation processes have to be considered. The first directly affects the ion oscillation, damping its motion and feeding it with thermal phonons. A master equation describes this process, which we have already analysed in detail in the CQED context (see in particular the discussion of a cavity field damping in Section 4.3.4). The oscillator was then a field mode, but the relaxation master equation is the same for a mechanical oscillator such as an ion in a trap. The relaxation can also be described, in a Monte Carlo approach, as a succession of quantum jumps in which phonons are created or annihilated.

The Q factor of this mechanical oscillation is very large. The quantum jumps are due to the interaction of the ion motion with fluctuating thermal fields and currents in the traps electrodes. Due to these perturbations, the ion motion relaxes slowly towards a highly excited thermal state, which generally contains millions of phonons. The typical time for an ion, initially in its motional ground state, to gain one phonon ranges from ~ 1 to ~ 100 ms, depending upon the trap size and, hence, upon the distance between the ion and the electrodes. This time is long compared to the duration of an ion–laser Rabi oscillation, whether on the carrier or on a motional sideband. We are thus justified to have neglected direct ion motion relaxation in the analysis of the experiments presented in the last section.[11]

The second relaxation mechanism is due to the excited qubit state spontaneous emission. Its rate is also extremely low in ion trap quantum information experiments. In the Ca^+ ion case, it is $\sim 1\ s^{-1}$ and in the Be^+ one it is completely negligible since $|e\rangle$ is then the upper state of a microwave transition with an extremely long spontaneous emission time. In a first approximation, we can thus also neglect this relaxation process during the coherent ion–laser coupling time. Of course, there are much more efficient spontaneous emission processes affecting the ion's excited states ($4P_{1/2}$ and $4P_{3/2}$ in Ca^+, $2P_{1/2}$ and $2P_{3/2}$ in Be^+) but they are not relevant for the evolution of the

---

[11]The thermalization of the trap is described by a rate equation similar to eqn. (4.79). It is a diffusive process resulting from the competition between phonon emission and absorption. Its characteristic time, of the order of the single-phonon leakage time multiplied by the average phonon number at equilibrium, is extremely long in a good trap (days or weeks).

ion qubit during the state engineering procedure since these states are excited only during the preliminary (laser pumping and cooling) or final (ion detection) phases of the experiments.

In the Be^+ case, the ions are also weakly coupled to the radiation environment by the Raman pulses used to manipulate them. A small fraction of the ion's excited $2P$ states is admixed by the non-resonant Raman light to the qubit states, resulting in a 10^{-3} to 10^{-2} probability of the ions scattering a photon during their coupling to the lasers. This process can be in first approximation ignored when analysing the few step ion manipulations considered in this chapter.

Damping in ion traps, which appears as a slow process as long as only a limited number of operations taking a relatively short time are concerned, becomes a much more serious issue when trying to extrapolate ion manipulations to big registers and large numbers of qubit gates. This question is directly related to the important problem of quantum error correction. We will not address it here, but rather try to answer a different question, which seems at first sight a bit provocative. Is damping always a bad thing? Can we envision situations in which relaxation, far from being a nuisance, could be useful to study specific quantum states? We will show in fact that spontaneous emission in an ion system can be cleverly used to *engineer* damping processes which prepare non-trivial equilibrium states and keep them protected from other unwanted relaxation mechanisms.

To achieve this environment engineering, we need first to enhance artificially the spontaneous emission rate from $|e\rangle$, creating an efficient irreversible damping channel for the ion's energy. When the ion is dressed with lasers of convenient amplitudes and frequencies, the spin–spring coupling transfers the damping from the spin to the ion motional degrees of freedom. This results in an effective relaxation of the ionic motion which can be tailored at will, by tuning the dressing lasers parameters. It becomes possible to engineer damping mechanisms with selected jump operators describing specific ionic motional relaxation processes. The ion can be efficiently driven in this way to the trap ground state. The final state of the relaxation process may also be a non-trivial quantum state, such as a Schrödinger cat.

To describe this reservoir engineering procedure (de Matos Filho and Vogel 1996; Poyatos *et al.* 1996), we will largely be inspired by the approach of Carvalho *et al.* (2001). We will first derive a master equation for the ion's enhanced spontaneous emission process, then achieve an adiabatic elimination of the spin degrees of freedom which will leave us with a master equation for the ionic vibration alone. The jump operators associated to this master equation will be analysed and we will conclude by giving a few examples of engineered reservoirs.

8.3.1   A master equation for the ion's motion in a trap

Enhanced spontaneous emission of the ion qubit

To be specific, we will restrict our analysis to the four-level system of Fig. 8.5, a good simplifying model for the Ca^+ ion. The $|e\rangle \rightarrow |g\rangle$ spontaneous emission rate can be increased at will by shining onto the ion a weak laser $\mathcal{L}_{ef}$, resonant on the transition between $|e\rangle$ and the short-lived $|f\rangle$ state. The coupling induced by $\mathcal{L}_{ef}$ admixes some of $|f\rangle$ into $|e\rangle$ and the spontaneous emission rate Γ_e from $|e\rangle$ is accordingly increased. This spontaneous rate enhancement is analogous to the Purcell effect discussed in

Chapter 4. There, the small spontaneous emission rate of an atomic excited state was enhanced by coupling it resonantly to a cavity mode of frequency ω_c and moderately large quality factor Q. The result was an enhanced rate, proportional to the square of the Rabi coupling between the atom and the cavity and inversely proportional to the cavity spectral width $\kappa = \omega_c/Q$ (eqn. 4.112, on page 206). Similarly, the enhanced rate for $|e\rangle$ is now:

$$\Gamma_e \approx \Omega_{ef}^2/\Gamma_{fg} \,, \tag{8.56}$$

where Γ_{fg} is the spontaneous emission rate from $|f\rangle$ to $|g\rangle$ (supposed to be much larger than the emission rate from $|f\rangle$ to $|e\rangle$) and Ω_{ef} the Rabi frequency induced by the $\mathcal{L}_{ef}$ laser.

Upon photon emission, the ion recoils due to momentum conservation. There is thus *a priori* a coupling between spontaneous emission and motion. When the Lamb–Dicke parameter associated to the $|f\rangle \to |g\rangle$ transition is small, the ion trapping is however so tight that the recoil energy is much smaller than a vibration quantum. Provided the number of phonons is not too large, a spontaneous emission jump is then unable to change appreciably the phonon number.[12]

We will assume here that this recoilless emission condition is fulfilled. The jump operator describing the spontaneous emission process then acts only on the ion's internal Hilbert space and can then be approximated as $L = \sqrt{\Gamma_e}\sigma_-$ where $\sigma_- = |g\rangle\langle e|$. The corresponding master equation, written in the Lindblad form, is:[13]

$$\frac{d\rho}{dt}\bigg|_{\text{se}} = \Gamma_e \sigma_- \rho \sigma_+ - \frac{\Gamma_e}{2}\left(\rho\sigma_+\sigma_- + \sigma_+\sigma_-\rho\right) \,. \tag{8.57}$$

Adiabatic elimination of the qubit states

In order to couple indirectly the ion's motion to the artificially enhanced qubit spontaneous emission, we must introduce an interaction between the spin and spring degrees of freedom admixing some of the spin relaxation rate into the spring's one. As we have seen in Section 8.1.4, this can be achieved by lasers $\mathcal{L}_{\pm q}$ at frequencies $\omega_{eg} \pm q\omega_z$ exciting resonantly the qth blue or red motional sideband. We can choose to realize this coupling with one laser, or with a combination of laser beams. To start with a simple situation, let us couple first the ion to a single laser resonant on the qth red sideband, with a Rabi frequency Ω_ℓ. The full density matrix evolves now (in interaction representation with respect to $H_0 = H_z + H_i$) under the joint action of the coupling Hamiltonian $H_{r;q}$ and of spin relaxation:

$$\frac{d\rho}{dt} = -\frac{i}{\hbar}[H_{r;q}, \rho] + \frac{d\rho}{dt}\bigg|_{\text{se}} \,. \tag{8.58}$$

Solving this complete master equation is in general difficult. We consider here only a simple limiting case, in which a master equation for the motional states alone can

---

[12]This has already been noticed in the discussion of sideband cooling: see note 7, page 469 where the recoilless emission of a trapped ion is linked to the Mösbauer effect.

[13]This is eqn. (4.71) of Chapter 4 written here in the interaction representation with respect to spin Hamiltonian.

be obtained. We assume that the sideband laser produces a weak excitation of the $|g\rangle \to |e\rangle$ transition, i.e. that:

$$|g_q| = \frac{\Omega_\ell \eta^q}{2q!} \ll \Gamma_e , \qquad (8.59)$$

(a condition that can be met only when the spontaneous emission rate of $|e\rangle$ is enhanced by the $\mathcal{L}_{ef}$ laser). The weakly excited ion remains then mostly in its ground state $|g\rangle$, a situation which will enable us to eliminate the spin degree of freedom.

Let us introduce the partial matrix elements of the density operator, $\rho_{ij} = \langle i| \rho |j\rangle$, where $i, j = g, e$. These matrix elements are still operators acting in the Hilbert space of the ion vibrational states. Taking into account eqns. (8.27), (8.57) and (8.58), we obtain the following evolution equations:

$$\frac{d\rho_{gg}}{dt} = -ig_q \left(D_q^\dagger \rho_{eg} - \rho_{ge} D_q \right) + \Gamma_e \rho_{ee} ; \qquad (8.60)$$

$$\frac{d\rho_{eg}}{dt} = -ig_q \left(D_q \rho_{gg} - \rho_{ee} D_q^\dagger \right) - \frac{\Gamma_e}{2} \rho_{eg} ; \qquad (8.61)$$

$$\frac{d\rho_{ee}}{dt} = -ig_q \left(D_q \rho_{ge} - \rho_{eg} D_q^\dagger \right) - \Gamma_e \rho_{ee} , \qquad (8.62)$$

the evolution of ρ_{ge} being obtained by hermitian conjugation of (8.61). The D_q operators are defined in eqn. (8.28). We should be careful to respect in these equations the order of the D_q and ρ_{ij} operators, which generally do not commute. In the weak excitation limit ($g_q \ll \Gamma_e$), we have:

$$|\rho_{ee}| \ll |\rho_{eg}| \ll |\rho_{gg}| , \qquad (8.63)$$

each term in this double inequality being $\sim |g_q/\Gamma_e|$ times smaller than the one at its right. The time derivatives of the small ρ_{eg} and ρ_{ee} terms can be neglected in eqns. (8.61) and (8.62), which become steady-state algebraic equations from which these terms can be obtained as a function of the dominant ρ_{gg} matrix element. In more physical terms, ρ_{eg} and ρ_{ee} adjust adiabatically to the slow variations of the dominant term and can thus be eliminated to yield a closed evolution equation for ρ_{gg}. Following this standard procedure of adiabatic elimination (already used in Section 4.5.2 when analysing the Purcell effect), we start by neglecting in eqn. (8.61) the ρ_{ee} term and we get an algebraic solution for ρ_{eg}:

$$\rho_{eg} = -\frac{2ig_q}{\Gamma_e} D_q \rho_{gg} . \qquad (8.64)$$

Inserting this solution in eqn. (8.62) and following a similar procedure, we obtain:

$$\rho_{ee} = \frac{4g_q^2}{\Gamma_e^2} D_q \rho_{gg} D_q^\dagger . \qquad (8.65)$$

Inserting these results into eqn. (8.60), we get a closed differential equation for ρ_{gg}. The motional density matrix ρ_z of the ion is the trace of ρ over the spin degree of freedom and nearly coincides with ρ_{gg}:

$$\rho_z = \rho_{gg} + \rho_{ee} \approx \rho_{gg} \ . \tag{8.66}$$

The closed evolution equation for ρ_{gg} leads thus to a master equation for the motional damping:

$$\frac{d\rho_z}{dt} = \Gamma_q D_q \rho_z D_q^\dagger - \frac{\Gamma_q}{2} \left(D_q^\dagger D_q \rho_z + \rho_z D_q^\dagger D_q \right) \ , \tag{8.67}$$

with:

$$\Gamma_q = \frac{4g_q^2}{\Gamma_e} = \frac{\Omega_\ell^2 \eta^{2q}}{(q!)^2 \Gamma_e} \ . \tag{8.68}$$

The evolution equation for the ion motion takes a Lindblad form, with a q-phonon jump rate $\Gamma_q \ll \Gamma_e$. The effective damping of the motion results from its weak Hamiltonian coupling to the spin, damped with rate Γ_e. This spin is in turn relaxing under the effect of its coupling to the excited state $|f\rangle$, which is damped with the rate $\Gamma_{fg} \gg \Gamma_e$. We have thus a cascading damping mechanism, in which a hierarchy of systems, less and less damped, acquire via successive laser couplings a smaller and smaller fraction of the damping rate of the most damped system (here the excited ionic state $|f\rangle$). The fraction of the relaxation rate transferred at each stage is of the order of the square of the ratio between the laser coupling and the damping rate which is being passed down. This mechanism can be summarized by the following chain of decreasing damping rates:

$$\Gamma_{fg} \to \Gamma_e \approx \Gamma_{fg} \Omega_{ef}^2 / \Gamma_{fg}^2 \to \Gamma_q \approx \Gamma_e (\Omega_\ell \eta^q)^2 / \Gamma_e^2 \ . \tag{8.69}$$

Tailored jump operators

The jump operator $L_{r;q}$ describing the ionic motion effective relaxation process is, according to eqns. (8.28), (8.67) and (8.68):

$$L_{r;q} = \sqrt{\Gamma_q} D_q = \frac{\Omega_\ell \eta^q}{\sqrt{\Gamma_e} q!} (ia)^q \ . \tag{8.70}$$

An excitation on the qth blue sideband corresponds similarly to the jump operator:

$$L_{b;q} = \frac{\Omega_\ell \eta^q}{\sqrt{\Gamma_e} q!} \left(ia^\dagger \right)^q = (-1)^q L_{r;q} \ . \tag{8.71}$$

Applying the same recipe to the carrier excitation, we find:

$$L_c = \frac{\Omega_\ell}{\sqrt{\Gamma_e}} \mathbb{1} \ . \tag{8.72}$$

This operator, being proportional to unity in the Hilbert space of the ion vibrational states, has by itself no action on ρ_z and the corresponding master equation reduces to $d\rho_z/dt = 0$ when the ion is subjected to a single laser tuned at the ω_{eg} frequency.

The jump operator L_c must however be retained to describe processes in which several lasers, including one driving the carrier transition, are simultaneously exciting the oscillating ion. In such a situation, the effective jump operator is the sum of the elementary operators associated to the different beams. A carrier excitation, even though it is ineffective by itself, has – as we will see below – an important interfering effect when other lasers simultaneously drive motional sidebands of the ion spectrum.

Since the various lasers act coherently on the ionic motion, their relative phases are important. More precisely, phase coherence between them should be maintained over times longer than the ion oscillation period. This phase does not appear explicitly in the above analysis, but it is nevertheless present in the formalism. The jump operators $L_{r;q}$, $L_{b;q}$ and L_c given by eqns. (8.70)–(8.72) are expressed in the interaction representation. In the usual representation in which the evolution due to the free oscillator Hamiltonian $H_z = \hbar\omega_z a^\dagger a$ is explicitly taken into account, we must replace $L_{r;q}$ and $L_{b;q}$ by $\exp(iq\omega_z t)L_{r;q}$ and $\exp(-iq\omega_z t)L_{b;q}$ respectively, while L_c is not modified. If the master equation involves a single jump operator $L_{b;q}$ or $L_{r;q}$, the transformation is purely formal and can be omitted, since the operator and its hermitian conjugate are combined in products in which the time-dependence introduced by the representation change is suppressed.

When the engineered jump involves an interference between different sideband and carrier laser excitations, the time-dependent phase terms become relevant in the representation in which H_z appears explicitly. In the case of a qth order red sideband interfering with a carrier laser excitation with Rabi frequency Ω_c, the jump operator becomes in the usual representation $\exp(iq\omega_z t)L_{r;q} + (\Omega_c/\sqrt{\Gamma_e})\mathbb{1}$. The phase change can equivalently be applied to the carrier contribution, the jump operators being expressed as $L_{r;q} + [\Omega_c \exp(-iq\omega_z t)/\sqrt{\Gamma_e}]\mathbb{1}$. The time-dependent term then describes the beating between the lasers used to engineer the reservoir. The time-dependence of these artificial Lindblad operators is a coherent feature distinguishing, in the usual representation, engineered environments from natural reservoirs.

In summary, environment engineering involves two kinds of lasers acting simultaneously on the ion: a set of phase-coherent, amplitude-controlled $\mathcal{L}_q$ lasers resonant with the carrier and sidebands of the $|g\rangle \to |e\rangle$ transition and the laser $\mathcal{L}_{ef}$ resonant on the $|e\rangle \to |f\rangle$ spectral line. The former produce artificial jump operators taking the form of arbitrary linear combinations of powers of a and $a^\dagger$. The strong $\mathcal{L}_{ef}$ laser, whose phase need not be stable, determines the artificial decay rate Γ_e of the $|e\rangle$ state (eqn. 8.56) and, hence, the time scale of the artificial relaxation process. By combining these lasers, one can build tailored environments generating, as we show below, a rich variety of relaxation dynamics.

8.3.2   Relaxation in an engineered environment

A radiative cryostat

As a first example of engineered environment, let us consider the artificial realization of a jump operator proportional to the phonon annihilation operator a. The corresponding reservoir is then equivalent to a $T = 0$ K cryostat, absorbing the phonons and bringing the ion's motion to its ground state. To realize it, we must use a laser beam $\mathcal{L}_1$ resonant on the first red sideband transition, while the $\mathcal{L}_{ef}$ laser is artificially broadening the upper qubit state $|e\rangle$. This recipe corresponds to the continuous version of the sideband cooling procedure already described in Section 8.1.7 and diagrammatically represented in Fig. 8.11: the laser $\mathcal{L}_1$ induces transitions between $|g, n\rangle$ and $|e, n - 1\rangle$, which decays towards $|g, n - 1\rangle$ under the action of $\mathcal{L}_{ef}$, and so on. We can analyse now this process in more quantitative terms. The jump operator is

$L_1 = \sqrt{\Gamma_1} a$ with $\Gamma_1 = \Omega_\ell^2 \eta^2 / \Gamma_e$. The corresponding master equation describes a relaxation towards the zero-phonon state with the rate Γ_1 depending upon the intensities of $\mathcal{L}_{ef}$ (which fixes Γ_e) and $\mathcal{L}_1$ (which determines Ω_ℓ).

The cooling rate obviously increases with the intensity of the sideband laser. A naive interpretation of eqn. (8.68) could lead to the impression that Γ_e, and hence the intensity of the $\mathcal{L}_{ef}$ laser, should on the contrary be decreased down to very small values, since it appears in the denominator of the cooling rate. This is deceptive though, since the adiabatic elimination requires Γ_e to be much larger than the coupling with the sideband laser (eqn. 8.59). Optimum sideband cooling thus requires a careful balance of the two lasers intensities, satisfying the set of restrictive conditions discussed above.

An environment absorbing phonons by pairs

Let us now irradiate the ion with a laser $\mathcal{L}_2$ resonant on the second red sideband. The corresponding jump operator is $L_2 = -\sqrt{\Gamma_2} a^2$ with $\Gamma_2 = \Omega_\ell^2 \eta^4 / 4\Gamma_e$. The elementary relaxation cycle corresponds to a $\mathcal{L}_2$-laser-induced transition from $|g, n\rangle$ to $|e, n - 2\rangle$, followed by the spontaneous decay of $|e, n - 2\rangle$ towards $|g, n - 2\rangle$ stimulated by the laser $\mathcal{L}_{ef}$. Two phonons are annihilated and the process goes on, suppressing vibration quanta by pairs. This is again an efficient cooling mechanism, which admits now two final stable states, the phonon states $|0\rangle$ and $|1\rangle$.[14]

The pair-phonon loss mechanism preserves the parity of the phonon number. It is thus interesting to study what happens, under this process, to eigenstates of the phonon number parity operator $\mathcal{P}$. Let us consider in particular the even or odd Schrödinger cat states of the ion vibration $|\Psi_\pm\rangle = (|\alpha\rangle \pm |-\alpha\rangle)/\sqrt{2}$ (see Chapter 7 for the description of the corresponding states in CQED). These states are eigenstates of $\mathcal{P}$ (eqn. 7.10), but also of the two-phonon jump operator, since they satisfy:

$$a^2 |\Psi_\pm\rangle = \alpha^2 |\Psi_\pm\rangle \ . \tag{8.73}$$

According to the discussion of Section 4.4.5, this feature makes them 'approximate pointer states' of the phonon-pair loss mechanism. The ion's motional density operator, supposed to be $\rho_z = |\Psi_\pm\rangle \langle \Psi_\pm|$ at the time $t = 0$ of the ion's coupling with the environment, satisfies then the condition $d(\mathrm{Tr}\rho_z^2)/dt = 0$ (eqn. 4.99). The linear entropy of the ion motional state, naught at $t = 0$, thus has a zero first-order time derivative. This means that the ion's entanglement with the engineered reservoir remains negligible over a time spanning several quantum jumps, while an ordinary relaxation process involving a one-phonon Lindblad operator would entangle it maximally with the environment after the characteristic time corresponding to a single jump.

These Schrödinger cats are not perfect pointer states because they do not satisfy the severe conditions discussed in Section 4.4.5. Being eigenstates of L_2 with a non-zero eigenvalue, they should also be eigenstates of the $[L_2, L_2^\dagger]$ commutator, equal to $\Gamma_2[a^2, a^{\dagger 2}] = \Gamma_2(4a^\dagger a + 2\mathbb{1})$. This is obviously not the case. As a result, their deterministic non-unitary evolution between jumps, ruled by the non-hermitian pseudo-Hamiltonian $-i\hbar L_2^\dagger L_2 / 2$, transforms them into states which are no longer eigenstates

---

[14]The two-dimensional space spanned by $|0\rangle$ and $|1\rangle$ is decoherence-free since these two states are eigenstates with zero eigenvalue of the Lindblad operator a^2 (see Section 4.4.5).

of a^2. We will show below that we can do better and tailor relaxation mechanisms which admit these motional cats as decoherence-free states remaining in principle indefinitely stable.

Tailoring the final state

Let us apply the general discussion of Section 4.4.5 to a master equation of the ion's motion, involving a single engineered Lindblad operator L. This equation is expressed here in the interaction representation with respect to the free ion oscillation. The Hamiltonian term is thus absent from it and we can adapt the discussion of Chapter 4 by merely assuming that the free Hamiltonian of the system is vanishing. We then conclude that an eigenstate $|\phi\rangle$ of the jump operator L with zero eigenvalue, satisfying the condition:

$$L|\phi\rangle = 0 , \tag{8.74}$$

is decoherence-free. This state is changed neither by quantum jumps (since $L|\phi\rangle = 0$) nor by the non-hermitian evolution between the jumps (since $L^\dagger L|\phi\rangle = 0$). Finally, its free evolution is frozen, since we are in the interaction representation. Such a state is time-invariant and does not couple to the environment. If $|\phi\rangle$ is a non-degenerate eigenstate of L, it is in fact the final equilibrium state of any Monte Carlo trajectory. The ion, prepared in an arbitrary initial motional state, will always end up in it.

These considerations determine a simple strategy for engineering reservoirs admitting specific quantum states of motion as decoherence-free states immune to entanglement and dissipation (Poyatos *et al.* 1996; Carvalho *et al.* 2001). We will say that these states are 'protected'. The principle is to tailor with appropriate laser beams, tuned on the carrier or motional sidebands of the ion, a Lindblad operator, linear combination of powers of a, $a^\dagger$ and the unity operator, which admit the quantum state to be protected as an eigenstate with zero eigenvalue. We will apply this strategy to analyse the 'protection' of two specific states, the coherent states and the even- and odd-parity Schrödinger cats.

Coherent state protection

Let us start by the protection of an arbitrary coherent state $|\alpha\rangle$. We have seen in Chapter 4 that an ordinary reservoir (with a jump operator proportional to a) leaves this state immune from entanglement during its continuous evolution towards the oscillator's ground state. The state remains a pointer state, but its amplitude changes and it is thus not protected from dissipation. It is in fact possible to improve on this situation and to find a reservoir which leaves the coherent state not only unentangled with the environment, but, in addition, invariant. We simply note that the state $|\alpha\rangle$ is the non-degenerate eigenstate with eigenvalue 0 of the operator $a - \alpha\mathbb{1}$. To protect it, we must thus engineer an artificial reservoir with a Lindblad term proportional to this operator. This is obtained by applying on the atom a laser on the carrier frequency (Rabi frequency Ω_c) and a laser on the first red sideband (Rabi frequency Ω_r). The jump operator is then:

$$L_1 = \frac{1}{\sqrt{\Gamma_e}}(\Omega_c\mathbb{1} + i\eta\Omega_r a) = \frac{i\eta\Omega_r}{\sqrt{\Gamma_e}}(a - \alpha) , \tag{8.75}$$

with:

Fig. 8.16 Transitions implied in the coherent state protection. The laser transitions are marked by solid line arrows and spontaneous decay by the dotted lines. The system's evolution halts when the $|g, n\rangle$ states fall into the 'dark' coherent superposition cancelling, by quantum interference, the ion excitation process.

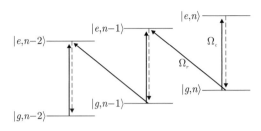

$$\alpha = i \frac{\Omega_c}{\Omega_r \eta} \ . \tag{8.76}$$

By properly tuning the ratio of the red sideband and carrier Rabi frequencies, we can protect any coherent state with an imaginary amplitude. Our calculation discarded the laser phases, all set to $\pi/2$. By adjusting the relative phase of the two beams, any coherent state can be protected. The combination of lasers realizing the jump operator proportional to $a - \alpha \mathbb{1}$ not only protects $|\alpha\rangle$, but also attracts any arbitrary initial state towards it. This state, being the non-degenerate eigenstate of $a - \alpha \mathbb{1}$ with eigenvalue 0 is indeed the final state of all the motional states Monte Carlo trajectories. It appears as stationary in the interaction picture.[15]

This coherent state stabilization can also be understood as a quantum interference process. Figure 8.16 presents the transitions between the spin–spring states involved in the evolution. The states $|g, n\rangle$ and $|e, n\rangle$ are coupled by the carrier frequency laser and by spontaneous decay. The red sideband couples $|g, n\rangle$ and $|e, n - 1\rangle$. These coupled transitions propagate phonon excitation up and down the motional states ladder. The evolution stops when the total probability amplitude of exciting the $|e, n - 1\rangle$ state from $|g, n - 1\rangle$ and $|g, n\rangle$ vanish. This cancellation can only result from a quantum interference process and, hence, requires a full coherence between motional states. Let c_n be the probability amplitude of $|g, n\rangle$. The transition amplitude to $|e, n - 1\rangle$ cancels when:

$$c_{n-1} \langle e, n - 1| H_c |g, n - 1\rangle + c_n \langle e, n - 1| H_{r;1} |g, n\rangle = 0 \ . \tag{8.77}$$

Using eqns. (8.26) and (8.30), this condition can also be written:

$$\eta \Omega_r \sqrt{n} c_n = i \Omega_c c_{n-1} \ , \tag{8.78}$$

a recurrence which leads, taking normalization into account, to:

$$c_n = e^{-|\alpha|^2/2} \frac{\alpha^n}{\sqrt{n!}} \ , \tag{8.79}$$

with α given by eqn. (8.76). We recognize here the complex probability amplitudes of a coherent state.

---

[15] If we analyse this relaxation process in the usual representation including the Hamiltonian evolution of the ion's vibration, we must describe it by the time-dependent Lindblad operator $a - \alpha \exp(-i\omega_z t)$. It admits the time-dependent state $|\alpha \exp(-i\omega_z t)\rangle$ as non-degenerate eigenstate with zero eigenvalue. We retrieve of course a freely evolving coherent state, corresponding to the 'frozen' state $|\alpha\rangle$ in the interaction picture.

In steady-state, quantum interference between two paths cancels the excitation probability. This situation is reminiscent of the 'dark states' observed in the coherent excitation of a three-level atom by a laser field (Alzetta *et al.* 1976). In these experiments, two nearly degenerate ground states are coupled by a laser field to an excited level. Under appropriate conditions, the atom does not scatter any photon, a surprising situation. One peculiar coherent superposition of the ground levels is then exactly decoupled from the laser beam, the two excitation probability amplitudes interfering destructively. Absorption–spontaneous emission cycles optically pump the atom in this dark state and the fluorescence stops after a short transient. This is exactly the same effect which is exploited here to stabilize coherent states under the competing effects of two lasers exciting the carrier and the red sideband of the ion spectrum.

Protecting Schrödinger cats

Let us now protect a highly non-classical state, the even Schrödinger cat $|\Psi_+\rangle = (|\alpha\rangle + |-\alpha\rangle)/\sqrt{2}$. It is an eigenstate with zero eigenvalue of $a^2 - \alpha^2 \mathbb{1}$. We can build such a jump operator by combining a carrier frequency laser with a second red sideband excitation. With obvious notation, the jump operator is then:

$$ L_2 = \frac{1}{\sqrt{\Gamma_e}} \left(\Omega_c \mathbb{1} - \frac{\Omega_r \eta^2}{2} a^2 \right) , \tag{8.80} $$

which stabilizes the desired cat state when:

$$ \frac{\Omega_c}{\Omega_r} = \frac{\alpha^2 \eta^2}{2} . \tag{8.81} $$

Let us remark that the same combination of lasers also protects the odd Schrödinger cat state $|\Psi_-\rangle = (|\alpha\rangle - |-\alpha\rangle)/\sqrt{2}$, since $(a^2 - \alpha^2 \mathbb{1})|\Psi_-\rangle = 0$.

The protection of the even (and odd) cats can also be seen as a 'dark state' effect. The combined effects of the laser exciting the carrier and the second red sideband induces phase relationships between motional states of the ion differing by even number of phonons. Starting from an arbitrary initial state developing on even phonon numbers, this optical pumping process will prepare the superposition of even $|g,n\rangle$ states which cancels by quantum interference the ion's excitation probability. The process comes then to a stop, which corresponds precisely to the generation of the even motional Schrödinger cat $|\Psi_+\rangle$. Similarly, an arbitrary initial state with odd photon numbers will get stabilized by the same optical pumping process in the odd cat state $|\Psi_-\rangle$. If we start with a general mixture of Fock states, the even and odd subsets of states will evolve independently, producing finally a mixture of the $|\Psi_+\rangle$ and $|\Psi_-\rangle$ cats. We have here an example of a relaxation process whose Lindblad operator has a degenerate zero eigenvalue. The final state of the relaxation process is in this case not unique. It can be any superposition state or mixture in the degenerate eigenspace and depends upon the initial conditions.

Other interesting quantum states can be generated or protected with this approach. The protection of an arbitrary qubit state $c_0 |0\rangle + c_1 |1\rangle$, for instance, has been discussed by Carvalho *et al.* (2001). Its implementation involves three lasers and will not be described here.

Robust protection

We have considered so far an ideal situation in which the ion is only submitted to a perfectly controlled engineered reservoir with a single Lindblad operator L. In the real world, it is also coupled to an unavoidable natural environment, which damps its motion by removing or adding phonons. This process is generally slow, but it has to be taken into account at long time scales. The ion relaxation results from the combined effects of the 'engineered jumps' described by L and occurring at a rate Γ_1 and 'spontaneous jumps' defined by the operators $\sqrt{\kappa(n_{\text{th}}+1)}\, a$ and $\sqrt{\kappa n_{\text{th}}}\, a^\dagger$ where n_{th} is the large number of thermal phonons at the temperature of the trap and κ the slow natural damping rate of the ion oscillator (Section 4.3.4). As discussed above, the natural rate κn_{th} of the phonon diffusion process is, in a good trap, of the order of $10\ \text{s}^{-1}$.

The evolution of the ion's motional state under these competing influences is in general complicated. It simplifies however if the engineered jump rate Γ_1 is much larger than the natural rate κn_{th}. The relatively fast engineered relaxation process is then dominant. It brings the ion's motion to a quasi-equilibrium before the slow natural evolution has had time to act efficiently. The effect of the natural jumps can then be described, on a long time scale, as a perturbation to the quasi-steady-state solution of the master equation of the engineered reservoir. This regime implies a hierarchy of time constants for the system's evolution, which is expressed by the cascading inequalities:

$$\kappa n_{\text{th}} \ll \Gamma_q \ll \Gamma_e \ll \Gamma_{fg} \ . \tag{8.82}$$

They can be all satisfied because of the enormous range between Γ_{fg} and κn_{th} which differ by seven orders of magnitude. It is for instance possible to tune $\mathcal{L}_{ef}$ and the lasers driving sideband or carrier transitions so that $\Gamma_e \sim 10^{-3}\Gamma_{fg} = 10^5\ \text{s}^{-1}$ and $\Gamma_q \sim 10^{-1}\Gamma_e = 10^4\ \text{s}^{-1}$. The damping rate of the engineered reservoir is then three orders of magnitude larger than the natural phonon diffusion rate. At the same time, the enhanced spontaneous emission rate Γ_e of the $|e\rangle$ state is much smaller than the ion's oscillation frequency ($\omega_z \sim 10^6\ \text{s}^{-1}$) so that all the approximations validating our analysis are satisfied.

Numerical solutions of the complete master equation performed under these conditions (Carvalho *et al.* 2001) lead us to distinguish two situations. If the state $|\phi\rangle$ protected by the engineered reservoir is a non-degenerate eigenstate of L (with zero eigenvalue), the natural relaxation has practically no effect. The protected state is an 'attractor' towards which the ion state always converges and the slow natural relaxation cannot displace the system from it. The state protection is then robust. This is the case of the protection of an arbitrary coherent state α, with an artificial reservoir realizing a jump operator proportional to $a - \alpha\mathbb{1}$.

When, on the contrary, $|\phi\rangle$ is a degenerate eigenstate of L, the perturbing effect of the natural reservoir mixes it with other states in the eigenspace with zero eigenvalue. The protection by the engineered reservoir is then, at long time scales, inefficient. This is the case of the even or odd cat states. They are degenerate eigenstates of $a^2 - \alpha^2\mathbb{1}$ and thus admixed in an uncontrollable way by the natural relaxation process. This engineered reservoir procedure cannot protect them from decoherence over a time longer than the decoherence time imposed by the natural ion vibration reservoir.

8.4 Quantum logic with trapped ions: individual qubit addressing

A variety of experiments have been realized to demonstrate gate operations and simple quantum logic procedures on linear strings of trapped ions. These experiments are now so numerous and diverse that it is impossible to present here a complete review. We will rather concentrate on two remarkable sets of studies, performed on Ca^+ by the Innsbruck group and on Be^+ by the Boulder team. In the first case, analysed in this section, the ions are addressed one by one, using simple entangling procedures which are straightforward to explain by analogy with the background of CQED physics developped in the previous chapters. In the Be^+ experiments, the ions are manipulated collectively and the physics involved is based on the clever engineering of a non-linear interaction between the ions which will be described in the next section.

8.4.1 An example of ion entanglement: a Bell state study

As a first illustration of ion entangling procedures, let us describe the preparation and detection of the four Bell states of two Ca^+ ions (Roos *et al.* 2004a). The ions are first initialized by optical pumping and sideband cooling in the state $|g_1, g_2; 0, 0\rangle$ (the symbols representing respectively the qubit states of the two ions and the phonon number in the CM and stretch modes). From then on, they are manipulated by laser beams addressing the ions individually, which are tuned either on the carrier or on the sideband frequencies of the stretch mode. We can thus disregard the unexcited CM mode and describe the system by three symbols only.

The first step of the Bell state preparation is realized by exciting the first ion on the first blue sideband with a $\pi/2$-pulse. The two-ion system then evolves into the state $[|g_1, (g_2); 0\rangle + |e_1, (g_2); 1\rangle]/\sqrt{2}$, which exhibits maximum entanglement between the first qubit and the collective motional state (we write between parentheses the ion which is not affected by the current operation). Next, the second ion is irradiated by a π-pulse on the carrier transition. This operation exchanges the qubit states of this ion, affecting neither the other ion nor the collective motional state. It brings the system into $[|(g_1), e_2; (0)\rangle + |(e_1), e_2; (1)\rangle]/\sqrt{2}$. Finally a blue sideband π-pulse is applied on the second ion to copy on it the vibrational state. The system ends up in $[|(g_1), e_2\rangle + |(e_1), g_2\rangle] |0\rangle /\sqrt{2}$, in which the qubit and motional states are separated and the initial zero-phonon state is retrieved. We prepare in this way the $|\psi^+\rangle$ Bell state. Combining these steps with obvious single-qubit operations (performed by applying on the ions carrier pulses) allows us to generate similarly the other Bell states, $|\psi^-\rangle$ and $|\phi^\pm\rangle$.

This entangling procedure is the direct transcription to ion trap physics of the CQED method described in Chapter 5 for preparing an EPR pair of atoms. Whereas the cavity field is in the latter case used as a catalyst to entangle two Rydberg atoms, this role is played here by the collective oscillation mode of the ion crystal. The ion experiment has, however, been carried further than the Rydberg atom one. The four Bell states have been mapped out by tomography, yielding the complete density matrix of the two-qubit system. The data have also been analysed to test Bell's inequalities.

The reconstruction of the Bell state density operator is realized according to the general method outlined in Section 8.1.5. The coefficients of the density matrix in eqn. (8.34) are obtained from separate measurements of the two ion's pseudo-spins, after proper rotations mapping the eigenvectors of σ_i ($i = X, Y, Z$) onto those of

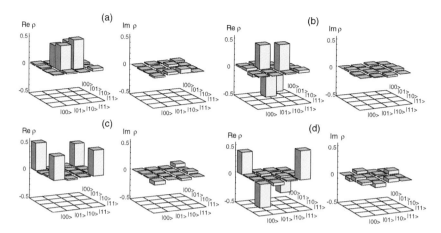

Fig. 8.17 Reconstruction of the two-ion density matrix after preparation of the four Bell states. (a)–(d) correspond respectively to $\left|\psi^{+}\right\rangle$, $\left|\psi^{-}\right\rangle$, $\left|\phi^{+}\right\rangle$ and $\left|\phi^{-}\right\rangle$ [see eqn. (2.54) for definition of these states]. The left-hand bar graph in each panel gives the real part of ρ, the imaginary part is on the right. Reprinted with permission from Roos *et al.* (2004a). © American Physical Society.

σ_{Z}, directly measured by laser-induced fluorescence. These rotations are realized by carrier frequency laser pulses. Altogether, nine different pulse settings are required to measure the sixteen coefficients of ρ.

Care should be taken that – due to noise – the reconstructed density matrix might be non-physical (it might have negative eigenvalues). Various regularization procedures have been proposed to cope with this situation. They are based on maximum likelihood (Banaszek *et al.* 1999a) or maximum entropy principle (Drobny and Buzek 2002). They produce a physically acceptable density matrix, which fits at best the experimental data. The reconstruction procedure can be fully automated. It can thus be transposed to larger quantum registers, in spite of the exponential increase in the number of required measurements.[16]

Figure 8.17 presents the reconstructed density matrices for the four Bell states (Roos *et al.* 2004a). Each state reconstruction involves about 2000 repetitions of the experiment. The imaginary parts of the elements of ρ are negligible, the real parts are close to the expected values. The state preparation fidelity, average value in ρ of the projector on the desired state, is about 0.9. It can be directly inferred from these measurements.

The two-ion entanglement results in a violation of the Bell inequality (eqn. 2.59). This non-locality test is performed by measuring two orthogonal spin components on each ion, the components measured on ion 1 being at 45 degrees of the ones measured on ion 2 (see the discussion on Bell's inequality in Chapter 2). Each component is obtained by rotating, prior to the fluorescence measurement, the pseudo-spin with a carrier pulse of appropriate duration and phase. When either the two ions or none

---

[16]The reconstruction of an eight-ion quantum state has been performed by the Innsbruck group (Häffner *et al.* 2005).

fluoresce, the corresponding correlation is counted as $+1$. It is registered as -1 if only one ion fluoresces. The experiment is repeated many times for each set of randomly selected spin polarizations. The statistical averages are combined to form the Bell sum Σ_b defined by eqn. (2.58). This sum, which should be smaller than 2 in any local hidden variable description, is here found to be 2.52.[17] This measurement, performed with a nearly hundred percent efficient atomic detection, does not suffer from the 'detection efficiency loophole' that affects optical experiments (Gisin and Gisin 1999). Nearly all prepared pairs are measured. It becomes impossible to sustain a far-fetched hidden-variable interpretation based on a selection process which would mysteriously bias the data by keeping un-detected a majority of pairs with $|\Sigma_b| < 2$. The distance between the two members of the non-local pair is however a few micrometres only and the 'light cone' loophole cannot be eliminated as well, since causal propagation between the two particles during detection remains possible. There is thus still a tiny and uncomfortable niche for hidden variables theories.

8.4.2 Principle of the Cirac–Zoller gate

The Bell state preparation procedure illustrates a central and general feature of logical operations on separately addressed ions. Local pulses applied on individual ions couple their qubit states to a collective vibration mode and inter-ion entanglement is achieved by a succession of such pulses which progressively 'knit' entanglement in a multi-qubit register. Cirac and Zoller (1995) have proposed to use this general principle to realize a universal control-not gate between two arbitrary ions in an $\mathcal{N}$-ion chain. This proposal has played an important role to trigger the whole field of quantum information. About ten years after it has been proposed, the Cirac–Zoller gate – or rather a closely related variant – has been demonstrated by the Innsbruck group (Schmidt-Kaler *et al.* 2003). We briefly recall here the principle of this gate, before describing in the next section its actual realization in the laboratory.

We want to realize a control-not gate between two arbitrary ions (not necessarily neighbours) of an $\mathcal{N}$-ion register. We call 1 the control ion and 2 the target ion. We use the CM collective mode to couple the ions and assume that this mode is initially cooled to its ground state. We also assume that the two ions have been prepared by carrier pulses into arbitrary qubit state superpositions of the form $c_g \left| g \right\rangle + c_e \left| e \right\rangle$ before the gate operation. The Cirac–Zoller (C–Z) proposal first realizes a control-phase gate, then transforms this gate into a control-not by sandwiching it between two $\pi/2$-carrier pulses applied to the target ion (the connexion between phase and control-not gates is discussed in Section 2.6.4).

The realization of the C–Z control-phase gate proceeds as follows. First, a π-area first red sideband pulse is applied to the control ion 1, achieving the transformation:

$$c_g \left| g_1; 0 \right\rangle + c_e \left| e_1; 0 \right\rangle \longrightarrow \left| g_1 \right\rangle \left(c_g \left| 0 \right\rangle + c_e \left| 1 \right\rangle \right), \qquad (8.83)$$

which copies the state of the control qubit on the motional state. Next, this state must be coupled to the ion 2 qubit to realize the core of the phase gate operation. The first idea of Cirac and Zoller was to apply to ion 2 a 2π-Rabi pulse which phase-shifts

---

[17]A similar violation was found in a Bell state experiment performed by the NIST group (Rowe *et al.* 2001).

by π the states involved in the process. This pulse, acting on the first red sideband, couples together – as in CQED – the combined states $|e_2; 0\rangle$ and $|g_2; 1\rangle$ and leaves unaffected the ground state $|g_2; 0\rangle$. After the 2π-rotation is completed, the following transformations are obtained:

$$|g_2; 0\rangle \rightarrow |g_2; 0\rangle \quad ; \quad |g_2; 1\rangle \rightarrow -|g_2; 1\rangle \ ;$$

$$|e_2; 0\rangle \rightarrow -|e_2; 0\rangle \quad ; \quad |e_2; 1\rangle \rightarrow \cos(\pi\sqrt{2}) |e_2, 1\rangle + \sin(\pi\sqrt{2}) |g_2; 2\rangle \ . \ (8.84)$$

The first three transformations have a form amenable to a gate operation, but there is clearly a problem with the last one. If ion 2 in the qubit state $|e\rangle$ is put in contact with the 'bus' mode in state $|1\rangle$, the transformation may generate a second phonon and bring the ion 2–bus system out of the computational two by two Hilbert space. This is because the Rabi frequency varies with the phonon number n as $\sqrt{n+1}$ (we assume here that the Lamb–Dicke parameter is small enough for this law to be valid). A closed-loop 2π-pulse between $|e; 0\rangle$ and $|g; 1\rangle$ becomes an open-loop $2\pi\sqrt{2}$ pulse between $|e; 1\rangle$ and $|g, 2\rangle$.

We have encountered a similar problem in CQED when attempting to describe the Rabi oscillation in the language of quantum information. When the atom–cavity system is initially in $|e; 1\rangle$, the Rabi flopping is mixing it with $|g, 2\rangle$ and the cavity field can no longer be treated as a qubit. The trick we used in CQED to realize a genuine phase gate was to introduce a third Rydberg state and to induce transitions in which this third state is only transiently excited, the system finally coming back to the computational space at the completion of the gate operation.

This is the idea that Cirac and Zoller described in their original proposal.[18] The specific solution they suggested is to make use of a third auxiliary state $|i\rangle$ to which the $|g\rangle$ qubit state is coupled by laser excitation on the first red sideband transition. This state is a Zeeman substate in the manifold of levels containing the excited qubit state.

To be specific, let us consider the case of Ca$^+$. The qubit state $|e\rangle$ is then the magnetic state with $m_J = -1/2$ in the $4D_{5/2}$ level, coupled to the qubit state $|g\rangle$ (also $m_J = -1/2$) via a π-polarized transition. We then choose for $|i\rangle$ the $m_J = +1/2$ Zeeman substate of $4D_{5/2}$, coupled to $|g\rangle$ by σ^+-polarized light. If we apply to the $|g_2, 1\rangle$ state a 2π-pulse with σ^+ polarization on the red sideband, the system will undergo a full loop via state $|i_2, 0\rangle$ and come back to the initial state, with a π-phase shift. The same σ^+-polarized pulse has, on the other hand, no effect on the initial state $|e_2; 1\rangle$ since there is no state with $m_j = -3/2$ in the Ca$^+$ $4S_{1/2}$ ground state. The same selection rule entails that $|e_2; 0\rangle$ is also unaffected. A similar recipe works also for the Be$^+$ ion, since the qubit states also belong to levels which have other substates among which the auxiliary qubit state $|i\rangle$ can be selected. In short, the σ^+-polarized pulse produces the transformations:

$$|g_2; 0\rangle \rightarrow |g_2, 0\rangle \ ; \quad |g_2; 1\rangle \rightarrow -|g_2; 1\rangle \ ; \quad |e_2; 0\rangle \rightarrow |e_2; 0\rangle \ ; \quad |e_2; 1\rangle \rightarrow |e_2; 1\rangle \ , \ (8.85)$$

which satisfy the truth table of a phase gate. The level $|i\rangle$ has been only momentarily excited and is not populated at the end of the operation.

---

[18]The C–Z proposal was actually anterior to the CQED phase gate (Rauschenbeutel *et al.* 1999) and 'rediscovered' independently by us on that occasion.

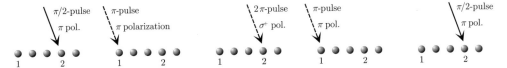

Fig. 8.18 The Cirac–Zoller gate consists of five pulses applied alternatively to the target and control ions (here the fourth and first ions in a linear crystal of five). The solid arrows depict a carrier pulse, the dashed ones a sideband pulse. The Rabi angle and polarization of the pulses are indicated.

Once the phase gate coupling ion 2 to the bus mode has been completed, we must finally copy back the motional state on the first ion to restore the first qubit in its initial state. This is again realized by a π-pulse on the first red sideband which realizes the transformation $|g_1; 1\rangle \rightarrow -|e_1; 0\rangle$. This last pulse, as the first one acting on ion 1, must be π-polarized in order to couple the two qubit states $|e\rangle$ and $|g\rangle$. The C–Z gate recipe thus uses two transitions with different polarizations, a π-polarized light beam to realize the first and last π-pulses acting on the control ion and a σ^+-polarized light beam to realize the intermediate 2π-pulse entangling ion 2 with the bus mode. It is straightforward to combine these three operations and to show that they realize the global phase gate transformation:

$$|g_1, g_2; 0\rangle \rightarrow |g_1, g_2; 0\rangle \quad ; \quad |g_1, e_2; 0\rangle \rightarrow |g_1, e_2; 0\rangle \quad ;$$
$$|e_1, g_2; 0\rangle \rightarrow |e_1, g_2; 0\rangle \quad ; \quad |e_1, e_2; 0\rangle \rightarrow -|e_1, e_2; 0\rangle \ . \tag{8.86}$$

In order to turn this phase gate into a control-not, two carrier $\pi/2$-area pulses with π-polarization are applied on ion 2, before and after the three pulses forming the phase gate. The complete gate operation thus consists of five pulses acting alternatively on the target and control ions, in a sequence which is schematized in Fig. 8.18.

8.4.3   Implementation of a variant of the Cirac–Zoller gate

Soon after this proposal was made, the Boulder team made a demonstration of the core of the C–Z gate operation on a Be^+ ion (Monroe *et al.* 1995a). Using as control bit a vibration mode and as target states two hyperfine Zeeman sublevels of the ion, this group has implemented the conditional dynamics of a control-not gate. Since it entangled the external and internal degrees of freedom of a single ion, this gate did not demonstrate interparticle coupling. After that early work, the Boulder team has progressively mastered the techniques of juggling with several ions at a time, but it did it by achieving multi-particle logic operations along routes different from the one pioneered by Cirac and Zoller. We discuss these alternative approaches in the next section.

The first complete demonstration of a gate *à la* Cirac–Zoller was made by the Innsbruck team on two Ca^+ ions in a linear trap (Schmidt-Kaler *et al.* 2003). The $|g\rangle$ and $|e\rangle$ states of the ions ($4S_{1/2}$, $m_J = -1/2$ and $4D_{5/2}$, $m_J = -1/2$ magnetic Zeeman sublevels) code respectively the $|0\rangle$ and $|1\rangle$ qubit logic states. As in the C–Z proposed version, the phase gate is realized in three steps. First, the internal state of the control ion 1 is copied onto the stretch vibration mode of the crystal. Then, a conditional dynamics is implemented between the target ion 2 and the vibration, realizing the actual gate operation. Finally, the motional state is copied back onto ion 1, replacing the control ion into its initial state and completing the process.

Principle of operation

The heart of this protocol – its second step – is, as suggested by C–Z, based on a 2π-Rabi pulse on a sideband excitation (the blue sideband being used instead of the red). But the trick employed to circumvent the difficulty of having the system leave the two by two computational space is solved in a different way. Instead of employing a transition coupling one of the qubit states to a third auxiliary level, the Innsbruck team resorts only to transitions between the two qubit states, but exploits to induce them a composite pulse scheme borrowed from nuclear magnetic resonance. The goal is to achieve, *with the same pulse sequence*, a 2π-Rabi flop *both* on the $|g_2;0\rangle \rightarrow |e_2;1\rangle$ and $|g_2;1\rangle \rightarrow |e_2;2\rangle$ transitions. This is obviously impossible if one uses a simple pulse, since the corresponding Rabi frequencies are different (in the ratio $\sqrt{2}$ for a small Lamb–Dicke parameter η). It becomes however possible if the Rabi flops are realized by combining four laser pulses.

This NMR trick exploits a general property of rotation products. One can visualise the operations by considering the pairs of states $\{|g_2;0\rangle, |e_2;1\rangle\}$ on the one hand, $\{|g_2;1\rangle, |e_2;2\rangle\}$ on the other hand, as representing two *composite* spins evolving on their respective Bloch spheres. Laser pulses resonant with the first blue sideband induce rotations on these spheres. A laser pulse of Rabi frequency Ω_ℓ, duration τ and phase φ produces on the first Bloch sphere a rotation of angle $\Omega_\ell\eta\tau$ around a direction of the OXY equatorial plane making an angle φ with the OY direction. The same laser pulse corresponds, on the second Bloch sphere, to a rotation of angle $\Omega_\ell\eta\tau\sqrt{2}$, around the same axis.[19]

Consider now the combination of laser pulses realizing on the first Bloch sphere the transformation R_1, product of four elementary rotations:

$$R_1 = R_Y(\phi)R_X(\pi)R_Y(\phi)R_X(\pi) . \tag{8.87}$$

These rotations are simply expressed in terms of the Pauli matrices as $R_X(\pi) = \exp(-i\pi\sigma_X/2) = \cos(\pi/2)\mathbb{1} - i\sin(\pi/2)\sigma_X = -i\sigma_X$ and $R_Y(\phi) = \exp(-i\phi\sigma_Y/2) = \cos(\phi/2)\mathbb{1} - i\sin(\phi/2)\sigma_Y$.[20] Making these replacements in eqn. (8.87), we can rewrite R_1 as:

$$R_1 = \left[\left(\cos\frac{\phi}{2}\mathbb{1} - i\sin\frac{\phi}{2}\sigma_Y\right)(-i\sigma_X)\right]^2 = -\mathbb{1} , \tag{8.88}$$

the last identity resulting from the simple commutation rules of the Pauli operators (eqn. 2.4). In other words, the square of the product of a rotation of angle π around OX by a rotation of *arbitrary* angle ϕ around OY is identical to a rotation by an angle 2π (around an arbitrary axis). It merely produces a π-phase shift (multiplication by

[19]The geometrical representation considered here should not be confused with the usual qubit Bloch sphere. We are dealing here with composite two-level systems whose evolution involves combined qubit and motional state transitions

[20]Similar identities have already been used in previous chapters. They result from a power series expansion of the exponential of the Pauli operators, with $\sigma_i^2 = \mathbb{1}$.

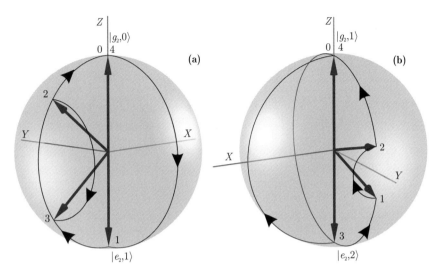

Fig. 8.19 Evolution of the combined Bloch vector realized by the composite pulse sequence. (a) $|g_2, 0\rangle \rightarrow |e_2, 1\rangle$ transition. (b) $|g_2, 1\rangle \rightarrow |e_2, 2\rangle$ transition. The combined Bloch vectors successive positions are marked by the arrows 0, 1, 2, 3 and 4 on both Bloch spheres.

−1) on any state of the first Bloch sphere. This remarkable result, which has been known for a long time by NMR spectroscopists, leaves us free to realize a 2π-pulse on the first Bloch sphere by alternating, on the blue sideband, π-pulses and pulses of arbitrary duration (the two kinds being phase-shifted by $\pi/2$ with respect to each other in order to realize rotations around OX and OY respectively).

Most remarkably, we can choose ϕ in order to produce a 2π-rotation on the second Bloch sphere as well. The same sequence of laser pulses produces on the $|g_2; 1\rangle$, $|e_2; 2\rangle$ composite spin a product of rotations whose angles are multiplied by $\sqrt{2}$:

$$R_2 = R_Y(\phi\sqrt{2})R_X(\pi\sqrt{2})R_Y(\phi\sqrt{2})R_X(\pi\sqrt{2}) .\tag{8.89}$$

Setting then $\phi = \pi/\sqrt{2}$, we obtain:

$$R_2 = R_Y(\pi)R_X(\pi\sqrt{2})R_Y(\pi)R_X(\pi\sqrt{2}) = \left[(-i\sigma_Y)\left(\cos\frac{\pi}{\sqrt{2}}\mathbb{1} - i\sin\frac{\pi}{\sqrt{2}}\sigma_X\right)\right]^2 = -\mathbb{1} .\tag{8.90}$$

The 'magic angle' $\phi = \pi/\sqrt{2}$ thus ensures that the second composite spin comes back, as the first one, to its initial position on its Bloch sphere, undergoing again a simple π-phase shift. Even if the system starting from $|e_2, 1\rangle$ transiently acquires two phonons, its motional state eventually comes back in the $\{0, 1\}$ phonon subspace and the computational space is conserved.

The evolutions of the two composite spins under the effect of the combined laser pulse sequence are represented on their respective Bloch spheres in Fig. 8.19(a) and (b). The initial states $|g_2, 0\rangle$ and $|g_2, 1\rangle$ are represented by vectors pointing to the north poles. The successive positions of the composite Bloch vectors are indicated by the labels 0, 1, 2, 3 and 4 (the first and last ones coinciding with the north pole). We see that both composite spins follow different trajectories but come back, at the end of their two 'circumnavigations', to their initial position.

The effect of the composite pulse sequence on the four tensor product states of the second ion with the stretch mode vibration is thus very simple. The three states $|g_2, 0\rangle$, $|e_2, 1\rangle$ and $|g_2, 1\rangle$ undergo a π-phase shift while $|e_2, 0\rangle$ (which is not coupled to any other state by the blue sideband excitation) remains invariant. Up to a global phase shift, this is equivalent to leaving all states unchanged, but $|e_2, 0\rangle$ which changes sign. This defines the operation of a quantum phase gate.

To turn this phase gate into a control-not, it is sandwiched between two $\pi/2$-pulses with opposite phases, exciting the carrier transition of ion 2. They mix $|g_2\rangle$ and $|e_2\rangle$ independently of the phonon number. When the stretch mode is in $|1\rangle$, the ionic state is unchanged by the phase gate. The $\pi/2$-pulses cancel their effects and ion 2 returns to its initial state. When the stretch mode is in $|0\rangle$, the $\pi/2$-carrier pulses combine in a π-pulse, due to the π-phase shift of $|e_2\rangle$ produced by the phase gate. The two levels $|g\rangle$ and $|e\rangle$ of ion 2 are then interchanged. We obtain a control-not gate, in which the control qubit is the vibration state (active in state $|0\rangle$) and the target ion 2.

At the beginning of the sequence, the internal state of ion 1 is mapped onto the vibration mode, initially in its ground state $|0\rangle$. This is realized with a π-pulse on the first blue sideband. When the ion is in $|e_1\rangle$, it is not affected and leaves the motional state in $|0\rangle$. When it is in $|g_1\rangle$, it is transferred to $|e_1\rangle$ and injects one phonon in the motion. The $|g\rangle$ state of ion 1 (which we take here as coding for $|0\rangle$), which is copied into the $|1\rangle$ state of the stretch mode is thus the inactive qubit state in the ion 1–ion 2 gate operation. After the phonon-controlled gate, the motional state is copied back onto the first ion's internal state, by another π-pulse on the blue sideband. Altogether, the scheme amounts to a control-not gate, in which the first ion is the control (active in its $|e\rangle$ state, coding here for $|1\rangle$), the second the target.

Experimental realization

In the experiment, the sequence of pulses, lasting altogether about 600 μs, can be interrupted at any time t. The probability $P_e(t)$ of finding each ion in $|e\rangle$ at time t is obtained from repeated realizations of the experimental sequence, using laser-induced fluorescence. The results of this 'quantum trace' procedure are presented in Fig. 8.20. Figure 8.20(a) shows the timing of the laser pulses. The pulses applied onto ion 1 perform the state-swap operation with the vibration mode. Between these pulses, ion 1 should always be found in $|e\rangle$. Six pulses are applied on ion 2, the four blue sideband pulses for the control-phase gate and the initial and final $\pi/2$-pulses on the carrier. The latter are shorter, since the laser–ion coupling is much larger for the carrier excitation than for the blue sideband, whose Rabi frequency is proportional to the small Lamb–Dicke parameter η.

The four panels in Fig.8.20(b)–(e) present the state evolution for the initial states $|g_1, g_2\rangle$, $|g_1, e_2\rangle$, $|e_1, g_2\rangle$ and $|e_1, e_2\rangle$ respectively. The initial state is prepared by a carrier excitation during the shaded time interval before the start of the gate sequence at $t = 0$. Ion 1 is then transferred to $|e\rangle$ (or left in this state) by the first state-swap. After this, the second ion undergoes a complex Rabi oscillation pattern under the action of the carrier and blue sideband pulses. The final state-swap is then performed between ion 1 and the vibration mode. The solid lines in Fig. 8.20(b)–(e) are computed

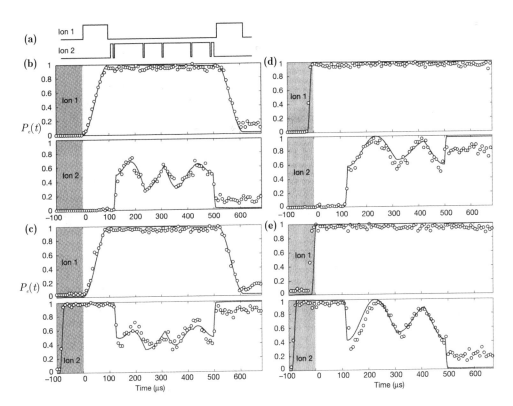

Fig. 8.20 State evolution of the two ions during the control-not gate operation. (a) Pulse sequence applied to the two ions. (b)–(e) State evolution for the four computational basis states. The $|e\rangle$ state probability, P_e is recorded in each case as a function of time during the gate operation. The initial state preparation takes place during the shaded period. The quantum gate operation begins at time $t = 0$. Reprinted by permission from Macmillan Publishers Ltd: Nature, Schmidt-Kaler *et al.* (2003).

from the expected Bloch vector evolution, without any adjustable parameters. They are in excellent agreement with the data. The final states can be read-out directly from these graphs. To a very good approximation, the gate produces the transformations $|g_1, g_2\rangle \rightarrow |g_1, g_2\rangle$, $|g_1, e_2\rangle \rightarrow |g_1, e_2\rangle$, $|e_1, g_2\rangle \rightarrow |e_1, e_2\rangle$ and $|e_1, e_2\rangle \rightarrow |e_1, g_2\rangle$.

Figure 8.21 presents, as a histogram, the measured gate's 'truth table'. The spurious elements are below 22%, corresponding to a final state fidelity of the order of 80%, mainly limited by laser frequency noise, resulting in uncontrolled phase modulations of the sideband pulses. The Innsbruck control-not gate has been used in more complex experiments on larger ion registers. To conclude this section, we briefly describe two experiments involving three ions.

8.4.4 A GHZ state of three ions

The following simple procedure realizes the entanglement of three ions in a GHZ state (Roos *et al.* 2004b). Starting from the initial state $|g, g, g; 0\rangle$, a $\pi/2$-blue sideband

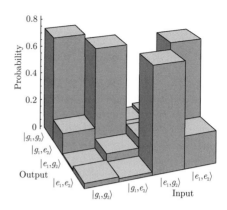

Fig. 8.21 Experimental truth table of the Ca$^+$ control-not gate: probabilities of the four output state for each input state in the logical basis. The maximum spurious element is 22%. Reprinted by permission from Macmillan Publishers Ltd: Nature, Schmidt-Kaler *et al.* (2003).

pulse applied to the first ion entangles it to the stretch vibration mode according to:[21]

$$|g, (g, g); 0\rangle \rightarrow [|g, (g, g); 0\rangle + |e, (g, g); 1\rangle]/\sqrt{2} \ . \tag{8.91}$$

Then, a π-carrier pulse, again applied to ion 1, exchanges its qubit states, leading to the state $[|e, (g, g); 0\rangle + |g, (g, g); 1\rangle]/\sqrt{2}$. Next, a control-not gate is operated between ion 2 and the stretch mode. This gate involves, as we have just shown above, six laser pulses applied on ion 2 (a composite excitation made of four blue sideband pulses, sandwiched by two $\pi/2$-carrier pulses). This operation switches the state of ion 2, conditioned to the stretch mode being in $|0\rangle$, according to:

$$[|(e), g, (g); 0\rangle + |(g), g, (g); 1\rangle] \rightarrow [|(e), e, (g); 0\rangle + |(g), g, (g); 1\rangle]/\sqrt{2} \ . \tag{8.92}$$

This transformation can be viewed as the unitary step in the quantum non-demolition measurement by ion 2 of the phonon number in the stretch mode. Finally, a π-carrier pulse on ion 3 exchanges $|g\rangle$ and $|e\rangle$, leading to $(|e, e, e; 0\rangle + |g, g, e; 1\rangle)/\sqrt{2}$ and a π-blue sideband pulse on that ion exchanges its internal excitation with the collective oscillation, bringing the ion string back to its ground state of motion:

$$\frac{1}{\sqrt{2}}(|e, e, e; 0\rangle + |g, g, e; 1\rangle) \rightarrow \frac{1}{\sqrt{2}}(|e, e, e\rangle + |g, g, g\rangle)|0\rangle = |\Psi_{\text{GHZ}}\rangle|0\rangle \ . \tag{8.93}$$

A GHZ state of three qubits is thus prepared according to a recipe which is strikingly similar to the method used to generate a GHZ state of three atoms in CQED (Section 6.3.2). We recognize the same entanglement-knitting procedure, involving $\pi/2$-, 2π- and π-Rabi pulses which entangle one by one the three spins to the spring system.

The Innsbruck group has applied this recipe to prepare a three-ion GHZ state and performed a complete tomography of its density operator ρ_{GHZ} (Roos *et al.* 2004b). Figure 8.22(a) displays the absolute values of the measured 64 matrix elements. The tomography has involved $3^3 = 27$ different settings of carrier pulses applied on the three ions prior to their independent fluorescence readings. The two large diagonal elements correspond to the probabilities (nearly equal to $1/2$) of finding

---

[21]To simplify notation, the ion states are no longer labelled by numbers; the states of the ions which are not affected by the current operation are still put between parentheses.

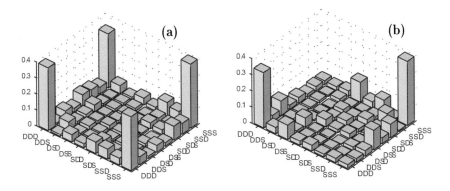

Fig. 8.22 (a) Absolute value of the elements of the density operator in the three-ion GHZ experiment, reconstructed by tomography (S stands for g and D for e). (b) Absolute values of the elements of the density operator after an unread measurement of ion 1. Reprinted with permission from Roos *et al.* (2004b), © AAAS.

the ion string in $|e, e, e\rangle$ and $|g, g, g\rangle$, while the two main non-diagonal terms, which have approximately the same magnitude, describe the coherence between these states. These four matrix elements largely dominate the background of the remaining 60 components, due to experimental imperfections. From this measurement, the fidelity $F_{\mathrm{GHZ}} = \langle \Psi_{\mathrm{GHZ}} | \rho_{\mathrm{GHZ}} | \Psi_{\mathrm{GHZ}} \rangle$ of the GHZ state is found to be 0.84.

Tracing a GHZ state over one of its qubits results in an incoherent mixture for the two remaining particles. We have analysed this property in detail in Section 6.3.2 and related it to the concepts of complementarity and decoherence. The three-ion GHZ state experiment beautifully illustrates these fundamental concepts. Figure 8.22(b) presents the tomography of the three-qubit system, after preparation in the $|\Psi_{\mathrm{GHZ}}\rangle$ state and a subsequent unread measurement on ion 1. This ion is subjected to the fluorescence cycle detection process, but the result of this procedure is not recorded. The three ions are then measured independently and the final density operator ρ_{GHZ}^{d} is reconstructed. The result of this decoherence experiment is clear: the two diagonal components of the density operator are conserved, while the two main off-diagonal terms have vanished. The GHZ state has been transformed into a statistical mixture with equal weights of $|g, g, g\rangle$ and $|e, e, e\rangle$.

The unread measurement has turned ion 1 into an 'environment' for the two other ions. The mere fact that information about the qubit state could be retrieved from the measurement records transforms the coherent superposition into a statistical mixture, a result which we have already commented in detail in the CQED context. An interesting aspect of this situation is that information stored in ion 1 can be erased if we decide to read the state of this qubit in a rotated basis different from $\{|e\rangle, |g\rangle\}$. We then expect to recover an entanglement between the two remaining ions, provided one detects them in correlation with the results of the measurement performed on ion 1 in the rotated basis. We have already analysed such an experiment realized on three Rydberg atoms in CQED (Section 6.3.2). A similar quantum eraser experiment, which we will not describe here, has been performed by the Innsbruck team on its three-Ca^{+} ion crystal (Roos *et al.* 2004b).

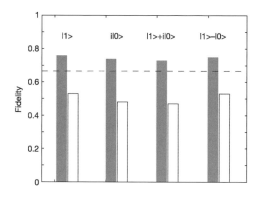

Fig. 8.23 Teleportation fidelity for four input states, $|1\rangle$, $|0\rangle$, $|1\rangle + i\,|0\rangle$ and $|1\rangle - i\,|0\rangle$. The gray bar gives the observed fidelity. The white bars present the fidelity when the final unitary operation is not performed by Bob. The dashed line gives the limit of a classical teleportation scheme. Reprinted by permission from Macmillan Publishers Ltd: Nature, Riebe *et al.* (2004).

8.4.5 Teleportation of an ion qubit

As a final illustration of the control-not gate operation, we briefly present a teleportation experiment which, as the GHZ one, has involved the manipulation of three ions (Riebe *et al.* 2004). The principle of teleportation has already been discussed in Section 2.6.3. Let us only recall here that the two quantum information partners, Alice and Bob, own a pair of particles, labelled a and b, in a maximally entangled Bell state. Alice has also at hand a particle, labelled u, in the unknown state $|\chi\rangle$, which should be transmitted to Bob. Alice first performs a complete measurement of the $u - a$ system in the Bell states basis. She obtains two bits of classical information that she transmits to Bob over a classical communication channel. Alice's measurement projects Bob's particle onto a state which is obtained from $|\chi\rangle$ by application of the Pauli or unity operators. Using the two bits provided by Alice, Bob determines which transformation should be performed on particle b to recover the initial state $|\chi\rangle$.

Implementing this protocol with three ions requires to combine various sets of operations achieved separately in other experiments already described in this section. The preparation of the ion a–ion b Bell state which serves as a teleportation 'resource' has been described in Section 8.4.1. The measurement of the ion u–ion a Bell state involves, as discussed in Chapter 2, the use of a control-not gate which turns the four Bell states into four qubit product states, subsequently detected by the fluorescence method. The gate used in this procedure was just described (Section 8.4.3). Once this measurement made, Bob's final operation is a mere identity or a single-qubit rotation realized by a carrier pulse whose phase depends on the outcome of Alice's measurement. The Innsbruck group has put together all these steps, including some variations and refinements which we will not describe here (Riebe *et al.* 2004).

Repeating the procedure many times on each of four test states $|\chi\rangle = |1\rangle$, $|0\rangle$, $|1\rangle + i\,|0\rangle$ and $|1\rangle - |0\rangle$, the density operator $\rho_{\exp}$ of the final state has been statistically reconstructed by tomography and compared to the theoretically expected state by extracting from the data the fidelity $F = \langle\chi|\,\rho_{\exp}\,|\chi\rangle$. Figure 8.23 presents as grey vertical bars the fidelities, close to 78%, obtained for the four test states.

The white bars show, for comparison, the fidelity obtained when no final operation is performed by Bob (no classical communication channel established with Alice). Qubit b is in this case described by a fully unpolarized density operator equal to $\mathbb{1}/2$, since it is maximally entangled with another qubit about which Bob has no information (see Chapter 2). The fidelity $\langle\chi|\,\mathbb{1}\,|\chi\rangle\,/2$ is then $1/2$ for any $|\chi\rangle$, a value very close to what is effectively obtained experimentally. If Alice and Bob have on the other hand established a classical channel without sharing entanglement, the best communication strategy consist for Alice to measure randomly one spin component of her qubit, to send her result to Bob who aligns the spin of his particle along this direction. It is easy to show that on average, the fidelity is in this case equal to $2/3$ (dotted horizontal line in Fig. 8.23). The grey bars are consistently above this classical limit, which demonstrates the essential role played by the quantum communication channel in this experiment.

8.5 Quantum logic with trapped ions: collective qubit addressing

The C–Z method illustrated by the Ca^+ experiments addresses one by one the ions of a linear crystal, using properly tailored laser beams whose waist is smaller than the qubit separation. This strategy has the advantage of obvious scalability. Any logical circuit can be realized by combining simple operations in which the successive control and target qubits are clearly identified. There is however a price to pay for this conceptual simplicity. Individual addressing requires inter-ion distances in the several micrometres range. This *loose binding* condition corresponds to relatively low oscillation frequencies, at most a few MHz. The characteristic time of the gate operations, which require resolved sideband spectra, must be at least an order of magnitude larger than the ion oscillation period, which restricts the speed of the device. In practice, the Ca^+ control-not gate, with its multiple-laser-pulse scheme, needs several hundreds of microseconds to operate.

The Boulder group, working on Be^+, has adopted an alternative strategy. In order to decrease the gate time, the non-local manipulations of the qubits are realized while the ions are tightly confined, with typical oscillation frequencies in the 10 MHz range. The ion separation is then too small for individual laser addressing. The qubits are irradiated together by light beams with large waists, which do not discriminate between the ions. Entangling phase gates, in which the two qubits play symmetrical roles, are very quickly realized in this way, within a time of a few tens of microseconds.

In order to perform single-qubit rotations, which have to be combined to two-qubit gates to achieve general 'computing' operations, the ions must be separated from each other before being submitted to focused carrier laser pulses. This strategy thus relies on a trap design in which ions can be moved, being separated or put in tight contact on demand. In this section, we discuss the physics of ion entanglement by collective laser addressing and we generalize the analysis to the collective manipulation of an ensemble of $\mathcal{N}$ ions. The architecture of a general-purpose computing machine operating with movable ions will be evoked in the last section of this chapter.

8.5.1 Geometric phase gate realized by collective laser addressing

Principle of geometric gate

The goal is to realize with a laser pulse addressing collectively two closely-spaced ions the symmetrical unitary transformation:

$$\mathcal{T}_{12} = e^{i(\pi/4)(\mathbb{1}-\sigma_{Z1}\sigma_{Z2})}\ , \tag{8.94}$$

which leaves the two-ion state unchanged if the qubits are in the same logical state and phase shifts it by $\pi/2$ if they are different, according to:[22]

$$T_{12}\,|e,e\rangle = |e,e\rangle \quad ; \quad T_{12}\,|g,g\rangle = |g,g\rangle \;\; ;$$
$$T_{12}\,|e,g\rangle = i\,|e,g\rangle \quad ; \quad T_{12}\,|g,e\rangle = i\,|g,e\rangle \;. \tag{8.95}$$

Under a simple disguise, T_{12} does implement a phase gate. Multiplying it by single-qubit operations which independently phase-shift the two qubits, we construct the transformation:

$$U_{12} = T_{12}\,e^{-i(\pi/4)(\sigma_{Z1}-\mathbb{1})}e^{-i(\pi/4)(\sigma_{Z2}-\mathbb{1})}\,, \tag{8.96}$$

whose truth table is that of the control-σ_Z phase gate (see eqn. 8.86):

$$U_{12}\,|e,e\rangle = -\,|e,e\rangle \quad ; \quad U_{12}\,|g,g\rangle = |g,g\rangle \;\; ;$$
$$U_{12}\,|e,g\rangle = |e,g\rangle \quad ; \quad U_{12}\,|g,e\rangle = |g,e\rangle \;. \tag{8.97}$$

In order to realize this simple transformation (Leibfried *et al.* 2003b), the Boulder group has exploited an idea proposed by Milburn *et al.* (2000). In a nutshell, it consists in pushing a collective mode of the ions oscillation around a closed loop in its phase space, bringing it finally back in its initial state. The qubit and motional states end up being separated, but the circumnavigation of the collective oscillation leaves a geometric phase imprint on the system's quantum state. If the pushing force is made to depend upon the internal states of the ions, one can achieve the phase-shifts described by eqn. (8.95) and thus realize a transformation which can easily be turned into a phase gate or a control-not.

The geometric phase shift accumulated by an oscillator undergoing a closed loop translation in its x,p phase space derives directly from the formula giving the composition of Glauber's displacement operators (eqn. 3.49, on page 117). Translating an oscillator in phase space by β then by α is identical to a direct translation by $\alpha+\beta$, up to a global phase factor $\exp(\alpha\beta^* - \alpha^*\beta)/2$. The exponent in this expression is equal to i times the external product of the vectors represented by α and β, i.e. i times the oriented area $S_{\alpha,\beta}$ of the triangle made by these two vectors. In other words, the closed circuit combining, in that order, the displacements β, α and $-(\alpha+\beta)$ brings the oscillator back to its initial state, with a phase shift equal to the oriented loop area. This result obviously generalizes to any closed polygon, which can be decomposed in elementary triangles.

Let us first analyse this geometric phase effect on a single trapped ion. We have shown in Section 8.2.2 that a displacement of the ion's oscillator is realized by momentarily applying to it a time-modulated force. For a Be^+ ion, this force can be produced by a pair of pulsed Raman lasers, oriented at right angles from each other and inclined at 45 degrees with respect to the oscillation direction. The frequency difference $\omega_{\ell1} - \omega_{\ell2}$ of these beams matches the oscillator's frequency ω_z and their phase difference φ defines the direction of the translation in the oscillator's phase space. The amplitude of the displacement is proportional to the Lamb–Dicke parameter η, to the Raman Rabi frequency Ω_R and to the duration τ of the pulse (eqn. 8.49). The

---

[22]Here $|g\rangle$ and $|e\rangle$ code for $|0\rangle$ and $|1\rangle$ respectively.

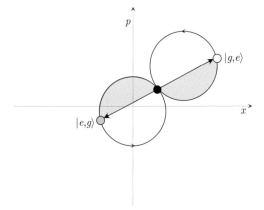

Fig. 8.24 Phase space representation of the stretch mode closed-loop trajectory for the two qubit configurations $|e, g\rangle$ and $|g, e\rangle$. The initial state is represented by the black circle and the current positions by the open and grey circles for $|g, e\rangle$ and $|e, g\rangle$ respectively. The phase accumulations at the current positions are identical, proportional to the shaded areas.

frequency Ω_R, proportional to the product of the two laser fields amplitudes, also depends upon the detunings of the lasers with the $2S_{1/2} \to 2P_{1/2}$ and $2S_{1/2} \to 2P_{3/2}$ transitions of the Be$^+$ ion. A triangular loop with a given area can in principle be achieved by combining three such Raman pushes, with proper adjustments of pulses durations and phases.

A more convenient way to realize a closed trajectory consists in applying to the ion a time-modulated pushing force at a frequency slightly detuned from the ion's frequency, realizing between the Raman lasers a frequency difference $\omega_{\ell 1} - \omega_{\ell 2} = \omega_z + \delta$ with $|\delta| \ll \omega_z$. During a short time interval, $[t, t + dt]$ with $dt \ll 1/\delta$, this force produces an elementary translation along a direction defined by the instantaneous phase δt. This direction changes as time evolves, resulting in a circular motion of the oscillator representative point in its phase space. After a time $2\pi/\delta$, the oscillator has travelled around a closed circle and comes back to its initial position. By adjusting the Raman pulse duration to this value, one realizes a circle whose area is proportional to $(\Omega_R \eta/\delta)^2$. A phase-shift equal to $\pi/2$ is achieved by proper setting of all the relevant parameters.

When two trapped ions are symmetrically irradiated in this way, they both experience a pushing force. The experimenters adjust the trap stiffness so that the ion separation at equilibrium is equal to an integer number of periods of the walking standing wave formed by the intersecting Raman beams. Both ions then experience exactly the same laser field. As a result, they are submitted to the same pushing force, provided they are in identical qubit states (configurations $|gg\rangle$ or $|ee\rangle$). The forces acting on the ions in unlike qubit states are on the contrary unbalanced.[23]

If the Raman forces are made nearly resonant with the anti-symmetrical stretch mode, we easily understand that they will have no net effect on the ion's motion when both ions are in the same qubit state. The equal forces then do not couple at all to the $z_1 - z_2$ collective coordinate. Only the CM mode [coordinate $(z_1 + z_2)/2$] could

---

[23] In the Schrödinger cat experiment (Section 8.2.3), one of these forces is set to 0. In the geometric gate experiment, the ratio of the forces is -2. This value is however not essential. The forces on the two qubit states must only be different. The Raman lasers must also produce the same light-shift on $|e\rangle$ and $|g\rangle$, in order to avoid an unwanted qubit dephasing during gate operation. The balancing of the light-shifts, and unbalancing of the resonant pushing forces on $|e\rangle$ and $|g\rangle$ are independently adjusted by setting properly the Raman laser detunings and polarizations (Wineland *et al.* 2003).

be excited in this case, but the force being modulated at a frequency far off-resonant with this mode, it remains unexcited as well. It is only when the ions are in the $|e, g\rangle$ and $|g, e\rangle$ two-qubit states that the stretch mode is excited. This excitation is tuned so as to produce a $\pi/2$-area closed loop on the collective ionic motion, resulting in the gate transformation described by eqn. (8.95).

The principle of this gate is illustrated in Fig. 8.24 where we show the trajectories of the stretch mode representative point in phase space for the two $|eg\rangle$ and $|ge\rangle$ configurations. The initial state of the ion's motion (black circle) needs not be the ground state. It can be any coherent state of motion, or even a non-classical or a mixed state. The phase geometric argument, which is a property of the transformations and not of the states themselves, applies to all these cases, provided the number of phonons remain small enough so that the system's evolution can be described within a model restricted to a first-order development in powers of η. In practice, this means that this gate can operate with a few thermal phonons, relaxing the strict requirement of preliminary sideband cooling to $T = 0$ K. We also see in Fig. 8.24 that the loops associated to the $|eg\rangle$ and $|ge\rangle$ states are symmetrical with respect to the oscillator initial position. They are both travelled in the same direction and correspond thus to the same phase shift, in sign as well as in magnitude.

Experimental realization

The Boulder group, following the procedure we have just analysed, has demonstrated a high-fidelity phase gate operating with two Be^+ ions (Leibfried *et al.* 2003b). In order to test its operation, the experimenters have coupled it to an interferometer design à *la* Ramsey. The effect of the gate by itself is a mere phase shift which has no effect on the ion qubit populations, the only physical quantities which are accessible to detection. By sandwiching the gate between single-qubit operations acting symmetrically on both ions, the Boulder team has turned the gate into a Bell-state generator and has been able in this way to analyse the gate operation and measure the fidelity of the states it produces.

Let us follow the successive steps of their experiment, which all address the two ions symmetrically. The first stage of the procedure, starting from the initial $|g, g\rangle$ state, prepares the qubits in superpositions of the logical states, by applying simultaneously to both of them a $\pi/2$-carrier pulse. This is achieved by a pair of co-propagating Raman laser beams, uniformly covering both ions. This operation, akin to an Hadamard operation, prepares the state $(|g, g\rangle + |g, e\rangle + |e, g\rangle + |e, e\rangle)/2$. Next, the pushing Raman lasers are applied to realize the phase gate operation, transforming the system's state into $[(|g, g\rangle + |e, e\rangle) + i(|g, e\rangle + |e, g\rangle)]/2$. Finally, a $3\pi/2$-carrier pulse is applied on both ions, which mixes independently again their qubit states. A simple calculation shows that the system ends up in state:

$$|\Psi_{12}\rangle = \frac{e^{i\pi/4}}{\sqrt{2}} |g, g\rangle + \frac{e^{-i\pi/4}}{\sqrt{2}} |e, e\rangle . \tag{8.98}$$

Within unimportant phase factors, which could be incorporated in the qubit state definitions, $|\Psi_{12}\rangle$ is a Bell state, exhibiting maximum entanglement between the two qubits.

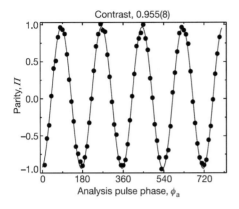

Contrast, 0.955(8)

Parity, Π

Analysis pulse phase, ϕ_a

Fig. 8.25 The parity signal after producing and analysing the maximally entangled state $|\Psi_{12}\rangle$. Reprinted by permission from Macmillan Publishers Ltd: Nature, Leibfried *et al.* (2003b).

In order to prepare this state with a high fidelity, it is important to reduce as much as possible the effects of ambient magnetic field fluctuations on the evolution of the qubit coherences. These fluctuations vary on time scales much longer than the experimental sequence, which must be repeated over and over to produce statistical results determining the fidelity of the Bell state preparation. The effect of these fluctuations on signals averaged over many realizations of the experiment can thus be described as an inhomogeneous perturbation. To suppress the corresponding noise, the $3\pi/2$-carrier pulse after the geometric phase gate is split into two parts. First a π-pulse is applied a time T after the first $\pi/2$-pulse of the sequence. Then, after another delay T, a final $\pi/2$-pulse completes the procedure. The dephasing of the qubit coherence during the first time interval T is compensated during the second one by this spin echo method (Section 7.4.3). This results, as we will see below, in an excellent Bell state fidelity. We can disregard this experimental subtlety in the theoretical analysis of an ideal experiment. In the discussion leading to eqn. (8.98), we have thus combined the π- and $\pi/2$-pulses in a single one, of angle $3\pi/2$.

The Bell state obtained by this procedure remains to be analysed. One way would be to perform state tomography, as described above in the Ca^+ case. This would require however individual ion addressing and reading, which could be done only after moving them away from each other. Instead of this, the Boulder group has chosen to leave the ions in place and to interrogate them collectively by an interferometric method.

A last $\pi/2$-carrier pulse with variable phase ϕ_a is applied to the system and the fluorescence level of both ions finally measured. We recall that each ion yields a signal only if in $|g\rangle$. We thus expect a large average signal, S_0, when the system is detected in $|g, g\rangle$, a signal $S_0/2$, when it is found in $|g, e\rangle$ or $|e, g\rangle$ and a zero signal when it is in $|e, e\rangle$. The whole experimental sequence is repeated many times for each ϕ_a and the statistics of the results yields the probabilities $P_{gg}(\phi_a)$, $P_{ge}(\phi_a) + P_{eg}(\phi_a)$ and $P_{ee}(\phi_a)$ of finding zero, one or two qubit excited.

From these data, the experimenters have reconstructed the expectation value of the atomic parity operator $\sigma_{Z1}\sigma_{Z2}$ which assumes the eigenvalue $+1$ if both ions are in the same state and -1 if they are in unlike states:

$$\langle \sigma_{Z1}\sigma_{Z2}(\phi_a)\rangle = P_{gg}(\phi_a) + P_{ee}(\phi_a) - [P_{ge}(\phi_a) + P_{eg}(\phi_a)] . \tag{8.99}$$

This parity signal recorded versus ϕ_a is shown in Fig. 8.25. It oscillates between extreme values very close to ± 1. If the experiment works according to theory, we expect that the pure parity states corresponding to these extreme values are maximally entangled because they are obtained from the Bell state $|\Psi_{12}\rangle$ (eqn. 8.98) by a one-qubit operation (the last $\pi/2$-carrier pulse) which cannot change the system's degree of entanglement. The state of parity $+1$ (at fringe maxima) must ideally be a superposition with equal probabilities of $|g, g\rangle$ and $|e, e\rangle$, while the state of parity -1 (at fringe minima) must be a superposition with equal probabilities of $|g, e\rangle$ and $|e, g\rangle$. We also notice that the signal oscillates versus ϕ_a at a rate which is the double of the one expected in one-particle Ramsey interferometry (the parity comes back to the same value after a phase increment of 180 degrees).

Let us give a theoretical derivation of this parity signal by expressing it in terms of the density operator ρ of the two-qubit system just before the last probing pulse is applied. If the gate has operated properly, we expect $\rho = |\Psi_{12}\rangle \langle \Psi_{12}|$, but we will not assume this result because we want to use the interferometric signal as a probe of the state prepared by the gate, without any presupposition about it. The $\pi/2$-carrier pulses phase is adjusted to produce on each ion a qubit rotation around an axis in the XOY plane of the Bloch sphere at an angle ϕ_a with OX, which is expressed by the operation $\exp(-i\pi\sigma_{\phi_a}/4)$ with $\sigma_{\phi_a} = \cos(\phi_a)\sigma_X + \sin(\phi_a)\sigma_Y$. With these definitions, the parity signal is:

$$\langle \sigma_{Z1}\sigma_{Z2} \rangle = \text{Tr}\left[\rho e^{i(\pi/4)(\sigma_{\phi_a 1} + \sigma_{\phi_a 2})} \sigma_{Z1}\sigma_{Z2} e^{-i(\pi/4)(\sigma_{\phi_a 1} + \sigma_{\phi_a 2})} \right] . \tag{8.100}$$

It is the expectation value in ρ of the product of single-qubit operators $\exp(i\pi\sigma_{\phi_a}/4)\sigma_Z$ $\exp(-i\pi\sigma_{\phi_a}/4)$, equal to $\cos(\phi_a)\sigma_Y - \sin(\phi_a)\sigma_X$.

Combining this expression with eqn. (8.100), we obtain:

$$\langle \sigma_{Z1}\sigma_{Z2} \rangle = \cos^2(\phi_a)\text{Tr}[\rho\sigma_{Y1}\sigma_{Y2}] + \sin^2(\phi_a)\text{Tr}[\rho\sigma_{X1}\sigma_{X2}]$$
$$- \sin(\phi_a)\cos(\phi_a)\text{Tr}[\rho(\sigma_{X1}\sigma_{Y2} + \sigma_{Y1}\sigma_{X2})] , \tag{8.101}$$

and after expliciting the trace operations and defining $\rho_{gg,ee}$ as $|\rho_{gg,ee}| \exp(i\psi_0)$:

$$\langle \sigma_{Z1}\sigma_{Z2} \rangle = -2|\rho_{gg,ee}| \cos(2\phi_a - \psi_0) + 2Re(\rho_{eg,ge}) . \tag{8.102}$$

The theoretical parity signal is expected to undergo two complete oscillations when the phase of the probing pulse makes one turn, in agreement with the observed data.[24] The contrast of this modulation measures the double of the absolute value of the matrix element between $|e, e\rangle$ and $|g, g\rangle$ in the state prepared by the geometric phase gate sandwiched between the two $\pi/2$- and $3\pi/2$-carrier pulses. From the experimentally measured contrast $C = 0.955$, we deduce $|\rho_{gg,ee}| = 0.477$, a result very close to the ideal $|\langle gg |\Psi_{12}\rangle \langle \Psi_{12}| ee\rangle| = 1/2$ value. More generally, the experimental data yield the fidelity of the Bell state preparation, $F_{12} = \langle \Psi_{12}| \rho |\Psi_{12}\rangle = (P_{ee} + P_{gg})/2 + C/2$.

---

[24] According to eqn. (8.102), the interference signal should be at a minimum for $\phi_a = \pi/4$ [since $\psi_0 = \pi/2$, as can be checked on eqn. (8.98)]. There is thus a $\pi/4$-phase offset between the theoretical prediction and the experimental signal shown in Fig. (8.25). This offset is accounted for by the phase-locking procedure used in the experiment, which added a non-compensated phase to the signal (D. J. Wineland and D. Leibfried, private communication).

From fluorescence measurements performed just after the Bell state preparation, the experimenters have obtained $P_{ee} + P_{gg} = 0.98$. Combining this value with the contrast C of the fringes, they get the remarkably good value $F_{12} = 0.97 \pm 0.02$.

8.5.2 GHZ state of up to six ions prepared by collective addressing

Principle of the GHZ state generation

Following the realization of this geometric two-qubit gate, the Boulder group has demonstrated the power of collective qubit addressing by generalizing the method to the manipulation of up to $\mathcal{N} = 6$ ions (Leibfried *et al.* 2005). It is convenient, in order to describe the qubit states in a symmetrical multi-ion system, to adopt the collective angular momentum point of view which we have introduced in Section 4.7 in the CQED context. The $\mathcal{N}$ two-level ions are initialized in a symmetrical state and they undergo laser-induced transformations which are invariant by ion permutation. They thus evolve within the Dicke subspace of states symmetrical by particle exchange, which is described as the Hilbert space of a $J = \mathcal{N}/2$ angular momentum. The $2J + 1$ $|J, M\rangle$, eigenstates of $J_Z = (\sum_i \sigma_{Zi})/2$ form a basis in this space $(-J \leq M \leq J)$. State $|J, M\rangle$ is the fully symmetrical state superposition in which $J + M$ ions are in state $|e\rangle$ (spins pointing up along OZ) and $J - M$ in state $|g\rangle$ (spins pointing down). The system's ground state is $|J, -J\rangle = |g, g, \ldots, g\rangle$ and the fully excited state is $|J, J\rangle = |e, e, \ldots, e\rangle$. The Boulder experiment has prepared and analysed GHZ states, linear superpositions with equal weights of $|J, -J\rangle$ and $|J, J\rangle$, with J up to 3.

The core of the GHZ state preparation is a collective geometric gate realizing the transformation:

$$\mathcal{T}_\mathcal{N} = e^{i(\pi/2)J_Z^2} = e^{i\pi\mathcal{N}/8} \exp\left[i\frac{\pi}{8}\sum_{i \neq j}\sigma_{Zi}\sigma_{Zj}\right], \tag{8.103}$$

which, up to a sign change and a global phase addition in the exponent, generalizes to $\mathcal{N} > 2$ the two-qubit transformation (8.94). To achieve this operation, a Raman pushing force is applied to all the ions. This force is now made nearly resonant with the collective CM mode of motion. It drives this mode on a circular loop in its phase space. The trap potential is adjusted so that the ions all sit at equivalent positions in the laser standing wave pattern.[25] The Raman lasers polarizations and frequencies are set so that the contribution to the pushing force takes opposite values when an ion is in $|g\rangle$ or $|e\rangle$. Under these conditions, the net driving force is proportional to the difference between the number of ions in $|e\rangle$ and $|g\rangle$, i.e. to $\langle J_Z \rangle$. The area of the loop undergone by the collective motional state is thus proportional to $\langle J_Z^2 \rangle$. The Ω_R, η and δ parameters are chosen so that the loop produces a $(\pi/2)\langle J_Z^2 \rangle$ phase-shift, thus achieving the transformation (8.103). This operation, which produces non-linear phase shifts of the qubit collective states, amounts to the action during an appropriate time of an effective Hamiltonian, symmetrical sum of the $\sigma_{Zi}\sigma_{Zj}$ inter-ion coupling terms. Note that, as in the $\mathcal{N} = 2$ case, the geometric phase shift is largely insensitive to the initial state of the ion's motion, so that the procedure does not need perfect cooling as a preliminary step.

---

[25] At equilibrium, the $\mathcal{N}$ ions are not equidistant, but their separations are set to correspond to integer values of the light-wave spatial period, so that they all see the same electric field.

The transformation $\mathcal{T}_\mathcal{N}$ acting alone on the initial $|J, -J\rangle = |g, g, \ldots, g\rangle$ state does not produce any effect, besides a simple phase-shift. In order to generate a GHZ state $\mathcal{T}_\mathcal{N}$ must be sandwiched between collective qubit rotations, following a recipe similar to that generating a Bell state in the $\mathcal{N} = 2$ case. The procedure depends upon the parity of the number $\mathcal{N}$ of ions. We will analyse it first for $\mathcal{N}$ even. A first global $\pi/2$-rotation of $|J, -J\rangle$ transforms this state into an un-entangled tensor product of $(|e_i\rangle + |g_i\rangle)/2$ 'transverse' spin states. The $\mathcal{T}_\mathcal{N}$ collective gate is then applied. It phase shifts the various terms of the state superpositions produced by the first pulse, generating entanglement. Finally, a global rotation, inverse of the first one, completes the procedure. The global transformation can be expressed as:

$$U_\mathcal{N}^{\text{even}} = e^{-i\pi J_X/2} e^{i\pi J_Z^2/2} e^{i\pi J_X/2} = e^{-i\pi J_Y^2/2} , \tag{8.104}$$

the last equality resulting from simple rotation properties of the angular momentum.

In order to compute $U_\mathcal{N}^{\text{even}} |J, -J\rangle$, we start by remarking that the states $|J, -J\rangle$ and $|J, J\rangle$ are related by the simple relation:

$$e^{-i\pi J_Y} |J, J\rangle = |J, -J\rangle , \tag{8.105}$$

which expresses that a rotation by an angle π of all the qubits around the OY axis of their Bloch spheres changes $|e, e, \ldots, e\rangle$ into $|g, g, \ldots, g\rangle$. This relation shows that the symmetrical and anti-symmetrical superpositions of $|J, J\rangle$ and $|J, -J\rangle$ can be expressed as:

$$|J, J\rangle \pm |J, -J\rangle = \left[1 \pm e^{-i\pi J_y}\right] |J, J\rangle . \tag{8.106}$$

Let us for a moment use as a basis the J_Y eigenstates (with eigenvalues M_Y). Since $\mathcal{N}$ is even (and J an integer), the phase factor $e^{-i\pi J_Y}$ in eqn. (8.106) assumes the value $(-1)^{M_Y}$ when applied to a state $|J, M_Y\rangle$. This shows that $|J, J\rangle + |J, -J\rangle$ and $|J, J\rangle - |J, -J\rangle$ contain respectively only even and odd states in M_Y. If M_Y is even (equal to $2p$ with p integer), the phase factor $\pi M_Y^2/2 = 2p^2\pi$ is a multiple of 2π. If M_Y is odd (equal to $2p+1$), we have on the other hand $\pi M_Y^2/2 = (2p^2+2p+1/2)\pi$, a phase factor equal to $\pi/2$ (modulo 2π). In other words, the operator $\exp[-i\pi J_Y^2/2]$ leaves the $|J, J\rangle + |J, -J\rangle$ state invariant and multiplies $|J, J\rangle - |J, -J\rangle$ by $e^{-i\pi/2} = -i$. We finally notice that $|J, -J\rangle$ can be written as half the difference between the symmetrical and anti-symmetrical superpositions of $|J, J\rangle$ and $|J, -J\rangle$, which immediately leads to:

$$U_\mathcal{N}^{\text{even}} |J, -J\rangle = \frac{1}{2}(|J, J\rangle + |J, -J\rangle) + \frac{i}{2}(|J, J\rangle - |J, -J\rangle) , \tag{8.107}$$

or, grouping the terms differently:

$$U_\mathcal{N}^{\text{even}} |J, -J\rangle = \frac{e^{i\pi/4}}{\sqrt{2}} |J, J\rangle + \frac{e^{-i\pi/4}}{\sqrt{2}} |J, -J\rangle . \tag{8.108}$$

For an odd number of ions, a slightly modified argument shows that a GHZ state of the same form is generated by the transformation:

$$U_\mathcal{N}^{\text{odd}} = e^{-i\pi J_X/2} e^{i\pi J_Z/2} e^{i\pi J_Z^2/2} e^{i\pi J_X/2} = e^{-i\pi J_Y(J_Y+\mathbb{1})/2} , \tag{8.109}$$

which requires the application of an additional rotation $\exp(i\pi J_Z/2)$ of the $\mathcal{N}$ ions between the collective geometric phase gate and the last $\exp(-i\pi J_X/2)$ operation. In this case, it is convenient to expand $|J, -J\rangle$ along the two states $|J, J\rangle + i |J, -J\rangle$ and $|J, J\rangle - i |J, -J\rangle$. Expressed in the J_Y basis, these states are respectively even and odd in $(M_Y - 1/2)$. The $\exp[-i\pi J_Y(J_Y + \mathbb{1})/2]$ operator multiplies the even state by $\exp(-3i\pi/8)$ and the odd one by $i \exp(-3i\pi/8)$. Summing up the two contributions, we find:

$$U_{\mathcal{N}}^{\mathrm{odd}} |J, -J\rangle = \frac{e^{-i\pi/8}}{\sqrt{2}} (|J, -J\rangle - |J, J\rangle) . \qquad (8.110)$$

The odd and even cases can be combined in a single formula by defining the operator $U_{\mathcal{N}}$ as:

$$U_{\mathcal{N}} = \exp\left[-i\frac{\pi}{2} J_X\right] \exp\left[i\frac{\xi_{\mathcal{N}}\pi}{2} J_Z\right] \exp\left[i\frac{\pi}{2} J_Z^2\right] \exp\left[i\frac{\pi}{2} J_X\right] , \qquad (8.111)$$

with $\xi_{\mathcal{N}} = \left[1 - (-1)^{\mathcal{N}}\right]/2$, equal to 0 for $\mathcal{N}$ even and to 1 for $\mathcal{N}$ odd. With this compact notation, the action of $U_{\mathcal{N}}$ on $|J, -J\rangle$ is:

$$U_{\mathcal{N}} |J, -J\rangle = \frac{e^{i\pi\xi_{\mathcal{N}}/8}}{\sqrt{2}} \left[e^{-i\frac{\pi}{4}} |J, -J\rangle + i^{\xi_{\mathcal{N}}} e^{i\frac{\pi}{4}} |J, J\rangle\right] . \qquad (8.112)$$

In the same way, $U_{\mathcal{N}}$ acting on $|J, J\rangle$ yields:

$$U_{\mathcal{N}} |J, J\rangle = \frac{e^{i\pi\xi_{\mathcal{N}}/8}}{\sqrt{2}} \left[(-i)^{\xi_{\mathcal{N}}} e^{i\pi/4} |J, -J\rangle + e^{-i\pi/4} |J, J\rangle\right] , \qquad (8.113)$$

the two states $U_{\mathcal{N}} |J, \pm J\rangle$ being obviously orthogonal. Starting from an initial state in which all the qubits are in the same level, $U_{\mathcal{N}}$ thus ideally prepares coherent superpositions of the two $|e, e, \dots, e\rangle$ and $|g, g, \dots, g\rangle$ states in which the $\mathcal{N}$ ions are all either 'up' or 'down'.

Analogy with the cat states produced by Kerr effect in optics

There is a striking analogy between this GHZ state preparation and the generation, described in Section 7.2, of photonic cat states by propagation of a coherent field in a Kerr medium. In both cases, a superposition of states with different classical attributes is due to the action of a non-linear effective Hamiltonian, affecting the phase relationships between the eigenstates of the system. In the optical case, the non-linearity results in the appearance, at specific 'magic' times, of state superpositions in the system. A two-component cat state appears in particular at the time $\pi/2\gamma_k$ corresponding to a $\pi/2$-accumulation of the Kerr phase distortion. We understand the generation of this cat by remarking that the initial coherent state $|\beta\rangle$ can be expressed as the half-sum of an even cat state $|\beta\rangle + |-\beta\rangle$ and an odd cat state $|\beta\rangle - |-\beta\rangle$. The first part, containing only even photon numbers, experiences no phase shift (modulo 2π) at the magic time $\pi/2\gamma_k$ while the second, which has only odd photon numbers, is phase-shifted by $\pi/2$. Regrouping the terms in a different way, we find the cat state given by eqn. (7.30) which looks conspicuously similar to the GHZ state of eqn. (8.112).

We have shown in Section 7.2 that, besides the two-component cat appearing at time $\pi/2\gamma_k$, other kinds of state superpositions, with more than two components, are also expected if the photon number is large enough. They appear in particular at the shorter 'magic' times $\tau = \pi/8\gamma_k, \pi/6\gamma_k, \pi/4\gamma_k, \ldots$ This suggests the possibility of generating similar multi-component state superpositions in large ensemble of ions, under the effect of collective geometric gate operations of the form $\exp(i\pi J_Z^2/q)$ with $q > 2$. To our knowledge, such states have not been studied so far.

Experimental generation and probing of GHZ ionic states with $\mathcal{N}$ up to 6

The Boulder group has prepared and analysed GHZ states of four, five and six Be^+ ions (Leibfried *et al.* 2005). The collective gate $T_\mathcal{N}$ was realized with pushing Raman laser beams according to the method described above. The qubit rotations sandwiching this gate were produced by carrier Raman pulses, with the frequency difference $\omega_{\ell 1} - \omega_{\ell 2}$ matching the qubit ω_{eg} frequency. The experimental procedure departed slightly from the recipe described above. Instead of ending with an $\exp(-i\pi J_X/2)$ rotation, the experimenters have, for practical convenience, applied finally an $\exp(i\pi J_X/2)$ pulse identical to the first rotation. This is equivalent to applying the $\exp(-i\pi J_X/2)$ rotation, followed by $\exp(i\pi J_X)$. In other words, the ion system was in fact subjected to the transformation[26] $U_\mathcal{N}^{\text{exp}} = \exp(i\pi J_X)U_\mathcal{N}$. Combining $U_\mathcal{N}$ with a rotation of angle π around OX amounts to a final exchange of the $|e\rangle$ and $|g\rangle$ qubits, so that the state ideally prepared in this experiment is:[27]

$$
\begin{aligned}
|\Psi_{\text{GHZ}}\rangle &= U_\mathcal{N}^{\text{exp}}|J, -J\rangle = e^{i\pi J_X} U_\mathcal{N}|J, -J\rangle \\
&= \frac{e^{-i\pi/4}}{\sqrt{2}}|e, e, \ldots, e\rangle + i^{\xi_\mathcal{N}}\frac{e^{i\pi/4}}{\sqrt{2}}|g, g, \ldots, g\rangle \,,
\end{aligned}
\tag{8.114}
$$

where we have discarded an irrelevant global phase factor in the last line.

This GHZ state is formally analogous to the 'high noon' states we have described in Chapter 7. The $U_\mathcal{N}^{\text{exp}}$ operation has put all the ions either in $|e\rangle$ or in $|g\rangle$, in the same way that a collective beam-splitter would dispatch the particles impinging on it either all in one output mode, or all in the other. We have already stressed that such a partition is fundamentally different from an ordinary beam-splitter operation, which leaves the particles 'choose' their output mode independently from each other. In the ion experiment context, such an ordinary beam-splitter is realized by a global rotation $\exp(i\pi J_X/2)$ or $\exp(i\pi J_Y/2)$ acting independently on all the ions. The $U_\mathcal{N}^{\text{exp}}$ operation is of a fundamentally different nature, since it involves non-local operations coupling the ions with each other. Here, as in the CQED case, realizing a collective beam-splitter requires a non-linear interaction.

To complete its GHZ experiment, the Boulder team has analysed the produced state by submitting it to a collective atomic interference test which is a beautiful realization of the thought experiment described in Chapter 7 (Fig. 7.43, on page 439). According to the textbook procedure, an adjustable phase shift φ must first be induced

---

[26]The $\pi/2$-rotations around OX were realized by proper adjustment of the Raman lasers phase difference. In the $\mathcal{N}$-odd case, the $\pi/2$-rotation around OZ combined with the last $\pi/2$-rotation around OX was replaced by an equivalent $\pi/2$-rotation around OY performed by a single Raman carrier pulse whose phase was offset by $\pi/2$ with respect to the phase producing the rotations around OX.

[27]Note the analogy with eqn. (8.98) giving the Bell's state prepared by an equivalent process in the $\mathcal{N} = 2$ case.

between the two output modes of the beam-splitter which has produced the GHZ state. These modes have then to be recombined by a second non-linear operation identical to the first one and the number of ions ending up in $|e\rangle$ and $|g\rangle$ must finally be measured as a function of φ.

Let us analyse in more detail how this sequence of operations has been actually realized in the experiment. The differential phase shifting between the two collective states is described by the transformation $\exp(i\varphi J_Z)$. It is then followed by a second application of the beam-splitting operation $U_{\mathcal{N}}^{\text{exp}}$ mixing again the states collectively. The complete decoding operation of the GHZ state can thus be expressed as:

$$
\begin{aligned}
U_{\mathcal{N}}^{\text{exp}} e^{i\varphi J_Z} &= e^{i\pi J_X/2} e^{i\pi \xi_{\mathcal{N}} J_Z/2} e^{i\pi J_Z^2/2} e^{i\pi J_X/2} e^{i\varphi J_Z} \\
&= e^{i\varphi J_Z} e^{-i\varphi J_Z} e^{i\pi J_X/2} e^{i\pi \xi_{\mathcal{N}} J_Z/2} e^{i\pi J_Z^2/2} e^{i\pi J_X/2} e^{i\varphi J_Z} \; . \quad (8.115)
\end{aligned}
$$

The $\exp(i\varphi J_Z)$ rotation around OZ in eqn. (8.115) is made explicit by introducing the component of the collective angular momentum $J_\varphi = \exp(-i\varphi J_Z) J_X \exp(i\varphi J_Z) = \cos(\varphi) J_X + \sin(\varphi) J_Y$, yielding finally:

$$
U_{\mathcal{N}}^{\text{exp}} e^{i\varphi J_Z} = e^{i\varphi J_Z} e^{i\pi J_\varphi/2} e^{i\pi \xi_{\mathcal{N}} J_Z/2} e^{i\pi J_Z^2/2} e^{i\pi J_\varphi/2} \; . \quad (8.116)
$$

The decoding operator $U_{\mathcal{N}}^{\text{exp}} \exp(i\varphi J_Z)$ can thus be realized – up to a last and undetectable phase shift $\exp(i\varphi J_Z)$ which we can disregard – by the same pulse sequence as the first collective beam-splitter operation, adding the offset φ to the phase of the carrier pulses. This phase adjustment implements the change $J_X \to J_\varphi$ transforming $U_{\mathcal{N}}^{\text{exp}}$ into $\exp(-i\varphi J_Z) U_{\mathcal{N}}^{\text{exp}} \exp(i\varphi J_Z)$. We rediscover here a property of Ramsey interferometers mentioned in Chapter 3. It is equivalent to introduce a variable phase shift between the two paths of the interferometer or to keep the phase difference fixed and build up instead a variable phase in the operation of the second beam splitter. For practical convenience, the experimenters have chosen here the second solution.

The first point of view (phase shifting of the paths, followed by the mixing of the levels) is however simpler to compute the expected signal. The phase shift operation transforms $|\Psi_{\text{GHZ}}\rangle$ into:

$$
e^{i\varphi J_Z} |\Psi_{\text{GHZ}}\rangle = \frac{e^{-i(\pi/4 - \mathcal{N}\varphi/2)}}{\sqrt{2}} |e, e, \ldots, e\rangle + i^{\xi_{\mathcal{N}}} \frac{e^{i(\pi/4 - \mathcal{N}\varphi/2)}}{\sqrt{2}} |g, g, \ldots, g\rangle \; . \quad (8.117)
$$

The $U_{\mathcal{N}}^{\text{exp}}$ second beam-splitter operator transforms the $|g, g, \ldots, g\rangle$ and $|e, e, \ldots, e\rangle$ components of eqn. (8.117) into superpositions of these states [which can be computed with the help of eqns. (8.112) and (8.113), taking into account the final $\exp(i\pi J_X)$ transformation in $U_{\mathcal{N}}^{\text{exp}}$], leading to the ideal final state before detection:

$$
\begin{aligned}
U_{\mathcal{N}}^{\text{exp}} e^{i\varphi J_Z} |\Psi_{\text{GHZ}}\rangle &= \cos\left(\frac{\xi_{\mathcal{N}}\pi}{2} - \frac{\mathcal{N}\varphi}{2}\right) |e, e, \ldots, e\rangle \\
&\quad - i^{\xi_{\mathcal{N}}} \sin\left(\frac{\xi_{\mathcal{N}}\pi}{2} - \frac{\mathcal{N}\varphi}{2}\right) |g, g, \ldots, g\rangle \; . \quad (8.118)
\end{aligned}
$$

This expression shows that the probabilities of finding the $\mathcal{N}$ ions all in one or the other qubit state undergo ideally π-out-of-phase oscillations between 0 and 1 as the

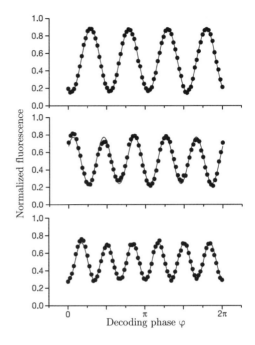

Fig. 8.26 The $\mathcal{N}$-ion collective interference fringes demonstrating the coherence of the GHZ states with $\mathcal{N} = 4, 5$ and 6 ions (from top to bottom). Reprinted by permission from Macmillan Publishers Ltd: Nature, Leibfried *et al.* (2005).

phase φ is swept, with $\mathcal{N}$ complete oscillations in the $[0, 2\pi]$ interval. Note also that the probability of finding the ions in $|e, e, \ldots, e\rangle$ at $\varphi = 0$ is equal to 1 for $\mathcal{N}$ even ($\xi_\mathcal{N} = 0$), to 0 for $\mathcal{N}$ odd ($\xi_\mathcal{N} = 1$).

The experimental interference signals are shown in Fig. 8.26, for $\mathcal{N} = 4$, 5 and 6. The total fluorescence rate of the ions is measured by repeating the experiment many times for each value of φ and accumulating data. The fluorescence signal is normalized to the difference of count rates when all the ions are in $|g\rangle$ (maximum fluorescence) and all in $|e\rangle$ (no fluorescence). As expected, $\mathcal{N}$ complete oscillations are observed when φ is varied between 0 and 2π. This is – as already analysed in Chapter 7 – a signature of collective $\mathcal{N}$-particle interference. The phase sensitivity of this collective interference is proportional to $\mathcal{N}$, as has already been noted in Section 7.6.3. The $1/\mathcal{N}$ variation of the interfringe spacing makes this interference with entangled states very promising for precision spectroscopy beyond the usual $\sqrt{\mathcal{N}}$ sensitivity (Leibfried *et al.* 2004). The phase of the observed fringes also agrees with the theoretical predictions of eqn. (8.118), with maximum fluorescence at $\varphi = 0$ for $\mathcal{N} = 5$ (minimum probability of finding the ions in $|e, e, \ldots, e\rangle$ for $\mathcal{N}$ odd) and a minimum fluorescence level for $\mathcal{N}$ even.

Ideally, the normalized signal should oscillate between +1 (all ions in $|g\rangle$) and 0 (all ions in $|e\rangle$). The reduced contrast is an indication of decoherence. Following a procedure similar to the one analysed above for $\mathcal{N} = 2$, one can deduce the fidelity of the GHZ state preparation from the contrast of these Ramsey fringes and from the independent measurements of the $P_{e,e,\ldots,e}$ and $P_{g,g,\ldots,g}$ probabilities in the state prepared by the first beam-splitting operation. The values obtained are 0.76, 0.60 and 0.51 for $\mathcal{N} = 4, 5$ and 6 respectively.

These decreasing values reflect once more an essential feature of the quantum-to-classical boundary. Here, as in CQED physics, it is increasingly difficult to prepare and maintain a superposition of states made of several independent or weakly bound particles when their number becomes large. The Boulder team has made a detailed analysis of its data, which proves the existence of genuine entanglement in the states they have produced up to $\mathcal{N} = 6$. The quantitative description of entanglement in multi-particle systems is beyond the scope of this book and we will not discuss this problem here.

Alternative collective entangling procedure via double sideband excitation

Let us conclude this section by mentioning another method of collective $\mathcal{N}$-ion entanglement which has also been demonstrated by the Boulder group on Be$^+$ (Sackett *et al.* 2000), following a procedure suggested by Mölmer and Sörensen (1999), and under a slightly different form by Solano *et al.* (1999). Instead of pushing the ions motional state along a loop which creates, then annihilates phonons in a real process, this alternative method employs virtual transitions in which no phonon is ever really emitted or absorbed. The excitation, which involves four Raman laser beams, submits the ions simultaneously to two slightly detuned sidebands, one on the blue side ($\omega_{\ell 1} - \omega_{\ell 2} = \omega_{eg} + \omega_z + \delta$), the other on the red side ($\omega'_{\ell 1} - \omega'_{\ell 2} = \omega_{eg} - \omega_z - \delta$) of the carrier frequency. Probability amplitudes in which a phonon is virtually created then annihilated interfere with opposite amplitudes in which a vibration quantum is first annihilated, then re-created. The net effect produces a coupling between the ions which contains terms of the form $\sum_{i \neq j} \sigma_{\pm i} \sigma_{\pm j}$ (which simultaneously excite or de-excite pairs of ions) and terms like $\sum_{i \neq j} \sigma_{\pm i} \sigma_{\mp j}$ (which describe transfers of excitations). When they are summed, all these terms add up to produce an effective Hamiltonian proportional to J_Y^2 (if the phases are well-chosen).

The action of this Hamiltonian for a proper time realizes the entangling transformation $\exp(-i\pi J_Y^2 / 2)$, preparing – as the geometric phase gate – GHZ states in the ion sample. Up to four ions have been entangled with this method, which presents strong analogies with the non-resonant collisional entangling procedure demonstrated with atoms in CQED (Section 5.4.6). The quantum interference at work in these Raman processes leads to quantum amplitude cancellations which have the remarkable effect to make the net coupling independent upon the initial phonon number (as long as the Lamb–Dicke parameter η remains small). As the collective geometric phase gate, the double sideband excitation gate thus has the attractive feature of remaining efficient in a relatively 'hot' environment. It does not require a careful cooling of the ion's vibrational motion.

8.6   Perspectives of ion traps for quantum information

8.6.1   A trap architecture with movable ions

The quantum information strategy based on collective qubit addressing requires traps in which ions are shuffled around. Before and after gate operations, the qubits need to be individually addressable to initialize their states, to realize single-qubit operations, and finally to perform measurements. This implies that a couple (or small set of ions) can be separated or brought in close contact on demand, in a fully deterministic way.

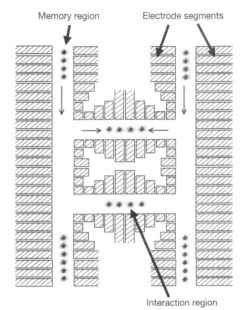

Fig. 8.27 A segmented trap architecture for scalable quantum information processing with trapped ions. Reprinted by permission from Macmillan Publishers Ltd: Nature, Kielpinski *et al.* (2002).

To achieve these tasks, the Boulder group has proposed a specific architecture sketched in Fig. 8.27 (Kielpinski *et al.* 2002).[28] The ions are stored along intersecting lines in a planar configuration. The trapping is produced by a layered structure. Electrodes above and below the plane of the figure (not shown), fed by a radio-frequency source, create a stiff potential holding tightly the ions in the third dimension. D.c. voltages, applied on segmented electrodes in the plane, create lines of micro-wells, each trapping a single ion.

By commuting the d.c. electrodes voltages, the ions can be moved around, while preserving their internal state coherence. The trap structure is divided in interaction and memory regions. Any two ions can be brought into adjacent traps in an interaction region, using the same procedure as for carriages in a marshalling yard. The barrier potential between these traps is then lowered and the two ions form a crystal in a single well, onto which a gate operation can be performed by collective ion addressing. The barrier is then raised again between the ions, separating them, in principle without loss of coherence. They are then shuffled back to their initial position. Single-qubit operations can be applied in the memory region or in the way to or from the interaction zone.

A sequence of such operations can in principle realize any quantum algorithm. The segmented ion configurations, the coherent transport (Rowe *et al.* 2002) of the ions and the merging and separation of ions in adjacent traps (Barrett *et al.* 2004; Chiaverini *et al.* 2005) have already been separately tested. A teleportation experiment in which three ions are shuffled around has successfully combined collective gate operations with single-qubit manipulation and detection (Barrett *et al.* 2004).

---

[28] Other schemes for scaling ion trap quantum processors are described in DeVoe (1998); Cirac and Zoller (2000); and Duan *et al.* (2004).

In the transport process, the ionic motion may progressively heat up. To combat this effect, the ions can be cooled down without affecting their internal state by sympathetic cooling in a dedicated region of the trap. The qubit ions are brought in contact with a 'cooling' ion of a different species. Sideband cooling of the common motion is then applied on the cooling ion transitions. The lasers involved are far off-resonance from the qubit transition and do not affect the qubit states. Finally, the two ion species are separated and the qubit ions are dragged back to the memory region.

The coherence between the qubit states are sensitive to decoherence processes, over time scales much smaller than the qubit natural lifetimes. Magnetic field fluctuations, for instance, produce a differential Zeeman shift of the qubit levels which is a cause of dephasing. Preserving the coherence over long processing times would require an unrealistic screening of stray magnetic fields. This limitation can be overcome by coding quantum information in 'decoherence-free subspaces' (Lidar *et al.* 1998; Kielpinski *et al.* 2001). In a simple case, the qubits are coded on a pair of adjacent ions, always moved together. The logical $|0\rangle$ and $|1\rangle$ qubit states are encoded onto $|e, g\rangle$ and $|g, e\rangle$ respectively. The two-ion states $|e\rangle$ and $|g\rangle$ are chosen to have opposite Zeeman shift in an homogeneous magnetic field. The dephasing between the logical qubits is then only due to field gradients, which can be made negligible at the scale of the intertrap spacing. The gate operations are then performed directly between pairs of ions.

Combining these techniques opens the way to the manipulation of large ensembles of ions, involving extended series of quantum operations. What will be the actual limits of these manipulations is an open question. Technical imperfections, classical noise and quantum decoherence still present formidable challenges. Ion trappers are applying their imagination and skills to overcome these immense difficulties. Theorists keep inventing clever procedures to increase the speed of the gate operations[29] and combat decoherence by various means, including the development of quantum error correcting schemes adapted to ion physics. Experimenters follow suite by putting these proposals to test in manipulations of steadily increasing complexity (Chiaverini *et al.* 2004). The experiments described in this chapter have shown how far they have been able to get, and give a feeling about the long way which lies ahead, on the road towards a true quantum computer juggling with thousands of ions or more.

8.6.2   Other avenues for ion trap physics

Even if this Holy Grail is out of reach in the foreseeable future, the quantum manipulation of a few tens of ions trapped at fixed positions in a linear or planar configuration holds remarkable promises for the quantum simulation of physics problems. As we have seen, the interplay of laser excitation with Coulomb interaction tailors interparticle couplings of the form $\sigma_{Zi}\sigma_{Zj}$ or $\sigma_{Xi}\sigma_{Yj}$. A wide range of Hamiltonians ruling the evolution of one- or two-dimensional spin lattices can be produced in this way on ensemble of ions. While the exact evolution of such a system in a 2^N dimension Hilbert space becomes rapidly intractable with classical computers, its simulation with an ion crystal would be very interesting to answer a variety of questions concerning quantum phase transitions in condensed matter physics (Porras and Cirac 2004).

---

[29]Among these, let us mention a fast gate proposal based on the application of counter-propagating pushing laser pulses, whose speed would not be limited by the ion oscillation frequency (Garcia-Ripoll *et al.* 2003).

Let us mention to conclude this chapter an interesting experimental convergence between ion trap and CQED physics. Preliminary attempts have been made to couple the ions in a trap to a high-quality optical cavity. In principle, such experiments could combine the best of two worlds, the exquisite control of the internal states and external motion of the ion, together with the strong coupling of this ion to an optical field mode. This architecture is very promising, for instance, for the realization of 'quantum repeaters', interfacing photonic flying qubits with static ions (Blinov *et al.* 2004).

The technical difficulties are serious, though. The dielectric mirrors of the cavities, for instance, get easily charged and perturb the electric trap. Significant progresses have nevertheless been made. An ion held in a cavity has produced a stream of individual photons (Keller *et al.* 2004), in the spirit of the optical CQED experiments with cold atoms described in Section 5.1.6. CQED effects, such as spontaneous emission modification and level shifts (Eschner *et al.* 2001; Wilson *et al.* 2003), have also been observed with a simpler arrangement involving an ion trapped in front of a mirror.

9
Entangling matter waves

Panurge... jette en pleine mer son mouton criant et bêlant. Tous les autres moutons, criant et bêlant en pareille intonation, commencèrent soi jeter et sauter en mer après à la file... Possible n'était les en garder comme vous savez être du mouton le naturel, toujours suivre le premier quelque part qu'il aille.

François Rabelais, Quart Livre

When bosonic atoms trapped in a box are cooled down to very low temperatures, their de Broglie wavelength increases until it becomes of the order of the mean inter-atomic distance. At this point, the system undergoes a phase transition, with a sizeable fraction of the atoms condensing into the trap lowest energy state. Eventually, when the temperature is lowered to still smaller values, the condensed phase extends to the whole sample. This very special form of matter has been predicted by Einstein in 1925. Its realization in the laboratory seventy years later has been one of the landmarks of the quantum century.

In trapped Bose–Einstein condensates the atoms all belong to a global wave function. They are quanta of this giant wave, in the same way as photons are quanta of the radiation field in a cavity. In this new form of matter, the wave–particle dualism, first introduced by Planck and Einstein at the beginning of last century, has come full circle. In 1905, light, classically viewed as a wave, was recognized as being also made of particles. Then, in the 1920s, atoms and atomic constituents, classically seen as particles were given a wavelike status through the work of de Broglie and Schrödinger. In this quantum synthesis, the particle aspect appears as a classical attribute of matter and a quantum property of radiation. Symmetrically, waviness is a classical feature for light, a quantum one for matter. This profound difference boils down to the fact that radiation is made of gregarious photonic bosons which naturally like to behave as collective waves, while matter is, deep down, a collection of fermions, spontaneously behaving as discrete individualistic entities.

Waviness becomes however an essential feature of matter when it is made of atoms in which fermions are bound together in even number. The pairing of fermions in composite bosons is essential to understand matter wave coherence, one of the most surprising manifestations of the quantum superposition principle. In this way, it is akin to superconductivity, produced by the pairing in phase space of conduction electrons in metals. Even-number fermion atoms have wave functions symmetrical under atom

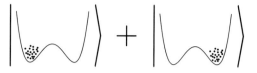

Fig. 9.1 A 'high noon' state of matter in a double potential well.

exchange. This leads, as discussed in Chapter 2, to their natural tendency to pack together in the same state below a critical temperature.

These bosonic atoms march in steps like photons in laser beams and exhibit coherence properties reminiscent of optics. By opening a leak in the trap holding the condensate, one forms directive and coherent matter wave beams quite similar to laser beams. They can be split and recombined in devices analogous to Young or Mach–Zehnder optical interferometers. Beams coming from independent condensates can also be made to interfere. Other manifestations of the matter wave coherence include superfluidity, which can be observed in various phenomena, notably the generation of vortices in rotating condensates.

Exploring the coherence of bosonic matter waves at the lowest temperatures encountered in Nature has become a very active field of experimental and theoretical research, attracting large numbers of atomic and condensed matter physicists. Among the new domains opening to their investigations, atom optics stands as especially important and promising. Here, coherent matter waves are reflected of walls made of electromagnetic fields, scattered by gratings made of periodic light patterns or focused by lenses made of inhomogeneous laser beams. In short, the roles of matter and fields are exchanged in the optics of matter waves. While in traditional optics, the interfering waves are made of photons and the reflective, dispersive and scattering devices made of atoms, here the waves are material and they are reflected, split and channelled by immaterial optical fields.

CQED finds a natural counterpart in atom optics. Instead of holding photons within reflecting metallic boundaries, experimentalists can now trap a giant matter wave in a potential well produced by a configuration of electromagnetic fields (magnetic bottle or optical trap). The atoms in this box, as the photons in the cavity, can be manipulated into mesoscopic superposition of states. The potential well can be split into two or more parts and the matter wave coherently channelled between them. Figure 9.1 represents a possible situation, bearing a strong analogy with the non-local photonic cat discussed in Chapter 7. Here, $\mathcal{N}$ atoms are suspended in a superposition of being all in the left well and all in the right well, a typical 'high noon' state situation.

In order to prepare such states in CQED, we had to rely on a strong non-linear interaction of the photons, either (theoretically) with a dispersive non-linear medium (Section 7.2) or with a single atom (Section 7.3). For matter waves, the non-linearity required to create the cat states does not need to be imposed externally. It turns out that the atomic bosons, unlike photons, interact weakly with each other when they collide. This interaction provides the non-linearity required to prepare mesoscopic state superpositions of the kind shown in Fig. 9.1.

More generally, the interactions between the matter wave quanta give us very convenient tools to engineer entanglement between them. While photonic states in CQED can be entangled only via their common interaction with atoms, atomic states in Bose–

Einstein condensates can be entangled without any external catalyst, by merely manipulating their intrinsic collisional coupling. This opens the way to interesting new routes for information processing. One can study condensates in which atoms have an internal degree of freedom akin to a spin (for instance they can be prepared in a superposition of two hyperfine sublevels of the atomic ground state). This internal state can be used to code a qubit and the collisions between the atoms lead to mechanisms for gate operations. A particularly simple situation occurs when the atoms of the condensate are trapped in a periodic lattice, making up a kind of register of qubits whose entanglement can be engineered by a proper arrangement of sequential or simultaneous collisional interactions.

In this chapter, we give a general introduction to this 'atomic CQED' physics. Our goal is to show how the basic methods developed in quantum optics can be exported in the domain of matter wave physics, opening new perspectives for fundamental research at the quantum–classical boundary. We will not present a general description of Bose–Einstein condensates which can be found in various textbooks (Pethick and Smith 2001; Meystre 2001; Pitaevskii and Stringari 2003), but introduce the minimal formalism required to understand the physics of mesoscopic state superposition and entanglement in matter waves.

We start in Section 9.1 by recalling the second quantization formalism which provides an elegant framework to describe ensembles of bosons in a potential well. This treatment introduces annihilation and creation operators for atoms in various energy eigenstates inside the confining box, which are analogous to the photon field operators used in the previous chapters. Observables involving one or two particles can be expressed in terms of these bosonic annihilation and creation operators.

We then give in Section 9.2 a qualitative description of the Bose–Einstein condensation process and of the experimental methods which are used to study it. Section 9.3 focuses on the simple situation of an ideal bosonic gas without interactions. We consider in particular the case of bimodal condensates, in which the giant matter wave has two components corresponding either to two different locations in space (double potential well) or to two different internal states of the atom (condensates made of spin-like particles). The coupling between these modes is realized either by a tunnel effect through the barrier between the wells or by applying an electromagnetic pulse coupling the spin-like states. These couplings are similar to that induced between two field modes by a beam-splitter, an analogy that we will develop in detail. We also introduce the concept of the phase of bimodal condensates and show that this phase is conjugate of the particle number difference between the two modes. This leads us to a discussion about matter wave interference and the visibility of fringes observed when two independent condensates overlap.

We next turn in Section 9.4 to the description of interatomic collisions in condensates. We recall the main features of these collisions, in particular their coherent nature which makes them very different from the stochastic collisions occurring in an ordinary thermal gas. We explain also briefly how the strength of these coherent collisions can be tuned via a magnetic effect, exploiting the Feshbach resonances (Feshbach 1962). These resonances give us a kind of knob to tune the interatomic coupling, which is very useful in atomic CQED or quantum information experiments. We show that the non-linearity introduced in the evolution of a bimodal condensate by the interatomic collisions produces a collapse and revival of its phase strikingly similar to the collapse and revival of a light field propagating in a Kerr medium. These effects present also analogies and differences with collapses and revivals in CQED physics (Section 3.4.3 and 7.4). We show that, as in optics, the system evolves into a Schrödinger-cat-like state in between the collapse and revival phases. A description of a possible experiment to observe this effect is presented.

In the next section (9.5), we consider a condensate trapped in a periodic potential, a generalization of the double-potential-well situation. We introduce the Bose–Hubbard Hamiltonian which rules the system's evolution. The condensate energy then contains, on top of the periodic potential energy, two additional and competing contributions, a kinetic 'hopping' term due to tunnelling through the interwell barriers and an intrawell collision energy term. When the number of particles per well is small, the condensate behaviour depends upon the relative strength of the hopping and collisional terms. The gas behaves either as a delocalized superfluid when interwell hopping dominates, or as an insulator with particles localized in each well when the collisional term is preponderant. The Mott transition between these two regimes has been observed in a pioneering experiment (Greiner *et al.* 2002a) whose principle and main results are briefly analysed. We also show that the phase of the condensate in the periodic structure exhibits collapses and revivals when the condensate, initially in the superfluid state, is suddenly turned into an insulator by increasing rapidly the barriers between the periodic wells. These collapse and revivals, similar to those predicted in the double-well case, have been observed in a beautiful experiment (Greiner *et al.* 2002b) which announces the production of Schrödinger cats in matter waves.

Finally, in Section 9.6 we describe simple entanglement experiments in Bose–Einstein condensates. We show how the coherent collisions between atoms trapped in potential wells can be controlled by merging the wells together and holding the atoms in contact for a well-defined time. The phase accumulation during the collision depends upon the internal states of the two atoms. The resulting dynamics is equivalent to the operation of a phase two-qubit gate. Experiments in which atoms in an optical lattice are entangled in this way are described (Mandel *et al.* 2003). Here again, as in the CQED experiments of previous chapters, the signature of the entanglement is obtained by studying interference effects, looking for instance at the fringe contrast of Ramsey experiments performed on the colliding atoms. Finally, the perspectives opened by these experiments will be briefly discussed.

9.1   Second quantization of matter waves

Consider a system of $\mathcal{N}$ bosons of mass m in a trap and assume first that the particles have no internal structure and do not interact with each other. We call $|\phi_\mu\rangle$ the single-particle states in the trap with energies ϵ_μ. The spatial dependence of the $\phi_\mu(\mathbf{r})$ wave functions and the ϵ_μ spectrum depend upon the shape of the trap potential. The single-particle ground state, assumed first to be non-degenerate, is labelled by $\mu = 0$. The collective quantum states of the $\mathcal{N}$ bosons could be written by assigning to each atom an index running from 1 to $\mathcal{N}$ and symmetrizing the global wave function with respect to particle exchange, according to the method outlined in Chapter 2. This procedure is very cumbersome. It is advantageously replaced by adopting the formalism of second quantization, which makes the connexion between matter wave physics and quantum optics transparent.

To each single-particle state $|\phi_\mu\rangle$ we associate an annihilation operator a_μ of one boson in this state. The adjunct operator $a_\mu^\dagger$ conversely creates a particle in state $|\phi_\mu\rangle$. These operators satisfy the canonical commutation rules of the creation and annihilation operators for a set of independent quantum oscillators:

$$[a_\mu, a_\nu^\dagger] = \delta_{\mu,\nu} . \tag{9.1}$$

We complete this notation by introducing the particle vacuum, $|0\rangle$. The global state $|\Psi_{n_0,n_1,\ldots,n_\mu,\ldots}\rangle$ representing n_0 bosons in state $|\phi_0\rangle$, n_1 particles in state $|\phi_1\rangle$ and so on... is:

$$|\Psi_{n_0,n_1,\ldots,n_\mu,\ldots}\rangle = \frac{1}{\sqrt{n_0!n_1!\ldots n_\mu!\ldots}}(a_0^\dagger)^{n_0}(a_1^\dagger)^{n_1}\ldots(a_\mu^\dagger)^{n_\mu}\ldots|0\rangle . \tag{9.2}$$

These states form a basis over which any physical state of the $\mathcal{N}$-boson ensemble can be expanded. Of particular importance is the state corresponding to $n_0 = \mathcal{N}$:

$$|\Psi_{n_0=\mathcal{N}}\rangle = \frac{1}{\sqrt{\mathcal{N}!}}(a_0^\dagger)^{\mathcal{N}}|0\rangle , \tag{9.3}$$

which describes all the particles in the ground state of the trap.

The a_μ and $a_\mu^\dagger$ operators can also be used to express any physical observable of the boson system. Let us first consider a one-atom observable O, sum of $\mathcal{N}$ identical single-particle operators $o_i = o$ acting on each boson: $O = \sum_i o_i$ $(i = 1, 2, \ldots, \mathcal{N})$. The potential energy V resulting from the action of an external force acting independently on each atom takes such a form, as well as the total kinetic energy $E_c = \sum_i p_i^2/2m$ of the bosonic gas. The single-particle observable o is completely defined by its dyadic expansion $o = \sum_{\mu,\nu} |\phi_\mu\rangle o_{\mu,\nu} \langle\phi_\nu|$ in terms of the matrix elements $o_{\mu,\nu} = \langle\phi_\mu|o|\phi_\nu\rangle$ of o in the $|\phi_\mu\rangle$ basis. From this expansion, we infer the second quantization expression of O:

$$O = \sum_{\mu,\nu} a_\mu^\dagger o_{\mu,\nu} a_\nu , \tag{9.4}$$

whose physical interpretation is clear. Under the action of o, a single particle is scattered from $|\phi_\nu\rangle$ to $|\phi_\mu\rangle$ with an amplitude $o_{\mu,\nu}$ and these elementary amplitudes must be summed on the distribution of initial and final available states. The second quantization expression (9.4) ensures that, in a sample of $\mathcal{N}$ identical bosons, the process occurs in a fully symmetrical way, making it impossible to distinguish the atoms undergoing single-particle scattering. This equation tells us that the indistinguishable bosons are annihilated in $|\phi_\nu\rangle$ and recreated in $|\phi_\mu\rangle$, with the amplitude corresponding to the single-particle process. The conservation of the number of bosons is ensured by the structure of O, a sum of quadratic terms involving each one annihilation and one creation operator.

The boson number operator $N = \sum_\mu a_\mu^\dagger a_\mu$ and the energy of the non-interacting gas (a diagonal single-particle operator in the $|\phi_\mu\rangle$ basis):

$$H = \sum_\mu a_\mu^\dagger a_\mu \epsilon_\mu , \tag{9.5}$$

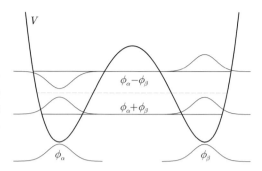

Fig. 9.2 A double potential well. The dotted horizontal line corresponds to the energy of the uncoupled states ϕ_α and ϕ_β (wave functions represented at the bottom). The solid lines correspond to the symmetrical (ground) and antisymmetrical states when tunnelling is taken into account.

are important examples of one-particle operators expressed in the second quantization formalism. The analogy with a light field whose photons belong to an ensemble of independent modes with different frequencies is conspicuous.

Two-atom observables can also be described simply in the second quantization point of view. Binary collisions in the gas are described by an interaction potential $W = \sum_{i \neq j} w(|\mathbf{r}_i - \mathbf{r}_j|)/2$, where $w(r)$ depends upon the distance r between two atoms. Following a procedure similar to the one applied above to single-particle operators, this interaction potential is expressed in the second quantization formalism as:

$$W = \frac{1}{2} \sum_{\mu, \mu'; \nu, \nu'} a_\mu^\dagger a_{\mu'}^\dagger w_{\mu\mu'; \nu\nu'} a_\nu a_{\nu'} , \tag{9.6}$$

where $w_{\mu\mu'; \nu\nu'} = \langle \phi_\mu, \phi_{\mu'} | w | \phi_\nu, \phi_{\nu'} \rangle$. Here again, the physical interpretation is obvious. The interaction which scatters pairs of atoms from the initial state $|\phi_\nu, \phi_{\nu'}\rangle$ into the final state $|\phi_\mu, \phi_{\mu'}\rangle$ with an amplitude $w_{\mu\mu'; \nu\nu'}$ is represented by a sum of processes in which pairs of indistinguishable bosons are annihilated in the initial state and recreated in the final one. The interaction is now a sum of quadruple products of bosonic operators conserving the total boson number. The W interaction plays, as we show below, an essential role in the physics of cold boson gases. Even if it is convenient to describe the gas in first approximation as a collection of independent non-interacting particles, the perturbing effect of the interatomic collisions described by eqn. (9.6) is essential to understand the dynamical properties of the Bose–Einstein condensate and, in first place, the mechanism leading to its formation.

The choice of the $|\phi_\mu\rangle$ basis which determines the form of the a_μ's is a matter of convenience, depending upon the physical situation to be described. Changing basis is achieved by a unitary transformation expressing the new basis states as linear combinations of the $|\phi_\mu\rangle$. New bosonic annihilation operators are obtained by linear combinations of the old ones, these combinations involving the same coefficients as those used in the transformation of the basis states. Let us consider a simple example. Assume that the trap potential is a symmetrical double well of the form shown in Fig. 9.2. Around the two minima, the potential is quasi-harmonic. When tunnelling is neglected, the single-particle ground states of the harmonic wells are $\phi_\alpha(\mathbf{r})$ and $\phi_\beta(\mathbf{r})$, the corresponding bosonic annihilation operators being a_α and a_β. When tunnelling is allowed, the degeneracy between these two states is lifted. The single-particle ground state is the symmetrical superposition of the two $|\phi_\alpha\rangle$ and $|\phi_\beta\rangle$ states and the lowest energy state of the $\mathcal{N}$ boson ensemble becomes:

$$|\Psi_\mathcal{N}\rangle = \frac{1}{\sqrt{2^\mathcal{N}\,\mathcal{N}!}} \left(a_\alpha^\dagger + a_\beta^\dagger \right)^\mathcal{N} |0\rangle \ . \tag{9.7}$$

The analogy with n photons in a superposition of two modes in the output arms of a beam-splitter is striking. Equation (9.7) describes a condensate state, in which each boson is in a superposition of being at the same time in the left and in the right well. Note that, as in the photonic counterpart, there is no entanglement if we view the system as $\mathcal{N}$ bosons in $(|\phi_\alpha\rangle+|\phi_\beta\rangle)/\sqrt{2}$ [see discussion following eqn. (3.113), on page 132]. This is very different from the high-noon state illustrated in Fig. 9.1 where all the particles are in the left well and *at the same time* in the right one, a highly entangled $\mathcal{N}$-particle situation which cannot be expressed as a tensor product of single-particle states.

Another example of basis change leads us to introduce the field bosonic operators $\Psi^\dagger(\mathbf{r})$ and $\Psi(\mathbf{r})$, which create or annihilate particles at a well-defined location. A single particle localized at point $\mathbf{r}$ is represented by a state expressed in the $|\phi_\mu\rangle$ basis as $|\mathbf{r}\rangle = \sum_\mu |\phi_\mu\rangle\langle\phi_\mu|\mathbf{r}\rangle = \sum_\mu \phi_\mu^*(\mathbf{r})|\phi_\mu\rangle$. In second quantization, this state thus becomes:

$$|1 : \mathbf{r}\rangle = \sum_\mu \phi_\mu^*(\mathbf{r}) a_\mu^\dagger |0\rangle \ , \tag{9.8}$$

which shows that the creation field operator $\Psi^\dagger(\mathbf{r})$ is defined as:

$$\Psi^\dagger(\mathbf{r}) = \sum_\mu \phi_\mu^*(\mathbf{r}) a_\mu^\dagger \ , \tag{9.9}$$

whereas its hermitian conjugate:

$$\Psi(\mathbf{r}) = \sum_\mu \phi_\mu(\mathbf{r}) a_\mu \ , \tag{9.10}$$

destroys a boson at point $\mathbf{r}$. These operators satisfy the canonical commutation relations of boson fields:

$$[\Psi(\mathbf{r}_1), \Psi^\dagger(\mathbf{r}_2)] = \delta(\mathbf{r}_1 - \mathbf{r}_2) \ , \tag{9.11}$$

which is easily deduced from eqn. (9.1) and the closure relationship over the $|\phi_\mu\rangle$'s. Let us note also that these field operators can be combined to define the second-quantization particle density operator $n(\mathbf{r})$:

$$n(\mathbf{r}) = \Psi^\dagger(\mathbf{r})\Psi(\mathbf{r}) \ . \tag{9.12}$$

There is a deep analogy between the annihilation and creation field operators and the positive and negative frequency parts of the electric field operator in quantum optics (eqn. 3.17). The particle density at point $\mathbf{r}$ is the counterpart of the photon counting rate operator, proportional to the product of the negative and positive frequency components of the electric field at the photodetector location.

9.2 Main features of Bose–Einstein condensation

The second quantization formalism is convenient to describe the general state of an ensemble of bosons and the operators acting on them. By itself it is not sufficient, though,

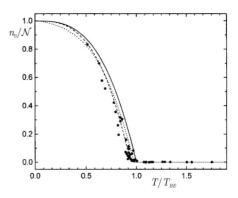

Fig. 9.3 Condensed fraction $n_0/\mathcal{N}$ versus temperature T in units of the transition temperature T_{BE} for a rubidium gas in an harmonic trap. The dots are experimental. The solid line corresponds to the ideal gas theory discussed in text. The dotted and dashed lines correspond to more refined fits. Reprinted with permission from Ensher *et al.* (1996). © American Physical Society.

to explain the condensation phenomenon. The quantum picture must be completed by a thermodynamical analysis which describes how, at a low enough temperature the system chooses to accumulate in the state described by eqn. (9.3) or (9.7). This analysis, whose principle was briefly outlined in Chapter 2, requires a book-keeping of the states available to the system, with their respective probabilities given by computing the corresponding Boltzmann factors. The evolution between these states is due to the binary atomic collisions which constantly reshuffle the system among its possible configurations. We will not carry out this analysis here in detail but only state its results and present a qualitative interpretation of the condensation phenomenon.

In a typical experiment, the condensate is formed out of a thermal gas of cold atoms confined in a magnetic trap. This trap is realized by a configuration of coils producing a gradient of magnetic fields. The action of these gradients on the magnetic moment of the atoms results in a force keeping them in a small volume. It is convenient to introduce first a few basic parameters which set the scale of time, length and temperature relevant for the discussion of Bose–Einstein condensation.

Let us first assume for simplicity that the trap realizes an isotropic harmonic potential well, entirely defined by the single-atom oscillation frequency ω_{ho}. The zero-point fluctuations of the position and energy of this harmonic oscillator in its ground state $|\phi_0\rangle$ sets the natural unit of length $a_{\mathrm{ho}} = (\hbar/m\omega_{\mathrm{ho}})^{1/2}$ and temperature $T_{\mathrm{ho}} = \hbar\omega_{\mathrm{ho}}/k_b$ of the condensation process. Typically, in a ^{87}Rb trap, $\omega_{\mathrm{ho}} \approx 10^3$ s^{-1}, which corresponds to $a_{\mathrm{ho}} \approx 1$ μm and $T_{\mathrm{ho}} \approx 5$ nK.

The thermodynamical book-keeping argument shows that, below a critical temperature T_{BE}, the $n_0/\mathcal{N}$ fraction of atoms in the trap ground state starts to increase significantly. The order of magnitude of T_{BE} can be estimated by equating the thermal de Broglie wavelength, $\lambda_{\mathrm{th}} = (2\pi\hbar^2/mk_bT)^{1/2}$, with the average interatomic distance in the trap, of the order of $(V/\mathcal{N})^{1/3}$ where V is the effective volume offered to the atoms. Roughly speaking, Bose–Einstein condensation occurs when $\lambda_{\mathrm{th}} \sim (V/\mathcal{N})^{1/3}$, meaning that the microscopic atomic wave packets, each extending over a de Broglie wavelength, start to overlap strongly with each other in the trap.

Let us estimate T_{BE} for an ideal Bose gas in an isotropic harmonic potential. At a temperature T, the radius a_T of the thermal cloud of atoms in the trap is of the order of $(2k_bT/m\omega_{\mathrm{ho}}^2)^{1/2}$, a value obtained by computing the amplitude of the harmonic motion whose energy is $k_b T$. The effective volume offered to the atoms is thus of the order of $(4\pi/3)a_T^3$ and the mean interatomic distance in the gas of the order of $(V/\mathcal{N})^{1/3} \approx (2k_b T/m\omega_{\mathrm{ho}}^2)^{1/2}(4\pi/3\mathcal{N})^{1/3}$. At the onset of condensation, we equate this distance with λ_{th}, and we find $T_{\mathrm{BE}} \approx 1.1\,\hbar\omega_{\mathrm{ho}}\mathcal{N}^{1/3}/k_b$. The precise thermodynamical calculation yields:

$$T_{\mathrm{BE}} = 0.94\,\hbar\omega_{\mathrm{ho}}\mathcal{N}^{1/3}/k_b = 0.94\,T_{\mathrm{ho}}\mathcal{N}^{1/3}\,, \tag{9.13}$$

which is in remarkable agreement with our back-of-the-envelope estimate. The thermodynamical argument giving T_{BE} can also be used to compute how the condensate fraction $n_0/\mathcal{N}$ increases when the temperature is reduced below T_{BE} (solid line in Fig. 9.3). The onset of condensation at $T = T_{\mathrm{BE}}$ is a typical phase transition phenomenon. We see that the condensed fraction rapidly increases to 1 when T/T_{BE} tends towards 0. For $T \ll T_{\mathrm{BE}}$, the boson gas is with a very good approximation condensed in the state described by eqn. (9.3).

The results we have recalled here are obtained within the ideal gas model. It is remarkable that they describe correctly real condensates, in spite of the existence of interatomic collisions, which play, as we will see shortly, a very important role in the dynamical behaviour of these systems. The critical temperature for a rubidium condensate containing $\mathcal{N} = 10^6$ atoms is $T_{\mathrm{BE}} \approx 100\,T_{\mathrm{ho}} \approx 500$ nK, in excellent agreement with the prediction of eqn. (9.13). The experimental points in Fig. 9.3, which correspond to a rubidium experiment (Ensher *et al.* 1996), show also that the variations of the condensed fraction versus T follow fairly well the predictions of the ideal gas model.

The submicrokelvin temperatures required for Bose–Einstein condensation correspond to atomic velocities in the range of centimetres per second. Such slow atoms cannot be directly obtained by standard laser cooling techniques. In a Bose–Einstein condensation experiment, the temperature of the gas, first reduced by laser cooling down to a few microkelvins, is further decreased by evaporative cooling. A radio-frequency field applied to the atoms induces the particles of highest energy to leave the trap, thus decreasing the mean energy of the remaining atoms. The collisions between the atoms continuously redistribute the energy between the particles. By sweeping the frequency of the radio-frequency fields, atoms of smaller and smaller energy are removed from the trap, resulting in a progressive cooling down of the system until Bose–Einstein condensation occurs. The final number $\mathcal{N}$ of bosons is smaller by about two orders of magnitude than the initial number of atoms in the trap before the start of evaporation. The signature of the transition is obtained by releasing suddenly the atoms and letting the gas expand. The spatial distribution of the expanded cloud is measured by detecting the shadow of an absorbing laser beam (Fig. 2.8, on page 50). For an ideal Bose–Einstein condensate without interactions, the spatial distribution of the expanded sample simply reflects the momentum distribution of the particles before their release. In the presence of interactions, this shape is modified, but remains markedly different from that of a thermal sample (Pitaevskii and Stringari 2003).

In fact, the interatomic collisions, even if they can be neglected in first approximation to evaluate the critical phase transition parameters, are essential to permit the dynamics of the condensation process. If these interactions were progressively switched off, it would take a longer and longer time for the gas to get into thermodynamical equilibrium. This equilibrium must be reached during the evaporation sequence before other adverse effects such as collisions on background gas have had time to perturb the system. It is thus very fortunate that the interatomic collisions have, in alkali atoms, the magnitude required to make the condensation process feasible. To understand this point better, we will have to wait Section 9.4 where we describe the collisions of the bosons in the trap.

Our description of ideal gas condensates, limited so far to simple harmonic oscillator wells, applies as well to other kinds of confining potentials, with critical temperatures of the same order of magnitude. Cigar-shaped condensates are realized with anisotropic magnetic traps. It is also possible to design double- or multiple-well systems. Evaporative cooling of bosons in a symmetrical double well leads, in the ideal gas limit, to the condensate state described by eqn. (9.7). Such double condensates can also be produced dynamically, starting with a condensate formed in one of the wells and lowering the barrier between them at a given time. This manipulation, which we describe in the next section, presents strong analogies with the use of beam-splitters in quantum optics.

9.3 The phase in Bose–Einstein condensate interference

The subtle concept of phase in interference experiments often leads to confusion and misconceptions. This is true for optical and for matter waves as well. There are in fact two very different kinds of possible experiments. In the first one, a single system is split into two (or more) components, and then recombined by some kind of beam-splitting device. In the second kind of interference, two independent sources of particles are used and interference is observed in the merged beams. All the interference processes we have analysed so far belong to the first category. We will see that they can also be observed with bosonic condensates, for which beam-splitting devices can be designed. Bose–Einstein condensates can also give rise to the second kind of interference, which presents more intriguing features.

Beam-splitter interference experiments are observed in a photonic Mach–Zehnder device even if the input field has no defined phase, if it is for instance a single photon or a field in a Fock state. What matters is the phase difference between the two arms of the device, which is well-defined, even if the phase in each path has no definite value. We will see that the same is true for matter waves. A bimodal condensate results from the splitting into two parts of a bosonic Fock state. It exhibits interference because the two parts acquire a well-defined phase difference, conjugated with a large fluctuation of the boson number distribution between the two modes. The phase difference accuracy is thus obtained at the expense of a loss of knowledge about the number of particles which have travelled along each path. These interference effects are observed statistically, by averaging the contribution of many particles. Whether these particles cross the interferometer together or one by one is irrelevant. The important point is that, in the end, we do not know how many have travelled in one way or in the other. These beam-splitting effects, whether photonic or atomic, are, deep down, single-particle phenomena.

Interference effects occurring with independent lasers or coherent matter waves are, on the other hand, never observed with single-particle sources or with thermal beams of photons or atoms. In their simplest form, this interference is observed with beams in Fock states, having initially a totally random relative phase. Before the first particle is detected, there is no phase information in the system and the particle number in each beam is perfectly known. The build-up of the phase in this interference, and the correlated loss of information about the particle number in each beam, is a very interesting process illustrating the nature of quantum measurement. The interference pattern is then revealed in a single realization of the experiment, which requires the existence of a very large number of particles in the same quantum state, a property specific to laser beams and boson condensates. In this way, this interference with independent source is a collective effect. We will, in this section, analyse in detail these two forms of interference, observed on matter waves.

9.3.1 Matter wave interference with beam-splitters

Let us start with the first kind of interference, which requires, as in optics, a splitting of the bosonic field into (at least) two parts, followed by their recombination. We consider here a boson gas with negligible interactions, leaving the description of important collisional effects for the next section. This ideal boson model is an approximation which will allow us to discuss in a very simple context the coherence features of the condensate and the relationship between its phase and the fluctuations of its particle number.

How can we, in the first place, prepare the bimodal condensate state described by eqn. (9.7)? One way, mentioned in the last section, is to perform evaporative cooling on a thermal ensemble of atoms trapped in a symmetrical double well, in the presence of tunnelling between the two wells. When the temperature is low enough, a symmetrical bimodal state is obtained. We can prepare the same state by following alternatively a dynamical route, suggested by quantum optics: first prepare $\mathcal{N}$ atoms in $|\phi_\alpha\rangle$, then coherently split the particles path, letting them stay in $|\phi_\alpha\rangle$ or end up in $|\phi_\beta\rangle$ with complementary amplitudes. This can be achieved by preparing first a condensate in one of the well, with an interwell barrier too high to allow for tunnelling. One then lowers for a time interval t the barrier between the two wells, inducing the atoms to tunnel coherently from one well to the other. The tunnelling Hamiltonian can be written as:

$$H_J = -i\hbar \frac{\mathcal{J}}{2} \left(a_\alpha^\dagger a_\beta - a_\beta^\dagger a_\alpha \right) , \qquad (9.14)$$

where $\mathcal{J}$ is a tunnelling parameter (real and positive without loss of generality) depending on the geometry of the lowered barrier. The analogy with the beam-splitter coupling between two field modes is obvious (see eqn. 3.101).[1] After a time t, the condensate has evolved into the bimodal state:

$$|\mathcal{N} : \theta\rangle = \frac{1}{\sqrt{\mathcal{N}!}} \left[\cos(\theta/2) a_\alpha^\dagger + \sin(\theta/2) a_\beta^\dagger \right]^{\mathcal{N}} |0\rangle , \qquad (9.15)$$

where we have made the notation change $\theta = \mathcal{J}t$. If the barrier is then suddenly raised again between the two wells, the evolution is frozen from time t on and the

---

[1]The tunnelling Hamiltonian of eqn. (9.14) is defined with a phase leading to real probability amplitudes in eqn. (9.15). The single-particle eigenstates of this Hamiltonian are $(a_\alpha^\dagger \pm i a_\beta^\dagger)|0\rangle / \sqrt{2}$. They are different from those represented in Fig. 9.2, where we assumed $H_J = -\hbar(\mathcal{J}/2)(a_\alpha^\dagger a_\beta + \text{h.c.})$. These different phase conventions have no consequences for physical predictions.

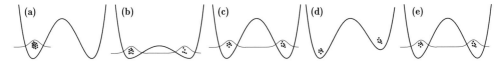

Fig. 9.4 Preparation of a bimodal condensate. (a) Initial situation: all atoms are in state $|\phi_\alpha\rangle$. (b) The potential barrier is lowered and tunnelling transfers atoms from one well to the other. (c) The potential barrier is raised again, stopping the tunnelling when the symmetrical state is reached. (d) A transient energy shift of one of the wells adjusts the relative phase of the wave function components. (e) The potential is reset to the configuration of (c).

bimodal condensate is stabilized. The case $\mathcal{J}t = \theta = \pi/2$ corresponds to a symmetrical splitting. The $\mathcal{N}$ particles are then accumulated in an equal weight superposition of the two wave functions $\phi_\alpha(\mathbf{r})$ and $\phi_\beta(\mathbf{r})$.

We can play a last trick to induce a phase difference between the two components of this bimodal superposition, a manipulation equivalent to the addition of a phase shifter in one arm of a Mach–Zehnder optical device. We offset transiently the right well, shifting its minimum by an amount ΔV for a small time interval τ, then restoring it to its initial value. The single-particle state $|\phi_\beta\rangle$ undergoes a phase shift $-\Delta V \tau/\hbar = -\phi$ and the boson system ends up in the 'bimodal phase state':

$$|\mathcal{N}:\theta,\phi\rangle = \frac{1}{\sqrt{\mathcal{N}!}}\left[\cos\left(\theta/2\right)a_\alpha^\dagger + e^{-i\phi}\sin\left(\theta/2\right)a_\beta^\dagger\right]^{\mathcal{N}}|0\rangle \ , \qquad (9.16)$$

entirely determined by the particle number $\mathcal{N}$ and the angles θ and ϕ, which are tuned by adjusting the parameters of the transient kicks applied to the trap potential. The sequence of operations performed on the condensate is sketched on Fig. 9.4(a)–(e), where we have represented the transformations applied to the double-well potential.

We have so far considered a bimodal situation in which the two components of the single-particle wave function are spatially separated. We can also consider a condensate in which the single-particle wave function, localized in a single well, is a superposition of two states with different internal atomic quantum numbers. The boson gas is then equivalent to a collection of atoms carrying each a fictitious spin. In alkali systems for instance, the atoms can be trapped in magnetic sublevels belonging to two different hyperfine ground states $|\phi_\alpha\rangle$ and $|\phi_\beta\rangle$ and these states can be coherently mixed by applying a classical microwave pulse resonant with the hyperfine transition. Starting with a condensate in which all the atoms are initially in one level, the pulse prepares a state of the form (9.15). The mixing between the modes is again of the form (9.14), where $\mathcal{J}$ is now the Rabi frequency of the classical pulse acting on the atomic fictitious spin.

By combining successively two such manipulations, one can perform a Mach–Zehnder type experiment (in the case of a spatial bimodal condensate) or a Ramsey type one (if a spin-like condensate is prepared). The Ramsey version of the bimodal condensate has been realized by Cornell *et al.* (1998). The condensate, a mixture of two rubidium hyperfine states was subjected to two successive $\pi/2$-pulses at frequency ω, nearly resonant with the hyperfine transition at frequency ω_{12} linking the two states and the number of atoms in these hyperfine levels was finally measured, as a function of the time interval T between the two pulses. The atomic detection was performed,

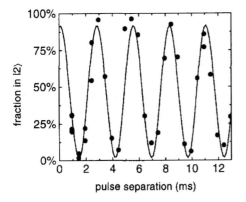

Fig. 9.5 Ramsey type oscillations in a bimodal hyperfine condensate. Population transfer from state $|\phi_\alpha\rangle = |1\rangle$ to state $|\phi_\beta\rangle = |2\rangle$ as a function of the time separation between the two $\pi/2$-pulses, proportional to the accumulated phase ϕ. From Cornell *et al.* (1998), with kind permission of Springer, Science and Business Media.

after letting the gas expand, by absorption of a probe laser beam, selectively sensitive to the two hyperfine levels. Figure 9.5 shows the atomic population transfer between $|\phi_\alpha\rangle$ and $|\phi_\beta\rangle$ as a function of T. The modulated signal, varying as $\cos(\omega - \omega_{12})T$, is typical of a Ramsey interferometer (eqn. 3.164, on page 150). The fringes are recorded on an ensemble of realizations of the experiment, the population of each level being averaged over many runs in which the condensate is prepared and probed under identical conditions.

The visibility of the matter wave fringes revealed by these experiments is related to the existence of large fluctuations in the number of particles in each of the condensate components. To make this clear, let us expand the bimodal phase state (eqn. 9.16) over a Fock state basis:

$$|\mathcal{N}; \theta, \phi\rangle = \sum_p \binom{\mathcal{N}}{p}^{1/2} \cos^{\mathcal{N}-p}(\theta/2) \sin^p(\theta/2) e^{-ip\phi} |\mathcal{N}-p\rangle_\alpha \otimes |p\rangle_\beta . \qquad (9.17)$$

This binomial partition of the bosons between the two modes is reminiscent of photon beam-splitting in optics (eqn. 3.116). The particle mean numbers in states $|\phi_\alpha\rangle$ and $|\phi_\beta\rangle$ and the dispersion of their difference are given by:

$$\mathcal{N}_\alpha = \mathcal{N} \cos^2(\theta/2) \; ; \; \mathcal{N}_\beta = \mathcal{N} \sin^2(\theta/2) \; ; \; \Delta(\mathcal{N}_\alpha - \mathcal{N}_\beta) = \sqrt{\mathcal{N}} \sin\theta . \qquad (9.18)$$

The fluctuation $\Delta(\mathcal{N}_\alpha - \mathcal{N}_\beta)$ is maximum, equal to $\sqrt{\mathcal{N}}$, for a symmetrical condensate ($\theta = \pi/2$) and zero for a mono-condensate ($\theta = 0$ or π). Let us consider now a symmetrical state. The coherence between Fock states with different particle numbers is:

$$\langle \mathcal{N} - p_1, p_1 \, | \mathcal{N}; \pi/2, \phi\rangle \, \langle \mathcal{N}; \pi/2, \phi \, | \mathcal{N} - p_2, p_2\rangle = \frac{1}{2^{\mathcal{N}}} \binom{\mathcal{N}}{p_1}^{1/2} \binom{\mathcal{N}}{p_2}^{1/2} e^{-i(p_1 - p_2)\phi} . \qquad (9.19)$$

Suppose now that ϕ is imperfectly known. This can happen if the phase-shifting pulse is noisy, with amplitude or time fluctuations. The distribution of ϕ is then defined by a probability law $P(\phi)$ and the bimodal condensate is described by its density operator:

$$\rho = \int_0^{2\pi} d\phi \, |\mathcal{N}; \pi/2, \phi\rangle \, P(\phi) \, \langle \mathcal{N}; \pi/2, \phi| . \qquad (9.20)$$

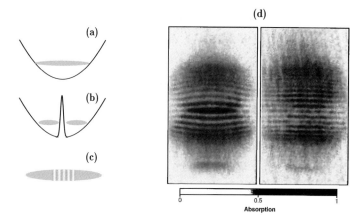

Fig. 9.6 Interference between two independent condensates. (a)–(c) Principle of the experiment. (a) Initial situation: a single condensate in a parabolic potential well. (b) A blue detuned laser creates a repulsive potential splitting the condensate into two parts. (c) All potentials are released and the two condensates freely expand. (d) Interference fringes for two initial separations. Reprinted with permission from Andrews *et al.* (1997), © AAAS.

The coherence between states with particle numbers p_1 and p_2 in state $|\phi_\beta\rangle$ then becomes:

$$\langle \mathcal{N} - p_1, p_1| \rho |\mathcal{N} - p_2, p_2\rangle = \frac{1}{2^\mathcal{N}} \binom{\mathcal{N}}{p_1}^{1/2} \binom{\mathcal{N}}{p_2}^{1/2} \times \int_0^{2\pi} d\phi \, P(\phi) \, e^{-i(p_1-p_2)\phi} . \quad (9.21)$$

This formula exhibits a conjugation relation between the phase ϕ and the difference of number of particles in two Fock states related to each other by a matrix element of ρ. When $P(\phi)$ is a Dirac-δ function, ρ has non-diagonal elements over a width $p_1 - p_2$ of order $\sqrt{\mathcal{N}}$. The wider $P(\phi)$ is, the narrower the non-diagonal distribution of ρ becomes. At the limit where $P(\phi) = 1/2\pi$ (undetermined phase), ρ is diagonal in $p_1 - p_2$. It is then an incoherent sum of diagonal operators $|\mathcal{N} - p, p\rangle\langle\mathcal{N} - p, p|$, each representing a projector associated to two uncorrelated condensates with perfectly fixed atom numbers. This 'uncertainty' relation between the phase fluctuation $\Delta\phi$ and the width $\Delta\mathcal{N}$ of the coherent partition of the particles between the two wells expresses a fundamental complementarity relation, already discussed in a different context in optics (Chapter 3).

Let us end this discussion by coming back to the issue we raised at the beginning. Is the bimodal interference signal of Fig. 9.5 specific of the condensed bosonic state? Would we observe it with thermal atoms? The answers to these questions are '*no*' and '*yes*', respectively. We have just shown that the phase relationship between the two components of the condensate is due to the build-up of particle number fluctuations in the two modes. These fluctuations are introduced by the beam-splitting mechanism, which is essentially a single-particle effect. The same signal would be observed by summing up $\mathcal{N}$ single-particle signals, exactly as in the optical Mach–Zehnder experiment described in Chapter 3. The phase is here a classical parameter, determined by the settings of the pulses which have manipulated the atoms. In short, this experiment does not require the atoms to be condensed. It is in no essential way different from the single-atom Ramsey experiments described in the previous chapters.

9.3.2   Interference of independent matter waves

We now consider a very different kind of matter wave interference, observed with independent condensates. The first experiment of this kind has been carried out by Andrews *et al.* (1997). Its principle is sketched in Fig. 9.6(a)–(c). A double-well trapping potential is obtained by superimposing a sharp barrier induced with laser light on top of the magnetic trap in which an initial condensate had been prepared (Fig. 9.6b). In this way, two independent condensates are generated, one on each side of the barrier. The height of the barrier is large enough so that tunnelling through it is negligible and the two condensates can thus be considered as completely independent.

The barrier is then suddenly switched off (Fig. 9.6c), as well as the trap holding the atoms and the samples are left free to expand and to overlap. The final density of the cloud is then measured, in a single shot, by absorption imaging. The final density exhibits clear fringes (Fig. 9.6d). This is at first sight surprising, since the condensates, contrary to the experiment described in Section 9.3.1, have no relative phase information imprinted on them.

We analyse this experiment following the argument of Castin and Dalibard (1997). We must first determine the initial state of the condensates. If the partition between the two wells is exactly symmetrical with $\mathcal{N}$ atoms in each well, the initial state of the double condensate is simply a tensor product of two independent condensates:

$$|\Psi\rangle = \frac{1}{\mathcal{N}!}(a_\alpha^\dagger)^\mathcal{N}(a_\beta^\dagger)^\mathcal{N}|0\rangle = |\mathcal{N},\mathcal{N}\rangle \ , \tag{9.22}$$

where the first and the second symbols in the ket refer to the numbers of bosons in modes ϕ_α and ϕ_β The exact partition between the two wells is in fact unknown. This statistical uncertainty can be accounted for by describing the initial state of the condensate by a density operator, incoherent mixture of projectors $|\mathcal{N}_1,\mathcal{N}_2\rangle\langle\mathcal{N}_1,\mathcal{N}_2|$ with $\mathcal{N}_1$ and $\mathcal{N}_2$ distributed around $\mathcal{N}$. This mathematical complication is not essential and the analysis of the interference experiment can be made by assuming that the number of particles initially in each well is perfectly known. We will come back to this point later.

After the release of the condensates, the single-particle wave functions $\phi_\alpha(\mathbf{r})$ and $\phi_\beta(\mathbf{r})$ expand, becoming two spatially overlapping functions $\overline{\phi_\alpha}(\mathbf{r})$ and $\overline{\phi_\beta}(\mathbf{r})$ at the detection time t_d. The expansion process being unitary, these final states are, as the initial ones, orthogonal. The bosonic field operator at time t_d can thus be expanded on a set of modes comprising $\overline{\phi_\alpha}$, $\overline{\phi_\beta}$ and other orthogonal modes we do not need to write explicitly:

$$\Psi(\mathbf{r},t_d) = \overline{\phi_\alpha}(\mathbf{r})a_\alpha + \overline{\phi_\beta}(\mathbf{r})a_\beta + \dots \ . \tag{9.23}$$

From this expression, we infer the average density of the expanded cloud:

$$\langle n(\mathbf{r})\rangle = \langle \mathcal{N},\mathcal{N}|\Psi^\dagger(\mathbf{r},t_d)\Psi(\mathbf{r},t_d)|\mathcal{N},\mathcal{N}\rangle = \mathcal{N}\left(|\overline{\phi_\alpha}(\mathbf{r})|^2 + |\overline{\phi_\beta}(\mathbf{r})|^2\right) \ , \tag{9.24}$$

which does not exhibit any spatial interference term. The bosonic state of the atomic cloud at the detection time is indeed diagonal in the Fock state basis. The operator products $a_\alpha a_\beta^\dagger$ and $a_\beta a_\alpha^\dagger$ describing the coherence between the two modes have a zero expectation value in this diagonal state. How can we then understand the experimental fringes shown in Fig. 9.6?

We must first notice that $\langle n(\mathbf{r}) \rangle$ is the mean value of the atomic density at point $\mathbf{r}$, averaged over a large number of realizations of the same experiment. What eqn. (9.24) really means is that a superposition of snapshots corresponding to many experiments repeated under the same condition will exhibit a flat image, without fringes. In other words, if a single atomic absorption image reveals an interference pattern, it must have a random phase and get completely washed out when the experiment is performed again and again.

The average density $\langle n(\mathbf{r}) \rangle$ is thus not the right quantity to compute in order to understand what happens in a single realization of the experiment. The double condensate is here a single quantum object, which is measured continuously, as the shadow of the atomic cloud appears on the detection screen. The real process is very fast, with many atoms detected almost instantaneously, but we can, for this analysis, assume that the process is progressive, with atoms detected one by one. We have learned in previous chapters that such a situation can be described very efficiently by a Monte Carlo approach. As photons from the probe laser beam get absorbed at various points, $\mathbf{r}_1, \mathbf{r}_2, \ldots, \mathbf{r}_{2\mathcal{N}}$, atoms are expelled from the condensate, a process equivalent to the action on the condensate of the annihilation operators $\Psi(\mathbf{r}_1), \Psi(\mathbf{r}_2), \ldots, \Psi(\mathbf{r}_{2\mathcal{N}})$. The bosonic field $\Psi(\mathbf{r})$ thus plays the role of a jump operator for the continuous measurement process.

Let us consider first what happens to the condensate when the first atom is detected. This occurs at a random point $\mathbf{r}_1$, within the density profile $\langle n(\mathbf{r}) \rangle$. Immediately after, the condensate is projected into the normalized state:

$$\frac{\Psi(\mathbf{r}_1)\,|\mathcal{N},\mathcal{N}\rangle}{\sqrt{\langle n(\mathbf{r}_1)\rangle}} = \sqrt{\frac{\mathcal{N}}{\langle n(\mathbf{r}_1)\rangle}}\left[\overline{\phi_\alpha}(\mathbf{r}_1)\,|\mathcal{N}-1,\mathcal{N}\rangle + \overline{\phi_\beta}(\mathbf{r}_1)\,|\mathcal{N},\mathcal{N}-1\rangle\right]. \qquad (9.25)$$

The detection process has thus a remarkable effect. It generates a state which presents fluctuations of the boson number between the two modes. This number is now either $\mathcal{N}$ or $\mathcal{N} - 1$. According to the discussion of the previous section, this fluctuation entails that the two modes have acquired a phase relationship. In other words, phase information appears spontaneously in the atomic ensemble, as a result of the first atom detection. It is also interesting to notice that this phase is produced by quantum interference in the detection process itself. When the first atom is detected, we have no way of knowing whether it comes from one mode or the other. This quantum ambiguity translates into the appearance of a quantum state superposition with different atom numbers, and introduces as we will see shortly, a privileged phase in the system.

Processing to the next stage of this Monte Carlo simulation, we obtain the (non-normalized) quantum state of the atomic cloud after detection of the second atom at $\mathbf{r}_2$:

$$\begin{aligned}
\Psi(\mathbf{r}_2)\Psi(\mathbf{r}_1)\,|\mathcal{N},\mathcal{N}\rangle =\ & \sqrt{\mathcal{N}(\mathcal{N}-1)}\,\overline{\phi_\alpha}(\mathbf{r}_2)\overline{\phi_\alpha}(\mathbf{r}_1)\,|\mathcal{N}-2,\mathcal{N}\rangle \\
& +\sqrt{\mathcal{N}(\mathcal{N}-1)}\,\overline{\phi_\beta}(\mathbf{r}_2)\overline{\phi_\beta}(\mathbf{r}_1)\,|\mathcal{N},\mathcal{N}-2\rangle \\
& +\mathcal{N}[\overline{\phi_\alpha}(\mathbf{r}_2)\overline{\phi_\beta}(\mathbf{r}_1) + \overline{\phi_\alpha}(\mathbf{r}_1)\overline{\phi_\beta}(\mathbf{r}_2)]\,|\mathcal{N}-1,\mathcal{N}-1\rangle \quad (9.26)
\end{aligned}$$

The fluctuation of the atomic numbers in the two components of the atomic cloud has increased, with the possibility for each one to contain $\mathcal{N}$, $\mathcal{N} - 1$ or $\mathcal{N} - 2$ particles. Most importantly, the last line of eqn. (9.26) shows that there are two ways to detect a couple of atoms at $\mathbf{r}_1$ and $\mathbf{r}_2$. The first particle can come from the left or the right well, and there is no way of telling. In both cases the system ends up in $|\mathcal{N} - 1, \mathcal{N} - 1\rangle$. This induces an interference pattern in the mean density of the atomic cloud at $\mathbf{r}_2$, after a first atom has been found at $\mathbf{r}_1$. This quantity, which we call $g(\mathbf{r}_2, \mathbf{r}_1)$, describes the distribution of positions where the second atom is likely to be found, once the first one has been detected at $\mathbf{r}_1$. It is equal to the expectation value in state (9.25) of the density operator $\Psi^{\dagger}(\mathbf{r}_2)\Psi(\mathbf{r}_2)$. It is thus a normalized fourth-order correlation function of the bosonic field operator:

$$g(\mathbf{r}_2, \mathbf{r}_1) = \frac{1}{\langle n(\mathbf{r}_1)\rangle} \langle \mathcal{N}, \mathcal{N}| \Psi^{\dagger}(\mathbf{r}_1)\Psi^{\dagger}(\mathbf{r}_2)\Psi(\mathbf{r}_2)\Psi(\mathbf{r}_1) |\mathcal{N}, \mathcal{N}\rangle \ . \tag{9.27}$$

To compute this correlation function, we remark that the expectation value of the product of field operators in eqn. (9.27) is the square of the norm of the state given by eqn. (9.26). After a straightforward rearrangement of terms, taking into account eqn. (9.24) and neglecting the difference between $\mathcal{N}$ and $\mathcal{N} - 1$ we get:

$$g(\mathbf{r}_2, \mathbf{r}_1) = \langle n(\mathbf{r}_2)\rangle + \frac{2\mathcal{N}^2}{\langle n(\mathbf{r}_1)\rangle} \mathrm{Re}\,[\overline{\phi}_{\alpha}^{*}(\mathbf{r}_2)\overline{\phi}_{\alpha}(\mathbf{r}_1)\overline{\phi}_{\beta}^{*}(\mathbf{r}_1)\overline{\phi}_{\beta}(\mathbf{r}_2)] \ . \tag{9.28}$$

The atomic correlation $g(\mathbf{r}_2, \mathbf{r}_1)$ exhibits an interference term, real part of a product of single-particle wave functions at $\mathbf{r}_1$ and $\mathbf{r}_2$. Depending upon the sign of this term, the conditional probability of detecting the second atom at $\mathbf{r}_2$ when the first has been found at $\mathbf{r}_1$ is larger or smaller than the unconditional probability, which corresponds to the $\langle n(\mathbf{r}_2)\rangle$ term in eqn. (9.28). The precise form of the interference pattern depends upon the geometrical parameters of the experiment. Let us assume that the two initial condensates are well-separated in point-like potential wells. The single-particle wave functions then expand towards each other over a long distance until they meet at midpoint. We can then approximate the wave functions by plane wave packets with opposite wave vectors $\mathbf{k}_{\alpha}$ and $\mathbf{k}_{\beta}$. Calling V the effective volume of the atomic cloud at detection time, we make the replacement $\langle n(\mathbf{r}_1)\rangle \approx \langle n(\mathbf{r}_2)\rangle \approx 2\mathcal{N}/V$, $\overline{\phi}_{\alpha}^{*}(\mathbf{r}_2)\overline{\phi}_{\alpha}(\mathbf{r}_1) = (1/V)\exp[i\mathbf{k}_{\alpha} \cdot (\mathbf{r}_1 - \mathbf{r}_2)]$ and $\overline{\phi}_{\beta}^{*}(\mathbf{r}_1)\overline{\phi}_{\beta}(\mathbf{r}_2) = (1/V)\exp[i\mathbf{k}_{\beta} \cdot (\mathbf{r}_2 - \mathbf{r}_1)]$. This yields the correlation function:

$$g(\mathbf{r}_2, \mathbf{r}_1) \approx \frac{2\mathcal{N}}{V}\left\{1 + \tfrac{1}{2}\cos\left[(\mathbf{k}_{\alpha} - \mathbf{k}_{\beta}) \cdot (\mathbf{r}_1 - \mathbf{r}_2)\right]\right\} \ . \tag{9.29}$$

This formula shows that the detection of pairs of atoms located along the axis joining the two initial condensates are most likely when they are separated by a distance $|\mathbf{r}_1 - \mathbf{r}_2| = 2p\pi/|\mathbf{k}_{\alpha} - \mathbf{k}_{\beta}|$ where p is an integer. There are three times less coincidences when $|\mathbf{r}_1 - \mathbf{r}_2| = (2p + 1)\pi/2|\mathbf{k}_{\alpha} - \mathbf{k}_{\beta}|$. The emergence of a clear interference pattern is conspicuous. This result shows that the first detection at $\mathbf{r}_1$ has already partly pinned-down the phase. The mere fact that an atom is detected at this point rules out the possibility of an interference pattern with a dark fringe at $\mathbf{r}_1$. We have already encountered this decimation argument in Chapter 6. Pursuing this Monte Carlo approach further, we could compute in a similar way the higher-order correlation functions corresponding to the detection of 3, 4 atoms and so on. The fringe contrast, already $1/2$ after a single atom has been detected, very rapidly increases towards unity.

If we resume the same experiment afresh, we find of course another $\mathbf{r}_1$ and the fringes build up with a different phase. This interference process, which requires the computation of high-order correlation functions of the field bosonic operator, is clearly not a single-particle phenomenon. It is quite different from the Mach–Zehnder or Ramsey type experiments with double condensates analysed above, in which the phase was a well-defined quantity, imprinted by a classical apparatus on the system. It is also very different from the Young double slit experiment described in Chapter 2. There, the phase of the fringes was defined classically by the geometry of the apparatus. All the particles, whether sent one by one or all together through the interferometer, were always falling on well-defined bright fringe positions. The situation is quite different now, the overall phase of the interference pattern being a random, inherently quantum quantity. The same analysis would apply to the interference of two independent laser beams, each being prepared in a Fock state.

The detection of the double condensate interference is reminiscent of other continuous measurement processes analysed in previous chapters. When, for instance, we measured the initially fluctuating energy of a field stored in a cavity, the photon number was pinned down progressively, through successive atomic detections, while the complementary observable, the field phase was continuously blurred. In an interference experiment involving two independent condensates, we process likewise to measure the relative phase between them, which is initially completely random. As we acquire progressively information, through the successive detection of atoms, the phase gets pinned down and we lose knowledge about the complementary quantity, the partition of the atoms between the two single-particle wave functions.

We can now remove the simplifying hypothesis made at the beginning of this analysis, namely that the initial partition between the two condensates is perfectly known. Starting with a statistical distribution of $|\mathcal{N}_1, \mathcal{N}_2\rangle$ states would obviously not change the results of our discussion. We would merely have to resume the same Monte Carlo calculation by starting from the different possible initial states and average over the initial state the results corresponding to a given sequence of detection events at $\mathbf{r}_1, \mathbf{r}_2, \ldots, \mathbf{r}_{\mathcal{N}_1 + \mathcal{N}_2}$. The first atomic detection at $\mathbf{r}_1$ would start to pin down to the same value the phase on all the trajectories starting from the different $|\mathcal{N}_1, \mathcal{N}_2\rangle$ initial states. An interference pattern would thus emerge in the same conditions as above. Of course, resuming the procedure anew with the same initial statistical mixture will yield a new sequence $\mathbf{r}_1, \mathbf{r}_2, \ldots, \mathbf{r}_{\mathcal{N}_1 + \mathcal{N}_2}$, so that an ensemble averaging completely washes out the interference. All the conclusions of the above analysis remain thus true.

9.4   Coherent collisions and cat-state generation

We have so far assumed that the boson condensate is an ideal gas, neglecting in the treatment of interference phenomena the effect of elastic collisions between the atoms. We show in this section that these interactions may affect strongly the evolution of a bimodal condensate in a trap and lead to the generation of atomic cat states which are closely related to the photonic cats encountered in CQED. We will see that these

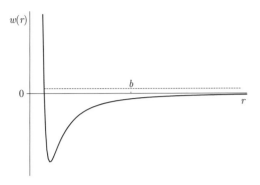

Fig. 9.7 A typical interatomic potential $w(r)$, with a range b. The horizontal dashed line corresponds to the collision kinetic energy.

states, as their photonic cousins, are very fragile. We need first to recall the main features of interatomic collisions in a cold atom gas.

9.4.1 A brief summary of cold atom collisions

The interaction energy between two atoms in the condensate is described by a potential $w(\mathbf{r})$ depending on their separation $\mathbf{r} = \mathbf{r}_1 - \mathbf{r}_2$. For ground state alkali atoms, this potential varies typically as shown in Fig. 9.7. Strongly repulsive at short range, it presents a minimum for an interatomic distance of a few angströms and falls off as $1/r^6$ at large distance. We define – loosely – the range b of this potential as the interatomic distance beyond which it can be truncated with a negligible effect on the description of the collision process. This range is typically of the order of a few nanometres. The relative kinetic energy of the atom pair for $r > b$, indicated by the horizontal line on Fig. 9.7, is exceedingly small compared to the range of variation of $w(\mathbf{r})$.

This potential admits bound states, with negative energies of the order of a fraction of an electron-volt, corresponding to stable molecular dimers. The dissociation temperature of these molecules is in the several hundred degree range, ten orders of magnitude larger than the relative kinetic temperature of the cold atom pairs. The association of a single pair thus releases an energy larger than the total kinetic energy of the whole boson cloud, made of millions of atoms. The molecule formation is, however, largely prevented by energy and momentum conservation during the binary elastic collisions.

The only way for the atoms in the condensate to bind into molecules is to undergo three-body collisions. Two atoms recombine and the molecule and the third partner (belonging to the condensate or to the background gas) recoil in opposite directions, escaping rapidly from the condensate. Three-body collisions are, however, rare at the very low density of the atomic cloud, their rate being typically in the subhertz range. The condensates remain thus metastable over a few seconds. During the short duration of a typical experiment, three-body processes are negligible and the atomic interactions are dominated by the elastic two-body collisions ruled by the potential $w(\mathbf{r})$.

Remarkably, these elastic collisions in the cold atomic gas can be described with a single parameter, the collision scattering length a_s. A brief reminder about formal scattering theory is needed to introduce this fundamental quantity. Let us start by analysing the evolution, in its rest frame, of a pair of distinguishable atoms undergoing a collision process. We define the stationary scattering states $\psi_{\mathbf{k}}^+(\mathbf{r})$ of the relative spatial coordinate $\mathbf{r}$ of the pair. They are obtained as eigenstates of the atomic Hamiltonian when one switches on adiabatically the potential $w(\mathbf{r})$ on initial plane waves $\psi_0(\mathbf{r}) = \exp(i\mathbf{k} \cdot \mathbf{r})$. Once these scattering states are known, the outgoing wave functions corresponding to any initial state are obtained by expanding it on a plane-wave basis and computing the corresponding superpositions of outgoing states. Provided the potential decays at large distances faster than $1/r^3$ (which is the case for alkali–alkali collisions), the scattering wave function, in the direction defined by the unit vector $\mathbf{n}$ (polar angles θ, ϕ) takes for $r \gg b$ the asymptotic form:

$$\Psi^+(\mathbf{r}) = \exp(i\mathbf{k} \cdot \mathbf{r}) + f_k(\theta, \phi)\frac{e^{ikr}}{r} , \tag{9.30}$$

with $k = |\mathbf{k}|$. It is the sum of a transmitted plane wave, identical to the ingoing one, and an outgoing wave function, determined by the scattering amplitude, $f_k(\theta, \phi)$:

$$f_k(\theta, \phi) = -\frac{m}{4\pi\hbar^2} \int d^3\mathbf{r}' \, e^{-ik\mathbf{n}\cdot\mathbf{r}'} w(\mathbf{r}')\psi_{\mathbf{k}}^+(\mathbf{r}') . \tag{9.31}$$

The square of the modulus of this amplitude is the differential scattering cross-section per unit solid angle in the direction θ, ϕ, whose integral over all directions yields the total scattering cross-section:

$$\sigma = \int |f_k(\theta, \phi)|^2 \, \sin\theta \, d\theta d\phi . \tag{9.32}$$

The definition (9.31) of $f_k(\theta, \phi)$ is implicit. It connects the behaviour of the scattered wave function at large distances to its values within the range of the potential. It takes a much simpler explicit form if one can perform the Born approximation, a perturbative development to first order in w. This approximation amounts to replacing $\Psi_{\mathbf{k}}^+(\mathbf{r}')$ by $\psi_0(\mathbf{r}')$ in eqn. (9.31). The scattering amplitude in a given direction then reduces to the Fourier transform of the potential, evaluated for the transfer of momentum $\mathbf{k} - k\mathbf{n}$:

$$f_k^{\text{Born}}(\theta, \phi) = -\frac{m}{4\pi\hbar^2} \int d^3\mathbf{r}' \, e^{i(\mathbf{k}-k\mathbf{n})\cdot\mathbf{r}'} w(\mathbf{r}') . \tag{9.33}$$

This simple result shows that a measurement of the angular dependence of the scattered wave provides, by inverse Fourier transform, the detailed shape of the scattering potential, provided its effect is weak enough to be treated within the Born approximation. This property is widely used, in quantum physics as well as in classical electromagnetism to determine the form factors of scattering objects of various kinds, when multiple scattering (corresponding to higher order terms in the Born expansion) is negligible.

Here, we want to describe the effect of the interatomic binary collisions on the behaviour of a gas made of identical bosons. We first give a qualitative picture, valid for a thermal sample above the Bose–Einstein critical temperature. Let us consider the wave function of an atom propagating in a gas made of other atoms, assumed first to be of a different kind. This wave function evolves into a superposition of the initial wave packet (for instance a plane wave) with the partial spherical waves resulting from the scattering by all the other gas particles. If the Born approximation holds, each scatterer contributes only up to first order and the forward scattering of the atomic wave function results in a phase delay. This delay, proportional to the propagation distance, can be described as a mean-field refractive index effect. In the Bose–Einstein condensate, this classical picture becomes inadequate, but we want to retain its main feature, the fact that an atom in the sample 'feels' the others as as a mean-field perturbation defined by a single parameter.

To build such a model, we face immediately a difficulty. The Born approximation is *a priori* not allowed in the description of alkali atom collisions, because the potential $w(\mathbf{r})$ has very strong attractive and repulsive parts (Fig. 9.7), making a linear perturbative treatment dubious. We will show however that we can replace the true potential $w(\mathbf{r})$ by a simple model potential $w_m(\mathbf{r})$ for which the Born approximation is valid and which has, on very cold atoms, the same effect as the real potential.

Fortunately, we can rely here on another kind of approximation, due to the extremely small energy of the colliding particles. The initial state of an atom pair expands on plane waves with very small $\mathbf{k}$ vectors, satisfying the condition $kb = 2\pi b/\lambda_{\text{th}} \ll 1$ (the atomic de Broglie wavelength is typically of the order of 0.1 to 1 μm, about 10^2 to 10^3 times larger than b). We can thus, within a very good approximation, replace $k\mathbf{n} \cdot \mathbf{r}'$ by 0 in the integral (9.31). It becomes then independent of $\mathbf{n}$. It is also easy to show, by symmetry arguments, that it is independent of the direction of $\mathbf{k}$. Hence, at the very low energy limit of the cold atom collisions, f_k reduces to a constant having the dimension of a length:

$$\lim_{k \to 0} f_k(\theta, \phi) = -a_s = -\frac{m}{4\pi\hbar^2} \lim_{k \to 0} \int d^3\mathbf{r}' \, w(\mathbf{r}')\psi_{\mathbf{k}}^+(\mathbf{r}') , \qquad (9.34)$$

and the scattered wave simply becomes asymptotically:

$$\Psi^+(\mathbf{r}) = \exp(i\mathbf{k}.\mathbf{r}) - a_s \frac{e^{i\,kr}}{r} , \qquad (9.35)$$

with a total scattering cross-section (for distinguishable particles):

$$\sigma = 4\pi a_s^2 . \qquad (9.36)$$

The scattering process has, at this limit, a spherical symmetry. It occurs in an S-wave carrying zero orbital angular momentum. This is a general feature of low-energy collisions, valid for a potential of arbitrary strength and spatial dependence, provided it decreases at large distance as $1/r^n$ with $n > 3$ (in particular, w needs not be isotropic).

We have introduced so far the concept of scattering length in the context of binary collisions of non-identical particles. For undistinguishable bosons, the cross section given by eqn. (9.36) is multiplied by a factor of two (see the discussion in Section 2.3.2) This extra factor helps the condensation process (note that in the case of fermions, the elastic cross-section in the S-wave vanishes because of the Pauli principle and direct evaporative cooling becomes impossible).

Since the collision process is fully defined at low energies by the single parameter a_s, any potential with the same scattering length has the same asymptotic scattering states as the real interatomic potential. We are thus free to replace w by a model

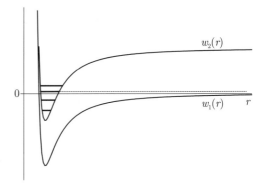

Fig. 9.8 Principle of Feshbach resonance. The incident atomic kinetic energy is given by the horizontal dotted line.

potential w_m, analytically tractable and amenable to a treatment within the Born approximation, and thus to a mean-field description. Before making this point more precise, let us give a few orders of magnitude.

The value of a_s is directly related to the phase shift δ_0 of the S-wave scattered by the potential from an impinging plane wave. At the limit of very low energy, $a_s = -\lim_{(k\to 0)} \delta_0/k$. We will not detail this calculation here, but only give orders of magnitude. For hydrogen, a_s is very small, of the order of the Bohr radius. For alkali atoms such as ^{87}Rb or ^{23}Na, a_s is of the order of a few nanometres, about a hundred times larger than the Bohr radius a_0, making the elastic scattering cross-section of these cold alkali atoms about four orders of magnitude larger than the atomic geometric cross-section. This is a very fortunate result for Bose–Einstein condensation of laser-cooled atoms, since, as we have seen above, these collisions are essential to ensure the thermalisation process during the evaporative cooling stage. This explains why achieving Bose–Einstein condensation is so much easier in alkali atoms than in hydrogen (Fried *et al.* 1998). The scattering length can have either sign (opposite to that of δ_0). Rubidium and sodium have a positive a_s, while it is negative for lithium atoms.

We assumed up to now that the colliding atoms can be treated as structureless particles, with an interatomic potential depending only upon the atom separation **r**. Collisions can however also involve several channels, corresponding to different asymptotic internal states of the atom at large distances. Figure 9.8 presents, in a very sketchy way, such a situation leading to an effect known as a Feshbach (1962) resonance. Two internal atomic states $|1\rangle$ and $|2\rangle$ with different magnetic moments correspond to collisional potentials w_1 and w_2 slightly offset with respect to each other. Two atoms in state $|1\rangle$ impinge on each other with a very small kinetic energy. This energy lies very close to that of a bound state in the potential w_2. If the $|1\rangle$ and $|2\rangle$ states are coupled by a small term of the interaction Hamiltonian, the scattering amplitude of atoms in state $|1\rangle$ is largely enhanced by the presence of this nearly resonant bound state in the other channel. The resonance can be tuned by applying a variable magnetic field B on the sample. This field shifts the relative energies of the two atomic states and makes it possible to sweep the bound state energy across the resonance which corresponds to a specific magnetic field value B_0. Theory shows, and experiments confirm (Inouye *et al.* 1998), that the scattering length varies as $1/(B - B_0)$, changing sign when the resonance is crossed and reaching high values on each side of B_0. Large scattering lengths, in the hundred of nanometres range can then be obtained. The application of a variable magnetic field to the atomic sample gives us a kind of 'knob' to adjust at will the interaction between the cold atoms in the condensate.

Which model potential can we choose to describe in a simple way the binary atomic interactions in the condensate? The simplest idea is to assume a contact interaction of the form $C\delta(\mathbf{r})$ and to adjust the constant C in order to get the desired scattering length. This naive approach has a problem since the action of the three-dimensional $\delta(\mathbf{r})$ distribution on a wave function diverging at the origin as $1/r$ does not yield a sensible result. This mathematical problem is solved by introducing instead a regularized contact distribution, the potential w_m being defined by its action on a two-body wave function $\psi(\mathbf{r})$ as (Dalibard 1999):

$$w_m(\mathbf{r})\psi(\mathbf{r}) = C_m\delta(\mathbf{r})\partial[r\psi(\mathbf{r})]/\partial r \ . \tag{9.37}$$

Solving the Schrödinger equation of the two-body system with this potential yields the exact solution for the scattering wave:

$$\Psi_{\mathbf{k}}^{+}(\mathbf{r}) = \exp(i\mathbf{k}\cdot\mathbf{r}) - \frac{a_m}{1 + ika_m}\frac{e^{ikr}}{r}, \tag{9.38}$$

with a_m defined as:

$$a_m = C_m m/4\pi\hbar^2 \ . \tag{9.39}$$

By choosing $C_m = 4\pi\hbar^2 a_s/m$, we adjust the model potential w_m to yield at the low-energy limit $ka_m \ll 1$ the *same* scattering length as the real interatomic interaction. Moreover, the low-energy scattering state, being linear in $a_m = a_s$ – hence in C_m – satisfies automatically the Born approximation. It renders thus the many-body interaction in the boson gas amenable to a mean-field description. The regularization term following the $\delta(\mathbf{r})$ distribution in eqn. (9.37) seems somewhat cumbersome. It will not bother us in the following since we will always apply the model potential to products of single-particle wave functions which are regular at $\mathbf{r}_1 = \mathbf{r}_2$. In this case, the model potential can simply be assimilated to a three-dimensional delta function.

Having identical asymptotic scattering states, the real and model potentials produce the same long-range correlation effects on the atomic gas, at interatomic distances $r > a_s$. The model and the real potential depart in their description of the gas correlations over smaller distances. This discrepancy is negligible, though, if the typical interatomic distance between particles is large compared to a_s. In other words, this modelization is possible if $\eta = (\mathcal{N}/V)^{1/3}a_s \ll 1$, which is usually satisfied in Na and Rb condensates ($\eta \sim 10^{-1}$–10^{-2}).

Injecting this model potential in eqn. (9.6), we get the second quantization expression of the interatomic energy:

$$W = \frac{2\pi\hbar^2 a_s}{m} \sum_{\mu,\mu',\nu,\nu'} \int d^3\mathbf{r}\ \phi_\mu^*(\mathbf{r})\phi_{\mu'}^*(\mathbf{r})\phi_\nu(\mathbf{r})\phi_{\nu'}(\mathbf{r})\ a_\mu^\dagger a_{\mu'}^\dagger a_\nu a_{\nu'} \ . \tag{9.40}$$

The effect of this interaction on an arbitrary boson state is to reshuffle the atoms among the various single-particle wave functions, destroying pairs of bosons in two populated states $\phi_\nu, \phi_{\nu'}$ to recreate them in two other states $\phi_\mu, \phi_{\mu'}$. The amplitude for each of these processes is equal to the product of the four corresponding single-particle wave functions, integrated over the volume of the gas. The expectation value of W in the ground state of the ideal boson gas condensate is:

$$\langle \mathcal{N} | W | \mathcal{N} \rangle = \frac{\hbar}{2} g \langle \mathcal{N} | (a_0^\dagger)^2 (a_0)^2 | \mathcal{N} \rangle = \frac{\hbar}{2} g \mathcal{N} (\mathcal{N} - 1) \,, \tag{9.41}$$

where we have introduced the collisional coupling g (homogeneous to a frequency) defined as:

$$g = \frac{4\pi \hbar a_s}{m} \int d^3 \mathbf{r} \, |\phi_0(\mathbf{r})|^4 \,. \tag{9.42}$$

To first order in perturbation theory, this expectation value describes the energy shift produced on the condensate ground state by the atomic interactions. We must remark, however, that a treatment which assumes that the system state is not changed and its energy is modified to first order, as expressed by eqn. (9.41), is valid only if the number of atoms is very small. We will come back to this point below. In large condensates, the atomic interactions change the system's ground state, which is no longer the harmonic oscillator Gaussian state $\phi_0(\mathbf{r})$, but becomes a new function $\psi_0(\mathbf{r})$. This function can be computed by a variational argument. The mean interaction energy, evaluated in state $\psi_0(\mathbf{r})$ adds to the trap potential energy $V_{\mathrm{trap}}(\mathbf{r})$ and to the atomic zero-point kinetic energy in this spatially modified state. To find out the form of the new ground state wave function, one must minimize the total energy with respect to small changes of the wave function around $\psi_0(\mathbf{r})$. Expressing this condition leads to solving an eigenvalue problem for a non-linear Schrödinger equation for ψ_0:

$$E\psi_0(\mathbf{r}) = \left[-\frac{\hbar^2}{2m} \Delta \psi_0(\mathbf{r}) + V_{\mathrm{trap}}(\mathbf{r}) + \frac{4\pi \hbar^2 a_s}{m} |\psi_0(\mathbf{r})|^2 \right] \psi_0(\mathbf{r}) \,. \tag{9.43}$$

This equation plays an important role in Bose–Einstein condensate studies. Its time-dependent version (replacing $E\psi_0$ by $i\hbar d\psi_0/dt$ in the left-hand side) is called the Gross–Pitaevskii equation (Pitaevskii and Stringari 2003). It describes in the mean-field approximation the evolution of the atom cloud under various conditions. We will not discuss here its validity and its limitations. Let us only say, qualitatively, that for a positive scattering length the gas stabilizes by expanding. It then decreases its positive interaction energy and its zero-point kinetic energy, at the expense of an increase of its potential energy in the trap. For negative scattering lengths, the opposite effect occurs. The cloud has a tendency to contract, which leads to an instability when the atom number reaches a critical value of the order of $a_{\mathrm{ho}}/|a_s|$ – about 1500 atoms in a lithium condensate (Bradley *et al.* 1997).

We restrict our discussion from now on to situations in which the effect of the interactions on the shape of the condensate is negligible. A simple back-of-the-envelope estimate of the rate of change of the three energy terms of the atomic cloud when its radius is varied around a_{ho} shows that we can assimilate $\psi_0(\mathbf{r})$ with $\phi_0(\mathbf{r})$ as long as $\mathcal{N} < a_{\mathrm{ho}}/|a_s|$. Typically $a_{\mathrm{ho}}/|a_s| \sim 10^2$–$10^3$, so that the effect of the interactions on the extension of the gas can be neglected in systems containing up to a few tens to a few hundred atoms. We can simply compute in this case the magnitude of the collisional coupling g which measures the strength of the interaction of an atom pair in the condensate. Noting that $|\phi_0|^4 = (1/\pi a_{\text{ho}}^2)^3 \exp(-2|\mathbf{r}|^2/a_{\text{ho}}^2)$, we perform the integration of eqn. (9.42) and we get:

$$g = \sqrt{\frac{2}{\pi}\frac{\hbar a_s}{m a_{\text{ho}}^3}} = \sqrt{\frac{2}{\pi}}\omega_{\text{ho}}\frac{a_s}{a_{\text{ho}}} \ . \tag{9.44}$$

A typical value of a_s is 5 nm. In a magnetic trap with $a_{\text{ho}} = 1$ μm and $\omega_{\text{ho}} = 10^3$ s^{-1}, we find $g \approx 4$ s^{-1}. The collisional interaction, proportional to $1/a_{\text{ho}}^3$ can be made much stronger by a tighter atomic confinement. In an optical lattice, ground state extensions a_{ho} of the order of 70 nm can be achieved, resulting in $g \approx 10^4$ s^{-1} (see next section).

We conclude this section by describing the effect of atomic interactions in a bi-modal condensate. If it is of the 'spin' type, ϕ_1 and ϕ_2 being two single-particle states associated to different internal quantum numbers within the same atomic species, we must introduce three different scattering lengths corresponding to the collisions within each species ($a_{s,1}$ and $a_{s,2}$) and to the interspecies collisions ($a_{s,12}$). The interaction energy then contains three terms:

$$\begin{aligned} W &= W_1 + W_2 + W_{12} \\ &= \frac{\hbar}{2}g_1 N_1(N_1 - 1) + \frac{\hbar}{2}g_2 N_2(N_2 - 1) + \hbar g_{12} N_1 N_2 \ , \end{aligned} \tag{9.45}$$

with N_i ($i = 1$, 2) being the boson number operators of each component and:

$$g_i = \frac{4\pi\hbar a_{s,i}}{m}\int d^3\mathbf{r}\,|\phi_i|^4 \quad (i = 1,\ 2) \ ; \quad g_{12} = \frac{4\pi\hbar a_{s,12}}{m}\int d^3\mathbf{r}\,|\phi_1|^2|\phi_2|^2 \ . \tag{9.46}$$

If the single-particle wave functions ϕ_1 and ϕ_2 are spatially non-overlapping states of the same atomic species confined in two different wells, the scattering lengths $a_{s,1}$ and $a_{s,2}$ are equal and the crossed coupling g_{12} vanishes, since the two species are never at the same place. The interaction energy W is restricted to the first two terms in eqn. (9.45). The coefficients g_1 and g_2 are in the ratio of the integrals $\int d^3\mathbf{r}\,|\phi_i|^4$, ($i = 1$, 2).

9.4.2 Collapse, revivals and cats in matter waves

The interactions described by eqns. (9.45) are reminiscent of the Kerr Hamiltonian in optics analysed in Section 7.2. Let us recall that the non-linear interaction of a light wave with a transparent medium contains a term proportional to the square of the light intensity $\hbar\gamma_k(a^\dagger a)^2$, where γ_k is a non-linear susceptibility. Its effect is to make the refractive index of the medium intensity-dependent and to give rise to a phase spreading of light beams propagating over long distances. It also leads, at least in theory, to phase revival effects and to the generation of multiple-components Schrödinger cat states of light. This cat generation procedure cannot be implemented in practice, due to too strong losses in the medium. The advent of Bose–Einstein condensation physics, which naturally introduces a Kerr-like coupling of matter waves, has led to a renewal of this early cat proposal which we now briefly present.

Spreading and refocusing of the phase in a bimodal condensate

We study now the effect of the atomic collisions on the phase of a bimodal condensate initially prepared in a symmetric phase state made up of two non-overlapping wave packets (Castin and Dalibard 1997). We assume that the tunnelling between the two wells is negligible (high interwell barrier). It is convenient to expand the initial phase state along the Fock states. The initial state (time $t = 0$) is obtained from eqn. (9.17), with $\theta = \pi/2$ and the notation change $p = \mathcal{N}/2 - \delta\mathcal{N}$:

$$|\psi(0)\rangle = |\mathcal{N}; \theta = \pi/2, \phi\rangle = \frac{e^{-i\mathcal{N}\phi/2}}{2^{\mathcal{N}/2}} \sum_{\delta\mathcal{N}} \sqrt{\frac{\mathcal{N}!}{(\mathcal{N}/2 + \delta\mathcal{N})!(\mathcal{N}/2 - \delta\mathcal{N})!}}$$
$$\times e^{i\delta\mathcal{N}\phi} |\mathcal{N}/2 + \delta\mathcal{N}, \mathcal{N}/2 - \delta\mathcal{N}\rangle . \qquad (9.47)$$

The quantity $\delta\mathcal{N}$ is a particle number fluctuation of the order of $\pm\sqrt{\mathcal{N}}$. For large $\mathcal{N}$'s, we can develop the factorials (using Stirling formula) and we get the Gaussian approximation:

$$|\psi(0)\rangle \approx (2/\pi\mathcal{N})^{1/4} e^{-i\mathcal{N}\phi/2} \sum_{\delta\mathcal{N}} e^{-\delta\mathcal{N}^2/\mathcal{N}} e^{i\delta\mathcal{N}\phi} |\mathcal{N}/2 + \delta\mathcal{N}, \mathcal{N}/2 - \delta\mathcal{N}\rangle . \qquad (9.48)$$

The sum in this expression, extending in principle over $\delta\mathcal{N}$ between $-\mathcal{N}/2$ and $+\mathcal{N}/2$, is in practice restricted to $-\sqrt{\mathcal{N}} < \delta\mathcal{N} < \sqrt{\mathcal{N}}$.

A simple calculation shows that a Fock state with given $\delta\mathcal{N}$ is an eigenstate of the Hamiltonian W, with the eigenenergy:

$$\langle W \rangle = \hbar g \frac{\mathcal{N}}{2} \left(\frac{\mathcal{N}}{2} - 1 \right) + \hbar g \delta\mathcal{N}^2 , \qquad (9.49)$$

(we assume that the collisional couplings g_1 and g_2 are equal and drop their index). This collisional energy varies quadratically with $\delta\mathcal{N}$. The minimum energy state (we consider here that the scattering length is positive) corresponds to the best possible equipartition of the particles between the two modes, $\delta\mathcal{N} = 0$ if $\mathcal{N}$ is even, $\delta\mathcal{N} = \pm 1/2$ if $\mathcal{N}$ is odd. Let us now study the effect of these collisions on the evolution of a bimodal phase state of the form (9.48). We apply to each Fock component its collisional dephasing, proportional to time and to its collisional energy (9.49). Up to a global phase factor, we get:

$$|\psi(t)\rangle \approx (2/\pi\mathcal{N})^{1/4} \sum_{\delta\mathcal{N}} e^{-\delta\mathcal{N}^2/\mathcal{N}} e^{i\delta\mathcal{N}\phi} e^{-ig\delta\mathcal{N}^2 t} \left| \frac{\mathcal{N}}{2} + \delta\mathcal{N}, \frac{\mathcal{N}}{2} - \delta\mathcal{N} \right\rangle . \qquad (9.50)$$

The phases in the Fock state expansion present a linear $\delta\mathcal{N}\phi$ term and a Kerr-like term, quadratic in $\delta\mathcal{N}$ describing the effect of the collisions. The phase coherence of this state is revealed by computing the expectation value of the interference term $\langle a_\beta^\dagger a_\alpha \rangle$:

$$\langle a_\beta^\dagger a_\alpha \rangle = (\mathcal{N}/2\pi)^{1/2} e^{i\phi} \sum_{\delta\mathcal{N}} e^{-2\delta\mathcal{N}^2/\mathcal{N}} e^{-ig(2\delta\mathcal{N}-1)t} . \qquad (9.51)$$

The phase spreading of the last exponential in eqn. (9.51) results in a collapse of this expectation value. The collapse is complete when the phase has fanned out over a full 2π-angle over the distribution of $\delta\mathcal{N}$, whose width is of the order of $2\sqrt{\mathcal{N}}$. This occurs within a time T_{collapse}:

$$T_{\text{collapse}} \approx \frac{\pi}{2g\sqrt{\mathcal{N}}} \ , \tag{9.52}$$

inversely proportional to the collision strength g and to the square root of the particle number. This effect is analogous to the spreading of the Gaussian wave packet of a free particle in a one-dimensional propagation problem, the phase ϕ and the fluctuation of the number of bosons $\delta\mathcal{N}$ being replaced by the particle's position x and momentum $\hbar k$ respectively. The wave packet of this fictitious particle is:

$$|\Psi(t)\rangle = \int dk\, e^{-k^2/\Delta^2 k} e^{ikx} e^{-i\hbar k^2 t/2m} |k\rangle \ , \tag{9.53}$$

which is very similar to eqn. (9.50).

There is, however, a big difference between the condensate phase and the particle position spreadings. In the latter case the sum over momentum is a continuous integral, whereas, in the condensate situation, the sum over $\delta\mathcal{N}$ is discrete, expressing the graininess of the matter wave. While the free-particle wave function spreading is an irreversible process, the phase condensate collapse is reversible. All the phases in eqn. (9.51) are refocused to the same value (modulo 2π) after a time T_{revival}:

$$T_{\text{revival}} = \pi/g \ , \tag{9.54}$$

as well as at all multiples of this time.[2] At these precise times, the coherence between the two modes of the condensates is fully restored to its initial value. The first revival time is $\sim \sqrt{\mathcal{N}}$ times larger than the collapse time, and is independent of the number of particles. It depends only upon the non-linearity g introduced by the collisions. Note the analogy of this effect with the collapse and revival of a coherent field propagating in a Kerr medium (Section 7.2).

A proposal for cat state generation in a bimodal condensate

The optical Kerr effect analogy suggests that multi-component matter-wave cats should appear at times which are rational fractions of T_{revival}. Let us describe here the matter-wave state at time $T_{\text{revival}}/2 = \pi/2g$. For the sake of simplicity, we will consider explicitly only the case of an even atom number $\mathcal{N}$ ($\delta\mathcal{N}$ is then an integer). The collisional phase appearing in eqn. (9.50) then takes the values:

$$g(\delta\mathcal{N})^2 T_{\text{revival}}/2 = \frac{\pi}{2}(\delta\mathcal{N})^2 = \begin{cases} 0 \text{ for } \delta\mathcal{N} \text{ even} \\ \\ \pi/2 \text{ for } \delta\mathcal{N} \text{ odd} \end{cases} \quad (\text{modulo } 2\pi) \ . \tag{9.55}$$

If we express the initial phase state as a sum of a component whose $\delta\mathcal{N}$'s are even and a component whose $\delta\mathcal{N}$'s are odd, the evolution produced by the collisions up to time $T_{\text{revival}}/2$ leaves the amplitude of the first component unaltered and multiplies

---

[2]The coherence revives for the first time with a phase equal or opposite to that of the initial state, depending upon the parity of $\mathcal{N}$.

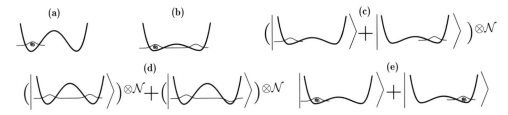

Fig. 9.9 Creation of a matter wave Schrödinger cat: (a)–(e) Successive stages of the non-local cat preparation.

by $e^{-i\pi/2}$ the amplitude of the second. A similar argument was used in Section 8.5.2 when discussing the generation of GHZ states of $\mathcal{N}$ ions in a trap. The separation between even and odd terms in $\delta\mathcal{N}$ is easy to perform analytically. Let us choose $\phi = 0$ and expand the phase bimodal state along the two states:

$$|\psi^{\pm}\rangle = \left[(a_{\alpha}^{\dagger} + a_{\beta}^{\dagger})^{\mathcal{N}} |0\rangle \pm (a_{\alpha}^{\dagger} - a_{\beta}^{\dagger})^{\mathcal{N}} |0\rangle\right] / \sqrt{\mathcal{N}!(2^{\mathcal{N}+1})}. \tag{9.56}$$

If $\mathcal{N}/2$ is an even integer, $|\psi^{+}\rangle$ contains only even powers of $\delta\mathcal{N}$ and $|\psi^{-}\rangle$ only odd powers. The $\delta\mathcal{N}$ parity of these states is exchanged if $\mathcal{N}/2$ is odd. Expressing the intial state as the sum of $|\psi^{+}\rangle$ and $|\psi^{-}\rangle$ and taking into account the evolution of the phase of the corresponding probability amplitudes under the effect of collisions up to the time $T_{\text{revival}}/2$ (factor 1 for the state with $\delta\mathcal{N}$ even, and $e^{-i\pi/2}$ for the state with $\delta\mathcal{N}$ odd), we find an expression which reminds us of eqn. (7.30) and (8.108):

$$|\psi(t = \pi/2g)\rangle = \frac{1}{2^{(\mathcal{N}+1)/2}} \frac{e^{-i\pi/4}}{\sqrt{\mathcal{N}!}} (a_{\alpha}^{\dagger} + a_{\beta}^{\dagger})^{\mathcal{N}} |0\rangle + \frac{(-1)^{(\mathcal{N}/2)}}{2^{(\mathcal{N}+1)/2}} \frac{e^{i\pi/4}}{\sqrt{\mathcal{N}!}} (a_{\alpha}^{\dagger} - a_{\beta}^{\dagger})^{\mathcal{N}} |0\rangle ,$$

$$\tag{9.57}$$

The system is now in a superposition of $\mathcal{N}$ particles all in the state $[|\phi_{\alpha}\rangle + |\phi_{\beta}\rangle]/\sqrt{2}$ with $\mathcal{N}$ particles all in the state $[|\phi_{\alpha}\rangle - |\phi_{\beta}\rangle]/\sqrt{2}$, obviously a Schrödinger-cat situation. As in the optical Kerr effect and trapped ion cases, a non-linear interaction, acting for a well-chosen time interval, prepares a mesoscopic superposition of states in a system containing several particles. When $\mathcal{N}$ is odd, a similar calculation leads to equivalent cat state expressions, provided we set $\phi = \pi/2$ in the initial bi-modal state (which amounts to redefining the relative phase between states $|\phi_{\alpha}\rangle$ and $|\phi_{\beta}\rangle$). The operators $a_{\alpha}^{\dagger} \pm a_{\beta}^{\dagger}$ in eqn. (9.57) are then replaced by $a_{\alpha}^{\dagger} \mp i a_{\beta}^{\dagger}$. Note that the exact expression of the cat state depends on the value of $\mathcal{N}$, a parameter hard to control in an experiment.

The recipe to prepare the $\mathcal{N}$-even state (9.57) could, in principle, follow the stages schematized in Fig. 9.9. We realize first a double potential well and start with a condensate with an even number of bosons prepared in the left well (α), separated from the right well (β) by a high barrier (Fig. 9.9a). The barrier is then suddenly lowered at a height such that $\mathcal{J} \gg g$ (Fig. 9.9b) and we wait a time such that $\mathcal{J}t = \pi/2$: the tunnel effect creates a phase state (Fig. 9.9c). This is not yet a cat, since all the particles occupy the same quantum state, but a mere tensor product of $\mathcal{N}$ particles in the same wave function. This is expressed by the $\mathcal{N}$ exponent at the right of the bracket around the sketch representing this state. The effect of collisions is negligible during this first stage. We then suddenly separate the two wells by raising the barrier. The tunnel effect is suppressed. The collisions become more efficient in each well, due to the extra confinement produced by deepening the wells. Enough time is left for the collisions to achieve the appropriate dephasing. At time $T_{\mathrm{revival}}/2$, a cat state is prepared (Fig. 9.9d): we have now $\mathcal{N}$ atoms, all occupying either one of two orthogonal wave functions $\phi_\alpha(\mathbf{r}) \pm \phi_\beta(\mathbf{r})$. The one-particle wave functions are delocalized in both wells, but they are still spatially overlapping. It is possible at this stage to separate them in the real space as well as in the Hilbert space. To achieve this, the tunnel effect is re-established by a sudden lowering of the barrier and the system is left to evolve for a quarter-period of the coherent oscillation between the two wells (collisions have a negligible effect during this time): the $\mathcal{N}$ atoms localize at left and at right, in a mesoscopic state superposition (Fig. 9.9e). The interwell barrier can then be finally raised, suppressing the tunnelling and freezing the system in a non-local state of the high-noon type:

$$|\Psi\rangle = \frac{1}{\sqrt{2(\mathcal{N}!)}} \left[e^{-i\pi/4}(a_\beta^\dagger)^{\mathcal{N}} + (-1)^{N/2} e^{i\pi/4}(a_\alpha^\dagger)^{\mathcal{N}} \right] |0\rangle \ . \qquad (9.58)$$

One way to demonstrate the coherence of the state (9.57) would be to leave the system alone after the time $T_{\mathrm{revival}}/2$, without lowering the interwell barrier and thus without separating spatially the two components. We would then wait until the two parts of the cat recombine at time T_{revival}. The recurrence of the bimodal state coherence at this time could be checked by releasing the condensate from the double-well trap and studying the interference term in the atomic density of the expanded cloud (see next section). Observing in this way the rephasing of the $\mathcal{N}$-particle matter wave would be a demonstration of the transient existence of a cat state in the system at half the revival time, even if we do not know its exact expression (in the probable case where we do not control exactly the value of $\mathcal{N}$ and its parity). In principle, the spatially separated cat state (9.58) could also be analysed in the same way. One would first overlap spatially via tunnelling the two components, then let them evolve under the effect of the binary collisions acting for an additional $T_{\mathrm{revival}}/2$ time interval, until the initial coherence is restored.

We have presented here a very general principle, without discussing the experimental limitations and the causes of decoherence of the system. The loss of a single atom from the condensate during this elaborate succession of operations would be lethal for the cat because the lost atom (or molecule formed by three-body recombination) would be in a quantum state entangled with the superposition and its detection would collapse it. The method seems thus to be restricted to the preparation and detection of mesoscopic systems made of a few particles. Their study would nevertheless be of great interest to investigate, on this new system, the quantum–classical boundary.

In the above proposals, the spatial separation between the two cat components is achieved by a sequence of operations, letting the $\mathcal{J}$ and g terms of the Hamiltonian act successively and separately on the atomic clouds. Situations in which tunnelling occurs in the presence of interatomic collisions are very interesting, but beyond the scope of this book (Albiez *et al.* 2005). They correspond to tunnelling phenomena analogous to the Josephson oscillations of superconducting physics (Barone and Paterno 1982). Let us just mention an interesting effect, predicted to occur in condensates with negative scattering length (Ho and Ciobanu 2000). The formation of a state of the form (9.58)

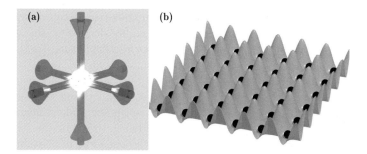

Fig. 9.10 Bosons in a periodic lattice. (a) Principle of the experiment. A cold atom sample is at the intersection of three pairs of counter-propagating detuned laser beams creating a periodic dipole optical potential. (b) The 'egg box potential', a two-dimensional cut of the optical lattice. The trapping sites are depicted by black spheres.

could then be obtained under steady-state operation, letting the system equilibrate in its natural ground state under the competitive effect of tunnelling and collisions, provided the condition $|g| \gg \mathcal{J}$ is satisfied and the temperature T small enough. The atoms have then a natural tendency to gather in states of maximum density, which minimize their interaction energy. This favours the accumulation of all the particles in one well or the other and we understand that, in this case, the system tends to end up in a state of the form (9.58). The quantum phase between the two parts of the state wave function is then pinned down by the small coupling between the two wells.[3]

9.5 Matter waves in periodical lattices

The double-well configuration discussed above can be extended to bosons occupying a periodic structure (Bloch 2005). An optical trap made of overlapping laser beams forming a three-dimensional standing wave is superimposed to the magnetic trap in which the condensate has been produced. These lasers are detuned with respect to the atomic transition, providing a dipolar force which tends to confine the atoms near the maxima (red detuning) or the minima (blue detuning) of the optical standing wave. The light forces form a periodic crystal-like structure of potential wells (Fig. 9.10). The evolution of the boson gas in this structure is determined by the competition between the collisional interactions and the tunnelling of the atoms across the potential interwell barriers.

The experimentalist can vary at will, and to some extent independently, the depth of the wells and hence the tunnelling rate (by changing the laser intensity), the collisional rate inside each well (e.g. by using Feshbach resonances) and the average occupation per lattice site (by changing the number of atoms in the condensate). A rich variety of effects has been observed in this system, under well-controlled conditions. One interesting aspect of these experiments is that they realize an ideal model of solid state physics. The behaviour of matter waves in periodic structures is a central problem in condensed matter physics. In metals, the matter wave is fermionic, with strong

---

[3]For other proposals of cat states and collapse–revival studies in Bose–Einstein condensates, see Wright *et al.* (1996); Cirac *et al.* (1998); Sinatra and Castin (1998); Gordon and Savage (1999); Dalvit *et al.* (2000); and Montina and Arecchi (2002).

constraints imposed by the Pauli principle. Well before the advent of Bose–Einstein condensation in cold atoms, theorists have wondered about what would happen to bosonic waves in crystals, looking for differences and similarities with electronic waves in metals and semiconductors. Bosons in lattice experiments opens a wonderful testing ground for their theories.[4]

In this section, we briefly analyse some important aspects of these experiments. We show in particular how the concept of complementarity between phase and particle number fluctuations can be used to reveal an essential feature of the system's behaviour, the transition between a superfluid and an insulating phase of the condensate. We also describe an experiment in which the collapse and revival of the condensate phase in a periodical structure has been observed, opening the way to Schrödinger cat generation in Bose–Einstein condensates. The quantum information aspect of these studies will be addressed in the next section.

The dynamics of an ensemble of $\mathcal{N}$ bosons in a periodic lattice made of $\mathcal{N}_w$ wells is ruled by the Bose–Hubbard Hamiltonian H_{BH} (Pitaevskii and Stringari 2003):

$$H_{\mathrm{BH}} = \sum_i \hbar\omega_{\mathrm{ho}} N_i - \hbar\mathcal{J} \sum_{<i,j>} a_i^\dagger a_j + \frac{\hbar g}{2} \sum_i N_i(N_i - 1) , \qquad (9.59)$$

with $N_i = a_i^\dagger a_i$. This second quantization Hamiltonian is expressed in terms of the bosonic operators a_i and $a_i^\dagger$ annihilating and creating bosons at each site i of the lattice. The single-particle wave function associated to each site is a Gaussian, corresponding to the zero-point motion of a particle around the corresponding potential minimum. The three terms in eqn. (9.59) describe respectively the external potential energy imposed by the lattice, the hopping kinetic energy between adjacent wells and the intrawell collisional interactions in the condensate. The $< i, j >$ symbol denotes a summation over pairs of first neighbours in the lattice. We suppose that the mean number of particles per well, $\mathcal{N}/\mathcal{N}_w$, is of the order of a few at most, a condition realized in the experiments which will be described below.

Three characteristic frequencies play an essential role in the system evolution: the oscillation frequency ω_{ho} in each well (supposed to be harmonic), the tunnelling rate $\mathcal{J}$ between adjacent wells and the collisional rate g in each well (we suppose $g > 0$, corresponding to a positive scattering length). The oscillation angular frequency ω_{ho}, typically in the 10 kHz range, is larger than $\mathcal{J}$ and g, while the ratio $\xi = g/\mathcal{J}$ can be varied at will. Depending upon ξ, the system is expected to behave in strikingly different ways (Fisher *et al.* 1989). Many-body theory predicts that, below a critical value ξ_c, the condensate is a single quantum system delocalized in the whole lattice. In this regime, defined as superfluid, the interwell hopping is dominant and the boson gas is characterized by a single giant wave function which is phase-coherent over the whole sample. At the critical value $\xi = \xi_c$, the system undergoes a phase transition corresponding to a sudden shrinking of the single-particle wave function correlation length. This phenomenon is very analogous to the Mott localization transition in condensed matter physics. In the solid state context, the phase transition corresponds to a cancellation of the conductivity, hence the name 'insulator' given to the system

---

[4]We should note also that these experiments have been extended to cold fermions in optical lattices, so that the behaviour of electrons can also be mimicked now in artificial structures.

when $\xi > \xi_c$. In this regime, the collisional intrawell interactions (defined by the rate g) are dominant over tunnelling (rate $\mathcal{J}$). The critical value of this phase transition is $\xi_c \sim 35$ when the mean occupation number per well $\mathcal{N}/\mathcal{N}_w$ is equal to 1.

The fluctuation of the number of particles in each lattice well undergoes a continuous change when ξ is varied across the transition value ξ_c. In the superfluid regime where tunnelling is dominant, the intrawell fluctuations are large and strong correlation exist between two arbitrary lattice sites. These correlations decrease as the critical point is crossed and they vanish for $\xi \gg \xi_c$, deep into the insulator regime. The number of particles in each well does not fluctuate any longer and there is no phase relation left between the wells. Before describing how this transition has been observed, let us analyse in more detail the coherence properties of the superfluid and insulator phases.

9.5.1 Superfluid regime in a lattice

When g is negligibly small, the tunnelling between the wells is dominant and the ground state wave function of the $\mathcal{N}$ bosons is the symmetric superposition of the Gaussian wave packets localized in all the wells. The corresponding bosonic creation operator is:

$$a_{\mathrm{sf}} = \frac{1}{\sqrt{\mathcal{N}_w}} \sum_i a_i \;, \tag{9.60}$$

and the $\mathcal{N}$ boson state is:

$$|\mathcal{N};\mathrm{sf}\,\rangle = \frac{1}{\sqrt{\mathcal{N}!}} (a_{\mathrm{sf}}^\dagger)^{\mathcal{N}} |0\rangle \;. \tag{9.61}$$

In order to analyse the fluctuations of the number of bosons among the wells, it is convenient to separate in the expression of $a_{\mathrm{sf}}^\dagger$ the contribution of one of the wells (labelled 1) from all the others:

$$a_{\mathrm{sf}}^\dagger = \frac{1}{\sqrt{\mathcal{N}_w}} a_1^\dagger + \sqrt{1 - \frac{1}{\mathcal{N}_w}} b_1^\dagger \;; \quad b_1^\dagger = \frac{1}{\sqrt{\mathcal{N}_w - 1}} \sum_{j=2,\ldots} a_j^\dagger \;. \tag{9.62}$$

This binomial partition of the $\mathcal{N}$ bosons puts, on average, $\mathcal{N}/\mathcal{N}_w$ particles in the first well, the reminder being in the state generated by the boson operator $b_1^\dagger$. The expectation values of the three bosonic number operators are thus:

$$\langle a_{\mathrm{sf}}^\dagger a_{\mathrm{sf}} \rangle = \mathcal{N} \;; \quad \langle a_1^\dagger a_1 \rangle = \langle N_1 \rangle = \frac{\mathcal{N}}{\mathcal{N}_w} \;; \quad \langle b_1^\dagger b_1 \rangle = \mathcal{N} \left(1 - \frac{1}{\mathcal{N}_w} \right) \;. \tag{9.63}$$

The binomial partition law yields immediately the fluctuation ΔN_1 of the number of bosons in well 1 (of course identical to the fluctuation in any well):

$$\Delta N_1 = \sqrt{\frac{\mathcal{N}}{\mathcal{N}_w} \left(1 - \frac{1}{\mathcal{N}_w} \right)} \approx \sqrt{\frac{\mathcal{N}}{\mathcal{N}_w}} = \sqrt{\langle N_1 \rangle} \;. \tag{9.64}$$

This fluctuation, approximately equal to the square root of the mean occupation number, is large in relative value when the occupation number per well is small. From the last three equations, we deduce the mean correlation between the bosonic operators of two arbitrary wells in the lattice. The expectation values of the occupation numbers obey the following relation:

$$\langle a_{\mathrm{sf}}^\dagger a_{\mathrm{sf}} \rangle = \frac{1}{\mathcal{N}_w}\langle a_1^\dagger a_1 \rangle + \left(1 - \frac{1}{\mathcal{N}_w}\right)\langle b_1^\dagger b_1 \rangle + \frac{\sqrt{\mathcal{N}_w - 1}}{\mathcal{N}_w}\langle a_1^\dagger b_1 + a_1 b_1^\dagger \rangle , \qquad (9.65)$$

which, taking into account (9.63), yields:

$$\langle a_1^\dagger b_1 + a_1 b_1^\dagger \rangle = 2\mathcal{N}\sqrt{\mathcal{N}_w - 1}/\mathcal{N}_w . \qquad (9.66)$$

We finally remark that $\langle a_1^\dagger b_1 + a_1 b_1^\dagger \rangle$ is the sum of $\mathcal{N}_w - 1$ terms $2\langle a_1^\dagger a_j \rangle / \sqrt{\mathcal{N}_w - 1}$ which, by symmetry, are all equal. Hence:

$$\langle a_1^\dagger a_j \rangle_{sf} = \langle a_1^\dagger a_1 \rangle_{sf} = \mathcal{N}/\mathcal{N}_w . \qquad (9.67)$$

We conclude from this simple book-keeping argument that the ground state of the condensate exhibits, in the absence of collisions, large correlations between two arbitrary lattice sites. The expectation value of the cross product of bosonic creation and annihilation operators is, for any pair of wells, equal to the average number of particles per site. As in the double-well case, this correlation is related to the coherence property of the bosonic ensemble. We will see that the partial matter waves originating from the various lattice sites interfere with each other when the condensate is released from its trap. This interference is a signature of the strong superfluid interwell correlations. Before analysing this interference, let us go to the other limiting case and study the boson distribution deep into the insulator phase.

9.5.2 The insulator regime

In the opposite situation of negligible tunnelling and large interactions between the atoms, the system evolution is predominantly ruled by the collisional term:

$$H_{\mathrm{coll}} = \frac{\hbar g}{2}\sum_i N_i(N_i - 1) , \qquad (9.68)$$

which admits as eigenstates tensor products of boson number states in the individual wells. As in the double-well case, it is convenient to define the fluctuation of the boson number in each well:

$$\delta N_i = N_i - \frac{\mathcal{N}}{\mathcal{N}_w} ; \qquad \sum_i \delta N_i = 0 , \qquad (9.69)$$

and the interaction term becomes:

$$H_{\mathrm{coll}} = \hbar g\frac{\mathcal{N}}{2}\left(\frac{\mathcal{N}}{\mathcal{N}_w} - 1\right) + \frac{\hbar}{2}g\sum_i(\delta N_i)^2 , \qquad (9.70)$$

an expression which, in operator form, generalizes eqn. (9.49), established for $\mathcal{N}_w = 2$.

The ground state of the $\mathcal{N}$-boson gas in the lattice corresponds to the minimum of the average fluctuation $\langle \sum_i(\delta N_i)^2 \rangle$, i.e. to the best possible equipartition of bosons among the lattice sites. The particles distribute themselves among the sites in order to minimize their local density and hence their interaction energy. If the average number of particles per well, $\mathcal{N}/\mathcal{N}_w$, is an integer n_0, the ground state corresponds to a perfect equipartition with $\langle \delta N_i \rangle = 0$ for all is and the ground state is:

$$|\mathcal{N}; \text{ins}\,\rangle = \left(\frac{1}{\sqrt{n_0!}}\right)^{\mathcal{N}_w} \prod_i (a_i^\dagger)^{n_0} |0\rangle \ . \tag{9.71}$$

For a non-integer mean particle number per well, we can write $\mathcal{N}/\mathcal{N}_w = n_0 + p$ where n_0 is an integer and p a fraction comprised between 0 and 1. The ground state corresponds to a fraction $(1 - p)$ of the sites with n_0 particles and a fraction p with $n_0 + 1$ particles. In a uniform lattice, this ground state is highly degenerate because the system can choose, without any energy cost, which well contains an extra particle. When the effect of the superimposed harmonic magnetic trap is taken into account, the bosons must distribute themselves among the wells to minimize their total energy, including their trapping potential energy. The non-uniformity of the system leads then to an organization of the particles in concentric layers with different atom numbers per well (Jaksch *et al.* 1998). In order to simplify the discussion, we will assume in the following that the number of particle per site is an integer and that the atom distribution is uniform in the lattice.

The first excited state of the $\mathcal{N}$-boson system is obtained, from the uniform ground state, by moving a single particle from one well to another. This costs an energy equal to $\hbar g$. If the same operation is repeated on two sites, the energy is increased by $2\hbar g$ and so on. The excitation spectrum of H_{coll} is thus made of equidistant levels, with a spacing $\hbar g$ between them. The existence of a gap between the ground and first excited state entails that, even if $\mathcal{J}$ is not vanishingly small, the ground state remains stable for $g/\mathcal{J}$ large enough. For $\xi \gg \xi_c$, the bosons are frozen in their wells, in the state described by eqn. (9.71). There are no correlations between different sites in this state:

$$\langle a_i^\dagger a_j \rangle_{\text{ins}} = n_0 \delta_{ij} \ , \tag{9.72}$$

which is a situation very different from the $\xi \ll \xi_c$ one (compare with eqn. 9.67). Note that the insulator regime cannot be accounted for by a mean-field approach. Deep into this regime, the boson gas is no longer described by a single giant wave function, but is broken in independent micro-condensates associated to the different wells, each having a well-defined particle number. There is then no phase relationship left between these micro-condensates. We will now see how these two regimes can be distinguished in an experiment.

9.5.3   Imaging the condensate: complementarity revisited again

The superfluid and insulator phases produce very different absorption images after release and expansion of the condensate. The principle of the experiment is very similar to the interference experiment described in Section 9.3. Once the condensate has reached its equilibrium, the magnetic trap and the optical lattice are suddenly switched off and the condensate is left free to expand during a time t, after which it is illuminated by an absorbing laser beam whose shadow on a screen is recorded. The partial waves coming from the different lattice sites overlap and may interfere with each other. Let us call $\phi_i(\mathbf{r}, t)$ the ground state wave function associated to site i, after expansion during the time t. The mean density at point $\mathbf{r}$ of the expanded cloud is given by:

$$\langle n(\mathbf{r}, t) \rangle = \langle \Psi^\dagger(\mathbf{r}, t)\Psi(\mathbf{r}, t) \rangle = \sum_{i,j} \phi_i^*(\mathbf{r}, t)\phi_j(\mathbf{r}, t)\langle a_i^\dagger a_j \rangle . \tag{9.73}$$

Well into the superfluid phase, all the $\langle a_i^\dagger a_j \rangle$ correlations are equal to $\mathcal{N}/\mathcal{N}_w = n_0$, so that they factor out in the expression of the atomic density which becomes:

$$\langle n(\mathbf{r}, t) \rangle_{\text{sf}} = n_0 \left| \sum_i \phi(\mathbf{r}, t) \right|^2 . \tag{9.74}$$

The situation is quite different deep into the insulator phase, in which all the cross-correlations with $i \neq j$ vanish. We get in this case:

$$\langle n(\mathbf{r}, t) \rangle_{\text{ins}} = n_0 \sum_i |\phi(\mathbf{r}, t)|^2 . \tag{9.75}$$

In the superfluid regime, the atomic density is proportional to the square of a sum of amplitudes and exhibits interference (eqn. 9.74). Deep into the insulator regime, it is proportional to a sum of squared amplitudes in which all interference effects have disappeared (eqn. 9.75).

The coherence between the partial waves originating from the different lattice sites is well-described in momentum space. Let us call $\overline{\phi_0}(\mathbf{p})$ the single-particle momentum wave function associated to the lattice well located at the origin of spatial coordinates. It is the Fourier transform of the wave function in the $\mathbf{r}$-representation, $\phi_0(\mathbf{r}, t = 0)$. The single-particle wave function $\phi_i(\mathbf{r}, 0)$ associated to the well centred at r_i is deduced from $\phi_0(\mathbf{r}, 0)$ by a translation in space, which corresponds to a multiplication by the phase factor $\exp(-i\mathbf{p} \cdot \mathbf{r}_i/\hbar)$ in the momentum representation. The propagation over time t adds the phase factor $\exp(-i|\mathbf{p}|^2 t/2m\hbar)$ to the $\mathbf{p}$-wave function (we neglect the effect of interatomic collisions during the expansion). Finally, the sum of amplitudes in eqn. (9.74) can be expressed as:

$$\sum_i \phi_i(\mathbf{r}, t) = \sum_i \int d^3\mathbf{p} \, \overline{\phi_0}(\mathbf{p}) e^{i\mathbf{p} \cdot (\mathbf{r} - \mathbf{r}_i)/\hbar} e^{-i|\mathbf{p}|^2 t/2m\hbar} . \tag{9.76}$$

This sum is reminiscent of a matter wave coherently scattered by a periodic lattice irradiated by an ingoing plane wave. It could describe for instance the wave function of a neutron elastically scattered by a crystal in the Born approximation (each lattice site scatters the wave once). The $\mathbf{p}$ vector represents then the momentum transfer during the scattering process and $\overline{\phi_0}(\mathbf{p})$ is the form factor describing the angular dependence of the scattering by a single lattice site. The partial waves originating from the different sites interfere with each other, as long as it is impossible to distinguish in the crystal which nucleus had scattered the neutron. After recording many single-neutron events, one obtains a Bragg scattering pattern of bright and dark spots corresponding to the directions where the interference is respectively constructive or destructive. The diffraction pattern can be used to reconstruct by Fourier transform the spatial distribution of nuclei in the scattering crystal.

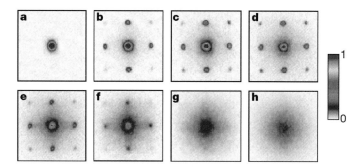

Fig. 9.11 Bosons in a periodic lattice: observation of the Mott transition. Absorption image of the atomic cloud after a free expansion of the sample. The depth of the optical lattice increases from (a) to (h). Reprinted by permission from Macmillan Publishers Ltd: Nature, Greiner *et al.* (2002a).

Figure 9.11(a)–(h) shows in different snapshots this pattern reminiscent of Bragg scattering, observed for increasing interwell separations, in an experiment performed on a ^{87}Rb condensate trapped in an optical lattice (Greiner *et al.* 2002a). Raising the barrier between the sites is realized by increasing the intensity of the lasers producing the optical lattice. This results in a decrease of the tunnelling rate $\mathcal{J}$, and to an increase of the interatomic coupling g, as the individual wells are made steeper and better confining for the single-atom wave functions. The ratio $\xi = g/\mathcal{J}$ is thus progressively increased from left to right and from top to bottom in Fig. 9.11. For each value of this ratio, the condensate is left to reach equilibrium, then released and observed by laser absorption after an expansion delay. The Bragg-like pattern becomes first more and more visible as the periodic structure is progressively established on the condensed cloud. It reaches a maximum contrast and then, around a value of $g/\mathcal{J}$ close to 35, it becomes progressively washed out, being replaced by a diffuse scattering spot (Fig. 9.11g). The vanishing of the Bragg pattern signals the change of the scattering process which evolves from a coherent regime described by eqn. (9.74) into an incoherent one, described by eqn. (9.75). This is a very spectacular signature of the Mott transition.

This evolution is reversible. If the interwell barriers are raised to a value larger than the critical one, then lowered, the sharp diffraction pattern reappears. The transition can thus be crossed adiabatically in both ways, which is a clear demonstration of the unitary evolution of the system. The passage from the superfluid to the insulator state is an adiabatic transformation, the system remaining at all times in the pure quantum ground state of a Hamiltonian whose parameters are continuously changed. The localization process occurring here is a quantum phase transition, different from a classical transition transforming a statistical system from a thermodynamic state into another one.

The diffuse image, characteristic feature of the condensate deep into the insulator phase, is also reminiscent of a scattering process, but an incoherent one. Equation (9.75) could describe the intensity of a neutron beam scattered in a non-elastic process by the nuclei in a crystal. Suppose for instance that the scattering event is accompanied by a spin-flip of the scattering nucleus. It would then be possible, at least in principle, to tell from which site each neutron has been scattered, by analysing the spin states left behind in the crystal. The existence of this which-path information, even if it is not actually read, is enough to destroy the interference. The complementarity argument we have developed in previous chapters when analysing interferometry experiments in quantum optics or cavity QED, apply to scattering theory and imply a fundamental distinction between coherent and incoherent scattering.

Complementarity can also be invoked in the condensate experiment. In the superfluid phase, the large atomic fluctuations make it impossible to find out, even in principle, from which well was originating a detected atom. Hence, the amplitudes associated to the different wells must interfere. Deep into the insulator phase, the atomic fluctuations are frozen. Nothing prevents us, in principle, to count the atoms in each well before and after an atom is detected. From these counts, one could infer from which well each atom was coming and, thus, the different amplitudes cannot interfere. Probabilities, not amplitudes, must be added. We understand again the connexion between phase definition on the one hand and fluctuations in the atom number partition on the other hand.

This experimental study of a condensed atomic cloud trapped in a lattice confirms in a spectacular way the Bose–Hubbard model. To make the comparison between experiment and theory complete, we should add a final remark. The atomic cloud density given by eqn. (9.73) refers to an ensemble average, whereas in the experiment, each snapshot is recorded in a single shot. Generalizing the double condensate discussion of Section 9.3, we could expect that each picture of the condensate in the insulator regime would exhibit a Bragg structure with resolved peaks, changing randomly form one shot to the next. The density averaged over many shots would then be in agreement with eqn. (9.75). This is not what is observed in Fig. 9.11 in which each image well above the Mott transition is diffuse. We thus still have to understand why, contrary to what happens in the interference of two independent condensates, the phase between the micro-condensates in the insulator regime does not build up during the atomic detection process.

We will not enter here in a detailed discussion of this interesting question. Let us just mention that several effects are expected to contribute to the structureless single shot condensate images, deep into the insulator phase. The detections of the first atoms should pin down the phase between microcondensates on different sites, generating a random speckle-type pattern for each plane of the expanded lattice normal to the detection laser. The complete washing out effect observed in Fig. 9.11 well above the Mott transition results from an averaging over the random speckle patterns corresponding to different planes. Similar experiments carried out on a one-dimensional array of 30 independent condensates (Hadzibabic *et al.* 2004) have shown that single-shot interference is clearly observed and well-understood with a simple model generalizing the analysis of the interference of two condensates made in Section 9.3.2.

9.5.4   Collapse and revival of Bragg scattering: towards cats again

In the Mott transition experiment, the atomic ensemble adjusts adiabatically to the change of the optical lattice. At the time when the condensate is released, it is in equilibrium in the ground state of the Bose–Hubbard Hamiltonian corresponding to a

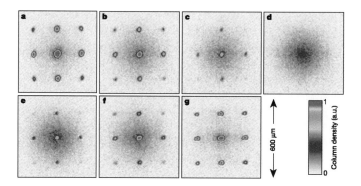

Fig. 9.12 Bosons in a periodic lattice: observation of collapses and revivals of the Bragg pattern. The evolution time increases from (a) to (g). Reprinted by permission from Macmillan Publishers Ltd: Nature, Greiner *et al.* (2002b).

well-defined $\xi = g/\mathcal{J}$ value. In a related experiment (Greiner *et al.* 2002b), a condensate initially prepared in the superfluid phase has been exposed to a sudden change of the lattice configuration. Starting from a situation in which $\xi < \xi_c$, the standing laser waves have been rapidly raised to a value corresponding to $\xi > \xi_c$ in a time so short that the condensate could not evolve. A superfluid state was in this way submitted suddenly to a Hamiltonian admitting insulator eigenstates. We have already encountered this situation in Section 9.4 where we dealt with a double-well configuration. To study the system's evolution, we must expand the initial superfluid state along the basis of the insulator states, which are the system's eigenstates in the final lattice configuration. Each of these Fock states evolves at a Bohr frequency multiple of g. The beating between these frequencies produces a phase spreading which leads first to a disappearance of the interwell correlations $\langle a_i^\dagger a_j \rangle$. The coherence collapse time $T_{\text{collapse}}^{\text{sf}}$ can be estimated as in the double-well case. A similar analysis yields:

$$T_{\text{collapse}}^{\text{sf}} \approx \frac{\pi}{2g\sqrt{2n_0}} \ . \tag{9.77}$$

The mean number of atoms per lattice site pair, $2n_0 = 2\mathcal{N}/\mathcal{N}_w$ merely replaces the number of atoms $\mathcal{N}$ in the double-condensate situation (compare eqns. 9.77 and 9.52).

As in the two-well case, the interwell coherence is expected to revive periodically. Whereas the first revival time was π/g for two wells (eqn. 9.54), it is now:

$$T_{\text{revival}}^{\text{sf}} = 2\pi/g \ . \tag{9.78}$$

To explain this difference, we write the coherence $\langle a_i^\dagger a_j \rangle$ between two adjacent wells, in the interaction picture, as:

$$\langle a_i^\dagger a_j \rangle = \sum_{\mathcal{N}_i, \mathcal{N}_j} C_{\mathcal{N}_i, \mathcal{N}_j} e^{ig(\mathcal{N}_i - \mathcal{N}_j + 1)t} \ , \tag{9.79}$$

where $\mathcal{N}_i$ and $\mathcal{N}_j$ are the fluctuating numbers of atoms in the wells and $C_{\mathcal{N}_i, \mathcal{N}_j}$ are stationary coefficients that we need not precise. Since we have a large number of sites

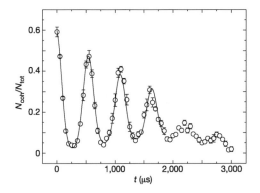

Fig. 9.13 Bosons in a periodic lattice: contrast of the Bragg pattern versus holding time in the insulator phase. Reprinted by permission from Macmillan Publishers Ltd: Nature, Greiner *et al.* (2002b).

with quasi-uncorrelated fluctuations of the atom number in adjacent wells, there are in the sum odd and even $\mathcal{N}_i - \mathcal{N}_j$ values. The coherence thus revives when $gt = 2\pi$. For $\mathcal{N}_w = 2$, this argument does not apply. The difference $\mathcal{N}_j - \mathcal{N}_i = 2\delta\mathcal{N}$ has a given parity (equal to that of $\mathcal{N}$), which reflects maximal correlation between the wells populations. In this case, the coherence revives for $gt = \pi$. This explains the factor of two between T_{revival} and $T_{\text{revival}}^{\text{sf}}$.

The periodic revival of the superfluid phase has been observed (Greiner *et al.* 2002b). After increasing suddenly the optical lattice intensity, the condensate is left to evolve freely for a time t_{hold}, then rapidly released and the resulting atomic cloud is imaged after a fixed expansion time. The time t_{hold} is then changed and a new picture is taken under the same conditions. Figure 9.12(a)–(g) shows a succession of images corresponding to increasing t_{hold} values. For short holding times, the images reveal a distinct Bragg pattern which progressively vanishes, being replaced by a diffuse image. At longer time, the Bragg pattern reappears, to vanish again and so on. Figure 9.13 shows the periodic variation of the Bragg pattern contrast as t_{hold} varies, exhibiting clearly the succession of collapses and revivals of the superfluid phase. The frequency of the revivals, $g/2\pi$, is about $2 \cdot 10^3$ Hz. Its order of magnitude is in good agreement with the evaluation of the collisional interaction in the tight confinement configuration of an optical lattice (Section 9.4). Note that the experiment is performed with a number of atoms per site of the order of 2. The small value of n_0 makes the collapse and revival stages not well-resolved in time. Instead of the sudden revival bursts predicted by theory when the number of atoms per well is large, we observe rather a continuous evolution of the Bragg contrast.

Following the analysis of Section 9.4, we also expect that, in between a collapse and a revival of the Bragg pattern, we should get Schrödinger-cat-like states in our condensate out of equilibrium. The situation, which we will not analyse here, is however more complex than in the double-well case, since it involves an entanglement between states pertaining to all the wells of the lattice.

This experiment demonstrates in a spectacular way that ultra-cold atom collisions are coherent phenomena which can be controlled to manipulate the evolution of a large ensemble of atoms. We show in the final section of this chapter that these collisions can be employed to achieve specific tasks in quantum information.

9.6 Entangling collisions in a Bose–Einstein condensate

Collisions between ultra-cold atoms have remarkable properties. They induce, as we have seen, reproducible and controllable phase shifts on atomic quantum states. Very soon after the birth of Bose–Einstein condensation physics, it was proposed (Jaksch *et al.* 1999) to exploit these collisional shifts in order to build quantum gates and to engineer entanglement in condensates trapped in potential wells. In this respect, cold atom collisions can be compared to the Rydberg atom collisions mediated by a cavity discussed in Chapter 5. In both cases, a process which is usually considered as generating randomness is made coherent and controllable by imposing constraining boundary conditions. It can then be harnessed to perform conditional dynamics and quantum logic operations. Cold collisions in optical lattices have an advantage over cavity-mediated collisions. They can be immediately generalized to entangle in a single operation a large ensemble of qubits. The possibilities opened by registers made of cold two-level atoms confined in one-, two- or three-dimensional optical lattices have raised a lot of interest in quantum information science. Proposals have been made to realize massive entanglement or to emulate complex quantum systems of interest in condensed matter physics with ensemble of cold bosons sitting in a periodic well structure and manipulated by collision-induced logical gates (Briegel *et al.* 2000).

It has not taken long for these ideas to be implemented in the laboratory. The set-up used to study the Mott transition in a lattice have been turned in entanglement factories for ensemble of cold atoms (Mandel *et al.* 2003). In addition to being promising for quantum information processing, these beautiful experiments open the way to fundamental physics studies in which entanglement can be used to perform interferometric studies and precision measurements of atomic scattering properties. We devote this last section to a brief description of this fascinating field of atom optics. This discussion will be greatly helped by analogies with the physics discussed in previous chapters. We will see that these matter-wave-in-a-box experiments share many conceptual ideas with the photon-in-a-box experiments.

9.6.1 A cold atom two-qubit collisional gate

Let us consider a pair of two-level bosonic atoms trapped in two identical potential wells which can be translated independently, superimposed or separated in a controlled way. To be specific, this system can be realized with two ^{87}Rb atoms in a one-dimensional optical lattice. The internal states of each atom, noted $|0\rangle$ and $|1\rangle$, to adopt explicitly a quantum information language, are two magnetic hyperfine sublevels. We will call them 'spin states' or 'qubit states' and need not specify them any more here. We will consider situations in which the atoms are brought into close contact in a single well only if they are in different spin states. The interaction between them is then measured by the cross-collisional rate g_{01} proportional to the scattering length $a_{s,01}$ associated to collisions between atoms in $|0\rangle$ and $|1\rangle$ states. This scattering length is of the same order as that corresponding to collisions between atoms in the same state. For a tight confinement ($a_{\text{ho}} \approx 70$ nm), we have typically $g_{01} \approx 10^4$ s^{-1}.

It is essential in these experiments that the potentials felt by the two spin states could be translated in space independently from each other. A simple way to achieve this is to play with the relative polarizations of the two counter-propagating beams making up the standing wave. We rely on the fact that the optical force exerted by a

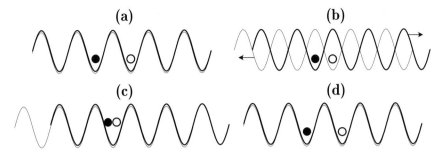

(a) (b)

(c) (d)

Fig. 9.14 Scheme showing the motion of the spin-dependent lattices in the gate operation when the initial qubit state is $|1,0\rangle$. The solid circle and the thick line represent the first atom in $|1\rangle$ and its lattice. The open circle and the thin line represent the second atom in $|0\rangle$ and the corresponding lattice. (a) Initial situation. (b) The two atoms are moved in opposite directions towards each other (to the right for the first atom, to the left for the second). (c) The atoms interact in the same lattice site after a $\lambda/2$ relative translation. (d) The atoms are returned to their original positions.

polarized laser beam on an atom is level-dependent. The optical transitions exciting the atom from the two spin states have different polarizations and couple thus with different strengths to the polarized standing wave. We assume that the 0 and 1 spin states are selectively sensitive to left- and right-circularly polarized light and that the standing wave is made of two linearly polarized beams propagating along Oz in opposite directions, with polarizations making an angle θ. The total field of the light wave is then a superposition of a right- and a left-circularly polarized standing wave which can be shifted with respect to each other by changing θ. This results immediately from:

$$\mathbf{u}_x e^{ikz} + (\mathbf{u}_x \cos\theta + \mathbf{u}_y \sin\theta)e^{-ikz} =$$
$$e^{-i\theta/2}(\mathbf{u}_x + i\mathbf{u}_y)\cos(kz + \theta/2) + e^{i\theta/2}(\mathbf{u}_x - i\mathbf{u}_y)\cos(kz - \theta/2) \ , \ (9.80)$$

where $\mathbf{u}_x$ and $\mathbf{u}_y$ are the unit vectors along Ox and Oy and $k = 2\pi/\lambda$ the wave number of the laser beams whose wavelength is λ. The potentials seen by the two spins are respectively proportional to the intensities of the two circularly polarized components, $\cos^2(kz \pm \theta/2)$. These potentials can be superimposed ($\theta = 0$) or separated at will by rotating the polarization of one running wave while keeping the other fixed. The relative translation of one standing wave with respect to the other is $\theta\lambda/2\pi$. After half a turn ($\theta = \pi$), the two periodic potentials are spatially shifted by one interwell separation $\lambda/2$.

Once this movable configuration of standing waves has been set, the two-qubit gate is achieved by the following set of operations (Fig. 9.14). We start by placing the two atoms in tightly confining adjacent wells, with negligible tunnelling between them. The spin-dependent 0- and 1-lattices are then rapidly translated by one interwell interval $\lambda/2$ with respect to each other, making them coincide again. We assume that the 0-lattice moves to the left and the 1-lattice to the right. The lattices are held in this configuration for an adjustable time t_{hold}, then shifted back to their initial position. The displacements of the lattices occur in a time short compared to g^{-1}, but long

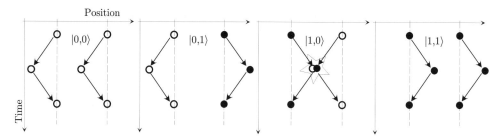

Fig. 9.15 Relative atomic motions for the four computational basis states of the two atoms.

compared to ω_{ho}^{-1}. The atoms thus follow adiabatically the lattice motion, without changing their internal or vibrational state. This adiabatic evolution produces, in addition to the spin-dependent collisional phase shifts which interest us primarily here, kinetic phases shifts affecting globally the atomic wave functions of the manipulated atoms (Jaksch *et al.* 1999). These additional shifts play no role in the entanglement process and will be disregarded in the following.

Figure 9.14 represents the two-atom evolution in only one of their spin states configurations. Conditional dynamics is achieved because this evolution is spin-dependent. We show in Fig. 9.15 the atomic trajectories corresponding to the four possible initial qubit states $|0,0\rangle$, $|0,1\rangle$, $|1,0\rangle$ and $|1,1\rangle$. If they are both in $|0\rangle$ or in $|1\rangle$, they move in the same direction, staying at a constant distance $\lambda/2$ from each other. If they are in $|0,1\rangle$ they move away from each other, their mutual distance increasing to λ. It is only if they are in the $|1,0\rangle$ state – the situation depicted in Fig. 9.14 – that they are moved towards each other and merged into the same well. During the holding time t_{hold} this state acquires in this case a quantum phase $\phi = g_{01}t_{\text{hold}}$ which reaches π for a typical t_{hold} of a few hundred μs. We thus have:

$$|00\rangle \to |00\rangle \, ; \quad |01\rangle \to |01\rangle \, ; \quad |10\rangle \to e^{-i\phi}|10\rangle \, ; \quad |11\rangle \to |11\rangle \, . \qquad (9.81)$$

These transformations describe a control-U_ϕ gate in which the left and right qubits are respectively the control and the target and U_ϕ is the single-qubit unitary transformation defined as:

$$U_\phi = \frac{e^{-i\phi}+1}{2}\mathbb{1} + \frac{e^{-i\phi}-1}{2}\sigma_Z \, . \qquad (9.82)$$

For $\phi = 2p\pi$ (p integer), the gate leaves the system invariant. For $\phi = (2p+1)\pi$ the gate performs the transformation $|\psi\rangle \to -\sigma_Z|\psi\rangle$ on the target, provided the control is in state $|1\rangle$. This is – up to a sign change – the control-π phase gate.

Entanglement is realized by preparing initially the qubits in a state superposition. This can be achieved, after initialization in $|0,0\rangle$, by submitting them to microwave pulses coherently mixing the hyperfine atomic spin states. We restrict our discussion here to situations in which the two qubits are simultaneously manipulated by the same pulse. This can be easily realized, whereas a single-qubit addressing is impractical when the two atoms are so closely spaced. After applying a $\pi/2$-pulse transforming $|0\rangle$ into $(|0\rangle+|1\rangle)/\sqrt{2}$ and operating the collisional gate, the two atoms have evolved according to:

$$|0,0\rangle \to \left|\Psi_\phi^{2\,\text{bit}}\right\rangle = \frac{1}{2}\left[|0\rangle\left(|0\rangle+|1\rangle\right) + |1\rangle\left(e^{-i\phi}|0\rangle+|1\rangle\right)\right] \, , \qquad (9.83)$$

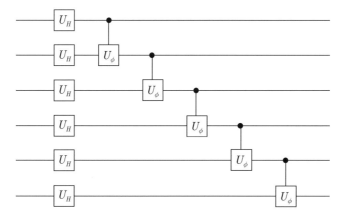

Fig. 9.16 Quantum circuit symbolizing the operation applied to an array of qubits undergoing controlled collisions in a one-dimensional lattice. U_H is the Hadamard gate preparing each qubit in the $(|0\rangle + |1\rangle)/\sqrt{2}$ state.

a state which can be expressed as the sum of a separate state of two qubits and a maximally entangled Bell type state:

$$\left|\Psi_\phi^{2\,\text{bit}}\right\rangle = \left(\frac{1 + e^{-i\phi}}{2}\right)\left(\frac{|0\rangle + |1\rangle}{\sqrt{2}}\right)\left(\frac{|0\rangle + |1\rangle}{\sqrt{2}}\right) + \left(\frac{1 - e^{-i\phi}}{2}\right)|\Psi_{\text{Bell}}\rangle \,, \qquad (9.84)$$

with:

$$|\Psi_{\text{Bell}}\rangle = \tfrac{1}{2}\left[|0\rangle\,(|0\rangle + |1\rangle) - |1\rangle\,(|0\rangle - |1\rangle)\right] \,. \qquad (9.85)$$

As the collisional time t_{hold} is increased, the system evolves periodically into a separate state (for $\phi = 0, 2\pi, 4\pi, \ldots$) and a maximally entangled Bell state (for $\phi = \pi, 3\pi, \ldots$). We will see below how this entanglement oscillation can be observed in an atomic interferometry experiment. Let us consider before how this entangling procedure generalizes to more qubits.

9.6.2 Multiple gates in a one-dimensional lattice

Assume now that the periodical lattice with spin-dependent translation capability is loaded with one atom per lattice site. The system then looks like a register made of $\mathcal{N}_w$ equidistant qubits. After submitting each atom initially in state $|0\rangle$ to a $\pi/2$-pulse, we perform the sequence of spin-selective operations described above. We put transiently in contact each atom with the one at its right, provided the pair is in $|1, 0\rangle$. This amounts to operating an ensemble of $\mathcal{N}_w - 1$ cascading control-U_ϕ gates in which each atom acts as a control for the atom located immediately at its right. Figure 9.16 shows a quantum circuit representation of the system evolution, with the convention that the left atom is the top qubit in the register.

This circuit realizes in general a massive entanglement of the $\mathcal{N}_w$ bits in a complex 'cluster' state (Raussendorf and Briegel 2001). We will not give the general expression of this state, but consider the simple $\mathcal{N}_w = 3$ case. A straightforward calculation shows that the three atoms evolve as $|0, 0, 0\rangle \rightarrow \left|\Psi_\phi^{(3\,\text{bit})}\right\rangle$, with:

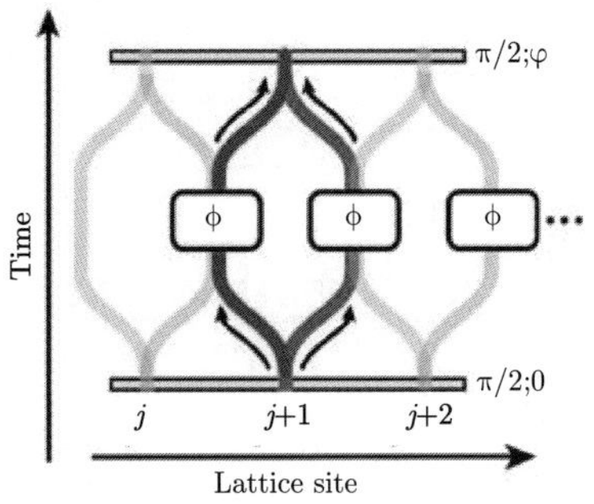

Fig. 9.17 Space–time representation of the Ramsey interferometer used to reveal the collisional qubit entanglement. Reprinted by permission from Macmillan Publishers Ltd: Nature, Mandel *et al.* (2003).

$$\left|\Psi_\phi^{(3\text{ bit})}\right\rangle = \left(\frac{1+e^{-i\phi}}{2}\right)\left(\frac{|0\rangle+|1\rangle}{\sqrt{2}}\right)\left(\frac{|0\rangle+|1\rangle}{\sqrt{2}}\right)\left(\frac{|0\rangle+|1\rangle}{\sqrt{2}}\right)$$
$$+ \left(\frac{1-e^{-i\phi}}{2}\right)|\Psi_{\text{GHZ}}\rangle \;, \tag{9.86}$$

and:

$$|\Psi_{\text{GHZ}}\rangle = \frac{1}{2\sqrt{2}}\left[(|0\rangle-|1\rangle)\,|0\rangle\,(|0\rangle+|1\rangle) - (|0\rangle+|1\rangle)\,|1\rangle\,(|0\rangle-|1\rangle)\right] \;. \tag{9.87}$$

These formulas generalize to three particles the results given for a pair of bits by eqns. (9.84) and (9.85). The state $|\Psi_{\text{GHZ}}\rangle$ is a three-particle entangled state of the Greenberger–Horne–Zeilinger (GHZ) type which we have already encountered in CQED and ion trap physics. The three-qubit system oscillates, when ϕ is increased, between a separate state ($\phi = 2p\pi$) and a GHZ state [$\phi = (2p+1)\pi$]. Note that when the atoms are in the GHZ state, each qubit is maximally entangled with the system made of the two others. Without entering into a detailed description of the $\mathcal{N}_w > 3$ registers, we will only notice that the cluster states share these important features with the $\mathcal{N}_w = 2$ and $\mathcal{N}_w = 3$ cases. When $\phi = 2p\pi$, all the qubits are in separate states while when $\phi = (2p+1)\pi$, each qubit is maximally entangled with the ensemble consisting of all the others. We leave as an exercise the inductive proof of this simple result which will be useful to remember in the following.

9.6.3 Collisional entanglement revealed by interferometry

In order to probe the entangled state produced in the linear qubit register, it is convenient to perform a Ramsey interference experiment on the cold atom cloud. Let us analyse it first in the case of two qubits. After the sequence of events leading to the state described by eqn. (9.84), we apply a last $\pi/2$-microwave pulse with a variable phase φ and we finally detect the atoms selectively in states $|0\rangle$ and $|1\rangle$. In the absence of entangling collisions, the probability of finding each atom in one of these levels undergoes oscillations versus φ, which are due to an interference between two paths in which each atom is either in $|0\rangle$ or in $|1\rangle$ between the two pulses (see Section 3.3.3).

In the presence of atomic interactions, the contrast of this interference signal depends crucially upon the amount of entanglement produced by the collisions between the two atoms. If there is no entanglement, the Ramsey signal has, ideally, a 100%

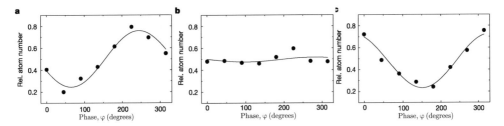

Fig. 9.18 Ramsey fringes observed for three different holding times. Reprinted by permission from Macmillan Publishers Ltd: Nature, Mandel *et al.* (2003).

contrast. In the case of maximal entanglement the contrast vanishes. The simple interpretation is that which-path information about the state of one atom is stored in the other. We retrieve again the very general complementarity argument which relates fringe visibility to the amount of entanglement between the system being investigated in an interference experiment and its environment. This argument obviously generalizes to multi-particle entanglement in a linear register of qubits. Ramsey fringes with a large contrast are expected if each atom is separable from all the others, which occurs when ϕ is a multiple of 2π. The fringes should vanish when each qubit is maximally entangled with all the others, which occurs when ϕ is an odd multiple of π.

This experiment has been carried out as a follow-up of the Mott transition and Bragg scattering collapse and revival experiments described above. We will not repeat the description of the set-up, and not discuss its many tricks, inessential to understand the basic features of the experiment. Details can be found in Mandel *et al.* (2003). At the start of the experiment, a three-dimensional lattice is loaded with a Bose–Einstein condensate containing, in the centre, about 1 atom per site. The polarization-dependent standing waves of this lattice are moved back and forth in one direction, realizing a large number of parallel registers which are simultaneously and massively entangled. The schematics of the multiple gate sequences applied to each of this linear registers is shown, as a space–time diagram, in Fig. 9.17. The splitting of the atomic trajectories by the spin-dependent transport and their recombination is indicated by the arrows. The contacts between the wave packets with different spin states, responsible for the conditional dynamics, are symbolized by the boxes. Two horizontal double lines represent the $\pi/2$-pulses, with phases zero and φ, making up the Ramsey interferometer. The detection process is finally performed by releasing the gas from the trap, letting it expand for a few milliseconds, then absorbing it by a laser beam selectively sensitive to the two-qubit states. The total atomic population in state $|0\rangle$ and $|1\rangle$ is recorded, without spatial resolution of the atomic cloud.

Figure 9.18 shows the relative population in one of the two states, versus the phase φ of the second Ramsey pulse. This signal corresponds, from left to right, to three increasing collisional times t_{hold}. The variation of the fringe contrast reveals the oscillation of the entanglement produced by the controlled collisions in the linear qubit registers. The first vanishing of the fringes is observed for $t_{\mathrm{hold}} = 210~\mu$s, in good agreement with the expected time π/g_{01}. Alternatively, the observation of these entanglement oscillations provide an elegant way to access to g_{01} and, knowing a_{ho} and ω_{ho}, to measure directly the scattering length a_{s01}. In a subsequent experiment,

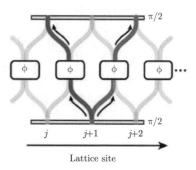

Fig. 9.19 Space–time representation of the Young spatial interferometer used to reveal the collisional qubit entanglement. Reprinted by permission from Macmillan Publishers Ltd: Nature, Mandel *et al.* (2003).

the same group has used this method to study the variations of the scattering length as a function of an applied magnetic field around a Feshbach resonance (Widera *et al.* 2004).

Instead of recording interference fringes versus time, it is also possible to perform a spatial interference experiment which reveals, in a very striking way, the entanglement dynamics in this qubit system (Mandel *et al.* 2003). In a variant of the sequence of operations described above, the experimenters have delocalized further the atoms after the collision time, moving the spin-dependent lattices in the same direction as before the collision, instead of bringing them back to the initial position. This operation is schematized in Fig. 9.19. Each atom, originating from a given lattice site, is now split into two wave packets finally separated by λ. After applying the second $\pi/2$-pulse, it is impossible to find out, even in principle, if the atom was in state $|0\rangle$ or $|1\rangle$ in between the two pulses and hence, whether it is coming from the right or from the left. The situation is reminiscent of a Young double slit interferometer where a single-particle wave function is split into two spatially separated wave packets whose trajectories cannot be determined.

In the final stage of the experiment, the condensate is released from the trap and expands until it is absorbed by a probe laser beam imaging its shadow on a screen, in a single shot. The wave packets coming from sites separated by λ recombine and give rise to an interference pattern quite similar to a double-slit interference signal. This interference results from the combination of all the single-particle waves which have been spatially split and recombined in the same way. The phase of the fringes is opposite for atoms detected in $|0\rangle$ and $|1\rangle$, which means that the final imaging must be state-selective. The interferograms obtained for increasing collision times t_{hold} are shown in Fig. 9.20. We again observe an oscillation of the fringe contrast, with maximum visibility when ϕ is close to 0 and 2π and vanishing contrast for $\phi = \pi$. Here again, complementarity is the key to understand the effect. The Young interference is, in essence, even in this multi-atom system, a single-particle phenomenon. Each single-particle wave packet interferes with itself. If an arbitrary atom in the ensemble is maximally entangled with all the others, they constitute an environment storing unambiguous information about the atom path and there is no fringes.

These beautiful experiments prove the existence of a controlled entanglement in the system, but are not specifically sensitive to the many-particle nature of this entanglement. As we have seen, the signals are well described by a two-atom model.

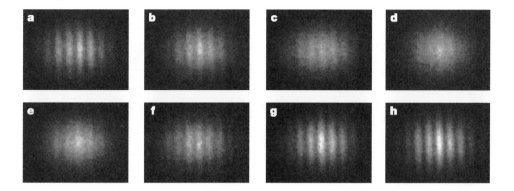

Fig. 9.20 Young type interferograms revealing the entanglement oscillations. Frames (a) to (h) correspond to increasing collision times. Reprinted by permission from Macmillan Publishers Ltd: Nature, Mandel *et al.* (2003).

Experiments to test the high degree of entanglement in the cluster qubit states would require the capacity to address individual lattice sites and perform correlated measurements on them.

Note also that this test of collisional entanglement is not particularly sensitive to the homogeneity of the lattice well filling. If there are vacancies in the lattice, subsets of directly connected qubits will contribute separately to the signal, making it impossible to recognize whether the filling of the qubit sites is uniform or not. We should remark however that the near vanishing of the interference for $\phi = \pi$ allows us to set an upper limit to the fraction of lattice sites having no direct neighbours. These isolated atoms, which cannot be entangled, produce a full contrast Ramsey signal, contributing to a non-cancellable fringe background for $\phi = \pi$. The small value of this background indicates that no more than $\approx 5\%$ of the atoms are isolated in the lattice. Multiple atom occupation of a single lattice site are also rare in this experiment. If a fraction of the sites has a double occupancy, these atoms remain permanently in collisional contact. The differences between the same spin and different spin scattering lengths produces in this case a loss of fidelity of the gate operation which we will not discuss here.

This brief account of controlled collisional entanglement in boson gases provides only a glimpse of the many possibilities offered by the manipulation of cold atoms in optical lattices, for quantum information or fundamental studies as well.[5] Very ingenious schemes have been proposed to use entanglement-sensitive signals in order to prepare by a filtering process qubit registers with homogeneous occupation throughout the lattice. Moving an individual atom in the lattice as a quantum reading and printing head coupled successively to different atoms could be used to build up quantum algorithms. Entanglement of atoms with clusters of qubits in their neighbourhood could implement error correction schemes. For a review, see for instance Briegel *et al.*

---

[5]We discussed here a 'top-down' approach in which the lattice is filled by a large number of atoms. A 'botton up' approach, in which atoms are inserted one by one in a multiple-well structure is also actively investigated (Kuhr *et al.* 2003; Beugnon *et al.* 2006).

(2000). Even without addressing individually the qubits, collective collisional effects could be exploited to simulate a wide class of quantum Hamiltonians relevant to condensed matter physics problems (Garcia-Ripoll *et al.* 2004). The variety of situations offered to the imagination of theorists and challenging the experimentalists is mind-bogging, especially if one includes the recent extension of these experiments to cold fermion systems.

10
Conclusion

Prediction is very difficult, especially about the future
attributed to N. Bohr

Juggling with atoms and photons in cavities has led us into a detailed exploration of the quantum laws. It has been a precious guide to understand fascinating experiments performed with trapped ions and cold atoms. In various disguises, all these studies feature spins and oscillators in interaction. These systems, with their simple Hamiltonians, lend themselves to analytical calculations. Their mathematics remains elementary, and yet their physics exhibits a wealth of subtle effects which illustrate the logic of the microscopic world.

The photon and atom boxes of this book are the actual realizations of the elaborate contraptions that the fathers of the quantum theory kept imagining in their thought experiments. Einstein, Bohr and their famous colleagues of the historical Solvay meetings were able to describe with accuracy what would happen if single particles were placed in such unusual situations, but they were far from believing that these experiments could be carried out one day in a laboratory. As we have seen above, Schrödinger has even alluded to the 'ridiculous' consequences that such experiments would entail.

This is indeed one of the fascinating paradoxes of quantum physics. Its laws have been discovered from very indirect hints at a time when the microscopic world was veiled. The founding fathers of the theory would certainly have guessed right – if reluctantly for some – the outcome of all the experiments we have described in these pages. They would probably have found highly improbable that such experiments could one day be carried out.

This paradox illustrates a constant and deep feature of scientific research. As we have noticed in the first chapter, the realization of thought experiments has been a consequence of technological advances that only the knowledge of the quantum laws has made possible. But these advances could not be foreseen by the founders of the theory, whatever bright and perceptive they were. Fundamental science is driven by curiosity alone, not by the search for applications, that one is even unable to imagine. For Schrödinger to foresee that atoms could be manipulated one at a time would have meant to envision the transistor, the computer and the laser as direct consequences of his famous equation. And he should have recognized that a combination of these devices could be used to achieve the intricate juggling games described in the previous chapters. This is obviously not the way scientific knowledge is progressing. And this is certainly better that way. There is something exhilarating in thinking that what we are doing today in the laboratory would be a surprise to Einstein, Schrödinger and, if we are lucky, even to Bohr!

Having stressed the difficulty of long-term predictions puts us in an awkward position to discuss the future of this physics. History tells us that it is dangerous to claim that some things are impossible, not to say 'ridiculous'. This does not mean, on the contrary, that everything is possible and will happen, given enough time and money. Let us simply mention interesting studies related to those described in this book and try to extrapolate them into the near future. Beyond this, physics is like weather, subjected to unpredictable events, which makes a physicist's life so exciting and full of surprises.

The physics of two-level systems and quantum oscillators finds a new and fascinating domain of application in mesoscopic physics (Imry 1997), an expanding field of research for quantum information which we have not covered in this volume, but which is worth mentioning when trying to get a glimpse at the future. Artificial atoms can now be made by confining electrons in quantum dots which are produced on solid substrates by lithographic techniques. These dots have discrete energy levels between which optical transitions can be driven. These systems are considered as qubit candidates in various laboratories (Elzerman *et al.* 2004; Kroutvar *et al.* 2004; Petta *et al.* 2005; Atatüre *et al.* 2006).

Another line of research studies the properties of microscopic superconducting circuits including Josephson junctions. The charge, magnetic flux and current in these devices are quantum variables whose evolution is ruled by a Hamiltonian exhibiting a discrete spectrum. Clever adjustments of critical parameters make it possible to isolate two states in these systems and to drive them in coherent superpositions. Single-qubit operations and two qubit gates have been realized in this way (Vion *et al.* 2002; Chiorescu *et al.* 2003; Yamamoto *et al.* 2003; McDermott *et al.* 2005). Superpositions of currents made of large electron numbers and circulating in opposite directions in small circuits have been observed (van der Wal *et al.* 2000).

In a remarkable development, a superconducting qubit has been coupled to a stripline made of parallel conductors, realizing a lump circuit equivalent to a cavity for microwave photons in the few GHz range. This circuit–QED experiment, as it has been dubbed, presents striking similarities with the CQED studies described in this book. Rabi splitting (Wallraff *et al.* 2004) and level shifts (Schuster *et al.* 2005) of the qubit induced by discrete photons in the cavity have been observed in experiments which hold very interesting promises for the development of exquisitely sensitive microwave detectors.

These miniaturized condensed-matter physics systems are very attractive for quantum information since one can apply to them the fabrication procedures which have been developed so successfully for integrated circuits in microelectronics. Building an array of Josephson junctions coupled to each others requires proper masks in the lithographic process and is not much more difficult than realizing only one or two circuits. We should however keep in mind that such man-made atoms lack a very important property of actual atoms or ions. They are not identical, their level structure depending on continuously variable parameters linked to the size and shape of the circuits. Tuning an array of such qubits for collective coherent operations will certainly not be a trivial task.

Decoherence is also a very important issue, limiting the performances of quantum information with these systems. They are *a priori* less well-isolated from their environment than atoms or ions in quantum optics experiments. The coupling of quantum dots to phonons (Vasanelli *et al.* 2002) and of Josephson circuits to stray electric and magnetic fields limits the time over which qubits built with such systems can be operated to a few microseconds at best. Many studies are aiming at elucidating these decoherence processes and at controlling their effects.

In other promising experiments, mesoscopic physics devices are now being combined with single-particle physics to build hybrid systems exploring quantum properties in novel directions. It is for instance proposed to couple very high Q cantilever oscillators of nanometric size either to a single molecule, or to an electron spin or even to a photon bouncing off a tiny mirror placed at the end of the lever. These devices could be used as very sensitive probes of single spins (Rugar *et al.* 2004) or of tiny mechanical forces. They could even ultimately be exploited to prepare and study superposition of states involving different spatial positions of a mesoscopic object made of a very large number of atoms (Marshall *et al.* 2003; Kleckner *et al.* 2006). Here again, decoherence will be a very challenging effect to overcome.

In a related domain, many experiments are now trapping small samples of laser-cooled atoms in the magnetic field of microchip circuits deposited on a substrate (Folman *et al.* 2002). Evaporative cooling of such samples has led to the realization of micro-Bose–Einstein condensate samples (Hänsel *et al.* 2001) which could be used for the Schrödinger cat studies proposed in the last chapter. These atom-chip experiments could also lead to the development of interferometers in which atoms with large de Broglie wavelengths will be coherently channelled along several interfering paths.

It has also been proposed to turn these microchip devices into rf electric traps for single circular Rydberg atoms (Hyafil *et al.* 2004; Mozley *et al.* 2005). In this case, the set-up needs be cooled at subkelvin temperatures to avoid blackbody radiation and the chip is superconducting. Two Rydberg atoms in nearby traps could then interact with each other via their coupling to a stripline cavity. CQED would then be realized in yet another configuration.

Where does the quantum computer fit in this otherwise bright future? Despite the efforts of theorists and experimenters, it is much too early to claim that this fabulous machine is just for engineers to build, following the blueprint of scientists who would have solved the main fundamental problems. Quantum simulators alluded to at the end of Chapters 8 and 9 will certainly be built in the near future and teach us a lot about quantum phase transitions in condensed matter physics. We are however still very far from the stage of a quantum computer 'Manhattan project'. First, because the number of quantum algorithms for which such a machine would be useful is very limited. And second and foremost because the critical issue of fidelity (reaching the level at which error corrections could make the machine fault-tolerant) is far from being resolved. Trapped ions are currently holding the fidelity record, but this record is still way off the mark. Even its most enthusiastic believers claim that a useful computing machine is still tens of years away. At such a time scale it is most probable that unexpected events and discoveries will chart the path of this research in unforeseen directions.

The issues raised by quantum information research go far beyond the practical advantages of building a powerful computer. These questions address the fundamental problem of the quantum-classical frontier which has haunted physicists since the birth of quantum physics and which we have discussed at length in this book. We have described how to generate and observe Schrödinger cats made of several atoms or photons and we have described ways to delocalize these systems in several places at once. Nobody knows how fat these cats will one day become and which tricks we will eventually be able to teach them to produce the wonderful technology of the future.

Appendix

Quantum states in phase space

Phase space distributions are fundamental tools in classical statistical physics. For a particle undergoing one-dimensional motion, the phase space is a plane, with the position x and conjugate momentum p as coordinates. A point in this plane defines the mechanical state of the particle. In the case of a cavity mode, equivalent to a one-dimensional oscillator, two orthogonal field quadratures play the role of position and momentum, and a point in the Fresnel plane defines a classical field state (Chapter 3). Statistical uncertainties or lack of knowledge about the system are accounted for by replacing the Dirac distribution associated to this point by a positive and normalized density of probability, $f(x, p)$, which takes non-vanishing values in a limited region of phase space. The statistical average $\overline{o}$ of any observable quantity, $o(x, p)$, is given by:

$$\overline{o} = \int f(x, p) o(x, p) \, dx dp \ . \tag{A.1}$$

The system's evolution due to external forces and to the coupling with the environment is described by a diffusion equation for f, relating its temporal and spatial derivatives.

The extension of this phase space representation to quantum states was discussed for the first time by Wigner (1932). A difficulty has immediately arisen because Heisenberg uncertainty relations forbid, even in the absence of any statistical indeterminacy, the precise and simultaneous determination of conjugate variables. In other words, a fundamental quantum blurring adds to the classical uncertainties. In spite of this problem, it remains possible to define a real phase space function for a quantum particle, which retains some of the essential characteristic features of the classical probability distribution. This quantum distribution is called the Wigner function W. Its definition and main properties are recalled in this appendix in the context of the physics of a harmonic oscillator or a field mode.

We will see that W is naturally defined from its Fourier transform, the system's symmetrical characteristic function $C_s(\lambda)$. Introducing C_s will lead us to consider two simply related characteristic functions, $C_n(\lambda)$ and $C_{an}(\lambda)$, and we will see that their Fourier transforms define two phase space distributions different from W, the Husimi-Q function and the Glauber–Sudarshan P distribution. The latter, which is highly singular, is of difficult mathematical use and we will not say much about it. The W and Q functions, which are ubiquitous tools in quantum optics, are used time and again in this book to give pictorial representations of oscillator and field states. Here, we will discuss their connexion, compare them and explain why W is more useful than Q for the analysis of the quantum features of a system. We will also study the main properties of C_n and C_s, two functions which turn out to be very convenient to describe the dynamics of a quantum oscillator undergoing a relaxation process.

This appendix, while presenting the main results in a self-consistent way, is intended as a brief reminder.[1] We start by introducing (Section A.1) the characteristic functions from which the W, Q and P functions are derived. We then describe the Wigner function (Section A.2), give its main properties and a few useful examples. The next section (A.3) is devoted to the Q distribution. The final section (A.4) deals with the evolution of the characteristic functions and W distribution under relaxation. We present a solution of the decoherence master equation for a Schrödinger cat state which is an instructive complement to the calculations presented in Chapter 7.

A.1   Characteristic functions

The phase space density functions W, Q and P are the Fourier transforms of characteristic functions, linked to the quantum average of the displacement operator $D(\lambda) = \exp(\lambda a^\dagger - \lambda^* a)$, in a state described by the density matrix ρ [see Section 3.1.3 for the main properties of $D(\lambda)$]. When operators appear as arguments of a function, we must address the issue of their ordering. A very detailed discussion of this question is found in Cahill and Glauber (1969). Let us mention only here that three simple orderings can be used when expanding in power series a function of the annihilation and creation operators a and $a^\dagger$. In the 'normal' order, all creation operators are placed to the left of the annihilation ones (the photon number operator $N = a^\dagger a$ is in normal order). The 'anti-normal' order puts all the a's on the left. Finally, the 'symmetric' order produces expressions in which all products of operators are symmetrized. For instance, the direct power series expansion of the displacement operator:

$$D(\lambda) = e^{\lambda a^\dagger - \lambda^* a} = \mathbb{1} + \lambda a^\dagger - \lambda^* a + \lambda^2 a^{\dagger 2} - \lambda \lambda^* (a^\dagger a + a a^\dagger) + \lambda^{*2} a^2 + \dots , \quad (A.2)$$

is naturally in symmetric order.

We now introduce three characteristic functions of the state of the system corresponding to these orders. The symmetric-order characteristic function, $C_s^{[\rho]}(\lambda)$, is the average of $D(\lambda)$ in the state described by the density operator ρ:

$$C_s^{[\rho]}(\lambda) = \langle D(\lambda) \rangle = \mathrm{Tr}\left[\rho e^{\lambda a^\dagger - \lambda^* a} \right] . \quad (A.3)$$

The operator $D(\lambda)$ being unitary, all its eigenvalues have a unit modulus. The modulus of its average is thus always bounded by one:

$$|C_s^{[\rho]}(\lambda)| \le 1 . \quad (A.4)$$

Note also the identity:

$$C_s^{[\rho]}(0) = \mathrm{Tr}(\rho) = 1 . \quad (A.5)$$

The upper bound of $C_s^{[\rho]}$ is thus reached at the origin. In addition, $C_s^{[\rho]}(\lambda)$ obeys the conjugation relation:

---

[1]More detailed discussions can be found in quantum optics textbooks: Walls and Milburn (1995); Barnett and Radmore (1997); Scully and Zubairy (1997); Schleich (2001); and Vogel *et al.* (2001).

$$C_s^{[\rho]}(-\lambda) = C_s^{[\rho]*}(\lambda) \ . \tag{A.6}$$

Let us finally give the expression of the symmetric characteristic function for a pure state $|\Psi\rangle$:

$$C_s^{[|\Psi\rangle\langle\Psi|]} = \langle\Psi| D(\lambda) |\Psi\rangle \ . \tag{A.7}$$

It is merely the overlap between $|\Psi\rangle$ and the state obtained by translating $|\Psi\rangle$ in phase space by $D(\lambda)$. It is thus an autocorrelation function of the quantum state in phase space.

The normal- and anti-normal-order characteristic functions are likewise defined as:

$$C_n^{[\rho]}(\lambda) = \mathrm{Tr}\left[\rho e^{\lambda a^\dagger} e^{-\lambda^* a}\right] \ , \tag{A.8}$$

and:

$$C_{an}^{[\rho]}(\lambda) = \mathrm{Tr}\left[\rho e^{-\lambda^* a} e^{\lambda a^\dagger}\right] \ . \tag{A.9}$$

These functions can be related to each other with the help of the Glauber identity:

$$e^A e^B = e^{A+B} e^{[A,B]/2} \quad \text{or} \quad e^{A+B} = e^A e^B e^{-[A,B]/2} \ , \tag{A.10}$$

valid when both A and B commute with $[A, B]$ (eqn. 3.39, on page 115). We obtain immediately:

$$C_n^{[\rho]}(\lambda) = e^{|\lambda|^2/2} C_s^{[\rho]}(\lambda) \ ; \quad C_{an}^{[\rho]}(\lambda) = e^{-|\lambda|^2/2} C_s^{[\rho]}(\lambda) \ . \tag{A.11}$$

Any couple of characteristic functions can thus be determined from the knowledge of the third one by a mere multiplication by $e^{\pm|\lambda|^2/2}$ or $e^{\pm|\lambda|^2}$.

As a simple example, let us determine the characteristic functions for a coherent state $|\alpha\rangle$. We have:

$$C_s^{[|\alpha\rangle\langle\alpha|]}(\lambda) = \langle\alpha| D(\lambda) |\alpha\rangle . \tag{A.12}$$

Recalling that:

$$D(\lambda) |\alpha\rangle = \exp[(\lambda\alpha^* - \lambda^*\alpha)/2] |\alpha + \lambda\rangle \ , \tag{A.13}$$

we get:

$$C_s^{[|\alpha\rangle\langle\alpha|]}(\lambda) = e^{(\lambda\alpha^* - \lambda^*\alpha)/2} \langle\alpha |\alpha + \lambda\rangle = e^{-|\lambda|^2/2} e^{\alpha^*\lambda - \lambda^*\alpha} \ , \tag{A.14}$$

a complex Gaussian function from which we derive:

$$C_n^{[|\alpha\rangle\langle\alpha|]} = e^{\alpha^*\lambda - \lambda^*\alpha} \ , \tag{A.15}$$

and:

$$C_{an}^{[|\alpha\rangle\langle\alpha|]} = e^{-|\lambda|^2} e^{\alpha^*\lambda - \lambda^*\alpha} \ . \tag{A.16}$$

The characteristic functions of a Fock state $|n\rangle$ are obtained in a similar way, by expanding the normal-ordered displacement operator in powers of $a^\dagger$ and a, retaining terms with the same number of creation and annihilation operators which are the only ones to have non-vanishing expectation values in an $|n\rangle$ state . The final result for C_s is (Barnett and Radmore 1997):

$$C_s^{[|n\rangle\langle|n\rangle|]} = e^{-|\lambda|^2/2}\mathcal{L}_n(|\lambda|^2) ,\qquad (A.17)$$

where $\mathcal{L}_n$ is the nth Laguerre polynomial:

$$\mathcal{L}_n(x) = \sum_{p=0}^{n} (-1)^p \frac{n!}{p!^2(n-p)!}x^p .\qquad (A.18)$$

Let us finally mention the symmetric-order characteristic function for a thermal field, obtained by summing the Fock state result over a thermal distribution:

$$\rho_{\text{th}} = \sum_{n} \frac{n_{\text{th}}^n}{(n_{\text{th}}+1)^{n+1}}\,|n\rangle\langle n| ,\qquad (A.19)$$

where n_{th} is the average number of thermal photons. We get a remarkably simple result:

$$C_s^{[\rho_{\text{th}}]}(\lambda) = e^{-(n_{\text{th}}+1/2)|\lambda|^2} .\qquad (A.20)$$

A.2   The Wigner distribution

The Wigner function is defined as the two-dimensional Fourier transform of the symmetric order characteristic function:[2]

$$W(\alpha) = \frac{1}{\pi^2}\int d^2\lambda\, C_s(\lambda)e^{\alpha\lambda^* - \alpha^*\lambda} .\qquad (A.21)$$

Using the notation λ', λ'' for the real and imaginary parts of λ and setting $\alpha = \alpha' + i\alpha'' = x + ip$ to emphasize the analogy between the field quadratures and the normalized dimensionless position and momentum of a particle, we can also write:

$$W(x,p) = \frac{1}{\pi^2}\int d^2\lambda\, C_s(\lambda)e^{2i(p\lambda' - x\lambda'')} .\qquad (A.22)$$

The Wigner function is real, as a direct consequence of the conjugate property of its Fourier transform (eqn. A.6). It is also normalized, which results from the identity:

$$\int d^2\alpha\, W(\alpha) = \frac{1}{\pi^2}\int d^2\alpha d^2\lambda\, C_s(\lambda)e^{\alpha\lambda^* - \alpha^*\lambda} = \int d^2\lambda\, C_s(\lambda)\delta(\lambda) = C_s(0) = 1 ,\qquad (A.23)$$

where we have used an integral definition of the two-dimensional Dirac distribution $\delta(\lambda)$:

$$\delta(\lambda) = \frac{1}{\pi^2}\int d^2\alpha\, e^{\alpha\lambda^* - \alpha^*\lambda}.\qquad (A.24)$$

Before exploring further the properties of W, we will write it in two other equivalent and useful forms.

---

[2]To simplify the notation, we omit for the time being, since no confusion is possible, the indication of the state in the expression of the C_s function and its Fourier transform. We write here $C_s(\lambda)$ instead of $C_s^{[\rho]}(\lambda)$ and $W(\alpha)$ instead of $W^{[\rho]}(\alpha)$. We come back later to the complete notation.

A.2.1   Two equivalent expressions of W

We first make explicit the trace operation in the definition of $C_s(\lambda)$, by using the continuous eigenbasis $\{|x\rangle\}$ of the quadrature operator $X_0 = (a + a^\dagger)/2$:

$$C_s(\lambda) = \int dx' \, \langle x'| \rho D(\lambda) |x'\rangle \; . \tag{A.25}$$

Using the Glauber identity, we can write:

$$D(\lambda) = e^{-i\lambda'\lambda''} e^{2i\lambda'' X_0} e^{-2i\lambda' P_0} \; , \tag{A.26}$$

where $P_0 = i(a^\dagger - a)/2$ is the field quadrature conjugate of X_0. It follows that:

$$D(\lambda) |x'\rangle = e^{i\lambda'\lambda''} e^{2i\lambda'' x'} |x' + \lambda'\rangle \; . \tag{A.27}$$

Within phase factors, the displacement operator translates the X_0 quadrature eigenstates by an amount λ'. Hence:

$$W(\alpha) = W(x, p) = \frac{1}{\pi^2} \int d\lambda' d\lambda'' dx' \, e^{2i(p\lambda' - x\lambda'')} e^{i\lambda'\lambda''} e^{2i\lambda'' x'} \, \langle x'| \rho |x' + \lambda'\rangle \; . \tag{A.28}$$

The integral over λ'' is a one-dimensional Dirac function:

$$\int d\lambda'' \, e^{i\lambda''(\lambda' + 2x' - 2x)} = 2\pi\delta(\lambda' + 2x' - 2x) \; . \tag{A.29}$$

The integration over λ' is then simple, leading to:

$$W(x, p) = \frac{2}{\pi} \int dx' \, e^{4ip(x - x')} \, \langle x'| \rho |2x - x'\rangle \; . \tag{A.30}$$

Setting finally $u = 2(x' - x)$, we obtain a well-known form for the Wigner function:

$$W(x, p) = \frac{1}{\pi} \int du \, e^{-2ipu} \, \langle x + u/2| \rho |x - u/2\rangle \; , \tag{A.31}$$

which is the Fourier transform of non-diagonal density matrix elements in the position eigenstates basis. The function W is clearly sensitive to quantum coherence in the field state. The Fourier transform can be inverted yielding the matrix elements of ρ in terms of W as:

$$\langle x + u/2| \rho |x - u/2\rangle = \int dp \, e^{2ipu} W(x, p) \; . \tag{A.32}$$

This equation shows that the knowledge of W is equivalent to that of the density operator, and hence that the Wigner function contains all information needed to compute the expectation value of any observable in the state of the system. We come back to this point below.

An even simpler expression for W can be derived from eqn. (A.31). We start from the translation relation:

$$\left| x - \frac{u}{2} \right\rangle = e^{-i(x-u)p} D(x + ip) \left| -\frac{u}{2} \right\rangle , \qquad (A.33)$$

[which follows directly from eqn. (A.27) in which we set $x' = -u/2$ and $\lambda = x + ip$] and from the conjugate relation, in which we change u into $-u$:

$$\left\langle x + \frac{u}{2} \right| = \left\langle \frac{u}{2} \right| D(-x - ip) e^{i(x+u)p} . \qquad (A.34)$$

We then replace $|x - u/2\rangle$ and $\langle x + u/2|$ by their expressions (A.33) and (A.34) in eqn. (A.31) and we introduce the hermitian parity operator $\mathcal{P}$ which performs a symmetry around the phase space origin according to:

$$\mathcal{P} |x\rangle = |-x\rangle ; \quad \mathcal{P} |p\rangle = |-p\rangle . \qquad (A.35)$$

We finally get:

$$W(x, p) = \frac{1}{\pi} \int du \left\langle \frac{u}{2} \right| D(-\alpha) \rho D(\alpha) \mathcal{P} \left| \frac{u}{2} \right\rangle = \frac{2}{\pi} \mathrm{Tr}[D(-\alpha)\rho D(\alpha)\mathcal{P}] . \qquad (A.36)$$

The Wigner function is thus the average value of $2\mathcal{P}/\pi$ in the state obtained by displacing the oscillator in phase space by the amount $-\alpha$, transforming its density operator according to $\rho \rightarrow D(-\alpha)\rho D(\alpha)$. It is easy to show that the $\mathcal{P}$ operator is the photon number parity observable introduced in Chapters 6 and 7. The $|n\rangle$ Fock state wave functions in the $|x\rangle$ and $|p\rangle$ representations is a product of an even parity Gaussian multiplied by a Hermite polynomial which has the parity of n (eqn. 3.13, on page 108). Reversing the sign of x or p in these functions thus amounts to a multiplication by $(-1)^n$, so that we can write:

$$\mathcal{P} = e^{i\pi a^\dagger a} . \qquad (A.37)$$

Important properties of this operator are summarized in Section 7.1.2.

Being the expectation value of an observable, W is a directly measurable quantity. We present in Section 6.5 a determination of W for the vacuum and the single-photon Fock state based on the definition given by eqn. (A.36). The eigenvalues of the parity operator being $+1$ and -1, its expectation lies between these values. It follows that the Wigner function is bounded:

$$-2/\pi \leq W(\alpha) \leq 2/\pi . \qquad (A.38)$$

We now use the equivalent definitions of W to establish its main properties.

A.2.2   Main properties

Average operator values

As noted above, the average value of any observable can be obtained from the knowledge of the Wigner distribution. There is in fact a very simple recipe to express this average by a simple integral, provided the observable, considered as a function of a

and $a^\dagger$, has been cast in the symmetric order. The relation (A.21) defining W can be written:

$$W(\alpha) = \frac{1}{\pi^2} \operatorname{Tr}\left[\rho \int d^2\lambda \, e^{\lambda^*(\alpha - a) - \lambda(\alpha^* - a^\dagger)}\right] , \qquad (A.39)$$

in which the integral is reminiscent of the two-dimensional Dirac distribution (see eqn. A.24). We can then formally write:

$$W(\alpha) = \operatorname{Tr}\left[\rho_c \delta(\alpha - a)\right] . \qquad (A.40)$$

The Wigner function is thus the average value of the operator-Dirac distribution of $\alpha - a$. Obviously, such a formal derivation has to be taken with care. We do not enter here in the mathematical justification of this expression [see for instance Cahill and Glauber (1969)], which is valid provided all quantum operators are put in the symmetric order.

Consider now an observable O of the field mode. It can be expanded as a power series of a and $a^\dagger$ and cast in the symmetric order by repeated commutations of these operators. Let us note $O_s(a, a^\dagger)$ the resulting expression. Formally, we can write:

$$O_s(a, a^\dagger) = \int d^2\alpha \, o_s(\alpha, \alpha^*) \delta(\alpha - a) , \qquad (A.41)$$

where $o_s(\alpha, \alpha^*)$ is the complex function of α obtained by replacing a by α and $a^\dagger$ by α^* in $O_s(a, a^\dagger)$ [the precise rules for this substitution can be found in Scully and Zubairy (1997)]. The average value of O_s is then:

$$\langle O_s \rangle = \operatorname{Tr}\left[\rho O_s\right] = \int d^2\alpha \, o_s(\alpha, \alpha^*) \operatorname{Tr}\left[\rho_c \delta(\alpha - a)\right] = \int d^2\alpha \, o_s(\alpha, \alpha^*) W(\alpha) . \quad (A.42)$$

This simple expression reminds us of eqn. (A.1) giving the average value of any classical observable as an integral over the probability density in phase space. The real and normalized Wigner function plays in this respect the role of $f(x, p)$ in classical physics. We will see however that, contrary to f, W can take negative values which are a signature of the oscillator's quantum behaviour.

Marginal distributions

The analogy between $f(x, p)$ and $W(x, p)$ can be pushed further by considering the marginal distributions of x and p, obtained by integrating the distribution over the conjugate variable. Setting $u = 0$ in eqn. (A.32), we get immediately the probability density of finding the value x of the oscillator's position (or field quadrature X_0) as:

$$P(x) = \langle x| \rho |x\rangle = \int dp \, W(x, p) . \qquad (A.43)$$

A simple calculation involving changes between the $|x\rangle$ and $|p\rangle$ bases shows that the integration of W over x yields likewise the probability density of finding the value p of the oscillator's momentum (or field quadrature $X_{\pi/2}$):

$$P(p) = \langle p| \rho |p\rangle = \int dx \, W(x, p) . \qquad (A.44)$$

More generally, the Wigner distribution can be expressed as a function of a any couple of orthogonal field quadratures x_ϕ and p_ϕ defined by:

$$x_\phi = x \cos \phi + p \sin \phi \; ; \quad p_\phi = -x \sin \phi + p \cos \phi \; , \tag{A.45}$$

and corresponding to axes rotated in the phase plane by an angle ϕ with respect to the x and p coordinate directions. The marginal distributions properties apply to any such couple of conjugate quadratures:

$$P(p_\phi) = \int dx_\phi \, W(x_\phi, p_\phi) \; . \tag{A.46}$$

It can be shown that W is the only quantum phase space distribution obeying this interesting property (Bertrand and Bertrand 1987), on which the 'quantum tomographic' methods for the reconstruction of W are based (Section 3.2.4).

For many quantum states, the probability distributions $P(x)$ or $P(p)$ present nodes. This is, for instance, the case for the x distribution of the excited Fock states (Fig. 3.2, on page 107). When $P(x_0) = 0$, it follows that:

$$\int dp \, W(x_0, p) = 0 \; . \tag{A.47}$$

Hence, W cannot be everywhere positive. When it takes negative values, it cannot be assimilated to a genuine probability distribution. In fact, as we now show, negativities in W are clear-cut indicators of the non-classical nature of a field state.

A.2.3 Examples of Wigner functions

We give in this paragraph the explicit expressions of W for a few quantum states classified in two categories: those whose Wigner function is everywhere positive, which we call 'quasi-classical' and those who present negativities, called 'non-classical'.

Quasi-classical states

Let us consider first a coherent state $|\beta\rangle$ (including obviously the vacuum state). Using the symmetric-order characteristic function (eqn. A.14, in which we replace α by β), the corresponding Wigner function, $W^{[|\beta\rangle\langle\beta|]}(\alpha)$ is:

$$
\begin{aligned}
W^{[|\beta\rangle\langle\beta|]}(\alpha) &= \frac{1}{\pi^2} \int d^2\lambda \, e^{-|\lambda|^2/2} e^{\lambda(\beta^* - \alpha^*) - \lambda^*(\beta - \alpha)} \\
&= \frac{2}{\pi} e^{-2|\beta - \alpha|^2} \; .
\end{aligned} \tag{A.48}
$$

It is a Gaussian, with a width $1/\sqrt{2}$, centred on the classical amplitude β and taking for $\alpha = \beta$ its maximum allowed value, $2/\pi$. Figure A.1(a) and (b) present the Wigner functions of the vacuum ($\beta = 0$) and of a coherent state with $\beta = \sqrt{5}$ ($\bar{n} = 5$ photons on the average).

The Wigner function of a thermal field (Fig. A.1c), with $n_{\rm th}$ photons on the average is derived also from the corresponding Gaussian characteristic function (eqn. A.20):

$$W^{[\rho_{\rm th}]}(\alpha) = \frac{2}{\pi} \frac{1}{2n_{\rm th} + 1} e^{-2|\alpha|^2/(2n_{\rm th}+1)} \; . \tag{A.49}$$

It is again a Gaussian, centred at the origin and represented in Fig. A.1(c) for $n_{\rm th} = 1$. The width is now $\sqrt{n_{\rm th} + 1/2}$ and the peak value is reduced to $1/\pi(n_{\rm th} + 1/2)$.

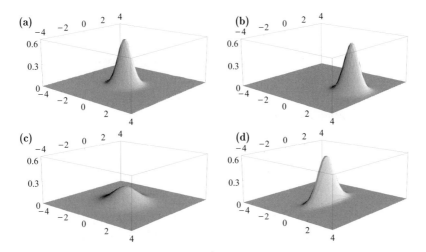

Fig. A.1 Classical state Wigner functions. (a) Vacuum state. (b) Coherent state with $\beta = \sqrt{5}$. (c) Thermal field with $n_{\text{th}} = 1$ photon on the average. (d) A squeezed vacuum state, with a squeezing parameter $\xi = 0.5$. The vertical scales are from 0 to $2/\pi$.

Let us finally examine the case of a squeezed vacuum state (Kimble 1992), often considered in quantum optics as 'non-classical'. It is obtained by the action on the vacuum of the 'squeezing' operator (Vogel *et al.* 2001):

$$S(\xi) = e^{(\xi^* a^2 - \xi a^{\dagger 2})/2} , \tag{A.50}$$

where we assume for the sake of this simple discussion that ξ is real. The squeezed states are minimal uncertainty states satisfying the Heisenberg uncertainty relation $\Delta X_0 \Delta P_0 = 1/4$. For a coherent state, the variances of X_0 and P_0 are equal. For a squeezed state the standard deviation of the X_0 quadrature is reduced to:

$$\Delta X_0 = \frac{1}{2} e^{-\xi} , \tag{A.51}$$

and the fluctuations of P_0 accordingly increased:

$$\Delta P_0 = \frac{1}{2} e^{\xi} . \tag{A.52}$$

Note that the limit of infinite squeezing corresponds to the position eigenstate $|x = 0\rangle$, with no position fluctuations and infinite momentum uncertainty. The Wigner function of a squeezed vacuum is a non-degenerate Gaussian:

$$W^{[sq,\xi]}(x,p) = \frac{2}{\pi} e^{-2\exp(2\xi)x^2} e^{-2\exp(-2\xi)p^2} , \tag{A.53}$$

represented in Fig. A.1(d) for $\xi = 0.5$. A general squeezed state corresponds to a complex ξ parameter and to a displacement towards a non-vanishing classical amplitude in phase space. Its Wigner function is also a non-degenerate Gaussian, centred on the classical amplitude, whose principal axis (minimum and maximum fluctuations) are tilted with respect to x and p.

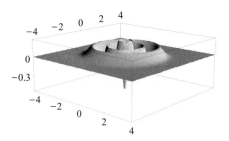

Fig. A.2 Wigner function of a five-photon Fock state.

For all states considered so far, including the squeezed ones, the Wigner function is definite and positive. It has all the properties of a classical probability distribution in phase space. For the computation of any observable eigenvalue, these states can be considered as classical fields with a stochastic complex amplitude whose probability distribution over phase space is given by W. This is not the case for the states we consider below.

Non-classical states

The Fock state Wigner function can be derived from the corresponding characteristic function (eqn. A.17):

$$W^{[|n\rangle\langle n|]}(\alpha) = \frac{2}{\pi}(-1)^n e^{-2|\alpha|^2} \mathcal{L}_n(4|\alpha|^2) . \tag{A.54}$$

Note that $W^{[|n\rangle\langle n|]}$ does not depend upon the phase of α, which reflects the complete phase indeterminacy in Fock states. Since $\mathcal{L}_n(0) = 1$:

$$W^{[|n\rangle\langle n|]}(0) = \frac{2}{\pi}(-1)^n . \tag{A.55}$$

The Wigner function of odd photon number states takes the minimal value $-2/\pi$ at the origin of phase space. It cannot thus be interpreted as a probability distribution of classical field fluctuations, as was the case for the states considered above. In fact, all photon number states, but $|0\rangle$, present negativities in some region of phase space. The single-photon Wigner function:

$$W^{[|1\rangle\langle 1|]}(\alpha) = -\frac{2}{\pi}(1 - 4|\alpha|^2)e^{-2|\alpha|^2} , \tag{A.56}$$

has a Mexican-hat shape. It is shown in Fig. 6.22(b), on page 349. As another example, we present in Fig. A.2 the five-photon state Wigner function. It presents circular rims around the origin, with alternating negative and positive values. Note that this representation does not coincide with the intuitive picture of a Fock state (a well-defined $|\alpha|$ value, corresponding to a single circular rim). We show below that the Husimi-Q function is closer to this simple picture. The inner rims in $W^{[|n\rangle\langle n|]}$ correspond to quantum interference features, revealing the non-classical nature of the Fock states.

Another example of non-classical state is given by the 'phase cat states', discussed in Chapter 7, which are superposition of two coherent fields with opposite phases:

$$|\Psi_{\text{cat}}^{\pm}\rangle = \frac{1}{\sqrt{\mathcal{N}_{\pm}}} (|\beta\rangle \pm |-\beta\rangle) , \tag{A.57}$$

where:

$$\mathcal{N}_\pm = 2\left(1 \pm e^{-2|\beta|^2}\right) , \tag{A.58}$$

is a normalization factor resulting from the finite overlap of $|\beta\rangle$ and $|-\beta\rangle$. The density matrix is $\rho_{\text{cat}}^\pm = |\Psi_{\text{cat}}^\pm\rangle\langle\Psi_{\text{cat}}^\pm|$ and the Wigner function:

$$
\begin{aligned}
W^{[\text{cat},\pm]} &= \frac{1}{\pi^2 \mathcal{N}_\pm} \int d^2\lambda\, e^{(\alpha\lambda^* - \alpha^*\lambda)} \Big[\langle\beta|\, D(\lambda)\, |\beta\rangle + \langle-\beta|\, D(\lambda)\, |-\beta\rangle \\
&\qquad \pm \langle\beta|\, D(\lambda)\, |-\beta\rangle \pm \langle-\beta|\, D(\lambda)\, |\beta\rangle \Big] .
\end{aligned}
\tag{A.59}
$$

The first line in this equation corresponds to the weighted sum of the Wigner functions of the coherent states $|\beta\rangle$ and $|-\beta\rangle$. The second line corresponds to non-diagonal terms involving the quantum coherence between the two components. These terms are absent in the case of a statistical mixture $(|\beta\rangle\langle\beta| + |\beta\rangle\langle\beta|)/2$. Assuming, for the sake of simplicity, that β is real we get:

$$\langle-\beta|\, D(\lambda)\, |\beta\rangle = e^{i\beta\lambda''}\langle-\beta\,|\beta+\lambda\rangle = e^{-2\beta(\beta+\lambda)-|\lambda|^2/2} . \tag{A.60}$$

The coherence terms also correspond to Gaussian integrations which can be performed explicitly, leading finally to:

$$W^{[\text{cat},\pm]}(\alpha) = \frac{2}{\pi(1 \pm e^{-2|\beta|^2})}\left[e^{-2|\alpha-\beta|^2} + e^{-2|\alpha+\beta|^2} \pm 2e^{-2|\alpha|^2}\cos(4\alpha''\beta) \right] . \tag{A.61}$$

These functions, represented in Fig. 7.4, on page 362, exhibit large negative values near the phase-space origin, a clear signature of non-classical behaviour which is analysed in detail in Chapter 7.

A.3 The Husimi-Q distribution

We have focused so far on the symmetrically ordered function C_s and its Fourier transform W. Let us consider now the anti-normal-order characteristic function C_{an} and the related Husimi-Q function. As we will see, the features of the Q function are very different from the ones of W. Being always positive, Q is in some cases more convenient to use for simple representations of quantum states such as Fock states. It is however less well-adapted than the Wigner function to exhibit explicitly the non-classical features of Schrödinger cats.

A.3.1 Definition and main properties

The Q distribution is the Fourier transform of the anti-normal-order characteristic function:

$$Q^{[\rho]}(\alpha) = \frac{1}{\pi^2}\int d^2\lambda\, e^{(\alpha\lambda^* - \alpha^*\lambda)} C_{an}^{[\rho]}(\lambda) . \tag{A.62}$$

This expression looks formidable, but in fact, $Q^{[\rho]}(\alpha)$ is simply related to the expectation value of the density operator ρ in state $|\alpha\rangle$. To show this, let us write Q as:

$$Q^{[\rho]}(\alpha) = \frac{1}{\pi^3}\text{Tr}\left[\rho\int d^2\lambda d^2\beta\, e^{(\alpha\lambda^* - \alpha^*\lambda)} e^{-\lambda^* a}\, |\beta\rangle\langle\beta|\, e^{\lambda a^\dagger}\right] , \tag{A.63}$$

where we have used the definition (A.9) of C_{an} and the closure relation on coherent states:

$$\frac{1}{\pi} \int d^2\beta \, |\beta\rangle \langle\beta| = \mathbb{1} \, . \tag{A.64}$$

The coherent state $|\beta\rangle$ being an eigenstate of $\exp(-\lambda^* a)$, Q reduces to:

$$Q^{[\rho]}(\alpha) = \frac{1}{\pi^3} \mathrm{Tr}\left[\rho \int d^2\lambda d^2\beta \, e^{\lambda^*(\alpha-\beta)-\lambda(\alpha^*-\beta^*)} \, |\beta\rangle \langle\beta|\right] \, , \tag{A.65}$$

and the integration over λ leads to a Dirac distribution $\delta(\alpha - \beta)$ (eqn. A.24). We thus finally get:

$$Q^{[\rho]}(\alpha) = \frac{1}{\pi} \mathrm{Tr}\left[\rho |\alpha\rangle \langle\alpha|\right] = \frac{1}{\pi} \langle\alpha| \rho |\alpha\rangle \, , \tag{A.66}$$

which can also be written as:

$$Q^{[\rho]}(\alpha) = \frac{1}{\pi} \langle 0| D(-\alpha)\rho D(\alpha) |0\rangle = \frac{1}{\pi}\mathrm{Tr}[|0\rangle \langle 0| D(-\alpha)\rho D(\alpha)] \, . \tag{A.67}$$

The Q function is thus the average of the projector onto the vacuum state, $|0\rangle \langle 0|$, in the field displaced in phase space by $-\alpha$. Being the expectation value of an observable, it is thus – as W – a directly measurable quantity. The Q function reconstruction method presented in Section 7.3.2 is based on this expression.

The Q distribution is positive, bounded by $1/\pi$ and normalized ($\int d^2\alpha \, Q(\alpha) = 1$), this last property resulting directly from eqn. (A.64). Using eqn. (A.66), it is easy to compute the Q functions of our favourite classical and non-classical states.

A.3.2 Examples of Q functions

The Q distribution of a coherent state $|\beta\rangle$ is:

$$Q^{[|\beta\rangle\langle\beta|]}(\alpha) = \frac{1}{\pi}| \langle\alpha |\beta\rangle |^2 = \frac{1}{\pi}e^{-|\alpha-\beta|^2} \, . \tag{A.68}$$

It is plotted in Fig. A.3(a) for $\beta = \sqrt{5}$. It is a Gaussian, with a width 1, reaching for $\alpha = \beta$ the maximum allowed value $1/\pi$. Besides their widths, the Q and W functions are thus very similar for a coherent state. Note that the pictorial representation of coherent states as an uncertainty disk superposed on a classical amplitude, used repeatedly from Chapter 3 onwards, corresponds to a cut at $1/e\pi$ in the $Q^{[|\beta\rangle\langle\beta|]}(\alpha)$ distribution.

Expanding $|\alpha\rangle$ on the Fock state basis, it is easy to find the Q function for an n-photon Fock state:

$$Q^{[|n\rangle\langle n|]}(\alpha) = \frac{1}{\pi} \frac{|\alpha|^{2n}}{n!}e^{-|\alpha|^2} \, . \tag{A.69}$$

Represented in Fig. A.3(b) for $n = 5$, it appears as a single circular rim around the origin, which peaks for $|\alpha|^2 = n$, an intuitive result. For Fock states, the Q function seems thus more 'reasonable' than the Wigner distribution.

Let us now turn to the phase-cat states $|\Psi^{\pm}_{\mathrm{cat}}\rangle$. Assuming again β real, and expanding coherent states scalar products, we get:

$$Q^{[\mathrm{cat},\pm]}(\alpha) = \frac{1}{\pi\mathcal{N}_{\pm}} \left[e^{-|\alpha-\beta|^2} + e^{-|\alpha+\beta|^2} \pm 2e^{-(|\alpha|^2+|\beta|^2)} \cos(2\beta\alpha'')\right] \, . \tag{A.70}$$

As for the Wigner function, the cat's Husimi distributions are made up of two Gaussians [first two terms in the right-hand side of eqn. (A.70)], corresponding to the

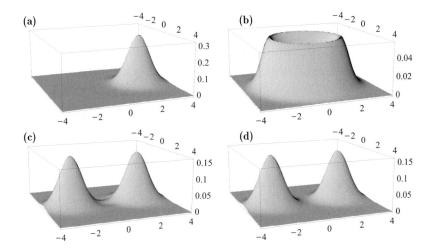

Fig. A.3 Husimi-Q distributions. (a) Coherent state $|\beta\rangle$, with $\beta = \sqrt{5}$. (b) Five-photon Fock state. (c) Schrödinger cat state, superposition of two coherent fields $|\pm\beta\rangle$, with $\beta = \sqrt{5}$. (d) Statistical mixture of the same coherent components.

field components and of an interference term, a Gaussian around the origin multiplied by a modulated cosine function. The maximum amplitude of the interference term, $\sim e^{-|\beta|^2}/\pi$ at $\alpha = 0$, is exponentially small for large β values. The Q function of the $|\Psi_{\text{cat}}^{+}\rangle$ cat state, plotted in Fig. A.3(c) for $\bar{n} = 5$, is thus almost identical to that of a statistical mixture, represented in Fig. A.3(d). The Q function is not able to display the quantum coherence of such a state superposition. Note that this exponential suppression of interference makes it experimentally difficult to reconstruct the field density matrix from Q (we would need to perform a measurement with an unrealistic precision).

The W and Q functions are connected in reciprocal space by the relation (A.11) between their Fourier transforms, $C_s(\lambda)$ being obtained from $C_{an}(\lambda)$ by a multiplication by $e^{|\lambda|^2/2}$. This relation explains why W is more sensitive than Q to a cat's state coherence. We have seen that this coherence is described by vanishingly small terms in Q around the phase space origin. The value of Q at $\alpha = 0$ is equal to the integral over the λ plane of its Fourier transform $C_{an}(\lambda)$. This integral is thus very small for a cat. The same integral, performed on the Fourier transform of W, yields a much larger result since $C_s(\lambda)/C_{an}(\lambda)$ diverges for $|\lambda| \to \infty$, entailing $W(0) \gg Q(0)$. The features of the phase space distribution around $\alpha = 0$, critical to the description of the coherence, are thus greatly amplified by the $e^{|\lambda|^2/2}$ multiplication transforming Q into W in reciprocal space.

We have defined W and Q as the Fourier transforms of C_s and C_{an}. What about C_n? Its Fourier transform is yet another phase space density function, the Glauber–Sudarshan-P distribution (Vogel *et al.* 2001). We do not use it here, since it turns out to be highly singular. For instance, the P distribution of the coherent state $|\beta\rangle$ is a δ-Dirac distribution centred at β and the P distribution of an $|n\rangle$ Fock state involves the n first derivatives of δ. In spite of their mathematical interest, these distributions do not provide a very practical representation of quantum states.[3]

A.4   Phase-space representations of relaxation

The phase space description of quantum states is well-suited to the description of relaxation mechanisms. The Lindblad equation introduced in Section 4.3.2 can be expressed as a differential equation for characteristic functions or phase space distributions. The solution of these equations provides an insightful view on relaxation, particularly in the case of Schrödinger cat states.

A.4.1   Relaxation of the normally-ordered characteristic function

We derive here first the differential equation for C_n, which takes a particularly simple form. The relations between characteristic functions (eqn. A.11) can then be used to obtain the corresponding equations for C_s or C_{an}. For the sake of simplicity, we examine only the $T = 0$ K case and we adopt the interaction representation with respect to the oscillator's free Hamiltonian. The Lindblad master equation is then simply (eqn. 7.112):

$$\frac{d\rho}{dt} = \frac{\kappa}{2}(2a\rho a^\dagger - a^\dagger a\rho - \rho a^\dagger a) \ . \tag{A.71}$$

Combining it with eqn. (A.8), we get the evolution equation for $C_n^{[\rho]}$:

$$\frac{dC_n^{[\rho]}}{dt} = \frac{\kappa}{2}\mathrm{Tr}[(2a\rho a^\dagger - a^\dagger a\rho - \rho a^\dagger a)e^{\lambda a^\dagger}e^{-\lambda^* a}] \ . \tag{A.72}$$

This equation involves, in the trace operation, a sum of products of ρ with a and $a^\dagger$. In order to deal with these products, it is convenient to extend the definition of C_n to these operator combinations. We start by generalizing eqn. (A.8) to an arbitrary operator U and define the normal order characteristic function of this operator as:

$$C_n^{[U]} = \mathrm{Tr}\left[Ue^{\lambda a^\dagger}e^{-\lambda^* a}\right] \ . \tag{A.73}$$

We next consider the action on $C_n^{[U]}$ of the differential operators $(\partial/\partial\lambda)$ and $(\partial/\partial\lambda^*)$, where λ and λ^* are formally taken as independent quantities when evaluating the derivatives. We get for instance:

$$\frac{\partial}{\partial\lambda}C_n^{[U]} = \mathrm{Tr}\left[Ua^\dagger e^{\lambda a^\dagger}e^{-\lambda^* a}\right] = C_n^{[Ua^\dagger]} \ . \tag{A.74}$$

Similar relations can be established for other products of U with a and $a^\dagger$ and we easily obtain simple correspondence rules linking partial derivatives of $C_n^{[U]}$ to the characteristic functions of $Ua^\dagger$, Ua, $a^\dagger U$ and aU:

$$C_n^{[Ua^\dagger]} = \frac{\partial}{\partial\lambda}C_n^{[U]} \quad ; \quad C_n^{[Ua]} = \left(-\frac{\partial}{\partial\lambda^*} + \lambda\right)C_n^{[U]} \ ;$$

---

[3]One can in fact consider a wide range of quasi-probability distributions, by defining the *s*-order of operators (*s* being a continuous parameter), whose symmetric, normal and anti-normal orders are limiting cases. None of these 'exotic' distributions is as useful as *W* or *Q*.

$$C_n^{[a^\dagger U]} = \left(\frac{\partial}{\partial\lambda} - \lambda^*\right)C_n^{[U]} \quad ; \quad C_n^{[aU]} = -\frac{\partial}{\partial\lambda^*}C_n^{[U]} \ . \tag{A.75}$$

Iterating these rules, all the terms in the right-hand side of eqn. (A.72) can be put in the form of products of differential operators acting on $C_n^{[\rho]}$ and we finally get an ordinary differential equation describing the evolution of this quantity:

$$\frac{dC_n^{[\rho]}}{dt}(\lambda,\lambda^*,t) = -\frac{\kappa}{2}\left[\lambda\frac{\partial}{\partial\lambda} + \lambda^*\frac{\partial}{\partial\lambda^*}\right]C_n^{[\rho]} \ . \tag{A.76}$$

This equation admits for solution the simple ansatz (Barnett and Radmore 1997):

$$C_n^{[\rho]}(\lambda,\lambda^*,t) = C_n^{[\rho]}(\lambda e^{-\kappa t/2},\lambda^* e^{-\kappa t/2},0) \ , \tag{A.77}$$

as can be checked by plugging it in eqn. (A.76). The relaxation of $C_n^{[\rho]}$ thus takes a simple form which can be used to derive the explicit expressions versus time of various quantum states.

The relaxation of the coherent state $|\alpha\rangle$ provides a simple application of this general solution. Its initial normal order characteristic function given by eqn. (A.15) evolves at time t into:

$$C_n^{[|\alpha\rangle\langle\alpha|]}(\lambda,\lambda^*,t) = e^{(\alpha^*\lambda - \alpha\lambda^*)\exp(-\kappa t/2)} \ , \tag{A.78}$$

which corresponds to the coherent state $|\alpha\exp(-\kappa t/2)\rangle$. We recover the simple result that a coherent state remains coherent in the relaxation process, with an exponentially damped amplitude. We give in Sections 3.2.5 and 4.4.4 detailed physical interpretations of this important result.

This simple formalism is also well-adapted to take into account a unitary evolution. Let us consider, for instance, the evolution of a field mode under the combined effects of its coupling to a classical source current and to an environment producing a relaxation described by eqn. (A.71). The additional term in the master equation due to the classical current is, using the notation of Section 3.1.3 [see in particular eqn. (3.45), on page 116]:

$$\left.\frac{\partial\rho_c}{\partial t}\right|_{\text{source}} = i\mathcal{S}[a + a^\dagger, \rho_c] \ , \tag{A.79}$$

where $\mathcal{S} = \mathcal{E}_0 J_1/\hbar\omega_c$ is the normalized current amplitude (assumed to be real without loss of generality). This commutator source term corresponds to an additional contribution to the right-hand side of eqn. (A.76), which is obtained by using again the correspondence rules (A.75). The final differential equation:

$$\frac{\partial C_n^{[\rho]}}{\partial t} = \left\{-i\mathcal{S}(\lambda + \lambda^*) - \frac{\kappa}{2}\left[\lambda\frac{\partial}{\partial\lambda} + \lambda^*\frac{\partial}{\partial\lambda^*}\right]\right\}C_n^{[\rho]} \ , \tag{A.80}$$

admits as an explicit solution (Barnett and Radmore 1997):

$$C_n^{[\rho]}(\lambda,t) = e^{(-2i\mathcal{S}/\kappa)(\lambda+\lambda^*)[1-\exp(-\kappa t/2)]}C_n^{[\rho]}(\lambda e^{-\kappa t/2},0) \ . \tag{A.81}$$

The steady state, obtained for $t \gg 1/\kappa$ is:

$$C_n^{[\rho]} = e^{-i(2\mathcal{S}/\kappa)(\lambda+\lambda^*)} \, , \tag{A.82}$$

corresponding to a coherent state with an amplitude:

$$\beta = 2i\,\mathcal{S}/\kappa = 2i\,\mathcal{E}_0 J_1/\hbar\omega_c\kappa \, . \tag{A.83}$$

This simple expression coincides with the steady-state solution of the classical differential equation for the field amplitude $\mathcal{E}$:

$$\frac{d\mathcal{E}}{dt} = i\mathcal{S} - \frac{\kappa}{2}\mathcal{E} \, . \tag{A.84}$$

A.4.2 Relaxation of the Wigner function

A similar approach can be used to express the Lindblad master equation as a differential equation for the Wigner distribution W. We start by defining, as above, the symmetric characteristic function $C_s^{[U]}$ associated to an arbitrary operator U and, by Fourier transform, the corresponding operator $W^{[U]}$. With this generalized notation, the time derivative of $W^{[\rho]}$ is:

$$\frac{dW^{[\rho]}}{dt} = \frac{\kappa}{2} \left[2W^{[a\rho a^\dagger]} - W^{[a^\dagger a\rho]} - W^{[\rho a^\dagger a]} \right] \, . \tag{A.85}$$

To express this equation in terms of the Wigner function $W^{[\rho]}$ alone, we need the algebraic correspondence rules between $W^{[U]}$ and the four quantities $W^{[Ua^\dagger]}$, $W^{[Ua]}$, $W^{[a^\dagger U]}$ and $W^{[aU]}$. These rules are:

$$W^{[Ua^\dagger]} = \left(\alpha^* + \frac{1}{2}\frac{\partial}{\partial\alpha} \right) W^{[U]} \quad ; \quad W^{[Ua]} = \left(\alpha - \frac{1}{2}\frac{\partial}{\partial\alpha^*} \right) W^{[U]} \, ;$$

$$W^{[a^\dagger U]} = \left(\alpha^* - \frac{1}{2}\frac{\partial}{\partial\alpha} \right) W^{[U]} \quad ; \quad W^{[aU]} = \left(\alpha + \frac{1}{2}\frac{\partial}{\partial\alpha^*} \right) W^{[U]} \, , \tag{A.86}$$

where α and α^* are to be considered as independent variables for the calculation of the derivatives. To obtain these equations, we establish first the correspondence rules linking $C_s^{[Ua^\dagger]}$, $C_s^{[Ua]}$, $C_s^{[a^\dagger U]}$ and $C_s^{[aU]}$ to $C_s^{[U]}$ (with the help of eqns. A.75 and A.11). These relations involve four differential operators $(\partial/\partial\lambda+\lambda^*/2)$, $(\partial/\partial\lambda^*-\lambda/2)$, $(\partial/\partial\lambda-\lambda^*/2)$ and $(\partial/\partial\lambda^*+\lambda/2)$ acting on $C_s^{[U]}$. In the Fourier correspondence between the C_s's and the W's, these differential operators are changed into the operators appearing in (A.86).

Iterating the rules (A.86), we transform eqn. (A.85) into a differential equation of the Fokker–Planck type for $W^{[\rho]}(\alpha)$, considered as a function of the two independent variables α and α^*:

$$\frac{d}{dt}W^{[\rho]}(\alpha,\alpha^*) = \frac{\kappa}{2}\left[\frac{\partial^2}{\partial\alpha\partial\alpha^*} + \frac{\partial}{\partial\alpha}\alpha + \frac{\partial}{\partial\alpha^*}\alpha^* \right] W^{[\rho]}(\alpha,\alpha^*) \, . \tag{A.87}$$

This equation describes the relaxation of the Wigner function of an oscillator coupled to a $T = 0$ K bath. Its right-hand side is the sum of two contributions. The first-order derivative terms produce a global drift of the distribution towards the origin and account for energy dissipation. The second-order derivative contribution represents a diffusion process which tends to broaden the distribution and prevents it from collapsing into a Dirac distribution at the origin, which would violate the Heisenberg uncertainty relations. This diffusion process blurs rapidly any small scale feature in the Wigner distribution (Habib *et al.* 1998). The evolution of a Schrödinger cat state illustrates this blurring in a striking way, as shown by the snapshots of W in Fig. 7.25, on page 406, where we observe the fast decay of the Wigner function interference feature. This decoherence phenomenon is analysed in detail in Chapter 7 and in the next section.

The action of an additional Hamiltonian interaction can be easily incorporated to the right-hand side of eqn. (A.87). For instance, in the presence of a classical source, it becomes:

$$\frac{d}{dt} W^{[\rho]}(\alpha, \alpha^*) = \left[\frac{\partial^2}{\partial\alpha\partial\alpha^*} + \frac{\partial}{\partial\alpha}(\alpha - \beta) + \frac{\partial}{\partial\alpha^*}(\alpha^* - \beta^*) \right] W^{[\rho]}(\alpha, \alpha^*) , \qquad \text{(A.88)}$$

with:

$$\beta = i\frac{2S}{\kappa} . \qquad \text{(A.89)}$$

Compared to eqn. (A.87), only the drift term has changed. The Wigner distribution maximum moves now towards the point β in phase space and the steady state of the process is the coherent state $|\beta\rangle$.

A.4.3 Relaxation of a Schrödinger cat state

We conclude this appendix by deriving the exact expression of the density matrix of a relaxing Schrödinger cat state. This result is useful for the analysis of cat state decoherence in Chapter 7. We consider initially a superposition of two coherent states with different phases:

$$|\Xi(t = 0)\rangle = \frac{1}{\mathcal{N}} \left[|\alpha e^{i\phi}\rangle + e^{-i\Psi} |\alpha e^{-i\phi}\rangle \right] , \qquad \text{(A.90)}$$

(where we assume that α is real) with:

$$\mathcal{N} = \sqrt{2\left\{1 + e^{-D^2/2} \cos\left[\Psi + D^2/(2\tan\phi)\right]\right\}} . \qquad \text{(A.91)}$$

This state generalizes the even and odd phase cats (eqn. A.57), which we recover for $\phi = \pi/2$ and $\Psi = 0$ or π. The distance between the two coherent components amplitude in phase space is:

$$D = 2|\alpha| \sin\phi = 2\sqrt{\bar{n}} \sin\phi . \qquad \text{(A.92)}$$

Setting $\beta = \alpha e^{i\phi}$, the initial density matrix is:

$$\rho = \frac{1}{\mathcal{N}^2} \left[|\beta\rangle\langle\beta| + |\beta^*\rangle\langle\beta^*| + e^{i\Psi} |\beta\rangle\langle\beta^*| + \text{h.c.} \right] , \qquad \text{(A.93)}$$

in which the first two terms describes the coherent components and the last ones their mutual coherence. We then deduce the initial normal-order characteristic function C_n:

$$C_n(\lambda, \lambda^*, t = 0) = \frac{1}{N^2} \left[e^{\lambda\beta^* - \lambda^*\beta} + e^{\lambda\beta - \lambda^*\beta^*} + e^{i\Psi} e^{iD^2/(2\tan\phi)} e^{-D^2/2} e^{(\lambda-\lambda^*)\beta} + \text{c.c.} \right].$$

$$(A.94)$$

The first two terms in the right-had side of this equation correspond to the characteristic functions of the two coherent components. The remaining complex terms correspond to the coherence in the initial field's density matrix.

For a relaxation in a $T = 0$ K bath, the characteristic function becomes at time t:

$$C_n(\lambda, \lambda^*, t) = C_n(\lambda\varepsilon, \lambda^*\varepsilon, t = 0), \qquad (A.95)$$

with:

$$\varepsilon = e^{-\kappa t/2}. \qquad (A.96)$$

It is easy to recognize that this characteristic function corresponds to the density operator:

$$\rho(t) = \frac{1}{N^2} \left[|\varepsilon\beta\rangle \langle\varepsilon\beta| + |\varepsilon\beta^*\rangle \langle\varepsilon\beta^*| \right.$$
$$\left. + e^{i\Psi} e^{iD^2(1-\varepsilon^2)/(2\tan\phi)} e^{-D^2(1-\varepsilon^2)/2} |\varepsilon\beta\rangle \langle\varepsilon\beta^*| + \text{h.c.} \right]. \qquad (A.97)$$

The first line describes two coherent components relaxing towards the origin. The damping time constant for the amplitude is $2/\kappa = 2T_c$. By comparing eqns. (A.93) and (A.97), we see that the non-diagonal coherence terms (last line in eqn. A.97) are strongly affected by relaxation. On the one hand, the quantum phase of the superposition rotates. Equal to Ψ at $t = 0$, it becomes $\Psi + \bar{n}\sin(2\phi)(1 - \varepsilon^2)$ at time t. On the other hand, the amplitude of the coherence terms decays rapidly, since it is multiplied by the factor:

$$\xi = e^{-2\bar{n}\sin^2\phi(1-\varepsilon^2)}. \qquad (A.98)$$

Let us consider times much shorter than T_c. The exponentials in the exponent of ξ can be expanded to first order, and we get:

$$\xi = e^{-D^2\kappa t/2}. \qquad (A.99)$$

The decoherence time T_D is thus:

$$T_D = \frac{T_c}{2\bar{n}\sin^2\phi}, \qquad (A.100)$$

reducing to $T_c/(2\bar{n})$ in the case of a π-phase cat with $\phi = \pi/2$. This time becomes very short for $\bar{n}$ large. The experiment presented in Section 7.5.3 has revealed the dynamics of the coherence damping and the quantum phase rotation of cat states corresponding to $\bar{n} \sim 3$–5 and $\phi \approx 50$ degrees.

The Wigner function of the decaying cat at time t can be derived fom eqn. (A.97). We find, in the simple case of the even and odd cats ($\phi = \pi/2$ and $\Psi = 0$ or π):

$$W^{[\text{cat}\pm]}(\alpha, t) = \frac{2}{\pi(1 + e^{-2|\beta|^2})} \left[e^{-2|\alpha - \beta\varepsilon|^2} + e^{-2|\alpha + \beta\varepsilon|^2} \right.$$
$$\left. \pm 2e^{-2|\alpha|^2} e^{-2|\beta|^2(1-\varepsilon^2)} \cos(4\alpha''\beta\varepsilon) \right]. \qquad (A.101)$$

This equation reduces obviously to eqn. (A.61) for $t = 0$, i.e. $\varepsilon = 1$.

Bibliography

Agarwal, G. S. (1974). *Quantum Statistical Theories of Spontaneous Emission and their Relations to Other Approaches*, Springer Tracts in Modern Physics, 70, Berlin.

Agarwal, G. S. and Gupta, S. D. (1998). *Phys. Rev. A*, **57**, 667.

Aharonov, Y. and Bohm, D. (1959). *Phys. Rev.*, **115**, 485.

Albiez, M., Gati, R., Fölling, J., Hunsmann, S., Cristiani, M. and Oberthaler, M. K. (2005). *Phys. Rev. Lett.*, **95**, 010402.

Alzetta, G., Gozzini, A., Moi, L. and Orriols, G. (1976). *Nuovo Cimento B*, **36**, 5.

Amoretti, M., Amsler, C., Bononi, G., Bouchta, A. and Athena collaboration (2002). *Nature (London)*, **419**, 456.

An, K., Childs, J. J., Dasari, R. R. and Feld, M. S. (1994). *Phys. Rev. Lett.*, **73**, 3375.

Anderson, M. H., Ensher, J. R., Matthews, M. R., Wieman, C. E. and Cornell, E. A. (1995). *Science*, **269**, 198.

Andrews, M. R., Townsend, C. G., Miesner, H.-J., Durfee, D. S., Kurn, D. M. and Ketterle, W. (1997). *Science*, **275**, 637.

Arnison, G., Ashbury, A., Aubert, B., Bacci, C. and Bauer, G. (1983). *Physics Letters B*, **122**, 103.

Aspect, A. (2004). *Bell's Theorem: The Naive View of an Experimentalist*, http://hal.ccsd.cnrs.fr/ccsd-00001079.

Aspect, A., Dalibard, J. and Roger, G. (1982a). *Phys. Rev. Lett.*, **49**, 1804.

Aspect, A., Grangier, P. and Roger, G. (1981). *Phys. Rev. Lett.*, **47**, 460.

Aspect, A., Grangier, P. and Roger, G. (1982b). *Phys. Rev. Lett.*, **49**, 91.

Atatüre, M., Dreiser, J., Badolato, A., Högele, A., Karrai, K. and Imamoglu, A. (2006). *Science*, **312**, 551.

Auffeves, A., Maioli, P., Meunier, T., Gleyzes, S., Nogues, G., Brune, M., Raimond, J.-M. and Haroche, S. (2003). *Phys. Rev. Lett.*, **91**, 230405.

Averbukh, I. S. (1992). *Phys. Rev. A*, **46**, R2205.

Averbukh, I. S. and Perel'Man, N. F. (1991). *Sov. Phys. Usp.*, **34**, 572.

Bachor, H.-A., Levenson, M. D., Walls, D. F., Perlmutter, S. H. and Shelby, R. M. (1988). *Phys. Rev. A*, **38**, 180.

Badurek, G., Rauch, H. and Tuppinger, D. (1986). *Phys. Rev. A*, **34**, 2600.

Banaszek, K., D'Ariano, G. M., Paris, M. G. A. and Sacchi, M. F. (1999a). *Phys. Rev. A*, **61**, 010304.

Banaszek, K., Radzewicz, C., Wodkiewicz, K. and Krasinski, J. S. (1999b). *Phys. Rev. A*, **60**, 674.

Banaszek, K. and Wodkiewicz, K. (1996). *Phys. Rev. Lett.*, **76**, 4344.

Banaszek, K. and Wodkiewicz, K. (1999). *Phys. Rev. Lett.*, **82**, 2009.

Barnett, S. M. and Radmore, P. M. (1997). *Methods in Theoretical Quantum Optics*, Oxford University Press, Oxford.

Barone, A. and Paterno, G. (1982). *Physics and Applications of the Josephson Effect*, Wiley, New York.

Barrett, M., Chiaverini, J., Schaetz, T., Britton, J., Itano, W., Jost, J., Knill, E., Langer, C., Leibfried, D., Ozeri, R. and Wineland, D. (2004). *Nature (London)*, **429**, 737.

Barton, G. (1994). P. Berman (ed.), *Advances in Atomic and Molecular Physics, Supplement 2*, p. 425, Academic Press, New York.

Bastard, G. (1991). *Wave Mechanics Applied to Semiconductor Heterostructures*, Wiley, New York.

Bauch, A. and Telle, H. R. (2002). *Rep. Prog. Phys*, **65**, 789.

Bayer, M., Reinecke, T. L., Weidner, F., Larionov, A., McDonald, A. and Forchel, A. (2001). *Phys. Rev. Lett.*, **86**, 3168.

Bell, J. S. (1964). *Phys. (N.Y.)*, **1**, 195.

Bell, J. S. (1966). *Rev. Mod. Phys.*, **38**, 447.

Bell, J. S. (1987). *Speakable and Unspeakable in Quantum Mechanics*, Cambridge University Press, Cambridge.

Bencheikh, K., Levenson, J. A., Grangier, P. and Lopez, O. (1995). *Phys. Rev. Lett.*, **75**, 3422.

Bencheikh, K., Simonneau, C. and Levenson, J. A. (1997). *Phys. Rev. Lett.*, **78**, 34.

Benkert, C., Scully, M. O., Bergou, J., Davidovich, L., Hillery, M. and Orszag, M. (1990). *Phys. Rev. A*, **41**, 2756.

Bennett, C. H. and Brassard, G. (1984). *Proceedings of IEEE International Conference on Computers, Systems and Signal Processing*, p. 175, IEEE, New York.

Bennett, C. H., Brassard, G., Crépeau, C., Jozsa, R., Peres, A. and Wootters, W. K. (1993). *Phys. Rev. Lett.*, **70**, 1895.

Bennett, C. H., Brassard, G. and Ekert, A. (1992). *Scientific American*, **267**, 50.

Benson, O., Raithel, G. and Walther, H. (1994). *Phys. Rev. Lett.*, **72**, 3506.

Bergou, J., Davidovich, L., Orszag, M., Benkert, C., Hillery, M. and Scully, M. O. (1989a). *Opt. Comm.*, **72**, 82.

Bergou, J., Davidovich, L., Orszag, M., Benkert, C., Hillery, M. and Scully, M. O. (1989b). *Phys. Rev. A*, **40**, 5073.

Bernardot, F., Nussenzveig, P., Brune, M., Raimond, J.-M. and Haroche, S. (1992). *Europhys. Lett.*, **17**, 33.

Bertet, P., Auffeves, A., Maioli, P., Osnaghi, S., Meunier, T., Brune, M., Raimond, J.-M. and Haroche, S. (2002). *Phys. Rev. Lett.*, **89**, 200402.

Bertet, P., Osnaghi, S., Rauschenbeutel, A., Nogues, G., Auffeves, A., Brune, M., Raimond, J.-M. and Haroche, S. (2001). *Nature (London)*, **411**, 166.

Bertrand, J. and Bertrand, P. (1987). *Found. Phys.*, **17**, 397.

Bethe, H. A. and Salpeter, E. E. (1977). *Quantum Mechanics of One- and Two-electron Atoms*, Plenum, New York.

Beugnon, J., Jones, M. P. A., Dingjan, J., Darquié, B., Messin, G., Browayes, A. and Grangier, P. (2006). *Nature (London)*, **440**, 779.

Birnbaum, K. N., Boca, A., Miller, R., Boozer, A. D., Northup, T. E. and Kimble, H. J. (2005). *Nature (London)*, **436**, 87.

Björk, G., Machida, S., Yamamoto, Y. and Igeta, K. (1991). *Phys. Rev. A*, **44**, 669.

Blatt, R. (2006). *Entangled World*, p. 235, Wiley-VCH, Weinheim.

Blinov, B. B., Moehring, D. L., Duan, L.-M. and Monroe, C. (2004). *Nature (London)*, **428**, 153.

Bloch, I. (2005). *Nature Physics*, **1**, 23.

Boca, A., Miller, R., Birnbaum, K. M., Boozer, A. D., McKeever, J. and Kimble, H. J. (2004). *Phys. Rev. Lett.*, **93**, 233603.

Bohm, D. J. (1951). *Quantum Theory*, Prentice-Hall, Englewood Cliffs, N.J.

Bohr, N. (1935). *Phys. Rev.*, **48**, 696.

Bohr, N. (1949). A. Schilpp (ed.), *Albert Einstein, Philosopher Scientist*, p. 201, Open Court, La Salle (IL).

Bonifacio, R. and Preparata, G. (1970). *Phys. Rev. A*, **2**, 336.

Bonifacio, R., Schwendimann, P. and Haake, F. (1971a). *Phys. Rev. A*, **4**, 302.

Bonifacio, R., Schwendimann, P. and Haake, F. (1971b). *Phys. Rev. A*, **4**, 854.

Bordé, C. J. (1999). *Proceeding of the ICOLS'99 Conference*, World Scientific.

Born, M. and Wolf, E. (1980). *Principles of Optics*, 6th edn., Pergamon Press, Oxford.

Boschi, D., Branca, S., de Martini, F., Hardy, L. and Popescu, S. (1998). *Phys. Rev. Lett.*, **80**, 1121.

Bouwmeester, D., Ekert, A. and Zeilinger, A. (2000). *The Physics of Quantum Information*, Springer, Berlin.

Bouwmeester, D., Pan, J.-W., Mattle, K., Eibl, M., Weinfurter, H. and Zeilinger, A. (1997). *Nature (London)*, **390**, 575.

Bradley, C. C., Sackett, C. A., Tolett, J. and Hulet, R. G. (1995). *Phys. Rev. Lett.*, **75**, 1687.

Bradley, C. C., Sackett, C. A., Tollett, J. J. and Hulet, R. G. (1997). *Phys. Rev. Lett.*, **79**, 1170.

Braginsky, V. B. and Khalili, F. Y. (1996). *Rev. Mod. Phys.*, **68**, 1.

Braginsky, V. B. and Vorontosov, Y. I. (1974). *Usp. Fiz. Nauk*, **114**, 41, [Sov. Phys. Usp., **17**, 644 (1975)].

Braginsky, V. B., Vorontosov, Y. I. and Khalili, F. Y. (1977). *Zh. Eksp. Teor. Fiz.*, **73**, 1340, [SPJETP, **46**, 705 (1977)].

Brattke, S., Varcoe, B. T. H. and Walther, H. (2001). *Phys. Rev. Lett.*, **86**, 3534.

Braun, D., Haake, F. and Strunz, W. T. (2001). *Phys. Rev. Lett.*, **86**, 2913.

Braunstein, S. L. and Kimble, H. J. (1998). *Phys. Rev. Lett.*, **80**, 869.

Braunstein, S. L., Mann, A. and Revzen, M. (1992). *Phys. Rev. Lett.*, **68**, 3259.

Brecha, R. J., Orozco, L. A., Raizen, M. G., Xiao, M. and Kimble, H. J. (1995). *J. Opt. Soc. Am. B*, **12**, 2329.

Brecha, R. J., Raithel, G., Wagner, C. and Walther, H. (1993). *Opt. Comm.*, **102**, 257.

Breitenbach, G., Schiller, S. and Mlynek, J. (1997). *Nature (London)*, **387**, 471.

Briegel, H.-J., Calarco, T., Jaksch, D., Cirac, J. I. and Zoller, P. (2000). *J. Mod. Opt.*, **47**, 415.

Bruckmeier, R., Hansen, H. and Schiller, S. (1997a). *Phys. Rev. Lett.*, **79**, 1463.

Bruckmeier, R., Schneider, K., Schiller, S. and Mlynek, J. (1997b). *Phys. Rev. Lett.*, **78**, 1243.

Brune, M., Hagley, E., Dreyer, J., Maître, X., Maali, A., Wunderlich, C., Raimond, J.-M. and Haroche, S. (1996a). *Phys. Rev. Lett.*, **77**, 4887.

Brune, M., Haroche, S., Lefèvre, V., Raimond, J.-M. and Zagury, N. (1990). *Phys. Rev. Lett.*, **65**, 976.

Brune, M., Haroche, S., Raimond, J.-M., Davidovich, L. and Zagury, N. (1992). *Phys. Rev. A*, **45**, 5193.

Brune, M., Nussenzveig, P., Schmidt-Kaler, F., Bernardot, F., Maali, A., Raimond, J.-M. and Haroche, S. (1994). *Phys. Rev. Lett.*, **72**, 3339.

Brune, M., Raimond, J.-M., Goy, P., Davidovich, L. and Haroche, S. (1987). *Phys. Rev. Lett.*, **59**, 1899.

Brune, M., Schmidt-Kaler, F., Maali, A., Dreyer, J., Hagley, E., Raimond, J.-M. and Haroche, S. (1996b). *Phys. Rev. Lett.*, **76**, 1800.

Bruss, D., Ekert, A. and Macchiavello, C. (1998). *Phys. Rev. Lett.*, **81**, 2598.

Buchleitner, A. and Mantegna, R. N. (1998). *Phys. Rev. Lett.*, **80**, 3932.

Buks, E., Schuster, R., Heiblum, M., Mahalu, D. and Umansky, V. (1998). *Nature (London)*, **391**, 871.

Burstein, E. and Weisbuch, C. (1995). *Confined Electrons and Photons*, Plenum, New York.

Buttler, W. T., Hughes, R. J., Kwiat, P. G., Lamoreaux, S. K., Luther, G. G., Morgan, G. L., Nordholt, J. E., Peterson, C. G. and Simmons, C. M. (1998). *Phys. Rev. Lett.*, **81**, 3283.

Buttler, W. T., Hughes, R. J., Lamoreaux, S. K., Morgan, G. L., Nordholt, J. E. and Peterson, C. G. (2000). *Phys. Rev. Lett.*, **84**, 5652.

Buzek, V. and Hillery, M. (1998). *Phys. Rev. Lett.*, **81**, 5003.

Buzek, V. and Knight, P. L. (1995). *Progress in Optics XXXIV*, vol. 34, p. 1, Elsevier, Amsterdam.

Buzek, V., Moya-Cessa, H., Knight, P. L. and Phoenix, S. J. D. (1992). *Phys. Rev. A*, **45**, 8190.

Cahill, K. E. and Glauber, R. J. (1969). *Phys. Rev.*, **177**, 1857.

Caldeira, A. O. and Leggett, A. J. (1983). *Physica*, **121A**, 587.

Caldeira, A. O. and Leggett, A. J. (1985). *Phys. Rev. A*, **31**, 1059.

Carmichael, H. (1993). *An Open-system Approach to Quantum Optics*, Springer, Berlin.

Carnal, O. and Mlynek, J. (1991). *Phys. Rev. Lett.*, **66**, 2689.

Carvalho, A. R. R., Milman, P., de Matos Filho, R. L. and Davidovich, L. (2001). *Phys. Rev. Lett.*, **86**, 4988.

Carvalho, C. R., Davidovich, L. and Zagury, N. (1989). *Opt. Commun.*, **72**, 306.

Casimir, H. B. G. (1948). *Proc. K. Ned. Akad. Wet.*, **51**, 793.

Casimir, H. B. G. and Polder, D. (1948). *Phys. Rev.*, **73**, 360.

Castin, Y. and Dalibard, J. (1997). *Phys. Rev. A*, **55**, 4330.

Caves, C. M. (1979). *Phys. Lett. B*, **80**, 323.

Caves, C. M., Thorne, K. S., Drever, R. W. P., Sandberg, V. D. and Zimmermann, M. (1980). *Rev. Mod. Phys.*, **52**, 341.

Chan, H. W., Black, A. T. and Vuletic, V. (2003). *Phys. Rev. Lett.*, **90**, 063003.

Chapman, M. S., Hammond, T. D., Lenef, A., Schmiedmayer, J., Rubenstein, R. A., Smith, E. and Pritchard, D. E. (1995). *Phys. Rev. Lett.*, **75**, 3783.

Chiaverini, J., Britton, J., Leibfried, D., Knill, E., Barrett, M., Blakestad, R., Itano, W., Jost, J., Langer, C., Ozeri, R., Schaetz, T. and Wineland, D. (2005). *Science*, **308**, 997.

Chiaverini, J., Leibfried, D., Schaetz, T., Barrett, M. D., Blakestad, R., Britton, J., Itano, W. M., Jost, J. D., Knill, E., Langer, C., Ozeri, R. and Wineland, D. J. (2004). *Nature (London)*, **432**, 602.

Chiorescu, I., Nakamura, Y., Harmans, C. J. P. M. and Mooij, J. E. (2003). *Science*, **299**, 1869.

Chuang, I. L. and Yamamoto, Y. (1996). *Proceedings of the 4th Workshop on Physics and Computation*, p. 82, New England Complex System Institute, Boston.

Cirac, J. I., Lewenstein, M., Molmer, K. and Zoller, P. (1998). *Phys. Rev. A*, **57**, 1208.

Cirac, J. I. and Zoller, P. (1995). *Phys. Rev. Lett.*, **74**, 4091.

Cirac, J. I. and Zoller, P. (2000). *Nature (London)*, **404**, 579.

Clauser, J. F., Horne, M. A., Shimony, A. and Holt, R. A. (1969). *Phys. Rev. Lett.*, **23**, 880.

Clauser, J. F. and Shimony, A. (1978). *Rep. Prog. Phys.*, **41**, 1881.

Cohadon, P. F., Heidmann, A. and Pinard, M. (1999). *Phys. Rev. Lett.*, **83**, 3174.

Cohen-Tannoudji, C. (1992). J. Dalibard, J.-M. Raimond and J. Zinn-Justin (eds.), *Fundamental Systems in Quantum Optics, Les Houches Summer School, Session LIII*, p. 1, North Holland, Amsterdam.

Cohen-Tannoudji, C. and Dalibard, J. (1986). *EuroPhys. Lett.*, **1**, 441.

Cohen-Tannoudji, C., Diu, B. and Laloë, F. (1977). *Quantum Mechanics*, Wiley, New York.

Cohen-Tannoudji, C., Dupont-Roc, J. and Grynberg, G. (1992a). *An Introduction to Quantum Electrodynamics*, Wiley, New York.

Cohen-Tannoudji, C., Dupont-Roc, J. and Grynberg, G. (1992b). *Photons and Atoms*, Wiley, New York.

Cornell, E. A., Hall, D. S., Matthews, M. R. and Wieman, C. E. (1998). *J. Low Temp. Phys.*, **113**, 151.

Crommie, M. F., Lutz, C. P. and Eigler, D. M. (1993). *Science*, **262**, 218.

Cundiff, S. T. and Ye, J. (2003). *Rev. Mod. Phys.*, **75**, 325.

Dalibard, J. (1999). *Proceedins of the international school Enrico Fermi*, p. 321, IOS Press, Amsterdam.

Dalibard, J., Castin, Y. and Mölmer, K. (1992). *Phys. Rev. Lett.*, **68**, 580.

Dalvit, D. A. R., Dziarmaga, J. and Zurek, W. H. (2000). *Phys. Rev. A*, **62**, 013607.

D'Ariano, G. M. and Yuen, H. P. (1996). *Phys. Rev. Lett.*, **76**, 2832.

Davidovich, L. (1996). *Rev. Mod. Phys.*, **68**, 127.

Davidovich, L., Brune, M., Raimond, J.-M. and Haroche, S. (1996). *Phys. Rev. A*, **53**, 1295.

Davidovich, L., Maali, A., Brune, M., Raimond, J.-M. and Haroche, S. (1993). *Phys. Rev. Lett*, **71**, 2360.

Davidovich, L., Raimond, J.-M., Brune, M. and Haroche, S. (1987). *Phys. Rev. A*, **36**, 3771.

Davidovich, L., Raimond, J.-M., Brune, M. and Haroche, S. (1988). N. Abraham, F. T. Arecchi and L. A. Lugiato (eds.), *Instabilities and Chaos in Quantum Optics II*, Plenum Press, New York.

Davis, K. B., Mewes, M.-O., Andrews, M. R., van Druten, N. J., Durfee, D. S., Kurn, D. M. and Ketterle, W. (1995). *Phys. Rev. Lett.*, **75**, 3969.

De Martini, F., Innocenti, G., Jacobovitz, G. R. and Mataloni, P. (1987). *Phys. Rev. Lett.*, **59**, 2955.

de Matos Filho, R. L. and Vogel, W. (1996). *Phys. Rev. Lett.*, **76**, 608.

Dehmelt, H. (1990). *Rev. Mod. Phys.*, **62**, 525.

Dekker, H. (1977). *Phys. Rev. A*, **16**, 2126.

Deutsch, D. and Josza, R. (1992). *Proc. Roy. Soc. London*, **A439**, 553.

DeVoe, R. G. (1998). *Phys. Rev. A*, **58**, 910.

Dicke, R. H. (1954). *Phys. Rev*, **93**, 99.

Diddams, S. A., Udem, T., Bergquist, J. C., Curtis, E. A., Drullinger, R. E., Hollberg, L., Itano, W. M., Lee, W. D., Oates, C. W., Vogel, K. R. and Wineland, D. J. (2001). *Science*, **293**, 825.

Diedrich, F., Bergquist, J. C., Itano, W. M. and Wineland, D. J. (1989). *Phys. Rev. Lett.*, **62**, 403.

Dirac, P. A. M. (1958). *The Principles of Quantum Mechanics*, 4th edn., Oxford University Press, Oxford.

Domokos, P. and Ritsch, H. (2002). *Phys. Rev. Lett.*, **89**, 253003.

Drexhage, K. H. (1974). E. Wolf (ed.), *Progress in Optics*, vol. XII, p. 163, North Holland, Amsterdam.

Drobny, G. and Buzek, V. (2002). *Phys. Rev. A*, **65**, 053410.

Duan, L. M., Blinov, B. B., Moering, D. L. and Monroe, C. (2004). *Quant. Inf. Comp.*, **4**, 165.

Ducloy, M. and Bloch, D. (1996). *Quantum Optics of Confined Systems*, Kluwer, Dordrecht.

Dürr, S., Nonn, T. and Rempe, G. (1998a). *Phys. Rev. Lett.*, **81**, 5705.

Dürr, S., Nonn, T. and Rempe, G. (1998b). *Nature (London)*, **395**, 33.

Eberly, J. H., Narozhny, N. B. and Sanchez-Mondragon, J. J. (1980). *Phys. Rev. Lett.*, **44**, 1323.

Edamatsu, K., Shimizu, R. and Itoh, T. (2002). *Phys. Rev. Lett.*, **89**, 213601.

Eichmann, U., Bergquist, J. C., Bollinger, J. J., Gilligan, J. M., Itano, W. M., Wineland, D. J. and Raizen, M. G. (1993). *Phys. Rev. Lett.*, **70**, 2359.

Eigler, D. M. and Schweizer, E. K. (1990). *Nature (London)*, **344**, 524.

Einstein, A. (1924). *Kgl. Preuss. Akad. Wiss.*, p. 261.

Einstein, A. (1936). *J. Franklin Inst.*, **221**, 349.

Einstein, A., Podolsky, B. and Rosen, N. (1935). *Phys. Rev.*, **47**, 777.

Eiselt, J. and Risken, H. (1989). *Opt. Comm.*, **72**, 351.

Ekert, A. (1991). *Phys. Rev. Lett.*, **67**, 661.

Ekert, A. and Josza, R. (1996). *Rev. Mod. Phys.*, **68**, 733.

Ekert, A. and Macchiavello, C. (1996). *Phys. Rev. Lett.*, **77**, 2585.

Ekert, A., Rarity, J. G., Tapster, P. R. and Palma, G. M. (1992). *Phys. Rev. Lett.*, **69**, 1293.

Elzerman, J. M., Hanson, R., Willems van Beveren, L. H., Wittkamp, B., Vandersypen, L. M. K. and Kouwenhoven, L. P. (2004). *Nature (London)*, **430**, 431.

Englert, B.-G. (1996). *Phys. Rev. Lett.*, **77**, 2154.

Englert, B. G., Schwinger, J., Barut, A. O. and Scully, M. O. (1991). *Europhys. Lett.*, **14**, 25.

Englert, B. G., Scully, M. O. and Walther, H. (1994). *Scientific American (december)*, **271**, 86.

Englert, B. G., Scully, M. O. and Walther, H. (1995). *Nature (London)*, **375**, 367.

Englert, B.-G., Walther, H. and Scully, M. O. (1992). *Appl. Phys. B*, **54**, 366.

Ensher, J. R., Jin, D. S., Matthews, M. R., Wieman, C. E. and Cornell, E. A. (1996). *Phys. Rev. Lett.*, **77**, 4984.

Ernst, R. R., Bodenhausen, G. and Wokaun, A. (1987). *Principles of Nuclear Magnetic Resonance in One and Two Dimensions*, Clarendon Press, Oxford.

Eschner, J., Morigi, G., Schmidt-Kaler, F. and Blatt, R. (2003). *J. Opt. Soc. Am. B*, **20**, 1003.

Eschner, J., Raab, C., Schimdt-Kaler, F. and Blatt, R. (2001). *Nature (London)*, **413**, 495.

Faist, A., Geneux, E., Meystre, P. and Quattropani, A. (1972). *Helv. Phys. Acta*, **45**, 956.

Fattal, D., Diamanti, E., Inoue, K. and Yamamoto, Y. (2004). *Phys. Rev. Lett.*, **92**, 037904.

Feher, G., Gordon, J. P., Buehler, E., Gere, E. A. and Thurmond, C. D. (1958). *Phys. Rev.*, **109**, 221.

Feshbach, H. (1962). *Ann. Phys.*, **19**, 287.

Feynman, R. (1985). *Opt. News*, **11**, 11.

Feynman, R. P. (1965). *Lectures on Physics*, Addison-Wesley, Reading MA.

Filipowicz, P., Javanainen, J. and Meystre, P. (1986). *Phys. Rev. A*, **34**, 3077.

Fischer, T., Maunz, P., Pinkse, P. W. H., Puppe, T. and Rempe, G. (2002). *Phys. Rev. Lett.*, **88**, 163002.

Fisher, M. P. A., Weichman, P. B., Grinstein, G. and Fisher, D. S. (1989). *Phys. Rev. B*, **40**, 546.

Fleischhauer, M. and Schleich, W. P. (1993). *Phys. Rev. A*, **47**, 4258.

Folman, R., Krüger, P., Schmiedmayer, J., Denschlag, J. and Henkel, C. (2002). *Advances in Atomic, Molecular and Optical Physics*, **48**, 263.

Fortunato, M., Raimond, J.-M., Tombesi, P. and Vitali, D. (1999). *Phys. Rev. A*, **60**, 1687.

Franson, J. D. and Ilves, H. (1994). *Appl. Optics*, **33**, 2949.

Friberg, S. R., Machida, S. and Yamamoto, Y. (1992). *Phys. Rev. Lett.*, **69**, 3165.

Fried, D. G., Killian, T. C., Willmann, L., Landhuis, D., Moss, S. C., Kleppner, D. and Greytak, T. J. (1998). *Phys. Rev. Lett.*, **81**, 3811.

Frolich, J. and Marchetti, P. A. (1989). *Comm. in Math. Phys.*, **121**, 177.

Furusawa, A., Sorensen, J. L., Braunstein, S. L., Fuchs, C. A., Kimble, H. J. and Polzik, E. S. (1998). *Science*, **282**, 706.

Gabrielse, G., Dehmelt, H. and Kells, W. (1985). *Phys. Rev. Lett.*, **54**, 537.

Gabrielse, G. and Dehmet, H. (1985). *Phys. Rev. Lett.*, **55**, 67.

Gallagher, T. F. (1994). *Rydberg Atoms*, Cambridge University Press, Cambridge.

Gallas, J. A. C., Leuchs, G., Walther, H. and Figger, H. (1985). D. Bates and B. Bederson (eds.), *Advances in Atomic and Molecular Physics*, vol. XX, Academic Press, New York.

Gangl, M. and Ritsch, H. (2000). *Euro. Phys. Journal D*, **8**, 29.

Garcia-Ripoll, J. J., Martin-Delgado, M. A. and Cirac, J. I. (2004). *Physical Review Letters*, **93**, 250405.

Garcia-Ripoll, J. J., Zoller, P. and Cirac, I. (2003). *Phys. Rev. Lett.*, **91**, 157901.

Gea-Banacloche, J. (1990). *Phys. Rev. Lett.*, **65**, 3385.

Gea-Banacloche, J. (1991). *Phys. Rev. A*, **44**, 5913.

Gérard, J.-M., Sermage, B., Gayral, B., Legrand, B., Costard, E. and Thierry-Mieg, V. (1998). *Phys. Rev. Lett.*, **81**, 1110.

Ghosh, P. K. (1995). *Ion Traps*, Oxford University Press, Oxford.

Ghosh, R. and Mandel, L. (1987). *Phys. Rev. Lett.*, **59**, 1903.

Gisin, N. (1989). *Helv. Physica Acta*, **62**, 363.

Gisin, N. and Gisin, B. (1999). *Phys. Lett. A*, **260**, 323.

Gisin, N. and Massar, S. (1997). *Phys. Rev. Lett.*, **79**, 2153.

Gisin, N., Ribordy, G., Tittel, W. and Zbinden, H. (2002). *Rev. Mod. Phys*, **74**, 145.

Giulini, D., Joos, E., Kiefer, C., Kupsch, J., Stamatescu, I.-O. and Zeh, H. D. (1996). *Decoherence and the Appearance of a Classical World in Quantum Theory*, Springer, Berlin.

Glauber, R. J. (1965). C. de Witt, A. Blandin and C. Cohen-Tannoudji (eds.), *Quantum Optics and Electronics, Les Houches Summer School*, Gordon and Breach, London.

Goldstein, E. V. and Meystre, P. (1997). *Phys. Rev. A*, **56**, 5135.

Goobar, E., Karlsson, A. and Björk, G. (1993a). *Phys. Rev. Lett.*, **71**, 2002.

Goobar, E., Karlsson, A., Björk, G. and Nilsson, O. (1993b). *Opt. Lett.*, **18**, 2138.

Gordon, D. and Savage, C. M. (1999). *Phys. Rev. A*, **59**, 4623.

Goy, P., Raimond, J.-M., Gross, M. and Haroche, S. (1983). *Phys. Rev. Lett.*, **50**, 1903.

Grangier, P., Courty, J.-M. and Reynaud, S. (1992). *Opt. Commun.*, **89**, 99.

Grangier, P., Levenson, J. A. and Poizat, J.-P. (1998). *Nature (London)*, **396**, 537.

Grangier, P., Roch, J. F. and Roger, G. (1991). *Phys. Rev. Lett.*, **66**, 1418.

Grangier, P., Roger, G. and Aspect, A. (1986). *Europhys. Lett.*, **1**, 173.

Greenberger, D. M., Horne, M., Shimony, A. and Zeilinger, A. (1990). *Am. J. Phys.*, **58**, 1131.

Greiner, M., Mandel, O., Esslinger, T., Hänsch, T. W. and Bloch, I. (2002a). *Nature (London)*, **415**, 39.

Greiner, M., Mandel, O., Hänsch, T. W. and Bloch, I. (2002b). *Nature (London)*, **419**, 51.

Griffiths, R. B. (2001). *Consistent Quantum Theory*, Cambridge University Press, Cambridge.

Gross, M. and Haroche, S. (1982). *Phys. Rep.*, **93**, 301.

Gross, M. and Liang, J. (1986). *Phys. Rev. Lett.*, **57**, 3160.

Grover, L. K. (1997). *Phys. Rev. Lett.*, **79**, 4709.

Grover, L. K. (1998). *Phys. Rev. Lett.*, **80**, 4329.

Gulde, S., Riebe, M., Lancaster, G. P. T., Becher, C., Eschner, J., Häffner, H., Schmidt-Kaler, F., Chuang, I. L. and Blatt, R. (2003). *Nature (London)*, **421**, 48.

Guthöhrlein, G. R., Keller, M., Hayasaka, K., Lange, W. and Walther, H. (2001). *Nature (London)*, **414**, 49.

Haake, F. (1973). *Statistical Treatment of Open Systems by Generalized Master Equation*, Springer, Berlin.

Haake, F. and Glauber, R. J. (1972). *Phys. Rev. A*, **5**, 1457.

Haake, F., King, H., Schröder, G., Haus, J. and Glauber, R. J. (1979). *Phys. Rev. A*, **20**, 2047.

Habib, S., Shizume, K. and Zurek, W. H. (1998). *Phys. Rev. Lett.*, **80**, 4361.

Hackermüller, L., Hornberger, K., Brezger, B., Zeilinger, A. and Arndt, M. (2004). *Nature (London)*, **427**, 711.

Hackermüller, L., Uttenthaler, S., Hornberger, K., Reiger, E., Brezger, B., Zeilinger, A. and Arndt, M. (2003). *Phys. Rev. Lett.*, **91**, 090408.

Hadzibabic, Z., Stock, S., Battelier, B., Bretin, V. and Dalibard, J. (2004). *Phys. Rev. Lett.*, **93**, 180403.

Häffner, H., Hänsel, W., Roos, C. F., Benhelm, J., Chek-al-kar, D., Chwalla, M., Körber, T., Rapol, U. D., Riebe, M., Schmidt, P. O., Bechera, C., Gühne, O., Dür, W. and Blatt, R. (2005). *Nature (London)*, **438**, 643.

Hagley, E., Maître, X., Nogues, G., Wunderlich, C., Brune, M., Raimond, J.-M. and Haroche, S. (1997). *Phys. Rev. Lett.*, **79**, 1.

Hänsch, T. W. and Schawlow, A. (1975). *Optics Comm.*, **13**, 68.

Hänsel, W., Hommelhoff, P., Hänsch, T. W. and Reichel, J. (2001). *Nature (London)*, **413**, 498.

Hare, J., Gross, M. and Goy, P. (1988). *Phys. Rev. Lett.*, **61**, 1938.

Haroche, S. (1984). G. Grynberg and R. Stora (eds.), *New Trends in Atomic Physics, Les Houches Summer School Session XXXVIII*, p. 347, North Holland, Amsterdam.

Haroche, S. (1992). J. Dalibard, J.-M. Raimond and J. Zinn-Justin (eds.), *Fundamental Systems in Quantum Optics, Les Houches Summer School, Session LIII*, p. 767, North Holland, Amsterdam.

Haroche, S., Brune, M. and Raimond, J.-M. (1991). *Europhys. Lett.*, **14**, 19.

Haroche, S., Brune, M. and Raimond, J.-M. (1992a). *Appl. Phys. B*, **54**, 355.

Haroche, S., Brune, M. and Raimond, J.-M. (1992b). *Journal de Physique II, Paris*, **2**, 659.

Haroche, S. and Raimond, J.-M. (1985). D. R. Bates and B. Bederson (eds.), *Advances in Atomic and Molecular Physics*, vol. XX, p. 347, Academic Press, New York.

Haroche, S. and Raimond, J.-M. (1994). P. Berman (ed.), *Advances in Atomic and Molecular Physics, Supplement 2*, p. 123, Academic Press, New York.

Haroche, S. and Raimond, J.-M. (1996). *Phys. Today*, **49**, 51.

Heinzen, D. J., Childs, J. J., Thomas, J. E. and Feld, M. S. (1987). *Phys. Rev. Lett*, **58**, 1320.

Heinzen, D. J. and Feld, M. S. (1987). *Phys. Rev. Lett.*, **59**, 2623.

Heinzen, D. J. and Wineland, D. J. (1990). *Phys. Rev. A*, **42**, 2977.

Hennrich, M., Kuhn, A. and Rempe, G. (2005). *Phys. Rev. Lett.*, **94**, 053604.

Hennrich, M., Legero, T., Kuhn, A. and Rempe, G. (2000). *Phys. Rev. Lett.*, **85**, 4872.

Herzog, T. J., Kwiat, P. G., Weinfurter, H. and Zeilinger, A. (1995). *Phys. Rev. Lett.*, **75**, 3034.

Hinds, E. A. (1990). D. Bates and B. Bederson (eds.), *Advances in Atomic and Molecular Physics*, vol. 28, p. 237, Academic Press, San Diego.

Hinds, E. A. (1994). P. Berman (ed.), *Advances in Atomic and Molecular Physics, Supplement 2*, p. 1, Academic Press, New York.

Hinds, E. A. and Sandoghdar, V. (1991). *Phys. Rev. A*, **43**, 398.

Ho, T. L. and Ciobanu, C. V. (2000). *Cond-Mat*, p. 0011095.

Höglund, B. and Metzger, P. B. (1965). *Science*, **150**, 339.

Holzwarth, R., Udem, T., Hänsch, T. W., Knight, J. C., Wadsworth, W. J. and Russel, P. S. J. (2000). *Phys. Rev. Lett.*, **85**, 2264.

Hong, C. K., Ou, Z. Y. and Mandel, L. (1987). *Phys. Rev. Lett.*, **59**, 2044.

Hood, C. J., Chapman, M. S., Lynn, T. W. and Kimble, H. J. (1998). *Phys. Rev. Lett.*, **80**, 4157.

Hood, C. J., Lynn, T. W., Doherty, A. C., Papkins, A. S. and Kimble, H. J. (2000). *Science*, **287**, 1447.

Horodecki, M., Horodecki, P. and Horodecki, R. (1996). *Phys. Lett. A*, **223**, 1.

Hulet, R. G., Hilfer, E. S. and Kleppner, D. (1985). *Phys. Rev. Lett.*, **55**, 2137.

Hulet, R. G. and Kleppner, D. (1983). *Phys. Rev. Lett.*, **51**, 1430.

Hyafil, P., Mozley, J., Perrin, A., Tailleur, J., Nogues, G., Brune, M., Raimond, J.-M. and Haroche, S. (2004). *Phys. Rev. Lett.*, **93**, 103001.

Imoto, N., Haus, H. A. and Yamamoto, Y. (1985). *Phys. Rev. A*, **32**, 2287.

Imry, Y. (1997). *Introduction to Mesoscopic Physics*, Oxford University Press, Oxford.

Inouye, S., Andrews, M. R., Stenger, J., Miesner, H.-J., Stamper-Kurn, D. M. and Ketterle, W. (1998). *Nature (London)*, **392**, 151.

Itzykson, C. and Zuber, J. B. (1980). *Quantum Field Theory*, Mc-Graw-Hill, New York.

Jackson, J. D. (1975). *Classical Electrodynamics*, 2nd edn., Wiley, New York.

Jacobson, J., Björk, G., Chuang, I. and Yamamoto, Y. (1995). *Phys. Rev. Lett.*, **74**, 4835.

Jaeger, G., Horne, M. A. and Shimony, A. (1993). *Phys. Rev. A*, **48**, 1023.

Jaekel, M.-T. and Reynaud, S. (1997). *Rep. Prog. Phys.*, **60**, 863.

Jaksch, D., Briegel, H.-J., Cirac, J. I., Gardiner, C. W. and Zoller, P. (1999). *Phys. Rev. Lett.*, **82**, 1975.

Jaksch, D., Bruder, C., Cirac, J. I., Gardiner, C. W. and Zoller, P. (1998). *Phys. Rev. Lett.*, **81**, 3108.

Jaksh, D., Cirac, J. I., Zoller, P., Rolston, S. L., Côté, R. and Lukin, M. D. (2000). *Phys. Rev. Lett.*, **85**, 2208.

James, D. F. V. (1998). *Appl. Phys. B*, **66**, 181.

Jaynes, E. T. and Cummings, F. W. (1963). *Proc. IEEE*, **51**, 89.

Jhe, W., Anderson, A., Hinds, E. A., Meschede, D., Moi, L. and Haroche, S. (1987a). *Phys. Rev. Lett.*, **58**, 666.

Jhe, W., Anderson, A., Hinds, E. A., Meschede, D., Moi, L. and Haroche, S. (1987b). *Phys. Rev. Lett.*, **58**, 1497.

Joos, E. and Zeh, H. D. (1985). *Z. Phys. B*, **59**, 223.

Kaluzny, Y., Goy, P., Gross, M., Raimond, J.-M. and Haroche, S. (1983). *Phys. Rev. Lett.*, **51**, 1175.

Keller, M., Lange, B., Hayasaka, K., Lange, W. and Walther, H. (2004). *Nature (London)*, **431**, 1075.

Kielpinski, D., Meyer, V., Rowe, M. A., Sackett, C. A., Itano, W. M., Monroe, C. and Wineland, D. J. (2001). *Science*, **291**, 1013.

Kielpinski, D., Monroe, C. and Wineland, D. J. (2002). *Nature (London)*, **417**, 709.

Kim, J. I., Fonseca-Romero, K. M., Horiguti, A. M., Davidovich, L., Nemes, M. C. and de Toledo Piza, A. F. R. (1999). *Phys. Rev. Lett.*, **82**, 4737.

Kim, Y.-H., Kulik, S. and Shih, Y. (2001). *Phys. Rev. Lett.*, **86**, 1370.

Kim, Y. H., Yu, R., Kulik, S. P. and Shih, Y. (2000). *Phys. Rev. Lett.*, **84**, 1.

Kimble, H. J. (1992). J. Dalibard, J.-M. Raimond and J. Zinn-Justin (eds.), *Fundamental Systems in Quantum Optics, Les Houches Summer School, Session LIII*, p. 545, North Holland, Amsterdam.

Kimble, H. J., Dagenais, M. and Mandel, L. (1977). *Phys. Rev. Lett.*, **39**, 691.

Kleckner, D., Marshall, W., de Dood, M. J. A., Dinyari, K. N., Pors, B.-J., Irvine, W. T. M. and Bouwmeester, D. (2006). *Physical Review Letters*, **96**, 173901.

Kleppner, D. (1981). *Phys. Rev. Lett.*, **47**, 233.

Knight, P. L. and Allen, L. (1983). *Concepts of Quantum Optics*, Pergamon, Oxford.

Knight, P. L. and Radmore, P. M. (1982). *Phys. Rev. A*, **26**, 676.

Knill, E., Laflamme, R. and Milburn, G. J. (2001). *Nature (London)*, **409**, 46.

Kobayashi, T., Zheng, Q. and Sekiguchi, T. (1995). *Phys. Rev. A*, **52**, 2835.

Kokorowski, D. A., Cronin, A. D., Roberts, T. D. and Pritchard, D. E. (2001). *Phys. Rev. Lett.*, **86**, 2191.

Kraus, K. (1983). *States, Effects and Operations*, Springer, Berlin.

Krause, J., Scully, M. O. and Walther, H. (1987). *Phys. Rev. A*, **36**, 4547.

Krause, J., Scully, M. O., Walther, T. and Walther, H. (1989). *Phys. Rev. A*, **39**, 1915.

Kroutvar, M., Ducommun, Y., Heiss, D., Bichler, M., Schuh, D., Abstreiter, G. and Finley, J. F. (2004). *Nature (London)*, **432**, 81.

Kuhn, A., Hennrich, M. and Rempe, G. (2002). *Phys. Rev. Lett.*, **89**, 067901.

Kuhr, S., Alt, W., Schrader, D., Dotsenko, I., Miroshnychenko, Y., Rosenfeld, W., Khudaverdyan, M., Gomer, V., Rauschenbeutel, A. and Meschede, D. (2003). *Phys. Rev. Lett.*, **91**, 213002.

Kurtsiefer, C., Pfau, T. and Mlynek, J. (1997). *Nature (London)*, **386**, 150.

Kurtsiefer, C., Zarda, P., Halder, M., Weinfurter, H., Gorman, P. M., Tapster, P. R. and Rarity, J. G. (2002). *Nature (London)*, **419**, 450.

Kwiat, P. G., Mattle, K., Weinfurter, H., Zeilinger, A., Sergienko, A. V. and Shih, Y. (1995). *Phys. Rev. Lett.*, **75**, 4337.

Kwiat, P. G., Steinberg, A. M. and Chiao, R. Y. (1992). *Phys. Rev. A*, **45**, 7729.

Laflamme, R., Miquel, C., Paz, J. P. and Zurek, W. H. (1996). *Phys. Rev. Lett.*, **77**, 198.

Lambrecht, A., Jaekel, M.-T. and Reynaud, S. (1996). *Phys. Rev. Lett.*, **77**, 615.

LaPorta, A., Slusher, R. E. and Yurke, B. (1989). *Phys. Rev. Lett.*, **62**, 28.

Legero, T., Wilk, T., Hennrich, M., Rempe, G. and Kuhn, A. (2004). *Phys. Rev. Lett*, **93**, 070503.

Leggett, A. J. (2001). *Rev. Mod. Phys.*, **73**, 307.

Leibfried, D., Barrett, M. D., Schaetz, T., Britton, J., Chiaverini, J., Itano, W. M., Jost, J. D., Langer, C. and Wineland, D. J. (2004). *Science*, **304**, 1476.

Leibfried, D., Blatt, R., Monroe, C. and Wineland, D. J. (2003a). *Rev. Mod. Phys.*, **75**, 281.

Leibfried, D., DeMarco, B., Meyer, V., Lucas, D., Barrett, M., Britton, J., Itano, W. M., Jelenkovic, B., Langer, C., Rosenband, T. and Wineland, D. J. (2003b). *Nature (London)*, **422**, 412.

Leibfried, D., Knill, E., Seidlin, S., Britton, J., Blakestad, R. B., Chiaverini, J., Hume, D. B., Itano, W. M., Jost, J. D., Langer, C., Ozeri, R., Reichle, R. and Wineland, D. J. (2005). *Nature (London)*, **438**, 639.

Leibfried, D., Meekhof, D. M., King, B. E., Monroe, C., Itano, W. M. and Wineland, D. J. (1996). *Phys. Rev. Lett.*, **77**, 4281.

Lemonde, P., Laurent, P., Santarelli, G., Abgrall, M., Kitching, J., Sortais, Y., Bize, S., Nicolas, C., Zhang, S., Schehr, G., Clairon, A., Dimarq, N., Petit, P., Mann, A., Luiten, A., Schang, S. and Salomon, C. (2000). A. Luiten (ed.), *Frequency Measurement and Control, Topics in Applied Physics*, p. 131, Springer Verlag, Heidelberg.

Lenstra, A. K. and Lenstra, H. W. (1993). *Lecture Notes in Mathematics*, vol. 1554, Springer, Berlin.

Leonhardt, U. (1997). *Measuring the Quantum State of Light*, Cambridge University Press, Cambridge.

Levenson, J. A., Abram, I., Rivera, T., Fayolle, P., Garreau, J.-C. and Grangier, P. (1993). *Phys. Rev. Lett.*, **70**, 267.

Levenson, M. D., Shelby, R. M., Reid, M. and Walls, D. F. (1986). *Phys. Rev. Lett.*, **57**, 2473.

Levitt, M. H. (2001). *Spin Dynamics: Basics of Nuclear Magnetic Resonance*, Wiley, Chichester.

Lhuillier, C. and Laloe, F. (1982). *J. de Physique Paris*, **43**, 197.

Lidar, D. A., Chuang, I. L. and Wharley, K. B. (1998). *Phys. Rev. Lett.*, **81**, 2594.

Lindblad, G. (1976). *Commun. Math. Phys.*, **48**, 119.

Lombardi, E., Sciarrino, F., Popescu, S. and de Martini, F. (2002). *Phys. Rev. Lett.*, **88**, 070402.

Loudon, R. (1983). *The Quantum Theory of Light*, Oxford University Press, Oxford.

Louisell, W. H. (1973). *Quantum Statistical Properties of Radiation*, Wiley, New York.

Lugiato, L. A., Scully, M. O. and Walther, H. (1987). *Phys. Rev. A*, **36**, 740.

Lutterbach, L. G. and Davidovich, L. (1997). *Phys. Rev. Lett.*, **78**, 2547.

Lvovsky, A. I., Hansen, H., Aichele, T., Benson, O., Mlynek, J. and Schiller, S. (2001). *Phys. Rev. Lett.*, **87**, 050402.

Lvovsky, A. I. and Mlynek, J. (2002). *Phys. Rev. Lett.*, **88**, 250401.

Mabuchi, H. and Doherty, A. C. (2002). *Science*, **298**, 1372.

Maia-Neto, P. A., Davidovich, L. and Raimond, J.-M. (1991). *Phys. Rev. A*, **43**, 5073.

Maioli, P., Meunier, T., Gleyzes, S., Auffeves, A., Nogues, G., Brune, M., Raimond, J.-M. and Haroche, S. (2005). *Phys. Rev. Lett.*, **94**, 113601.

Maître, X., Hagley, E., Nogues, G., Wunderlich, C., Goy, P., Brune, M., Raimond, J.-M. and Haroche, S. (1997). *Phys. Rev. Lett.*, **79**, 769.

Mandel, O., Greiner, M., Widera, A., Rom, T., Hänsch, T. W. and Bloch, I. (2003). *Nature (London)*, **425**, 937.

Margalit, N. M., Babic, D. I., Streubel, K., Mirin, R. P., Mars, D. E., Brower, J. E. and Hu, E. L. (1996). *Appl. Phys. Lett.*, **69**, 471.

Marrocco, M., Weidinger, M., Sang, R. T. and Walther, H. (1998). *Phys. Rev. Lett.*, **81**, 5784.

Marshall, W., Simon, C., Penrose, R. and Bouwmeester, D. (2003). *Phys. Rev. Lett.*, **91**, 130401.

Martorell, J. and Lawandy, N. M. (1990). *Phys. Rev. Lett.*, **65**, 1877.

Maunz, P., Puppe, T., Schuster, I., Syassen, N., Pinkse, P. W. H. and Rempe, G. (2004). *Nature (London)*, **428**, 50.

Maunz, P., Puppe, T., Schuster, I., Syassen, N., Pinkse, P. W. H. and Rempe, G. (2005). *Phys. Rev. Lett.*, **94**, 033002.

McDermott, R., Simmonds, R. W., Steffen, M., Cooper, K. B., Cicak, K., Osborn, K. D., Oh, S., Pappas, D. P. and Martinis, J. M. (2005). *Science*, **307**, 1299.

McEvoy, J. and Zarate, O. (1999). *Introducing quantum theory*, Icon Books, Duxford, UK.

McGillivray, J. C. and Feld, M. S. (1976). *Phys. Rev. A*, **14**, 1169.

McKeever, J., Boca, A., Boozer, A. D., Buck, J. R. and Kimble, H. J. (2003a). *Nature (London)*, **425**, 268.

McKeever, J., Buck, J. R., Boozer, A. D., Kuzmich, A., Nägerl, H.-C., Stamper-Kurn, D. M. and Kimble, H. J. (2003b). *Phys. Rev. Lett.*, **90**, 133602.

Meekhof, D. M., Monroe, C., King, B. E., Itano, W. M. and Wineland, D. J. (1996). *Phys. Rev. Lett.*, **76**, 1796.

Mermin, N. D. (1990a). *Phys. Rev. Lett.*, **65**, 1838.

Mermin, N. D. (1990b). *Phys. Today*, p. 9.

Meschede, D., Walther, H. and Müller, G. (1985). *Phys. Rev. Lett.*, **54**, 551.

Messiah, A. (1961). *Quantum Mechanics*, North Holland, Amsterdam.

Meunier, T., Gleyzes, S., Maioli, P., Auffeves, A., Nogues, G., Brune, M., Raimond, J.-M. and Haroche, S. (2005). *Phys. Rev. Lett.*, **94**, 010401.

Meystre, P. (2001). *Atom Optics*, Springer, Berlin.

Meystre, P., Rempe, G. and Walther, H. (1988). *Opt. Lett.*, **13**, 1078.

Meystre, P. and Sargent, M. (1999). *Elements of Quantum Optics*, 3rd edn., Springer, Heidelberg.

Meystre, P. and Scully, M. O. (1983). *Quantum Optics, Experimental Gravity and Measurement Theory*, Plenum, New York.

Milburn, G. J., Schneider, G. and James, D. F. (2000). *Fortschr. Physik*, **48**, 801.

Millburn, G. J. and Walls, D. F. (1983). *Phys. Rev. A*, **28**, 2065.

Miller, C. A., Hilsenbeck, J. and Risken, H. (1992). *Phys. Rev. A*, **46**, 4323.

Milman, P., Auffeves, A., Yamaguchi, F., Brune, M., Raimond, J.-M. and Haroche, S. (2005). *Eur. Phys. J. D*, **32**, 233.

Milman, P., Castin, Y. and Davidovich, L. (2000). *Phys. Rev. A*, **61**, 063803.

Misra, B. and Sudarshan, E. C. G. (1977). *J. Math. Phys.*, **18**, 756.

Mitchell, M. W., Lundeen, J. S. and Steinberg, A. M. (2004). *Nature (London)*, **429**, 161.

Mölmer, K. and Sörensen, A. (1999). *Phys. Rev. Lett.*, **82**, 1835.

Monroe, C., Meekhof, D. M., King, B. E., Itano, W. M. and Wineland, D. J. (1995a). *Phys. Rev. Lett.*, **75**, 4714.

Monroe, C., Meekhof, D. M., King, B. E., Jefferts, S. R., Itano, W. M., Wineland, D. J. and Gould, P. (1995b). *Phys. Rev. Lett.*, **75**, 4011.

Monroe, C., Meekhof, D. M., King, B. E. and Wineland, D. J. (1996). *Science*, **272**, 1131.

Montina, A. and Arecchi, F. T. (2002). *Phys. Rev. A*, **66**, 013605.

Morigi, G., Solano, E., Englert, B.-G. and Walther, H. (2002). *Phys. Rev. A*, **65**, 040102.

Morin, S. E., Yu, C. C. and Mossberg, T. W. (1994). *Phys. Rev. Lett.*, **73**, 1489.

Mozley, J., Hyafil, P., Nogues, G., Brune, M., Raimond, J.-M. and Haroche, S. (2005). *Eur. Phys. J. D*, **35**, 45.

Münstermann, P., Fischer, T., Maunz, P., Pinkse, P. W. H. and Rempe, G. (1999). *Phys. Rev. Lett.*, **82**, 3791.

Münstermann, P., Fischer, T., Maunz, P., Pinkse, P. W. H. and Rempe, G. (2000). *Phys. Rev. Lett.*, **84**, 4068.

Myatt, Q. J., King, B. E., Turchette, Q. A., Sackett, C. A., Kielpinski, D., Itano, W. M., Monroe, C. and Wineland, D. J. (2000). *Nature (London)*, **403**, 269.

Nägerl, H. C., Bechter, W., Eschner, J., Schmidt-Kaler, F. and Blatt, R. (1998a). *Appl. Phys. B*, **66**, 603.

Nägerl, H. C., Leibfried, D., Rohde, H., Thalhammer, G., Eschner, J., Schmidt-Kaler, F. and Blatt, R. (1999). *Phys. Rev. A*, **60**, 145.

Nägerl, H. C., Leibfried, D., Schmidt-Kaler, F., Eschner, J. and Blatt, R. (1998b). *Optics express*, **3**, 89.

Nagourney, W., Sandberg, J. and Dehmelt, H. (1986). *Phys. Rev. Lett.*, **56**, 2797.

Nielsen, M. A. and Chuang, I. L. (2000). *Quantum Computation and Quantum Information*, Cambridge University Press, Cambridge.

Nielsen, M. A., Knill, E. and Laflamme, R. (1998). *Nature (London)*, **396**, 52.

Nogues, G., Rauschenbeutel, A., Osnaghi, S., Brune, M., Raimond, J.-M. and Haroche, S. (1999). *Nature (London)*, **400**, 239.

Nussenzveig, A., Hare, J., Steinberg, A. M., Moi, L., Gross, M. and Haroche, S. (1991). *Europhys. Lett.*, **14**, 755.

Nussenzveig, P., Bernardot, F., Brune, M., Hare, J., Raimond, J.-M., Haroche, S. and Gawlik, W. (1993). *Phys. Rev. A*, **48**, 3991.

Nussmann, S., Murr, K., Hijlkema, M., Weber, B., Kuhn, A. and Rempe, G. (2005). *Nature Physics*, **1**, 122.

Omnès, R. (1994). *The Interpretation of Quantum Mechanics*, Princeton University Press, Princeton.

Osnaghi, S., Bertet, P., Auffeves, A., Maioli, P., Brune, M., Raimond, J.-M. and Haroche, S. (2001). *Phys. Rev. Lett.*, **87**, 037902.

Ou, Z. Y. and Mandel, L. (1988). *Phys. Rev. Lett.*, **61**, 50.

Peil, S. and Gabrielse, G. (1999). *Phys. Rev. Lett.*, **83**, 1287.

Pereira, S. F., Ou, Z. Y. and Kimble, H. J. (1994). *Phys. Rev. Lett.*, **72**, 214.

Peres, A. (1995). *Quantum theory: concepts and methods*, Kluwer, Dordrecht.

Peres, A. (1996). *Phys. Rev. Lett.*, **77**, 1413.

Perina, J. (1991). *Quantum Statistics of Linear and Non-linear Optical Phenomena*, 2nd edn., Springer, Heidelberg.

Peter, E., Senellart, P., Martrou, D., Lemaître, A., Hours, J., Gérard, J.-M. and Bloch, J. (2005). *Phys. Rev. Lett.*, **95**, 067401.

Pethick, C. and Smith, H. (2001). *Bose–Einstein condensation in dilute gases*, Cambridge University Press, Cambridge.

Petta, J. R., Johnson, A. C., Taylor, J. M., Laird, E. A., Yacoby, A., Lukin, M. D., Markus, C. M., Hanson, M. P. and Gossard, A. C. (2005). *Science*, **309**, 2180.

Pfau, T., Spälter, S., Kurtsiefer, C., Ekstrom, C. R. and Mlynek, J. (1994). *Phys. Rev. Lett.*, **73**, 1223.

Pinard, M., Fabre, C. and Heidmann, A. (1995). *Phys. Rev. A*, **51**, 2443.

Pitaevskii, L. P. and Stringari, S. (2003). *Bose–Einstein condensation*, Oxford University Press, Oxford.

Planck, M. (1900). *Verhandlungen der Deutschen Physikalischen Gesellschaft*, **2**, 237.

Plenio, M. B. and Knight, P. L. (1998). *Rev. Mod. Phys.*, **70**, 101.

Poizat, J.-P. and Grangier, P. (1993). *Phys. Rev. Lett.*, **70**, 271.

Poizat, J.-P., Roch, J.-F. and Grangier, P. (1994). *Ann. Phys. Fr.*, **19**, 265.

Poole, C. P., Farach, M. A. and Creswick, R. J. (1995). *Superconductivity*, Academic Press, San Diego.

Porras, D. and Cirac, I. (2004). *Phys. Rev. Lett.*, **92**, 207901.

Poyatos, J. F., Cirac, J. I. and Zoller, P. (1996). *Phys. Rev. Lett.*, **77**, 4728.

Preskill, J. (2005). *Quantum computation*, http://www.theory.caltech.edu/~ preskill/ph229/.

Protsenko, I. E., Reymond, G., Schlosser, N. and Grangier, P. (2002). *Phys. Rev. A*, **65**, 052301.

Purcell, E. M. (1946). *Phys. Rev.*, **69**, 681.

Raimond, J.-M., Brune, M. and Haroche, S. (2001). *Rev. Mod. Phys.*, **73**, 565.

Raimond, J.-M., Goy, P., Gross, M., Fabre, C. and Haroche, S. (1982). *Phys. Rev. Lett.*, **49**, 1924.

Raimond, J.-M., Meunier, T., Bertet, P., Gleyzes, S., Maioli, P., Auffeves, A., Nogues, G., Brune, M. and Haroche, S. (2005). *J. Phys. B*, **38**, S535.

Rainville, S., Thompson, J. K. and Pritchard, D. E. (2004). *Science*, **303**, 334.

Raithel, G., Wagner, C. and Walther, H. (1994). P. Berman (ed.), *Advances in Atomic and Molecular Physics, Supplement 2*, p. 57, Academic Press, New York.

Raizen, M. G., Thompson, R. J., Brecha, R. J., Kimble, H. J. and Carmichael, H. J. (1989). *Phys. Rev. Lett.*, **63**, 240.

Ramsey, N. F. (1985). *Molecular Beams*, International Series of Monographs on Physics, Oxford University Press, Oxford.

Rangaswamy, M., Porejsz, B., Ardekani, B. A., Choi, S. J., Tanabe, J. L., Lim, K. O. and Begleiter, H. (2004). *Neuroimage*, **21**, 329.

Rarity, J. G., Owens, P. C. M. and Tapster, P. R. (1994). *J. Mod. Opt.*, **41**, 2435.

Rarity, J. G. and Tapster, P. R. (1990). *Phys. Rev. Lett.*, **64**, 2495.

Rauch, H. and Werner, S. A. (2000). *Neutron Interferometry*, Oxford University Press, Oxford.

Rauch, H., Zeilinger, A., Badurek, G., Wilfing, A., Bauspiess, W. and Bonse, U. (1975). *Phys. Lett. A*, **54**, 425.

Rauschenbeutel, A., Bertet, P., Osnaghi, S., Nogues, G., Brune, M., Raimond, J.-M. and Haroche, S. (2001). *Phys. Rev. A*, **64**, 050301.

Rauschenbeutel, A., Nogues, G., Osnaghi, S., Bertet, P., Brune, M., Raimond, J.-M. and Haroche, S. (1999). *Phys. Rev. Lett.*, **83**, 5166.

Rauschenbeutel, A., Nogues, G., Osnaghi, S., Bertet, P., Brune, M., Raimond, J.-M. and Haroche, S. (2000). *Science*, **288**, 2024.

Raussendorf, R. and Briegel, H. J. (2001). *Phys. Rev. Lett.*, **86**, 5188.

Reithmaier, J. P., Sek, G., Löffler, A., Hofmann, G., Kuhn, S., Reitzenstein, S., Keldysh, L. V., Kulakovskii, V. D., Reinecke, T. L. and Forchel, A. (2004). *Nature (London)*, **432**, 197.

Rempe, G., Schmidt-Kaler, F. and Walther, H. (1990). *Phys. Rev. Lett.*, **64**, 2783.

Rempe, G., Scully, M. O. and Walther, H. (1991). *Physica Scripta*, **T34**, 5.

Rempe, G. and Walther, H. (1990). *Phys. Rev. A*, **42**, 1650.

Rempe, G., Walther, H. and Klein, N. (1987). *Phys. Rev. Lett.*, **58**, 353.

Reynaud, S., Lamine, B., Lambrecht, A., Maia Neto, P. and Jaekel, M.-T. (2002). *Int. J. Mod. Phys. A*, **17**, 1003.

Riebe, M., Häffner, H., Roos, C. F., Hänsel, W., Benhelm, J., Lancaster, G. P. T., Körber, T. W., Becher, C., Schmidt-Kaler, F., James, D. F. V. and Blatt, R. (2004). *Nature (London)*, **429**, 734.

Roch, J. F., Poizat, J. P. and Grangier, P. (1993). *Phys. Rev. Lett.*, **71**, 2006.

Roch, J.-F., Vigneron, K., Grelu, P., Sinatra, A., Poizat, J.-P. and Grangier, P. (1997). *Phys. Rev. Lett.*, **78**, 634.

Roos, C., Zeiger, T., Rohde, H., Nägerl, H. C., Eschner, J., Leibfried, D., Schmidt-Kaler, F. and Blatt, R. (1999). *Phys. Rev. Lett.*, **83**, 4713.

Roos, C. F., Lancaster, G. P. T., Riebe, M., Häffner, H., Hänsel, W., Gulde, S., Becher, C., Eschner, J., Schmidt-Kaler, F. and Blatt, R. (2004a). *Phys. Rev. Lett.*, **92**, 220402.

Roos, C. F., Leibfried, D., Mundt, A., Schmidt-Kaler, F., Eschner, J. and Blatt, R. (2000). *Phys. Rev. Lett.*, **85**, 5547.

Roos, C. F., Riebe, M., Häffner, H., Hänsel, W., Benhelm, J., Lancaster, G. P. T., Becher, C., Schmidt-Kaler, F. and Blatt, R. (2004b). *Science*, **304**, 1478.

Rowe, M. A., Ben-Kish, A., Demarco, B., Leibfried, D., Meyer, V., Beall, J., Britton, J., Hughes, J., Itano, W. M., Jelenkovic, B., Langer, C., Rosenband, T. and Wineland, D. J. (2002). *Quantum Infor. Comput.*, **2**, 257.

Rowe, M. A., Kielpinski, D., Meyer, V., Sackett, C. A., Itano, W. M., Monroe, C. and Wineland, D. J. (2001). *Nature (London)*, **409**, 791.

Rugar, D., Budakian, B., Mamin, H. J. and Chui, B. W. (2004). *Nature (London)*, **430**, 329.

Sackett, C. A., Kielpinski, D., King, B. E., Langer, C., Meyer, V., Myatt, C. J., Rowe, M., Turchette, Q. A., Itano, W. M., Wineland, D. J. and Monroe, C. (2000). *Nature (London)*, **404**, 256.

Sakurai, J. J. (1994). *Modern Quantum Mechanics*, Addison Wesley, New York.

Sandoghdar, V., Sukenik, C., Haroche, S. and Hinds, E. A. (1996). *Phys. Rev. A*, **53**, 1919.

Sandoghdar, V., Sukenik, C. I., Hinds, E. A. and Haroche, S. (1992). *Phys. Rev. Lett.*, **68**, 3432.

Santori, C., Fattal, D., Vukovic, J., Solomon, G. S. and Yamamoto, Y. (2002). *Nature (London)*, **419**, 594.

Scharf, G. (1970). *Helv. Physica Acta*, **43**, 806.

Schleich, W. P. (2001). *Quantum Optics in Phase Space*, Wiley, Berlin.

Schmidt-Kaler, F., Häffner, H., Riebe, M., Gulde, S., Lancaster, G. P. T., Deuschle, T., Becher, C., Roos, C. F., Eschner, J. and Blatt, R. (2003). *Nature (London)*, **422**, 408.

Schrödinger, E. (1935). *Naturwissenschaften*, **23**, 807, 823, 844.

Schuster, D. I., Wallraff, A., Blais, A., Frunzio, L., Huang, R.-S., Majer, J., Girvin, S. M. and Schoelkopf, R. J. (2005). *Phys. Rev. Lett.*, **94**, 123602.

Schwinger, J., Scully, M. O. and Englert, B.-G. (1988). *Z. Phys. D*, **10**, 135.

Scully, M. O. and Drühl, K. (1982). *Phys. Rev. A*, **25**, 2208.

Scully, M. O., Englert, B.-G. and Schwinger, J. (1989). *Phys. Rev. A*, **40**, 1775.

Scully, M.-O., Englert, B.-G. and Walther, H. (1991). *Nature (London)*, **351**, 111.

Scully, M. O. and Walther, H. (1989). *Phys. Rev. A*, **39**, 5229.

Scully, M. O. and Zubairy, M. S. (1997). *Quantum Optics*, Cambridge University Press, Cambridge.

Shimizu, F., Shimizu, K. and Takuma, H. (1992). *Phys. Rev. A*, **46**, R17.

Shor, P. W. (1994). S. Goldwasser (ed.), *Proceedings of the 35th Annual Symposium on the Foundations of Computer Science*, p. 124, IEEE Computer Society Press, Los Alamitos, CA.

Shor, P. W. (1995). *Phys. Rev. A*, **52**, R2493.

Simon, D. R. (1994). S. Goldwasser (ed.), *Proceedings of the 35th Annual Symposium on the Foundations of Computer Science*, p. 116, IEEE Computer Society Press, Los Alamitos, CA.

Sinatra, A. and Castin, Y. (1998). *Eur. Phys. J. D*, **4**, 247.

Sinatra, A., Roch, J. F., Vigneron, K., Grelu, P., Poizat, J.-P., Wang, K. and Grangier, P. (1998). *Phys. Rev. A*, **57**, 2980.

Smithey, D. T., Beck, M., Raymer, M. G. and Faridani, A. (1993). *Phys. Rev. Lett.*, **70**, 1244.

Solano, E., de Matos Filho, R. L. and Zagury, N. (1999). *Phys. Rev. A*, **59**, R2539.

Spiller, T. (1994). *Phys. Lett. A*, **192**, 163.

Steane, A. (1996a). *Proc. Roy. Soc. London*, **452**, 2551.

Steane, A. M. (1996b). *Phys. Rev. Lett.*, **77**, 793.

Storey, P., Tan, S., Collet, M. and Walls, D. (1995). *Nature (London)*, **375**, 368.

Storry, C. H., Speck, A., Sage, D. L., Guise, N., Gabrielse, G., Grzonka, D., Oelert, W., Schepers, G., Sefzick, T., Pittner, H., Herrmann, M., Walz, J., Hänsch, T. W., Comeau, D., Hessels, E. A. and ATRAP Collaboration (2004). *Phys. Rev. Lett.*, **93**, 263401.

Sukenik, C. I., Boshier, M. G., Cho, D., Sandoghdar, V. and Hinds, E. A. (1993). *Phys. Rev. Lett.*, **70**, 560.

Tanaka, K., Nakamura, T., Takamatsu, W., Yamanishi, M., Lee, Y. and Ishihara, T. (1995). *Phys. Rev. Lett.*, **74**, 3380.

Tavis, M. and Cummings, F. W. (1969). *Phys. Rev.*, **188**, 692.

Teich, M. C. and Saleh, B. E. A. (1989). *Quantum Opt.*, **1**, 153.

Thompson, R. J., Rempe, G. and Kimble, H. J. (1992). *Phys. Rev. Lett.*, **68**, 1132.

Thorne, K. S., Drever, R. W. P., Caves, C. M., Zimmermann, M. and Sandberg, V. D. (1978). *Phys. Rev. Lett.*, **40**, 667.

Tittel, W., Brendel, J., Zbinden, H. and Gisin, N. (1998). *Phys. Rev. Lett.*, **81**, 3563.

Tittel, W., Brendel, J., Zbinden, H. and Gisin, N. (2000). *Phys. Rev. Lett.*, **84**, 4737.

Tonomura, A., Endo, J., Matsuda, T., Kawasaki, T. and Ezawa, H. (1989). *Am. J. Phys.*, **57**, 117.

Unruh, W. G. (1978). *Phys. Rev. D*, **18**, 1764.

Unruh, W. G. and Zurek, W. H. (1989). *Phys. Rev. D*, **40**, 1071.

Vahala, K. J. (2003). *Nature (London)*, **424**, 839.

Vaidyanathan, A. G., Spencer, W. P. and Kleppner, D. (1981). *Phys. Rev. Lett.*, **47**, 1592.

van der Wal, C. H., ter Haar, A. C. J., Wilhelm, F. K., Schouten, R. N., Harmans, C. J. P. M., Orlando, T. P., Lloyd, S. and Moij, J. E. (2000). *Science*, **290**, 773.

van Dyck, R. S., Schwinberg, P. B. and Dehmelt, H. G. (1977). *Phys. Rev. Lett.*, **38**, 310.

van Dyck, R. S., Schwinberg, P. B. and Dehmelt, H. G. (1987). *Phys. Rev. Lett.*, **59**, 26.

Vandersypen, L. M. K., Steffen, M., Breyta, G., Yannoni, C. S., Sherwood, M. H. and Chuang, I. L. (2001). *Nature (London)*, **414**, 883.

Vandersypen, L. M. K., Yannoni, C. S., Sherwood, M. H. and Chuang, I. L. (1999). *Phys. Rev. Lett.*, **83**, 3085.

Varcoe, B. T. H., Brattke, S., Weidinger, M. and Walther, H. (2000). *Nature (London)*, **403**, 743.

Vasanelli, A., Ferreira, R. and Bastard, G. (2002). *Phys. Rev. Lett.*, **89**, 216804.

Vion, D., Aassime, A., Cottet, A., Joyez, P., Pothier, H., Urbina, C., Esteve, D. and Devoret, M. H. (2002). *Science*, **296**, 886.

Vogel, W., Welsch, D. G. and Wallentowitz, S. (2001). *Quantum Optics: an Introduction*, 2nd edn., Wiley, Berlin.

Vuletic, V. and Chu, S. (2000). *Phys. Rev. Lett.*, **84**, 3787.

Wallraff, A., Schuster, D. I., Blais, A., Frunzio, L., Huang, R. S., Majer, J., Kumar, S., Girvin, S. M. and Schoelkopf, R. J. (2004). *Nature (London)*, **431**, 162.

Walls, D. F. and Milburn, G. J. (1995). *Quantum Optics*, Springer Verlag, New York.

Walther, H. (1988). *Phys. Scripta*, **T 23**, 165.

Walther, H. (1992). J. Dalibard, J.-M. Raimond and J. Zinn-Justin (eds.), *Fundamental Systems in Quantum Optics, Les Houches Summer School, Session LIII*, p. 211, North Holland, Amsterdam.

Walther, H. (1993). H. Walther, T. Hänsch and B. Neizert (eds.), *Atomic Physics*, vol. 13, AIP Press, New York.

Walther, P., Pan, J. W., Aspelmeyer, M., Ursin, R., Gasparoni, S. and Zeilinger, A. (2004). *Nature (London)*, **429**, 158.

Weidinger, M., Varcoe, B. T. H., Heerlein, R. and Walther, H. (1999). *Phys. Rev. Lett.*, **82**, 3795.

Weihs, G., Jennewein, T., Simon, C., Weinfurter, H. and Zeilinger, A. (1998). *Phys. Rev. Lett.*, **81**, 5039.

Werner, S. A., Collella, R., Overhauser, A. W. and Eagen, C. F. (1975). *Phys. Rev. Lett.*, **35**, 1053.

White, A. G., James, D. F. V., Eberhard, P. H. and Kwiat, P. G. (1999). *Phys. Rev. Lett.*, **83**, 3103.

Widera, A., Mandel, O., Greiner, M., Kreim, S., Hänsch, T. W. and Bloch, I. (2004). *Phys. Rev. Lett.*, **92**, 160406.

Wigner, E. P. (1932). *Phys. Rev.*, **40**, 749.

Wilson, M. A., Bushev, P., Eschner, J., Schmidt-Kaler, F., Becher, C., Blatt, R. and Dorner, U. (2003). *Phys. Rev. Lett.*, **91**, 213602.

Wineland, D. J., Barett, M., Britton, J., Chiaverini, J., DeMarco, B., Itano, W. M., Jelenkovic, B., Langer, C., Leibfried, D., Meyer, V., Rosenband, T. and Schätz, T. (2003). *Phil. Trans. Roy. Soc.*, **361**, 1349.

Wineland, D. J. and Dehlmelt, H. (1975). *Bull. Am. Phys. Soc.*, **20**, 637.

Wineland, D. J., Drullinger, R. E. and Walls, F. L. (1978). *Phys. Rev. Lett.*, **40**, 1639.

Wineland, D. J. and Itano, W. M. (1979). *Phys. Rev. A*, **20**, 1521.

Wootters, W. K. and Zurek, W. H. (1982). *Nature (London)*, **299**, 802.

Wright, E. M., Walls, D. F. and Garrison, J. C. (1996). *Phys. Rev. Lett.*, **77**, 2158.

Wuerker, R. F., Shelton, H. and Langmuir, R. V. (1959). *J. Appl. Phys.*, **30**, 342.

Yablonovich, E., Gmitter, T. J. and Bhat, R. (1988). *Phys. Rev. Lett.*, **61**, 2546.

Yamamoto, T., Pashkin, Y. A., Astafiev, O., Nakamura, Y. and Tsai, J. S. (2003). *Nature (London)*, **425**, 941.

Yamamoto, Y. and Slusher, R. E. (1993). *Phys. Today*, **46**, 66.

Yang, G. J., Zobay, O. and Meystre, P. (1999). *Phys. Rev. A*, **59**, 4012.

Ye, J., Vernooy, D. W. and Kimble, H. J. (1999). *Phys. Rev. Lett.*, **83**, 4987.

Yokoyama, H., Nishi, K., Anan, T., Yamada, H., Brorson, S. D. and Ippen, E. P. (1990). *Appl. Phys. Lett.*, **57**, 2814.

Yokoyama, H., Suzuki, M. and Nambu, Y. (1991). *Appl. Phys. Lett.*, **58**, 2598.

Yoshie, T., Scherer, A., Hendrickson, J., Khitrova, G., Gibbs, H. M., Rupper, G., Ell, C., Shchekin, O. B. and Deppe, D. G. (2004). *Nature (London)*, **432**, 200.

Yurke, B. and Stoler, D. (1986). *Phys. Rev. Lett.*, **57**, 13.

Zeilinger, A. (1999). *Rev. Mod. Phys.*, **71**, S288.

Zheng, S. B. and Guo, G. C. (2000). *Phys. Rev. Lett.*, **85**, 2392.

Zippilli, S., Vitali, D., Tombesi, P. and Raimond, J.-M. (2003). *Phys. Rev. A*, **67**, 052101.

Zurek, W. H. (1981). *Phys. Rev. D*, **24**, 1516.

Zurek, W. H. (1991). *Phys. Today*, **44**, 36.

Zurek, W. H. (2003). *Rev. Mod. Phys.*, **75**, 715.

Zurek, W. H., Habib, S. and Paz, J. P. (1993). *Phys. Rev. Lett.*, **70**, 1187.

Index